COMPUTER MODELING FOR ENVIRONMENTAL MANAGEMENT SERIES

PUBLISHED TITLES

Computer Generated Physical Properties
Stan Bumble

Computer Simulated Plant Design for Waste Minimization/Pollution Prevention
Stan Bumble

FORTHCOMING TITLES

Computer Modeling and Environmental Management
William C. Miller

COMPUTER MODELING FOR ENVIRONMENTAL MANAGEMENT SERIES

COMPUTER SIMULATED PLANT DESIGN for WASTE MINIMIZATION/POLLUTION PREVENTION

COMPUTER MODELING FOR ENVIRONMENTAL MANAGEMENT SERIES

COMPUTER SIMULATED PLANT DESIGN for WASTE MINIMIZATION/POLLUTION PREVENTION

Stan Bumble, Ph.D.

LEWIS PUBLISHERS
Boca Raton London New York Washington, D.C.

Library of Congress Cataloging-in-Publication Data

Bumble, Stan.
Computer simulated plant design for waste minimization/pollution prevention / Stan Bumble.
p. cm. -- (Computer modeling for environmental management series)
Includes bibliographical references and index.
ISBN 1-56670-352-2 (alk. paper)
1. Chemical plants--Design and construction--Computer simulation. 2. Chemical plants--Environmental aspects--Computer simulation. 3. Waste minimization--Computer simulation. 4. Pollution--Computer simulation. I. Title. II. Series.

TP155.5.B823 2000
660′.28′ 0286—dc21 99-057318

Direct all inquiries to CRC Press LLC, 2000 N.W. Corporate Blvd., Boca Raton, Florida 33431.

Lewis Publishers is an imprint of CRC Press LLC

International Standard Book Number 1-56670-352-2
Library of Congress Card Number 99-057318
Printed in the United States of America 1 2 3 4 5 6 7 8 9 0
Printed on acid-free paper

Preface

When I asked an EPA repository of information for any references on the subject of this book, I was given a very swift and professional reply: "There isn't any." This was, of course, counter to my experience of years working on this subject and collecting huge numbers of papers and referrals that detailed progress and enthusiasm for my attempts. A summary of these findings is in this book.

I think it true that the kind of person who will be successful in finding results or creating results in *Computer Simulated Plant Design for Waste Minimization/Pollution Prevention* is not the average kind of scientist or engineer one finds today. Indeed, the proper person for this work is a multidisciplined computer scientist, chemical engineer, chemist, mathematician, etc. There are not many people like that today, particularly creative ones. However, you will meet some in this book.

The book is divided into five parts and each part has a number of sections. The title of the parts describes the main theme of the part but not all of the included matter.

The first part is entitled Pollution Prevention and Waste Minimization. It begins with descriptions of process flowsheets and block flow diagrams. It then describes pollution prevention, cost, and energy. It describes control of exhausts from processes or, in other words, reduction of emissions. There is then a very brief description of the design or simulation of a plant so the reader can get the flavor of it before pollution prevention is discussed more thoroughly. Reaction systems and separation systems appropriate for waste minimization are then introduced. Continuing in this manner, computer simulation as it pertains to pollution prevention is introduced. The Inorganic Chemical Industry Notebook Section from EPA is then shown as an example. The important introduction to models is introduced next and this is systematized with process models and simulation. Process information and waste minimization are tied together. The very important cost factors are discussed with waste minimization and Department of Energy (DOE) processes. A number of sections on pollution prevention then occur and a discussion proceeds on tools for P2.

A discussion of the redesign of products and processes follows. A very proper set of results for the environment, health, and safety in the early design phases of a process is presented. An interesting article is summarized that correlates the size of plants and the exposure to pollution. The work on the motivation for pollution prevention among top executives in the company is very educational. This is also true of the article on why the reason for pollution prevention has not been more favorably received publicly. A description of a graduate student's work on a plantwide controllability and flowsheet structure for complex continuous plants is shown. A 3D Design, 3D chemical plant program is described. A computer-aided flowsheet design and analysis for nuclear fuel reprocessing is also described.

Conceptual designs of "clean processes" are shown as well as the development of tools to facilitate the design of plants that generate as little pollution as possible. Computer Simulated Plant Design for Waste Minimization/Pollution Prevention and flowsheet tools for spreadsheets are shown. Integrated synthesis and analysis of chemical process designs using heuristics in the context of pollution prevention are studied. Also presented are model-based environmental sensitivity analysis for designing a clean process plant. Ways to reduce gas emissions in utility plants and elsewhere are shown. Upsizing or inputting the waste of one plant into another is strongly urged. This is further discussed for zero emissions where plants are clustered together.

Permix is a reactor design, from SRI, which helps pollution prevention. Batch chromatography is a technique that can help develop optimum processes. There are P2 opportunities that can be identified from the various sectors mentioned before. Excerpts on waste minimization are included from the latest Federal Register. The definitions of bioaccumulation, persistence, and toxicity are discussed as they will be used to spotlight the worst chemical compounds.

The ATSDR section concentrates on health. There is a chapter on OSHA software. The idea of having communities monitor toxic compounds is discussed (EMPACT). The very fine work of the EDF (Environmental Defense Fund) in matters of health and Scorecard is reviewed. Screening for endocrine disruptors is discussed. A paper on reducing risk for man and animals is included. Risk is then discussed as a "human science." The IPPS (industrial pollution projection system) is a way to compare pollution country by country.

Part II begins with a sequential set of chapters that prepares the reader for chapters on mathematical methods considered or used in computer programs for pollution prevention and waste minimization. They are in order: Linear Programming, The Simplex Model, Quadratic Programming, Dynamic Programming, Combinatorial Optimization, Elements of Graph Theory, Organisms and Graphs, Trees and Searching, Network Algorithms, Extremal Programs, Traveling Salesman Problem, Optimization Subject to Diophantine Constraints, Integer Programming, MINLP (Mixed Integer Nonlinear Programming), Clustering Methods, Simulated Annealing, Tree Annealing, Global Optimization Methods, Genetic Programming, Molecular Phylogenetic Studies, and Adaptive Search Techniques.

It is to be noted that Organisms and Graphs is included in Part II, Mathematical Methods, although it is a little different than the other methods cited. It refers to processes in living organisms that are to be compared to processes or flowsheets in chemical plants.

Advanced mathematical techniques are used in RISC-Lenz work and also the work of Drs. Friedler and Fan. Scheduling of processes for waste minimization is for batch and semicontinuous processes. Multisimplex can optimize 15 controls and responses at once. Extremal optimization provides high quality solutions to hard optimization problems,. Petri nets and Synprops compare two processes and show the graph model and concurrent processing together. Petri net-digraph models are for automating HAZOP analyses of batch process plants. DuPont CRADA is a description of neural network controllers for chemical process plants. KBDS is about design history to support chemical plant design, and dependency-directed backtracking helps when objects, assumptions, or external factors have changed previously in a design. Interactive collaborative environments allow different people at far removed places to work on the same drawings. The control kit for O-matrix is a control system without the need for programming, the clean process advisory system (CPAS) is a system of software tools for design information on clean techniques for pollution prevention to conceptual process and product designers when needed. Finally, nuclear applications are discussed. Also, it is important to have a process for viewing of the environmental impact at the beginning of the design process. There are tools to accomplish this such as OPPEE (Optimization for Pollution Prevention, and Energy and Environment) as well as CPASTM. Following is a discussion of computers, as they are very important in this work. The future will lead to better computers for doing the work needed for pollution prevention and waste minimization.

Part III is entitled Computer Programs for Pollution Prevention and/or Waste Minimization. It first discusses such programs as HYSYS, ICPET, and HYSIS. Then a discussion of Green Design describes environmentally benign products. There is then a study of chemicals and materials from renewable resources. One of the software companies into simulation software by the name of Simulation Sciences is then discussed. Two federal agencies, NFS and EPA, are interested in providing funds for deserving applied research for environmentally benign methods in industrial processes, design, synthetic processes, and products used in manufacturing processes. BDK is then discussed, and is an integrated batch development. An ingenious and very useful program called Process Synthesis is then introduced. It optimizes the structure of a process system, while minimizing cost and maximizing profit and will be discussed further later. Synphony is the commercial name for the process synthesis program that is now available. It determines all possible flowsheets from all possible operating units and raw materials for a given product and ranks these. The following programs are then discussed: Aspen, CAPD (Computer-Aided Process Design), work at CMU, Silicon Graphics/Cray Research, work by Floudas, etc. Work on robust self-assembly using highly designable structure and self-organizing systems are then described. The work of El-Hawagi and Spriggs on Mass Integration is then given prominence. The synthesis of mass energy integration for waste minimization via in-plant modification then follows naturally. A very clever scheme for the whole picture of environmentally acceptable reactions follows. Work concerning pollution prevention by reactor network synthesis is outlined. LSENS is the NASA program for chemical kinetics. It was the first of its kind and DOE's program followed. Chemkin was developed at Sandia and is used by many people. It was instrumental in the application to NOx chemistry and has a huge library of thermodynamic and kinetic data, but uses the NASA format. There follows a discussion of what Chemkin can do. Multiobjective Optimization is a continuous optimizer and performs waste minimization. Risk Reduction through Waste Minimizing Process Synthesis follows. It combines process design integration, risk reduction, waste minimization and Chemkin. Kineticus is a program written by a graduate student at Drexel University. It can perform similar operations to Chemkin. SWAMI (Strategic Waste Minimization) from EPA enhances process analysis techniques and identifies waste minimization techniques. Super Pro is a program that designs manufacturing processes with environmental constraints. P2-Edge software helps engineers and designers incorporate pollution prevention into the design stage. CWRT is a program for aqueous effluent stream pollution prevention design options. The OLI program ESP (Environmental Simulation Program) enhances the productivity of engineers and scientists (it is a steady state program). Process Flowsheeting and Control has multiple recycles and control loops. Environmental Hazard Assessment for Computer-Generated Alternative Syntheses is the general Syngen program for generation of shortest and least costly synthesis paths. The computer generated wastewater minimum program in a dairy plant is described. A LCA (Life Cycle Analysis) Program is described. Minimization of free energy (for chemical equilibrium) and free radicals are discussed. A pollution prevention process modification using on-line optimization is described. Genetic algorithms for generation of molecules is outlined. Finally, coding theory, cellular optimization, Envirochemkin, and the chemical equilibrium program are used together as the best among alternatives.

Part IV is entitled Computer Programs for the Best Raw Materials and Products of Clean Processes. The first section describes how regression is used with much data to predict physical properties. Later this is extended to Risk Based Concentrations. The properties are predicted from chemical groups. This method is used in a spreadsheet and is tied in with an optimization scheme, and the whole program is called SYNPROPS and used to replace toxic solvents with benign solvents with the same physical properties. There is toxic ignorance for almost 75% of the top-volume chemicals in use. However, SYNPROPS (from groups) can yield MCL, tap water, ambient air, and commercial/industrial/residential soil risk based concentrations. There is then a study of drug design followed by a discussion of a source of pollution: aerosols. A program called Computer-Aided Molecular Design (CAMD) is discussed. An applied case is described; Texaco Chemical Company plans to reduce HAP emissions through an early pressure reduction program by vent recovery system. The work of Drs. Fan and Friedler is introduced with a description of the design of molecules with desired properties by combinatorial analysis. Some of the extensive mathematical background needed for this follows. There then follows another method which is called Automatic Molecular Design Using Evolutionary Techniques. This uses genetic software techniques to automatically design molecules under control of a fitness function within the realm of nanotechnology. Algorithmic generation of feasible partitions returns us to the method of Fan and Friedler. Testsmart promotes faster, cheaper, and more humane lab tests without cruelty to animals and also uses SAR techniques to obtain toxicity data. European Cleaner Technology Research, Cleaner Manufacturing in the European Union involving substitution, minimization, etc. is described and Cleaner Synthesis is discussed. This finds an alternate, cleaner synthesis rather than dealing with after-effects. THERM is introduced. This is a very useful program that derives thermodynamic functions from groups, puts them in NASA format for use in Chemkin and LSENS, and also obtains thermodynamic functions for reactions. Design trade-offs for pollution prevention are then discussed, as is the shift of responsibility to industry with pollution product defects. Programming waste minimization within a process simulation program aims at eliminating pollution at the source. The discussion leads to product and process design tradeoffs for pollution prevention. This entails integrating multiobjective design optimization with statistical quality control and lifecycle analysis. Incorporating pollution prevention in the U.S. Department of Energy Design Projects is next. This raises awareness and provides specific examples of pollution prevention design opportunities. A description of PMN (Pre Manufacturing Notice) within TSCA follows. There is then a short article on why pollution prevention founders. ICPET (Institute for Chemical Process and Environmental Technology) is described as supplying innovative computer modeling and numerical techniques. The programs HYSYS, IVPET, and HYSIS are then discussed. Cost effective optimization is highlighted. Pinch technology as part of process integration and the effective use of heat is described. The Geographic Information System is shown as important to many parts of environmental work. Chronic environmental effects are included in the Health chapter. The EDF Scorecard, which tracks pollution and its causes in many geographies has had large impact. Also, HAZOP and process safety identifies hazards in a plant and what causes it. Safer by Design is a study about making plants safer by design. Design theory and methodology includes three parts: product and process design tradeoffs for pollution prevention, pollution prevention and control, and integration of environmental impacts into product design.

Part V is entitled Pathways to Prevention. It opens with a similarity between the Grand Partition Function of Statistical Mechanics and the mass and energy balance of chemical engineering. Then part of the data for mechanisms from the Department of Chemistry from the University of Leeds is shown. Blurock's extensive Reaction program is then described. R & D concerning catalytic reaction technology controlling the efficiency of energy and material conversion processes under friendly and environmental measures is shown. An article for building the shortest synthesis route is included. A description of how DuPont controls greenhouse emissions is given (for at least one plant). Another article describes how software simulations lead to better assembly lines. A theoretical connection between equations of state and connected irreducible integrals as well as the mathematics of generating functions is shown. An article on ORDKIN, a model of order and kinetics for the chemical potential of cancer cells is reproduced. Another article shows what chemical engineers can learn from nature as to isolation versus interaction in research. There is also a description of design synthesis using adaptive search techniques and multicriteria decision analysis. The Path Probability method is shown with application to environmental problems. The method of steepest descents is shown. The Risk Reduction Laboratory/ Pollution Prevention Branch Research (RREL/PPRB) is discussed. The PPRB is a project that develops and demonstrates cleaner production technologies, cleaner products and innovative approaches to reducing the generation of pollutants in all media.

The Author

Stan Bumble, Ph.D., has guided research, development, and engineering at DuPont and Dow Corning with computer programs that optimized the best products and properties. He has used computer programs for assisting the U.S. government with the development of their missile program and with the recovery of disaster victims. He has helped (with the assistance of computers) the U.S. Department of Justice and the Environmental Protection Agency at many hazardous sites such as Love Canal.

Table of Contents

Part I. Pollution Prevention and Waste Minimization

Part II. Mathematical Methods

Part III. Computer Programs for Pollution Prevention and/or Waste Minimization

Part V. Pathways to Prevention

List of Figures

Part I. Pollution Prevention and Waste Minimization

1.1 Chemical Process Structures and Information Flow

Systematic study of structural problems is of relatively recent origin in chemical engineering. One of the first areas to receive such attention is process flowsheet calculations. These calculations typically occur in process design.

Process design may be perceived as a series of distinct tasks. Starting with a market need or a business opportunity, a number of process alternatives are created or synthesized. The task of creating these alternatives is sometimes referred to as process synthesis. The outcome of process synthesis is usually expressed in terms of process flowsheets. The best solution is arrived at by systematically evaluating each of these alternatives. This quantitative evaluation usually begins with the material and energy balances, followed by equipment size and costing and culminates in an analysis of the economic merits of the process. As the initial choice of the process is not expected to be optimal, it is usually possible to improve the process by a different choice of process flows and conditions. This is called parameter optimization. Some of these decided variables may be continuous, others may be discrete such as stages or size of equipment.

A process can be improved by a different choice of processing units and interconnections. The task of identifying such improvements is termed structural optimization. While some structural improvements are but minor modifications of the same process, others give rise to different processes.

The above description is of course a gross simplification of the reality. In practice, these tasks are not always neatly partitioned, nor are they carried out in sequence, nor to completion. This evaluation or optimization may be truncated once the outcome is apparent, or its purpose is fulfilled. However, it is an iterative nature of process design activities and the central role of process flowsheet calculations and the heart of process evaluation and optimization. Because the calculations are so repetitive, efficiency, reliability, and accuracy of the solution procedure deserve special attention.

Though the first computer calculations to process design were limited to design calculations involving a single unit such as a heat exchanger or a flash separator, it did not take very long before chemical engineers recognized the far greater potential of a process flowsheet simulator. In the years since the first such program was reported, process flowsheeting programs have become the accepted workhorse of many a process design organization. One feature of such a program is its capability to input and modify the process flowsheet configuration and to perform design calculations involving a process flowsheet. Because of the need to enhance material and energy utilization, a chemical process is typically highly integrated. Unconverted reactants and unwanted byproducts arising from incomplete chemical conversion are typically recycled after they are first separated from the desired products. The recycle enhances the overall chemical conversion and yield. Also, the reaction or separation may have to be carried out at a high temperature. In order to minimize energy requirements, a feed-effluent heat exchanger may be introduced to recover waste heat and to preheat the feed. The ideal design structure of a process flowsheet is a tree from the viewpoint of design calculations. Then the calculations can proceed sequentially. This is never ideal from the viewpoint of material and energy utilization. The introduction of recycle streams and heat exchangers creates more cyclic structures in a process flowsheet and makes it more difficult to determine an appropriate calculation sequence.

1.2 Analysis Synthesis & Design of Chemical Processes

Three principal diagrams for a chemical process are the block flow diagram (BFD), process flow diagram (PFD) and the piping & instrumentation diagram, (P&ID). Design is an evolutionary process which can be represented by the sequence of process diagrams describing it. To begin, an input-output diagram may be sketched out. One can then break down the process into its basic functional elements such as the reaction and separation sections. One could also identify recycle streams and additional unit operations in order to reach desired temperature and pressure conditions. These basic elements lead to a generic process block flow diagram, which can be drawn after estimates of process flows and material and heat balances are made. After preliminary equipment specifications, a process flow diagram is made. Finally, as the mechanical and instrumentation details are considered, the piping and instrumentation diagram is created.

Other parts of the plant must be included. These are:

Engineering Economic Analysis of Chemical Processes

- Estimates of Capital Cost
- Estimation of Manufacturing Costs
- Engineering Economic Analysis
- Profitability Analysis

Technical Analysis of a Chemical Process

- Structure of Chemical Process Flow Diagrams
- Tracing Chemicals Through the Process Flow Diagram
- Understanding Process Conditions
- Utilizing Experience-Based Principles to Confirm the Suitability of a Process Design

Analysis of System Performance

- Process Input/Output Models
- Tools for Evaluating System Performance
- Performance Curves for Individual Unit Operations
- Multiple Unit Performance
- Reactor Performance
- Regulating Process Conditions
- Process Troubleshooting

Synthesis and Optimization of a Process Flow Diagram

- Synthesis of the PFD from the Generic Block Flow Process Diagram
- Synthesis of a Process Using a Simulator and Simulator Troubleshooting
- Process Optimization

The Professional Engineer, The Environment, and Communications

- Ethics and Professionalism
- Health, Safety, and the Environment
- Written and Oral Communications
- The Written Report

1.3 Strategy and Control of Exhausts

Limits for exhaust emissions from industry, transportation, power generation, and other sources are increasingly legislated. One of the principal factors driving research and development in the petroleum and chemical processing industries in the 1990s is control of industrial exhaust releases. Much of the growth of environmental control technology is expected to come from new or improved products that reduce such air pollutants as carbon monoxide (CO), volatile organic compounds (VOCs), nitrogen oxides (NOx), or other hazardous air pollutants. The mandates set forth in the 1990 amendments to the Clean Air Act (CAA) push pollution control methodology well beyond what, as of this writing, is in general practice, stimulating research in many areas associated with exhaust system control. In all, these amendments set specific limits for VOCs, nitrogen oxides, and the so-called criteria pollutants. An estimated 40,000 facilities, including establishments as diverse as bakeries and chemical plants are affected by the CAA.

There are 10 potential sources of industrial exhaust pollutants which may be generated in a production facility:

1. Unreacted raw materials
2. Impurities in the reactants
3. Undesirable by-products
4. Spent auxiliary materials such as catalysts, oils, solvents, etc.
5. Off spec product
6. Maintenance
7. Exhausts generated during start-up or shutdown
8. Exhausts generated from process upsets and spills
9. Exhausts generated from product and waste handling, sampling storage, and treatment
10. Fugitive sources

Exhaust streams generally fall into two general categories, intrinsic and extrinsic. The intrinsic wastes represent impurities present in the reactants, by-products, co-products, and residues as well as residues used as part of the process, i.e., sources 1-5. These materials must be removed from the system if the process is to continue to operate safely. Extrinsic wastes are generated during operation of the unit, but are more functional in nature. These are generic to the process industries overall and not necessarily inherent to a specific process configuration, i.e., sources 6-10. Waste generation may occur as a result of unit upsets, selection of auxiliary equipment, fugitive leaks, process shutdown, sample collection and handling, solvent selection, or waste handling practices.

Control Strategy Evaluation

There are two broad strategies for reducing volatile organic compound (VOC) emissions from a production facility:

1. Altering the design, operation, maintenance, or manufacturing strategy so as to reduce the quantity or toxicity of air emissions produced.
2. Installing after-treatment controls to destroy the pollutants in the air emission stream.

The most widely used approach to exhaust emission control is the application of add-on control devices. For organic vapors, these devices can be one of two types, combustion or capture. Applicable combustion devices include thermal incinerators, i.e., rotary kilns, liquid injection combustors, fixed hearths, and fluidized bed combustors; catalytic oxidation devices; flares or boilers/process heaters. Primary applicable capture devices include condensers, adsorbers, and absorbers, although such techniques as precipitation and membrane filtration are finding increased application.

The most desirable of the control alternatives is capture of the emitted materials followed by recycle back into the process. However, the removal efficiencies of the capture techniques generally depend strongly on the physical and chemical characteristics of the exhaust gas and the pollutants considered. Combustion devices are the more commonly applied control devices, because these are capable of a high level of removal efficiencies, i.e., destruction for a variety of chemical compounds under a range of conditions. Although installation of emission control devices requires capital expenditures, they may generate useful materials and be net consumers or producers of energy. The selection of an emission control technology is affected by nine interrelated parameters:

1. Temperature, T, of the inlet stream to be treated
2. Residence time
3. Process exhaust flow rate
4. Auxiliary fuel needs
5. Optimum energy use
6. Primary chemical composition of exhaust stream
7. Regulations governing destruction requirements
8. The gas stream's explosive properties or heat of combustion
9. Impurities in the gas stream

Given the many factors involved, an economic analysis is often needed to select the best control option for a given application.

Capture devices are discussed extensively elsewhere. Oxidation devices are either thermal units that heat alone or catalytic units in which the exhaust gas is passed over a catalyst usually at an elevated temperature. The latter speed oxidation and are able to operate at temperatures well below those of thermal systems.

Oxidation Devices

Thermal Oxidation

Thermal oxidation is one of the best known methods for industrial waste gas disposal. Unlike capture methods such as carbon adsorption, thermal oxidation is an ultimate disposal method destroying the objectionable combustible compounds in the waste gas rather than collecting them. There is no solvent or adsorbent to dispose or regenerate. On the other hand, there is no product to recover. A primary advantage of thermal oxidation is that virtually any gaseous organic stream can be safely and cleanly incinerated, provided proper engineering design is used.

A thermal oxidizer is a chemical reactor in which the reaction is activated by heat and is characterized by a specific rate of reactant consumption. There are at least two chemical reactants, an oxidizing agent and a reducing agent. The rate of reaction is related both to the nature and to the concentration of reactants, and to the conditions of activation, i.e., the temperature (activation), turbulence (mixing of reactants), and time of interaction.

Some of the problems associated with thermal oxidizers have been attributed to the necessary coupling of the mixing, the reaction chemistry, and the heat release in the burning zone of the mixing. These limitations can reportedly be avoided by using a packed-bed flameless thermal oxidizer, which is under development.

Catalytic Oxidation

A principal technology for the control of exhaust gas pollutants is the catalyzed conversion of these substances into innocuous chemical species, such as water and carbon dioxide. This is typically a thermally activated process commonly called catalytic oxidation, and is a proven method for reducing VOC concentrations to the levels mandated by the CAA. Catalytic oxidation is also used for treatment of industrial exhausts containing halogenated compounds.

As an exhaust control technology, catalytic oxidation enjoys some significant advantages over thermal oxidation. The former often occurs at temperatures that are less than half those required for the latter, consequently saving fuel and maintenance costs. Lower temperatures allow use of exhaust stream heat exchangers of a low grade stainless steel rather than the expensive high temperature alloy steels. Furthermore, these lower temperatures tend to avoid the emissions problems arising from the thermal oxidation processes.

Critical factors that need to be considered when selecting an oxidation system include:

1. Waste stream heating values and explosive properties. Low heating values resulting from low VOC concentration make catalytic systems more attractive, because low concentrations increase fuel usage in thermal systems.
2. Waste gas performance that might affect catalyst performance. Catalyst formulations have overcome many problems owing to contaminants, and a guard bed can be used in catalytic systems to protect the catalyst.
3. The type of fuel available and optimum energy use. Natural gas and No. 2 fuel oil can work well in catalytic systems, although sulfur in the fuel oil may be a problem in some applications. Other fuels should be evaluated on a case-by-case basis.
4. Space and weight limitations on the control technology. Catalysts are favored for small light systems.

There are situations where thermal oxidation may be preferred over catalytic oxidation. For exhaust streams that contain significant amounts of catalyst poisons and/or fouling agents, thermal oxidation may be the only mechanically feasible control. Where extremely high VOC destruction efficiencies of difficult to control VOC species are required, thermal oxidation may attain higher performance. Also, for relatively rich waste gas streams, i.e., having ±20 to 25% lower explosive limits (LEL), the gas stream's explosive properties and the potential for catalyst overheating may require the addition of dilution air to the waste gas system.

Catalysts — For VOC oxidation a catalyst decreases the temperature, or time required for oxidation, and hence also decreases the capital, maintenance, and operating costs of the system.

Catalysts vary both in terms of compositional material and physical structure. The catalyst basically consists of the catalyst itself, which is a finely divided metal; a high surface area carrier; and a support structure. Three types of conventional metal catalysts are used for oxidation reactions: single- or mixed-metal oxides, noble (precious) metals, or a combination of the two.

Exhaust Control Technologies

In addition to VOCs, specific industrial exhaust control technologies are available for nitrogen oxides, NOx, carbon monoxide, CO, Halogenated hydrocarbon, and sulfur and sulfur oxides, SOx.

Nitrogen Oxides

The production of nitrogen oxides can be controlled to some degree by reducing formation in the combustion system. The rate of NOx formation for any given fuel and combustor design is controlled by the local oxygen concentration, temperature, and time history of the combustion products. Techniques employed to reduce NOx formation are collectively referred to as combustion controls and U. S. power plants have shown that furnace modifications can be a cost-effective approach to reducing NOx emissions. Combustion control technologies include operational modifications, such as low excess air, biased firing, and burners-out-of-service, which can achieve 20 to 30% NOx reduction; and equipment modifications such as low NOx burners, overfire air, and reburning, which can achieve a 40 to 60% reduction. As of this writing, approximately 600 boilers having 10,000 MW of capacity use combustion modifications to comply with the New Source Performance Standards (NSPS) for NOx emissions.

When NOx destruction efficiencies approaching 90% are required, some form of post-combustion technology applied downstream of the combustion zone is needed to reduce the NOx formed during the combustion process. Three post-combustion NOx control technologies are utilized: selective catalytic reduction (SCR); nonselective catalytic reduction (NCR); and selective noncatalytic reduction (SNCR).

Carbon Monoxide

Carbon monoxide is emitted by gas turbine power plants, reciprocating engines, and coal-fired boilers and heaters. CO can be controlled by a precious-metal oxidation catalyst on a ceramic or metal honeycomb. The catalyst promotes reaction of the gas with oxygen to form CO2 at efficiencies that can exceed 95%. CO oxidation catalyst technology is broadening to applications requiring better catalyst durability, such as the combustion of heavy oil, coal, municipal solid waste, and wood. Research is under way to help cope with particulates and contaminants, such as fly ash and lubricating oil, in gases generated by these fuels.

Halogenated Hydrocarbons

Destruction of halogenated hydrocarbons presents unique challenges to a catalytic oxidation system. The first steps in any control strategy for halogenated hydrocarbons are recovery and recycling. However, even with full implementation of economic recovery steps, significant hydrocarbons are present as impurities in the exhaust stream. Impurity sources are often intermittent and dispersed.

The principal advantage of a catalytic oxidation system for halogenated hydrocarbons is operating cost savings. Catalytically stabilized combustors improve the incineration conditions, but still must employ very high temperatures as compared to VOC combustors.

Uses

Catalytic oxidation of exhaust streams is increasingly used in those industries involved in surface coatings: printing inks, solvent usage, chemical and petroleum processes, engines, cross media transfer, and a number of other industrial and commercial processes.

1.4 Chemical Process Simulation Guide

The following is a very brief account of a rough draft. It is a description of a process simulation without pollution prevention or waste minimization as essential parts. The structure consists of four parts:

1. User Interface
2. Executive Program
3. Thermodynamic Unit Operations
4. Constants, Database, and Equations

(See Figure 78). The part the user sees is the user interface. (This is where the user enters data (e.g., stream temperature, pressure and composition and design parameters such as the distillation column number of stages). The second part (executive program) takes the user input and follows the instructions to control such things as calculation sequence and convergence routines. It finds a solution in which all the recycle loops have converged and all the user specifications have been met. In the third part, the chemical, physical, and thermodynamic properties can be calculated. Here the thermodynamics constant database, the correlation constants, and the limits of the correlations and the equations are stored. The fourth part is the unit operations modules. They perform the engineering calculations, such as the pressure drop in a pipe, based on the pipe diameter and the Reynolds number.

You must satisfy the degrees of freedom and supply all needed information to the simulator. This includes all compositional data as well as all data to satisfy the Gibbs Phase Rule. This must be done for all equipment, whether it is a pump or a flash drum.

There are two simulator types: sequential modular and simultaneous equation. Sequential modular simulators are more common. There are also hybrid systems. The sequential modular approach sequentially calculates modules. It takes the process feeds and performs the unit operation calculation to which it is fed. The output is the conditions of the outlet stream(s) along with information on the unit operation. This outlet stream(s) are fed to subsequent unit operations and the calculations proceed sequentially. If recycle streams are present in the chemical process, these streams are "torn" (i.e., the user is asked to supply an estimate of the stream specification or the program responds with an initial zero flow). The simulator calculates around the loop(s), revising the input tear stream values, until the input and output tear streams match. This is called converging the recycle; often this is the major time requirement and cause of simulator failure.

Below is an overview of a process simulator's capabilities:

1. Steady state process simulation is not the right tool for every process problem; it is effective when vapor-liquid equilibrium is important, for evaluating the steady state effect of process changes, and for preliminary equipment sizing.
2. The engineer should always perform short-cut calculations to estimate the solution; this allows him to evaluate the process simulation results and to speed-up and successfully complete recycle convergence problems.
3. The thermodynamics property correlation is at the heart of any process simulation; if it is wrong, all the simulation results are wrong.
4. Most commercial process simulators are sequential modular; thus, they converge individual unit operation modules sequentially and then seek to converge recycle loops. Thus, useful information can sometimes be obtained from an "unconverged" simulation.
5. Of the four parts of a typical process simulator, problems usually occur in the executive program being unable to converge the program to meet the specifications, in the thermodynamics equations because the wrong thermodynamic correlation is chosen by the user or adequate thermodynamic data is unavailable, and in unit operations modules again because user specifications cannot be met.
6. The process simulator forces the user to satisfy the degrees of freedom before it will simulate the process.

Component Separation via Flash and Distillation

Although the chemical reactor is the heart of the process, the separation system is often the most expensive. Making good product and avoiding co-product production is economically significant; this may make the difference between an economical and an uneconomical process. However, the product must meet purity specifications before it can be sold. We must deal with separations where the components move between the liquid-vapor or liquid-liquid phases. This includes flashing (also called flash distillation, decanting), distillation, and absorption. Distillation accomplishes the component distillation based upon the difference in boiling point or vapor pressure where absorption is based on the gas solubility difference. Since the trade-off between operating and capital cost determines the equipment design, estimating these costs is included. Extraction and leaching use similar equipment and the design issue is again solubility or mass transfer from one phase to another (i.e., liquid to liquid and solid to liquid, resp.).

The design of all this equipment is based on the phase approaching equilibrium. An equilibrium stage involves two steps: first is the perfect mixing of the two phases such that equilibrium is reached, and the second is perfect separation between the phases (e.g., vapor and liquid, and liquid and liquid).

Phase Separation: Flash Drums and Decanters

Phase separation can be a very cost effective separation method. Flash drums are very popular with cost conscious chemical engineers. It should be noted that the product purity from a flash drum is limited for it acts as a single equilibrium stage and thus there must be significant differences in the component boiling points to obtain relatively pure products.

Column Design: Objective

Tower operating costs are investigated based upon operating cost. These costs and the column design are initially based upon short-cut calculations. Using the short-cut results and some initial specifications, the column can be simulated. Assuming the simulation converges, the column simulation can be improved by changing the specifications.

Selecting Column Pressure Based Upon Operating Cost (See Figure 82)

Energy is what drives the separation in a distillation cost. The operating costs of a distillation are the energy input in the reboiler and the energy remover in the condenser. Refrigeration costs more than steam per BTU transferred. A large portion of the cost is the compression (both the associated capital and operating costs). So to avoid refrigeration costs, it is often economical to operate at higher pressure. A pump is used rather than a compressor, to pump the feed to the column. In this way cooling water can be used for cooling. The exceptions are for very high pressures and when the high temperature in the bottom of the column leads to product degradation. For the first exception, the high pressure leads to high capital cost (thick walled vessels) and hazard considerations (e.g., mechanical explosion).

When we have a reasonable operating line pressure we need to find the number of equilibrium stages. The distillation module in the process simulator will not calculate the required number of equilibrium stages. It can be done by below bounds found via short-cut calculations. The stream compositions and column diameters found using short-cut calculations are only approximations. They may be sufficient to eliminate this design option, but are not necessarily good enough to use to design the column. It is the rigorous tower simulation that gives real answers. Unfortunately they are not always easy to converge. Therefore a step wise approach is advocated. The first step is the short-cut calculations. The second is a simple rigorous simulation. The next steps refine the rigorous simulation specifications, and the last step is to optimize the column design using the well-specified rigorous simulation.

The process simulator can easily calculate these bounds. They also can estimate from the Gilliland correlation, the column reflux ratio, and the number of stages for a range of actual to minimum reflux ratio values. The calculations are typically based upon key component recoveries. Usually one specifies the light-key component recovered in the distillate product and the heavy-key component recovered in the bottom product. These are close to 100%. Calculations rate existing equipment by comparing them to ideal operation. In this case one could calculate the predicted number of equilibrium stages and compare this to the number of trays to calculate tray efficiency. The short-cut calculations can be performed in a rating mode; however, it is more typical to perform a rigorous simulation with actual feed compositions, duties, and reflux ratio and then to manipulate the number of equilibrium stages until the product compositions are matched.

1.5 Integrated Design of Reaction and Separation Systems for Waste Minimization

Pollution prevention is one of the most serious challenges that is currently facing the industry. With increasingly stringent environmental regulations, there is a growing need for cost and energy efficient pollution prevention. In the 1970s the main focus of environmental pollution was end of pipe treatment. In the 1980s the main environmental activity of chemical processes was in implementing recycle/

reuse policies in which the pollutants can be recovered from terminal streams and reused. The current approach towards pollution prevention is source reduction in addition to end of pipe treatment and recycle/reuse. Source reduction pertains to any step that limits the extent of waste generated at the source. It focuses on in-plant activities that reduce the amount of hazardous species entering any waste stream. The objective can be achieved through changes in design/operating conditions that alter the flow rate/composition of pollutant-laden streams. The measures such as process modifications (temperature/pressure changes, etc.) and unit replacement and feedstock substitution, and reactor/separation network design can be manipulated to achieve cost-effective waste minimization. A systematic pollution prevention methodology has been developed, taking into account the fundamental understanding of the global insights of the process. The problem is formulated as an optimization program and solved to identify the optimum operating conditions in various units, reaction schemes, system design, optimum selection of feedstocks, separating agents, etc. for a fixed product throughput.

1.6 A Review of Computer Process Simulation in Industrial Pollution Prevention

EPA report 600R94128 discusses process simulator needs as a tool for P2. Most state of the art simulators provide many features that make them powerful tools for the analysis of P2 alternatives in a wide range of industrial processes. They have extensive libraries of unit operation models, physical property data, ability to incorporate user-supplied models and data, and they can perform sensitivity analyses and set design specifications using any process variable. They include other important features such as process optimization.

They are now very user friendly. They can significantly contribute to U.S. Industrial P2 efforts because they can easily model and analyze waste water streams. Industrial waste water is the largest volume of hazardous waste in the U.S., and waste water treatment is probably the largest application of process simulation.

Current measurement obstacles of data collection and data quality are overcome by the accurate and reliable waste generation data provided by simulation models. The obstacle of material balance closure is also overcome with the material balance done by these simulators.

Although possessing many features that make them powerful and convenient tools for process design and analysis, current process simulators lack many critical aspects needed for P2. Some are general, yet some are specific to P2. Some of these needs are:

Fugitive emissions estimations
P2 technology databases
Access to public domain data
Life cycle and ancillary operation analysis
Combustion byproduct estimation
Biological process modeling

Process synthesis could help determine alternative chemical reaction pathways and catalysts, determine alternative chemical separation sequences and efficiently incorporate waste treatment units into a process design. Process simulation tools could be helpful in dilute streams as the hazardous components in chemical process streams are present in trace amounts and the simulation could evaluate alternative reaction pathways to prevent these troublesome byproducts.

Improved models are needed for dynamic simulation of process transients such as start-ups or shutdowns, stochastic modeling to deal with non-routine events such as accidents, upsets and spills and large-scale modeling to understand the environmental conditions that result from interactions among unit operations. Process simulators need to handle various non-equilibrium phenomena (reaction kinetics, sorption, transport) impacting waste generation.

The following list contains some more capabilities that would be desirable in process simulators for P2 purposes:

1. Fugitive emissions estimation. It is possible to include emission factors into simulation architecture, application of deterministic emissions correlations, and application of equipment failure analysis.
2. P2 Technology databases. P2 case studies have revealed a series of effective equipment and process modifications. They can be organized by chemical, process, or unit operation, and can be made available in the form of an expert system for the process simulator user.
3. Access to public domain data. The TRI, RCRA biennial survey, CMA waste data bank, and a number of other sources of data could be useful to the process simulator user in benchmarking process configurations. Process simulators could query these data banks.
4. Life cycle and ancillary operation analysis. Simulation tools could be useful in evaluating the upstream and downstream impacts of alternative process designs and modifications, as well as the impacts of process ancillary operations such as maintenance, cleaning, and storage.

5. Combustion and byproduct estimation. Stack air emissions from incinerators and combustors may contain products of incomplete combustion such as chlorinated dioxins and furans and unburned principle organic hazardous constituents. They may be difficult to predict and measure. Process simulators, without the data support to model these trace species, now have the potential to do so.
6. Biological process modeling. These are increasingly being applied for the treatment, remediation and separation of hazardous wastes in air emissions, waste waters, sludges, soils, and sediments. Few simulators currently contain unit operation models for these processes.

Waste minimization and pollution prevention via source reduction of a chemical process involves modifying or replacing conventional chemical production processes. The impact of these activities upon process economics may be unclear, as increasing treatment and disposal costs and a changing regulatory environment make the cost of waste production difficult to quantify.

There are some basic strategies for reducing process wastes at their source. The flowrate of a purge stream can be reduced by decreasing the purge fraction, by using a higher purity feedstock, or by adding a separation device to the purge or recycle stream that will remove the inert impurity. Reaction byproduct production can be reduced by using a different reaction path, by improving catalyst selectivity, or by recycling byproducts back to the reactor so that they accumulate to equilibrium levels. Solvent wastes can be reduced by recovering and recycling the spent solvent, replacing the system with a solventless process, or replacing the existing solvent with a less toxic or more easily recovered solvent.

Previous work in source reduction has focused upon generating alternatives. Hierarchical approaches to identify clean processes and the industrial viability of solvent substitutions have been explored. Waste minimization via alternative reactor conditions and parameters has also been explored.

Integrating environmental concerns into the design and operation of chemical manufacturing facilities has become a necessity. Product and process design with environment as an objective and not just as a constraint on operations can lead to design alternatives that improve both the environmental and economic performance.

The usual way to reduce pollutant emissions has been to add control technology to bring the process into compliance with discharge standards. This has led to the allocation of large amounts of capital to the installation and operation of environmental control equipment. There has been little operational guidance about how to do better.

Design is not an easy activity. The input can be an abstract description of an organization and the result a detailed description of a concrete product, process, or system capable of satisfying those desires. It is a decision process with many decision makers and multiple levels of detail. After the design is specified, methods for generating alternatives are used, but because the time for completing a design is limited, the number of alternatives and the level of detail with which they can be analyzed is often compromised. The analysis of alternatives using engineering analysis (usually starting with mass and energy balances) is applied to each alternative to make predictions of the expected performance of the system. Inputs and outputs of the process, flow rates, compositions, pressure, temperature and physical state of material streams, energy consumption rate, stock of materials in the process, and sizing of the equipment units are listed and analyzed.

The information for each alternative is then summarized into indicators of performance to assess whether the requirements specified during the objective formulation have been met. These objectives include economic indicators (capital investment and operating cost) and should include indicators of safety and environmental performance. The alternatives can then be ranked.

Process design is iterative. Results are evaluated to identify opportunities for improvement before returning to the beginning of the design cycle. When the design team concludes that there are no opportunities for improvement, then the work stops.

The goal of proper design generation should be that the design (1) have high economic potential, (2) have high conversion of raw materials into desired products, (3) use energy efficiently, and (4) avoid the release of hazardous substances to the environment.

Pollution from a chemical process can be viewed as the use of the environment as a sink for unwanted by-products and unrecovered materials. Thus, design alternatives that increase the use of process units and streams as material sources and sinks could have lower environmental impact. Energy integration techniques can reduce utilities consumption by using process streams as sources and sinks of heat. The use of processing task integration in reactive distillation processes can reduce costs, energy use and emissions.

The mathematical programming approach to process synthesis usually uses a reducible superstructure that is optimized to find the best combination of process units that achieve the design task. A common feature is the use of cost minimization as the objective function in the optimization. As the value of recovered materials is not included, opportunities to improve economic performance of the networks involved by increasing material recovery beyond targets specified in the original optimization problem may be overlooked.

Huang and Edgar generate waste minimization alternatives with knowledge-based expert systems and fuzzy logic as attractive tools for designers. This is knowledge intensive as it requires knowledge from many disciplines.

Huang and Fan developed a hybrid intelligent design that improves the controllability of heat and mass exchanger networks by choosing stream matches that improve an index of controllability while keeping the operating cost of the network at its minimum. The system combines pinch analysis for the generation of targets with an expert system, fuzzy logic, and neural networks to assign stream matches. This addresses the fact that highly integrated processes are difficult to control.

Computer-assisted systems for the rapid generation of alternative synthesis paths to a desired chemical such as SYNGEN and LHASA are available. They can support pollution prevention processes.

EnviroCAD is an extension of BioDesigner, a program for the design and evaluation of integrated biochemical processes. Input data consists of waste streams and the system recommends alternatives for waste recovery, recycling, treatment, and disposal based on three knowledge bases. An expert system for generating feasible treatment trains for waste streams has also been embedded in the Process_Assessor module of the BatchDesign_Kit under development at M. I. T. The expert system is based on heuristic rules containing the knowledge of regulations and treatment technologies.

Some environmental impacts of design are not normally generated in the analysis stage. Such impacts include fugitive emissions and selectivity losses in reactors. In the latter case, estimation of individual by-products is usually not required. Frequently economic performance is the only criterion. Mass and energy balances, relevant for estimating the pollutant emissions from a process, are not included in the standard flow sheets used during process design. Environmental concentrations of released pollutants may be necessary for a proper evaluation of the potential environmental impact of a design.

Commercial process simulators are frequently deficient in predicting species concentration in dilute process effluent or waste streams. Unit operation models for innovative separation technologies (e.g., membrane separations) and waste treatment equipment are not included in commercial process simulators and are therefore usually not included in conceptual process designs.

Difficulties in evaluating environmental performance, needed for summarizing flow-sheet information, include (1) relevant properties of chemicals (toxicity, environmental degradation constants) are not readily available to chemical engineers in process simulators, chemical process design handbooks, etc.; (2) location-specific knowledge is needed to estimate potential environmental impacts; and (3) people differ in the importance they assign to various environmental impacts.

When the emission of a single pollutant is the most important environmental concern affecting a design, then the mass of that pollutant released into the environment can be used as an indicator of environmental impact. This was used to study the trade-off between control cost and emissions of nitrogen oxides from a power plant and a refinery. When more than one chemical is a source of environmental concern, environmental evaluation becomes more complicated.

Dozens of different ranking and scoring schemes have been proposed to evaluate chemicals based on measures of toxicity or measures of toxicity and exposure. Grossman and coworkers multiplied the material flows in a chemical process by the inverse of the 50% lethal dose of each material and added the resulting figures to obtain a toxicity index. Fathi Afshar and Yang divided material flows by their threshold limit values (TLVs) and multiplied them by their vapor pressure (assuming that fugitive emissions are proportional to vapor pressure).

Selection and refinement of a final design is a multiobjective decision problem, where economic, environmental, and safety concerns may be in conflict. Improving one objective may worsen another. For example, decreasing solvent emissions by increased separations may lead to increased emissions of combustion gases from energy generation.

In decision problems with multiple objectives, the set of nondominated alternatives must be identified. Each dominated alternative has at least one win-win alternative that can be attained without sacrificing achievement in any of the design objectives. The set of nondominated alternatives remains after the removal of all the dominated alternatives. The "best compromise" alternative is selected from the set of nondominated alternatives and this requires input about the values and preferences of the people responsible for making the decision.

Multiobjective goal programming is a technique that has also been used to solve chemical process design problems without specifying weighting factors to trade off one objective against another. The procedure involves stating goals for each objective of the design, ranking the objectives in order of importance, and choosing the alternative that minimizes lexicographically the vector of deviations from the aspiration levels. This allows the decision-maker to make trade-offs implicitly by specifying the aspiration levels. The aspiration levels will be case specific. This technique does not attempt to balance conflicting objectives. A marginal improvement in a highly ranked goal is preferred to large improvements in many goals.

Sensitivity analysis determines whether the best alternative identified advances the design objectives sufficiently, given the levels of uncertainty, to make further search unnecessary. The aspects of design that are driving the environmental impact and the trade-offs associated with the modifications of the aspects of the design driving the impacts must be identified and understood.

In December of 1992 the Center for Waste Reduction of the AICHE, the U.S. EPA and the U.S. DOE sponsored a workshop to identify requirements for improving process simulation and design tools with respect to the incorporation of environmental considerations in the simulation and design of chemical processes. Most are still present today. Such needs are:

Generation of Alternatives

1. Increase the integration of process chemistry into the generation of design alternatives.
2. Develop tools to identify new reaction pathways and catalysts.
3. Extend alternate generation methods to include unconventional unit operations.
4. Develop methods that allow the rapid identification of opportunities to integrate processes.
5. Develop methods to recognize opportunities to match waste streams with feed streams and to prescribe the operations needed to transform a waste stream into a usable feed stream.

Analysis of Alternatives

1. Predict generation of undesired by-products.
2. Improve prediction of reaction rates.
3. Predict fugitive emissions and emissions from nonroutine operations (e.g. start-up).
4. Improve characterization of non-equilibrium phenomena.
5. Include waste-treatment unit operations in process simulators.
6. Increase the ability of process simulators to track dilute species.
7. Improve stochastic modeling and optimization.
8. Link process and environmental models.
9. Build databases of properties relevant to environmental characterization of process and link them to process simulators.
10. Include information about uncertainties in databases.
11. Create databases with typical mass and energy balances (including trace components of environmental significance) for widely used raw materials in the chemistry industry to facilitate the characterization of upstream processes.
12. Develop guidelines to match the level of detail used in process models with the accuracy needed to make decisions.

Evaluation of Alternatives

1. Develop the accounting rules to allocate environmental impacts to specific processes and products in complex plants.
2. Develop environmental impact indices that are able to combine data of different quality while preserving their information content.
3. Develop screening indicators.
4. Develop frameworks that facilitate the elicitation of preferences needed as input to multiobjective optimization.

Sensitivity Analysis

1. Incorporate sensitivity analysis as a standard element in papers and books related to chemical process design.
2. Develop indicator frameworks that allow rapid identification of the features of a design that drive its environmental impact.

In the language of economists, zero emissions sets the objective of maximizing value added per unit resource input. This is equivalent to maximizing resource productivity, rather than simply minimizing wastes or pollution associated with a given product. It emphasizes seven objectives:

1. Minimize the material intensity of goods and services.
2. Minimize the energy intensity of goods and services.
3. Minimize the toxic dispersion.
4. Enhance ability of material to be recycled.
5. Maximize sustainable use of renewable sources.
6. Extend product durability.
7. Increase the service intensity of goods and services.

From the management standpoint there seem to be four elements. They are identified as follows:

1. Providing real services based on the customer needs.
2. Assuring economic viability for the firm.
3. Adopting a systems (life-cycle) viewpoint with respect to processes and products.

4. Recognizing at the firm's policy level that the environment is finite, that the "carrying capacity" of the Earth is limited, and that the firm bears some responsibility regarding the environment.

1.7 EPA Inorganic Chemical Industry Notebook Section V

The best way to reduce pollution is to prevent it in the first place. Here are some ways to promote pollution prevention.

Substitute raw materials
Improve reactor efficiencies
Improve catalyst
Optimize processes
Reduce heat exchanger wastes and inefficiencies
Improve wastewater treatment and recycling
Prevent leaks and spills
Improve inventory management and storage

The best way to reduce pollution is to prevent it in the first place. Here are some problems that may occur, which can cause pollution problems. They must be prevented from occurrence. In most cases the way to do that is obvious. The following is a list of factors contributing to pollution in plant design that can be averted: byproducts, coproducts, heavy metals in the catalysts, spent catalysts, catalyzed reaction has by-product formation, incomplete conversion and less than perfect yield, intermediate reaction products, trace levels of toxic constituents, high heat exchange tube temperatures, high localized temperatures, higher operating temperatures, fugitive emissions, seal linkage, higher gas pressures, corrosion, waste generation from corrosion inhibitors or neutralization, vent gas lost during batch fill, high conversion with low yield, non-regenerative treatment systems, insufficient R & D into alternative reaction pathways missing opportunities for eliminating waste, raw material and/or product have/has bad environmental impact, impurities, high vapor pressures, low odor threshold materials, toxic or nonbiodegradable materials that are water soluble, community and worker safety, large inventory, unknown characteristics and sources of waste streams, unknown fate and waste properties, unknown treatment and management of hazardous and toxic waste, leaks, leaks to groundwater, waste and releases from shutdown and startup, furnace emissions, inadequate mixing, waste discharge from jets, tank breathing, frequent relief, discharge to environment from over pressure, injection of seal flush fluid into process stream, fugitive emissions from shaft seal leaks, releases when cleaning or purging lines, leaks and emissions during cleaning, contaminated water condensate from steam stripping, etc.

1.8 Models

Model usage requires:

1. Problem recognition
2. Acceptance of responsibility to deal with the problem
3. A sufficient incentive for action
4. A belief that the possibility of finding a solution exists

Models are used as partial substitutes for their prototypes to assist in designing, understanding, predicting the behavior of or controlling the prototype. They must represent significant characteristics of their prototype.

Steady state simulation is used for process design and optimization through generation of mass and/or energy balances.

Dynamic simulation is used for process control, start-up and shut-down.

General process simulators model the behavior of a wide variety of processes.

A specific case simulator is designed to predict the behavior of a particular process.

A local simulator is intended to look at a specific part of a process.

A whole process simulator is designed to be able to consider a complete process flowsheet.

General Simulation Packages
ASPEN PLUS
PROCESS/PROII
HYSIM
DESIGN II
CHEMCAD
METSIM (general metallurgy)
SIMMET (mineral processing)

Local Simulation Packages
F*A*C*T
GTT-ChenSage, ChemApps
Thermochemistry
MTDATA
HSC

Specific Case Simulation Packages
MAPPS (pulp and paper)
SULSIM (Claus process)
BENEFITS

When used properly, process simulation has the following benefits:

cost reduction
improved organizational effectiveness
reduction of capital cost by better design
reduction of time for design, commissioning and start-up
reduction of pilot plant cost, size and complexity
material and energy optimization
improved productivity and efficiency
provision of training for new personnel
provision of screening model for projects
provision of repository for technical knowledge
definition of gaps, deficiencies, and inconsistencies in process knowledge

Optimization
- optimize flowsheet
- optimize unit operation
- optimize operation economically
- evaluate alternative raw materials
- optimize location and flow of recycle streams

In New Processes Design

- optimize flowsheet
- optimize operation economically
- optimize unit
- determine process sensitivities

Evaluation
- predict effect on whole system
- aid feasibility studies
- guide scale-up
- evaluate alternatives
- estimate possible effluents
- guide further research

Other
- equipment design
- sensitivity testing
- operating strategy evaluation
- process and production control
- energy conservation
- management information
- business planning
- training and teaching
- improve communication, reproducibility and accuracy

Limitations of Process Simulation
- requires discipline with respect to record keeping
- high initial training cost
- results are only as good as the models and plant data available and hence may be given undue credibility
- costly for small, simple, one-time problems
- many process units do not have equivalent simulator models
- the properties of many substances are not in simulator data banks

1.9 Process Simulation Seen as Pivotal in Corporate Information Flow

Process modeling and simulation go hand in hand. A process model is a mathematical representation of a production process. Simulation is the use of the model to predict a plant's performance and its economics. Until recently, their application has been a rather specialized design province of chemical engineering.

Because of profitability, companies generally don't have the large engineering staffs they used to have, or their staffs have been cut back significantly. So, while the workload has not changed, there are fewer people to do the work. Thus a greater percentage of the engineering staff needs access to the results and benefits that come from simulation software.

At one time, DuPont adopted an eight-point approach to becoming "best in the industry." Each point has a modeling or simulation component.

The first six points lead to construction of a plant: discovery of a superior product that can be manufactured with world-class chemistry; understanding of the process; translation of that understanding into a viable, dynamic process model; confirmation of the model in pilot facilities, as necessary; development of on-line analyzers to get compositional and physical property data in real time; and design of the most effective manufacturing facilities. The process model developed in step two or three, refined in steps four and five, and used for design in step six will then be used to run the plant — step seven — perhaps with model-predictive control.

The final point deals with operations optimization, the kind of modeling needed to make a business run exceptionally well. Manufacturing plants must get the raw materials they need when they need them. Products must be manufactured so that customers get what they want when they want it. Production facilities must be maintained in a way that ensures their reliability, safety, and operability.

This leads to the desire that we run the plant with less maintenance, higher yield, less energy, quality at least as good or better than was produced before, and at the same time increase productivity, which means fewer people and less capital.

We know that A plus B makes C, but we would like to know: the reaction mechanism; the transient intermediates that A and B go through in producing C; the reaction mechanisms of all the by-product reactions; which of all the steps in the reaction mechanism are kinetically controlled, which mass-transfer controlled, and which heat-transfer controlled; if the reaction is homogeneous, then what takes place at every point in the reactor at every point in time; and if the reaction is heterogeneous, the diffusion characteristics of raw materials to the catalyst surface or into the catalyst, as well as the reaction, reaction mechanism and by-product reactions within the catalyst, the diffusion characteristics of products away from the catalyst, and the nature of the heat transfer around the catalyst particle.

The optimization of the plant would be either technical supervision, not solely upon the expertise of the operators and plant personnel, but upon the knowledge of all the people who have worked on the development of the chemistry and the process design. It would involve development of a model of the entire plant by using rigorous chemical engineering techniques. The chemistry of the primary reactions and by-product reactions would be modeled.

It would be possible to run the plant as a model on a computer and test out operating scenarios — higher rates, different feedstocks, modified operating conditions — before they are tried on the actual plant. The model could also be used for operator training and to test plant start-ups and shut-downs. The model would run in real time, parallel to the plant, for model-predictor control.

"Optimization of the site" would eliminate control rooms for individual plants. When there is a control system that permits hands-off operation of a plant, and there is an expert system to coach the operating personnel through unusual events, then a centralized control room serving many plants is certainly possible.

Goals change. On one day, for example, the business decision might be to optimize plant output because of a need for more product. Another day, or week, the decision might be to optimize the costs from the plant, for example, by selecting some alternate feedstock. Optimization of energy consumption or minimization of undesirable effluents from the plant, or some combination of such factors, might be other goals.

1.10 Model-Based Environmental Sensitivity Analysis for Designing a Clean Process Plant

Process integration is being employed for the reduction of energy and material costs. More recently it has been used for the minimization of waste. Process streams are heavily interacted among units in an integrated process plant. Thus, various operational problems are due to severe disturbance propagation in the system. If these disturbance variables are environmentally sensitive, then the operational problems will become environmental problems. Such environmental problems may not be solved by control design. Thus, they should be prevented during process design. This means the design must be environmentally benign, i.e., effectively rejecting environmentally sensitive disturbance propagation through the process system, as well as cost effective. Thus we propose a model-based system sensitivity analysis approach for developing an environmentally benign process. The important elements of such a system are the waste generation/propagation models for different types of process operations, such as reaction, distillation, extraction, adsorption, and heat exchange operations. They represent system responses to environmentally sensitive disturbances and fluctuations. Although first-principles based, the models are simplified in linear form. Therefore, they can be easily embedded into a process design procedure. Also an introduction is made of a model-based waste minimization index for evaluating the cleanness of a process design. Now the design decisions will be evaluated not only by cost constraints, but also by waste reduction requirements. This allows process designers to perform effective sensitivity analysis of process alternatives, to identify any inappropriate connections among process units, and to quantify the severity of environmental impact of the design. It will also facilitate designers to derive improved process designs. This approach will apply to solve real-world problems by analyzing industrial processes where reactors, distillations, extraction, and heat integration occur. The resulting process will reduce waste by 10% through restructuring the topology of the system, and by introducing appropriate recycling.

1.11 Pollution Prevention in Design: Site Level Implementation Strategy For DOE

As the true cost of waste management has been realized, pollution prevention has received increased interest in recent years. Costs associated with pollution control and environmental compliance are continuing to rise as environmental regulations become more stringent so that DOE must develop or buy technologies to meet stricter emissions and disposal standards. Pollution prevention has become not only an environmentally friendly policy, but also a prudent business practice. It can be used as a powerful tool for cutting the rising costs associated with waste management while it helps to create superior products, services, and facilities.

The best time to incorporate pollution prevention strategies is up front, in the design phase of products, services, and facilities, because this is where and when the process that reduces or generates waste is defined and the materials that create or eliminate waste is chosen. As much as 70% of a product's or facility's life cycle costs are fixed during the design phase. Instead of waiting until the operational phase to examine pollution permission opportunities, large efficiency gains can be achieved by incorporating pollution prevention up front during design where they can be planned into the full life cycle of the project, including construction, operation, and decommissioning. It is easier to change a drawing than to retrofit an entire facility. In addition, identifying pollution prevention opportunities during design can reduce or eliminate environmental compliance and liability concerns before they are even created.

Pollution prevention is any practice that eliminates or minimizes waste generation. EPA defines it as source reduction, meaning reducing waste or pollutants before they are created, prior to recycling, treatment, or disposal.

The P2 by Design project has provided an excellent opportunity for collaboration among facilities, combining pollution prevention expertise to assist operations offices throughout the U.S. Department of Energy (DOE) Complex since fiscal year 1993.

There have been nine barriers to implementation of pollution prevention in design. They are:

Pollution prevention is a separate program rather than a routine part of the design process.
Widespread application of pollution prevention requires a paradigm shift for designers and managers otherwise accustomed to pollution control.
Lack of definitive pollution prevention criteria.
Pollution prevention requests tend to conflict with budget/schedule requirements.
Perception that pollution prevention and environmentally sound practices are more expensive and less efficient.
Project managers and designers need incentives to make pollution prevention a routine part of design.
A large majority of design work is non-project work, which falls out of the scope of many DOE requirements.
Resource/energy management, safety and industrial health and pollution prevention are each viewed as separate programs.
Engineers often do not receive feedback on performance of equipment, specified materials and processes, and actual facility operating and maintenance costs.

The pollution prevention conceptual design reports (CDR) should be sufficiently detailed to ensure that:

1. the concepts are feasible
2. risks are identified and addressed, and
3. life project cost estimates can be prepared.

Areas to address in the CDR are:

Anticipated waste streams during construction, operation and decommissioning
Management methods to prevent or minimize anticipated waste streams
Methods to eliminate and/or minimize use of hazardous materials
Methods to eliminate and/or minimize energy and resource-intensive processes
Decontamination and disposal requirements
Methods to conserve resources.

1.12 Pollution Prevention in Process Development and Design

Incorporating pollution prevention into process development and design means cost effectiveness. Early decisions in the development process determine later development activities, such as laboratory and pilot plant studies, equipment and materials selection and project economic analysis. Unforseen technical, regulatory and economic consequences of design choices can be anticipated. Thus the technical and economic risk associated with environmental risks is reduced. There can also be quicker time-to-market, process innovation, improved quality of products and increased efficiency when there is early consideration of environmental design.

Some questions to be raised in the process development cycle include:

Exploration of the toxicity of the product
What is the basis for product purity specifications? (Useful for design of separation and recycle systems).
What related products are anticipated? (Can cleaning related wastes be minimized).
How will the product be packaged?

Process Development should include the following:

Bench scale testing to validate process chemistry
Conceptual design to determine economic feasibility
Pilot-scale testing to determine engineering issues for process scale-P

Preliminary engineering: specifications for detailed (pre-construction and construction-phase) engineering

Pollution prevention data from bench-scale testing should include corrosion rates of candidate construction materials, screen for catalytic effects of candidate materials, corrosion products, feed impurities, etc. It is also essential to obtain vapor pressure data for products and intermediates, vapor-liquid equilibrium data for potential entrainers, diluents, and trace compounds. Also needed are loading capacity and regenerative properties of adsorbents. The reactor will need data for reaction stoichiometry, equilibrium yield, catalyst activity, activity and lifetime, the identity and characterization of reaction byproducts, the kinetics of major side-reactions, and the effects of recycle.

The effect of reactor mixing and feed distribution on byproduct formation, fouling rates in heat exchange equipment, corrosion studies, and sedimentation rates and product stability all will be necessary.

It is important to allow easy access to storage tanks, reactors, etc. for cleaning. Tanks and vessels should drain completely. Piping design should allow recovery of waste streams separately, should have minimal lengths of piping runs, minimal valves and flanges, drains, vents and relief lines should go to recovery or treatment and valves should be bellow-seal or zero-emission.

In-line process analyzers are to be used. Closed-loop (purge style) sampling ports, preventative maintenance monitoring equipment, real-time monitoring of foulage, leaking of heat exchangers and mode-based control are all emphasized.

Foul-resistant materials (e.g., Teflon) on heat exchanger surfaces need frequent cleaning and so do glass or polymer lined vessels.

Use "hidden waste costs" in cost equations and "penalty functions" for releases based on environmental objectives.

Some of the process design heuristics for pollution prevention are:

Seek to minimize the number of process steps
Minimize potential for leaks
Maximize process selectivity at each unit operation
Minimize process utility requirements
Segregate process streams where possible
Design for operability

In regard to wastes, one must be vigilant for:

wastes related to production and extraction (i.e., mining) of raw materials and intermediate products
emissions resulting from production of energy for the process (including off-site generation)
wastes resulting from packaging, storage, and transportation of raw materials and products
wastes from decommissioning of process facilities
cleaning, maintenance, start-up and shut-down wastes
non-point source emissions including contamination of storm water, trash, and soils in processing areas
secondary wastes generated during waste treatment operations (ash, sludges, biosolids, spent adsorbents, etc.)
direct release of product to the environment during use

Appropriate screening vectors might include:

Persistent or bioaccumulative toxins (heavy metals, dioxins, etc.)
Acute toxins or other materials requiring special handling
Specifically-regulated materials (e.g., Toxics Release inventory chemicals)
Greenhouse gases
Ozone-depleting chemicals
Materials specifically identified in existing or anticipated permits

1.13 Pollution Prevention

Pollution Prevention (P2) opportunity can be identified by the industry sector, product, or production process that is related to your business. Pollution Prevention can be accomplished by any of the five following methods.

1. Product Design/Material Selection — Product design is a process of synthesis in which product attributes such as cost, performance, manufacturability, safety, and consumer appeal are considered together. Product design for P2 incorporates environmental objectives with minimum loss to the product's performance, useful life, or functionality.
2. Process Design — Process designers consider production attributes such as cost, productivity, end-part manufacturability, and operator safety when designing a production process. Process design P2 incorporates environmental objectives with minimum loss to the production process, stability and productivity in particular.
3. Process Improvement/Material Substitution — Production Process improvements for P2 are considered after the process equipment is already in place. Although this continuous method of improvement yields varying degrees of suc-

cess, it is a viable option to incorporate pollution prevention into an already existing process.

4. Energy Conservation—Energy conservation minimizes power plant emissions through the efficient use of energy in the production process by minimizing pollution offsite at the power plant.
5. Environmental Management Systems—Environmental management systems identify pollution prevention opportunities in order to implement one of the above P2 methods.

1.14 Pollution Prevention Research Strategy

The four objectives of this strategy are to: deliver broadly applicable tools and methodologies for P2 and sustainability; develop and transfer P2 technology approaches; verify selected P2 technologies; and conduct research to address economic, social, and behavioral research and P2.

The next advances will represent more fundamental changes in individual lifestyle, industrial process design, consumer products, and land use so that future research must focus on quantum leaps instead of incremental improvements. They require a commitment by the public and private sectors to support long-term research that can, if carefully planned, produce the needed technology and tools that take pollution prevention to the next level.

Some of the goals and program emphases will:

develop process simulation tools.
support fundamental engineering research in addressing green chemistry
develop and test improved synthesis pathways
continually develop process feedback techniques for pollution prevention.
develop intelligent controls for process operations.
understand organizational decision related to human health and environmental protection.

Also, ORD (Office of Research and Development) will use electronic technology (e.g., Internet home pages, distance learning) to the maximum extent possible as a means of engagement with stakeholders. The research products developed by ORD will be designed to be available electronically, and ORD intends to be a major provider of pollution prevention research and development products via the Internet.

ORD will only be able to contribute meaningfully to the future direction if it concentrates on longer term research which will produce a new generation of technologies to move pollution prevention beyond the "low hanging fruit". This can be achieved with a commitment from the public and private sectors.

Protecting human health and the environment means we must look beyond the TRI-listed chemicals. A number of industries already see the need for a more holistic approach with the design for environment (DfE) and industrial ecology, product stewardship (DfE, life cycle assessment) and clean technology.

The SAB chose the high priority human health and environmental risks to be

High-priority human health risks
Ambient air pollutants
Worker exposure to chemicals in industry and agriculture
Indoor air pollution
Pollutants in drinking water
High-priority risks to natural ecology and human welfare
Habitat alteration and destruction
Species extinction and overall loss of biological diversity
Stratospheric ozone depletion
Global climate change

Pollution prevention approaches are needed for targeted industries in the industrial sector. (In most cases, these were aligned with specific regulatory programs or agency initiatives.)

A sector-based approach was used to organize and evaluate recent research and development activities already occurring for pollution prevention. An economic sector can be defined as a grouping of enterprises that produce similar goods and services. The sectors identified by SAB were: industrial, agricultural, consumer, energy, and transportation.

The criteria for choosing topical areas addressing high-risk human health or environmental problems did not exclude a problem based solely on the lack of available data indicating high risk. The major funders of pollution prevention research for the manufacturing sector are DOE, DOD, DOC-NIST, The National Science Foundation (NSF), the EPA, and DOA.

ORD's goals are

I. ORD will deliver broadly applicable tools and methodologies for pollution prevention and sustainability.
II. ORD will develop and transfer pollution prevention technologies and approaches.
III. ORD will verify selected pollution prevention technologies.
IV. ORD will conduct research to address economic, social, and behavioral research for pollution prevention.

The goal of the research effort is to improve existing design practices by developing more environmentally benign chemical syntheses and safer commercial substances. This encompasses all types and aspects of chemical processes (e.g., synthesis, catalysis, analysis, monitoring, separations and reaction conditions). Emphasis will be on (1) an extramurally focused program on green chemistry and (2) an in-house program on improved oxidation pathways.

Both continuous and discrete engineering approaches are being pursued to prevent and reduce pollution including equipment and technology modifications, reformulation or redesign of products, substitution of alternative materials, and in-process changes. ORD will support the potential areas of improved reactor, catalyst, or process designs in order to reduce unwanted products.

Engineering will rely on technologies that allow at or near "zero releases" of waste. Recycling will be very important, preventing or minimizing releases of toxic metals and organics. Separation technologies, such as adsorption, membranes, filtration, distillation, and combinations of these will be used.

ORD will design approaches for predicting the performance of intelligent controls (IC) in pollution prevention applications. Such approaches as fuzzy logic, neural networks, and genetic algorithms will play a part.

Pollution Prevention in Design: Site-Level Implementation Strategy for DOE

Costs associated with pollution control and environmental compliance are continuing to rise as environmental regulations become more stringent, requiring DOE to develop or buy technologies to meet ever-stricter emissions and disposal standards. Pollution prevention has become not only an environmentally friendly policy, but also a prudent business practice. It can and should be used as a powerful tool for cutting the rising costs associated with waste management while it helps to create superior products, services, and facilities.

The most effective time to incorporate pollution prevention strategies is up front, in the design phase of products, services, and facilities because this is where and when the processes generate or reduce waste are defined and the materials that create or eliminate waste are chosen. Up to 70% of a product's or a facility's life cycle costs are fixed during the design phase. By changing the traditional practice of waiting until the operational phase to examine pollution prevention opportunities, large efficiency gains can be achieved by incorporating pollution prevention up front during design where they can be planned into the full life cycle of the project, including construction, operation and decommissioning. It is easier to change a drawing than to retrofit an entire facility. Also identifying pollution prevention opportunities during design can reduce or eliminate environmental compliance and liability concerns before they are even created.

Moving pollution prevention into the design phase supports a desired paradigm shift from pollution control to prevention. The end result of such a shift in thinking will be designs that result in less emissions to the environment, less waste being shipped to hazardous waste landfills, less burial of radioactive and/or mixed waste, fewer compliance obligations for DOE, reduced liability concerns, improved public perception, improved worker safety, and ultimately, less cost to DOE.

1.15 Pollution Prevention Through Innovative Technologies and Process Design at UCLA's Center for Clean Technology

The redesign of products and processes to prevent waste is becoming more attractive than the retrofitting and disposal strategies needed to handle waste, if only based on cost alone. The CCT pollution prevention program reduces the generation of waste and their educational program focuses on developing innovative technologies and understanding the flow of materials.

Process flowsheet analysis aims to identify process configurations that minimize waste. Reaction engineering research enables the prediction of trace level pollutant formation, which will aid in the design of clean synthesis and safe reactor technologies. Investigators strive for advances in separation technologies which will allow by-products to be effectively concentrated and recycled.

The flow of materials, from the acquisition of raw materials to the disposal of products and wastes is essential for pollution prevention. This can help to identify whether wastes in one industrial sector can be viewed as raw materials in another. Detailed study of material flows can also reveal the types of processes and products responsible for toxic waste generation.

There is a complex interdependency of many products and processes. Two related approaches are employed to further study the complex systems used to convert raw materials to products. One is known as Industrial Ecology. It examines how wastes can be converted into raw materials. Another approach, called Life Cycle Assessment, starts with a particular product and identifies the precursors required for its manufacture and use, and then examines the impacts of its ultimate disposal.

Process reaction engineering has always stressed the search for pathways between raw materials or reactants and high yields of desirable end products. Now, the new process reaction engineering required for source reduction must employ design methods capable of minimizing production of trace byproducts. Instead of following the concentration of components present at percent-level concentrations, we must follow species in the low ppm concentration levels. Thus, two levels of fundamental research in the analysis of reaction pathways are currently being examined.

Detailed Chemical Kinetic Models (DCKMs) involving thousands of elementary reactions and fluid mechanical modeling are being developed to describe industrial processes such as combustion, petroleum cracking, and chemical vapor deposition. Generic software tools for these kinetic models are being developed to identify process conditions that minimize either the total amount of waste generated or the formation of particular species, such as the formation of air toxics in combustion systems.

At another level, both ab initio and semi-empirical quantum mechanical methods are being used to estimate the thermochemistry and rate parameters associated with the formation of key pollutants. Combined, both molecular and atomic modeling approaches provide powerful new tools for the selection and control of chemical pathways.

Overview of the Combustion Air Toxics Program

Combustion of fossil fuels is both the major source of energy production today as well as the principal source of air pollution. It is imperative that combustion devices be designed not only to operate at peak thermal efficiencies, but also to result in the emission of lowest possible levels of toxic by-products.

The development of "clean" combustion devices requires a better understanding of the fundamental chemical and physical processes that are responsible for the formation of toxic by-products and how these by-products are related to fuel structure, operating conditions, and device design.

The experimental and computational program is for the chemical kinetic processes responsible for formation and destruction of trace toxic combustion by-products such as PAHs, in hydrocarbon flames.

The parallel theoretical work involves the Detailed Chemical Kinetic Models (DCKM) to account for experimental measurements and to predict the formation of toxic formation by-products over a range of conditions from first principles. DCKMs are then coupled with transport models to simulate the behavior of laboratory flames for mechanism validation and to predict the fate of PAHs as a function of fuel structure and operating conditions. Ultimately these mechanisms will be used to predict the emission behavior of large-scale combustors for the optimal design and control of these devices.

Formation of Aerosols in Combustion

This program is to describe the formation and dynamics of aerosols, including those containing metals, in combustion processes and the generalized models for predicting the scavenging rate of submicron aerosols by larger ones. New measurements are underway to establish the size of distributions of metal-containing aerosols formed in flames and to assess the extent of partition toxic species on different aerosol particle sizes formed in combustion, as control technologies needed to manage aerosol emissions are highly dependent on particle sizes.

Interaction of Fluid Dynamics and Chemical Kinetics

This is directed towards the interaction of chemistry and fluid mechanics in influencing the emissions from combustion devices. The effect of hydrodynamic strain on mixing, flame ignition, extinction, and burning rate are investigated. Strain rates large enough to cause flame intermittency are typically found in highly turbulent flames but can also be introduced by aerodynamic and acoustic phenomena; this can lead to incomplete burning in combustors and incinerators. Consequently, research is underway to induce high strain rates to enhance mixing and burning rates while controlling the levels of emissions of toxic hydrocarbons and NOx at a minimum level.

Minimizing the Wastes from the Manufacture of Vinyl Chloride

Eleven billion pounds of vinyl chloride monomer (VCM) are produced annually in the United States by the thermal cracking of 1,2-C2H4Cl2 (ethylene dichloride to EDC). The yields are better at atmospheric pressure, the decomposition of EDC to VCM commercially is at pressures of 10 to 30 atm and at temperatures of 500 to 600 C to reduce equipment size, to improve heat transfer and to separate better HCL from the product VCM. To maintain high VCM selectivities, EDC conversions are generally kept at 50 to 70%. Higher temperatures and longer reaction times lead to the production of undesirable by-products that include C2H2, CH3Cl, CCL4, CHCl3, C4H4, and C6H6 as well as higher molecular products and tars. These light and heavy by-products, some of which are potentially toxic, must be further processed, incinerated, or disposed of by other means. This extends the use of detailed studies of elementary reactions to VCM production. Further, through determining by semi-empirical quantum mechanical calculations the rate parameters for these reactions, the study tries to demonstrate the link between DCKM and the atomic modeling of rate parameters.

Mass Exchange Networks for Pollution Prevention

The general Mass Exchange Network (MEN) synthesis problem is stated as: Given a set of pollutant rich process streams and a set of pollutant lean streams, synthesize a network of mass exchange units that can transfer polluting species from the rich streams to the lean streams at minimum cost. The goal of the synthesis is to identify the set of exchangers and the configuration of streams that optimize the transfer. An optimal network of mass exchangers may achieve the desired separation at minimum capital cost, minimum oparing cost, or some combination of the two. A key feature of this approach is that it combines thermodynamic and driving force constraints into the optimization. A recently developed linear programming formulation of the variable target MEN problems allows the computation of the minimum utility cost for large scale problems. A novel perturbation technique is being employed to establish structural properties of the optimal solutions of the nonisothermal MEN synthesis problems.

1.16 Assessment of Chemical Processes with Regard to Environmental, Health, and Safety Aspects in Early Design Phases

In the early phases of designing a new fine chemical process (reaction path synthesis, conceptual flowsheet synthesis) many important decisions must be made. Selection of the optimal chemical synthesis pathway to the desired product fixes a large amount of costs and development time of the process. Thus, such decisions should include all available information, such as chemical and economic knowledge (yield, selectivity, raw material prices) as well as technological and environment, health, and safety (EHS) aspects. Any EHS problem that is not identified and considered at early phases can lead to wrong decisions and large problems during the later design process.

Here a tool is proposed which allows fast evaluation of a chemical reaction or a basic flowsheet of a chemical process in order to identify major problems in the fields of environment, health, and safety. The identified problems are quantified according to its relevance to the design process, i.e., according to the effort to handle an EHS problem. The results can be displayed both to user (chemist or chemical engineer) who has only general knowledge in these fields and to an expert in EHS-questions. The decision process to find the optimal chemical reaction alternative can be supported. The tool is applied to a six-stage batch process to 8a-Amino-2,6-Dimethylergolin, an intermediate of the pharmaceutical industry (Novartis). The results of the assessment of this process are compared to a detailed risk analysis done by industry.

The tool consists of:

- Database of EHS-data: Uing an interface to public databases (e.g., ECDIN, CHRIS), all relevant information is automatically collected and checked
- Interface to batch process simulation software: Material balance for process assessment are obtained from the Batch Plus or Batch Design Kit
- New assignment method: First all EHS-information of the process is collected in 11 effect categories (e.g., fire, acute toxicity, aquatic toxicity). Then the information is aggregated according to clearly stated principles (e.g., scientific, expert fixed) and finally yields in indices which represent the size of the EHS-problem (how cheap and fast can a problem be managed during process development). As the assessment is fully automatized, results of changes in input data or assessment method can be checked easily.
- Method library: Besides the new assessment methods, other methods from the literature can be chosen for process assessment for comparison (e.g., Waste Reduction, Inherent Environmental Hazard)
- User interface: The relevant EHS-information of processes can be displayed in a graphical way depending on the degree of detail chosen by the user. Simple overall indices of reaction route alternatives can be displayed as well as in depth studies of the thermal risk of a certain stage or details of the aquatic toxicity of a certain substance.

A tool which delivers such information could be a valuable method to support the process of comparing alternative reaction pathways. It also improves the transparency of considering EHS-aspects during the decision process. Since decisions can be reproduced and documented easily, communicating the design decision based on EHS-effects should be easier. Therefore, this tool could be a useful completion to existing chemical, technological, and economic methods for process design.

1.17 Small Plants, Pollution and Poverty: New Evidence from Brazil and Mexico

New data from Mexico and Brazil analyze relationships to linking economic development, the size distribution of manufacturing plants, and exposure to industrial pollution. This study addresses air pollution and small plants (with 1 to 20 employees) as well as medium and large plants. The study shows that small plants are more pollution-intensive than large facilities. It also shows that small plants dominate poor regions and are a relatively small source of employment in high-income areas. Industry is also more pollution intensive in low-income regions, at least for Brazil. A standard of six dirty industries document a decline of 40% in the "dirty sector" share of industrial activity form the poorest to the richest municipalities. Yet, poor areas do not suffer more from industrial pollution. The risk of mortality from industrial air pollution is much higher in the top two income deciles among Brazil's municipalities. The great majority of projected deaths in the high-income areas are attributable to emissions from large plants. The scale of large-plant emissions dominates all other factors. So lower-income areas suffer much less from industrial air pollution in Brazil, despite a higher dirty-sector share and greater prevalence of emissions-intensive small plants.

1.18 When Pollution Meets the Bottom Line

If a manufacturer learned that there were untapped opportunities to reduce waste and emissions within a plant that would also significantly cut costs, one would think that the company would seize on such opportunities and implement them.

Experience dictates, however, that these opportunities are not always taken. These findings were found in a collaborative study by the National Resources Defense Council (NRDC), an environmental advocacy group, Dow Chemical, Monsanto, Amoco, and Rayonier Paper.The study participants were all interested in pollution prevention in a real life setting and they wanted to know the reason for the lack of widespread reliance on promising pollution prevention techniques.

The project found that once pollution prevention opportunities were found, corporate business priorities and decision making structures posed formidable barriers to implementing those opportunities. Most environmental professionals outside of industry incorrectly assume that a pollution prevention plan that actually saves money and is good for the environment will be quickly seized upon by U.S. business. This work shows that such opportunities may not be sufficiently compelling as a business matter to ensure their voluntary implementation.

1.19 Pollution Prevention as Corporate Entrepreneurship

Since pollution prevention can produce significant and quantifiable corporate gains, one wonders why it would not be most likely developed and implemented as such activities benefit the strategic and financial position of the corporation. The research challenge is to understand why pollution prevention has not received greater attention and action by corporations.

First, corporations do not widely view pollution prevention as its benefits are rarely recognized. Corporations also do not identify the factors that prevent or allow the marshalling of resources to exploit this potential opportunity. However, the data from corporate environmental reports are an imperfect mirror of corporate pollution activities and these conclusions must be interpreted with some care.

Further research is needed to better understand why pollution prevention is not recognized as an opportunity. Perhaps it is due to the fact that pollution prevention is an activity whose gains are produced by generating less of something, in this case pollution, and this contrasts with the norm of growth and greater production. In addition, tools and methods to measure the strategic benefits of pollution prevention, such as environmental accounting, are greatly lacking. Some research also indicates that there is a managerial bias towards threat avoidance. This by itself means that opportunities such as pollution prevention might not receive as much recognition as their potential benefits would imply. One implication for managers wishing to implement pollution prevention is that pollution prevention's potential as a threat avoidance tool, such as decreasing Superfund disposal liability, may prove more powerful than its opportunity characteristics.

The lack of opportunity recognition may result in corporations failing to link pollution prevention and strategic management. A model has been constructed that relates corporate entrepreneurship and strategic management. One is external environment, including competitive, technological, social, and political factors. A second is strategic leaders, values/beliefs, and behavior. A third is organization conduct/form involving strategy, structure, process, and core values/beliefs. A final influence is organization performance, including effectiveness, efficiency, and stakeholder satisfaction. The above analysis of the corporate reports indicates that these variables rarely exist in a form which promotes pollution prevention entrepreneurship. Without a tie to strategic management, pollution prevention will remain at best an add-on activity.

One of the greatest potential gains from pollution prevention recognizes its potential as not only individual ventures but as a form of corporate self-renewal. There pollution prevention can be a corporate change agent. Implementing pollution prevention requires the reconceptualization of the whole corporate approach to materials, energy, and water use as well as to the central manufacturing processes. By focusing on what enters and moves through the corporate enterprise, managers improve the efficiency and outcomes of these processes. By drawing on the lessons of entrepreneurship, the corporation can move towards realizing those potential significant gains.

1.20 Plantwide Controllability and Flowsheet Structure of Complex Continuous Process Plants

A. J. Groenendijk, a Ph.D. student at the University of Amsterdam, states that the material balance for main components and impurities in a complex plant is related to the plantwide properties of the components inventory. His study is of the recycle loops in the complex continuous plant. Steady-state and dynamic simulation with controllability tools characterize interactions between recycles to evaluate plantwide control properties of different flowsheet alternatives. He intends to do further work developing an algorithm for the design of an optimal flowsheet and control structure where connectivity is a degree of freedom. For this mixed integer non-linear programming problem optimization techniques will be used.

3D Design, 3D Chemical Plant Design

GB MM has produced 3D Chemical Plant Visualization. Their 3D graphics have the ability to vary the transparency of plant equipment to look inside (variable transparency), view an object through 360 degrees (multiple viewpoints), analyze a pump working from the outside to the inner workings (layer by layer), etc.

1.21 Development of COMPAS

A computer-aided flowsheet design and analysis system, COMPAS has been developed in order to carry out the flowsheet calculation on the process flow diagram of nuclear fuel reprocessing. All of the equipment: dissolver, mixer-settler, etc. in the process flowsheet diagram are graphically visualized as icon on a bitmap display of a UNIX workstation. The flowsheet drawing was easily carried out by mouse operation. Two examples show that the COMPAS is applicable for deciding operating conditions of the Purex process and for analyzing the extraction behavior in a mixer-settler extractor.

1.22 Computer-Aided Design of Clean Processes

There are a very large number of possible alternative processes for converting a defined raw material to a desired product. There is vigorous research in the development of computer-aids that can assist the design process. Some of them can now contribute effectively to assist the design process and contribute effectively to the generation of clean economic processes. The computer-aided synthesis can have optimization criteria (e.g., cost of effluent) and the optimization can be constrained to forbid effluent fluxes exceeding pre-defined levels. We are concerned with the conceptual design rather than the detailed mechanical design of individual components.

Design space is divided into four areas: Overall Design, Discrete Synthesis, Superstructure Optimization and Subsystem Synthesis.

Overall Design is the space of all possible designs. There cannot be a design method that can provide a mathematical guarantee of optimality. We cannot define all possible designs because we do not fully know the objectives or the constraints. In this way, one of the design objectives is to relax the constraints by discovering new catalysts, new reactants, new extractants, new pieces of equipment, and new products with the same or improved functionality. These discoveries are dominated by human innovation and have the greatest potential for generating radically improved designs. Some synthesis methods aim to stimulate innovation but this area is the most difficult to automate.

Discrete Synthesis combines currently known chemistry and operations in the most effective manner. Generating novel designs necessitates going beyond assembling a kit of standard operations. It must look at the chemical and physical processes that occur within the units to generate new operations that combine the known processes in cleaner, more economic packages. The number of combinations, even of the known unit operations that can be presented to a synthesis program is very large. It is greater than the bounds of any currently conceivable technique to guarantee generating an optimal combination, which also incorporates fully optimized operating conditions. The computer methods do take short cuts, while covering a very wide area, but give a reasonable assurance that the results will be much better than can be achieved by incremental evolution of previous plant designs. Methods applicable include implicit enumeration (Dynamics, Programming, Branch and Bound, etc.), psuedo-random methods (genetic algorithms, simulated annealing and mixed methods), Artificial Intelligence (AI), Fuzzy AI, and incremental AI.

The problem is discretized, i.e., a finite number of component flowrates, equipment sizes and operating conditions. Recursive formulation leads to a Dynamic Programming approach. Every initial decision decomposes the overall process synthesis into smaller problems each of which is similar to the original in having defined inputs and outputs. A finite number of possible streams, solutions to these immediate problems can be recorded to give an efficient overall optimization procedure. AI methods use heuristic (provable) rules to enable design decisions to be made sequentially without recursion. Proper choice of the hierarchical sequence for decisions minimizes coupling between successive decisions, thus minimizing the penalty of omitting a recursion. An alternative enhancement to AI is fuzzy AI. There the AI rules are ranked by degree of belief. The belief weightings are revised by matching against detailed evaluation of the process synthesized and the overall synthesis repeated until convergence of ranking is achieved. There is also an evolutionary application of AI in which an initially simple process is augmented until the cost of further refinement gives no further benefit.

Discrete Synthesis Designs can give rise to radical departures from current practice so that experimental work may be necessary to validate the designs.

Superstructure Optimization is applicable when the number of discrete alternatives is small. Then the alternatives in the superstructure can be optimized by rigorous methods such as Mixed-Integer Non-Linear Programming (MINLP). Floudas and Grossman make MINLP practical by tackling two of the major aspects the limit its application, binary variables and non-convexity. Binary variables cover the selection/non-selection of units and the connections of streams between them; non-convexity gives rise to multiple local optima and arises, for example, as a consequence of the characteristic "economy of scale" cost relationships.

Subsystem Synthesis is the optimal design of parts of a process where the major features of the process flowsheet are fixed. The fixed flowsheet limits the options, yet significant benefits can be achieved. For the heat exchanger networks, the pinch effect is used and results in energy savings. Computerized versions of the method are available, e.g., Advent (from Aspen Technology, Inc.) and Super target (from Linhoff March, Ltd). Alternative approaches using Mathematical Programming are also under development, e.g., Johns and Williams. Also, saving energy, on its own, has substantial environmental benefits (less CO2 released, etc.).

Douglas and Stephanopoulos describe how their AI structure can incorporate MINLP to give mutually beneficial performance. They provide a human interface to stimulate the introduction of human innovation into the AI synthesis. Similarly, it has been shown how implicit enumeration can allow an impossible step (i.e., not seen in the database) which, if it radically improves the overall process, can stimulate the designer to devise a practical implementation of the step. Innovations can be incorporated into any of the computer-based methods, to reduce manual effort and give a greater assurance that the innovations are being deployed most effectively. Efficient methods of predicting the optimal performance of subsystems can also improve the efficiency of either computer-based or manual methods of whole-process synthesis.

All the computer-aids reviewed can incorporate environmental constraints and criteria related to the computed release rates of potentially harmful materials. There are only limited ways of incorporating other environmentally important criteria into such quantitative design objectives and other methods are required to handle, for example, safety, start-up and shut-down, and controllability. However, a potential exists to automate a significant part of the design process within the area that is automated, there is a better guarantee of optimality. Significant human design time can be released to concentrate on the parts of the process that can only be tackled by human insight and ingenuity.

Progress is possible in design under uncertainty. Here many environmentally important factors are unknown (e.g., properties of trace components). Robust designs must operate reliably whatever the actual values of these uncertain parameters may be. In such cases, conventional design criteria have been modified on a statistical basis. Uncertain outcomes have been discretized through quadrature, though neither allows the optimization to synthesize radically different flowsheets.

Computer-aided procedures are in the development stage. Some, however, are routinely used. When a new process design, variant, or modification is conceived, its performance needs to be evaluated. Commercial tools for steady-state and dynamic process flowsheet simulation exist. Their use in verifying the performance of novel, potentially beneficial processes pays even greater benefit than in studying more traditional designs.

1.23 Computer-Aided Chemical Process Design for P2

Chemical engineers do not always have tools to facilitate the tasks of designing plants to generate as little pollution as possible. The object of research is to develop computer simulation tools for the design and development of less polluting chemical manufacturing processes. Such tools can be used to minimize the potential environmental impact of emissions by designing and building entirely new plants, modifying existing facilities or altering operating conditions. This is to be done while keeping capital and operating costs from increasing. Sophisticated computer aided design methodologies for chemical process simulators have progressed while simultaneously optimizing operating variables and cost. It was stated, however, by the Environmental Chemical Engineering Lab at Seoul National University, that no currently available computer simulation tool is designed for minimizing pollution impact while containing costs.

1.24 LIMN-The Flowsheet Processor

LIMN adds a set of flowsheeting tools to the Microsoft Excel 7 spreadsheet for Windows 95, greatly enhancing the value of spreadsheets to metallurgists and process engineers. The process flowsheets built into a spreadsheet has the ability to sketch rapidly report quality flowsheets using a "flowsheet aware" drawing package. It has an extensive process unit icon library, block diagram option, easy addition of user drawn custom icons. The data is stored internally within the spreadsheet. It has an WYSIWYG presentation.

1.25 Integrated Synthesis and Analysis of Chemical Process Designs Using Heuristics in the Context of Pollution Prevention

Process modification identification and comparison is not consistently practiced for pollution prevention, particularly during conceptual design. P2TCP (Pollution Prevention Tool for Continuous Processes) is a computer-based system developed to help designers identify pollution prevention opportunities. It can be used for continuous chemical processes as well as conceptual and retrofit design and can help in the development of cleaner processes. Case studies are used to validate P2TCP and to identify further extensions, not at the principal knowledge source. P2TCP is a novel design approach. It is not like hierarchical or step-wise design techniques; heuristics (knowledge based rules) are used to analyze each system of a chemical process (reaction and separation) independently for potential alternatives. Effects associated with the interacting streams, i.e., streams leaving the interacting streams and potential recycles, are then taken into consideration to further reduce the number of options requiring consideration. The effectiveness of this heuristic approach has been demonstrated in a number of pollution prevention case studies. Unlike hierarchical techniques, it is theoretically possible to consider all alternatives. Furthermore, the case studies demonstrate that the number of design alternatives requiring consideration is not prohibitive.

1.26 Model-Based Environmental Sensitivity Analysis for Designing a Clean Process Plant

Process integration is now more employed for the reduction of energy and material costs, and more recently, the minimization of waste. Process streams are heavily interacted among units in an integrated process plant. This has led to various operational problems due to newly introduced severe disturbance propagation in the system. Industrial practice has also shown that if the disturbance variables are environmentally sensitive, then the operational problems will become environmental problems. These environmental problems usually cannot be resolved through control design. Needless to say, they should be prevented during process design. This renders the design to be not only cost-effective, but also environmentally benign in terms of effectively rejecting environmentally sensitive disturbance propagation through a process system. We propose a mode-based system sensitivity analysis approach for developing an environmentally benign process. One of the fundamental elements of the approach is the waste generation/propagation models for different types of process operations, such as reaction, distillation, extraction, adsorption, and heat exchange operations. Those models characterize system responses to environmentally sensitive disturbances and fluctuations. The models are first principles based, but simplified in linear form. Thus they can be easily embedded into a process design procedure. In addition, a model-based waste minimization index for evaluating the cleanness of a process design was introduced. Thus, every design decision will be evaluated not only by cost constraints, but also by waste reduction requirements. The approach provides a useful tool for process designers to perform effective sensitivity analysis of process alternatives, to identify any inappropriate connections among process units, to quantify the severity of environmental impact of the design. This will greatly facilitate designers to derive improved process systems. The applicability of the approach to solve real-world problems is demonstrated by analyzing an industrial process where reactors, distillations, extraction, and heat integration occur. The resultant process will reduce waste by 10% by restructuring the topology of the system, and by introducing appropriate recycling.

1.27 Achievement of Emission Limits Using Physical Insights and Mathematical Modeling

Gaseous emissions are produced in all chemical processes from the generation of utilities. There are many ways to minimize gaseous emissions. They are changes to heat recovery in process, changes in the configuration of the existing steam turbine network, process changes, use of different fuel used in the boilers/furnaces, integration of gas turbine, installation of low NOx burners, and EOP techniques. Here we show a method to integrate the different flue gas minimization techniques for an existing site. The objective of this problem is to minimize the capital investment to achieve the required emission limits. The approach to the problem is divided into three steps: setting targets, screening of options, and optimization. The first step utilizes the difference between the limits and current operation of turbines and boilers. Depending on the efficiency of the equipment, savings in utilities are converted into a targeting curve using different paths. The screening of options is done in two parts. The first part consists of generation of different options: heat recovery, fuel switch, gas turbine integration, and change in the process. The second part eliminates the uneconomic options using the targeting curve generated in the first step. The options left after screening the existing system are then formulated in a maximal superstructure of a MILP. The structure is then subjected to optimization. This gives the maximum capital required to achieve the emission limits. The proposed hierarchical method is a simultaneous approach to the problem. Physical insights are used higher in the hierarchy to understand the problem. This generates a smaller size of the superstructure. The understanding of the problem also gives bounds on different variables, which reduces the solution space. Thus any problem can be easily solved using existing optimization techniques. Moreover it has been shown by several case studies that it is possible to satisfy emission limits and make annual savings at the same time.

1.28 Fritjof Capra's Foreword to *Upsizing*

Upsizing is the latest book by Gunther Pauli, and Fritjof Capra has written a foreword to it. It states that we must redesign our businesses and industries so that the waste of one industry will be a resource for the next. These industries need to be clustered geographically so that they are embedded in an "ecology of organizations" in which the waste of any one organization would be the resource of another. In such a sustainable industrial system, the total outflow of each organization — its products and wastes — would be perceived and treated as resources cycling through the system.

Ecological clusters of industries exist in several parts of the world under auspices of ZERI — the Zero Emissions Research Initiative. Recently new mathematics has helped with the complexity of living systems and the understanding of the basic characteristics of life. The simplest living bacterial cell is a highly intricate network involving literally thousands of interdependent chemical reactions. A characteristic of the new mathematics is that it is non-linear. Now powerful computers have helped us understand that the surprising patterns underneath the seemingly chaotic behavior of non-linear systems. They show an underlying order beneath the seeming chaos. Chaos theory is really a theory of order, but a new kind of order that is revealed by the mathematics of complexity. The emerging science of complexity has brought many new insights into the patterns and processes of organization of living systems, which are crucial to understanding the principles of ecology and to building sustainable human communities. The emerging theory of living systems, including the new science of complexity, is relevant to three distinct but gradually coalescing areas of concern: (1) the endeavor of creating and nurturing the sustainable human communities; (2) the tasks of understanding our present technological complexities and of redesigning our technologies so as to make them ecologically sustainable; and (3) the challenge of carrying out the profound organizational changes required by the first two areas of concern.

1.29 ZERI Theory

Zero emissions represents a shift in our concept of industry away from that in which wastes are considered the norm, to integrated systems in which it has its use. Zero emissions envisages all industrial inputs being used in the final product converted into value-added inputs for other industries or processes.

Here industries will reorganize into "clusters" such that each industry's wastes/products are fully matched with others' input requirements, and the total set of factories produces no waste of any kind. This will amount to a standard of efficiency with Total Quality Management (zero defects). The methodology can be applied to any industry and can be summarized as follows:

1. Total throughput.
2. Output-Input Models: An inventory of all "wastes" not consumed in the final product or its process of manufacture. Then an identification of industries that can use the outputs or modifications of them.
3. Identification of potential candidates for clustering, optimized as to size and number of participating industries.
4. Where economic coupling is difficult, and research into proper system design.
5. Design of appropriate government policies.
6. Additional information channel for ZERI design with global dialogues.

1.30 SRI's Novel Chemical Reactor — PERMIX

This reactor improves the ratio of desired product to waste product by a factor of 20 over comparable technology. In addition to reducing or eliminating waste disposal and increasing product quality, the technology also increases yields and thus decreases the cost of raw materials. This is an exothermic process with porous means to control reaction rate and exothermic heat (U.S. Patent No. 5,583,240). The new reactor uses the progressive addition of one reactant permeating all along the reactor and mixing in the entire volume of the reactor to minimize or eliminate local high concentration gradients and hot spots as well as to control the ratio of reactants as the reaction proceeds. The mixing elements, catalyzed or inert, are a key to the improved performance of the new reactor. In the liquid phase where the flows are laminar, particulate mixing elements change the mass transport from molecular diffusion to convective diffusion, increasing it by a factor of 100,000. In the gas phase, the transport is increased by a factor of 100 and the flow is highly turbulent, which reduces the scale of mixing by turbulent eddies to a much smaller scale than the mixing element dimensions. These order-of-magnitude improvements in transport can be used to control the ratios of reactants and products and therefore decrease waste products and increase yields. The most economic method for heat of reaction removal is adiabatic reactor operation for incremental conversion followed by heat removal in a conventional heat exchanger.

1.31 Process Simulation Widens the Appeal of Batch Chromatography

Batch chromatography and SMB (Simulated Moving Bed) chromatography has selective value for purifying or recovering certain high-value biomolecules, and in processing fine chemicals and foodstuffs. Developing optimal processing schemes, however, tends to be timely and expensive because elaborate testing is necessary. Process simulation technology is now available to significantly expedite the development of new applications, or the optimization of existing ones. It consists of injection of the feed mixture to be separated into a packed column of adsorbent particles, through which there is a continuous flow of a mobile phase. Chromatographic separation involves a lower use of energy than other separation techniques, such as distillation. Furthermore, liquid chromatography is often performed at room temperature, thus preventing loss of activity of heat-sensitive components such as occurs in some industries.

At DuPont, in the late Fifties and early Sixties, I conceived of using the Wenograd apparatus, (a condenser discharging into a hypodermic needle, containing a test chemical mixture, that discharged its contents into a chromatograph). Dr. Wenograd used it to test hazardous materials such as explosives, thus to replace such tests as "the Drop Test." I thought it might help test for hazardous pollution by-products, but it was never actuated.

1.32 About Pollution Prevention

P2 opportunities can be identified by the industry sector, product, or production process that is related to your business. It can be accomplished by any of the five methods below:

1. Product Design/Material Selection — Product design is a process of synthesis in which product attributes such as cost, performance, manufacturability, safety, and consumer appeal are considered together. Product design for P2 incorporates environmental objectives with minimum loss to the product's performance, useful life, or functionality.
2. Process Design — Process designers consider production attributes such as cost productivity, end-part manufacturability, and operator safety when designing a production process. Process design for P2 incorporates environmental objectives with minimum loss to the production process, stability, and productivity in particular.
3. Process Improvement/Material Substitution — Production process improvements for P2 are

considered after the process equipment is already in place. Although this continuous method of improvement yields varying degrees of success, it is a viable option to incorporate P2 into an already existing process.

4. Energy Conservation — Energy Conservation minimizes poor plant emissions through the efficient use of energy in the production process by minimizing pollution offsite at the power plant.
5. Environmental Management System — Environmental management systems identify P2 opportunities in order to implement one of the above P2 methods.

November 16, 1994, the United States Environmental Protection Agency released a report entitled "The Waste Minimization National Plan". This plan establishes three goals.

1. To reduce, as a nation, the presence of the most persistent, bioaccumulative, and toxic constituents by 25% by the year 2000 and by 50% by the year 2005.
2. To avoid transferring these constituents across environmental media.
3. To ensure that these constituents are reduced at their source whenever possible, or, when not possible, that they are recycled in an environmentally sound manner.

1.33 Federal Register/Vol. 62, No. 120/Monday, June 23, 1997/ Notices/33868

This plan presented a combination of five objectives of voluntary, regulatory, and institutional mechanisms to achieve these objectives. They were:

1. Develop a framework for setting national priorities; develop a flexible screening tool for identifying priorities at individual facilities; identify constituents of concern.
2. Promote multimedia environmental benefits and prevent cross-media transfers.
3. Demonstrate a strong preference for source reduction; shift attention to the nation's hazardous waste generators to reduce hazardous waste generation at its source.
4. Clearly define and track progress; promote accountability for EPA, states, and industry.
5. Involve citizens in waste minimization implementation decisions.

EPA promised to help in such ways as using the results from the prototype screening approach to set priorities for metals and proposing guidance to encourage the implementation of multimedia pollution prevention programs at all facilities. In addition EPA would implement several voluntary mechanisms, including:

1. Promoting focused technical assistance to small- and medium-sized generators of constituents of concern.
2. Developing outreach and communication mechanisms.
3. Providing guidance to states on incorporating waste minimization in hazardous waste management planning.

EPA would implement several mechanisms within the RCRA REGULATORY framework including:

1. Developing a program for working with generators to promote waste minimization.
2. Issuing revised guidance on the use of Supplemental Environmental Projects (SEPs).
3. Working with EPA Regions and states to provide waste minimization training for inspectors, permit writers, and enforcement officials.

There are also a number of Institutional Mechanisms that are not cited here. However, the report ends with the statement that EPA will publish guidance to regions, states, and industry, identifying when and how waste minimization information should be made available to the public during the permit process.

1.34 EPA Environmental Fact Sheet, EPA Releases RCRA Waste Minimization PBT Chemical List

States, industry environmental groups, and citizens advised EPA in 1994 that waste minimization should consist of the following:

Reduce as a nation the presence of the most persistent, bioaccumulative, and toxic chemicals in industrial hazardous wastes by 25 percent by the year 2000 and by 50 percent by the year 2005.

Avoid transferring these chemicals across environmental media.

Ensure that these chemicals are reduced at their source whenever possible, or, when not possible, that they are recycled in an environmentally sound manner.

To address these recommendations, EPA first developed the Waste Minimization Prioritization Tool, which scores thousands of chemicals based on their mass generated, persistence, bioaccumulation, and toxicity. EPA then identified the chemicals of greatest concern to the RCRA program on a national basis — those chemicals that are very persistent, bioaccumulative, and toxic; are generated in largest volumes or by many facilities; are present in soils and sediments; and are hard to manage, clean up, or pose other RCRA concerns. The proposed RCRA PBT List contains 53 chemicals that ranked highest for these factors from a national perspective. EPA recognizes that other PBT chemicals may be identified as priorities by regional, state, or local organizations or companies, and encourages coordinated efforts to address the reduction of those chemicals as well.

The Clinton Administration, the Environmental Defense Fund (EDF), and the Chemical Manufacturers Association (CMA) jointly announced a six-year program to test 2,800 compounds, major industrial chemicals, for their health and environmental effects.

The unprecedented, cooperative program covers U.S. high-production chemicals, each produced or imported in a volume of more than 1 million pounds per year. All tests are to be completed by the year 2004. Industry's estimated cost of the tested program is between $500 million and $700 million.

Under the announced program, chemical manufacturers will have 13 months to volunteer their products for testing, after which EPA will order tests for the chemicals that have not been volunteered. EDF will monitor testing process and provide free on-line information to the public via the Internet, on a chemical-by-chemical and company-by-company basis.

1.35 ATSDR

The Agency for Toxic Substances and Disease Registry is an agency of the U.S. Department of Health and Human Services. It has the mission to prevent exposure and adverse human health effects and diminished quality of life associated with exposure to hazardous substances from waste sites, unplanned releases, and other sources of pollution present in the environment. Directed by congressional mandate, it is to perform specific functions concerning the effect on public health of hazardous substances in the environment. These functions include public health assessments of waste sites, health consultations concerning specific hazardous substances, health surveillance and registries, response to emergency releases of hazardous substances, applied research in support of public health assessments, information development and dissemination, and education and training concerning hazardous substances.

1.36 OSHA Software/Advisors

Some software available from OSHA includes:

Hazard Awareness
Lead in Construction
Logging Technical
Safety Pays
Silica Technical
Respiratory Protection Technical
Asbestos 2.0
Confined Spaces 1.1
Online Confined Spaces 1.1
Fire Safety
GOCAD
Best I. T. Practices in the Federal Government
elaws

1.37 Environmental Monitoring for Public Access and Community Tracking

Introduction

The purpose is to solicit applications under the Environmental Monitoring for Public Access and Community Tracking (EMPACT) Grants Program sponsored by the US EPA.

EPA has a competition for grants in 1999. The goal of EMPACT is to assist communities to provide sustainable public access to environmental monitoring data and information that are clearly communicated, time relevant, useful, and accurate in the largest U.S. metropolitan areas. Environmental monitoring consists of the systematic measurement, evaluation, and communication of physical, chemical, and/or biological information intended to give insight into environmental conditions. EMPACT seeks to assist the American public in day-to-day decision-making about their health and the environment.

Pilot programs will be established in a limited number of eligible cities with grant awards. The pilot programs will emphasize using advanced and innovative technologies to monitor environmental conditions and provide and communicate environmental information to citizens. The pilots also require effective partnerships between local and state governments, research institutions non-governmental organizations, the private sector, and/or the Federal Government to provide timely environmental information to the public. It is essential that data and information derived from EMPACT monitoring activities be disseminated using terminology and format that are clearly understandable, relevant, and credible to the lay public.

1.38 Health: The Scorecard that Hit a Home Run

The EDF's Chemical Scorecard web site yields, instantly, information about the health effects of chemical emissions from 17,000 industrial facilities. The Scorecard was developed in consultation with grassroots groups who will use the information to monitor and improve their local environments.

When a user types his or her zip code, neighborhood maps appear on the screen, with schools and industrial facilities marked. Users see what chemicals are released and can find out which are the most toxic. Because the Scorecard puts a spotlight on toxic emissions, it encourages companies to expedite emissions reductions.

1.39 Screening and Testing for Endocrine Disrupters

Reproductive abnormalities in wildlife, increasing breast cancer in women, and decreasing sperm counts in men may have a common link. Many believe that pesticides and other chemicals that disrupt the endocrine system are the underlying thread. However, analytical methods for testing the endocrine disrupters are scarce. EDSTAC (the Endocrine Screening and Testing Advisory Committee) chartered to help EPA is underway. QSAR work is also being used to aid the process. There is a compilation of searchable, up-to-date inventory of research by the federal government that is also available.

1.40 Reducing Risk

An analog formulation, in its most simple form, is used to express toxicity for six mammals and eight chemical species. General control theory is discussed and the system transfer function is shown to be similar to the analogue toxicity equation. Also general kinetic equations of Lotka are of this nature. Electrical network equations can be solved for LC50/100 for man and animals in a more complex system by the network systems model of the environment. By analogy then, the system can be controlled by feedback control or any of a dozen methods to reduce the overall LC50/100 of the ecological population, at any site, by reducing the emissions of specific chemicals to a site whose ecological nature is known.

Introduction

While studying risk and toxicology of small mammals for the ecological impact on various Superfund sites, it was discovered that the use of simple analogue techniques led to an acceptable description of the data for several species. Encouraged by this, further general electrical (and mechanical) analogues revealed specific exponential and algebraic equations, with parameters that could be fitted to laboratory and field data and whose general form occurs in books and papers with actual experimental ecological field data and did fit the data quite well. When the analog method was joined with network theory [1], control theory [1-5], system theory [6], chemical kinetics [7], operational (e.g., LaPlace transform) methods [1-6], order-disorder techniques [10,11], system techniques used in medicine [10], and catastrophe theory [11], then powerful techniques for planning the control of pollution, as well as describing its impact and the exact nature of its character to man as well as animals became evident.

Mathematical Ecotoxicity

This section introduces the broad outline of the concepts of characterization of the mathematical character of ecotoxicity and its control with several descriptions of applications.

The simple electrical analog first used is

$$I(t) = (1 - esp - (R/L)t) \qquad (1)$$

where E is e. m. f., R is resistance, L is inductance and t is time. An analogue was used by setting R equivalent to k, L equivalent to m and E equivalent to p', where k is the toxicological reference concentration, m is the mass, and p' is the physiological factor for the organism in question. This equation is also found in mechanical or viscoelastic behavior [12]. Also, I(t) = LC50/100. Then,

$$LC50/100 = p'/k(1 - \exp - (kmt)) = p'/k(f(t)). \qquad (2)$$

In practice, when t is very large, the function of t, (1-exp(k/mt)), is equivalent to unity and an additive constant, C, is added so that

$$LC50/100 = p'/k + C. \qquad (3)$$

Then each animal has a different physiological factor p', each chemical has a different value of k, and when LC50/100 = y is plotted vs. 1/k = x, then the experimental data fall on a straight line with slope p' and intercept C (Figure 1). The eight chemicals used to plot Figure 1 were vinyl chloride, xylene, acrolein, formaldehyde, benzene, styrene, epichlorohydrin, and chlorobenzene. The data was extended to the horse with data from the University of Pennsylvania Veterinary School and found to fit logically on the simple graphs.

The more rigorous derivation leads to an expression with hyperbolic trigonometric functions as well as exponential functions and is non-linear, bending towards the abscissa at higher values of $1/k$. It is this expression that should be used for very toxic substances, such as "dioxin".

Control Theory

Then, describing control [3] as

Initial Conditions
input(u) → System → output(y)

and P(D) . y(t) = Q(D) . u(t), and calling

$$g(s) = Q(s)/P(s)$$

the system transfer function,

$$u(s) \rightarrow g(s) \rightarrow y(s) \quad (4)$$

where s is a complex variable, D = d/dt and P(D) and Q(D) are polynomials. It is then true that the parallel, series, and feedback configurations can be represented as equations 5, 6, and 7, respectively.

$$g(s) = g_1(s) + g_2(s) + \text{—} \; g_m(s) \quad (5)$$

$$g(s) = g_1(s)g_2(s) + \text{—} \; g_m(s) \quad (6)$$

$$g(s) = g_1(s)/(1 + g_1(s) + g_2(s)) \quad (7)$$

and Figures 2, 3, and 4, respectively.

The Transfer Function

The Laplace transform equation (4) is an algebraic function and can be expanded in a partial fraction expansion. The zeros of the P(s) are the poles of g(s) and may be real or complex conjugate pairs. They may be simple or multiple. The general form of the expansion [1,3,6] is

$$g(s) = Q(s)/P(s) = a_{0+}a_1s + as^2 + \text{—} + A_1/(s - s_1) + \text{—} + a_2/(s - s_2) + A_{p2}/(s - s_{p2})^2 + \text{—}A^{pr}/(s - s_{pr})^r. \quad (8)$$

The inverse transform is then

$$g(t) = a_0u_1(t) + a_1u_2(t) + a_2u_3(t) + \text{—} + A_1\exp(s_1t) + A_2\exp(s_2t) + \text{—} + A_{pr}/(r\text{-}1)!t^{r\text{-}1}\exp(spt)(-1)^{r\text{-}1} \quad (9)$$

where u is the unit step function. In this connection, it is noted in a text on ecotoxicity that a function such as

$$LC50/100 = a\exp(kt) + a\exp(kt) + \text{—} \quad (10)$$

fits the data for selected chemicals and animals very well.

In Bumble and Honig [8] and Bumble [9], various applications of order-disorder theory to various physical phenomena, a lattice approach and states of occupation of the vertices and neighbor itneractions were proposed. It was found that the ratio of probabilities of the unoccupied basic figure could be expressed as the ratio of polynomials, and if the lateral interactions were very small, as a series in one variable which would rapidly converge when the variable was very small.

$$g(s) = (a_0p_m + a_1p_{m-1} + \text{—} \; a_m)/(b_0p_n + b_1p_{n-1} + \text{—} \; b_n) \quad (11)$$

This leads to [8] and [9] in many cases. It also indicates that there may be a critical value for the transfer function of value in controlling pollution cases by analogy to order-disorder work. The Grand Canonical Ensemble of a space-time lattice in three dimensions in its various states of occupation can then be thought of as a model for the toxic components traveling into the biological system.

A Boolean, or an on-off idealization captures the main features of many continuous dynamic systems. Many cellular and biochemical processes exhibit a response which follows an S-shaped, or sigmoidal curve as a function of altered levels of some molecular input. Even a function whose maximum slope is less than vertical (e.g., coupled systems governed by such signoidal functions) are often properly idealized by on-off systems. Over iterations, feedback of signals through a sigmoidal function tends to sharpen to an all-or-nothing response. A biological rate equation dependent on the Hill function law can lead to

$$x = (k_1/k_{-1})(1-\exp(-(k_{-1})(t))) \quad (12)$$

where k_1 is the rate equation in the forward reaction and k_{-1} is that in the reverse directgion. This regulation endows biological systems with the possibility to choose between two or more well distinct states of regime. This, characterized by multiple steady states, is epigenetic. It is ensured by feedback loops.

Networks and Ecosystems

In [14], Lotka proposes general kinetic equations that lead to similar results (17) as (9) or (9a). If X is mass, and excess x_i of each mass X over corresponding equilibrium values is

$$x_i = X_i - C_i \quad (13)$$

$$dX_i/dt = F_i(X_1, X_2, \text{—} X_n, P, Q) \qquad (14)$$

where F is equal to growth of any component, dependent on the others and the parameters P and Q. Then

$$dxi/dt = f(x_1, x_2, \text{—} x_n), \qquad (15)$$

and using Taylor's theorem, we obtain

$$\begin{aligned} dx_1/dt &= a_{11}x_1 + a_{12}x_2 + \text{—}\ a_{1n}x_n + a_{111}x_1^2 + a_{112}x_1x_2 + a_{122}x_2^2 + \text{—} \\ dx_2/dt &= a_{21}x_1 + a_{22}x_2 + \text{—}\ a_{2n}x_n + a_{211}x_1^2 + a_{212}x_1x_2 + a_{222}x_2^2 + \text{—} \\ &\cdots \qquad \cdots \\ dx_n/dt &= a_{n1}x_1 + a_{n2}x_2 + \text{—}\ a_{nn}x_n + a_{n11}x_1^2 + a_{n12}x_1x_2 + a_{n22}x_2^2 + \text{—} \end{aligned} \qquad (16)$$

A general solution is

$$\begin{aligned} x_1 &= G_{11}\exp(m_1t) + G_{12}\exp(m_2t) + \text{—}\ G_{1n}\exp(m_nt) + G_{111}\exp(2m_1t) + \text{—} \\ x_2 &= G_{21}\exp(m_1t) + G_{22}\exp(m_2t) + \text{—}\ G_{2n}\exp(m_nt) + G_{221}\exp(2m_1t) + \text{—} \\ &\cdots \qquad \cdots \\ x_n &= G_{n1}\exp(m_1t) + G_{n2}\exp(m_2t) + \text{—}\ G_{nn}\exp(m_nt) + G_{n11}\exp(2m_1t) + \text{—} \end{aligned} \qquad (17)$$

Now the Gs are constants (n arbitrary) and m1—m_n are n roots of the equation for m.

$$\begin{vmatrix} a_{11} - m & a_{12} & \text{—}\ a_{1n} \\ a_{21} & a_{22} - m & \text{—} a_{2n} \\ a_{n1} & a_{n2} & \text{—}\ a_{nn} - m \end{vmatrix} = \Delta = 0 \qquad (18)$$

If some of G are positive and some are negative, then oscillations may occur. If the roots m are complex, exp (a + ib)t = exp(at)(cosbt + isinbt), and there are damped oscillations about equilibrium.

Consider a general type of network [15] made up of m independent circuits. Each circuit contains an e.m.f., resistances, inductances and capacitances, then each circuit current will flow. Letting $z_{jk}(p)$ be the operator acting on I_k in the jth equation

$$z_{jk} = L_{jk}p + R_{jk} + 1/(C_{jk}p), \text{ then} \qquad (19)$$

$$\begin{aligned} z_{11}(p)I_1 + z_{12}(p)I_2 + \text{—} + z_{1m}(p)I_m &= p(L_{11}I_{10} + L_{12}I_{20} + \text{—} + L_{1m}I_{m0}) + E_{11} \\ z_{21}(p)I_1 + z_{22}(p)I_2 + \text{—} + z_{2m}(p)I_m &= p(L_{21}I_{10} + L_{22}I_{20} + \text{—} + L_{2m}I_{m0}) + E_{22} \\ \cdots \qquad \cdots \\ z_{m1}(p)I_1 + z_{m2}(p)I_2 + \text{—} + z_{mm}(p)I_m &= p(L_{m1}I_{10} + L_{m2}I_{20} + \text{—} + L_{mm}I_{m0}) + E_{mm} \end{aligned} \qquad (20)$$

At t = 0, I_k = 0, (k = 1, 2, — m) and the right parentheses are all zero, and all E on the right are zero except E_{11} = E(t), as a single e.m.f. will produce the same effect as if all others were operating. Then,

$$I_k = (M_1K(p)/\Delta(p))E(t) \qquad (21)$$

where p is the determinant, $M_1K(p)$ is the cofactor $(-1)^{k-1}$ times the minor of $z_{ik}(p)$.

$$\begin{vmatrix} z_{11}(p) & z_{12}(p) & z_{1m}(p) \\ z_{21}(p) & z_{22}(p) & z_{2m}(p) \\ z_{m1}(p) & z_{m2}(p) & z_{mm}(p) \end{vmatrix} \qquad (22)$$

If E(t) is a function of t,

$$I_k(t) = E(0)A_1k(t) + \int_0^t A_1K(t-2)E'(z)dz. \qquad (23)$$

$$\text{Consider } LdA/dt + RA = 1 \qquad (24)$$

$$A(t) = 1/R(1 - \exp(-Rt/L)). \qquad (25)$$

$$\text{Then } 1 = 1/R\ (1\text{-}\exp(R(t\text{-}z)/L))E_0 w\cos wz dz \qquad (26)$$

$$1 = (E_0/(R^2+L^2w^2))(R\sin wt - Lw\cos wt + Lw\exp(-Rt/L)). \qquad (27)$$

Solution of the equation,

$$LdI/dt+RI = E_0\sin wt \qquad (28)$$

would also lead to (27(for I(o) = 0)). (Figure 5 shows a simple series circuit).

It is now seen that if we designate R = k, L = m, E = p', etc., by analogy, we can solve (18) and (21) for an ecological or anthropomorphic system and we will obtain LC50/100 for man and other mammals in a more complex system than shown in (2) by the network systems model of the environment.

Feedback Loops and the Environment

There have been applications in medicine [10] that are similar to the analogue results, such as the equation relating the rate of erythropoietin release, concentration in the blood plasma, etc. Also, models of respiratory control do correspond as examples.

The polynomials of the functions form minimums that are equilibrium states and are a surface (manifold) in n + k dimensional space, where n is the dimension of the state and k is the dimension of the control state. The behavior of the system is described by a trajectory on the manifold. The canonical polynomial form

$$f(x) = x^m+u_1x^{m-2}+u_2x^{m-3}+—u(n-2)x$$

represents the universal unfolding of singularities. Regions of more than one solution represent catastrophic [11] separation and are seen as jumps to another branch.

Optimizing Environmental Quality

Once the analog has been made between the physical and biological cases and a mathematical fit for the ecotoxic function is present, it can be ascertained whether the toxic limit is exceeded for man [16] and every other animal that may be present in the area of assessment. When remediation is necessary, control steps can be used for the system and the control design can alleviate the environmental impact on man and the ecology [17]. These steps can be made prior to the building of the plant, or afterwards as a retrofit, if necessary.

Various kinds of control, such as cascade, feedforward, adaptive, proportional, integral or reset, derivative or rate, bang bang, epidemic, impulsive, singular and profit are examples and various combinations of these controls are used today. It is realized that the control systems can be both characterized and optimized by mathematical techniques to yield optimal feedback systems for man and the ecology. There are two types of feedback control systems. The first type is called a regulator, whose function is to keep the output or controlled variable constant in the face of load variations, parameter changes, etc. The second type is the servomechanism, whose inputs are time-varying. The function of a servomechanism is to provide a one-to-one correspondence between input and output. The system speed of response is another important consideration in feedback control systems. Information about response time can be gained from the frequency-response characteristic of the system.

Stability of the system is important. Thus, if any perturbation of the system occurs, the effect on the system should be transitory, with the system ultimately coming to rest. Nyquist has made use of contour mapping methods in the analysis of stability and this has been used for many years as an important technique in the analysis of stability of linear servomechanisms and regulating systems. Satche [19] extended the use of contour integration to systems with simple time delays. It is procedures such as these that can lead to the prevention of such pollution catastrophes as did occur at Bhopal.

Now that the system is characterized, analyzed, analogized and [20,21,22] solved mathematically, it can be optimized using calculus of variations, filter, control and predictor methods, dynamic programming methods, automatic control and the Pontryagin [21,22] Principle, linear and non-linear Prediction Theory, in addition to newer methods [22-27]. Then the system can be converted back to the original ecosystem to plan or direct the prevention or minimization of pollution to the ecosystem. Sensitivity analysis [21] can also be used to determine the specific influence of the parameters on the system process, so that steps can be taken to bring it into synchronization with the goals of reducing risk optimally.

Applications

The following smattering of examples merely indicate applications of control to pollution problems.

1. The theory of the optimal control of systems with Markov jump disturbances to the control of industrial processes with stages subject to breakdown leads to explicit analytic solutions (Figures 5a and 5b, where U = processors). In more complicated examples where an analytic solution is not tractable, the Markov Decision Process approach can provide an attractive method of solution [26a].
2. Multipass systems describe processes in which the material or workpiece involved is processed by a sequence of passes of the processing tool, e.g., longwall coal cutting. Sufficient conditions can be obtained so that an arbitrary initial start on the first pass can be steered to within an arbitrary small neighborhood of a desired state on some subsequent pass [26b].
3. Based on the state equations of bilinear distributed parameter systems, one can use independent linear and bilinear control variables. This method applied to the control of a simplified model of a continuous kiln (Figure 6), reduces rise time, overshoot and settling time of the reference response. This has also been obtained

in various types of disturbances acting on the plant [26c].

where xe = [1-exp(-a/(Ve)^2)]Ue, U and V are control vectors, a is a constant, and e is equilibrium.

All of the above examples serve not only as process control illustrations but also as pollution control illustrations and merely scratch the surface of a plethora of applications to pollution minimization in many industries.

Reference [29] shows many cases for applications. Three examples are described below. Evolution strategy [29a] starts with a set of realizations of the system which mutate and lead to a new set. Only the best members of both sets, measured by a certain quality criterion, survive as the new generation. Reference [29b] tells of a hybrid computer and determination of drug dosage regimens in individual patients. The system consists of an analog computer and a digital computer, linked by three interfaces. Reference [29c] (designated and demonstrated) uses low grade thermal energy for commercial power production facilities in a hybrid methane generation plant.

Each of the above systems, or all of them together, can be used in the overall system design outlined in this chapter.

Computer Simulation, Modeling and Control of Environmental Quality

A powerful program called "Envirochemkin" is being used (for pollution abatement services together with its subsidiary programs (such as "Therm", etc.) which will aid and abet the program described as a controller to bring systems or plants into the optimum mode for pollution prevention or minimization. Self-optimizing or adaptive control systems can be developed now. These consist of three parts: the definition of optimum conditions of operation (or performance), the comparison of the actual performance with the desired performance and the adjustment of system parameters by closed-loop operation to drive the actual performance toward the desired performance [30]. The first definition will be made through a regulatory agency requiring compliance; the latter two by a program such as Envirochemkin. Further developments that are now in force include learning systems as well as adaptive systems. The adaptive system modifies itself in the face of a new environment so as to optimize performance. A learning system is, however, designed to recognize familiar features and patterns in a situation and then, from its past experience or learned behavior, reacts in an optimum manner. Thus the former emphasizes reacting to a new situation and the latter emphasizes remembering and recognizing old situations. Both attributes are contained in the "mechanism" of Envirochemkin.

Envirochemkin can also use the Artificial Intelligence technique of "backward chaining" to control chemical processes to prevent pollution while maximizing profit during computation. To do this time is considered as negative in the computation and the computations are made backward in time to see what former conditions should be in order to reach the present desired stage of minimum pollution and maximum profit. Then the amount of each starting species, the choice of each starting species, the process temperature and pressure and the mode of the process (adiabatic, isothermal, fixed temperature profile with time, etc.) and associated chemical reaction equations (mechanism) are chosen so as to minimize pollution and maximize profit. Pilot runs have already been performed to measure the success of this procedure.

Conclusions

A simple analytic expression is derived for toxicity for ecosystems using analogues between systems found in physics and engineering and data for man. This is compared with the literature. Control theory from Electrical Engineering is discussed as important to Environmental Systems. The Transfer System, Networks, and Feedback Loops are some of the more important concepts to understand in the applications to the Environment. Methods for optimizing Environmental Quality are described and some applications that are very prone to applying Simulation and Control to Environmental Systems are discussed. A particular Computer Simulation Modeling System with Control Applications to the environment and Ecosystem is described briefly.

1.41 Risk: A Human Science

On February 1, 1994, a flammable, corrosive, and toxic solution escaped from a corroded pump connection at Associated Octel's ethyl chloride manufacturing plant at Ellesmere Poet. A dense cloud of toxic gas formed, which first enveloped the plant and then began to move away from the site. Despite the efforts of the on-site emergency services to isolate the leak and to stop the gas from spreading, the ethyl chloride collected into a pool and then caught fire.

One employee and 17 firefighters were injured in the blaze, which destroyed the plant and led to Octel being prosecuted under Sections 2 and 3 of the Health and Safety at Work etc. Act 1974. The company was fined 150,000 pounds for failing to ensure the safety of employees and others. The Health and

Safety Executive (HSE) concluded in its report that the incident might have been prevented if a more detailed assessment of the hazards and risks of the operation on site had been carried out by the company beforehand. Octel has since rebuilt the plant incorporating improved safety features; it has also introduced better standards for health and safety management, particularly for maintenance.

A Risky Business

Assessing and managing risks and hazards is an essential part of working with chemicals, but the range of models and theories that try to explain how accidents are caused within organizations, and how best to manage risks, can be almost baffling. Most chemical processes are inherently unsafe, which means that the safe operation of these processes depends on engineered safety devices and good operating processes. There are three main factors contributing to accident causation and risk assessment: hardware; people; and systems and cultures. Important lessons could be missed in accounts of accidents that emphasize one of these factors rather than taking a balanced view. Sometimes human error is the direct cause of an incident, but more often human error is an indirect or contributory factor, resulting from poor safety management or poor safety culture. So-called hardware failures can also take place on more than one level. Besides failures such as structural collapse, others involve underlying causes like poor design. Hardware failure was emphasized in the HSE's report into the Allied Colloids Bradford warehouse fire in 1992. The incident took place when kegs of a self-reactive substance ruptured, although it was not clear at the time whether this was a result of operator error or close proximity to a manufacturing steam heating system.

There is a strong human dimension to risk assessment because people are involved throughout the risk assessment process. People estimate and evaluate risks, are implicated in the cause of accidents and have to live with or accept risks every day. Because economic and political considerations represent human dimensions too, these are also important factors in risk assessment. The clean-up costs or threat of fines are additional factors that need to be weighed up when installing safety features to prevent pollution in rivers, for example.

Policy, legislation, economics, and public opinion are all human factors that come into play in deciding the best course of action. Shell, for instance, learned to its cost the importance of public opinion in assessing risks during the recent Brent Spar debacle. Even after assessing a number of options for the spar, deep-sea disposal — Shell's initial solution — proved to be one of the two most acceptable proposals in terms of cost, safety and environmental impact. The public's hostile reaction to deep-sea disposal cost Shell £20 million.

If the human dimension is taken into account, then the concept of risk assessment as a purely objective, scientific activity needs to be abandoned. The way that society deals with risk is deeply cultural, and therefore cannot be improved by simply applying more, or better science. The two ends of the spectrum in the 'subjective/objective debate' are represented by the so-called 'cultural relativists', who define risk as subjective and socially constructed, and the 'naive positivists' who think that risk is an objectively measurable reality. But both of these extreme positions oversimplify risk. Risk assessment can be objective in that it can be the subject of rational dispute and criticism, is dependent on probabilities affected by empirical events and can be criticized in terms of how well the scientific goals of explanation and prediction are served. Whilst choosing the middle ground, the strengths of both sides should be recognized rather than simply watered down. Both scientific and ethical components have an important part to play in forming a strategy to deal with risk.

Safe Science

Despite the theories on risk being proffered by both scientists and social scientists, companies need to assess and manage risks on a day to day basis. Practical advice and assessment are available from a number of sources that have the experience and expertise to set industry on the right track.

One such source is the HSE's Health and Safety Laboratory (HSL), which undertakes research into risks to health and safety, and how they can be controlled. As well as providing scientific and technological advice, HSL also carries out forensic investigations when workplace incidents take place.

HSL's Sheffield site houses biomedical, occupational hygiene, and environmental equipment as well as engineering, risk assessment, and safety management facilities. Large scale experiments involving fires and explosions are carried out at the HSL's remote 550-acre site in Buxton, Derbyshire, which includes facilities such as open air test ranges, tunnels, giant rigs, and scaffolding as well as a pilot chemical reactor plant. These enable scientists to study fire, smoke, and toxic and flammable liquids and disaster scenarios such as gas explosions.

HSL provides scientific and technological advice to both private and public sector organizations, and undertakes R&D work. As well as providing the expertise to solve customer problems, the laboratory also provides independent and impartial scientific expertise on a national and international scale. Other services include rapidly assembling multi-disciplinary teams for accident and investigative work. All areas of worker safety and health are covered, including those dealing with meeting European and international standards.

Another source of guidance for those who regularly deal with hazardous chemicals is the Chemical Industries Association (CIA), which represents about 75% of British chemical manufacturers. The CIA publishes 'green books' — guidelines that have been devised by risk managers and loss prevention specialists from CIA member companies and which identify methods of quantifying risks and identifying methods of risk control. These guides cover areas such as employer and public liability, material damage, and product liability, and come under the umbrella of CIA's responsible care program. Other publications cover communicating and comparing risks, assessing risks associated with substances, and risk-benefit analysis. The CIA has also published "Risk — Its Assessment, Control, and Management", a guide aimed at the general public which gives information on how the chemical industry goes about assessing risks that arise from its products and processes, and their effects on the public and the environment.

One Step Ahead

The HSE's Chemical and Hazardous Installations Division (CHID), which began operating in April, 1996, is responsible for all of the HSE's on-shore operational functions including the manufacture, processing, storage, and road transportation of all chemicals and explosives.

When CHID was set up, Paul Davies, head of the newly formed division, anticipated that the new regulations covering high hazard industries would come into force as a result of new European directives. CHID's role would be to help industry to prepare for the changes that lay ahead especially in providing effective management systems to insure the safety of chemical installations at all stages-from design through to decommissioning.

The Control of Major Accident Hazards (COMAH) directive is an EU directive that came into force in February 1999 and is a development of existing Cimah regulations. According to Peter Sumption, operational strategy manager at CHID, the Health and Safety Commission (HSC) is currently consulting with stakeholders, such as the CIA and the UK Petrochemical Industries Association (PIA), in order to publish draft regulations later this year. Guidelines will also be published at the same time, highlighting contentious issues in order to register any comments. Sumption thinks that one issue that will be raised is whether CHID should have powers to stop construction of a hazardous installation if it is not satisfied with the site's safety provisions. Overall, CHID's holistic approach takes into account underlying design and organization problems as well as more obvious technical considerations. The new directive places much greater emphasis on safety management systems says Sumption. The HSC will report to ministers in the autumn on completion of the HSC report.

The proposed regulations will cover safety report requirements, major accident prevention policy, and on- and off-site emergency plans. CHID will test its procedure for evaluating these requirements during a trial run with four unnamed companies. Local authorities and emergency services take part in testing companies' emergency plans, and local people surrounding the plant will be informed about hazards and emergency measures. In order to help industry comply with these requirements, CHID plans to publish a booklet dealing with emergency planning, as well as its criteria for accepting safety reports.

The Future

The chemical industry may learn from its own past experience, as well as from experiments like those carried out at HSL on how to assess and manage risk more effectively. But with the advances in computer technology now becoming available, it is not always essential to have hands-on experience of a 'real-life' chemical plant in order to assess its risk potential. A recent collaboration between Imperial College, Cadcentre, and Silicon Graphics/Cray Research has resulted in a virtual plant, a three-dimensional computer-generated plant, which the viewer can 'walk through'. The plant is designed to provide training for operators, maintenance staff, and hazard prevention specialists before a real-life plant is even built, and also to assist in safe design. The plant is useful for training because information about the physical processes taking place inside the reactors and typical hazards taught during a plant inspection can be experienced and understood in a safe, virtual environment. There are also design advantages; the time taken from design to completion of the plant could be reduced to just six months. Currently it takes up to three years.

Companies can now learn from each other's mistakes without 'washing their dirty linen in public', thanks to a new database launched by the Institution of Chemical Engineers (ICHEME). The database contains information on over 8000 industrial and chemical accidents, including accident accounts and the lessons that were learned. The aim is to help other companies to prevent accidents and loss of production time, and to save resources and people's lives. The database is available in CD-ROM format.

Sound Precautions

So how can the risk of incidents like the Octel fire be assessed? Unfortunately, says Hurst, accidents like these serve to confirm the public's suspicion that chemical processes are inherently unsafe. However, by learning lessons from such incidents, by understanding the human and technical dimensions of risk, and with the guidance of organizations like CHID, HSL, and the CIA, sound preventative measures can be put in place. Communicating and comparing risks and implementing tougher European directives will help both plant managers and the neighbors to sleep more soundly.

1.42 IPPS

The Industrial Pollution Projection System (IPPS) is a modeling system which can use industry data to estimate comprehensive profiles of industrial pollution for countries, regions, urban areas or proposed new projects. IPPS has been developed to exploit the fact that industrial pollution is heavily affected by the scale of industrial activity, its sectoral composition, and the process technologies which are employed in production. The U.S. prototype has a database for 200,000 facilities in all regions of the U.S. IPPS spans approximately 1,500 product categories, all operating technologies and hundreds of pollutants. It can separately project air, water, and solid waste emissions, and incorporates a range of risk factors for human toxic and ecotoxic effects. Since it has been developed from a database of unprecedented size and depth, it is undoubtedly the most comprehensive system of its kind in the world. When applying the U.S.-based estimates to other economies, patterns of sectoral intensity are likely to be similar, but the present goal is to expand the applicability of IPPS by incorporating data from developing countries. This paper provides a brief assessment of the available databases, describes methods for estimating pollution intensities by combining U.S. Manufacturing Census data with the U.S. EPA's pollution databases, focuses on estimation of toxic pollution intensities weighted by human and ecological risk factors, and describes the media-specific pollution intensities developed for the U.S. EPA's criteria air pollutants, major water pollutants, and toxic releases by medium (air/water/land). Results are critically assessed and the complete set of IPPS intensities is then available. The World Bank's technical assistance work with new environmental protection institutions (EPIs) can then stress cost-effective regulation with implementation of market-based pollution control instruments.

Part II. Mathematical Methods

2.1 Linear Programming

Linear Programming (LP) is a procedure for optimizing an objective function subject to inequality constraints and non-negativity restrictions. In a linear program, the objective function as well as the inequality constraints are all linear functions. LP is a procedure that has found practical application in almost all facets of business, from advertising to production planning. Transportation, distribution, and aggregate production planning problems are the most typical objects of LP analysis. The petroleum industry seems to be the most intensive user of LP. Large oil companies may spend 10% of the computer time on the processing of LP and LP-like models.

2.2 The Simplex Model

LP problems are generally solved via the Simplex model. The standard Solver uses a straightforward implementation of the Simplex method to solve LP problems, when the Assume Linear Model Box is checked in the Solver Option dialog. If the Simplex or LP/Quadratic is chosen in the Solver Parameters dialog, the Premium and Quadratic Solvers use an improved implementation of the Simplex method. The Large-Scale LP Solver uses a specialized implementation of the Simplex method, which fully exploits sparsity in the LP model to save time and memory. It uses automatic scaling, matrix factorization, etc. These same techniques often result in much faster solution times, making it practical to solve LP problems with thousands of variables and constraints.

2.3 Quadratic Programming

Quadratic programming problems are more complex than LP problems, but simpler than general NLP problems. Such problems have one feasible region with "flat faces" on its surface, but the optimal solution may be found anywhere within the region or on its surface. Large QP problems are subject to many of the same considerations as large LP problems. In a straightforward or "dense" representation, the amount of memory increases with the number of variables times the number of constraints, regardless of the model's sparsity. Numerical instabilities can arise in QP problems and may cause more difficulty than in similar size LP problems.

2.4 Dynamic Programming

In dynamic programming one thinks about what one should do at the end. Then one examines the next to last step, etc. This way of tackling a program backward is known as dynamic programming. Dynamic programming was the brainchild of an American mathematician Richard Bellman, who described the way of solving problems where you need to find the best decisions one after another. The uses and applications of dynamic programming have increased enormously.

2.5 Combinatorial Optimization

Optimization just means "finding the best", and the word "combinatorial" is just a six syllable way of saying that the problem involves discrete choices, unlike the older and better known kind of optimization which seeks to find numerical values. Underlying almost all the ills is a combinatorial explosion of possibilities and the lack of adequate techniques for reducing the size of the search space. Technology based on combinatorial optimization theory can provide ways around the problems. It turns out that the "assignment problem" or "bipartite matching problem" is quite approachable — computationally intensive, but still approachable. There are good algorithms for solving it.

2.6 Elements of Graph Theory

Graphs have proven to be an extremely useful tool for analyzing situations involving a set of elements in which various pairs of elements are related by some property. Most obvious are sets with physical links, such as electrical networks, where electrical components are the vertices and the connecting wires are the edges. Road maps, oil pipelines, telephone connecting systems, and subway systems are other examples. Another natural form of graphs are sets with logical or hierarchical sequencing, such as computer flow charts, where the instructions are the vertices and the logical flow from one instruction to possible successor instruction(s) defines the edges. Another example is an organizational chart where the people are the vertices and if person A is the immediate superior of person B then there is an edge (A,B). Computer data structures, evolutionary trees in biology, and the scheduling of tasks in a complex project are other examples.

2.7 Organisms and Graphs

I will discuss the use of graphs to describe processes in living organisms. Later we will review graphs for processes in chemical plants commonly known as flowsheets. Ingestion f_1 (Figure 7) is followed by digestion f_2, which leads on one hand to excretion f_3 and on the other to absorption f_4. The absorbed materials are then transported via f_{4T5} to the sites of synthetic processes f_5. Then the synthesis of digestive enzymes, represented by f_6 , follows via transport f_{5T6}. These enzymes are transported via f_{6T7} to the site of secretion, represented by f_7, and digestion f_2 again follows.

On the other hand, some of the synthesized products are transported via f_{5T8} to the site of the catabolic processes, which are represented by f_8 . Products of catabolism are transported via f_{8T9} to the site of elimination of waste products, and there elimination, represented by f_9, takes place. Catabolic processes result in the liberation of energy, represented by f_{10}, which in turn provides the possibility of transport f_T. On the other hand, after a transport f_{8T11}, the catabolic reactions give rise to the production f_{11} of CO2, and the latter is transported within the cell via f_{11T12}. This eventually results in the elimination of CO2, represented by f_{12}.

The intake of O2 from the outside, represented by f_{13}, results in a transport of O2 to the sites of different reactions involved in catabolic processes. Liberation of energy combined with anaprocesses as well as other biological properties result in the process of multiplication, which is not intended in the figure to simplify the latter.

2.8 Trees and Searching

The most widely used special type of graph is a tree. A tree is a graph with a designated vertex called a root such that there is a unique path from the root to any other vertex in the tree. Trees can be used to decompose and systematize the analysis of various search problems. They are also useful for graph connectivity algorithms based on trees. One can also analyze several common sorting techniques in terms of their underlying tree structure.

2.9 Network Algorithms

Network algorithms are used for the solution of several network optimization problems. By a network, we mean a graph with a positive integer assigned to each edge. The integer will typically represent the length of an edge, time, cost, capacity, etc. Optimization problems are standard in operations research and have many practical applications. Thus good systematic procedures for their solution on a computer are essential. The flow optimization algorithm can also be used to prove several important combinatorial theorems.

2.10 Extremal Problems

Extremal problems or optimization problems may be regarded abstractly in terms of sets and transformations of sets. The usual problem is to find, for a specified domain of a transformation, a maximal element of the range set. Problems involving discrete optimization and methods for determining such values, whether exactly, approximately, or assymtotically are studied here. We seek upper and lower bounds and maximum and minimum values of a function given in explicit form.

2.11 Traveling Salesman Problem (TSP)-Combinatorial Optimization

Problems in combinatorial optimization involve a large number of discrete variables and a single "cost" function to be minimized, subject to constraints on these variables. A classic example is the traveling salesman problem: given N cities, find the minimum length of a path connecting all the cities and returning to its point or origin. Computer scientists classify such a problem as NP-hard; most likely there exists no algorithm that can consistently find the optimum in an amount of time polynomial in N. From the point of view of statistical physics, however, optimizing the cost function is analogous to finding the ground-state energy in a frustrated, disordered system. Theoretical and numerical approaches developed by physicists can consequently be of much relevance to combinatorial optimization.

2.12 Optimization Subject to Diophantine Constraints

A Diophantine equation is a polynomial equation in several variables whose coefficients are rational and for which a solution in integers is desirable. The equations are equivalent to an equation with integer coefficients. A system of Diophantine equations consists of a system of polynomial equations, with rational coefficients, whose simultaneous solution in integers is desired. The solution of a linear Diophantine equation is closely related to the problem of finding the number of partitions of a positive integer N into parts from a set S whose elements are positive integers. Often, a Diophantine equation or a system of such equations may occur as a set of constraints of an optimization problem.

2.13 Integer Programming

Optimization problems frequently read: Find a vector x of nonnegative components in E, which maximizes the objective function subject to the constraints. Geometrically one seeks a lattice point in the region that satisfies the constraints and minimizes the objective function. Integer programming is central to Diophantine optimization. Some problems require that only some of the components of x be integers. A requirement of the other components may be that they be rational. This case is called mixed-integer programming.

2.14 MINLP

Mixed Integer Nonlinear Programming (MINLP) refers to mathematical programming algorithms that can optimize both continuous and integer variables, in a context of nonlinearities in the objective function and/or constraints. MINLP problems are NP-complete and until recently have been considered extremely difficult. Major algorithms for solving the MINLP problem include: branch and bound, generalized Benders decomposition (GBD), and outer approximation (OA). The branch and bound method of solution is an extension of B&B for mixed integer programming. The method starts by relaxing the integrality requirements, forming an NLP problem. Then a tree enumeration, having a subset of the integer variables is fixed successively at each node. Solution of the NLP at each node gives a lower bound for the optimal MINLP objective function value. The lower bound directs the search by expanding nodes in a breadth first or depth first enumeration. A disadvantage of the B&B method is that it may require a large number of NLP subproblems. Subproblems optimize the continuous variables and provide an upper bound to the MINLP solutions, while the MINLP master problems have the role of predicting a new lower bound for the MINLP solution, as well as new variables for each iteration. The search terminates when the predicted lower bound equals or exceeds the current upper bound.

MINLP problems involve the simultaneous optimization of discrete and continuous variables. These problems often arise in engineering domains, where one is trying simultaneously to optimize the system structure and parameters. This is difficult. Engineering design "synthesis" problems are a major application of MINLP algorithms. One has to determine which components integrate the system and also how they should be connected and also determine the sizes and parameters of the components. In the case of process flowsheets in chemical engineering, the formulation of the synthesis problem requires a superstructure that has all the possible alternatives that are a candidate for a feasible design embedded in it. The discrete variables are the decision variables for the components in the superstructure to include in the optimal structure, and the continuous variables are the values of the parameters of the included components.

2.15 Clustering Methods

Clustering methods have been used in various fields as a tool for organizing (into sub-networks or astronomical bodies) data. An exhaustive search of all possible clusterings is a near impossible task, and so several different sub-optimal techniques have been proposed. Generally, these techniques can be classified into hierarchical, partitional, and interactive techniques. Some of the methods of validating the structure of the clustered data have been discussed as well as some of the problems that clustering techniques have to overcome in order to work effectively.

2.16 Simulated Annealing

Simulated annealing is a generalization of a Monte Carlo method for examining the equations of state and frozen states of n-body systems. The concept is based on the manner in which liquids freeze or metals recrystallize in the process of annealing. In that process a melt, initially at high temperature and disordered is slowly cooled so that the system at any time is almost in thermodynamic equilibrium and as cooling proceeds, becomes more disordered and approaches a frozen ground state at T = 0. It is as if the system adiabatically approaches the lowest energy state. By analogy the generalization of this Monte Carlo approach to the combinatorial approach is straightforward. The energy equation of the thermodynamic system is analogous to an objective function, and the ground state is analogous to the global minimum.

If the initial temperature of the system is too low or cooling is done insufficiently slowly, the system may become quenched forming defects or freezing out in metastable states (i.e., trapped in a local minimum energy state). By analogy the generalization of this Monte Carlo approach to combinatorial problems is straightforward.

2.17 A Tree Annealing

Simulated annealing was designed for combinatorial optimization (assuming the decision variables are discrete). Tree annealing is a variation developed to globally minimize continuous functions. Tree annealing stores information in a binary tree to keep track of which subintervals have been explored. Each node in the tree represents one of two subintervals defined by the parent node. Initially the tree consists of one parent and two child nodes. As better intervals are found, the path down the tree that leads to these intervals gets deeper and the nodes along these paths define smaller and smaller subspaces.

2.18 Global Optimization Methods

This section surveys general techniques applicable to a wide variety of combinatorial and continuous optimization problems. The techniques involved below are:

- Branch and Bound
- Mixed Integer Programming
- Interval Methods
- Clustering Methods
- Evolutionary Algorithms
- Hybrid Methods
- Simulated Annealing
- Statistical Methods
- Tabu Search

Global optimization is the task of finding the absolutely best set of parameters to optimize an objective function. In general, there can be solutions that can be locally optimal but not globally optimal. Thus global optimization problems are quite difficult to solve exactly; in the context of combinatorial problems, they are often NP-hard. Global optimization problems fall within the broader class of nonlinear programming (NLP). Some of the most important classes of global optimization problems are differential convex optimization, complementary problems, minimax problems, bilinear and biconvex programming, continuous global optimization, and quadratic programming.

Combinatorial Problems have a linear or nonlinear function defined over a set of solutions that is finite but very large. These include network problems, scheduling, and transportation. If the function is piecewise linear, the combinatorial problem can be solved exactly with a mixed integer program method, which uses branch and bound. Heuristic methods like simulated annealing, tabu search, and genetic algorithms have also been used for approximate solutions.

General unconstrained problems have a nonlinear function over reals that is unconstrained (or have simple bound constraints). Partitioning strategies have been proposed for their exact solution. One must know how rapidly the function can vary or an analytic formulation of the objective function (e.g., interval methods). Statistical methods can also partition to decompose the search space but one must know how the objective function can be modeled. Simulated annealing, genetic algorithms, clustering methods and continuation methods can solve these problems inexactly.

General constrained problems have a nonlinear function over reals that is constrained. These problems have not been as well used; however, many of the methods for unconstrained problems have been adapted to handle constraints.

Branch and Bound is a general search method. The method starts by considering the original problem with the complete feasible region, which is called the root problem. A tree is generated of subproblems. If an optimal solution is found to a subproblem, it is a feasible solution to the full problem, but not necessarily globally optimal. The search proceeds until all nodes have been solved or pruned, or until some specified threshold is met between the best solution found and the lower bounds on all unsolved subproblems.

A mixed-integer program is the minimization or maximization of a linear function subject to linear constraints. If all the variables can be rational, this is a linear programming problem, which can be solved in polynomial time. In practice linear programs can be solved efficiently for reasonably sized problems. However, when some or all of the variables must be integer, corresponding to pure integer or mixed integer programming, respectively, the problem becomes NP-complete (formally intractable).

Global optimization methods that use interval techniques provide rigorous guarantees that a global maximizer is found. Interval techniques are used to compute global information about functions over large regions (box-shaped), e.g., bounds on function values, Lipschitz constants, or higher derivatives.

Most global optimization methods using interval techniques employ a branch and bound strategy. These algorithms decompose the search domain into a collection of boxes for which the lower bound on the objective function is calculated by an interval technique.

Statistical Global Optimization Algorithms employ a statistical model of the objective function to bias the selection of new sample points. These methods are justified with Bayesian arguments that suppose that the particular objective function that is being optimized comes from a class of functions that is modeled by a particular stochastic function. Information from previous samples of the objective function can be used to estimate parameters of the stochastic function, and this refined model can subsequently be used to bias the selection of points in the search domain.

This framework is designed to cover average conditions of optimization. One of the challenges of using statistical methods is the verification that the statistical model is appropriate for the class of problems to which they are applied. Additionally, it has proved difficult to devise computationally interesting version of these algorithms for high dimensional optimization problems.

Virtually all statistical methods have been developed for objective functions defined over the reals. Statistical methods generally assume that the objective function is sufficiently expensive so that it is reasonable for the optimization method to perform some nontrivial analysis of the points that have been previously sampled. Many statistical methods rely on dividing the search region into partitions. In practice, this limits these methods to problems with a moderate number of dimensions. Statistical global optimization algorithms have been applied to some challenging problems. However, their application has been limited due to the complexity of the mathematical software needed to implement them.

Clustering global optimization methods can be viewed as a modified form of the standard multistart procedure, which performs a local search from several points distributed over the entire search domain. A drawback is that when many starting points are used, the same local minimum may be identified several times, thereby leading to an inefficient global search. Clustering methods attempt to avoid this inefficiency by carefully selecting points at which the local search is initiated.

Evolutionary Algorithms (EAs) are search methods that take their inspiration from natural selection and survival of the fittest in the biological world. EAs differ from more traditional optimization techniques in that they involve a search from a "population" of solutions, not from a single point. Each iteration of an EA involves a competitive selection that weeds out poor solutions. The solutions with high "fitness" are "recombined" with other solutions by swapping parts of a solution with another. Solutions are also "mutated" by making a small change to a single element of the solution. Recombination and mutation are used to generate new solutions that are biased towards regions of the space for which good solutions have already been seen.

Mixed Integer Nonlinear Programming (MINLP) is a hybrid method and refers to mathematical programming algorithms that can optimize both continuous and integer variables, in a context of non-linearities in the objective and/or constraints. Engineering design problems often are MINLP problems, since they involve the selection of a configuration or topology as well as the design parameters of those components. MINLP problems are NP-complete and until recently have been considered extremely difficult. However, with current problem structuring methods and computer technology, they are now solvable. Major algorithms for solving the MINLP problem can include branch and bound or other methods. The branch and bound method of solution is an extension of B&B for mixed integer programming.

Simulated annealing was designed for combinatorial optimization, usually implying that the decision variables are discrete. A variant of simulated annealing called tree annealing was developed to globally minimize continuous functions. These problems involve fitting parameters to noisy data, and often it is difficult to find an optimal set of parameters via conventional means.

The basic concept of Tabu Search is a meta-heuristic superimposed on another heuristic. The over all approach is to avoid entrainment in cycles by forbidding or penalizing moves which take the solution, in the next iteration, to points in the solution space previously visited (hence tabu).

2.19 Genetic Programming

Genetic algorithms are models of machine learning that uses a genetic/evolutionary metaphor. Fixed-length character strings represent their genetic information.

Genetic Programming is genetic algorithms applied to programs.

Crossover is the genetic process by which genetic material is exchanged between individuals in the population.

Reproduction is the genetic operation which causes an exact copy of the genetic representation of an individual to be made in the population.

Generation is an iteration of the measurement of fitness and the creation of a new population by means of genetic operations.

A function set is the set of operators used in GP. They label the internal (non-leaf) points of the parse trees that represent the programs in the population.

The terminal set is the set of terminal (leaf) nodes in the parse trees representing the programs in the population.

2.20 Molecular Phylogeny Studies

These methods allow, from a given set of aligned sequences, the suggestion of phylogenetic trees which aim at reconstructing the history of successive divergence which took place during the evolution, between the considered sequences and their common ancestor.

One proceeds by

1. Considering the set of sequences to analyze.
2. Aligning these sequences properly.
3. Applying phylogenetic making tree methods.
4. Evaluating statistically the obtained phylogenetic tree.

2.21 Adaptive Search Techniques

After generating a set of alternative solutions by manipulating the values of tasks that form the control services and assuming we can evaluate the characteristics of these solutions, via a fitness function, we can use automated help to search the alternative solutions. The investigation of the impact of design decisions on nonfunctional as well as functional aspects of the system allows more informed decisions to be made at an earlier stage in the design process.

Building an adaptive search for the synthesis of a topology requires the following elements:

1. How an alternative topology is to be represented.
2. The set of potential topologies.
3. A fitness function to order topologies.
4. Select function to determine the set of alternatives to change in a given iteration of the search.
5. Create function to produce new topologies.
6. Merge function to determine which alternatives are to survive each iteration.
7. Stopping criteria.

Genetic Algorithms offer the best ability to consider a range of solutions and to choose between them. GAs are a population based approach in which a set of solutions are produced. We intend to apply a tournament selection process. In tournament solution a number of selections are compared and the solution with the smallest penalty value is chosen. The selected solutions are combined to form a new set of solutions. Both intensification (crossover) and diversification (mutation) operators are employed as part of a create function. The existing and new solutions are then compared using a merge function that employs a best fit criterion. The search continues until a stopping criterion, such as n iterations after a new best solution is found.

If these activities and an appropriate search engine is applied, automated searching can be an aid to the designer for a subset of design issues. The aim is to assist the designer not prescribe a topology. Repeated running of such a tool as a design and more information emergence is necessary

2.22 Advanced Mathematical Techniques

This section merely serves to point out The Research Institute for Symbolic Computation (RISC-LINZ). This Austrian independent unit is in close contact with the departments of the Institute of Mathematics and the Institute of Computer Science at Johannes Kepler University in Linz. RISC-LINZ is located in the Castle of Hagenberg and some 70 staff members are working at research and development projects. Many of the projects seem like pure mathematics but really have important connection to the projects mentioned here. As an example, Edward Blurock has developed computer-aided molecular synthesis. Here algorithms for the problem of synthesizing chemical molecules from information in initial molecules and chemical reactions are investigated. Several mathematical subproblems have to be solved. The algorithms are embedded into a new software system for molecular synthesis. As a subproblem, the automated classification of reactions is studied. Some advanced techniques for hierarchical construction of expert systems have been developed. This work is mentioned elsewhere in this book. He is also involved in a project called Symbolic Modeling in Chemistry, which solves problems related to chemical structures.

A remarkable man also is Head of the Department of Computer Science in Vesprem, Hungary. Ferenc Friedler has been mentioned before in this book for his work on Process Synthesis, Design of Molecules with Desired Properties by Combinatorial Analysis, and Reaction Pathway Analysis by a Network Synthesis Technique.

2.23 Scheduling of Processes for Waste Minimization

The high value of specialty products has increased interest in batch and semicontinuous processes. Products include specialty chemicals, pharmaceuticals, biochemicals, and processed foods. Because of the small quantities, batch plants offer the producing of several products in one plant by sharing the available production time between units. The order or schedule for processing products in each unit of the plant is to optimize economic or system performance criterion. A mathematical programming model for scheduling batch and semicontinuous processes, minimizing waste and abiding to environmental constraints is necessary. Schedules also include equipment cleaning and maximum reuse of raw materials and recovery of solvents.

2.24 Multisimplex

Multisimplex can optimize almost any technical system in a quick and easy way. It can optimize up to 15 control and response variables simultaneously. Its main variables include continuous multivariate on-line optimization, handling unlimited number of control variables, handling unlimited number of response variables and constraints, multiple optimization sessions, fuzzy set membership functions, etc. It is a Windows-based software for experimental design and optimization. Only one property or measure seldom defines the production process or the quality of a manufactured product. In optimization, more than one response variable must be considered simultaneously. Multisimplex uses the approach of fuzzy set theory, with membership functions, to form a realistic description of the optimization objectives. Different response variables, with separate scales and optimization objectives, can then be combined into a joint measure called the aggregated value of membership.

2.25 Extremal Optimization (EO)

Extremal Optimization is a general-purpose method for finding high-quality solutions to hard optimization problems, inspired by self-organizing processes found in nature. It successively eliminates extremely undesirable components of sub-optimal solutions. Using models that simulate far-from equilibrium dynamics, it complements approximation methods inspired by equilibrium statistical physics, such as simulated annealing. Using only one adjustable parameter, its performance proves competitive with, and often superior to, more elaborate stochastic optimization procedures.

In nature, highly specialized, complex structures often emerge when their most efficient components are selectively driven to extinction. Evolution, for example, progresses by selecting against the few most poorly adapted species, rather than by expressly breeding those species best adapted to their environment. To describe the dynamics of systems with emergent complexity, the concept of "self-organized criticality" (SOC) has been proposed. Models of SOC often rely on "extremal" processes, where the least fit components are progressively eliminated. The extremal optimization proposed here is a dynamic optimization approach free of selection parameters.

2.26 Petri Nets and SYNPROPS

Petri Nets are graph models of concurrent processing and can be a method for studying concurrent processing. A Petri Net is a bipartite graph where the two classes of vertices are called places and transitions. In modeling, the places represent conditions, the transitions represent events, and the presence of at least one token in a place (condition) indicates that that condition is met. In a Petri Net, if an edge is directed from place p to transition t, we say p is in an input place for transition t. An output place is defined similarly. If every input place for a transition t has at least one token, we say that t is enabled. A firing of an enabled transition removes one token from each input place and adds one token to each output place. Not only do Petri Nets have relations to SYNPROPS but also to chemical reactions and Flowsheet Synthesis methods such as SYNPHONY.

2.27 Petri Net-Digraph Models for Automating HAZOP Analysis of Batch Process Plants

Hazard and Operability (HAZOP) analysis is the study of systematically identifying every conceivable deviation, all the possible causes for such deviation, and the adverse hazardous consequences of that deviation in a chemical plant. It is a labor-and time intensive process that would gain by automation.

Previous work automating HAZOP analysis for continuous chemical plants has been successful; however, it does not work for batch and semi-continuous plants because they have two additional sources of complexity. One is the role of operating procedures and operator actions in plant operation, and the other is the discrete-event character of batch processes. The batch operations characteristics are represented by high-level Petri Nets with timed transitions and colored tokens. Causal relationships between process variables are represented with subtask digraphs. Such a Petri Net-Gigraph model based framework has been implemented for a pharmaceutical batch process case study.

Various strategies have been proposed to automate process independent and items common to many chemical plants. Most of these handle the problem of automating HAZOP analysis for batch plants. The issues involved in automating HAZOP analysis for batch processes are different from those for continuous plants.

Recently, the use of digraph based model methods was proposed for hazard identification. This was the emphasis for continuous plants in steady state operation. The digraph model of a plant represents the balance and confluence equations of each unit in a qualitative form thus giving the relationships between the process state variables. The relationships stay the same for the continuous plant operating under steady-state conditions. However, in a batch process, operations associated with production are performed in a sequence of steps called subtasks. Discontinuities occur due to start and stop of these individual processing steps. The relationships between the process variables are different in different subtasks. As the plant evolves over time, different tasks are performed and the interrelationships between the process variables change. A digraph model cannot represent these dynamic changes and discontinuities. So, the digraph based HAZOP analysis and other methods proposed for continuous mode operation of the plant cannot be applied to batch or semi-continuous plants and unsteady operation of continuous plants. In batch plants, an additional degree of complexity is introduced by the operator's role in the running of the plant. The operator can cause several deviations in plant operation which cannot occur in continuous plants. The HAZOP procedure has to be extended to handle these situations in batch processes.

Batch plant HAZOP analysis has two parts: analysis of process variable deviation and analysis of plant maloperation. In continuous mode operation hazards are due only to process variable deviations. In continuous operation, the operator plays no role in the individual processing steps. However, in batch operation the operator plays a major role in the processing steps. Subtask initiation and termination usually requires the participation of the operator. Hazards can arise in batch plants by inadvertent acts of omission by the plant operator. Such hazards are said to be due to plant maloperation.

The detailed description of how each elementary processing step is implemented to obtain a product is called the product recipe. The sequence of tasks associated with the processing of a product constitutes a task network. Each subtask has a beginning and an end. The end of a subtask is signaled by a subtask termination logic. The subtask termination logic is either a state event or a time event. A state event occurs when a state variable reaches a particular value. When the duration of a subtask is fixed *a priori*, its end is flagged by a time event. A time event causes a discontinuity in processing whose time of occurrence is known *a priori*.

A framework for knowledge required for HAZOP analysis of batch processes has been proposed. High level nets with timed transitions and colored tokens represent the sequence of subtasks to be performed in each unit. Each transition in a TPN represents a subtask and each place indicates the state of the equipment. Colored tokens represent chemical species. The properties of chemical species pertinent to HAZOP analysis; Name, Composition, Temperature, and Pressure were the attributes with colored tokens.

In classical Petri Nets, an enabled transition fires immediately, and tokens appear in the output places the instant the transition fires. When used for representing batch processes, this would mean that each subtask occurs instantaneously and all temporal information about the subtask is lost. Hazards often occur in chemical plants when an operation is carried out for either longer or shorter periods than dictated by the recipe. It is therefore necessary to model the duration for which each subtask is performed. For this, an optimum, representing the duration for which the subtask occurs, was associated with each transition in the task Petri Net. The numerical value of op-time is not needed to perform HAZOP analysis since only deviations like HIGH and LOW in the op-time are to be considered. A deadtime was also associated with each transition to represent the time between when a subtask is enabled and when operation of the subtask actually starts. This is required for HAZOP analysis because a subtask may not be started when it should have been. This may cause the contents of the vessel to sit around instead of the next subtask being performed, which can result in hazardous reactions.

Recipe Petri Nets represent the sequence of tasks to be performed during a campaign. They have timed transitions and the associated tokens are the colored chemical entity tokens. Each transition in these Petri Nets represent a task. The places represent the state of the entire plant. Associated with each transition in the recipe Petri Net is a task Petri Net.

In batch operations, material transfer occurs during filling and emptying subtasks. During other subtasks, operations are performed on the material already present in the unit. However, the amount of the substance already present in the unit may change during the course of other subtasks due to reaction and phase change. Similarly, the heat content of materials can also undergo changes due to heat transfer operations. Therefore, digraph nodes representing amount of material which enters the subtask, amount of material which leaves the subtask, amount of heat entering the subtask, and the amount of heat leaving the subtasks are needed in each subtask digraph.

Using the framework above, a model based system for automating HAZOP analysis of batch chemical processes, called Batch HAZOP Expert, has been implemented in the object-oriented architecture of Gensym's real-time expert system G2. Given the plant description, the product recipe in the form of tasks and subtasks and process material properties, Batch HAZOPExpert can automatically perform HAZOP analysis for the plant maloperation and process variable deviation scenarios generated by the user.

2.28 DuPont CRADA

DuPont directs a multidisciplinary Los Alamos team in developing a neural network controller for chemical processing plants. These plants produce polymers, household and industrial chemicals, and petroleum products that are very complex and diverse and where no models of the systems exist.

Improved control of these processes is essential to reduce energy consumption and waste and to improve quality and quantity. DuPont estimates its yearly savings could be $500 million with a 1% improvement in process efficiency. For example, industrial distillation consumes 3% of the entire U.S. energy budget. Energy savings of 10% through better control of distillation columns would be significant.

The team has constructed a neural network that models the highly bimodal characteristics of a specific chemical process, an exothermic Continuously Stirred Tank Reactor (CSTR). A CSTR is essentially a big beaker containing a uniformly mixed solution. The beaker is heated by an adjustable heat source to convert a reactant into a product. As the reaction begins to give off heat, several conversion efficiencies can exist for the same control temperature. The trick is to control the conversion by using history data of both the solution and the control temperatures.

The LANL neural network, trained with simple plant simulation data, has been able to control the simulated CSTR. The network is instructed to bring the CSTR to a solution temperature in the middle of the multivalued regime and later to temperature on the edge of the regime. Examining the control sequence from one temperature target to the next shows the neural network has implicitly learned the dynamics of the plant. The next step is to increase the complexity of the numerical plant by adding time delays into the control variable with a time scale exceeding that of the reactor kinetics. In a future step, data required to train the network will be obtained directly from an actual DuPont plant.

The DuPont CRADA team has also begun a parallel effort to identify and control distillation columns using neural network tools. This area is rich in nonlinear control applications.

2.29 KBDS-(Using Design History to Support Chemical Plant Design)

The use of design rationale information to support design has been outlined. This information can be used to improve the documentation of the design process, verify the design methodology used and the design itself, and provide support for analysis and explanation of the design process. KBDS is able to do this by recording the design artifact specification, the history of its evolution and the designer's rationale in a prescriptive form.

KBDS is a prototype computer-based support system for conceptual, integrated, and cooperative chemical processes design. KBDS is based on a representation that accounts for the evolutionary, cooperative and exploratory nature of the design process, covering design alternatives, constraints, rationale and models in an integrated manner. The design process is represented in KBDS by means of three interrelated networks that evolve through time: one for design alternatives, another for models of these alternatives, and a third for design constraints and specifications. Design rationale is recorded within IBIS network. Design rationale can be used to achieve dependency-directed backtracing in the event of a change to an external factor affecting the design. This suggests the potential advantages derived from the maintenance and further use of design rationale in the design process.

The change in design objectives, assumptions, or external factors is used as an example for an HDA plant. The effect on initial-phase-split, separations, etc. is shown as an effect of such changes. The effect of the change in the price of oil affects treatment-of lights, recycle-light-ends, good-use-of-raw-materials, vent/flare lights, lights-are-cheap-as-fuel, etc.

The use of design rationale information to support design can be used to improve the documentation of the design process, verify the design methodology used and the design itself, and provide support for analysis and explanation of the design process. KBDS is able to do this by recording the design artifact specification, the history of its evolution, and the designer's rationale in a prescriptive form.

2.30 Dependency-Directed Backtracking

Design objectives, assumptions, or external factors often change during the course of a design. Such changes may affect the validity of decisions previously made and thus require that the design is reviewed. If a change occurs the Intent Tool allows the designer to automatically check whether all issues have the most promising positions selected and thus determine from what point in the design history the review should take place. The decisions made for each issue where the currently selected position is not the most promising position should be reviewed.

The evolution of design alternatives for the separation section of the HDA plant is chosen as an example. An example of a change to a previous design decision (because the composition of the reactor effluent has changed) due to an alteration to the reactor operating conditions is another example. Also, the price of oil is an example of an external factor that affects the design.

2.31 Best Practice: Interactive Collaborative Environments

The computer scientists at Sandia National Laboratories developed a concurrent engineering tool that will allow project team members physically isolated from one another to simultaneously work on the same drawings. This technology is called Interactive Collaborative Environments (ICE). It is a software program and networking architecture supporting interaction of multiple X-Windows servers on the same program being executed on a client workstation. The application program executing in the X-Windows environment on a master computer can be simultaneously displayed, accessed and manipulated by other interconnected computers as if the program were being run locally on each computer. The ICE acts as both a client and a server. It is a server to the X-Windows client program that is being shared, and a client to the X-Servers that are participants in the collaboration.

Designers, production engineers, and the other groups can simultaneously sit at up to 20 different workstations at different geographic locations and work on the same drawing since all participants see the same menu-driven display. Any and all of the participants, if given permission by the master/client workstation, may edit the drawing or point to a feature with a mouse, and all work station pointers are all simultaneously displayed. Changes are immediately seen by everyone.

2.32 The Control Kit for O-Matrix

This is an ideal tool for a "classical" control system without the need for programming. It has a user friendly Graphical User Interface (GUI) with push buttons, radio buttons, etc. The user has many options to change the analysis, plot range, input format, etc., through a series of dialog boxes. The system is single input-single output and shows the main display when the program is invoked consisting of transfer functions (pushbuttons) and other operations (pulldown menus). The individual transfer functions may be entered as a ratio of s-polynomials, which allows for a very natural way of writing Laplace transfer functions.

Once the model has been entered, various control functions may be invoked. These are:

- Bode Plot
- Nyquist Plot
- Inverse Nyquist Plot
- Root Locus
- Step Response
- Impulse Response
- Routh Table and Stability
- Gain and Phase Margins

A number of facilities are available to the user regarding the way plots are displayed. These include:

- Possibility to obtain curves of the responses of both the compensated and uncompensated systems of the same plot, using different colors.
- Bode plot: The magnitude and phase plots may be displayed in the same window but if the user wishes to display them separately (to enhance the readability for example), it is also possible to do this sequentially in the same window.
- Nyquist plot: When the system is lightly damped, the magnitude becomes large for certain values of the frequency; in this case, ATAN Nyquist plots may be obtained which will lie in a unit circle for all frequencies. Again, both ordinary and ATAN Nyquist plots may be displayed in the same window.
- Individual points may be marked and their values displayed with the use of the cursor (for example the gain on the root locus or the frequency, magnitude, and phase in the Bode diagram).

The user can easily change the system parameters during the session by using dialog boxes. Models and plots may be saved and recalled.

1997 Progress Report: Development and Testing of Pollution Prevention Design Aids for Process Analysis and Decision Testing

This project is to create the evaluation and analysis module which will serve as the engine for design comparison in the CPAS Focus Area. The current title for this module is the Design Options Ranking Tool or DORT.

Through the use of case studies, it will be intended to demonstrate the use of the Dort module as the analysis engine for a variety of cost and non-cost measures which are being developed under CPAS or elsewhere. For example, the CPAS Environmental Fate and Risk Assessment Tool (EFRAT) and Safety Tool (Dow Indices Tools) are index generators that can be used to rank the processes with respect to environmental fate and safety. These process attributes can then be combined with cost or other performance measures to provide an overall rank of process options based on user-supplied index weightings. Ideally this information will be provided to the designer incrementally as the conceptual process design is being developed.

2.33 The Clean Process Advisory System: Building Pollution Into Design

CPAS is a system of software tools for efficiently delivering design information on clean technologies and pollution prevention methodologies to conceptual process and product designers on an as-needed basis. The conceptual process and process design step is where the potential to accomplish cost effective waste reduction is greatest. The goals of CPAS include:

reduce or prevent pollution
reduce cost of production
reduce costs of compliance
enhance U.S. global competitiveness
attain sustainable environmental performance

The attributes of CPAS include:

CPAS is a customizable, computer-based suite of design tools capable of easy expansion.

The tools are not intended to evaluate which underlying methodologies are correct or best, but rather to ensure all design options are presented and considered.

Tools that can be used as stand-alone or as an integrated system should be used to ensure that product and process designers will not have to wait until the entire system is released before using individual tools.

Each tool will interface with others and with commercial process simulators. The system will operate on a personal computer/workstation platform with access on the World Wide Web for some tools.

Nuclear Applications
Development of COMPAS, Computer-Aided Process Flowsheet Design and Analysis System of Nuclear-Fuel Reprocessing

A computer aided process flowsheet design and analysis system, COMPAS has been developed in order to carry out the flowsheet calculation on the process flow diagram of nuclear fuel reprocessing. All of the equipment in the process flowsheet diagram are graphically visualized as icons on a bitmap display of UNIX workstation. Drawing of the flowsheet can be carried out easily by the mouse operation. Specifications of the equipment and the concentrations of the components in the stream are displayed as tables and can be edited by a computer user. Results of calculations can also be displayed graphically. Two examples show that the COMPAS is applicable to decide operating conditions of the Purex process and to analyze extraction behavior in a mixer-settler extractor.

2.34 Nuclear Facility Design Considerations That Incorporate WM/P2 Lessons Learned

Many of the nuclear facilities that have been decommissioned or which are currently undergoing decommissioning have numerous structural features that do not facilitate implementation of waste minimization and pollution prevention (WM/P2) during decommissioning. Many were either “one of a kind” or “first of a kind” facilities at the time of their design and construction. They provide excellent opportunities for future nuclear facility designers to learn about methods of incorporating features in future nuclear facility designs that will facilitate WM/P2 during the eventual decommissioning of these next-generation nuclear facilities. Costs and the time for many of the decommissioning activities can then be reduced as well as risk to the workers. Some typical design features that can be incorporated include: improved plant layout design, reducing activation products in materials, reducing contamination levels in the plant, and implementing a system to insure that archival samples of various materials as well as actual “as built” and operating records are maintained.

Computer based systems are increasingly being used to control applications that can fail catastrophically, leading to either loss of life, injury, or significant economic harm. Such systems have hard timing constraints and are referred to as Safety Critical–Real Time (SC-RT) systems. Examples are flight control systems and nuclear reactor trip systems. The designer has to both provide functionality and minimize the risk associated with deploying a system. Adaptive search Techniques and Multi-Criteria Decision Analysis (MCDA) can be employed to support the designers of such systems. The Analysis-Synthesis-Evaluation (ASE) is used in software engineering. In this iterative technique the Synthesis element is concentrated on with "what-if" games and alternative solutions. In addition, in one example, architectural topology is used in ordering alternatives having a fitness function with adaptive search techniques.

2.35 Pollution Prevention Process Simulator

Conceptual design and pollution control traditionally have been performed at different stages in the development of a process. However, if the designer was given the tools to view a process's environmental impact at the very beginning of the design process, emphasis could be placed on pollution prevention and the selection of the environmentally sound alternatives. This could help eliminate total pollution as well as reduce the costs of the end-of-the-pipe treatment that is currently done. The Optimizer for Pollution Prevention, Energy, and Economics (OPPEE) started the development of such tools.

The concept of pollution prevention at the design stage started by OPPEE has grown into a much broader project called the Clean Process Advisory System (CPAS). CPAS has a number of complementary components that comprise a tool group:

The Incremental Economic and Environmental Analysis Tool which compares a process's pollution, energy requirements, and economics

An information-based Separation Technologies Database

Environmental Fate Modeling Tool

Pollution Prevention Process Simulator activities have been merged into the CPAS Design Comparison Tool Group.

2.36 Reckoning on Chemical Computers

(Dennis Rouvray, professor of chemistry in Dept of Chemistry at the University of Georgia, Athens, GA 30602-2556)

The days of squeezing ever more transistors onto silicon chips are numbered. The Chemical computer is one new technology that could be poised to take over, says Dennis Rouvray, but how will it perform?

The growth in the use of the electronic computer during the latter half of the 20th century has brought in its wake some dramatic changes. Computers started out as being rather forbidding mainframe machines operated by white-coated experts behind closed doors. In more recent times, however, and especially since the advent of PCs, attitudes have changed and we have become increasingly reliant on computers. The remarkable benefits conferred by the computer have left us thirsting for more: more computers and more powerful systems. Computer power is already astonishing. State-of-the-art computers are now able to compute at rates exceeding 109 calculations per second, and in a few years we should be able to perform at the rate of 1012 calculations per second. To the surprise of many, such incredible achievements have been accomplished against a backdrop of steadily falling prices. It has even been claimed that we are currently on the threshold of the era of the ubiquitous computer, an age when the computer will have invaded virtually every corner of our existence. But, the seemingly unstoppable progress being made in this area could be curtailed if a number of increasingly intractable problems are not satisfactorily solved.

Let us take a look at these problems and consider what our options might be. The breathtaking pace of computer development to date has been possible only because astounding human ingenuity has enabled us to go on producing ever more sophisticated silicon chips. These chips consist of tiny slivers of silicon on which are mounted highly complex arrays of interconnected electronic components, notably transistors. A single transistor (or group of transistors that performs some logic function) is referred to as a logic gate. Progress in achieving greater computer power ultimately depends on our ability to squeeze ever more logic gates on to each chip. It could be argued that the technology employed in fabricating very large-scale integrated (VLSI) chips is the most ingenious of all our modern technologies. By the end of the year 2000 we can confidently expect that it will be possible to cram as many as 1017 transistors into 1 cm^3 of chip. An oft-quoted but only semi-serious scientific law, known as Moore's Law, suggests that the number of transistors that can be accommodated on a single chip doubles every year. Until the mid-1970s this law appears to hold; since then the doubling period has gradually lengthened and is now closer to 18 months. This means that processor speed, storage capacity, and transmission rates are growing at an annual rate of about 60%, a situation that cannot be expected to continue into the indefinite future.

Current Status

Clearly, Moore's Law will eventually experience a major breakdown. Why will this occur and when might we expect it to happen? We are currently in a position to give fairly precise answers to both these questions. The breakdown will occur because the natural limits for the systems involved have been reached. When electronic components are miniaturized and packed extremely closely together they begin to interfere with one another. For example, the heat generated by operating them becomes very difficult to dissipate and so overheating occurs. Moreover, quantum tunneling by the electrons in the system assumes intolerable proportions. At present, the electronic components mounted on chips are not less than 0.5 um in size. By the year 2010, however, this size could well have shrunk to 0.1 um or less. Components of such dimensions will operate effectively only if several increasingly urgent issues have been resolved by that time. These include the heat dissipation problem, constructing suitable potential energy barriers to confine the electrons to their prescribed pathways, and developing new optical lithography techniques for etching the chips. Although imaginative new procedures will continue to be introduced to overcome these problems, the consensus is that by the year 2012 at the latest we will have exhausted all the tricks of the trade with our current technology. A transition to some new technology will then be imperative.

What then are our options for a brand new technology? The battle lines have already been drawn and it is now clear that two technologies will be competing to take over. These will be based either on a molecular computing system or a quantum computing system. At this stage it is not clear which of these will eventually become established, and it is even possible that some combination of the two will be adopted. What is clear is that the molecular computer has a good chance in the short term, because it offers a number of advantages. For example, the problems associated with its implementation tend to be mild in comparison with those for the quantum computer. Accordingly, it is believed that the molecular computer is likely to have replaced our present technology by the year 2025. This technology could, in turn, be replaced by the quantum computer. However, the arrival on the scene of the latter could well be delayed by another quarter of a century, thus making it the dominant technology around the year 2050. Both of these new technologies have the common feature that they depend on manipulating and controlling molecular systems. This implies, of course, that both will be operating in the domain in which quantum effects are paramount.

Future Prospects

The control of matter in the quantum domain and the exploitation of quantum mechanics present significant challenges. In the case of the quantum computer, for example, the whole operation is based on establishing, manipulating, and measuring pure quantum states of matter than can evolve coherently. This is difficult to achieve in practice and represents a field that is currently at the cutting edge of quantum technology. The molecular or chemical computer on the other hand gives rise to far fewer fundamental problems of this kind and is, at least in principle, quite feasible to set up. The primary difficulties lie rather in more practical areas, such as integrating the various component parts of the computer. The differences between the two technologies are well illustrated in the distinctive ways in which the various component parts are interconnected together in the two computers.

In quantum computers, connections are established between the components by means of optical communication, which involves using complex sequences of electromagnetic radiation, normally in the radio frequency range. A system that functions reliably is difficult to set up on the nanoscale envisaged. But for the molecular computer there are already several proven methods available for interconnecting the components. There is, for example, the possibility of using so-called quantum wires, an unfortunate misnomer because these have nothing to do with quantum computers. Research on quantum wires has been so extensive that there are now many options open to us. The most promising at present are made of carbon and are based on either single- or multi-walled carbon nanotubes. Single-walled nanotubes offer many advantages, including their chemical stability, their structural rigidity, and their remarkably consistent electrical behavior. In fact, they exhibit essentially metallic behavior and conduct via well separated electronic states. These states remain coherent over quite long distances, and certainly over the ca. 150 nm that is required to interconnect the various components.

Other possible starting materials for fabricating quantum wires include gallium arsenide and a variety of conducting polymers, such as polyacetylene, polyaniline, or polyacrylonitrile. When electrical insulation of these wires is necessary, molecular hoops can be threaded on to them to produce rotaxane-type structures.

Connecting the components of a chemical computer together is one thing. Having all the components at hand in suitable form to construct a working chemical computer is quite another. Can we claim that all the necessary components are currently available, at least in embryonic form? This would be going too far, though there are many signs that it should be feasible to prepare all these components in the not too distant future. Consider, for example, how close we are now to producing a molecular version of the key computer component, the transistor. A transistor is really no more than a glorified on/off switch. In traditional, silicon chip-based computers this device is more correctly referred to as a metal oxide semiconductor field effect transistor or Mosfet. The charge carriers (electrons in this case) enter at the source electrode, travel through two n-type regions (where the charge carriers are electrons) and one p-type channel (where the charge carriers are positive holes), and exit at the drain electrode. The Mosfet channel either permits or forbids the flow of the charge carriers depending on the voltage applied across the channel. Currently, a gap of 250 nm is used between the electrodes, but if this distance were reduced to below 10 nm, the charges could jump between the electrodes and render the transistor useless.

Chemical Computers

In the chemical computer this problem does not arise because the switching is carried out by individual molecules. The switching function is based on the reversible changing of some feature of the molecule. One could envisage that the relevant molecules are packed together in a thin molecular film and that each molecule is addressed independently by using a metallic probe of the kind used in scanning tunneling microscopy. Switching would thus be an integral feature of the molecular film and would exploit some aspect of the molecular structure of the species making up the film. The noting of molecules performing electronic functions is not new. As long ago as 1974, a proposal was put forward for a molecular rectifier that could function as a semiconductor p/n junction. Since then, researchers have synthesized a variety of molecular electronic switches.

The precise manner in which the different layers of molecular switches and other molecular components might be positioned in a chemical computer remains to be resolved. Moreover, the chemical techniques to be adopted in producing such complicated, three-dimensional arrays of molecules have yet to be worked out. Things are still at a rudimentary stage, though considerable experience has been amassed over the past two decades. In the case of one-dimensional structures, we now have at our disposal the well-known Merrifield polypeptide synthesis technique. This allows us to synthesize high yields of polypeptide chains in which the amino acids are linked together in some predetermined sequence. For two-dimensional structures, our extensive experience with Langmuir-Blodgett films makes it possible to build up arrays by successively depositing monolayers on to substrates while at the same time controlling the film thickness and spatial orientation of the individual species in the layers. More recent work on molecular assemblies constructed from covalent species demonstrates that the judicious use of surface modification along with appropriate self-assembly techniques should render it possible to construct ordered assemblies of bistable, photo-responsive molecules.

Looking Ahead

The chemical computer may as yet be little more than a glint in the eye of futurists. But substantial progress, especially over the past decade, has already been made toward its realization. As our need for radical new computer technology becomes increasingly urgent during the next decade, it seems likely that human ingenuity will see us through. Most of the components of a molecular computer, such as quantum wires and molecular switches, are already in existence. Several of the other molecular components could be used to replace our current silicon chip-based technology. Moreover, our rapidly accruing experience in manipulating the solid state, and knowledge of the self-assembly of complex arrays, should stand us in good stead for the tasks ahead. When the new technology will begin to take over is still uncertain, though few now doubt that it can be much more than a decade away. Rather than bursting on the scene with dramatic suddenness, however, this transition is likely to be gradual. Initially, for example, we might see the incorporation of some kind of molecular switch into existing silicon chip technology, which would increase switching speeds by several orders of magnitude. This could rely on pulses of electromagnetic radiation to initiate switching. Clearly, things are moving fast and some exciting challenges still lie ahead. But, if our past ingenuity does not fail us, it cannot be long before some type of molecular computer sees the light of day. Always assuming, of course, that unexpected breakthroughs in quantum technology do not allow the quantum computer to pip it to the post.

Part III. Computer Programs for Pollution Prevention and/or Waste Minimization

3.1 Pollution Prevention Using Chemical Process Simulation

Chemical process simulation techniques are being investigated as tools for providing process design and developing clean technology for pollution prevention and waste reduction.

HYSYS, commercially available process simulation software, is used as the basic design tool. ICPET is developing customized software, particularly for reactor design, as well as custom databases for the physical and chemical properties of pollutants, that can be integrated with HYSYS. Using these capabilities, studies are being carried out to verify reported emissions of toxic chemicals under "voluntary-action" initiatives and to compare the performance of novel technology for treating municipal solid waste with commercially available technology based on incineration processes.

3.2 Introduction to the Green Design

"Green Design" is intended to develop more environmentally benign products and processes. Some examples of practices include:

Solvent substitution in which single use of a toxic solvent is replaced with a more benign alternative, such as biodegradable solvents or non-toxic solvents. Water based solvents are preferable to organic based solvents. Technology change such as more energy efficient semiconductors or motor vehicle engines. For example, the Energy Star program specifies maximum energy consumption standards for computers, printers, and other electronic devices. Products in compliance can be labeled with the "Energy Star." Similarly, "Green Lights" is a program that seeks more light from less electricity.

Recycling of toxic wastes can avoid dissipation of the materials into the environment and avoid new production. For example, rechargeable nickel-cadmium batteries can be recycled to recover both cadmium and nickel for other uses. Inmetco Corporation in Pennsylvania and West Germany are routinely recycling such batteries using pyrometallurgical distillation.

Three goals for green design are:

Reduce or minimize the use of non-renewable resources;

Manage renewable resources to ensure sustainability and;

Reduce, with the ultimate goal of eliminating toxic and otherwise toxic harmful emissions to the environment, including emissions contributing to global warming.

The object of green design is to pursue these goals in the most cost-effective fashion. A green product or process is not defined in any absolute sense, but only in comparison with other alternatives of similar function. For example, a product could be entirely made of renewable materials, use renewable energy, and decay completely at the end of its life. However, this product would not be green if, for example, a substitute product uses fewer resources during production and uses or results in the release of fewer hazardous materials.

Green products imply more efficient resource use, reduced emission, and reduced waste, lowering the social cost of pollution control and environmental protection. Greener products promise greater profits to companies by reducing costs (reduced material requirements, reduced disposal fees, and reduced environmental cleanup fees) and raising revenues through greater sales and exports.

How can an analyst compare a pound of mercury dumped into the environment with a pound of dioxin? Green indices or ranking systems attempt to summarize various environmental impacts into a simple scale. The designer or decision maker can then compare the green score of alternatives (materials, processes, etc.) and choose the one with minimal environmental impacts. This would contribute to products with reduced environmental impacts.

Following are some guiding principles for materials selection:

Choose abundant, non-toxic materials where possible.

Choose materials familiar to nature (e.g., celluose), rather than man-made materials (e.g., chlorinated aromatics).

Minimize the number of materials used in a product or process.

Try to use materials that have an existing recycling infrastructure.

Use recycled materials where possible.

Companies need management information systems that reveal the cost to the company of decisions about materials, products, and manufacturing processes. This sort of system is called a "Full cost accounting" system. For example, when an engineer is choosing between protecting a bolt from corrosion by plating it with cadmium vs. choosing a stainless steel bolt, a full cost accounting system could provide information about the purchase price of two bolts and the additional costs to the company of choosing a toxic material such as cadmium.

Green Design is the attempt to make new products and processes more environmentally benign by making changes in the design phase.

3.3 Chemicals and Materials from Renewable Resources

Renewable carbon is produced at a huge annual rate in the biosphere and has been regarded as a valuable source of useful chemicals, intermediates, and new products. The use of renewable feedstocks will progressively move toward a CO_2 neutral system of chemical production. A "biomass refinery" describes a process for converting renewable carbon into these materials. The petrochemical industry, however, has a significant lead in technology for selectively converting their primary raw material into products. The scope of methodology for conversion of biomass is much smaller and the list of products available from biomass is much shorter than for petrochemicals.

Tools are needed to transform selectively nontraditional feedstocks into small molecules (non-fuel applications) and discrete building blocks from renewables. Feedstocks include monosaccharides, polysaccharides (celluose, hemicelluose, and starch), extractives, lignin, lipids, and proteinaceous compounds. New transformations of these feedstocks using homogeneous and heterogeneous catalysis are needed as are new biochemical transformations. Sessions on synthesis and use of levuinic acid and levoglucosan, as well as sessions on new transformations and new building blocks from renewables are necessary.

3.4 Simulation Sciences

Commercial software packages allow engineers to quickly and easily evaluate a wide range of process alternatives for batch plants. To reduce costs for specialty chemical and pharmaceutical plants manufacturing high-value products requires many hours of engineering time or the use of process simulation. Commercial simulator packages have replaced in house tools over the last 10 to 15 years. They are also much improved. They can address waste minimization. Following are several examples.

Solvents can either be sent to waste disposal or recovered. Since recovery is preferred, simulation can be used to answer the questions:

Batch or continuous distillation?
What equipment is available?
Are there enough trays?
What should the reflux ratio be?
Where should the feed go?

One can optimize a simple flash off the reactor, determine cut points at various purity levels, etc.

A simulator can also remove bad actors from waste streams with liquid extraction. The questions of how many theoretical stages are needed and which solvents are best can be determined. Some reactive components are unstable and hazardous, so disposal may not be recommended by a carrier, etc.

Simulators may help with controlling vapor emissions. Absorbers may be designed with the right number of stages, the right number of vapor/liquid ratios. Pilot work can be cut down. The simulator can help to find the right diameter, etc., also ensuring minimum cost.

Simulators can help with distillation, crystallization, and flash performance, ensuring proper solvents and process development work.

They can evaluate whether the most cost-effective solids removal procedure is in place.

They also have improved greatly in their physical generation capability so important in developing process systems.

Simulators are very useful in evaporative emissions reports, and are important for government reporting records.

They are very important for a plant's emergency relief capabilities, needed for both safety and process capability.

They can help tell whether the vapor above a stored liquid is flammable.

3.5 EPA/NSF Partnership for Environmental Research

Research proposals were invited that advance the development and use of innovative technologies and approaches directed at avoiding or minimizing the generation of pollutants at the source. The opening date was November 18, 1997 and the closing date was February 17, 1998.

NSF and EPA are providing funds for fundamental and applied research in the physical sciences and engineering that will lead to the discovery, development, and evaluation of advanced and novel environmentally benign methods for industrial processing and manufacturing. The competition addresses technological environmental issues of design, synthesis, processing and production, and use of products in continuous and discrete manufacturing industries. The long-range goal of this program activity is to develop safer commercial substances and environmentally friendly chemical syntheses to reduce risks posed by existing practices. Pollution prevention has become the preferred strategy for reducing the risks posed by the design, manufacture, and use of commercial chemicals. Pollution Prevention at the source involves the design of chemicals and alternative chemical syntheses that do not utilize toxic feedstocks, reagents, or solvents, or do not produce toxic by-products or co-products. Investigations include:

> Development of innovative synthetic methods by means of catalysis and biocatalysis; photochemical, electrochemical or biomimetric synthesis; and use of starting materials which are innocuous or renewable.
>
> Development of alternative and creative reaction conditions, such as using solvents which have a reduced impact on health and the environment, or increasing reaction selectivity thus reducing wastes and emissions.
>
> Design and redesign of useful chemicals and materials such that they are less toxic to health and the environment or safer with regard to accident potential.

The aim of this activity is to develop new engineering approaches for preventing or reducing pollution from industrial manufacturing and processing activities, both for continuous and discrete processes. The scope includes: technology and equipment modifications, reformulation or redesign of products, substitution of alternative materials, and in-process changes. Although these methods are thought of in the chemical, biochemical, and materials process industries, they are appropriate in other industries as well, such as semiconductor manufacturing systems. Areas of research include:

- Biological Applications: Includes bioengineering techniques such as metabolic engineering and bioprocessing to prevent pollution. Examples are conversion of waste biomass to useful products, genetic engineering to produce more specific biocatalysts, increase of energy efficiency, decreased use of hazardous reactants or byproducts, or development of more cost effective methods of producing environmentally benign products.
- Fluid and Thermal Systems: Includes improved manufacturing systems that employ novel thermal or fluid and/or multiphase/particulate systems resulting in significantly lower hazardous effluent production. Examples are novel refrigeration cycles using safe and environmentally benign working fluids to replace halogenated hydrocarbons hazardous to upper atmosphere ozone levels; improved automobile combustion process design for reduced pollutant production.
- Interfacial Transport and Separations: Includes materials substitutions and process alternatives which prevent or reduce environmental harm, such as change of raw materials or the use of less hazardous solvents, organic coatings, and metal plating systems where the primary focus is on non-reactive diffusional and interfacial phenomena. Examples include: use of special surfactant systems for surface cleaning and reactions; novel, cost-effective methods for the highly efficient in-process separation of useful materials from the components of process waste streams (for example, field enhanced and hybrid separation processes); novel processes for molecularly chemical and materials synthesis of thin films and membranes.
- Design, Manufacturing, and Industrial Innovations: Includes: (a) New and improved manufacturing processes that reduce production of hazardous effluents at the source. Examples include: machining without the use of cutting fluids that currently require disposal after they are contaminated; eliminating toxic electroplating solutions by replacing them with ion or plasma-

based dry plating techniques; new bulk materials and coatings with durability and long life; and other desirable engineering properties that can be manufactured with reduced environmental impact. (b) Optimization of existing discrete parts manufacturing operations to prevent, reduce, or eliminate waste. Concepts include: increased in-process or in-plant recycling and improved and intelligent process control and sensing capabilities; in-process techniques that minimize generation of pollutants in industrial waste incineration processes.

- Chemical Processes and Reaction Engineering: Includes improved reactor, catalyst, or chemical process design in order to increase product yield, improve selectivity, or reduce unwanted by-products. Approaches include novel reactors such as reactor-separator combinations that provide for product separation during the reaction, alternative energy sources for reaction initiation, and integrated chemical process design and operation, including control. Other approaches are: new multifunctional catalysts that reduce the number of process stages; novel heterogeneous catalysts that replace state-of-the -art homogeneous ones; new photo- or electro catalysts that operate at low temperatures with high selectivity; novel catalysts for currently uncatalyzed reactions; processes that use renewable resources in place of synthetic intermediates as feedstocks; novel processes for molecularly controlled materials synthesis and modification.

3.6 BDK-Integrated Batch Development

This program is an integrated system of software and is advertised as capable of streamlining product development, reducing development costs, and accelerating the time it takes to market the products. It is said to allow a rapid selection of the optimum chemical synthesis and manufacturing routes with consideration of scale-up implications, have a seamless transfer of documentation throughout the process and a smoother path to regulatory compliance and a optimized supply chain, waste processing, equipment allocation and facility utilization costs.

Furthermore, it identifies the optimum synthetic route and obtains advice on raw material costs, yields, and conversion and scale-up; finds the smoothest path to comply with environmental, safety, and health regulations; uses equipment selection expert systems to draw on in-depth knowledge of the unit operations used in batch processing; increases efficiency in the allocation and utilization of facilities; enables product development chemists and process development engineers to share a common frame of reference that supports effective communication, information access, and sharing throughout the project, and captures the corporate product development experience and shares this among future product development teams. There are other claims for this program that were developed by Dr. Stephanopoulos and co-workers at MIT.

3.7 Process Synthesis

Process Synthesis is the preliminary step of process design that determines the optimal structure of a process system (cost minimized or profit maximized). This essential step in chemical engineering practice has traditionally relied on experience-based and heuristic or rule-of-thumb type methods to evaluate some feasible process designs. Mathematical algorithms have then been used to find the optimal solution from these manually determined feasible process design options. The fault in this process is that it is virtually impossible to manually define all of the feasible process system options for systems comprising more than a few operating units. This can result in optimizing a set of process system design options that do not even contain the global optimal design.

For example, if a process has over 30 operating units available to produce desired end products, there are about one billion possible combinations available. Now, a systematic, mathematical software method to solve for the optimal solution defining all of the feasible solutions from a set of feasible operating units has been developed, and this software method performs well on standard desktop computers. A discussion of the mathematical basis and cost estimation methods along with a glimpse of this new software is presented.

Friedler and Fan have discovered a method for process synthesis. It is an extremely versatile, innovative and highly efficient method that has been developed to synthesize process systems based on both graph theory and combinatorial techniques. Its purpose is to cope with the specificities of a process system. The method depicts the structure of any process system by a unique bipartite graph, or P-graph in brief, wherein both the syntactic and semantic contents of the system are captured. An axiom system underlying the method has been established to define exactly the combinatorial feasible process structures. The method is capable of rigorously generating the maximal structure comprising every feasible possible structure or flowsheet for manufacturing desired products from given raw materials provided that all plausible operating units are given and the corresponding intermediates are known. The method is also capable of generating the optimal and some near-optimal structures or flowsheets from the maximal structure in terms of either a linear or non-linear cost function. The task is extremely difficult or impossible to perform by any available process synthesis method. Naturally the optimal and near-optimal flowsheets can be automatically forwarded to an available simulation program for detailed analysis, evaluation, and final selection. Such effective integration between synthesis and analysis is rendered by adhering to the combinatorial techniques in establishing the method. The maximal structure may be construed as the rigorously constructed superstructure with minimal complexity. The superstructure as traditionally generated in the MINLP (Mixed Integer Non-linear Programming) or MILP (Mixed Integer Linear Programming) approach, has never been mathematically defined; therefore, it is impossible to derive it algorithmically.

The method has been implemented on PCs with Microsoft Windows because the search space is drastically reduced by a set of axioms forming the foundation of the method and also because the procedure is vastly sped up by the accelerated branch and bound algorithm incorporated in the method. To date, a substantial number of process systems have been successfully synthesized, some of which are industrial scale containing more than 30 pieces of processing equipment, i.e., operating units. Nevertheless, the times required to complete the syntheses never exceeded several minutes on the PCs; in fact, they are often in the order of a couple of minutes or less. Unlike other process-synthesis methods, the need for supercomputers, main-frame computers, or even high-capacity workstations is indeed remote when the present method is applied to commercial settings. Intensive and exhaustive efforts are ongoing to solidify the mathematical and logical foundation, extend the capabilities, and improve the efficiency of the present method. Some of these efforts are being carried out in close collaboration with Friedler and Fan and others are being undertaken independently. In addition, the method has been applied to diverse processes or situations such as separation processes, azeotropic distillation, processes with integrated waste treatment, processes with minimum or no waste discharges, waste-water treatment processes, chemical reactions in networks of reactors, biochemical processes, time-staged development of industrial complexes or plants, and retrofitting existing processes. Many of these applications have been successfully completed.

A new approach, based on both graph theory and combinatorial techniques, has been used to facilitate the synthesis of a process system. This method copes with the specifics of a process sytem using a unique bipartite graph (called a P-graph) and captures both the syntactic and semantic contents of the process system. There is an axiom system underlying the approach and it has been constructed to define the combinatorial feasible process structures. This axiom system is based on a set of specifications for the process system problem. They include the types of operating units and the raw materials, products, by-products, and a variety of waste associated with the operating units. All feasible structures of the process system are embedded in the maximal structure, from which individual solution-structures can be extracted subject to various technical, environmental, economic, and societal constraints. Various theorems have been derived from the axiom system to ensure that this approach is mathematically rigorous, so that it is possible to develop efficient process synthesis methods on the basis of a rigorous mathematical foundation.

Analysis of the combinatorial properties of process synthesis has revealed some efficient combinatorial algorithms. Algorithm MSG generates the maximal structure (super-structure) of a process synthesis problem and can also be the basic algorithm in generating a mathematical programming model for this problem. This algorithm can also synthesize a large industrial process since its complexity grows merely polynomially with the size of the synthesized process. Another algorithm, SSG, generates the set of feasible process structures from the maximal structure; it leads to additional combinatorial algorithms of process synthesis including those for decomposition and for accelerating branch and bound search. These algorithms have also proved themselves to be efficient in solving large industrial synthesis problems.

Process synthesis has both combinatorial and continuous aspects; its complexity is mainly due to the combinatorial or integer variable involved in the mixed integer-nonlinear programming (MINLP) model of the synthesis. The combinatorial variables of the model affect the objective or cost function more profoundly than the continuous variable of this model. Thus, a combinatorial technique for a class of process synthesis problems has been developed and it is based on directed bipartite graphs and an axiom system. These results have been extended to a more general class of process design problems.

A large set of decisions is required for the determination of the continuous or discrete parameters when designing a chemical process. This is especially true if waste minimization is taken into account in the design. Though the optimal values of the continuous variables can usually be determined by any of the available simulation or design programs, those of the discrete parameters cannot be readily evaluated. A computer program has been developed to facilitate the design decisions on the discrete parameters. The program is based on both the analysis of the combinatorial properties of process structures and the combinatorial algorithms of process synthesis.

The very complex decisions of process synthesis occurs because the decisions are concerned with specifications or identification of highly connected systems such as process structures containing many recycling loops. Now, a new mathematical notion, decision mapping, has been introduced. This allows us to make consistent and complete decisions in process design and synthesis. The terminologies necessary for decision-mappings have been defined based on rigorous set theoretic formalism, and the important properties of decision-mappings.

Process network synthesis (PNS) has enormous practical impact; however, its mixed integer programming (MIP) is tedious to solve because it usually involves a large number of binary variables. The recently proposed branch-and-bound algorithm exploits the unique feature of the MIP model of PNS. Implementation of the algorithm is based on the so-called decision-mapping that consistently organizes the system of complex decisions. The accelerated branch-and-bound algorithm of PNS reduces both the number and size of the partial problems.

3.8 Synphony

Synphony provides the ability to determine all feasible flowsheets from a given set of operating units and raw materials to produce a given product and then ranks these by investment and operating costs. Synphony has been proven to significantly reduce investment and operating costs by minimizing by-products and identifying the best overall design. The software analyzes both new process designs and retrofits of existing operations to generate all feasible solutions and ranks the flowsheets based on investment and operating costs.

The program called Synphony is commercially available. A case study using Synphony at a manufacturing facility demonstrated a 40% reduction in waste water and a 10% reduction in operating costs. The software analyzes both new process designs and retrofits of existing operations to generate all feasible solutions and ranks the flowsheets based on investment and operating costs. Synphony is the first flowsheet synthesis software program to rigorously define all feasible flowsheet structures from a set of feasible unit operations and to rank flowsheets according to the lowest combined investments and operating costs. All feasible flowsheets are determined from a set of operating units and raw materials to produce a given product and then these flowsheets are ranked by investment and operating costs. Each solution can be viewed numerically or graphically from the automatically generated flowsheets. A significant advantage of Synphony is that it generates all feasible flowsheet solutions while not relying on previous knowledge or heuristic methods. If the objective is to minimize waste, Synphony has been proven to achieve significant reductions while also reducing operating costs.

3.9 Process Design and Simulations

Aspen is a tool that can be used to develop models of any type of process for which there is a flow of materials and energy from one processing unit to the next. It has modeled processes in chemical and petrochemical industries, petroleum refining, oil and gas processing, synthetic fuels, power generation, metals and minerals, pulp and paper, food, pharmaceuticals, and biotechnology. It was developed at the Department of Chemical Engineering and Energy Laboratory of the Massachusetts Institute of Technology under contract to the United States Department of Energy (DOE). Its main purpose under that contract was the study of coal energy conversion.

Aspen is a set of programs which are useful for modeling, simulating, and analyzing chemical processes. These processes are represented by mathematical models, which consist of systems of equations to be solved. To accomplish the process analysis, the user specifies the interconnection and the operating conditions for process equipment. Given values of certain known quantities, Aspen solves for the unknown variables. Documentation is available and the ASPEN PLUS Physical Properties Manual is very important.

Aspen Tech's Smart Manufacturing Systems (SMS) provides model-centric solutions to vertical and horizontal integrated management systems. These embody Aspen Tech's technology in the area of modeling, simulation, design, advanced control, on-line optimization, information systems, production management, operator training, and planning and scheduling. This strategy is enabled by integrating the technology through a Design-Operate-Manage continuous improvement paradigm.

The consortium in Computer-Aided Process Design (CAPD) is an industrial body within the Department of Chemical Engineering at CMU that deals with the development of methodologies and computer tools for the process industries. Directed by

Professors Biegler, Grossmann, and Westerberg, the work includes process synthesis, process optimization, process control, modeling and simulation, artificial intelligence, and scheduling and planning.

Unique software from Silicon Graphics/Cray Research allows virtual plant, computational fluid dynamics analysis, and complex simulations. The CFD analysis solution focuses on analyzing the fluid flows and associated physical phenomena occurring as fluids mix in a stirred tank or fluidized bed, providing new levels of insight that were not possible through physical experimentation.

Advances in computational fluid dynamics (CFD) software have started to impact the design and analysis processes in the CPI. Watch for them.

Floudas at Princeton has discussed the computational framework/tool MINOPT that allows for the efficient solution of mixed-integer, nonlinear optimization (MINLP) methodologies and their applications in Process Synthesis and Design with algebraic and/or dynamic constraints. Such applications as the areas of energy recovery, synthesis of complex reactor networks, and nonideal azeotropic distillation systems demonstrate the capabilities of MINOPT.

Paul Matthias has stated that the inorganic-chemical, metals, and minerals processing industries have derived less benefit from process modeling than the organic-chemical and refining industries mainly due to the unique complexity of the processes and the lack of focused and flexible simulation solutions. He highlighted tools needed (i.e., thermodynamic and transport properties, chemical kinetics, unit operations), new data and models that are needed, how models can be used in day-to-day operations, and most important, the characteristics of the simulation solutions that will deliver business value in such industries.

The industrial perspective of applying new, mostly graphical tools for the synthesis and design of nonideal distillation systems reveals the sensitivity of design options to the choice of physical properties representation in a more transparent way than simulation, and such properties are very useful in conjunction with simulation.

Barton discusses three classes of dynamic optimization problems with discontinuities: path constrained problems, hybrid discrete/continuous problems, and mixed-integer dynamic optimization problems.

3.10 Robust Self-Assembly Using Highly Designable Structures and Self-Organizing Systems

Through a statistical exploration of many possibilities, self-assembly creates structures. These explorations may give rise to some highly designable structures that can be formed in many different ways. If one uses such structures for self-assembly tasks, a general approach to improving their reliability will be realized.

Manufacturing builds objects from their components by placing them in prescribed arrangements. This technique requires knowledge of the precise structure needed to serve a desired function, the ability to create the components with the necessary tolerances, and the ability to place each component in its proper location in the final structure.

If such requirements are not met, self-assembly offers another approach to building structures from components. This method involves a statistical exploration of many possible structures before settling into a possible one. The particular structure produced from given components is determined by biases in the exploration, given by component interactions. These may arise when the strength of the interactions depends on their relative locations in the structure. These interactions can reflect constraints on the desirability of a component being near its neighbors in the final structure. For each possible structure the intersections combine to give a measure of the extent to which the constraints are violated, which can be viewed as a cost of energy for that structure. Through the biased statistical exploration of structures, each set of components tends to assemble into that structure with the minimum energy for that set. Thus, self-assembly can be viewed as a process using a local specification, in terms of the components and their interactions, to produce a resulting global structure. The local specification is, in effect, a set of instructions that implicitly describes the resulting structure.

We describe here some characteristics of the statistical distributions of self-assembled structures.

Self-assembly can form structures beyond the current capacity of direct manufacturing. The most straightforward technique for designing self-assembly is to examine with a computer simulation the neighbors of each component in the desired global structure, and then choose the interactions between components to encourage these neighbors to be close together.

A difficulty in designing the self-assembly process is the indirect or emergent connection between the interactions and the properties of resulting global structures. There is a possibility of errors due to defective components or environmental noise. To address this problem, it would be useful to arrange the self-assembly so the desired structure can be formed in many ways, increasing the likelihood they will be correctly constructed even with some unexpected changes in the components or their interactions. That is, the resulting global structure should not be too sensitive to errors that may occur in the local specification.

A given assembly can then be characterized by the number of different component configurations producing a given global designability. Self-assembly processes with skewed distributions of designability can also produce relatively large energy gaps for the highly designable structures. A large energy gap with small changes in the energies of all the global structures do not change the one with the minimum energy, but small changes with a small gap are likely to change the minimum energy structure. If there are several structures that adjust reasonably well to the frustrated constraints in different ways, the energy differences among these local minima will determine the gap.

Self-assembly of highly designable structures is particularly robust, both with respect to errors in the specification of the components and environmental noise. Thus we have a general design principle for robust self-assembly: select the components, interactions and possible global structures so the types of structures desired for a particular application are highly designable.

Applying this principle requires two capabilities. The first is finding processes leading to highly designable structures of the desired forms. The second requirement is the ability to create the necessary interactions among the components

Achieving a general understanding of the conditions that give rise to highly designable structures is largely a computational problem that can be addressed before actual implementations become possible. Thus developing this principle for self-assembly design is particularly appropriate in situations where explorations of design possibilities take place well ahead of the necessary technnological capabilities. Even after the development of precise fabrication technologies, principles of robust self-assembly will remain useful for designing and programming structures that robustly adjust to changes in their environments or task requirements.

3.11 Self-Organizing Systems

Some mechanisms and preconditions are needed for systems to self-organize. The system must be exchanging energy and/or mass with its environment. A system must be thermodynamically open because otherwise it would use up all the available usable energy in the system (and maximize its entropy) and reach thermodynamic equilibrium

If a system is not at or near equilibrium, then it is dynamic. One of the most basic kinds of change for SOS is to import usable energy from its environment and export entropy back to it. Exporting entropy is another way to say that the system is not violating the second law of thermodynamics because it can be seen as a larger system-environment unit. This entropy-exporting dynamic is the fundamental feature of what chemists and physicists call dissipative structures. Dissipation is the defining feature of SOS.

The magic of self-organization lies in the connections, interactions, and feedback loops between the parts of the system; it is clear that SOS must have a large number of parts. These parts are often called agents because they have the basic properties of information transfer, storage, and processing.

The theory of emergence says the whole is greater than the sum of the parts, and the whole exhibits patterns and structures that arise spontaneously from the parts. Emergence indicates there is no code for a higher-level dynamic in the constituent, lower-level parts.

Emergence also points to the multiscale interactions and effects in self-organized systems. The small-scale interactions produce large-scale structures, which then modify the activities at the small scales. For instance, specific chemicals and neurons in the immune system can create organism-wide bodily sensations which might then have a huge effect on the chemicals and neurons. Prigogine has argued that micro-scale emergent order is a way for a system to dissipate micro-scale entropy creation caused by energy flux, but this is still not theoretically supported.

Even knowing that self-organization can occur in systems with these qualities, it's not inevitable, and it's still not clear why it sometime does. In other words, no one yet knows the necessary and sufficient conditions for self-organization.

3.12 Mass Integration

An industrial process has two important dimensions: (1) Mass which involves the creation and routing of chemical species. These operations are performed in the reaction, separation, and by-product/waste processing systems. These constitute the core of the process and define the company's technology base. (2) Energy which is processed in the supporting energy systems to convert purchased fuel and electric power into the forms of energy actually used by the process, for example, heat and shaft work. Design, part science and part art, commands a detailed understanding of the unit operation building blocks. They must be arranged to form a complete system which performs desired functions. It starts with a previous design and uses experience-based rules and know-how along with their creativity to evolve a better design. They are aided by computer-based tools such as process simulators and unit operation design programs.

These designs have scope for improvement, frequently large and expensive. Now engineers realize it is just as important to assemble the building blocks correctly as it is to select and design them correctly as individual components. This led to integrated process design or process integration which is a holistic approach to design that emphasizes the unity of the whole process. Pinch analysis was an example of this. It is the definitive way to design heat recovery networks, to select process wide utility heating and cooling levels to establish the energy/capital tradeoff for heat recovery equipment. Mass flow is more recent in process integration. It is similar to energy integration but tackles the core of the process and has a consequence of more direct and significant impact on process performance. It addresses the conversion, routing, and separation of mass and deals directly with the reaction, separation, and byproduct/waste processing systems. It guides designers in routing all species to their most desirable destinations and allows them to establish mass-related cost tradeoffs. Mass integration also defines the heating, cooling, and shaft work requirements of the process. It also provides insight into other design issues such as providing resources (e.g., fuel and water) to break up bottlenecks in the utility systems and selecting the catalysts and other material utilities.

3.13 Synthesis of Mass Energy Integration Networks for Waste Minimization via In-Plant Modification

In recent years academia and industry envisioned the development of the transshipment of a commodity (the pollutant) from a set of sources to a set of sinks to address pollution prevention. Some of the design tools developed on the basis of this approach are: Mass Exchange Networks (MENs), Reactive Mass Exchange Networks (REAMENs), Combined Heat and Reactive Mass Exchange Networks (CHARMENs), Heat Induced Separation Networks (HISENs), and Energy Induced Separation Networks (EISENs). These designs are systems based (rather than unit based) and trade off the thermodynamic, economic, and environmental constraints on the system. They answer the questions: (1) What is the minimum cost required to achieve a specified waste reduction task, and (2) What are the optional technologies required to achieve the specified waste reduction task? They are applicable only towards the optimal designed end-of-pipe waste reduction systems. However, source reduction is better because of regulatory agencies and also for economic incentives. This is attributed to the fact that unit cost of separation increases significantly with dilution (i.e., lower costs for concentrated streams that are within the process, and higher costs for dilute, end-of -pipe streams). Thus, it is important that systematic design techniques target waste minimization from a source reduction perspective.

3.14 Process Design

Process design uses molecular properties extensively and it is a very important part of such work. El-Halwagi uses the concept of integrated process design or process integration which is a holistic approach to design that emphasizes the unity of the whole process. He states that powerful tools now exist for treating industrial processes and sites as integrated systems. These are used together with a problem solving philosophy that involves addressing the big picture first, using fundamental principles, and dealing with details only after the major structural decisions are made. In further work, two approaches are developed: graphical and algorithmic. In the graphical approach, a new representation is developed to provide a global tracking for the various species of interest. The graphical approach provides a global understanding of optimum flow, separation, and conversion of mass throughout the process. It also provides a conceptual flowsheet that has the least number of processing stages. In the algorithmic approach, the problem is formulated as an optimization program and solved to identify the optimum flowsheet configuration along with the optimum operating conditions.

A systematic tool is developed to screen reaction alternatives without enumerating them. This task of synthesizing Environmentally Acceptable Reactions is a mixed-integer non-linear optimization program that examines overall reactions occurring in a single reactor to produce a specified product. It is designed to maximize the economic potential of the reaction subject to a series of stoichiometric, thermodynamic and environmental constraints. It is a screening tool, so additional laboratory investigation, path synthesis, kinetics, and reactor design may be needed, but it is an excellent starting point to plan experimental work.

3.15 Pollution Prevention by Reactor Network Synthesis

Chemical Reactor Synthesis is the task of identifying the reactor or network of reactors which transform raw materials to products at optimum cost. Given a set of chemical reactions with stoichiometry and kinetics, the goal is to find the type, arrangements, and operating conditions of reactors which meet design constraints. Reactor network synthesis is a powerful tool since it gives the optimum reactor flowsheet while minimizing cost. However, reactor synthesis is difficult to achieve. Recently, a geometric approach has shown promise as a method of reactor network synthesis. The strategy is to construct a region defining all possible species concentrations which are attainable by any combination of chemical reaction and/or stream mixing; this is called the Attainable Region (AR). The two types of chemical reactors considered in this work are the Plug Flow Reactor (PFR) and the Continuously Stirred Reactor (CSTR). Once the AR is defined, the reactor network optimization is essentially solved. The synthesis of the optimum reactor network coincides with the construction of the AR. An algorithm for generating candidate attainable regions is available.

3.16 LSENS

LSENS, from the NASA Lewis Research Center, has been developed for solving complex, homogeneous, gas-phase, chemical kinetics problems. It was motivated by the interest in developing detailed chemical reaction mechanisms for complex reactions such as the combustion of fuels and pollutant formation and destruction. Mathematical descriptions of chemical kinetics problems constitute sets of coupled, nonlinear, first-order ordinary differential equations (ODEs). The number of ODEs can be very large because of the numerous chemical species involved in the reaction mechanism. Further complicating the situation are the many simultaneous reactions needed to describe the chemical kinetics of practical fuels. For example, the mechanism describing the oxidation of the simplest hydrocarbon fuel, methane, involves over 25 species participating in nearly 100 elementary reaction steps. Validating a chemical reaction mechanism requires repetitive solutions of the governing ODEs for a variety of reaction conditions. Consequently, there is a need for fast and reliable numerical solution techniques for chemical kinetics problems. In addition to solving the ODEs describing chemical kinetics, it is often necessary to know what effects variations in either initial condition values or chemical reaction parameters have on the solution. Such a need arises in the development of reaction mechanisms from experimental data. The rate coefficients are often not known with great precision and in general, the experimental data are not sufficiently detailed to accurately estimate the rate coefficient parameters. The development of reaction mechanism is facilitated by a systematic sensitivity analysis which provides the relationships between the predictions of a kinetics model and the input parameters of the problem.

3.17 Chemkin

Complex chemically reacting flow simulations are commonly employed to develop a quantitative understanding and to optimize reaction conditions in systems such as combustion, catalysis, chemical vapor deposition, and plasma processing. They all share the need for accurate, detailed descriptions of the chemical kinetics occurring in the gas-phase or on reactive surfaces. The Chemkin suite of codes broadly consists of three packages for dealing with gas-phase reaction kinetics, heterogeneous reaction kinetics, and species transport properties. The Chemkin software was developed to aid the incorporation of complex gas-phase reaction mechanisms into numerical simulations. Currently, there are a number of numerical codes based on Chemkin which solve chemically reacting flows. The Chemkin interface allows the user to specify the necessary input through a high-level symbolic interpreter, which parses the information and passes it to a Chemkin application code. To specify the needed information, the user writes an input file declaring the chemical elements in the problem, the name of each chemical species, thermochemical information about each chemical species, a list of chemical reactions (written in the same fashion a chemist would write them), and rate constant information, in the form of modified Arrhenius coefficients. The thermochemical information is entered in a very compact form as a series of coefficients describing the species entropy (S), enthalpy (H), and heat capacity (Cp) as a function of temperature. The thermochemical database is in a form compatible with the widely used NASA chemical equilibrium code. Because all of the information about the reaction mechanism is parsed and summarized by the chemical interpreter, if the user desires to modify the reaction mechanism by adding species or deleting a reaction, for instance, they only change the interpreter input file and the Chemkin application code does not have to be altered. The modular approach of separating the description of the chemistry from the set-up and solution of the reacting flow problem allows the software designer great flexibility in writing chemical-mechanism-independent code. Moreover, the same mechanism can be used in different chemically reacting flow codes without alteration.

Once the nature of the desired or substituted product or intermediate or reactant is known, we wish to describe how it and the other species change with time, while obeying thermodynamic laws. In order to do this we use another program called

Envirochemkin, which is derived from a program called Chemkin.

Chemkin is a package of FORTRAN programs which are designed to facilitate a chemist's interaction with the computer in modeling chemical kinetics. The modeling process requires that the chemist formulate an applicable reaction mechanism (with rate constants) and that he formulate and solve an appropriate system of governing equations.

The reaction mechanism may involve any number of chemical reactions that concern the selected named species. The reactions may be reversible or irreversible, they may be three body reactions with an arbitrary third body, including the effects of enhanced third body efficiencies, and they may involve photon radiation as either a reactant or product.

The program was used by Bumble for air pollution, water pollution, biogenic pollution, stationary sources, moving sources, remedies for Superfund sites, environmental forensic engineering, the stratospheric ozone problem, the tropospheric ozone problem, smog, combustion problems, global warming, and many other problems. It was found to function well with room temperature reactions, working well with free radicals, etc.

In order to describe Envirochemkin, a simplified case is shown involving the cracking of ethane, to convert it to a less toxic and profitable species. First, create the reaction file called ethane.dat. Then create the input file: ethane.sam.

The output Spec.out file is shown in the appendix, from which we can plot 2 and 3 dimensional graphs.

In the ethane.dat file, we first type the word ELEMENTS, then all the chemical elements in the problem, then END, then the word SPECIES, then all the chemical formulas, then the word END, then all the chemical equations and next to each three constants a, b, and c from the rate constant for the equation that comes from the literature: $k = aT^{-b}exp(-c/RT)$. Finally, at the end of the reactions, which may number 100 in some problems, we type END. The program can solve for 50 unknowns (species) and 100 differential equations and such problems are often run.

In the ethane.sam file we first type 0 for the isothermal problems, where T and P are constant, then the temperature in degrees K, and the pressure in atm. next to it. Other modes for running problems are 1 for constant H and P, 2 for constant U and V, 3 for T varies with time with V constant and 4 for T varies with time with P constant. Below the numbers 3 and 4 are the coefficients for $dT/d\ t= c1exp(-c2T) + c3 + c4T + c5T^2$, displayed as 3000. 1000. 0.0 0.0 0.0. Below that we put 0.0d-6, then the residence time in microseconds in the form shown (which is 100000 usec. or 0.1 sec.) and then the interval in microsec. between times for calculation and display. Then all the chemical formulas for the species and below each one the initial concentration in mole fractions (shown as 1.0 or 0.0) and a three digit code consisting of 0 or 1. The ones indicate that a sensitivity analysis calculation is wanted and is placed in the 2nd position or that data is needed to make the plotting of results simple.

The spec.out file presents the mole fraction of each chemical species as a function of time, temperature, and pressure as indicated. The mole fraction of each species is presented in a matrix in the same position as the chemical formulas at the top.

The program is originally in FORTRAN and will not run if there is the slightest error. Now if the program refuses to run, type intp.out and it will indicate what the errors are, so you can correct them and run the program again. Intp.out reveals your errors with an uncanny sort of artificial intelligence that will appear at the appropriate equation shown below the last statement.

In order to run, the thermodynamic data for each species is needed and is contained in either the file sandia.dat or chemlib.exe\thermdat. The data that is used is

$$Cpi/R = a1 + a2iT + a3iT^2 + a4iT^3 + a5iT^4 \quad (1)$$

$$Hi/RT = a1I + a2i/2T + a3i/3aT^2 + a4i/4T^3 + a5i/5T^4 + a6i/T \quad (2)$$

$$Si/R = a1ilnT + a2iT + a3i/2T^2 + a4i/3T^3 + a5i/4T^4 + a7i \quad (3)$$

There are seven constants for each species (a1...a7) and each species is fitted over two temperature ranges, so there are fourteen constants for each species in all.

Other information imbedded in the NASA code are name, formula, date of creation, physical state, temperature range of validity, and temperature at which the two temperature ranges fit smoothly together.

Now to run the program type (in the order shown below where ethane.dat is the reaction file):

```
c:\ckin\intp
ethane.dat
C:\ckin\sandia.dat
```

then wait a few moments and type

```
ckin\ckin
[enter] [enter]
ethane.sam          [program will run]
```

After the 1st, 2nd, and 3rd line press enter. After the fourth line press enter twice.

If the thermodynamic data is in chemlib.exe\ thermdat, substitute that for sandia.dat.

In every run the file rates.out will be created.

The above is repeated for every instant of time for which there is a calculation in the forward and reverse direction and then the equations are ranked according to the speed of their reaction. Note that when the sign is negative in the fifth column, the reaction proceeds in the reverse direction. Also note that these data are very important in determining the mechanism of the reaction.

Another file, sense.out, is created when the code in ethane.sam indicates that it is desired for up to five species.

There is often a great deal of uncertainty in the rate constants for some reaction mechanisms. It is, therefore, desirable to have an ability to quantify the effect of an uncertain parameter on the solution to a problem. A sensitivity analysis is a technique which is used to help achieve this end. Applying sensitivity analysis to a chemical rate mechanism requires partial derivatives of the production rates of the species with respect to parameters in the rate constants for the chemical reactions. This file shows the partial derivatives and how the increase or decrease of each species changes the speed or velocity of each reaction shown for every interval of time and like the rates.out file is very important in determining the mechanism and optimizing the reactions.

3.18 Computer Simulation, Modeling and Control of Environmental Quality

Programs such as Envirochemkin and Therm discussed later can help controllers bring systems or plants into the optimum mode for pollution prevention or minimization. Self-optimizing or adaptive control systems can be developed now. These consist of three parts: the definition of optimum conditions of operation (or performance), the comparison of the actual performance with the desired performance, and the adjustment of system parameters by closed-loop operation to drive the actual performance toward the desired performance. The first definition will be made through a Regulatory Agency requiring compliance; the latter two by a program such as Envirochemkin. Further developments that are now in force include learning systems as well as adaptive systems. The adaptive system modifies itself in the face of a new environment so as to optimize performance. A learning system is, however, designed to recognize familiar features and patterns in a situation and then, from its past experience or learned behavior, reacts in an optimum manner. Thus, the former emphasizes reacting to a new situation and the latter emphasizes remembering and recognizes old situations. Both attributes are contained in the "mechanism" of Envirochemkin.

Envirochemkin can also use the Artificial Intelligence technique of "backward chaining" to control chemical processes to prevent pollution while maximizing profit during computation. "Backward Chaining" is a method whereby the distance between the nth step and the goal is reduced while the distance between the n-1th step and the nth step is reduced and so on down to the current state. To do this, time is considered as negative in the computation and the computations are made backward in time to see what former conditions should be in order to reach the present desired stage of minimum pollution and maximum profit. This has been applied in forensic work, where people were sickened by hazardous material, not present when the analytical chemistry was performed at a later date. However, the computer kinetics detected the hazardous material during the reaction of the starting material. Then the amount of each starting species, the choice of each starting species, the process temperature and pressure and the mode of the process (adiabatic, isothermal, fixed temperature profile with time, etc.) and associated chemical reaction equations (mechanism) are chosen such as to minimize pollution and maximize profit.

3.19 Multiobjective Optimization

In the early 1990s, A. R. Ciric sent me a paper by Ciric and Jia entitled "Economic Sensitivity Analysis of Waste Treatment Costs in Source Reduction Projects: Continuous Optimization Problems", University of Cincinnati, Department. of Chemical Engineering, October 1992. This fine paper was really the first I had seen that treated waste minimization within a process simulation program. Waste Minimization and pollution prevention via source reduction of a chemical process involves modifying or replacing chemical production processes. The impact of these activities upon process economics is unclear, as increasing treatment and disposal costs and a changing regulatory environment make the cost of waste production difficult to quantify.

There are two ways to address treatment costs. One way is to solve a parametric optimization problem that determines the sensitivity of the maximum net profit to waste treatment costs. The other way is to formulate the problem as a multiobjective optimization problem that seeks to maximize profits and minimize wastes simultaneously.

If waste treatment costs are well defined, source reduction projects can be addressed with conventional process synthesis and optimization techniques that determine the process structure by maximizing the net profit. However, future waste treatment and disposal costs are often not well defined, but may be highly uncertain. Since the treatment costs are rapidly increasing, the uncertainty in treatment costs will make the results of conventional optimization models very unreliable. Systematic techniques for taking care of this critical feature have not been developed.

The parametric method referred to above (treating waste treatment as a parameter in the optimization study) will lead to a sensitivity of the maximum profit determined by solving numerous optimization problems and a plot of the maximum net profit as a function of the waste treatment cost. Alternately, the process reduction source reduction can be formulated as a multiobjective optimization problem. There one would not try to place a cost on waste treatment. Instead, one would seek to simultaneously minimize waste generation and to maximize profits before treatment costs. If both of these objectives can be achieved in a single design, multiobjective optimization will identify it. If these objectives cannot be achieved simultaneously, multiobjective optimization will identify a set of noninferior designs. Fundamentally this design contains all designs where profits cannot be increased without increasing waste production. A plot of this set gives the trade-off curve between waste production and profitability. Each element of the noninferior set corresponds to a design where profits have been maximized for a fixed level of waste production. The entire trade-off curve (or noninferior set) can be generated by parametrically varying waste production. In both approaches the final choice of the best design is left to the decision maker capable of weighing potential profits against risks.

3.20 Risk Reduction Through Waste Minimizing Process Synthesis

Waste minimization may be accomplished by source reduction, recycling, waste separation, waste concentration, and waste exchange but these all depend on the structure of the process. However, these all need different waste treatment systems even when generating the same product. Also, the risk depends on the structure of the process. Conventionally, the design of facilities for waste minimization and risk reduction cannot be isolated from that of the process for product generation. Process design and waste minimization and risk reduction should be integrated into one consistent method. This, however, makes the already complex tasks involved in process synthesis very cumbersome. The work in this section establishes a highly efficient and mathematically rigorous technique to overcome this difficulty.

A special directed bipartite graph, a process graph or P-graph in short, has been conceived for analyzing a process structure. In a P-graph, an operating unit is represented on a process flowsheet by a horizontal bar, and a material by a circle. If material is an input to an operating unit, the vertex representing this material is connected by an arc to the vertex representing the operating unit. If a material is an output from an operating unit, the vertex representing the material is connected by an arc to the vertex representing the material. In Figures 17 and 18 the conventional and P-graph representations of a reactor and distillation column are shown.

All materials in the process being synthesized are divided into five disjoint classes: raw materials, required products, potential products, disposable materials, and intermediates. The intermediates are similar to a disposable material; nevertheless, unlike the disposed material, the intermediate must be fed to some operating units for treatment or consumption. The intermediate would be a waste which may induce detrimental effects if discharged to the environment or marketed as a by-product, and an intermediate can be fed to some operating units of the process, if produced. The potential product and the production of the disposable product need not occur. The operating units that generate a product or treat an undesirable output can also produce the disposable materials. A raw material, a required product, a potential product, or a disposable material can be fed to operating units. The intermediate is like the disposable material, but unlike the disposable material, it must be fed to some operating units for treatment or consumption. It needs to be treated or consumed within the process.

Specific symbols are assigned to the different classes of materials in their graphical representations. For illustration, a process yielding product H, potential product G, and disposable material D, from raw materials A, B, and C by operating units 1, 2, 3 is shown in Figure 83. The method is founded on an axiom system, describing the self-evident fundamental properties of combinatorially feasible process structures and combinatorics. In the conventional synthesis of a process, the design for the product generation and that for the waste minimization or treatment are performed separately. This frequently yields a locally optimum process. Now we will integrate these two design steps into a single method for process synthesis.

This truly integrated approach is based on an accelerated branch and bound algorithm. The product generation and waste treatment are considered simultaneously in synthesizing the process. This means the optimal structure can be generated in theory. (The enumeration tree for the conventional branch-and-bound algorithm which generates 75 subproblems in the worst case is shown in Figure 84). The cost-optimal structure corresponds to node # 14, and it consists of operating units 2, 8, 9, 10, 15, 20, 25, and 26, as shown in Figure 35. Risk is yet to be considered in this version of process synthesis.

The same product(s) can be manufactured by various structurally different processes, each of which may generate disposable materials besides the product(s). Often, materials participating in structurally different processes can pose different risks. Also, if a material produced by any process can be safely disposed in an environmentally benign manner, the risk associated with it is not always negligible.

Materials possessing risk may be raw materials, intermediates, or final products. These risks can be reduced with additional expenditure for designing and constructing the process. The extent of reduction depends on the interplay between economic, environmental, toxicological or health-related factors. However, here we consider only the cost, waste generation, and risk factors. Cost is defined as the objective function to be minimized subject to the additional constraints on both the second and third factors.

Two types of risk indices are: (1) internal risk associated with a material consumed within the process, e.g., a raw material or intermediate, and (2) an external risk index associated with a material discharged to the environment, e.g., a disposable material; both defined on the basis of unit amount of material. The overall risk of a process is the sum of the risk of all materials in the process. Each is obtained as the sum of its internal and external risks, and each is obtained by multiplying the amount of the material and the corresponding risk index.

The branch-and-bound algorithm of process synthesis incorporating integrated in-plant waste treatment has been extended to include the consideration of risk.

The first example has been revisited for risk consideration. The enumeration tree of branch-and-bound algorithm remains the same for the worst case (Figure 15). The optimal solution with the integrated in-plant waste treatment, resulting from the subproblem corresponding to node # 14 does not satisfy the constraint on risk; instead, subproblem corresponding to node #17 gives rise to the optimal solution of the problem (Figure 16). Although the cost of this solution is higher than that obtained from the subproblem corresponding to node # 14, it has the minimal cost among the solutions satisfying the constraint on risk; the resultant structure is given in Figure 16.

This algorithm generates the cost optimal solution of synthesis problem, satisfying the constraints on both waste generation and risk. It has been demonstrated with an industrial process synthesis problem that the process optimal structure synthesized by taking into account risk can be substantially different from that by disregarding it.

Determining continuous parameters and discrete parameters are decisions needed in designing a process. They have different effects on production cost and waste generation. The highest levels of the EPA waste reduction hierarchy depend on the discrete parameters, i.e., on its structure. While optimal values of the continuous parameters can be determined by almost any simulation program, the values of the discrete parameters cannot be readily optimized because of the large number of alternatives involved. It is not possible to optimize the discrete parameters of an industrial process incorporating waste minimization. Thus, it is often done heuristically based on the designers experience. As the decisions needed are interdependent, a systematic method is required to carry them out consistently and completely as shown below.

Suppose material A is produced from raw materials D, F, and G by a process consisting of 5 operating units shown in Figure 19 and 20. Operating units are represented in Figures (a) and (b) for reactive separator. The graph representation, a vertex for the material, is different for that for an operating unit. Thus, the graph is bipartite. The graphs for all of the candidate operating units of the examples are shown in Figure 25. These operating units can be linked through an available algorithm, i.e., algorithm MSG (Maximal Structure Generation — Figure 85 and Figure 86), to generate the so called maximal structure of the process being designed. The maximal structure contains all candidate process structures capable of generating the product.

The set of feasible process structures can be generated by an available algorithm, algorithm SSG (Solution Structure Generation — Figure 87), from the maximal structure. It is difficult to optimize individually the process structures because of the very large number of the structures involved.

Material in the maximal structure are

a. Materials that can not be produced by any operating unit (purchased raw materials).

b. Material that can be produced by only one operating unit.
c. Materials that can be produced by two or more alternative operating units.

Only case c above requires a decision. Here we must select the operating unit(s) to be included in the process producing this material. When designing a process, decisions should not be made simultaneously for the entire set of materials in c because the decisions may be interdependent. When the maximal structure has been decided by algorithm MSG, the major steps for designing the structure of the process are

1. Determine set g of materials in class c.
2. Generate the feasible process structures by algorithm SSG, optimize them individually by an available process simulation program, select the best among them, and stop if set g is empty or it has been decided that no decision is to be made for producing any material in this set. Otherwise, proceed to step 3.
3. Select one material from set g and identify the set of operating units producing it.
4. Decide which operating unit or units should produce the selected material.
5. Update set g and return to step 2.

By applying the general stepwise procedure outlined above, this example has been solved as presented. In step 1, set g is determined as g = {A, A-E}. If it is decided that no decision is to be made with regard to material A or A-E, all feasible process structures given in Figure (A) through (g) are generated by algorithm SSG. These structures can be evaluated by a process simulation program.

If the number of feasible structures is to be reduced, a decision is needed whether to produce A or A-E. This is done by selecting the former in step 3. Operating units 1 and 2 can yield this material. In step 4, operating unit 1 is selected from heuristic rules or the knowledge base. Then set g, updated in step 5, has only one element, material A-E and returning to step 2 and knowing that no additional decisions need to be made on the process structures illustrated in Figures 25 (a) and (b). The structures in Figure 25 (c) are generated by algorithm SSG.

To reduce the number of generated structures further, additional decisions must be made on the production of an element in set g. Since material A-E is now the only material in set g, this material is selected in set 3. Material A-E can be produced by operating units 3 and 4: see later Figures. Suppose that the decision in step 4, again based on heuristics or knowledge bases, is to produce material A-E by operating unit 4. After executing step 5 and eventually returning to step 2, set g is found to be empty. As a result only one process structure is generated by algorithm SSG. This process structure is to be evaluated (See Figures 20, 22, and 25).

In the design of an industrial process, the number of possible structures is 3465 in this real example for producing material A61 (Folpet) with the operating units listed in Figure 88 and the maximal structure listed as Figure 85. Materials A5, A14, A16, A22, A24, A25, A48, and A61 belong to class c. If operating unit 23 is selected for producing material A14, then 584 different structures remain. With an additional decision on material A61, the number of structures is reduced to 9. This number is small enough so that all the structures can be evaluated by an available simulation or design program.

The solution-structures of an industrial process synthesis problem with a set of M materials has 65 elements, M={A1, A2,.....A65}, where, R = {A1, A2. A3, A4, A6, A7, A8, A11, A15, A17, A18, A19, A20, A23, A27, A28, A29, A30, A34, A43, A47, A49, A52, A54} is the set of raw materials. Moreover, 35 operating units are available for producing the product, material A61. The solution structure of the problem is given in Figure 87. The structure of Synphony is outlined in Figure 89 as outlined by Dr. L. T. Fan.

An algorithm and a computer program were developed to facilitate the design decisions of the discrete parameters of a complex chemical process to reduce the number of processes to be optimized by a simulation program. They are highly effective for both hypothetical and real examples.

3.21 Kintecus

After encountering James Ianni's work on Kintecus on the Internet, I arranged an interview with him at Drexel University. He is a graduate student in Metallurgical Engineering. The program models the reactions of chemical, biological, nuclear, and atmospheric processes. It is extremely fast and can model over 4,000 reactions in less than 8 megabytes of RAM running in pure high speed 32-bit under DOS. It has full output of normalized sensitivity coefficients that are selectable at any specified time. They are used in accurate mechanism reduction, determining which reactions are the main sources and sinks, which reactions require accurate rate constants, and which ones can have guessed rate constants. The program can use concentration profiles of any wave pattern for any species or laser profile for any hv. A powerful parser with mass and charge balance checker is present for those reactions that the OCR or operator entered incorrectly, yet the model is yielding incorrect results or divergent results. The operator can also create an optional name file containing common names for species and their mass representation. The latter can be used for biological and nuclear reactions. It is also possible to have fractional coefficients for species. It can quickly and easily hold one or more concentrations of any species at a constant level. It has support for photochemical reactions involving hv and Loseschmidt's number. It can model reactions from fermtoseconds to years. It automatically generates the spreadsheet file using the reaction spreadsheet file. It can do reactions in a Continuous Stirred Tank Reactor (CSTR) with multiple inlets and outlets. It can compute all internal Jacobians analytically. This is very useful for simulating very large kinetic mechanisms (more than 1,000).

3.22 SWAMI

The Strategic Waste Minimization Initiative (SWAMI) software program is a user friendly computer tool for enhancing process analysis techniques to identify waste minimization opportunities within an industrial setting. It is involved in promoting waste reduction and pollution prevention at the source.

The software program assists the user in:

Simplifying the highly complex task of process analysis of hazardous materials use, identification, and tracking.

Storing process information for any future reassessment and evaluation of pollution prevention opportunities due to changes in process design.

Simulating the effect of waste stream analysis based on process changes in promoting pollution prevention alternatives.

Developing mass balance calculations for the entire process and for unit operation by total mass, individual chemical compounds, and special chemical elements.

Performing cost benefit studies for one or more feasible waste reduction or pollution prevention solutions.

Prioritizing opportunity points by a cost of treatment and disposal or volume of hazardous waste generated.

Developing flow diagrams of material inputs, process sequencing, and waste output streams.

Identifying pollution prevention strategies and concepts.

Consolidating pollution prevention and waste information reports for in-house use and meeting pollution prevention toxic material inventory report requirements.

Interfacing with other EPA pollution prevention tools including the "Waste Minimization Opportunity Assessment Manual," the Pollution Prevention Clearinghouse On-Line Bulletin Board (PPIC), and the Pollution Prevention Economic Software Program.

3.23 SuperPro Designer

Waste generation from process manufacturing facilities is best accomplished when systematic pollution prevention thinking is incorporated in the design and development of such processes. To help, Intelligen, Inc., has developed SuperPro Designer, a comprehensive waste minimization tool for designing manufacturing processes within environmental constraints. SuperPro enables engineers to model on the computer integrated manufacturing processes, characterize waste streams, assess the overall environmental impact, and readily evaluate a large number of pollution prevention options.

3.24 P2-EDGE Software

Pollution Prevention Environmental Design Guide for Engineers (P2-EDGE) is a software tool designed to help engineers and designers incorporate pollution prevention into the design stage of new products, processes and facilities to reduce life cycle costs and increase materials and energy efficiency. P2-EDGE is a project related software tool that provides more than 200 opportunities to incorporate pollution prevention into projects during the design phase. Each opportunity is supported by examples, pictures, and references to help evaluate the applicability and potential benefits to the project. Built-in filters narrow the focus to only the opportunities that apply, based on project size and design stage.

P2-EDGE displays a qualitative matrix to compare the opportunities based on implementation difficulty and potential cost savings. The program indicates which stage of the project will realize pollution prevention benefits (engineering/procurement, construction, startup, normal operations, offnormal operations, or decommissioning) and who will benefit (the project, the site, the region, or global). If a technology is recommended, P2-EDGE shows whether that technology is currently available off the the shelf or is still in development.

Flowsheeting on the World Wide Web

This preliminary work describes a 'flowsheeting', i.e., mass and energy balance computation, tool running across the WWW. The system will generate:

A 3-D flowsheet
A hypertext document describing the process
A mass balance model in a spreadsheet
A set of physical property models in the spreadsheet

The prototype system does not have the first two and last two features fully integrated, but all features have been integrated. The illustration of the prototype is with the Douglas HDA process.

1. Process description
2. Hypertext description and flowsheet, as generated
3. Spreadsheet
4. Physical property data

Process Description

feed2, feed2,3 and 4 mix react and are separated into 5 and 6.

5 split into 3,7.

6 separates to 8,9. Separates to 4 11.

Process Flowsheet
Click here for flowsheet
Node List
Process contains 14 nodes:

Node 1 (feed)
Node 2 (feed)
Node 3 (mixer)
Node 4 (mixer)
Node 5 (mixer)
Node 6 (reactor)
Node 7 (reactor)
Node 8 (separator)
Node 9 (splitter)
Node 10 (separator)
Node 11 (separator)
Node 12 (product)
Node 13 (product)
Node 14 (product)

Node Information

Node 1 is a feed
It is an input to the process. It has 1 output stream:
stream 1 to node node 3 (mixer)
Node 2 is a feed
It is an input to the process. It has 1 output stream:
stream 2 to node 3 (mixer)
Node 3 is a mixer
It has 2 input streams:
Stream 1 from node 1 (feed)
Stream 2 from node 2 (feed)
It has 1 output stream
Stream 12 to node 4 (mixer)
Node 4 is a mixer
It has 2 input streams
Stream 12 from node 3 (mixer)
Stream 3 from node 9 (splitter)
It has 1 output stream:
Stream 13 to node 5 (mixer)
Node 5 is a mixer
It has 2 input streams
Stream 13 from node 4(mixer)
Stream 4 from node 11 (separator)
It has 1 output stream
Stream 14 to node 6 (reactor)
Node 6 is a reactor
It has 1 input stream:
Stream 14 from node 5 (mixer)
It has 1 output stream:
Stream 15 to node 7 (reactor)
Node 7 is a reactor
It has 1 output stream:
Stream 15 from node 6 (reactor)
It has 1 output stream:
Stream 26 to node 8 (separator)
Node 6 is a separator
It has 1 input stream:
Stream 16 from node 7 (reactor)
It has 2 output streams:
Stream 5 to node 9 (splitter)
Stream 6 to node 10 (separator)
Node 9 is a splitter
It has 1 input stream:
Stream 5 from node 8 (separator)
It has output streams:
Stream 3 to node 4 (mixer)
Stream 7 to node 12 (product)
Node 10 is a separator
It has 1 input stream
Steam 6 from node 8 (separator)
It has 2 output streams:
Stream 8 to node 13 (product)
Stream 9 to node 11 (separator)
Node 11 is a separator
It has 1 input stream:
Stream 9 from node 10 (separator)

It has 2 output streams:
Stream 4 to node 5 (mixer)
Stream 11 to node 14 (product)
Node 12 is a product
It has 1 input stream:
Stream 7 from node 9 (splitter)
It is an output from the process
Node 13 is a product
It has 1 input stream
Stream 8 from node 10 (separator)
It is an output from the process
Node 14 is a product
It has 1 input stream:
Stream 11 from node 11 (separator)
It is an output from the process
Stream Information
Stream 2 from 1 (feed) to 3 (mixer)
Stream 2 from 2 (feed) to 3 (mixer)
Stream 3 from 9 (splitter) to 4 (mixer)
Stream 4 from 11 (separator) to 5 (mixer)
Stream 5 from 8 (separator) to 9 (splitter)
Stream 6 from 8 (separator) to 10 (separator)
Stream 7 from 9 (splitter) to 12 (product)
Stream 8 from 10 (separator) to 13 (product)
Stream 9 from 10 (separator) to 11 (separator)
Stream 11 from 11 (separator) to 14 (product)
Stream 12 from 3 (mixer) to 4 (mixer)
Stream 13 from 4 (mixer) to 5 (mixer)
Stream 14 from 5 (mixer) to 6 (reactor)
Stream 15 from 6 (reactor) to 7 (reactor)
Stream 16 from 7 (reactor) to 8 (separator)
Process contains 15 streams

A very simple language has been developed to describe the topology of a process. It consists of verbs which are processing operation names and nouns which are stream numbers. Observe the HDA plant description provided below.

"Feed 1, feed 2, 3 and 4 mix then react twice and are separated into 5 and 6.
5 splits into 3, 7.
6 separates to 8 9, which separates to 4 11."

3.25 CWRT Aqueous Stream Pollution Prevention Design Options Tool

This tool will contain a compilation of applied or considered source reduction design option information from industry that deals with aqueous effluent streams. The information will include simple to complex technologies and techniques, and specific technologies and combinations of technologies applied to result in a reduced waste generation profile from the facility or plant involved.

3.26 OLI Environmental Simulation Program (ESP)

The Environmental Simulation Program (ESP) is a steady state process simulator with a proven record in enhancing the productivity of engineers and scientists. It has applications industry-wide and the software is not only applied to environmental applications but to any aqueous chemical process.

A wide range of conventional and environmental unit operations are available:

Mix	Precipitator	Feedforward
Split	Extractor	Crystallizer
Separate	Component Split	Clarifier
Neutralizer	Incinerator	Sensitivity
Absorber	Compressor	Membrane (UF, RO)
Stripper	Bioreactor	Electrodialysis
Reactor	Manipulate	Saturator
Exchanger	Controller	Dehydrator

ESP provides the engineer or scientist accurate answers to questions involving complete aqueous systems. Design, debottlenecking, retrofitting, troubleshooting, and optimizing of existing or new processes is easy with ESP. Upstream waste minimization, as well as the waste treatment itself, is possible with ESP. The dynamic response of a process can be studied using the dynamic simulation program, DynaChem, to examine control strategy, potential upsets, scheduled waste streams, controller tuning, and startup/shutdown studies.

3.27 Process Flowsheeting and Control

Process flowsheeting with multiple recycles and control loops are allowed. Feedforward and feedback Controllers and Manipulate blocks help to achieve process specifications.

Rigorous Biotreatment Modeling

Heterotrophic and autotrophic biological integration is integrated with rigorous aqueous chemistry. Single or multiple substrates are allowed. Substrates may be specific molecules from the Databank or characterized by ThOD, MW, or statistical stoichiometry. Simultaneous physical (e.g., air stripping) and chemical (e.g., pH, trace components) effects are applied. ESP provides for flexible configuration of biotreatment processes, including sequential batch reactors and clarifiers with multiple recycles.

Sensitivity Analysis

The sensitivity block allows the user to determine easily the sensitivity of output results to changes in Block Parameters and physical constants.

Dynamic Simulation with Control

Discrete dynamic simulation of processes with control can be accomplished and is numerically stable using DynaChem. Studies of pH and compositional control, batch treatment interactions, multistage startup and shutdown, controller tuning, Multicascade, and adaptive control are all possible.

Access to OLI Thermodynamic Framework and Databank

All ESP computations utilize the OLI predictive thermodynamic model and have access to the large inplace databank.

Access to OLI Toolkit

The Toolkit, including the Water Analyzer and OLI Express, provides flexible stream definition and easy single-case (e.g., bubble point) and parametric case (e.g., pH sweep) calculations. This tool allows the user to investigate and understand the stream chemistry, as well as develop treatment ideas before embarking on process flowsheet simulation. The Toolkit also allows direct transfer of stream information to other simulation tools for parallel studies.

3.28 Environmental Hazard Assessment for Computer-Generated Alternative Syntheses

The purpose of this project is to provide a fully operational version of the SYNGEN program for the rapid generation of all the shortest and least costly synthetic routes to any organic compound of interest. The final version will include a retrieval from literature databases of all precedents for the reactions generated. The intent of the program is to allow all such alternative syntheses for commercial chemicals to be assessed. Once this program is ready it will be equipped with environmental hazard indicators, such as toxicity, carcinogenicity, etc., for all the involved chemicals in each synthesis, to make possible a choice of alternative routes of less environmental hazard than any synthesis currently in use.

3.29 Process Design for Environmentally and Economically Sustainable Dairy Plant

Major difficulties in improving economics of current food production industry such as dairy plants originate from problems of waste reduction and energy conservation. A potential solution is a zero discharge or a dry floor process which can provide a favorable production environment. In order to achieve such an ideal system, we developed a computer-aided wastewater minimization program to identify the waste problem and to obtain an optimized process. This method can coordinate the estimation procedure of water and energy distribution of a dairy process, MILP (Mixed Integer Linear Programming) formulation, and process network optimization. The program can specify the waste and energy quantities of the process streams by analyzing audit data of the plant. It can show profiles of water and energy demand and wastewater generation, which are normally functions of the production amount and the process sequence. Based on characterized streams in the plant, wastewater storage tanks and membrane separation units have been included in the waste minimization problem to search for cost-effective process based on MILP models. The economic study shows that cost of an optimized network is related to wastewater and energy charges, profit from by-products, and equipment investments.

3.30 Life Cycle Analysis (LCA)

Industry needs to know the environmental effect of its processes and products. Life Cycle Analysis (LCA) provides some of the data necessary to judge environmental impact. An environmental LCA is a means of quantifying how much energy and raw material are used and how much (solid, liquid and gaseous) waste is generated at each stage of a product's life.

```
                |-------------|—waste heat
Raw
materials  ----|             |—solid waste
                |             |—emission to air
                |             |—emissions to water
Energy/
fuels  ---------|             |—usable products
                |-------------|
```

The main purpose of an LCA is to identify where improvements can be made to reduce the environmental impact of a product or process in terms of energy and raw materials used and wastes produced. It can also be used to guide the development of new products.

It is important to distinguish between life cycle analysis and life cycle assessment. Analysis is the collection of the data. It produces an inventory; assessment goes one step further and adds an evaluation of the inventory.

Environmentalists contend that zero chlorine input to the industrial base means zero chorinated toxins discharged to the environment. Industry experts claim that such a far reaching program is unnecessary and will have large socioeconomic impacts. Environmentalists have responded with the argument that overall socioeconomic impacts will be small since there are adequate substitutes for many of the products that currently contain chlorine.

3.31 Computer Programs

Free radicals are important intermediates in natural processes involved in cytotoxicity, control of vascular, tone, neurotransmission. The chemical kinetics of free-radical reactions control the importance of competing pathways. Equilibria involving protons often influence the reaction kinetics of free radicals important in biology. Free radicals are very important in atmospheric chemistry and mechanisms. Yet, little is known about their physical or biological properties.

In 1958, White, Johnson, and Dantzig (at Rand) published an article entitled "Chemical Equilibrium in Complex Mixtures." It was a method that calculated chemical equilibrium by the method of the minimization of free energy. It was an optimization problem in non-linear programming and was used in industry and in defense work on main frame computers. PCs were not available at that time. Also, environmental matters were not as much of a concern as they are now.

The literature and computer sites on Geographic Information Sytems (GIS) are rife with a tremendous amount of information. The number of such maps are increasing greatly every day as exploration, assessment, and remediation proceeds across the world wherever environmental work is taking place.

There are many software programs for geotechnical and geo-environmental and environmental modeling. They are in the category of contaminant modeling. Most of them are in the DOS platform and are public domain.

Massively parallel computing systems provide an avenue for overcoming the computational requirements in the study of atmospheric chemical dynamics. The central challenge in developing a parallel air pollution model is implementing the chemistry and transport operators used to solve the atmospheric reaction-diffusion equation. The chemistry operator is generally the most computationally intensive step in atmospheric air quality models. The transport operator (advection equation) is the most challenging to solve numerically. Both of these have been improved in the work of Dabdub and Seinfeld at Cal. Tech. and have been improved in the next generation of urban and regional-scale air quality models. HPCC (High Performance Computing and Communications) provides the tools essential to develop our understanding of air pollution further.

EPA has three main goals for its HPCC Program activities:

> Advance the capability of environmental assessment tools by adapting them to a distributed heterogeneous computing environment that includes scalable massively parallel achitectures.
>
> Provide more effective solutions to complex environmental problems by developing the capability to perform multipollutant and multimedia pollutant assessments.
>
> Provide a computational and decision support environment that is easy to use and responsive to environmental problem solving needs to key federal, state and industrial policy-making organizations.

Thus, EPA participates in the NREN, ASTA, IITA, and BRHR components of the HPCC Program, where: NREN: increasing access to a heterogeneous computing environment, ASTA: environmental assessment grand challenges, IITA: enhancing user access to environmental data and systems, BRHR: broadening the user community tools by adapting them to a distributed heterogeneous computing environment that includes scalable massively parallel architectures.

Environmental modeling of the atmosphere is most frequently performed on supercomputers. UAMGUIDES is an interface to the Urban Airshed Model (UAM). An ozone-compliance simulator is required by the Clean Air Act of 1990, so that modeling groups across the United States have asked the North Carolina Supercomputing Center (NCSC) to develop a portable version. NCSC's Environmental Programs Group used the CRAY Y-MP system, a previous-generation parallel vector system from Cray Research to develop UAMGUIDES as a labor-saving interface to UAM. Running UAM is very complex. The Cray supercomputers have, since then, been upgraded. Computational requirements for modeling air quality have increased significantly as models have incorporated increased functionality, covered multi-day effects and changed from urban scale to regional scale. In addition, the complexity has grown to accommodate increases in the number of chemical species and chemical reactions, the effects of chemical particle emissions on air quality, the effect of physical phenomena, and to extend the geographical region covered by the models.

The effects of coal quality on utility boiler performance are difficult to predict using conventional methods. As a result of environmental concerns, more utilities are blending and selecting coals that are not the design coals for their units. This has led to a wide range of problems, from grindability and moisture concerns to fly ash collection. To help utilities predict the impacts of changing coal quality, the Electric Power Research Institute (EPRI) and the U.S. Department of Energy (DOE) have initiated a program called Coal Quality Expert (CQE). The program is undertaken to quantify coal quality impacts using data generated in field-, pilot-, and laboratory-scale investigations. As a result, FOULER is a mechanistic model placed into a computer code that predicts the coal ash deposition in a utility boiler and SLAGGO is a computer model that predicts the effects of furnace slagging in a coal-fired boiler.

In Europe, particularly Prof. Mike Pilling and Dr. Sam Saunders at the Department of Chemistry at the University of Leeds, England, have worked on tropospheric chemistry modeling and have had a large measure of success. They have devised the MCM (Master Chemical Mechanism), a computer system for handling large systems of chemical equations, and were responsible for quantifying the potential that each VOC exhibits to the development of the Photochemical Ozone Creation Potential (POCP) concept. The goal is to improve and extend the Photochemical Trajectory Model for the description of the roles of VOC and NOx in regional scale photo-oxidant formation over Europe. In their work they use Burcat's "Thermochemical Data for Combustion Calculations" in the NASA format.

Statistical methods, pattern recognition methods, neural networks, genetic algorithms and graphics programming are being used for reaction prediction, synthesis design, acquisition of knowledge on chemical reactions, interpretation and simulation of mass spectra analysis and simulation of infrared spectra, analysis and modeling of biological activity, finding new lead structures, generation of 3D-dimensional molecular models, assessing molecular similarity, prediction of physical, chemical, and biological properties, and databases of algorithms and electronic publishing. Examples include the course of a chemical reaction and its products for given starting materials using EROS (Elaboration of Reactions for Organic Synthesis) where the knowledge base and the problem solving techniques are clearly separated. Another case includes methods for defining appropriate, easily obtainable starting materials for the synthesis of desired product. This includes the individual reaction steps of the entire synthesis plan. It includes methods to derive the definition of structural similarities between the target structure and available starting materials, finding strategic bonds in a given target, and rating functions to assign merit values to starting materials. Such methods are integrated into the WODCA system (Workbench for the Organization of Data for Chemical Application).

In 1992 the National Science Foundation was already looking to support work for CBA (Computational Biology Activities); software for empirical analysis and/or simulation of neurons or networks of neurons; for modeling macromolecular structure and dynamics using x-ray, NMR or other data; for simulating ecological dynamics and analyzing spatial and temporal environmental data; for improvement of instrument operation; for estimation of parameters in genetic linkage maps; for phylogenetic analysis of molecular data; and for visual display of biological data. They were looking for algorithm development for string searches; multiple alignments; image reconstruction involving various forms of microscopic, x-ray, or NMR data; techniques for aggregation and simplification in large-scale ecological models; optimization methods in molecular mechanics and molecular dynamics, such as in the application to protein folding; and spatial statistical optimization.

They sought new tools and approaches such as computational, mathematical, or theoretical approaches to subjects like neural systems and circuitry analysis, molecular evolution, regulatory networks of gene expression in development, ecological dynamics, physiological processes, artificial life, or ion channel mechanisms.

There has been constructive cross-fertilization between the mathematical sciences and chemistry. Usually in QSAR methods, multiple linear or non linear regression, classical multivariate statistical techniques were used. Then discriminant analysis, principal components regression, factor analysis, and neural networks were used. More recently partial least squares (PLS), originally developed by a statistician for use in econometrics, has been used and this has prompted additional statistical research to improve its speed and its ability to forecast the properties of new compounds and to provide mechanisms to include nonlinear relations in the equations. QSAR workers need a new method to analyze matrices with thousands of correlated predictors, some of which are irrelevant to the end point. A new company was formed called Arris with a close collaboration of mathematicians and chemists that produced QSAR software that examines the three-dimensional properties of molecules using techniques from artificial intelligence.

Historically, mathematical scientists have worked more closely with engineers and physicists than with chemists, but recently many fields of mathematics such as numerical linear algebra, geometric topology, distance geometry, and symbolic computation have begun to play roles in chemical studies. Many problems in computational chemistry require a concise description of the large-scale geometry and topology of a high-dimensional potential surface. Usually, such a compact description will be statistical, and many questions arise as to the appropriate ways of characterizing such a surface. Often such concise descriptions are not what is sought; rather, one seeks a way of fairly sampling the surface and uncovering a few representative examples of simulations on the surface that are relevant to the appropriate chemistry. An example is a snapshot or typical configuration or movie of a kinetic pathway. Several chemical problems demand the solution of mathematical problems connected with the geometry of the potential surface. Such a global understanding is needed to be able to picture long time scale complex events in chemical systems. This includes the understanding of the conformation transitions of biological molecules. The regulation of biological molecules is quite precise and relies on sometimes rather complicated motions of a biological molecule. The most well studied of these is the so-called allosteric transition in hemoglobin, but indeed, the regulation of most genes also relies on these phenomena. These regulation events involve rather long time scales from the molecular viewpoint. Their understanding requires navigating through the complete navigation space. Another such long-time scale process that involves complex organization in the configuration space is bimolecular folding itself.

Similarly, specific kinetic pathways are important. Some work has been done on how the specific pathways can emerge on a statistical energy landscape. These ideas are, however, based on the quasi-equilibrium statistical mechanics of such systems, and there are many questions about the rigor of this approach. Similarly, a good deal of work has been carried out to characterize computationally pathways on complicated realistic potential energy surfaces. Techniques based on path integrals have been used to good effect in studying the recombination of ligands in biomolecules and in the folding events involved in the formation of a small helix from a coiled polypeptide. These techniques tend to focus on individual optimal pathways, but it is also clear that sets of pathways are very important in such problems. How these pathways are related to each other and how to discover them and count them is still an open computational challenge.

The weak point in the whole scenario of new drug discovery has been identification of the "lead." There may not be a good lead in a company's collection. The wrong choice can doom a project to never finding compounds that merit advanced testing. Using only literature data to derive the lead may mean that the company abandons the project because it cannot patent the compounds found. These concerns have led the industry to focus on the importance of molecular diversity as a key ingredient in the search for a lead. Compared to just 10 years ago, orders of magnitude more compounds can be designed, synthesized, and tested with newly developed strategies. These changes present an opportunity for the imaginative application of mathematics.

There are three aspects to the problem of selecting samples from large collections of molecules: First, what molecular properties will be used to describe the compounds? Second, how will the similarity of these properties between pairs of molecules be quantified? Third, how will the molecules be paired or quantified?

For naturally occurring biomolecules, one of the most important approaches is the understanding of the evolutionary relationships between macromolecules. The study of the evolutionary relationship between biomolecules has given rise to a variety of mathematical questions in probability theory and sequence analysis. Biological macromolecules can be related to each other by various similarity measures, and at least in simple models of molecular evolution, these similarity measures give rise to an ultrametric organization of the proteins. A good deal of work has gone into developing algorithms that take the known sequences and infer from these a parsimonious model of their biological descent.

An emerging technology is the use of multiple rounds of mutation, recombination, and selection to obtain interesting macromolecules or combinatorial covalent structures. Very little is known as yet about the mathematical constraints on finding molecules in this way, but the mathematics of such artificial evolution approaches should be quite challenging. Understanding the navigational problems in a high-dimensional sequence space may also have great relevance to understanding natural evolution. Is it punctuated or is it gradual as many have claimed in the past? Artificial evolution may obviate the need to completely understand and design biological molecules, but there will be a large number of interesting mathematical problems connected with the design.

Drug leads binding to a receptor target can be directly visualized using X-ray crystallography. There is physical complexity because the change in free energy is complex as it involves a multiplicity of factors including changes in ligand bonding (with both solvent water and the target protein), changes in ligand conformation or flexibility, changes in ligand polarization, as well as corresponding changes in in the target protein.

Now structural-property refinement uses parallel synthesis to meet geometric requirements of a target receptor binding site. Custom chemical scaffolds are directed to fit receptor binding sites synthetically elaborated through combinatorial reactions. This may lead to thousands to millions of members, while parallel automated synthesis is capable of synthesizing libraries containing of the order of a hundred discrete compounds. Structure property relationships are then supplied to refine the selection of sub-libraries. 3D structural models, SAR bioavailability and toxicology are also used in such searches. Additional 3D target-ligand structure determinations are used to iteratively refine molecular properties using more traditional SAR methods.

In the Laboratory for Applied Thermodynamics and Phase Equilibria Research, an account of a Computer Aided Design of Technical Fluids is given. The environmental, safety, and health restrictions impose limitations on the choice of fluids for separation and energy processes. Group contribution methods and computer programs can assist in the design of desired compounds. These compounds and mixtures have to fulfill requirements from an integrated point of view. The research program includes both the design of the components and the experimental verification of the results.

The Molecular Research Institute (MRI) is working in many specific areas, among which are Interdisciplinary Computer-Aided Design of Bioactive Agents and Computer-Aided Risk Assessment and Predictive Toxicology, and all kinds of models for complicated biological molecules. The first area, cited above, designs diverse families of bioactive agents. It is based on a synergistic partnership between computational chemistry and experimental pharmacology allowing a more rapid and effective design of bioactive agents. It can be adapted to apply to knowledge of the mechanisms of action and to many types of active systems. It is being used for the design of CNS active therapeutic agents, particularly opioid narcotics, tranquilizers, novel anesthetics, and the design of peptidomimetics. In Computer-Aided Risk Assessment they have produced strategies for the evaluation of toxic product formation by chemical and biochemical transformations of the parent compound, modelling of interactions of putative toxic agents with their target biomacromolecules, determination of properties leading to toxic response, and use of these properties to screen untested compounds for toxicity.

3.32 Pollution Prevention by Process Modification Using On-Line Optimization

Process modification and on-line optimization have been used to reduce discharge of hazardous materials from chemical and refinery processes. Research has been conducted at three chemical plants and a petroleum refinery that have large waste discharges. The research has been done where development of process modification methodology for source reduction has been accomplished. The objective is to combine these two important methods for pollution prevention and have them share process information to efficiently accomplish both tasks.

Process modification research requires that an accurate process model be used to predict the performance of the plant and evaluate changes proposed to modify the plant to reduce waste discharges. The process model requires precise plant data to validate that the model accurately describes the performance of the plant. This precise data is obtained from the gross error detection system of the plant. In addition, the economic model from the process optimization step is used to determine the rate of return for the proposed process modifications. Consequently, a synergism from the two methods for pollution prevention and Process Modification have selected important processes for their application. Moreover, cooperation of companies has been obtained to apply these methods to actual processes rather than to simulated generic plants.

3.33 A Genetic Algorithm for the Automated Generation of Molecules Within Constraints

A genetic algorithm has been designed which generates molecular structures within constraints. The constraints may be any useful function such as molecular size.

3.34 WMCAPS

A system is herein proposed that uses coding theory, cellular automata, and both the computing power of Envirochemkin and a program that computes chemical equilibrium using the minimization of the chemical potential. The program starts with the input ingredients defined as the number of gram-atoms of each chemical element as

$$b_i, \; i = 1, 2, 3, 4. \ldots$$

Now if a_{ij} is the number of gram-atoms of i in the jth chemical compound and x_j is the number of moles of the jth chemical compound we have two equations or constraints

$$\Sigma^{n}_{j=1} a_{ij}x_j = b_i \qquad i = 1,.....,m$$

$$x_j > 0 \qquad j = 1,......,n$$

with n >= m. Subject to these constraints it is desired to minimize the total Gibbs free energy of the system.

$$\Sigma_{j}{}^{n}{}_{=1}\, c_j\, x_j + \Sigma^{n}_{j=1}\, x \log (x_j / \Sigma^{n}_{i=1} x_i)$$

where $c_j = F_j/RT + \log P$
F_j = Gibbs energy per mole of jth gas at temperature T and unit atmospheric pressure
R = universal gas constant

My experience is that this method works like a charm on a digital computer and is very fast.

Now we have the equilibrium composition at the given temperature and pressure in our design for our industrial plant. This is a very important first step. However our products must go through a series of other operations at different conditions. Also, our products are at their equilibrium values and they may not be allowed to reach their true values for the residence time in the reactor. This is where Envirochemkin comes in. Starting with the equilibrium values of each compound, it has rate constants for each reaction in the reactor and again at the proper temperature and pressure will calculate the concentration of each compound in the mixture. This will be a deviation from the equilibrium value for most cases.

It is important to note that both the above program and Envirochemkin come with a very large data file of thermodynamic values for many species. The values that are given are standard enthalpy and entropy and also heat capacity over a wide range. This allows the program to take care of phase change over the many unit operations that compose an industrial plant.

There is a third program used and that is SYNPROPS. Let us say that we have a reaction in our plant that leads to what we want except that one percent of the product is a noxious, toxic, and hazardous compound that we wish to eliminate. We then set many of the properties (especially the toxic properties) of a molecule that is virtual equal to our unwanted species and also set the stoichiometric formula of this virtual molecule also equal to that of the unwanted molecule. This data is put into the SYNPROPS spreadsheet to find the kin of the unwanted molecule that is benign.

A fourth program is then used called THERM. We use it to show whether the reaction of the mix in the reactor to form the benign substitution is thermodynamically of sufficient magnitude to create the benign molecule and decrease the concentration of the unwanted molecule to below a value that will not cause any risk to be above that of significance.

The industrial plant may be composed of many different unit operations connected in any particular sequence. However, particular sequences favor better efficacy and waste minimization and the optimum sequence, of course, is the best. In order to find the best among the alternatives we have used a hierarchical tree and in order to depict the flowsheet we use CA (cellular automata).

Part IV. Computer Programs for the Best Raw Materials and Products of Clean Processes

4.1 Cramer's Data and the Birth of Synprops

Cramer's data (Figures 43 and 44) is in the table of group properties. Results so obtained were from extensive regressions on experimental data from handbooks and were tested and statistically analyzed. The data was used to predict physical properties for other compounds than those used to derive the data. In this work, optimization procedures are combined with the Cramer data (in an extended spreadsheet), and applied for Pollution Prevention and Process Optimization. In addition, Risk Based Concentration Tables from Smith, etc., are included as constraints to ensure that the resulting composite structures are environmentally benign.

During the course of many years, scientists have recognized the relationship between chemical structure and activity. Pioneering work has been done by Hammett in the 1930s, Taft in the 1950s, and Hansch in the 1960s. Brown also recognized the relation between steric effects and both properties and reactions. QSAR methodologies were developed and used in the areas of drug, pesticide, and herbicide research. In the 1970s, spurred by the increasing number of chemicals being released to the environment, QSAR methods began to be applied to environmental technology.

Meanwhile, the hardware and software for personal computers have been developing very rapidly. Thus the treatment of many molecules through their composite groups and the connection with their properties becomes an exercise of obtaining good data to work with. A Compaq 486 Presario PC with a Quattro Pro (version 5.0) program was available. In the "Tools" part of the program is an Optimizer program, which was used in this work. The technology of the modern PC was matched with the power of mathematics to obtain the following results. The values of the parameters B, C, D, E, and F for thirty-six compounds are shown in Figure 41 and used to obtain physical properties and Risk Based Concentrations.

4.2 Physical Properties form Groups

It has also been known that a wide range of properties can be derived using The Principle of Corresponding States which used polynomial equations in reduced temperature and pressure. In order to obtain the critical properties needed for the reduced temperature and reduced pressure, the critical constants are derived from the parameters for the groups of which the molecules are composed.

Thus, the treatment of many molecules through their composite groups and the connection with their properties becomes an exercise of obtaining good data to work with. This is particularly difficult for drug and ecological properties that are not in the public domain.

Cramer's method consisted of applying regressions to data from handbooks, such as the *Handbook of Chemistry and Physics*, etc., to fit the physical properties of molecules with the groups comprising their structures. The results considered about 35 groups and were used in the Linear-Constitutive Model and a similar number of groups (but of a different nature) were used in the Hierarchical Additive-Constitutive Model. Statistically a good fit was found and the prediction capabilities for new compounds were found to be excellent.

Twenty-one physical properties were fitted to the structures. The Properties (together with their dimensions) were Log activity coefficient and Log partition coefficient (both dimensionless), Molar refractivity (cm^3/mol), Boiling point (degrees C.), Molar volume (cm^3/mol), Heat of vaporization (kcal./mol), Magnetic susceptibility (cgs molar), Critical temperature (degrees C.), Van der Waals $A^{1/2}$ ($L\ atm^{1/2}/mol$), Van der Waals B (L/mol), Log dielectric constant (dimensionless), Solubility parameter (cal/cm^3), Critical pressure (atm.), Surface Tension (dynes/cm), Thermal Conductivity ($10^4 \times (cals^{-1}cm^{-2}(cal/cm)^{-1})$), Log viscosity (dimensionless), Isothermal ($m^2/mol \times 10^{10}$), Dipole moment (Debye units), Melting point (degrees C), and Molecular weight (g./mol). Later the equations for molar volume (Bondi scheme) and molar refractivity (Vogel scheme) were included as were equations for the Log Concentration X/Water, where X was ether, cyclohexane, chloroform, oils, benzene and ethyl alcohol, respectively. Risk-Based Concentrations and Biological Activity equations were also included. The units of the molar volume by the Bondi technique is 22 cm^3/mole and the other newer equations have dimensionless units.

The Hierarchical Model (Figure 43), shows the parameters for the groups in five columns. This was set up in a spreadsheet and the structure of each molecule was inserted as the number of each of the groups that comprised the molecule. The sum of each column then being called B, C, D, E, and F after the parameters in each column multiplies the number of appropriate groups. In Figures 43 and 44, the column B contains the variables, which are the number of each of the groups denoted in column A, and these can be manually set to find the values of the parameters B, C, D, E, and F, or determined automatically by the optimizer program. Columns N and O essentially repeat columns A and B, respectively, except near the bottom where there are equations to determine the number of gram-atoms of each chemical element for the molecule whose groups are currently displayed in column B. The top and bottom of column O and all of column Q have embedded in them formulas for physical properties, activities or Risk Based Concentrations in the general linear combination equation

$$P_{ij} = a_i + b_iB_j + c_iC_j + d_iD_j + e_iE_j + f_iF_j$$

The i subscripts stand for different properties and the j subscripts indicate different molecules. The values for B, C, D, E, and F are found in cells D111, F111, H111, J111, and L111, respectively, and are linear equations in terms of all the group entries in column B.

It is seen that the spreadsheets (Figures 42 and 43) are like the blueprints of a molecule whose structure is the composite of the numbers in column B and whose properties are given in column O and Q. The quantities B...F are the conversion factors of the numbers in column B to the properties in columns O and Q. In this manner they are analogous to the genes (5 in this case) in living systems. Values for B, C, D, E, and F are shown for thirty-six of the most hazardous compounds found on Superfund sites in Figure 41.

Linear graphs were drawn that show how the parameters B, C, and D vary with the molecular groups. Also constructed were graphs of how the parameters B, C, D, E, and F vary with the groups on spiral or special radar graphs. This was collated for all the parameters and all the groups on one spiral graph. Also the values for all the hazardous compound were shown on a linear graph. A regression fits the plot of the parameter B versus the groups on a spiral plot. A good fit was also obtained for the parameters C, D, E, and F as well.

The Linear Model Spreadsheet is shown in Figure 44. It is exactly similar to another table called the Hierarchical Model except that it uses groups that are different. The Hierarchical Model Spreadsheet is shown in Table II.

4.3 Examples of SYNPROPS Optimization and Substitution

Some of the results for the Linear Model (using 21 groups) are indicated below:

1. Substitutes for Freon-13 can be CF3CL (a redundancy) or CHBRFCH3,
2. Substitutes for Freon-12 can be CF2CL2 (a redundancy) or CHF2CL.
3. Substitutes for alanine can be: C(NH2)3CN or CH(CONH2)2CN or CH(CCF3)BR or CH(CF3)CONH2,
4. A substitute for CH3CL3 can be CF3I,
5. Substitutes for 1,1-dichloroethylene can be CH2=CHOH and CH2=CHNO2.

If these substitute compounds do not fit exactly to the desired properties, they can serve as the starting point or as precursors to the desired compounds.

Skeleton compounds were used to find the best functional groups for each property. As examples the Linear Model and 21 groups were used with the >C< skeleton (4 groups allowed) and the constraints:

1. Tc is a maximum: C(-NAPTH)2(CONH2)2,
2. Critical pressure smaller or equal to 60, Boiling Point greater or equal to 125, Solubility Parameter greater or equal to 15: CF2(OH)2,
3. Heat of Vaporization a maximum: C(CONH2)4,
4. Heat of Vaporization a minimum: CH4,
5. Log Activity Coefficient greater or equal to 6, Log Partition Coefficient smaller or equal to -2, Critical Pressure equal to 100:, C(CN)2NO2CONH2,
6. Minimum Cost: CH4,
7. Maximum Cost: C(NAPTH)4,
8. Maximum Cost with Critical Temperature greater or equal to 600, Critical Pressure greater or equal to 100: C(NAPTH)2I(CONH2),
9. Minimum Cost with Critical Temperature greater or equal to 600, Critical Pressure equal to 60: CH(OH)(CN)2.

Results for some of the runs made to ascertain which groups confer maximum and/or minimum properties to a substance follow, using the >C< skeleton. They show COOH for maximum magnetic susceptibility, minimum Activity Coefficient, maximum Log Partition Coefficient, maximum Heat of Vaporization, maximum Surface Tension, and Viscosity. NH2 conferred minimum Critical Pressure and maximum Activity Coefficient. C=O occurred for minimum Dipole Moment, minimum Log Partition Coefficient, and minimum Viscosity; NO2 occurred for minimum Critical Temperature and minimum Surface Tension; CL appeared for maximum Dielectric Constant; CONH2 appeared for minimum Critical Temperature; OH appeared for minimum Boiling Point; and F for minimum Heat of Vaporization.

An optimization leading to a most desired structure with non-integer values showed 8.67 hydrogen atoms, 1.88 cyclohexane groups, and 5.41 >C< groups. This is a string of >C< groups attached to each other with a proper number of cyclohexane rings and hydrogens attached. This was rounded off to 8 hydrogens, 2 cyclohexane rings, and 5 >C<s. Results show that a resulting molecule, cyclopentane with 8 hydrogens and 2 cyclohexane groups appended, satisfies most of the desired physical properties very well.

The hierarchical model was used to find the best substitution model for methyl chloroform or 1,1,1-trichloroethane. It was CH3CH2.8(NO2)0.2, if the melting point, boiling point, and log of the ratio of the equilibrium concentration of methyl chloroform in octanol relative to its concentration in water are taken from the literature. The result was also obtained by constraining the molecule to be C-C surrounded by six bonds. This is a hydrocarbon in accord with practical results and that of S.F. Naser. In the same way, TCE's substitute (constrained to be C=C surrounded by four bonds) was C=CI0.383(CONH2)1.138(NO2)2.480 and PCE's substitute was C=CI3COOH.

The precision goal of the fit between predicted and actual Risk Based Concentrations was for adequate internal program control or constraint purposes. Comparisons between predicted and actual Risk Based Concentrations for air are shown in a Figure contained in a previous book, Computer Generated Physical Properties by the author. Tapwater, soils, and MCL results are similarly contained.

A SYNPROPS run that searched for a substitute for carbon tetrachloride, CCL4, is shown in Table VI. Also tables for freons are shown in the freon tables for (CF3CH2F), R125 (CF3CHF2), HFC-338mccq (CF3CF2CF2CH2F), R32 (CH2F2), and hfc-245fa (CF3CH2CHF2) One can print out the intermediate results of a search, where the first result indicates that for the original compound, CCl4, the second that for an intermediate SYNPROP result sheet on the way to an answer and the last the result that was the one the process found closest to the final answer. This last one cited for a substitute for CCl4 was a compound with about 9 -CH=CH- groups and about 8 -CH=CH2 endcap groups indicating a highly olefinic molecule with the Air-Risk Concentration rising from about 1.7 to 94000 with the solubility parameter remaining constant at 10.2.

A similar run with 1,1,1-trichloroethane is shown in three Tables, where the Air-Risk Concentration rose from 3.2 to 164, while the solubility parameter remained fairly constant, changing from 8.75 to 8.96. The molecule that was formed had 2 -CH=CH- groups and 2-CH=CH2 groups similar to the above but had to add a small amount of the naphthyl group. The molecule C(NAPTH)4 had an Air Risk-Concentration of 26000 and when the unlikely molecule, C14(NAPTH)30, was inserted in SYNPROPS, the Air Risk-Concentration 2.6 E+27 was predicted indicating that this group needs a revision of data.

The tables in Figures 43 and 44 show that a compound such as CCL2=CCL2 can be formed from the molecule in the Linear spreadsheet Mode by taking 1 >C=CH2 and -2 for -H and 4 -CL groups. Thus one can use negative numbers when the need arises. Notice that the Air Risk -Concentration here is 0.17 and the solubility parameter is 12.5.

4.4 Toxic Ignorance

For most of the important chemicals in American commerce, the simplest, safest facts still cannot be found. Environmental Defense Fund research indicates that, today, even the most basic toxicity testing results cannot be found in the public record for nearly 75% of the top-volume chemicals in commercial use.

The public cannot tell if a large majority of the highest-use chemicals in the United States pose health hazards or not — much less how serious the risks might be, or whether those chemicals are actually under control. These include chemicals that we are likely to breathe or drink, that build up in our bodies, that are in consumer products, and that are being released from industrial facilities into our backyards, streets, forests, and streams.

In 1980, the National Academy of Science National Research Council completed a four-year study and found that 78% of the chemicals in highest-volume commercial use had not even "minimal" toxicity testing. No improvement was noted 13 years later. Congress promised 20 years ago that the risk of toxic chemicals in our environment would be identified and controlled. That promise is now meaningless.

The chemical manufacturing industry itself must now take direct responsibility in solving the chemical ignorance problem.

The first steps are simple screening tests that manufacturers of chemicals can easily do. All high-volume chemicals in the U.S. should have been subjected to at least preliminary health-effects screening with the results publicly available. A model definition of what should be included in preliminary screening tests for high-volume chemicals was developed and agreed on in 1990 by the U.S. and the other member nations of the Organization for Economic Cooperation and Development, with extensive participation from the U.S. Chemical Manufacturing industry.

4.5 Toxic Properties from Groups

The equation derived was

$$-LN(X) = a + bB + cC + dD + eE = fF$$

which can also be written as

$$X = exp(-a).exp(-bB).exp(-cC).exp(-dD).exp(-eE).exp(-fF)$$

where X is MCL (mg/L), or tap water (ug/L), or ambient air (ug/m^3), or commercial/industrial soil (mg/kg), or residential soil (mg/kg).

Graphs for the Risk-Based Concentration for tap water, air, commercial soil, residential soil, and MCL for the hazardous compounds from superfund sites can be found in *Computer Generated Physical Properties* (Bumble, S., CRC Press, 1999).

4.6 Rapid Responses

The first serious excursions by the pharmaceutical industry into designing protease inhibitors as drugs began over 30 years ago. However, although the angiotensin converting enzyme (ACE) inhibitors such as Captopril and Enalapril emerged as blockbuster drugs, interest waned when the difficulties of designing selective, bioavailable inhibitors became apparent, and efforts to design bioavailable throm and renin inhibitors were not so successful.

The resurgence of interest in protease research has been kindled by the continual discovery of new mammalian proteases arising from the human genome project. At present, researchers have characterized only a few hundred mammalian proteases but extrapolating the current human genome data suggests that we will eventually identify over 2000. Recent advances in molecular biology have helped us to identify and unravel the different physiological roles of each mammalian protease. In summary, we can now predict with more confidence what the consequences of inhibiting a particular protease might be, and therefore make informed decisions on whether it will be a valid target for drug intervention. Further, we know that select protease inhibition can be the Achilles heel of a vast number of pathogenic organisms, including viruses such as HIV, bacteria, and parasites.

Better by Design

Knowledge-based drug design is an approach that uses an understanding of the target protein, or protein-ligand interaction, to design enzyme inhibitors, and agonists or antagonists of receptors. Researchers have recently made substantial inroads into this area, thanks to the developments in X-ray crystallography, NMR, and computer-aided conversion of gene sequences into protein tertiary structures.

In addition to these physical approaches, Peptide Therapeutics, Cambridge, Massachusetts developed a complementary, empirical method, which uses the power of combinatorial chemistry to generate arrays of structurally related compounds to probe the catalytic site and examine the molecular recognition patterns of the binding pockets of enzymes. The system that was patented can be adapted to generate structure-activity relationships (SAR) data for any protein-ligand interaction. In the first instance, however, it was demonstrated that this strategy using proteases as the enzyme target and termed this section of the platform technology RAPID (rational approach to protease inhibitor design).

The conversion of peptide substrates into potent non-peptide inhibitors of proteases possessing the correct pharmokinetic and pharmacodynamic properties is difficult but has some precedents, for example, in designing inhibitors of aspartyl protease such as HIV protease and the matrix metalloproteases. Further, recent work by groups from Merck, SmithKline Beecham, Zeneca, and Pfizer on the cysteinyl proteases Ice and cathepsin K, and the serine proteases elastase and thrombin also opened up new strategies for designing potent reversible and bioavailable inhibitors starting from peptide motifs.

A RaPiD Approach

One of the Peptide Therapeutics' initial objectives was to synthesize selective inhibitors of Der pl, the cysteinyl protease that is considered to be the most allergenic component secreted by the house dust mite.

The house dust mite lives in warm moisture-rich environments such as the soft furnishings of sofas and beds. To feed itself, the mite secretes small particles containing a number of proteins, including Der pl, to degrade the otherwise indigestible proteins that are continuously being shed by its human hosts. When these proteins have been sufficiently tenderized by the protease, the mite returns to its meal. It is a slightly discomforting thought that most of the 'house dust' that can be seen on polished furniture originates from shed human skin. The problems arise when humans, especially young children with developing immune systems, inhale Der pl-containing particles into the small airways of the lung, because the highly active protease can destroy surface proteins in the lung and cause epithelial cell shedding. Further, there is evidence to suggest that the protease also interferes with immune cell function, which leads directly to a greatly accentuated allergic response to foreign antigens.

To test the concept that the Der pl inhibitors will be effective in treating house dust mite related atopic asthma, first we needed to synthesize a selective and potent compound that could be used for *in vivo* studies and would not inhibit other proteases. We set as our criteria that an effective, topically active compound should be 1000 times more selective for Der pl than for cathepsin B, an important intercellular cysteinyl protease.

To map the protease and so to understand the molecular recognition requirements, the binding pockets that surround the catalytic site, we designed and synthesized fluoresence resonanance energy transfer (Fret) library. Four residues, A, B, C, and D were connected via amide bonds in a combinatorial series of compounds of the type A10-B10-C8-D8 which represent 6400 compounds. The central part of each molecule, A-B-C-D, was flanked by a fluorescer (aminobenzoic azid) and quench (3-nitrotyrosine) pair. No fluorescence was detected while the pair remained within 50A of one another, but on proteolytic cleavage of the substrate the quencher was no longer there and fluorescence was generated in direct proportion to the affinity of the substrate (1/Km where Km is the Michaelis constant for the protease and its subsequent turnover (k_{ca}).

The combinatorial mapping approach lends itself readily to the inclusion of non-peptides and peptidomimetic compounds, because all that is required is the cleavage in the substrate of one bond between the fluorescer-quencher pair. The sissile bond is usually a peptidic amide bond, but in the case of weakly active proteases we have successfully incorporated the more reactive ester bond.

We synthesized and then screened the resulting library of 6400 compounds against Der pl and cathepsin B using an 80-well format, where each well contains 20 compounds. Each library was built twice, but the compounds were laid out differently so that we could easily identify the synergistic relationships between the four residues A-D, and decipher immediately the structure-activity relationships that emerged.

At the beginning of our work we could analyze the amount of SAR data that was produced using pencil and paper. However, as the Fret libraries approached 100,000 compounds, the amount of data generated made SAR analysis extremely difficult and time consuming. Therefore, we developed a unique software and automated the SAR analysis, so that the RAPiD is now a powerful decision making tool for the medicinal chemist, who can who can quickly analyze the SAR data in fine detail.

Using clear SAR patterns, medicinal chemists can select a variety of compounds from the Fret library for resynthesis, and obtain full kinetic data on the k_{cat} and Km values. We used the SAR data that we obtained for Der pl and cathe B to convert the most selactive and active motifs into an extremely potent and >1 fold selective inhibitor PTL11031, which we are currenly evaluating in vivo and are currently adapting it for designing selective protein inhibitors. It is important to note that the initial output from this modular approach is genuine SAR patents, which can be quickly converted into SAR data. More than a year after we patented the RAPiD concept, Merck also published a spatially addressable mixture approach using larger mixtures of compounds. This described a similar system for discovering a 1-adrenergic receptor agonists, and independently evaluated the point of this approach for generating quickly large amounts of SAR data for understanding the synergies involved in protein-ligand interactions.

We think that the RAPiD system will allow the medicinal chemist to make knowledge-based drug design decisions for designing protease inhibitors, and can easily be extended by changing the assay readout, to generating useful SAR or other protein-ligand interactions.

4.7 Aerosols Exposed

Research into the pathways by which aerosols are deposited on skin or inhaled is shedding light on how to minimize the risk of exposure, says Miriam Byrne, a research fellow at the Imperial College Centre for Environmental Technology in London.

Among the most enduring TV images of 1997 must be those of hospital waiting rooms in Southeast Asia, crowded with infants fighting for breath and wearing disposable respirators. Last autumn, many countries in the region suffered from unprecedented air pollution levels in particle (aerosol) form, caused by forest fires and exacerbated by low rainfall and unusual wind patterns associated with El Niño. At the time, the director general of the World Wide Fund for Nature spoke of a "planetary disaster: the sky in Southeast Asia has turned yellow and people are dying." In Sumatra and Borneo, more than 32,000 people suffered respiratory problems during the episode, and air pollution was directly linked to many deaths in Indonesia.

In such dramatic situations, we do not need scientific studies to demonstrate the association between pollutant aerosol and ill health: the effects are immediately obvious. However, we are developing a more gradual awareness of the adverse health effects associated with urban air pollution levels, which are now commonplace enough to be considered "normal." Air pollution studies throughout the world, most notably the Six Cities study conducted by researchers at Harvard University, U.S., have demonstrated a strong association between urban aerosol concentrations and deaths from respiratory diseases. Although researchers have yet to confirm exactly how particles affect the lungs, and whether it is particle chemistry, or simply particle number that is important, the evidence linking air pollution to increased death rates is so strong that few scientists doubt the association.

Hospital reports indicate that excess deaths due to air pollution are most common in the elderly and infirm section of the population, and the U.K. Department of the Environment (now the DETR) Expert Panel on Air Quality Standards concluded that particulate pollution episodes are most likely to exert their effects on mortality by accelerating death in people who are already ill (although it is also possible prolonged exposure to air pollution may contribute to disease development). One might think that the elderly could be unlikely victims, since they spend a great deal of their time indoors, where they should be shielded from outdoor aerosol. Unfortunately, aerosol particles readily penetrate buildings through doors, windows, and cracks in building structures, especially in domestic dwellings, which in the UK are naturally ventilated. Combined with indoor particle sources, from tobacco smoke and animal mite excreta, for example, the occupants of buildings are continuously exposed to a wide range of pollutants in aerosol form.

Exposure Routes

So if particles are generated in buildings, and infiltrate from outdoors anyway, is there any point in advising people to stay indoors, as the Filipino health department did during last autumn's forest fires? In fact, staying indoors during a pollutant episode is good practice: airborne particles often occur at lower levels indoors, not because they do not leak in, but because they deposit on indoor surfaces.

The ability of particles to deposit is one of the key features that distinguishes this behavior from that of gases. Although some reactive gases, SO_2 for example, absorbed onto surfaces, the surface gas interaction is primarily a chemical one in the case of aerosol particles; their physical characteristics govern transport adherence to surfaces. Particles greater than a few um in size are strongly influenced by gravity and settle readily on horizontal surfaces, whereas smaller particles have a greater tendency to move by diffusion. In everyday life, we encounter particles in a wide range of size distributions.

There is another important factor that distinguishes pollutant particles from gases. "If you don't breathe it in, you don't have a problem" is a philosophy that we might be tempted to apply to aerosol pollution. But this is by no means true in all cases; unlike gases, aerosol particles may have more than one route of exposure, and are not only a hazard while airborne. There are three major routes by which pollutant particles can interact with the human body: inhalation, deposition, and ingestion on the skin. Even the process of inhaling particles is complex, relative to gases, because particles occur in a wide range of size distributions and their size determines their fate in the respiratory system. When entering the nose, some particles may be too large to penetrate the passages between nasal hairs or negotiate the bends in the upper respiratory tract, and may deposit early in their journey, whereas smaller particles may penetrate deep in the alveolar region of the lung, and if soluble, may have a toxic effect on the body.

The second route by which particles intercept the body is by depositing on the skin, but this tends to be more serious for specialized occupational workers — notably those involved in glass fiber and cement manufacture — than for the general public. In an average adult, the skin covers an area of about $2m^2$, and while much of this is normally protected by clothing, there is still considerable potential for exposure. In the U.K., the Health and Safety Executive estimates that 4 working days per year are lost through occupational dermatitis — although not all of these cases arise from pollutant particle deposition; liquid splashing and direct skin contact with contaminated surfaces are also contributors. It is not only the skin itself that is at risk from particle deposition. It is now almost 100 years since A. Schwenkenbacher discovered that skin is selectively permeable to chemicals; the toxicity of agricultural pesticides, deposited on the skin as an aerosol or by direct contact with contaminated surfaces, is an issue of major current concern.

Particle Deposition

The third human exposure pathway for pollutant particles is by ingestion. Unwittingly, we all consume particles that have deposited on foodstuffs, as well as picking up particles on our fingertips through contact with contaminated indoor surfaces, and later ingesting them. Toxic house dust is a particular menace to small children, who play on floors, crawl on carpets, and regularly put their fingers in their mouths. Research by the environmental geochemistry group at Imperial College, London, has shown that for small children, hand-to-mouth transfer is the major mechanism by which children are exposed to lead and other metals, which arise indoors from infiltrated vehicle and industrial emissions and also from painted indoor surfaces.

Of the three exposure routes, particle deposition dictates which one dominates any given situation: while particles are airborne, inhalation is possible, but when they are deposited on building or body surfaces, skin exposure and ingestion exposures result. And the route of exposure may make all the difference: some chemicals may be metabolically converted into more toxic forms by digestive organs and are therefore more hazardous by ingestion than by inhalation or skin penetration. Therefore, to predict how chemicals in aerosol form influence our health, we must first understand how we become exposed. A sensible first step in trying to make comprehensive exposure assessments, and developing strategies for reducing exposure, is to understand the factors influencing indoor aerosol deposition, for a representative range of particle sizes. We can then apply this knowledge to predicting exposure for chemicals that occur as aerosols in these various size ranges.

At the Imperial College, together with colleagues from Riso National Laboratory, Denmark, we have dedicated more than a decade of research to understanding factors that control indoor aerosol deposition and which, in turn, modify exposure routes. Motivated by the Chernobyl incident, and in an effort to discover any possible benefits of staying indoors during radioactive particulate cloud passage, we measured, as a starting point, aerosol deposition rates in test chambers and single rooms of houses for a range of particle sizes and indoor environmental conditions. We use these detailed data to formulate relationships for the aerosol surface interaction, and use computational models to make predictions for more complex building geometries, such as a whole house.

Precise Locations

Using the tracer aerosol particles for deposition experiments in UK and Danish houses, we have found that aerosol deposition on indoor surfaces occurs most readily for larger particles, and in furnished and heavily occupied rooms. This probably comes as no surprise: as mentioned before, gravity encourages deposition of larger particles, and furnishings provide extra surface area on which particles can deposit. What may be surprising, though, are our supplementary measurements, which compare aerosol deposition on the walls and floor of a room-sized aluminum test chamber. We can see, for the smallest particle size examined (0.7 um), that total wall deposition becomes comparable to floor deposition. We found that adding textured materials to the walls enhances aerosol deposition rate by at least a factor of 10, even for particles that we might expect to be large enough to show preferential floor deposition.

What are the implications of these observations? The predicted steady-state indoor/outdoor aerosol concentrations, from an outdoor source, generated using our measured indoor aerosol deposition rates in a simple compartmental model, indicates that indoor aerosol deposition is an important factor in lowering indoor concentrations of aerosols from outdoor sources, particularly in buildings with low air exchange rates. However, encouraging particles to deposit on surfaces is only a short-lived solution to inhalation exposure control, because the particles can be readily resuspended by disturbing the surfaces on which they have deposited. It is prudent to clean not only floors regularly but also accessible walls, and particularly vertical soft furnishings such as curtains which are likely to attract particles and are also subject to frequent agitation. The same cleaning strategies can also be applied to minimizing house-dust ingestion by small children: in this case, surface contact is the key factor.

We have seen that carpets and wallpaper can be readily sampled for tracer particles by NAA; so too can the surface of the human body. While there are relatively few skin contaminants in the normal urban indoor environment, there are many in the workplace, and data for indoor aerosol deposition rates on skin are important for occupational risk assessment. In addition, such data are relevant in the nuclear accident context: after the Chernobyl incident, calculations by Arthur Jones at the National Radiological Protection Board suggested that substantial radiation doses could arise from particles deposited on the skin, and that the particle deposition rate on skin was a critical factor in determining the significance of this dose.

Susceptible Skin

In an ongoing study, we are using our tracer particles to measure aerosol deposition rates on the skin of several volunteers engaged in various sedentary activities in a test room. Following aerosol deposition, we wipe the volunteers' skin with moistened cotton swabs according to a well-validated protocol, and collect hair and clothing samples. We then use NAA to detect tracer particles deposited on the wipes, hair and clothing. The most striking finding so far is that particle deposition rates on skin are more than an order of magnitude higher than deposition rates on inert surfaces such as walls. We think that there are several factors contributing to this result, including the fact that humans move, breathe, and have temperature profiles that lead to complex air flows around the body.

As well as providing occupational and radiological risk assessment data, our work on skin deposition may raise some issues concerning indoor aerosol inhalation, because it provides information on particle behavior close to the human body, i.e., where inhalation occurs. In the urban environment, personal exposure estimates for particulate pollutants are often derived from stationary indoor monitoring, but some researchers, notably those working in the University of California at Riverside, have noted elevated particle levels on personal monitors positioned around the nose and mouth. These workers conclude that this is due to the stirring up of "personal clouds," i.e., particles generated by shedding skin and clothing fragments, and by dust resuspended by the body as it moves. This may well be the case, but our tracer particle measurements on sedentary volunteers do not show up human-generated particles; however, they are still sufficiently high to suggest that particles are actually being drawn into the region surrounding a person. While questions remain about how stationary particle monitoring relates to personal exposure, and until we understand whether it is particle number, mass, pattern of exposure, or a combination of all of these that contributes to respiratory ill health, we are left with a complex and challenging research topic.

4.8 The Optimizer Program

The Quattro Pro Program (version 5.0 or 7.0) contains the optimizer program under the Tools menu. This has been used to optimize structure in terms of a plethora of recipes of desired physical and toxicological properties. Such a program can be used for substitution for original process chemicals that may be toxic pollutants in the environment and also for drugs in medicine that need more efficacy and fewer side effects. These studies can be made while ensuring minimum cost. In order to do this, the computer is instructed as to what the constraints are (= or >= or <=) in the equations, what the variables are, what the constants are, and which variables are constrained to be integers. Conditions are also set up to constrain the number and types of bonds, if desired.

When the Optimizer is called up, a template appears, in which you are to name the solution cell, say whether you want a maximum, minimum, or none (neither), name a Target value, and assign the Variable Cells in the spreadsheet. Finally, the constraints are added. These may look like Q1..Q1 Integer, Q2..Q2<=5, etc. You may Add, Change or Delete (to) any constraint. The difficulty is only in understanding what terms such as Solution Cell, Variable Cell, etc., mean. There is also an Options choice in the Optimizer Box. In it you can fix the maximum time, maximum iterations, and the precision and tolerance of the runs. It also allows choices for estimates: tangent or quadratic, derivatives: forward or central, and search methods: Newton and conjugate. It allows options for showing iteration results and assumptions of linear and automatic scaling. You can save the model, load the model, and also have a report. The system can use nonlinear as well as linear models. Before proceeding, it is well to set up your variable cells, constraint cells and solution cells on your spreadsheet. This normally utilizes only a small part of your spreadsheet and the solution will appear within this small part of your spreadsheet that is set aside.

4.9 Computer Aided Molecular Design (CAMD): Designing Better Chemical Products

A new class of molecular design, oriented towards chemical engineering problems, has developed over the last several years. This class of CAMD software focuses on three major design steps:

1. Identifying target physical property constraints. If the chemical must be a liquid at certain temperatures we can develop constraints on melting and boiling points. If the chemical must solvate a particular solute we can develop constraints on activity coefficients.
2. Automatically generating molecular structures. Using structural groups as building blocks, CAMD software generates all feasible molecular structures. During this step we can restrict the types of chemicals designed. We could eliminate all structural groups which contain chlorine or we may require that an ether group always be included.
3. Estimating physical properties. Using structural groups as our building blocks enables us to use group contribution estimation techniques to predict the properties of all generated structures. Using group contribution estimation tech-

niques enables CAMD software to evaluate new compounds.

As an example we design an extraction solvent for removing phenol from an aqueous stream other than toluene which is strongly regulated. The extraction substitute for toluene must satisfy three property constraints: the selectivity and capacity for the solute must be high, the density should be significantly different from the parent liquor to facilitate phase separation, and the vapor-liquid equilibrium with the solute should promote easy solvent recovery.

To satisfy these property constraints it is often easy to simply specify that the substitute should have the same properties as the original solvent. We will find a new chemical that has the same selectivity for extracting phenol from water as does toluene.

To quantify selectivity we can use activity coefficients, infinite dilution activity coefficients or solubility parameters. We use the latter and our target is Spd = 16.4, Spp = 8.0 and Sph = 1.6, where they are the dispersive, polar, and hydrogen-bonding solubility parameters in units of $MPa^{1/2}$. We add a small tolerance to each value.

Next we generate structural groups. Halogenated groups were not allowed because of environmental concerns. Acidic groups were not allowed because of corrosion concerns. Molecules can be represented as simple connected graphs. Such graphs must satisfy the following constraint:

$$b/2 = n + r - 1$$

where b is the number of bonds, n is the number of groups, and r is the number of rings in the resulting molecule. Our case has b = 6, n = 3, and r = 0. For this particular example one of the CAMD denerated solvents, butyl acetate, matched the solvent chosen as the toluene substitute in the plant.

4.10 Reduce Emissions and Operating Costs with Appropriate Glycol Selection

BTEX emissions from glycol dehydration units have become a major concern and some form of control is necessary. One method of reducing BTEX emissions that is often overlooked is in the selection of the proper dehydrating agent. BTEX compounds are less soluble in diethylene glycol (DEG) than triethylene glycol (TEG) and considerably less soluble in ethylene glycol (EG). If the use of DEG or EG achieves the required gas dew point in cases where BTEX emissions are a concern, a significant savings in both operating costs and the cost of treatment of still vent gases may be achieved. The paper described here compares plant operations using TEG, DEG, and EG from the viewpoint of BTEX emissions, circulation rates, utilities, and dehydration capabilities.

4.11 Texaco Chemical Company Plans to Reduce HAP Emissions Through Early Reduction Program by Vent Recovery System

For the purposes of the Early Reduction Program, the source identification includes a group of stationary air emission locations within the plant's butadiene purification operations. The process includes loading and unloading and storing of crude butadiene, transferring the butadiene to unit and initial pretreatment; solvent extraction; butadiene purification; recycle operations; and hydrocarbon recovery from wastewater. The emissions location include: process vents, point sources, loading operations, equipment leaks, and volatiles from water sources. To reduce HAP emissions Texaco Chemical plans to control point source emissions by recovering process vent gases in a vent gas recovery system. The vent recovery system first involves compression of vent gases from several process units. The compressed vent gases go through cooling. Next, the gases go to a knockout drum for butadiene condensate removal. The liquid butadiene is again run through the process. Some of the overhead vapors route to a Sponge Oil tower which uses circulating wash oil to absorb the remaining hydrocarbons. The remaining overhead vapors burn in the plant boilers.

4.12 Design of Molecules with Desired Properties by Combinatorial Analysis

Suppose that a set of groups to be considered and the intervals of values of the desired properties of the molecule to be designed are given. Then, the desired properties constitute constraints on the integer variables assigned to the groups. The feasible region defined by these constraints is determined by an algorithm involving a branching strategy. The algorithm generates those collections of the groups that can constitute structurally feasible molecules satisfying the constraints on the given properties. The molecular structures can be generated for any collection of the functional groups.

The proposed combinatorial approach considers only the feasible partial problems and solutions in the procedure, thereby resulting in a substantial reduction in search space. Available methods exist in two classes. One composes structures exhaustively, randomly, or heuristically, by resorting to expert systems, from a given set of groups; the resultant molecule is examined to determine if it is endowed with the specified properties. This "generate-and-test" strategy is usually capable of taking into account only a small subset of feasible molecular structures of the compound of interest. It yields promising results in some applications, but the chance of reaching the target structure by this strategy can be small for any complex problem, e.g., that involving a large number of groups. In the second class, a mathematical programming method is applied to a problem in which the objective function expresses the "distance" to the target. The results of this assessment may be precarious since the method for estimating the properties of the structure generated, e.g., group contributions, is not sufficiently precise. The work here is a combinatorial approach for generating all feasible molecular structures, determined by group contributions, in the given intervals. The final selection of the best structure or structures are to be performed by further analysis of these candidate structures with available techniques.

4.13 Mathematical Background I

Given:

a. Set G of n groups of which a molecular structure can be composed,
b. The lower bounds, p_j's and the upper bounds, P_j's of the properties to be satisfied, where j=1, 2, ..., m;
c. Upper limit L_i (i=1, 2, ...,n) for the number of appearances of group i in a molecular structure to be determined; and
d. Function f_k *(k=1, 2, ..., m) representing the value of property k which is estimated by the group contribution method as*

$$f_k(x_1,x_2,\ldots,x_n).$$

In the above expression, x_1, x_2, ..., x_n are, respectively, the number of groups #1, #2, ..., and #n contained in the molecular structure or compound.

The problem can now be formulated as follows:

Suppose that f_k (k = 1, 2, ...m_i) is an invertible function on the linear combinations of coefficients a_{ki} (i = 1, 2, ..., n), on S $a_{ki}x_i$. Furthermore, assume that function f_k (k = m_i + 1, m_i + 2, ..., m_2) has a sharp linear outer approximation, i.e., there are coefficients a_{ki} and a'_{ki} such that

$$\Sigma a_{ki}x_i \le f_k(x_1,x_{2,\ldots,}x_n) \le \Sigma a'_{ki}x_i \quad (k = m_i + 1,\ m_i + 2,\ldots,\ m_2) \tag{1}$$

We are to search for all the molecular structures formed from given groups, #1, #2,... and #n, whose numbers are x_1, $x_{2,}$ ... x_n, respectively, under the condition that the so-called property constraints given below are satisfied.

$$p_j \le f_j(x_1,x_2,\ldots,x_n) \le P_j \quad (j = 1,2\ldots,m).$$

Throughout this paper, the constraints imposed by the molecular structure on feasible spatial configurations are relaxed, and the molecular structures are expressed by simple connected graphs whose vertices and edges represent, respectively, the functional groups from the G set and the associated bonds. Thus, the set of such connected graphs need be generated from the set of functional groups G that satisfies the property constraints admitting multiple appearances of the functional groups.

In the conventional generate-and-test approach, all or some of the connected graphs, i.e., structurally feasible graphs, are generated from the available functional groups and are tested against the property constraints. This usually yields an unnecessarily large number of graphs. To illustrate the inefficiency of this approach, let the structurally feasible graphs be partitioned according to the set of functional groups of which they are composed; in other words, two graphs are in the same partition if they contain the same groups with identical multiplicities. Naturally, all the elements in one partition are either feasible or infeasible under the property constraints. Moreover, the graph generation algorithm of this approach may produce all elements of the partition, even if an element of this partition has been found to be infeasible earlier under the property constraints: obviously this is highly inefficient.

4.14 Automatic Molecular Design Using Evolutionary Techniques

Molecular nanotechnology is the precise three-dimensional control of materials and devices at the atomic scale. An important part of the nanotechnology is the design of molecules for specific purposes. This draft paper describes early results using genetic software techniques to automatically design molecules under the control of a fitness function. The software begins by generating a population of random molecules. The population is then evolved towards greater fitness by randomly combining parts of the better individuals to create new molecules. These new molecules then replace some of the worst molecules in the population. The approach here is genetic crossover to molecules represented by graphs. Evidence is presented that suggests that crossover alone, operating on graphs, can evolve any possible molecule given an appropriate fitness function and a population containing both rings and chains. Prior work evolved strings or trees that were subsequently processed to generate molecular graphs. In principle, genetic graph software should be able to evolve other graph representable systems such as circuits, transportation networks, metabolic pathways, computer networks, etc.

4.15 Algorithmic Generation of Feasible Partitions

The feasible partitions can easily be generated for the problem defined in the preceding section by a tree search algorithm similar to the branch-and-bound framework.

A novel approach is proposed in the present work which is substantially different from the generate-and-test approach. It first identifies the feasible partitions satisfying the property constraints as well as the structural constraints; this is followed by the generation of the different molecular structures for each of the resultant partitions. The proposed approach is more effective than the generate-and-test approach because each partition need be considered only once, and the algorithm for generating molecular structures is performed only for a feasible partition. In addition, the approach can be conveniently implemented by means of a tree search.

Suppose that the values of variables $x_1, x_2, \ldots, x_k$ (k n-1) are fixed *a priori* as $l_1, l_2, \ldots, l_k$ at an intermediate phase of the procedure; then, the problem is branched to L_{k+1} partial problems for $l_{k+1} = 0, 1, 2, \ldots, L_{k+1}$ according to the following two cases.

Case 1. $k \le n-2$:

$$p'_j \le \Sigma\, a_{ji}l_i + \Sigma\, a_{ji}x_i \le P'_j \qquad (j = 1, 2, \ldots, m_i)$$

$$p_j \le \Sigma\, a'_{ji}l_i + \Sigma\, a'_{ji}xi \qquad (j = m_i{+}1, m_i + 2, \ldots, m_2)$$

$$\Sigma\, a_{ji}l_i + \Sigma\, a_{ji}x_i \le P_j \qquad (j = m_i + 1, m_i + 2, \ldots, m_2)$$

$$0 \le x_i \le M_i \qquad (i = k + 2, k + 3. \ldots, n)$$

$$\Sigma x_i - \Sigma\, x_i \ge -2$$

where the values of variable x's are extended from integers to real values; p'_j and P'_j denote $f^{-1}(p'_j)$ and $f^{-1}(P'_j)$ ($j = 1, 2, \ldots, m_1$), respectively; constraint ( ) expresses a necessary condition to have a connected graph in a partition; and i_1 indicates the indices of branching functional groups while i_2 indicates the indices of terminator groups.

The feasibilities of partial problems generated above must be tested; for example, this can be done by the first phase of the simplex algorithm. If any of the problems pass the test, it needs to be branched further.

Case 2. $k = n-1$:

A test must be performed by simple substitution to determine if the constraints and condition below are satisfied for $l_1, l_2, \ldots, l_n$.

$$p_j \le f_j(l_1, l_2, \ldots, l_n) \le P_j \qquad (j = 1,2, \ldots, m)$$

$\Sigma li - \Sigma l_i$ is an even number not less than -2

Condition 1. If the partition given by $l_1, l_2, \ldots, l_n$ includes functional groups with different types of bonds (e.g., single and double bonds), then there must be a group included in the partition, which has at least two different types of bonds, with each type belonging to at least one group of the partition containing other types of bonds.

After the feasible partitions are generated, the feasible molecular structures can be generated by an available computer program or a combinatorial algorithm. The present procedure is most advantageous when applied to problems involving large numbers of constraints on the predicted properties, especially if most of them are linear or can be sharply bounded by linear functions.

4.16 Testsmart Project to Promote Faster, Cheaper, More Humane Lab Tests

The Environmental Defense Fund (EDF), John Hopkins University, the University of Pittsburgh, and Carnegie-Mellon University announced the launch of TestSmart, a project to find more efficient and humane methods of conducting a preliminary toxicity screening test on chemicals. The four institutions will explore new testing methods that minimize the use of laboratory animals and produce reliable results faster and for less money than in the past. The search to gather basic information on the health and environmental effects of nearly 3000 high-production volume industrial chemicals is under way.

In October 1998, Al Gore announced a cooperative agreement among EDF, the U.S. EPA and the Chemical Manufacturers Association (CMA) to test thousands of industrial chemicals that are used in the U.S. in volumes of more than one million pounds each year. The agreement to test came after separate studies by EDF, EPA, and CMA all concluded that basic health effects information is not publicly available for most major industrial chemicals.

The high-production volume chemicals will be tested over the next five years using screening methods as defined through the Organization for Economic Cooperation and Development's international consensus process. Some of the test procedures now call for testing on laboratory rodents, fish, and insects. The TestSmart project will explore alternative testing and evaluation techniques.

We need assurance that chemicals in our economy are not causing unknown harm to our health and environment. Animal-based studies are used by virtually every regulatory agency and we must evaluate emerging techniques and identify areas where further targeted research is needed to develop new approaches.

A key component of the initiative will involve the use of Structure-Activity Relationship (SAR) analysis. Using SAR, it is possible in appropriate circumstances to extrapolate on the health and environmental effects of classes of chemicals to structurally related agents that have not yet been tested. SAR is well established for certain endpoints and we will be evaluating its application in a wider variety of contexts.

The project team will help review proposals under the voluntary testing initiative to group similar chemicals into categories. Selected members of the categories would be tested and results interpolated to other members. While an important mechanism for enhancing the efficiency of testing and minimizing use of test animals, proper definition of scientifically robust categories is essential to the success of this approach.

4.17 European Cleaner Technology Research

Cleaner manufacturing is an increasingly important goal for industry in Europe for both existing and new facilities. For such work, significant research funding is made available by governments and industry in Europe to achieve cleaner process and product technologies. European funding goals are to improve industrial competitiveness and technology transfer to industry. Topics include plastic and polymer recycling, expansion of products with renewable materials, recycle of an increasing number of chemicals, diverse CFC replacement, carbon dioxide utilization, heavy metal minimization, and reduced chemical use.

European Community

Program on Industrial and Materials Technologies (BRITE-EURAM II)

The EC has 15 research and technology development (RTD) areas. The following descriptions highlight the clean technology topics covered under BRITE/EURAM

- Friendly Polymers
  - Biodegradability
  - polyamide polymers
  - bacterial polyesters
  - blends of thermoplastic resins
- Polymer Coatings Research Efforts
  - high-solids, isocyanate-free paint
  - a water-based improvement
- Ultimate Recyclability of Heterogeneous Materials
  - composite materials
- Substitutes for PVC in various applications
- Use of Wastes in Paving Materials
- Use of Gypsum in the Building Industry
- Clean Technology in the Leather Industry.
  - nonchrome alternatives
- Acid Recovery
- Catalyst Recycle and Recovery

Germany

Low Emission Processes in Industry

The projects in this category focus on process development utilizing chemical routes or achievement of greater efficiency through process understanding. Though the projects focus on specific processes or plants, there are significant advances in the knowledge underlying each process involved. This is a transferrable element that benefits the overall issue of cleaner manufacturing.

- Low Emission Products
  - Halogen-free fire retardants for plastics in electronic equipment
  - recyclable lawn mowers
  - television parts and assemblies for ultimate reuse
  - biologically degradable lubricants
- CFC Replacement
- Chlorinated Hydrocarbon Replacement
- Reduction of Volatile Emissions
  - reduction of non chlorinated hydrocarbons
  - volatile emissions from two-layer enamels for industrial coatings
  - development of powder coating alternatives in automobiles
- Plastics Recycling
  - expand the type of plastic products that can be recycled
  - recycling process, for pure or impure or mixed plastics.
- Pyrolysis and hydrogenation techniques.

DECHEMA

- Research Focus Areas
  - new materials
  - principles of catalysis
  - basics of recycling, and
  - Renewable resources
- Recycling of Plastics and Metals/Inorganics
  - plastics recycling by thermal methods

degradation of polymers to monomers and oligomers

Renewable Resources

use of olochemical surfactants on a broader basis

renewable resources for biologically degradable lubricants and hydraulic oils

derivatives from natural products such as starch, glucose, protein, celluose, lignin and fat combined to give new possible applications

use of natural polymers and derivatives

new materials using renewable resources

intermediates, special, and fine chemicals

biotechnology for use of renewable resources

Plant Protection and Resistance

Carbon Dioxide Utilization

Fraunhofer Institute for Food Technology and Packaging

quality assurance in Food and Packaging

technical microbiology

process engineering and biotechnology

packaging technology

Reduction of Chemicals in Paper Products

Biopreservation of Food

Solvent-Free Extraction

Packaging Materials from Renewable Resources

CFC Elimination

Biopreservation of Food

Solvent-Free Extraction

Packaging Materials from Renewable Resources

CFC Elimination

Energy Efficient Organic Vapor Removal from Air Streams

Metal and Acid Recovery in the Copper Plating Industry

Improvement of Membrane Properties

Italy

Membrane Systems

emphasis on membrane material characterization and process modeling

ceramic membrane and nanofiltration projects included

Photocatalysis

Photocatalytic treatment of aqueous materials

Recycling of Membranes and Catalysts

Switzerland

Reaction Engineering:Effects of Mass Transfer on Secondary Product Formation (ETH).

Chemical Substitution in the Textile Industry

Solar Energy

Recycling of Building and Pavement Materials (ETH).

Catalysts for Carbon Dioxide Coversion (ETH)

United Kingdom

Clean Synthesis of Effect Chemicals

Light Harvesting

Farming as an Engineering Process

Cities and Sustainability

Fuel Cells

Clean Combustion

Analysis and Measurment

Life Cycle Analysis (LCA)

Hub technologies include the most significant environmental problems, work currently undertaken by industry and universities and areas of cleaner technologies

process flow sheeting

process control software

reactor engineering

membrane processing

super- and sub-critical fluids

photo(electro)chemistry

ultrasonics and sonoelectrochemistry

ohmic processes

induction heating

4.18 Cleaner Synthesis

Chemical companies are coming under increasing pressure to clean up their activities by finding alternative cleaner syntheses rather than by dealing with the after-effects, says Tim Lester. When the technical adviser for cleaner synthesis on the Science and Engineering Council's (now EPSRC's) Clean Technology Programme was appointed four years ago, one of the first things he did was to carry out some literature searches. Using the keyword 'clean' yielded almost nothing, but I suspect that the position would be very different today. The terms clean technology and clean (or cleaner) synthesis are heard much more frequently nowadays at gatherings of industrial chemists. Journal publishers have also seized on the opportunity — the *Journal of Chemical Technology and Biotechnology* now boldly states on its front cover that its coverage encompasses clean technology, while the *Journal of Cleaner Production* was launched in 1993. Clean technology forestalls pollution by circumventing waste production and minimizing the use of energy — avoiding the problem in the first place rather than treating the effluents. So where does cleaner synthesis fit in?

Cleaner synthesis involves making changes to the chemistry, biology, or engineering of the original process. It is just one of several waste minimization options and may not always be the most appropriate; another strategy might be to change the product to one that serves the same purpose, but is cleaner to manufacture. Alternatively, carefully implementing "good practice" can lead to substantial reductions in discharges to water and land, as demonstrated by a recent project to clean up the rivers Aire and Calder.

For those involved in manufacturing chemicals, cleaner synthesis is particularly satisfying — making a product by a novel route that is both commercially advantageous and less of a burden on the environment. Many sub-disciplines of chemistry and process engineering can contribute to the technology at various levels of sophistication — from leading edge science to the innovative application of existing knowledge.

Chemical manufacturing processes vary enormously in their waste production characteristics. Commodity chemicals made from petroleum feedstocks are often produced by using catalytic processes with very high yields and selectivities. Byproducts can frequently be used in other processes or, as a last resort, as fuels to help run large integrated petrochemical plants. Nevertheless, even the petrochemical and refining industries acknowledge certain challenges, notably finding a selective route to ethene synthesis; currently only ca 30% of the naphtha cracked product is ethene. It would also be useful to identify an alternative acid catalyst to replace HF or H_2SO_4 used for alkylating various gasoline components.

At the other end of the scale lies the synthesis of complicated biologically active molecules in the pharmaceutical and agrochemical industries, often requiring multistep procedures giving overall yields as low as 10%.

Roger Sheldon has categorized sectors of the chemical industry by their quantity of byproduct per kilogram of product (Table 1). Many of the byproducts that arise in these various industry sectors do so by stoichiometric reactions, side reactions, or further reactions.

Stoichiometric Reactions

$$A + B \rightarrow C + D$$

Unless by some fortunate chance both C and D are useful products, stoichiometric reactions of this type produce a molecule of a waste for each molecule of product. Acids and alkalis are often reagents, with salts as the unwanted byproducts.

Other reactions that fit this type of scheme are $AlCl_3$-catalyzed Friedel-Crafts alkylations and acylations, which generate large quantities of aluminum wastes. Despite considerable advances towards replacing $AlCl_3$ with other catalysts, such as zeolites and Envirocats, worldwide demand for $AlCl_3$ is ca. 75 000 t pa — much of it used to catalyze Friedel-Crafts reactions.

Researchers are also developing cleaner alternatives to oxidizing agents based on chromium and manganese, which can give toxic metal byproducts that are subject to regulatory pressures.

We can take a quantitative look at the waste produced by different reactions using the concept of atom utilization, calculated by dividing the molecular weight of the desired product by the sum of the molecular weights of all "products".

The manufacture of t-butylamine illustrates the concept. The conventional route is via the Ritter reaction:

$$(CH_3)_2C{=}CH_2 \xrightarrow{HCN\ +\ H_3O} (CH_3)_3CNHCHO \xrightarrow{H_3O}$$

$$(CH_3)_3CNH_2 + HCOOH$$

Based on these reactions the atom utilization is 61%, but the reaction suffers from salt formation during work up. However, BASF has developed a cleaner synthesis using zeolite catalysis, which demonstrates 100% atom utilization:

$$(CH_3)_2C = CH_2 + NH_3 \rightarrow (CH_3)_3CNH_2$$

Stoichiometric reactions are common in work-up procedures to neutralize an acidic or basic medium that was used in a previous reaction step. Product work up and isolation can pose as great a constraint on clean synthesis as the synthetic reaction itself and careful reappraisal of a longstanding work up procedure can sometimes realize significant improvements.

Side Reactions

$$A + B \rightarrow C \text{ (desired reaction)}$$

$$A + B \rightarrow D \text{ (side reaction)}$$

Aromatic substitutions are a well-known example of this type of reaction, with their competition between ortho, meta, and para sites. Nitrating systems that maximize the production of para-nitrotoluene are commercially attractive because the para-isomer costs about twice as much as ortho-nitrotoluene. Keith Smith at the University of Swansea recently reported producing ca. 80% of para-nitrotoluene using nitric acid and acetic anhydride to generate acetyl nitrate in the presence of zeolite beta.

Attempts to make particular enantiomers are also subject to side reactions as a consequence of imperfect stereoselectivity.

Further Reactions

$$A + B \rightarrow C \rightarrow D$$

In these reactions C is the desired product, but the reaction is difficult to stop cleanly at this stage. During the manufacture of ethene oxide by oxidizing ethene, ca 20% of the starting material undergoes further oxidation to carbon dioxide and water. Another example is the production of methylene chloride by chlorinating methyl chloride, which also results in the over-chlorinated byproducts chloroform and carbon tetrachloride.

Fate of Solvents

Besides avoiding the above three reaction types wherever possible, cleaner synthesis also involves finding the most appropriate solvents. Water is frequently thought of as the only solvent that can be readily discharged, but finding chemistry that works in water is not helpful unless the reactants and/ or products (plus by-products) are harmless or can be thoroughly removed. Supercritical carbon dioxide is one of very few other environmentally friendly solvents, but as yet its use is limited to a few specialist applications — extracting flavors, or caffeine from coffee beans, for example.

We can sometimes avoid conventional solvents by using an excess of one of the reagents, which circumvents the need to separate and contain another component. For example, polypropene is now prepared in propene, replacing solvents such as kerosene. In many cases, however, effective containment and recycling of solvents may be the best option.

The manufacture of lactic acid illustrates several different aspects of waste production. Two routes — chemical synthesis and fermentation — satisfy most of the worldwide demand for the acid, ca 40 000 tpa. In the chemical synthesis route, the first step involves reacting ethanal with HCN under base-catalyzed conditions at atmospheric pressure to give lactonitrile:

$$CH_3CHO + HCN \rightarrow CH_3CHOHCN \quad (1)$$

After purifying the product by distillation, the next step is to hydrolyse the lactonitrile using concentrated sulphuric (or hydrochloric) acid — a typical stoichiometric reaction resulting in the low value byproduct ammonium sulphate (or chloride):

$$CH_3HCOHCN + 2H_2O + HH_2SO_4 \rightarrow CH_3CHOHCOOH + H(NH_4)2SO_4 \quad (2)$$

The crude lactic acid is then esterified to give methyl lactate:

$$CH_3CHOHCOOH + CH_3OH \rightarrow CH_3CHOHCOOCH_3 + H_2O$$

which is purified and hydrolyzed under acid conditions to give lactic acid and methanol, for recycling back into the system.

The lactic acid molecule is 'built' in steps (1) and (2), with atom utilization efficiencies of 100% and ca. 60%, respectively, but producing a product of a suitably high quality also involves two distillations besides the esterification/hydrolysis steps and their associated energy requirements and wastes.

The second route to lactic acid manufacture relies on fermenting carbohydrates, such as molasses or corn syrup. The resulting acid is neutralized by adding calcium carbonate to yield calcium lactate. Fermentation takes four to six days and, to keep the calcium lactate in solution so that it can be separated by filtration, its concentration must be below ca. 10%, with inevitable consequences for plant cost and capacity. Work up proceeds by carbon treatment, evaporation, and acidification with sulphuric acid, to produce lactic acid plus stoichiometric quantities of calcium sulphate. The resulting lactic acid is only technical grade. If higher quality product is needed, we need to include esterification, distillation, and hydrolysis steps as for the previous route. Both processes therefore fall well short of the ideals of clean synthesis for a number of reasons.

Is It Cleaner?

Choosing the cleanest process from various different routes to make the same product is not always straightforward. One measure that we can look at is the E factor — the amount of waste produced for a given amount of product, taking into account yields and solvent losses as well as atom utilization. However, wastes vary in composition as well as quantity, and consequently may need to be discharged to different media — for example to landfill rather than to water. Finding ways to quantify the environmental damage done by different wastes is not always straightforward.

It is also important, when evaluating the options for cleaner synthesis, to include all the relevant factors, such as the implications of making the necessary reagents or the energy consumed during product separation. For example, ozone looks like an attractive oxidizing agent because it does not leave any toxic residues behind, unlike metal-based reagents, but we need to weigh this against the fact that 90% of the electrical energy used in the electric discharge ozone generator results in heat rather than ozone.

What started out as a simple matter of judging the cleanest process may begin to look like a life cycle analysis. At least one pharmaceutical company has found that an innovative piece of chemistry required so much extra solvent compared to the existing reaction path that the perceived improvements turned out to be a mirage.

Moreover, while clean synthesis addresses environmental concerns, we must not overlook the implications of such changes for health and safety. In a recent review of inherently safer chemistry, Abe Mittelman and Daniel Lin point out that substituting HF for aluminum trichloride in Friedel-Crafts reactions may not be as good for employees as it is for the environment. Companies need to assess the consequences of any changes to processes fully before implementing them.

Industry Response

Manufacturers have been changing and refining their processes in ways that produce less waste since the early days of the chemical industry. Many changes have been driven by economics — by reduced costs for raw materials and energy, or improvements to produce purer products that command higher prices. Tighter regulations for waste disposal, leading to increased treatment or off-site disposal costs, have added to economic pressures — sometimes tipping the balance in favor of less polluting processes. For example, in the chlor-alkali industry, membrane cells, which avoid the use of mercury, are now preferred to the traditional mercury cathode cells.

Today, the literature contains many examples of industrially applied cleaner syntheses. A good example is the Hoechst-Celanese route to ibuprofen from isobutylbenzene, which involves three steps — two of them catalytic — and has an atom utilization of 100%, a marked improvement on the previous six-step synthesis.

Companies are continuing to conduct in-house studies on ways to improve their manufacturing processes, but they are also collaborating with each other, for example, through the Cefic (European Chemical Industry Council) Sustech R&D program, which is subdivided into: bio-Sustech; catalyst design and application; process intensification; safety and environmental management; process modeling, simulation, and control; contaminated land issues; separation technologies; particulate solids processing; and recovery, recycling, and reuse.

Government Activity

At the supra-national level the UN, through its environmental program UNEP, has identified a number of cleaner production industry sector groups, among them several sectors using chemical processes — the textile and pulp and paper industries for example. By organizing conferences, newsletters, and other forms of networking, UNEP aims to encourage better exchange of information between different sectors. A recent example concerns the transfer of "know-how" about clean production in the metal finishing industry between Australia and Hong Kong. When implemented, the resulting recommendations — based on good housekeeping, minimizing water use, and the recovery of excess metals — should reduce discharges to Hong Kong Harbor.

Industrial development work on clean processes is underpinned by government funded research programs in the U.K., U.S., and other countries. In the U.K., the Cleaner Synthesis program was launched in 1992, and in the U.S. the National Science Foundation is running a Benign Chemical Synthesis and Processing program with similar aims.[15] A Gordon research conference on Environmentally benign organic synthesis was held in July 1996, while in Italy a number of universities have established an Inter-university Consortium on Chemistry for the Environment. The EPSRC's financial commitment to cleaner synthesis through its funding of research in universities reached £15m, with the award of a further £1.5m for nine projects in June 1996. Topics that are attracting most support include various aspects of catalysis, supercritical fluids, chemistry in water, radical reactions that circumvent undesirable initiators, such as tin hydrides, electrochemical synthesis, and some novel reactor concepts.

The EPSRC's Clean Technology Program also has a related research target on Waste minimization through recycling, re-use, and recovery in industry, and there is a parallel Link program on this subject.

The past few years have seen much more attention being given to cleaner ways of making chemicals and there is a substantial commitment to research in both industry and universities. Industry is introducing cleaner processes, though confidentiality issues can delay their disclosure. However, to quote a speaker at a recent Royal Society meeting: "clean technology has to be cost effective, unlike some environmental projects". Clean synthesis is about a "win–win" approach — a "win" for the environment and a "win" for the manufacturer.

Dr. Tim Lester is EPSRC technical adviser for cleaner synthesis and may be contacted at Deft Technology and Design, 11 Nightingale Road, Hampton, Middlesex TW12 3HU. He would be pleased to hear from any readers wanting to discuss industry's needs or ideas for research projects to explore cleaner synthesis.

1 Better by Design

In October 1995, the antibiotic project team at Glaxo Wellcome's R&D division in Stevenage was engaged in trying to improve an existing route for preparing a promising antibiotic that is currently in clinical trials. One step of the synthesis that posed a particular challenge was the desilylation of the trinem to give the corresponding hydroxyester.

The original route involved deprotecting the trinem using excess tetrabutylammonium bromide, potassium fluoride, and acetic acid. In the laboratory these conditions gave us a yield of 75% of the hydroxyester after product isolation. On scaling up the reaction in a pilot plant to produce about 250 kg of hydroxyester, the yield fell to ca 68%. This was because of the instability of the product during the protracted work up, which involved numerous extractions to remove the acetic acid. Of greater concern, however, was the considerable amount of waste generated at this scale (ca 150 kg of waste per kg of product). The waste included 5 kg of quaternary amine byproducts per kg of product, which give off considerable amounts of nitrogen oxide gases during disposal by combustion.

Initially, our literature search for alternative desilylation procedures looked very promising. However, although many methods are available, conditions for deprotecting silylesters are often very harsh and completely degraded our starting compound and/or the required product. Eventually, however, we tried out a milder method described by Seigfried Hünig and coworkers at Würzburg University, using triethylamine trihydrogen fluoride in tetrahydrofuran. Switching to this method gave us just 50% of the unisolated product (still in solution), but we later increased solution yields to over 90% simply by changing the solvent to dimethylformamide or N-methylpyrrolidone. Because the new method no longer uses acetic acid, we did not have to carry out extensive extractions and we were able to isolate the product in 80% yield, irrespective of the scale. This reduced the amount of waste to 60 kg per kg of product, including just 0.8 kg per kg of amine byproducts.

It was an impressive result, but as usual there was a catch. Under the new reaction conditions the lactone impurity gradually forms, and is difficult to remove using the simpler isolation conditions. The amount of impurity that forms is critically dependent on when we stop the reaction — too early and conversion is incomplete, or too late and the level of impurity becomes unacceptable. Traditionally, chemists have used a "one factor at a time" approach to optimization — in other words assessing the effect of a particular factor by keeping all other conditions constant. But while this can be a quick method of improving a single response such as yield, it does not give us an overall picture of what is going on. If there are many conflicting goals, such as simultaneously maximizing yield and purity while minimizing material and waste, then it becomes very difficult to apply this approach efficiently.

Instead, we chose to use an experimental design software package called DesignExpert to generate a model of the reaction based on the statistical analysis of a series of 20 experiments in which we altered all of the factors simultaneously. From this we determined the optimum reaction conditions for various desired outcomes — maintaining high product yields, controlling quality, maximizing throughput, and minimizing wastes. After applying these conditions experimentally, we were reassured to find that the actual responses closely matched the expected outcomes predicted by the computer.

One drawback of this approach may be having to carry out 20 experiments; fortunately automation can help to remove the drudgery ... but that's another story.

Martin Owen is automation project leader in the process research and development division of Glaxo Wellcome at Gunnels Wood Road, Stevenage, Hertfordshire SG1 2NY.

TABLE 1 Sectors of the Chemical Industry by Quantity of By-Product/kg Product

Industry sector	Product tonnage	kg by-product/ kg product
Oil refining	10^6 – 10^8	c 0.1
Bulk chemicals	10^4 – 10^6	<1–5
Fine chemicals	10^2 – 10^4	5–50
Pharmaceuticals	10^1 – 10^3	25–100+

2 Catalysts for Change

In September 1995, Contract Chemical's Knowsley site became one of the first fine chemical companies in the UK to be accredited to the new standard for environmental management systems, BS7750. This recognizes the company's commitment to uphold the principles of responsible care, which aim to improve environmental performance. As part of the company's continuous improvement program, Contract Chemicals is developing environmentally benign chemistry for synthesizing a range of organic intermediates for the pharmaceutical, agrochemical, and related industries.

One area that has been investigated recently, in conjunction with researchers at the University of York, is the development of a range of supported-reagent catalysts called Envirocats, which demonstrate significant advantages over traditional catalysts in terms of processing efficiency, health, safety, and the environment. Important applications for Envirocats include replacing conventional catalysts such as para-toluenesulphonic acid and methane sulphonic acid in a number of high temperature esterifications. Using the solid Envirocat alternative — a strong Brönsted acid — avoids the problem of colored final products and reduces processing times without necessitating any plant modifications.

Envirocats also have the advantage of producing better quality esters for certain applications, using fewer aqueous washes, and do not require solvent or carbon treatment to purify the products.

We have also applied another Envirocat to synthesize the pharmaceutical intermediate para-chlorobenzophenone (see Scheme), which conventionally involves Friedel-Crafts acylation using stoichiometric quantities of the toxic and potentially hazardous catalyst $AlCl_3$. Unlike $AlCl_3$, the Envirocat doesn't complex with the final products, eliminating the need for an aqueous quench and the problem of acidic effluent. Besides reducing HCl emissions by as much as 75%, using the alternative non-toxic Envirocat reduces the weight of catalyst by 10-fold while yielding 70% of the desired para-isomer, with only minimal amounts of the ortho-isomer byproduct.

Keith Martin is a technical service specialist at Contract Chemicals, Penrhyn Road, Knowsley Industrial Park South, Prescot, Merseyside L34 9HY.

4.19 THERM

A set of subprograms collectively called THERM, written by Ritter and Bozelli, has proven very useful in the two programs whose descriptions follow (NASA's LSENS and DOE's Chemkin). THERM supplies the thermochemical data needed by these two programs based on the data files in Benson's book "Thermochemical Kinetics." This data in THERM is given for 330 groups. They are divided into three types of groups, BD-bond dissociation groups, CDOT-radical groups, and Regular groups — all other groups with no unbonded electrons. The last groups consist of HC-hydrocarbon groups, CYCH-ring corrections for hydrocarbon, oxygen, and nitrogen containing ring systems, INT-interaction groups/substituent effects, CLC-chlorine and halogen containing groups, HCO-oxygen containing groups, and HCN-nitrogen containing groups. Recipes guide the user into the various subprograms. ENTER/ESTMATE SPECIES asks for specie ID, # of groups in the species, # of different groups to be entered, elemental formula, groups contained in the species, and the species symmetry number. (The code for the groups in the species takes a while to get used to). The result when stored in Therm.lst is in the form heat of formation, entropy, and heat capacity at 300, 400, 500, 800, 1000, and 1500 degrees Kelvin. The whole Therm.lst file for many species can be converted to the Therm.dat file, which is exactly what is needed in the NASA format for LSENS and Chemkin. This is done with amazing speed by the subprogram Thermfit. Incidentally, literature values for H, S, and Cp can be converted to the NASA format with Thermfit as well. The Therm.dat file is also needed in Thermrxn which is a sub-program where a chemist writes a reaction which is immediately converted to a table of enthalpies entropies free energies, equilibrium constants, ratio of Arrhenius factors at a series of temperatures and in units of one's option. All of the information found by THERM can be sorted and stored in specific files easily recalled.

4.20 Design Trade-Offs for Pollution Prevention

Design for the environment poses special problems just as design for manufacturability does. There are no analytical tools to integrate these issues into the conventional engineering design analysis. Unavoidable tradeoffs must often be made between cost, performance, manufacturability, and customer satisfaction. Decisions must be made under a great deal of uncertainty and with input from multiple sources.

The current trend in environmental protection legislation shifts the financial responsibility for environmental mitigation of industrial impact to the industry carrying out the activity. Traditional manufacturing cost analyses do not reflect this total long-term cost. This project integrated design evaluation and optimization and lifecycle analysis in two ways: (1) statistical manufacturing process control which treats pollution as a product defect, and (2) the cost of compliance with regulations. The "internalization of externalities" will be analyzed with the same degree of mathematical rigor that engineers traditionally utilize only for models of physical systems.

4.21 Programming Pollution Prevention and Waste Minimization Within a Process Simulation Program

Now we will discuss the programming of a waste minimization/pollution prevention program within the process simulation program so that we have a truly holistic program which prevents pollution at the source, does not allow a production plan without disallowing any waste or pollution to ensue, and at the same time optimizes the profit and minimizes the cost of the plant including the operational costs.

In order to do this we have at least three subprograms and several data files for assistance. The programs used are Chemkin and Envirochemkin for chemical kinetics of at least 50 species in 100 equations, THERM and specifically Thermrxn, and SYNPROPS, the program that finds physical chemical and biological properties for each molecular structure and conversely finds molecular structure for specific properties. Data files will also be used, such as those of Alexander Burcat and Bonnie McBride, "1915 Ideal Gas Thermodynamic Data for Combustion and Air-Pollution Use," Technion Aerospace Engineering (TAE) Report #732 January, 1995. Also used is a simple table "Estimates Rate Constants for Reactions" and a large matrix of properties vs. chemical groups derived from Synprops.

Now it should be noticed that in Synprops a large number of variables can be constrained. Let us propose that our system is made up of the elements carbon, hydrogen, oxygen, and nitrogen only. Now we can constrain the gram-atoms or mass of each of these elements to be constant and equal to what they were on input. Thus if, let us say, the solubility parameter is maximized, under these circumstances, we will find a compound with maximum solubility parameter wth a stoichiometric formula corresponding to that we specified. Similarly, we can set up species to have toxicity in air, in water, in soil, etc. that deigns the specie as not toxic to man or animal and is considered non-toxic by the regulatory agencies. So now we have a case where a toxic chemical shows up in a process plant and we know that it proceeds to become one of several other species that are found to be benign, then we must be on our way to pollution prevention for our process.

Two programs then come to our aid. Thermrxn is a subprogram of THERM. If we have the thermodynamic functions for the species concerned, and we can write out the chemical equation, then Thermrxn will print out tables of the thermodynamic functions of the reaction involved over a range of temperatures of interest including the equilibrium constants, free energy, as well as the enthalpy, entropy, and the energy and also the ratio of the rate constant in the forwards direction as well as the rate constant in the backward direction. So now we know, mainly from an equilibrium point of view, how likely the reaction will proceed as planned. If the result is positive, then we insert the chemical mechanism into Chemkin or Envirochemkin. There as mentioned before we may have approximately 100 equations and 50 species. We may need to have rate constants for this dubious reaction. In that event we can use numbers from the table, Estimated Rate Constants for Reactions, which are approximate and based on the number of atoms constituting the molecules and whether they are linear or non-linear. Also, Chemkin can be run in several modes, i.e., isothermal, adiabatic, constant volume, constant pressure, etc. This allows further leeway.

After a mode is run in statistical fashion and we know the bounds on each species, we take the ith species and form the fraction $C_i/k_i = l_i$ where C is the concentration and k is the toxicity standard and we test whether the value of l_i is << then m_i. If this is so, we have achieved pollution prevention for this particular species. We may then have to test all other species (particularly byproducts and hazardous species) in the same way. This method is for pollution prevention. For Waste Minimization, we may proceed in a similar way except that the values of m_i for all modes, form a curve and the the minima of the curves for all species then would accomodate Waste Minimization.

We wish to be non biased as to the products of reaction we allow in Chemkin. Let us suppose the process only has four chemical elements. Then:

C H O N
C C2 CH CO CN
H CH H2 OH NH
O CO OH O2 NO
N CN NH NO N2

C3 C2H C2O C2N C2H CH2 HCOHCN C20 HCO CO2 NCO C2N HCN NCO CN2

C2H CH2 CHO HCN CH2 H3 H2O NH2 HCO H2O HO2 HNO HCN NH2 NHO N2H

C2O HCO CO2 NCO HCO H2O HO2 HNO CO2 HO2 O3 NO2 CO HNO NO2 N2O

C2N HCN NCO CN2 HCN NH2 HNO HN2 NCO HNO NO2 N2O CN2 HCN HNO N3

Notice that we have formed a tree in permuting the elements. In the first row we have all possible diatomic molecules and free radicals . In the second row we have all the tri atomic molecules or free radicals. They were generated by multiplying all the diatomics by the scalar quantities C, H, O, or N. We stopped the tree there. I have found all the molecules generated in significant quantity in my computer runs except for H3 and N3. Yet lately there was an article that N3 was important in certain circumstances. The very important atomic specie C, H, O, and N were left out in the tree above but they must be included. The tree was stopped in the tree above at the tiatomic level but in practice the next step involves multiplying each 4 × 4 matrix by C, H, O, and N, until we reach formulas CxHyOzNw, where the value of n = x + y + z + w reaches a level that yields a total number of compounds smaller or equal to 100. Notice, however that the number of compounds accessed already above is about three dozen. Notice also that the matrices are symmetrical and each compound is counted only once.

Now we may think of the matrix as three dimensional, i.e., as if it was like the inside of a PC. First we have the diatomic slice and behind it the four triatomic slices and so on. Now the elemental value for each compound changes as we compute to the latest value of the concentration of the molecule represented by that particular element. However, the element also contains the toxicological standard for that particular molecule so that when the calculation of the concentration divided by the standard is low enough that molecule can be eliminated from the computations. In order to have this happen the calculations should be run in all its modes, i.e., isothermal, adiabatic, constant temperature, constant volume, variable temperatures, and variable pressures.

4.22 Product and Process Design Tradeoffs for Pollution Prevention

There are significant parallels between design for manufacturability and design for the environment:

1. In the past, designers were not directly confronted with the effect of their designs on the manufacturing process or the environment;
2. Traditional analytic design procedures are not capable of dealing with these issues in a mathematically rigorous manner;
3. Both involve the harsh realities of unavoidable tradeoffs.
4. The interplay between product and process design is significant.

This project sought to develop a new design methodology that integrate current multiobjective design optimization with statistical process quality control and lifecycle analysis. The basis of the method is the internalization of previously external environmental impacts into concurrent engineering.

4.23 Incorporating Pollution Prevention into U.S. Department of Energy Design Projects

Pollution Prevention seeks to eliminate the release of all pollutants (hazardous and non-hazardous) to all media (land, air, and water). Beyond eliminating pollution prevention at the source, pollution prevention includes energy conservation, water conservation, and protection of natural resources. Therefore, pollution prevention addresses not only wastes exiting a process, but materials entering and being consumed by the process as well. Historically, pollution prevention activities within the U.S. Department of Energy (DOE) have focused on existing process waste streams — the Pollution Prevention Opportunity Assessment (P2OA) being the central tool for identifying and implementing pollution prevention opportunities.

The software used here was a good template for tracking pollution prevention features implemented, but the database behind the template seemed spotty. The software should be able to compute lifecycle costs and to have a clear treatment of operating and maintenance costs of selected opportunities. However, the Electronic Design Guideline (EDG) software is not intended to be computational. It is intended to raise awareness and provide specific examples of pollution prevention design opportunities. The user must then compute the cost and benefits of implementation specific to their project circumstances.

4.24 EPA Programs

EPA's new chemical program under the Toxic Substances Control Act (TSCA) plays an important role in preventing pollution. TSCA requires that EPA review new chemicals for the risks they may pose to human health and the environment before they enter the market. Anyone who plans to manufacture or import a new chemical substance must provide EPA with a pre-manufacture notice (PMN) at least 90 days prior to the activity. To determine whether a substance is "new", the company must consult EPA's Inventory of Chemical Substances (the TSCA Inventory). If the substance is not listed, it is a "new" chemical. EPA personnel are also evaluating computer-based software for synthesis design and new chemical substances that may be safer substitutes for chemicals currently used, or that will be created via pollution prevention processes. During the course of PMN, EPA has identified the following desirable criteria:

> Test data on the PMN substance itself (toxicity and fate).
>
> No reports of adverse effects.
>
> Safer substitute.
>
> Less toxic or fewer toxic associated chemicals.
>
> Safer pollution prevention, source reduction, or recycling processes that reduce exposure/releases.
>
> Successful implementation or a recommendation resulting from EPA's "Alternative Synthetic Pathway".
>
> The use of the chemical substance should be environmentally beneficial rather than harmful. Conservation of energy and water during manufacturing, processing, or use.

No TSCA enforcement actions filed against the PMN submitter within the past year.

4.25 Searching for the Profit in Pollution Prevention: Case Studies in the Corporate Evaluation of Environmental Opportunities

A perception of P2 is that, although it exists, it is too slow. Evidence contradicts the view that firms suffer from organizational weaknesses that makes them unable to appreciate the financial benefits of P2 investments. Instead, the projects foundered because of significant technical difficulties, marketing challenges, and regulatory barriers. In addition, there are environmental policy reforms likely to promote P2 innovation.

Of late, regulators and private industry are turning to the environmental strategies that target the causes, rather than the consequences, of polluting activities. P2 is a challenge for the private sector because it requires diverse forms of innovation. Innovation is difficult and often costly and inherently uncertain and firms must find new ways of integrating environmental concerns into the corporate planning process. There is an implication of evidence that companies fail to pursue P2 opportunities that would profit them.

Cases of five company examples were studied to shed light on this matter. The results say much about the way in which firms are regulated. Also, a fruitful approach is to focus on barriers to P2's profitability. Efforts to promote regulatory flexibility and innovations should be embraced. The regulations should be performance-based in the redesign of complex products and processes in ever-changing markets. Emission standards should be applied to broader categories of effluent rather than individual substances. Performance-based environmental permitting should be explored as a means to lower barriers and constraints. Such flexibility would foster better environmental accounting information and methods. This would help decision-making too. The technical identification of P2 opportunities may well be served by greater efforts at basic R & D and firm specific material accounting.

4.26 Chemical Process Simulation, Design, and Economics

NRC's (ICPET), Institute for Chemical Process and Environmental Technology, specializes in developing innovative computer modeling and numerical analysis techniques. Canadian industries use these to improve design and operation of chemical and manufacturing process systems and products.

Additional software-based support helps clients explore additional cost implications that result from the impact of industrial processes on the environment

ICPET also has related expertise in computational fluid dynamics and reactive flow modelling and environmental methodologies.

4.27 Pollution Prevention Using Process Simulation

Chemical process simulation techniques are being investigated as tools for improving process design and developing clean technology for pollution prevention and waste reduction.

HYSYS, commercially available process simulation software, is used as the basic design tool. ICPET is developed customized software, particularly for reactor design, as well as custom databases for the physical and chemical properties of pollutants, that can be integrated with HYSIS. Using these capabilities, studies are being carried out to verify reported emissions of toxic chemicals under "voluntary-action" initiatives and to compare the performance of novel technology for treating municipal solid waste with commercially available technology based on incineration processes.

4.28 Process Economics

ICPET's simulation capability includes process economics for pollution prevention. Research is focused on establishing methods that will estimate the cost of environmental management systems (EMS) including contingent environmental liability costs associated with a given design for a manufacturing or a chemical process. This capability complements the conventional process economics methodologies in HYSIS. Therefore, clients can access a very conprehensive package of process design and economic assessment tools. These tools allow them to evaluate cost effective approaches for optimizing chemical processes for pollution prevention within complex operating constraints.

4.29 Pinch Technology

Pinch technology is the analysis of the energy usage of plants and then the optimization of the plant process from the viewpoint of the qualitatively effective use of heat. Process analysis using pinch technology started in 1992 at an ethylene oxide plant and in 1993 at an ethylene plant. Energy-saving rationalization enabled by these analyzes was implemented stating in 1995.

Pinch technology was developed in 1978 by Linhoff of Lead University in the United Kingdom, then put into practical use in 1984, by Linhoff March Co. Pinch technology is a method of using the overall thermal balance of a process to theoretically predict minimum energy consumption, the construction costs of the plant using a heat recovery system and other information. These predictions are then employed as targets for the design or renovation of the process.

4.30 GIS

Geographic Information Systems (GIS) Computer programs manipulate and analyze spatial data. The environmental field continues to make up a large part of the GIS market. This includes site assessment and cleanup, wildlife management, pollution monitoring, risk analysis, vegetation mapping, and public information. Natural resource managers use or are familiar with GIS. Most people do not have the ability to carry an entire map in their head, and GIS gives you the ability to organize the information in a real-world manner. It allows you to put together maps more simply and they are produced faster. It allows you to put information together from different sources that you would probably not ever have a reason to put together, and make sense of it easily. Whatever the definition, GIS shares the ability to integrate and manipulate spatial data that have been "geocoded" or fixed in geographic space. These may include census data, ZIP codes, or digital photographs. Environmental researchers might use data on watershed boundaries, pesticide loading figures, EPA air quality readings, or locations of chemical factories. A GIS program typically deals with these various attributes in "layers," mixing and matching the information to reveal associations.

GIS is also changing the way data are collected. Inexpensive and portable global positioning system (GPS) receivers, which use satellite navigation to pinpoint a location, are now commonly used in field operations to provide accurate digital data for use in a GIS model. Lighweight "pen" computers and speech recognition software also can be used to produce digital field notes that can be easily plugged into GIS database. The database becomes the product of GIS, rather than its input. The next step in the evolution of GIS, is to merge 3-D GIS applications with computerized techniques for "scientific visualization" which adds the fourth dimension of time. This convergence, which has just begun, will eventually produce rich, 4-D animations showing complex environmental processes.

4.31 Health

Chronic environmental effects are normally considered to be a function of the time-integrated concentration to which a receptor is exposed, or dose, which is formally defined by

$$D=\int_0^t C(t)Qdt$$

Here D is the dose in milligrams of contaminant absorbed, inhaled, or ingested; C(t) is the time dependent concentration in the medium to which the individual is exposed; and Q is the volumetric rate of intake of that medium by the individual. The estimation of the intake dose involves translation of the exposure concentration into the rate of mass uptake by members of the at-risk population. The basic relationship required to estimate intake dose can be written

$$ID = C.CR.EF.ED/Wt.\tau$$

where

ID = Intake dose (mg.kg .day)
C = Exposed concentration (e.g., mg/L)
CR = Contact rate with medium containing con taminant (e.g., L/day)
EF = Exposure Frequency (days/year)
ED = Exposure duration (years)
Wt = Body weight (kg)
τ = Averaging time (days)

The Hazard Quotient is then

$$HQ = ID/RfD$$

where

HQ = Hazard Quotient
ID = Intake dose ($mg.kg^{-1}.day^{-1}$)
RfD = Reference Dose ($mg.kg^{-1}.day^{-1}$)

The Hazard Index is

$$HI = \Sigma_i (HQ)_i$$

where

HI = Hazard index
I = Contaminant and pathway index
HQ_i = Hazard quotient for contaminant or pathway i

Carcinogenic Risks

The lifetime cancer risk is assumed to be modeled by the equation

$$Risk = 1-\exp(-(ID.PF))$$

where

Risk = Risk of contracting cancer over a lifetime
ID = Intake dose ($mg.kg^{-1}.day^{-1}$)

PF = Cancer potency factor from the slope of the dose-response relationship ($mg.kg^{-1}.day^{-1}$)

For low levels of risk ($risk<10^{-3}$), this relationship is approximated by

$$Risk = ID.PF$$

The cumulative effect is again summed

$$Risk = S_i \ (Risk)_i$$

If the calculated total risk exceeds 10^{-4} - 10^{-6} excess lifetime cancer risk, then the risk is normally considered unacceptable.

4.32 Scorecard-Pollution Rankings

The Chemical Scorecard issued by the Environmental Defense Fund on the Internet has proved to be a very important method for ascertaining not only the pollution in a geographical part of the country but also who is responsible for causing it. It ranks states, counties, zip codes, and facilities by pounds of pollution. It can be customized by type of pollution (air, water, ground) and it's data is manipulated from the Toxic Release Inventory of the EPA.

Scorecard can also rank some large industrial plants by their environmental performance (e.g., pounds of pollution per barrel of oil produced). These rankings were only available for plants in five industrial sectors but this is growing.

The U.S. EPA's Toxic Release Inventory was established by Section 313 of the Emergency Planning and Community Right-To-Know Act of 1986 (EPCRA). Under EPCRA, manufacturing facilities in specific industries are required to report their environmental releases and chemical waste management annually to EPA. Covered facilities must disclose their releases of approximately 650 toxic chemicals and chemical categories to air, water, and land, as well as the quantities of chemicals they recycle, treat, burn, or otherwise dispose of on-site and off-site.

Scorecards 1996 TRI data is derived from the reports of over 21,500 manufacturing and federal facilities. Actually, scorecard provides environmental release profiles on an even larger set of facilities (over 23,000), because the system continues to provide access to archived reports on facilities with 1995 TRI data, even if they are no longer reporting to TRI.

Even so, there are limits to TRI data:

1. TRI does not cover all toxic chemicals that have the potential to adversely affect human health or the environment.
2. TRI does not require reporting from many major sources of pollution releases.
3. TRI does not require companies to report the quantities of toxic chemicals used or the amounts that remain in products.
4. TRI does not provide information about the exposures people may experience as a consequence of chemical use.

Scorecard provides two types of information about the potential health hazards of chemicals reported to the TRI. It identifies the kinds of health effects specific chemicals may cause, relying on the findings of authoritative scientific and regulatory organizations, and on scientific studies compiled in major toxicology databases. It also ranks chemicals by their potential to cause either cancer or noncancer health risks, relying on a sophisticated scoring system developed by scientists at EDF in collaboration with colleagues at the School of Public Health, University of California at Berkeley.

Scorecard identifies chemicals that can cause cancer, harm the immune system, contribute to birth defects, or lead to nine other types of health impacts. Chemicals with health hazards that are widely recognized by authoritative scientific organizations are listed separately from the chemicals whose hazards are only suspected on the basis of more limited data. Scorecard uses the State of California's official list of chemicals with known toxic properties as its source for chemicals that are recognized to cause cancer, reproductive toxicity, and/or developmental toxicity.

Scorecard can rank manufacturing facilities using raw TRI data and it can rank some industrial plants using pollution measures that have been normalized to take into account production levels. Production levels have an important impact on chemical releases and waste generation: large manufacturing operations generally use and release more chemicals than smaller operations. The different levels of production across plants should be taken into account.

While the overall volume of chemicals released are a concern to host communities, raw TRI data does not provide the best indicator of the chemical use efficiency of a plant. Scorecard can also rank plants by various measures that take into account production differences between plants. Normalized values for Scorecard's 40 ranking categories are calculated by dividing a plant's release data by its production data.

Normalized values are only used to compare plants within the same industrial sector, because the normalizing unit obviously varies across manufacturing sectors. Scorecard only generates production-normalized rankings for nine sectors:

Automobile Assembly
Iron and Steel Mills (Integrated)
Iron and Steel Mills (Mini)
Petroleum Refining
Pulp and Paper Mills
Nonferrous Metals (Aluminum)
Nonferrous Metals (Copper)
Nonferrous Metals (Lead)
Nonferrous Metals (Zinc)

Scorecard displays raw rankings side by side with normalized rankings. To find a plant that could potentially improve its environmental performance, look for one that ranks high in the "raw rank" column and high in the "normalized rank" column. This means that they are likely not doing as well as their counterparts at reducing waste per unit of production. To find plants that are doing better than others within the same sector, look for the plants that rank high in the "raw rank" column but low in the "normalized rank" column.

Large differences (e.g., differences greater than a factor of ten) in normalized rank probably are more important to pay attention to than small differences, as no two facilities are identical and there may be good reasons for small differences in seemingly similar plants.

Scorecard data on plant production capacities is obtained from the EPA's 1998 Sector Facility Indexing Project (SFIP). SFIP defines production data as an indicator of the overall production and a surrogate for the size and complexity of a plant's operations.

4.33 HAZOP and Process Safety

The method used for hazards analysis at a process facility involving P&IDs is a HAZOP analysis. This was developed to identify and evaluate the safety hazards in a process plant, and to identify operability problems which, although not hazardous, could compromise the plant's ability to achieve design productivity.

HAZOP is a simple, structured methodology for hazard identification. It is designed to inspire imaginative thinking (or brainstorming) by a team of experts to identify hazards and operational problems while examining a process or system in a thorough and systematic manner.

A HAZOP study involves a systematic, methodical examination of design documents that describe the facility. The study is performed by a multidisciplined team to identify safety hazards or operability problems by evaluating deviations from design intents.

During the initial stages of the HAZOP study, PLG establishes a risk ranking matrix with severity rankings and frequency rankings to allow for a qualitative assessment of the consequences of each hazard. This matrix allows the HAZOP team to screen each identified hazard and to assign a priority that will focus subsequent corrective action on significant potential hazards rather than on those that are insignificant, trivial, or operational concerns.

To accomplish this qualitative assessment of risk, each hazard event is ranked through a team consensus judgement in two ways: first, deciding the severity of the consequences of the event, and second, deciding the expected frequency or likelihood of the scenario resulting in the consequences identified for the hazard. The consequence severity and event likelihood used in ranking the hazards are found in HAZOP studies.

4.34 Safer by Design

Any conference on safety will have many papers on safety management systems — they are in fashion and are an undoubted improvement on the ad hoc methods of the past. The papers usually list the steps to be followed in dealing with hazards:

> Identify, using a systematic technique such as hazard and operability studies (Hazop); prevent, if possible by adding on protective equipment, if not, by procedures; control; and mitigate the consequences of the hazard.

This process is followed until the combination of probability and consequences is as low as reasonably practicable, and for major hazards quantitative methods are often used. Only rarely is the first step followed by 'avoid', yet it seems obvious that whenever possible we should do so. Consider a simple analogy: the most hazardous piece of equipment in our homes is the stairs; more people are killed or injured by falling down them than any other way. The traditional methods of prevention and control are to train people to use the handrails, to keep the stairs free from junk and to make sure the carpet is secure. The way to avoid the hazard — if we regard the risk of falling as too high — is to buy a bungalow. This is an inherently safer solution because it does not depend on equipment that might fail or procedures that might be ignored. You can't fall down stairs that aren't there.

Interest in inherently safer designs was stimulated by the explosion at Fixborough in 1974. The leak was so large (about 30 to 50t) and the explosion was so devastating because the inventory in the plant was large (several hundred tonnes). The inventory was so large because only about 6% of the raw material was reacted per pass; the rest got a free ride and had to be recovered and recycled. Reducing the inventory is not easy and the only company that tried to do so for this process soon abandoned the research because it saw no need for a new plant in the forseeable future.

Ten years later, interest in inherently safer designs was boosted by the disaster at Bhopal. Methyl isocyanate, the material that leaked and killed over 2000 people, was not a product or raw material, but an intermediate. It was convenient to store it, but not essential to do so, a point missed by most commentators. Yet within a year Union Carbide and other companies had reduced their stocks of methyl isocyanate and other hazardous intermediates by as much as 75%.

The most widely used method of inherently safer design is intensification: using so little of hazardous material that it hardly matters if it all leaks out. Today hazardous intermediates such as phosgene are increasingly manufactured at the point of use and the only stock is a few kilograms in a pipeline. Intensified designs are available for reactors, liquid-vapor contacting equipment, mixers, scrubbers, dryers, liquid-liquid extractors, heat pumps, etc. Inherently safer designs are usually cheaper as well as safer, because less added on protective equipment is needed. In addition, if we can intensify, the saving is cheaper because intensified equipment is smaller and therefore cheaper. To use the popular phrase, safety need not be a 'zero sum game'.

If we cannot intensify, an alternative is substitution: using safer materials. Thus non-flammable or less flammable, non-toxic or less toxic solvents, refrigerants and heat transfer can be used in place of flammable or toxic ones; water-based paints can replace solvent-based ones. Chlorofluorocarbons (CFCs), hailed as wonder refrigerants when they were introduced, are now known to affect the ozone layer and there has been a move back to liquefied petroleum gases (LPG) and ammonia, although hydrofluorocarbons (which do not damage the ozone layer) and hydrochlorofluorocarbons (which cause much less damage than CFCs) are now available. In some ethylene oxide plants the hundreds of tonnes of boiling paraffin used to cool the reactor tubes is a bigger hazard than the mixture of ethyne and oxygen in the tubes; more modern plants use water cooling.

Apart from substituting auxiliary materials we can also make changes in the chemistry. The product made at Bhopal, the insecticide carbaryl, is made from a-naphthol, methylamine, and phosgene. At Bhopal, the first two of these were reacted together to make methyl isocyanate (MIC). The MIC was then reacted with a-naphthol to make carbaryl (Scheme 1). In an alternative process the same three raw materials are used, but they are reacted in a different order. α-Naphthol and phosgene are reacted together to give a chloroformate ester, which is then reacted with methylamine. No MIC is produced.

Neither process is ideal because both involve the use of phosgene, but the second process at least avoids producing MIC. More fundamentally, instead of carbaryl could we make an alternative insecticide that is safer to produce, develop pest-resistant plants, or make use of natural predators? I am not suggesting that we should — these ideas have disadvantages — only that the question should be asked.

Design Constraints

If we cannot intensify or substitute, and we have to use or store a large amount of hazardous material, we should handle it in the least hazardous form. Thus, large quantities of ammonia, chlorine, and LPG are now usually stored refrigerated at low pressure below their boiling points rather than under pressure at atmospheric temperature; if a leak should occur, the driving force is low and evaporation is small.

Another route to inherently safer design is to limit the energy available. It is better to prevent overheating by limiting the temperature of the heating medium, such as steam, than to rely on interlocks, which may fail or be disconnected.

Simpler plants are safer than complex ones because there is less equipment to fail and fewer opportunities for human error.

Inherently safer design has been adopted far more slowly than HAZOP or quantitative risk assessment, both introduced only a few years earlier, for a number of reasons, some of which are common to all innovations:

Innovation requires more time than repeating an earlier design, time that is usually not available. We should, therefore, try to recognize the need for new plants sooner than we have done in the past. When designing a plant we are conscious of all the changes we would like to make, but cannot do so this time round, so despite the pressure on numbers of people we should give some thought to 'the plant after next'.

We like to follow the procedures — in plant design and everything else — that we have always followed. New processes or equipment may produce unforeseen problems that will delay or prevent the achievement of flowsheet output. Better to stick to designs that we know. There are good reasons for these fears. During the 1960s a new generation of plants was built, larger than those built before, operating under more extreme conditions. At the same time there was a demand for minimum capital cost. Many of these new plants required extensive modification and the expenditure of much money and effort before flowsheet output was achieved. The industry had burnt its fingers and tended to play safe for many years afterwards.

About 1980 a senior engineer in an international chemical company was asked to survey attitudes to innovation. He read through about 15 major expenditure proposals submitted to the head office for approval and found that all but a few claimed, as an advantage, that no innovation was involved. (In one or two cases he suspected that there was some innovation, but the originators concealed the heresy.)

In the chemical industry despite Flixborough and Bhopal, many managers still believe that they can handle large inventories safely. However, when we are dealing with hazardous materials, only very low failure rates — of people and equipment — are acceptable today and these rates are often beyond, or close to, the best that people and equipment can provide. We can keep up a tip-top performance for an hour or so while playing a game or a piece of music, but not everybody, everywhere, all the time.

The influence of licensers and equipment suppliers is firmly on the side of tradition. Why develop new processes and designs when there is a market for those we have already?

Not all innovators recognize the need to sell their ideas. Ideas do not sell themselves. Ramshaw has discussed the qualities necessary for innovation in a large company:

1. Exceptional tenacity; it takes a long time to persuade colleagues to accept innovations.
2. Allies and collaborators within the company who will receive credit for any success.
3. An ability to spot applications and demonstrate satisfactory and economic performance.

There are also constraints to which inherently safer designs are particulaly prone:

In asking for intensified and other inherently safer designs we are asking for more than better widgets; we are asking for a change in the design process. This is inevitably slower than a change in the technology. Many engineers are happier carrying out calculations than handling ideas. My 'how to' books sell more copies than my idea books. Universities tend to teach calculation methods rather than the innovative approaches such as intensification.

Companies today tend to be organized in business areas rather than functional departments such as research and design. With this sort of organization there may be no central department to sponsor developments that might benefit the company as a whole, but which no single business or project can afford.

Will intensification make control more difficult? Large inventories of equipment such as reactors or the bases of distillation columns have a damping effect, smoothing out the effect of minor changes in feed composition, heat input, and other variables. Control engineers are confident that overcoming reductions is not an insuperable problem, though in some cases faster responding instruments may be needed. This constraint is not real.

Large inventories between the section of a plant allow the rest of a plant to operate when part is shut down. The need for them, however, is a self-fulfilling prophecy: if they are there we use them and carry out repairs at leisure. If we do not have them — the Japanese philosophy — we carry out repairs more promptly.

Many intensified designs use moving parts whereas the traditional designs do not. This is often quoted as a disadvantage even though the mechanical engineering is well proven. As with control, we have to overcome a perception that may not be real, but is widely believed.

A study carried out for the Health and Safety Executive found that lack of constraint was a major constraint. It found that safety advisers are now familiar with the concept of inherently safer design, including intensification, but that there is still a lack of general awareness in design organizations and among senior managers. There is a need to raise awareness and to develop techniques and tools that will make it easier for designers to use intensified designs.

There is a lack of investigative tools, similar to hazard and operability studies, for examining designs and uncovering ways of introducing intensified and other inherently safer designs.

The Actions Needed

Most of the actions needed follow automatically from the descriptions of the constaints, but some are worth further discussion.

We will not get intensified and other inherently safer plants unless designers are convinced that they are possible and desirable. We have got to keep on spreading the message. One single publication of an example or a technique is not enough to catch the eyes of those who might make use of it. We need an 'orthodontic' approach: continuous steady pressure over the years.

However, designers do not live and work in a vacuum, but are influenced by the culture of their organization. If it is unsympathetic to innovation, they will produce traditional designs. Those at the top have to set the tone. Statements of policy carry little weight. The little things count for more. If a designer sees an expenditure proposal supported by a statement that no innovation is involved, put forward as a reason for satisfaction, s/he assumes that the innovation is not wanted.

Senior managers, for example, could ask for an annual report on the progress made in reducing inventories, and before authorizing the construction of a new plant they could ask how its inventory compares with that of the last one. A Japanese industrialist who joined the board of ICI was surprised to find that they did not discuss new technologies and the progress made in implementing them.

In the U.K., although not in other countries, all chemical engineers receive some training in safety and loss prevention, but inherently safer design (including intensification) is not always included. It should be. Students should be taught that safety is not — or should not be — a coat of paint that can be added to their design by a safety expert, but an integral part of the design for which they are responsible. Students who are not familiar with inherently safer designs have not been equipped for their future careers.

Inherently safer design has grown slowly, but it is an oak tree that will mature slowly and last a long time.

4.35 Design Theory and Methodology

D. L. Thurston has written a series on this topic. Three of them are reviewed here.

Product and Process Design Tradeoffs for Pollution Prevention

Significant parallels exist between design for manufacturability and design for the environment: (1) In the past designers were not directly confronted with the effect of their design on the manufacturing process or the environment; (2) traditional analytic design procedures are not capable of dealing with these matters in a mathematically rigorous manner; (3) both involve the harsh realities of unavoidable tradeoffs; and (4) the interplay between product and process design is significant. This project seeks to develop a new methodology that integrates concurrent multiobjective design optimization with statistical process quality control and analysis. The basis of the method is the internalization of previously external environmental impacts into concurrent engineering.

Pollution Prevention and Control

Upcoming amendments to the Clean Air Act require industry to install expensive pollution control systems. Pollution Prevention methods can decrease pollution control costs, but might also require tradeoffs between treatment cost, manufacturing cost, product quality, customer satisfaction, reliability, manufacturability, and other attributes. In addition, tradeoffs between environmental impacts that occur at different cycles of the life cycle might be necessary. This project does not presume to identify the "environmentally" correct design alternative. Rather, we are developing a structured framework within which engineers can rationally consider tradeoffs between environmental impacts, manufacturing cost, pollution control cost and product performance.

Integrating Environmental Impacts into Product Design

This project will develop a formal method for integrating impacts directly into the product engineering design process. This requires consideration of tradeoffs between traditional engineering design issues, quality, manufacturing cost, and environmental impact. A methodology by which engineers can rationally examine these tradeoffs and develop alternatives that are optimal in the sense of maximizing overall design utility will be developed. The benefit realized by this projects is that designers will be able to consider the environmental impact at the very beginning of the product design process, and will make decisions which have a less detrimental impact on the environment.

Part V. Pathways to Prevention

5.1 The Grand Partition Function

Upon studying such topics as mass integration of El-Hawagi et alia, certain matters seem to occur in the mind of a theoretical physicist. First of all, the difference between energy integration and energy plus mass integration seem similar to that between the canonical ensembles and the grand canonical ensembles of statistical mechanics in the minds of the theoretical chemist and physicist. Very briefly, the Grand Canonical Ensemble (G.P.F.) is defined as

$$(G.P.F.) = (P.F.)^N e^{Nu/kT} \lambda^N ,$$

where $\lambda^N = \exp(u/kT)$, and

$$(P.F.) = \Sigma_i \, \Omega_i e^{-E_i / kT}$$

and (P.F.) is used in the canonical ensemble.

Now u is the chemical potential that controls the movement of particles (hence mass) into or out of the system, whereas E denotes the movement of energy or heat out of the system.

On researching order-disorder or cooperative phenomena, it was found that probabilities of occurrence of particulate matter was denoted by a direct product of the factor x times the factor y, each raised to appropriate powers. Now if x denotes matter (or material) and y denotes energy or energy of interaction and if the x and y are very large numbers, we would have expressions quite similar to the G.P.F.

The successive filling of sites and the creation of bonds, starting with a completely empty figure of sites can be symbolized by mis. Then for each site filled, introduce a factor, x; for each short range bond formed, introduce a factor, y; for each long range interaction formed, introduce the factor, z. Each of these factors is to be raised to an appropriate power. The power of x is the number of sites of one type filled; the power of y is the number of short-range bonds of on type filled, and the power of z is the number of long-range bonds formed.

For example, if g sites become occupied and h of these sites are of the type a, then g-h are of type b. Suppose further that r short range bonds and t long range bonds are formed. The equilibrium constant for such a "reaction" will then be

$$mis = [x]^{g-h} [x']^h [y]^r [z]^t [y']^s$$

and it remains to find the values of x, x', y, and z. This can be done by setting up the various sets of equations among the various geometrical figures involved. Such equations are called consistency equations and normalizing equations.

5.2 A Small Part of the Mechanisms from the Department of Chemistry of Leeds University

```
*************************************************************************;
* INORGANIC CHEMISTRY ;
*************************************************************************;
* Ox/NOx CHEMISTRY ;
% J<1>: O3 = O1D    ;
% J<2>: O3 = O      ;
% 6.00D-34*O2*O2*((TEMP/300)@-2.8): O = O3;
% 5.60D-34*O2*N2*((TEMP/300)@-2.8): O = O3;
% 8.00D-12*EXP(-2060/TEMP): O + O3 =;
% KMT01      : O + NO = NO2;
% 6.50D-12*EXP(120/TEMP)  : O + NO2 = NO;
% KMT02      : O + NO2 = NO3;
% 3.20D-11*O2*EXP(70/TEMP) : O1D = O;
% 1.80D-11*N2*EXP(110/TEMP) : O1D = O;
% 1.80D-12*EXP(-1370/TEMP) : NO + O3 = NO2 ;
% 1.20D-13*EXP(-2450/TEMP) : NO2 + O3 = NO3;
% 3.30D-39*EXP(530/TEMP)*O2: NO + NO = NO2 + NO2;
% 1.67D-06*(HO/H): NO2 = HONO    ;
% 1.80D-11*EXP(110/TEMP): NO + NO3 = NO2 + NO2 ;
% 4.50D-14*EXP(-1260/TEMP) : NO2 + NO3 = NO + NO2 ;
% KMT03 % KMT04    : NO2 + NO3 = N2O5  ;
% 4.00D-04    : N2O5 = NA + NA ;
% J<4>: NO2 = NO + O ;
```

```
% J<5>: NO3 = NO    ;
% J<6>: NO3 = NO2 + O      ;
* HOx FORMATION, INTERCONVERSION AND RE-
MOVAL ;
% 2.20D-10*H2O       : O1D = OH + OH      ;
% 1.90D-12*EXP(-1000/TEMP) : OH + O3 = HO2 ;
% 7.70D-12*EXP(-2100/TEMP) : OH + H2 = HO2;
% 1.50D-13*KMT05     : OH + CO = HO2      ;
% 2.90D-12*EXP(-160/TEMP) : OH + H2O2 = HO2;
% 1.40D-14*EXP(-600/TEMP) : HO2 + O3 = OH ;
% 4.80D-11*EXP(250/TEMP)  : OH + HO2 = ;
% 2.20D-13*KMT06*EXP(600/TEMP): HO2 + HO2 =
H2O2  ;
% 1.90D-33*M*KMT06*EXP(980/TEMP) : HO2 + HO2
= H2O2           ;
% KMT07        : OH + NO = HONO    ;
% KMT08        : OH + NO2 = HNO3   ;
% 2.30D-11     : OH + NO3 = HO2 + NO2     ;
% 3.70D-12*EXP(240/TEMP): HO2 + NO = OH +
NO2   ;
% KMT09 % KMT10    : HO2 + NO2 = HO2NO2 ;
% 3.50D-12             : HO2 + NO3 = OH + NO2 ;
% 1.80D-11*EXP(-390/TEMP) :OH + HONO = NO2;
% KMT11        : OH + HNO3 = NO3   ;
% 1.50D-12*EXP(360/TEMP):OH + HO2NO2 = NO2;
% 6.00D-06      : HNO3 = NA  ;
% J<3>: H2O2 = OH + OH     ;
% J<7>: HONO = OH + NO     ;
% J<8>: HNO3 = OH + NO2    ;
* SOx CHEMISTRY ;
% 4.00D-32*EXP(-1000/TEMP)*M:O + SO2 = SO3;
% KMT12                 : OH + SO2 = HSO3    ;
% 1.30D-12*EXP(-330/TEMP)*O2      : HSO3 = HO2
+ SO3  ;
% 1.20D-15*H2O        : SO3 = SA              ;

*************************************************************************;
* ALKANES ;

*************************************************************************;
* METHANE ;
% 7.44D-18*TEMP@2*EXP(-1361/TEMP) : OH + CH4
= CH3O2          ;
% KRO2NO*0.999 : CH3O2 + NO = CH3O + NO2;
% KRO2NO*0.001 : CH3O2 + NO = CH3NO3 ;
% 7.20D-14*EXP(-1080/TEMP)*O2 : CH3O = HCHO
+ HO2  ;
% KMT13 % KMT14:CH3O2 + NO2 = CH3O2NO2 ;
% KRO2NO3 : CH3O2 + NO3 = CH3O + NO2;
% 4.10D-13*EXP(790/TEMP) : CH3O2 + HO2 =
CH3OOH          ;
% 1.82D-13*EXP(416/TEMP)*0.33*RO2 : CH3O2 =
CH3O  ;
% 1.82D-13*EXP(416/TEMP)*0.335*RO2 : CH3O2 =
HCHO  ;
% 1.82D-13*EXP(416/TEMP)*0.335*RO2 : CH3O2 =
CH3OH         ;
% 1.00D-14*EXP(1060/TEMP) : OH + CH3NO3 =
HCHO + NO2   ;
% J<51> : CH3NO3 = CH3O + NO2   ;
% 1.90D-12*EXP(190/TEMP) : OH + CH3OOH =
CH3O2;
% 1.00D-12*EXP(190/TEMP) : OH + CH3OOH =
HCHO + OH    ;
% J<41> : CH3OOH = CH3O + OH     ;
* ETHANE ;
% 1.51D-17*TEMP@2*EXP(-492/TEMP) : OH + C2H6
= C2H5O2         ;
% KRO2NO*0.991:C2H5O2 + NO = C2H5O + NO2;
% KRO2NO*0.009 : C2H5O2 + NO = C2H5NO3 ;
% 6.00D-14*EXP(-550/TEMP)*O2 : C2H5O =
CH3CHO + HO2           ;
% KRO2NO3 : C2H5O2 + NO3 = C2H5O + NO2 ;
% 7.50D-13*EXP(700/TEMP) : C2H5O2 + HO2 =
C2H5OOH        ;
% 3.10D-13*0.6*RO2 : C2H5O2 = C2H5O      ;
% 3.10D-13*0.2*RO2 : C2H5O2 = CH3CHO   ;
% 3.10D-13*0.2*RO2 : C2H5O2 = C2H5OH   ;

% 4.40D-14*EXP(720/TEMP) : OH + C2H5NO3 =
CH3CHO + NO2           ;
% J<52> : C2H5NO3 = C2H5O + NO2            ;
% 1.90D-12*EXP(190/TEMP) : OH + C2H5OOH =
C2H5O2          ;
% 1.00D-11 : OH + C2H5OOH = CH3CHO + OH;
% J<41> : C2H5OOH = C2H5O + OH;
* PROPANE ;
% 1.50D-17*TEMP@2*EXP(-44/TEMP)*0.307 : OH +
C3H8 = NC3H7O2        ;
% 1.50D-17*TEMP@2*EXP(-44/TEMP)*0.693 : OH +
C3H8 = IC3H7O2        ;
% KRO2NO*0.71*0.98 : NC3H7O2 + NO = NC3H7O
+ NO2  ;
% KRO2NO*0.71*0.02:NC3H7O2 + NO = NC3H7NO3
;
% 3.70D-14*EXP(-460/TEMP)*O2 : NC3H7O =
C2H5CHO + HO2          ;
% KRO2NO3 : NC3H7O2 + NO3 = NC3H7O + NO2
;
% KRO2HO2*0.64 : NC3H7O2 + HO2 = NC3H7OOH
;
% 6.00D-13*0.6*RO2 : NC3H7O2 = NC3H7O ;

% 6.00D-13*0.2*RO2 : NC3H7O2 = C2H5CHO;
% 6.00D-13*0.2*RO2 : NC3H7O2 = NPROPOL ;
% 7.30D-13 : OH + NC3H7NO3 = C2H5CHO + NO2
;
% J<53> : NC3H7NO3 = NC3H7O + NO2      ;
% 1.90D-12*EXP(190/TEMP) : OH + NC3H7OOH =
NC3H7O2         ;
% 1.53D-11 : OH + NC3H7OOH = C2H5CHO + OH;
```

% J<41> : NC3H7OOH = NC3H7O + OH ;
% KRO2NO*0.71*0.958 : IC3H7O2 + NO = IC3H7O + NO2 ;
% KRO2NO*0.71*0.042 : IC3H7O2 + NO = IC3H7NO3 ;
% 1.50D-14*EXP(-200/TEMP)*O2 : IC3H7O = CH3COCH3 + HO2 ;
% KRO2NO3 : IC3H7O2 + NO3 = IC3H7O + NO2 ;
% KRO2HO2*0.64 : IC3H7O2 + HO2 = IC3H7OOH ;
% 4.00D-14*0.6*RO2 : IC3H7O2 = IC3H7O ;
% 4.00D-14*0.2*RO2 : IC3H7O2 = CH3COCH3 ;
% 4.00D-14*0.2*RO2 : IC3H7O2 = IPROPOL ;
% 4.90D-13 : OH + IC3H7NO3 = CH3COCH3 + NO2 ;
% J<54> : IC3H7NO3 = IC3H7O + NO2 ;
% 1.90D-12*EXP(190/TEMP) : OH + IC3H7OOH = IC3H7O2 ;
% 2.42D-11 : OH + IC3H7OOH = CH3COCH3 + OH ;
% J<41> : IC3H7OOH = IC3H7O + OH ;
* BUTANE (N-BUTANE) ;
% 1.51D-17*TEMP@2*EXP(190/TEMP)*0.147 : OH + NC4H10 = NC4H9O2 ;
% 1.51D-17*TEMP@2*EXP(190/TEMP)*0.853 : OH + NC4H10 = SC4H9O2 ;
% KRO2NO*0.60*0.967 : NC4H9O2 + NO = NC4H9O + NO2 ;
% KRO2NO*0.60*0.033 : NC4H9O2 + NO = NC4H9NO3 ;
% 3.70D-14*EXP(-460/TEMP)*O2 : NC4H9O = C3H7CHO + HO2 ;
% 1.30D+11*EXP(-4127/TEMP) : NC4H9O = HO1C4O2 ;
% KRO2NO*0.60*0.987 : HO1C4O2 + NO = HO1C4O + NO2 ;
% KRO2NO*0.60*0.013 : HO1C4O2 + NO = HO1C4NO3 ;
% 8.4D+10*EXP(-3523/TEMP) : HO1C4O = HOC3H6CHO + HO2 ;
% KRO2NO3 : NC4H9O2 + NO3 = NC4H9O + NO2 ;
% KRO2HO2*0.74 : NC4H9O2 + HO2 = NC4H9OOH;
% 1.30D-12*0.6*RO2 : NC4H9O2 = NC4H9O ;
% 1.30D-12*0.2*RO2 : NC4H9O2 = C3H7CHO ;
% 1.30D-12*0.2*RO2 : NC4H9O2 = NBUTOL ;
% KRO2NO3 : HO1C4O2 + NO3 = HO1C4O + NO2 ;
% KRO2HO2*0.74 : HO1C4O2 + HO2 = HO1C4OOH ;
% 1.30D-12*0.6*RO2 : HO1C4O2 = HO1C4O;
% 1.30D-12*0.2*RO2 : HO1C4O2 = HOC3H6CHO ;
% 1.30D-12*0.2*RO2 : HO1C4O2 = HOC4H8OH ;
% 1.78D-12 : OH + NC4H9NO3 = C3H7CHO + NO2;
% J<53> : NC4H9NO3 = NC4H9O + NO2 ;
% 5.62D-12 : OH + HO1C4NO3 = HOC3H6CHO + NO2 ;
% J<53> : HO1C4NO3 = HO1C4O + NO2 ;
% 1.90D-12*EXP(190/TEMP) : OH + NC4H9OOH = NC4H9O2 ;
% 1.67D-11 : OH + NC4H9OOH = C3H7CHO + OH;
% J<41> : NC4H9OOH = NC4H9O + OH ;
% 1.90D-12*EXP(190/TEMP) : OH + HO1C4OOH = HO1C4O2 ;
% 2.06D-11 : OH + HO1C4OOH = HOC3H6CHO + OH ;
% J<41> : HO1C4OOH = HO1C4O + OH ;
% 1.02D-11 : OH + HOC4H8OH = HOC3H6CHO + HO2 ;
% J<15> : HOC3H6CHO = HO1C3O2 + HO2 + CO ;
% 3.04D-11 : OH + HOC3H6CHO = HOC3H6CO3 ;
% KNO3AL : NO3+HOC3H6CHO = HOC3H6CO3+HNO3 ;
% KRO2NO*2.7 : HOC3H6CO3 + NO = HO1C3O2 + NO2 ;
% KRO2NO*0.71*0.981 : HO1C3O2 + NO = HO1C3O + NO2 ;
% KRO2NO*0.71*0.019 : HO1C3O2 + NO = HO1C3NO3 ;
% 3.70D-14*EXP(-460/TEMP)*O2 : HO1C3O = HOC2H4CHO + HO2 ;
% KFPAN % KBPAN : HOC3H6CO3 + NO2 = C4PAN1;
% KRO2NO3 : HOC3H6CO3 + NO3 = HO1C3O2 + NO2 ;
% KAPHO2*0.71 : HOC3H6CO3+HO2 = HOC3H6CO3H;
% KAPHO2*0.29 : HOC3H6CO3+HO2 = HOC3H6CO2H + O3 ;
% 5.00D-12*0.7*RO2 : HOC3H6CO3 = HO1C3O2 ;
% 5.00D-12*0.3*RO2 : HOC3H6CO3 = HOC3H6CO2H;
% KRO2NO3 : HO1C3O2 + NO3 = HO1C3O + NO2 ;
% KRO2HO2*0.64 : HO1C3O2 + HO2 = HO1C3OOH;
% 6.00D-13*0.6*RO2 : HO1C3O2 = HO1C3O;
% 6.00D-13*0.2*RO2 : HO1C3O2 = HOC2H4CHO ;
% 6.00D-13*0.2*RO2 : HO1C3O2 = HOC3H6OH ;
% 9.60D-12 : OH + C4PAN1 = HOC3H6CO3 + NO2;
% 4.23D-12 : OH + HO1C3NO3 = HOC2H4CHO + NO2 ;
% J<53> : HO1C3NO3 = HO1C3O + NO2 ;
% 1.32D-11 : OH + HOC3H6CO3H = HOC3H6CO3 ;
% J<41> : HOC3H6CO3H = HO1C3O2 + OH ;
% 1.04D-11 : OH + HOC3H6CO2H = HO1C3O2 ;
% 1.90D-12*EXP(190/TEMP) : OH + HO1C3OOH = HO1C3O2 ;
% 1.92D-11 : OH + HO1C3OOH = HOC2H4CHO + OH ;
% J<41> : HO1C3OOH = HO1C3O + OH ;
% 9.10D-12 : OH + HOC3H6OH = HOC2H4CHO + HO2 ;
% J<15> : HOC2H4CHO = HOCH2CH2O2+HO2+CO;
% 3.50D-11 : OH + HOC2H4CHO = HOC2H4CO3 ;
% KNO3AL : NO3+HOC2H4CHO = HOC2H4CO3+HNO3 ;

% KRO2NO*2.7 : HOC2H4CO3+NO = HOCH2CH2O2 + NO2 ;
% KFPAN % KBPAN : HOC2H4CO3 + NO2 = C3PAN1;
% KRO2NO3 : HOC2H4CO3+NO3 = HOCH2CH2O2+NO2 ;
% KAPHO2*0.71 : HOC2H4CO3+HO2 = HOC2H4CO3H;
% KAPHO2*0.29 : HOC2H4CO3+HO2 = HOC2H4CO2H + O3 ;
% 5.00D-12*0.7*RO2 : HOC2H4CO3 = HOCH2CH2O2;
% 5.00D-12*0.3*RO2 : HOC2H4CO3 = HOC2H4CO2H;
% 1.42D-11 : OH + C3PAN1 = HOC2H4CO3 + NO2;
% 1.78D-11 : OH + HOC2H4CO3H = HOC2H4CO3 ;
% J<41> : HOC2H4CO3H = HOCH2CH2O2 + OH ;
% 1.50D-11 : OH + HOC2H4CO2H = HOCH2CH2O2;
% KRO2NO*0.60*0.91 : SC4H9O2 + NO = SC4H9O + NO2 ;
% KRO2NO*0.60*0.09 : SC4H9O2 + NO = SC4H9NO3 ;
% 1.80D-14*EXP(-260/TEMP)*O2 : SC4H9O = MEK + HO2 ;
% 2.70D+14*EXP(-7398/TEMP) : SC4H9O = CH3CHO + C2H5O2 ;
% KRO2NO3 : SC4H9O2 + NO3 = SC4H9O + NO2 ;
% KRO2HO2*0.74 : SC4H9O2 + HO2 = SC4H9OOH;
% 2.50D-13*0.6*RO2 : SC4H9O2 = SC4H9O ;
% 2.50D-13*0.2*RO2 : SC4H9O2 = MEK ;
% 2.50D-13*0.2*RO2 : SC4H9O2 = BUT2OL ;
% 9.20D-13 : OH + SC4H9NO3 = MEK + NO2 ;
% J<54> : SC3H9NO3 = SC4H9O + NO2 ;
% 1.90D-12*EXP(190/TEMP) : OH + SC4H9OOH = SC4H9O2 ;
% 3.21D-11 : OH + SC4H9OOH = MEK + OH ;
% J<41> : SC4H9OOH = SC4H9O + OH ;
* 2-METHYL PROPANE (I-BUTANE) ;
% 1.11D-17*TEMP@2*EXP(256/TEMP)*0.233 : OH + IC4H10 = IC4H9O2 ;
% 1.11D-17*TEMP@2*EXP(256/TEMP)*0.767 : OH + IC4H10 = TC4H9O2 ;
% KRO2NO*0.60*0.967 : IC4H9O2 + NO = IC4H9O + NO2 ;
% KRO2NO*0.60*0.033 : IC4H9O2 + NO = IC4H9NO3;
% 3.70D-14*EXP(-460/TEMP)*O2 : IC4H9O = IPRCHO + HO2 ;
% KRO2NO3 : IC4H9O2 + NO3 = IC4H9O + NO2 ;
% KRO2HO2*0.74 : IC4H9O2 + HO2 = IC4H9OOH ;
% 1.30D-12*0.6*RO2 : IC4H9O2 = IC4H9O ;
% 1.30D-12*0.2*RO2 : IC4H9O2 = IPRCHO ;
% 1.30D-12*0.2*RO2 : IC4H9O2 = IBUTOL ;
% 1.50D-12 : OH + IC4H9NO3 = IPRCHO + NO2 ;
% J<53> : IC3H7NO3 = IC4H9O + NO2 ;
% 1.90D-12*EXP(190/TEMP) : OH + IC4H9OOH = IC4H9O2 ;
% 1.68D-11 : OH + IC4H9OOH = IPRCHO + OH ;
% J<41> : IC4H9OOH = IC4H9O + OH ;
% KRO2NO*0.60*0.975 : TC4H9O2 + NO = TC4H9O + NO2 ;
% KRO2NO*0.60*0.025 : TC4H9O2 + NO = TC4H9NO3 ;
% 2.70D+14*EXP(-8052/TEMP) : TC4H9O = CH3COCH3 + CH3O2 ;
% KRO2NO3 : TC4H9O2 + NO3 = TC4H9O + NO2 ;
% KRO2HO2*0.74 : TC4H9O2 + HO2 = TC4H9OOH;
% 6.70D-15*0.7*RO2 : TC4H9O2 = TC4H9O ;
% 6.70D-15*0.3*RO2 : TC4H9O2 = TBUTOL ;
% 1.67D-13 : OH + TC4H9NO3 = CH3COCH3+HCHO+NO2 ;
% J<55> : TC4H9NO3 = TC4H9O + NO2 ;
% 2.20D-12*EXP(190/TEMP) : OH + TC4H9OOH = TC4H9O2 ;
% J<41> : TC4H9OOH = TC4H9O + OH ;

5.3 REACTION: Modeling Complex Reaction Mechanisms

Dr. Edward S. Blurock, who was at the RISC-Linz at Johannes Kepler University in Austria, has done some valuable work with computers. His program REACTION is an expert system for the generation, manipulation, and analysis of molecular and reaction information. The goal of the system is to assist in the modeling of complex chemical processes such as combustion. REACTION enables both "numeric" and "symbolic" analysis of mechanisms. The major portion of the numeric analysis results from an interface to the CHEMKIN system where the reaction and molecule data is either generated automatically or taken from a database. The symbolic methods involve graph theoretical and network analysis techniques. The main use of this tool is to analyze and compare the chemistry within mechanisms of molecules of different structure. Current studies involve comparing hydrocarbons with up to ten carbons and the influence of the structure on autoignition ('knocking phenomenon and octane number').

In 1995 Dr. Blurock taught a course at Johannes Kepler University at the Research Institute for Symbolic Computation. It was called Methods of Computer Aided Synthesis. It included Symbolic Methods in Chemistry

CAOS
REACCS
LHASA
SYNGEN
EROS/WODCA
EROS

Daylight

Daylight is a program providing computer algorithms for chemical information processing. Their manual

(Daylight Theory Manual, Daylight 4.51) consists of the following sections: Smiles, Smarts, Chuckles, Chortles and Charts, Thor, Merlin, and Reactions. Smiles is a language that specifies molecular structure and is a chemical nomenclature system. Smarts is a substructure searching and similarity tool. It reveals the principles of substructure searching, NP-complete problems and screening, structural keys, fingerprinting and similarity metrics. Chuckles, Chortles and Charts are for mixtures. They are languages representing combinatorial libraries which are regular mixtures of large numbers of compounds. Thor is a chemical information database system consisting of fundamental chemical nomenclature, chemical identifiers, etc. Merlin is a chemical information database exploration system and a pool of memory-resident information and Reactions has all of the features for reaction processing.

5.4 Environmentally Friendly Catalytic Reaction Technology

The establishment of a clean energy acquisition/ utilization system and environmentally friendly industrial system is necessary to lower global pollution and reach a higher level of human life. Here we aim to conduct systematic and basic R & D concerning catalytic reaction technology controlling the efficiency of energy and material conversion processes under friendly and environmental measures. Basic technology development for the molecular design of a catalyst using computer aided chemical design will be combined with the development of new catalysts on the strength of wide-choice/normal-temperature and pressure reaction technologies.

The basic steps are:

1. Preparing model catalytic substances ideal in making controllable various catalytic properties, including absorption, reaction, diffusion and desorption will be studied using thin film preparation technology, a process to synthesize materials on a nanometer scale.
2. This will be done through studies on computer-aided high functioning catalyst design, surface analyzed instruments-based catalyst properties evaluations, etc.
3. The acquisition and utilization of clean energy leads to development that is aimed at a new photo-catalyst that can quite efficiently decompose water not only with ultraviolet but also with rays of sunlight, thus generating hydrogen.
4. Search for and develop methods for facilitated operation of elementary catalytic reaction processes, including light excitation, electric charge separation, oxidation, and reduction reactions that will be integrated and optimized for the creation of a hybrid-type photo-catalyst.

In order to efficiently manufacture useful chemical materials such as liquid fuels that emit less CO2 using natural gas and other gaseous hydrocarbons as raw materials, a new high performance catalyst acting under mild reaction conditions is to be developed.

5.5 Enabling Science

Building the Shortest Synthesis Route

The goal is to make the target compound in the fewest steps possible, thus avoiding wasteful yield losses and minimizing synthesis time.

R&D laboratories synthesize many new compounds every year, yet there seems to be no clear protocol for designing acceptable and efficient routes to target molecules. Indeed, there must be millions of ways to do it. Some years ago, in an effort to use the power of the computer to generate all the best and shortest routes to any compound, a group at Brandeis began to develop the SYNGEN program.

The task is huge, even for the computer. Imagine a graph that traces the process of building up a target molecule; we call it a synthesis tree. The starting materials for the possible synthesis routes are molecules we can easily obtain. As the routes progress, new starting materials are added from time to time until the target is obtained. Each line represents a reaction step, or level, from one intermediate to another, and each step decreases the yield. Two of many possible routes are traced in Figure 50.

To find these routes, we presume to start with the target structure and a catalog of all possible starting materials. Then, the computer generates all the points (intermediates) and lines (reactions) of the graph. If the computer has been programmed with an extensive knowledge of chemical reactions, it could do this by generating all possible reactions backward one step from the target structure to the intermediate structures, then repeating this on each intermediate as many times as necessary to return to the available starting materials.

At this stage, the problem gets too big. Suppose there are 20 possible last reactions to the target (level 1) and that each of these reactions also has 20 possible reactions back to level 2. Going back only five levels will generate 20^5 (3.2 million) routes. How do we select only one to try in the laboratory?

This generation of reactions and intermediates is a brute-force approach; clearly, it must be focused and simplified with some stringent logic. The central criterion should be economy — that is, to make the target in the fewest steps possible, thus avoiding wasteful yield losses and minimizing synthesis time.

A Protocol for Synthesis Generation

The key to finding the shortest path seems to be to join the fewest possible starting materials and those that are closest to the target on the graph. The starting material skeletons are usually smaller than the target skeleton, so joining them to assemble the target will always require reactions that construct skeletal bonds. This underlying skeleton is revealed by deleting all the functional group bonds on a structure and leaving only the framework, usually just C-C σ-bonds.

The central feature of any synthesis is the assembly of the target skeleton from the skeletons of the starting material. Looking for all the possible ways of cutting the target skeleton into the skeletons of available starting materials represents a major focus for examining the synthesis tree.

We illustrate this task by looking at the steroid skeleton of estrone and cutting it in two at different points in the structure (Figure 52). Each cut creates two intermediate skeletons, and each skeleton is then cut in two again to obtain four skeletons. This procedure creates a convergent synthesis, and convergent routes are the most efficient (4). With four starting skeletons, we will need to construct only 6 (or fewer) of the 21 target skeleton bonds. We could keep dividing each skeleton until we ultimately arrive at a set of one-carbon skeletons, but it is not necessary to go that far, that is, to a "total synthesis".

With our four starting skeletons, each skeleton represents a family of many compounds with different functional groups placed on the same skeleton. Suppose that we find a set in which all four skeletons are represented by real compounds in an available library of starting materials; this set could form the basis of a synthesis route with no more than six construction steps to the steroid if the functional groups are right. The skeletal bonds we cut, which must be constructed in the synthesis route, are called a **bondset**, and these bondsets are a basis for generating the shortest syntheses. Each skeletal bondset represents a whole family of potential syntheses.

The Ideal Synthesis

There are two kinds of reactions: construction reactions, which build the target skeletal bonds (usually C-C bonds), and refunctionalization (ΔFG) reactions, which alter the functional groups without changing the skeleton. Any synthesis must do construction reactions, because the starting materials are smaller than the target, but must a synthesis route have any ΔFG reactions?

Imagine a synthesis route with its set of starting materials chosen so that their functional groups are correct to initiate the first construction, leave a product correctly functionalized for the second construction, and so on, continuing to construct skeletal bonds until the target skeleton is built. This is the ideal synthesis in that it must have the fewest steps possible. It requires no FG reactions to get from one construction product to the next.

In a survey of many syntheses, we found that

- the average nonaromatic starting material has a skeleton of only three carbons
- one skeletal bond in three of the targets is constructed
- there are twice as many FG reactions as constructions

Therefore, for an average synthesis, the number of steps equals the number of target skeleton bonds.

We think we can do better. Building the shortest, most economical syntheses requires first finding those skeletal dissection bondsets with the fewest bonds, to minimize construction reactions. It also requires no more than four correctly functionalized starting materials, to minimize FG reactions. Common targets have 20 or fewer carbons, which implies an average starting material of 5 carbons. In our experience with catalogs of starting materials, functional diversity on the skeletons is ample up through five carbons but decreases sharply with larger molecules.

Generating the Chemistry

Once we find the four commercially available starting materials, we need to make a second pass, down from the target through the ordered designated bonds of the bondset. This process generates the actual construction reactions we require, in reverse. So, we need a method of generalizing structures and reactions to quickly find the reactions appropriate to the functional groups present.

Any carbon in a structure can have four general kinds of bonds, as summarized in Figure 53: skeletal bonds to other carbons (R); Π-bonds to adjacent carbons (Π); bonds to heteroatoms that are electronegative (Z); and bonds to heteroatoms that are electropositive (H). The numbers of bonds are referred to as σ, z, and h, respectively. If we know the values of and obtain h by subtraction from 4, only two digits (z and Π) are needed to describe each carbon. This description is summarized in Figure 53, where each carbon is marked in the example structure with its z value. This digitalized general description of the structure is easy for the computer.

In Figure 53, a reaction change at each carbon is just a simple exchange of one bond type for another.

This change may be designated by the two letters for the bond made and the bond lost. Thus, reaction HZ indicates making a bond to hydrogen by loss of a bond to heteroatom — that is, a reduction. The 16 possible combinations are shown and described with general reaction families in Figure 53.

Using this system, we can generate all possible generalized reactions, forward or backward, from any structure. No routes are missed, and we can find all the best routes back from the target to real starting materials. Relatively few generalized reactions are created, and we refine the abstract into real chemistry only at the end. When starting materials are generated through successive applications of these reaction families, we can look them up in the catalog, where they are indexed by skeleton and by generalized z lists of the functionality on each skeletal carbon.

The SYNGEN Program

We have applied this approach in our SYNGEN program, an earlier version of which found its way into laboratories at Glaxo-Wellcome, Wyeth-Ayerst, and SmithKline Beecham, but is currently being improved significantly. The two phases of the generation are summarized in Figures 50 through 55 for one particular result, the Wyeth estrone synthesis. In the first phase (Figure 54, left side), we see the skeletal dissection down to four starting skeletons, all found in the catalog; in fact, the intermediate skeleton B also was found, so further dissection to E and F may not be needed.

In the second phase (Figure 54, right side), this ordered bondset is followed, one bond at a time, generating the construction reactions for an ideal synthesis until all of the functional groups have been generated. These actual starting materials are found in the catalog, so a full synthesis route can be written from them that goes up the right side in a quick, constructions-only ideal synthesis of the target. This three-step synthesis of a target structure can be converted to estrone in two more steps. The prediction for an average synthesis would have been much longer.

The catalog for the current version of SYNGEN has about 6000 starting materials, but it is being expanded from available chemicals directories. After the target is drawn on the screen, the program generates the best routes in <1 min. It displays the bondsets, the starting materials used, and the actual routes, which are ordered by their calculated overall cost.

The output screen from SYNGEN for the example analyzed in Figures 50 through 55 is shown in Figure 55. Two other sample outputs, from a different bondset of the same target, are shown in Figures 56 and 57. The notations on the arrows use abbreviations to describe the nature of the reaction; explanations are available on a help screen. The routes shown are still in a generalized form and require further elaboration of chemical detail by the user. Literature precedents, however, are being added to the program, as described later.

The Future of SYNGEN

Three developments are currently under way on the SYNGEN program. The first and perhaps most important improvement is creating a graphical output presentation that is easy for a chemist to read and navigate; this work is nearing completion. The second deals with the problem of validating the generated reactions with real chemistry. The third development, currently supported by the U.S. Environmental Protection Agency (EPA), is to assign starting material indexes of environmental hazard — such as toxicity and carcinogenicity — so that the routes generated may be flagged for environmental concern when these starting materials are involved.

The second development deals with a major problem in previous versions of SYNGEN: The program generated too many reactions that chemists saw as clearly nonviable. Such results tended to destroy their confidence in the program as a whole. We now have a way to validate the generated reactions from the literature, eliminating many of these nonviable reactions.

The generalizing procedure for describing structures and reactions in SYNGEN also was applied to create an index-and-retrieval system to find matches for any input query reaction from a large database of published reactions. This program, RECOGNOS, has been applied to an archive of 400,000 reactions originally published between 1975 and 1992 and packaged as a single CD-ROM that allows instant access to matching precedents in that archive. The RECOGNOS program is available on CD-ROM from InfoChem GmbH, Munich, Germany, combined with their ChemReact database of 370,000 reactions and renamed "ChemReact for Macintosh".

This archive of literature reactions, now almost double the original size, has been distilled to more than 100,000 construction reactions. These reactions, in turn, have been converted into a look-up table for use by the SYNGEN program. With this tool, SYNGEN can validate any reaction it generates by searching for matches in the archive and determining the average yield. Unprecedented reactions are therefore set aside, and a realistic yield can be estimated for each reaction to be used in the overall cost accounting.

We believe that SYNGEN has considerable potential for discovering new alternatives for creating organic chemicals in the most economical way possible. Even when the program does not yield a directly usable synthesis, it often starts the chemist thinking about different approaches previously not considered. No chemist can think of all the possible routes to the target, but SYNGEN does this quickly. It also provides a powerful and focused output of the possibilities.

5.6 Greenhouse Emissions

Two years ago, the nations of the world gathered in Kyoto to hammer out a plan to curb those man-made gases that are believed to be raising the temperature of the planet.

When the Kyoto Protocol reached Washington, however, it was pronounced too expensive, too unworkable. It was dead on arrival.

But on the Texas–Louisiana border a DuPont chemical plant is doing what Washington politicians and bureaucrats have been unable and unwilling to do — cutting greenhouse gases.

DuPont's Orange plant sits on 400 acres amid wetlands and waterfowl on the Texas-Louisiana border. It makes the chemicals used to make nylon. It also has been emitting tons of nitrous oxide — a greenhouse gas. "Our aim was to control those emissions. The problem was there was no known technology to do it," said the plant manager.

So DuPont invented its own. Today, the fumes from the plant run through a building packed with a catalytic filtering unit that breaks the nitrous oxide into harmless nitrogen and oxygen.

Another closely watched corporate experiment was launched last year by oil giant BP Amoco. It set a goal of reducing 1990 greenhouse gas emission levels by 10% by 2010-regardless of sales growth.

BP Amoco's strategy involves an emissions trading program under which each BP refinery and plant is given a reduction target. Those plants that can do better than their target can "sell" their excess reduction to other facilities.

There have been five "trades" among BP facilities for 50,000 tons of carbon dioxide, with a ton of carbon dioxide "valued" at $17 to $22.

It is widely agreed that some sort of country-to-country emissions trading will have to be part of any accord on climate change. So, BP Amoco's experiment is viewed by many as a valuable test case.

5.7 Software Simulations Lead to Better Assembly Lines

Many years ago this author inspected the Budd auto parts (frames) plant in Kitchener, Ontario. The frames (for all American cars) moved in and out of a station while hanging from a conveyor belt. An instrument measured a point on this frame and compared it to the blueprint in a computer. If there was no match that frame and a number of succeeding ones on the belt were scrapped. It was regarded as a technological marvel.

Engineers now boot up programs that let them tinker, in three dimensions, with every permutation and combination of a product's design. Now engineers aim their computers at designing and refining the assembly lines on which those products are made.

As an example, Dow Chemical Co., now uses computers to simulate its methods for making plastics, running what-if scenarios to fine-tune the temperatures, pressures and rates at which it feeds in raw materials. Dow can now switch pressures and rates at which it feeds in raw materials. Dow can switch production among 15 different grades of plastics in minutes, with almost no wasted material. Before computer modeling, the process took two hours and yielded lots of useless by-products.

Production engineers in industries as diverse as chemicals, automobiles, and aluminum smelting are manipulating virtual pictures of their plants and processes to see whether moving a clamp or adding a new ingredient will make existing equipment more productive, or will enable the same assembly line to skip freely from product to product. Some are even testing out a new virtual reality program that enables engineers wearing special goggles to detect problems by "walking through and around" a three-dimensional model of their factory designs.

The entire relationship between product design and production engineering is being turned on its ear. No longer is it enough for designers to create products that can be made and maintained efficiently. Increasingly, management is asking them whether the products can be manufactured with a minimum of retooling or work stoppages — and if not, whether it is worth giving up a particular product feature in order to wring time and money from the manufacturing routine.

The software is letting manufacturing influence design, not just the other way around.

Real-life examples of modeling's efficiency are mounting. Ford says that one of its plants now uses the same assembly line to make compact and mid-sized models.

Computer simulations of the tread-etching process has enabled tire makers like Goodyear Tire and Rubber to switch production from one type of tire to another in about an hour — a process which previously took an entire work shift. Simulations have shown cookie companies like Nabisco how to use the same packaging machines to make 5-pound bags for price clubs, 1-pound bags for groceries and 6-cookie packs for vending machines.

Various forces are driving the trend towards computer modeling. For one thing, computer technology has finally caught up with manufacturing pipe dreams. Recently computers have been powerful enough to quickly simulate what happens if you change something in a chemical reactor.

Consumers have grown increasingly picky and expect to be able to choose among myriad colors, sizes, and shapes for almost any product. This means that the manufacturers must mix and match parts as the orders come in. And that, in turn, means having tools that can respond to electronic commands to switch paint wells, move clamps, or change packaging and labels.

Already, the modeling procedure has led to development of a conveyor belt that can sense what model is in production and instruct robotic arms to pluck the right hood or other part from a storage bin and have it ready to meet the truck chassis as it moves down the line.

When a conveyor system was too slow, a computer helped us figure out why. "Freightliner" now uses work-flow simulation to figure out how to keep trucks moving evenly from worker to worker, when some models need more handling than others — putting 12 bolts into a wheel well, for example, instead of four.

When a worker lost time on a difficult truck, he should make it up on an easy one. Do not wait until the line is set up to find out you could have prevented bottlenecks.

5.8 Cumulants

There is a relationship between the equation of state and a set of irreducible integrals. These irreducible integrals have a graphical representation in which each point (or molecule) is connected by a bond (f_{ij}) to at least two other points. By the introduction of irreducible integrals a great economy is achieved in accounting for all possible interactions. An important property of cumulants which makes them useful in the treatment of interacting systems is the following: a cumulant can be explicitly represented only by the lower (not higher) moments, and vice versa.

5.9 Generating Functions

Consider a function F(x,t) which has a formal (it need not converge) power series expansion in t:

$$F(x,t) = \Sigma^{\infty}_{n=0} f_n(x)t^n$$

The coefficient of t is, in general, a function of x. We say that the expansion of F(x,t) has generated the set f_n (x) and that F(x,t) is a generating function for the $f_n(x)$.

Examples of generating functions are: Bessel functions, Gegenbauer polynomials, Hemite polynomials, Laguerre polynomials, Legendre functions of the second kind, Legendre polynomials, semidiagonal kernels, trigonometric functions, etc.

5.10 ORDKIN a Model of Order and Kinetics for the Chemical Potential of Cancer Cells

A method for deriving the chemical potential of particles adsorbed on a two dimensional surface has previously been derived for lateral and next nearest neighbor interactions of the particles. In order to do so a parameter called K was used and $K = ((\exp((\mu-\varepsilon)/kT))(\theta/1-\theta))^{1/Z}$. It was found from a series of normalizing, consistency, and equilibrium relations shown in papers by Hijmans and DeBoer (1) and used by Bumble and Honig (2) in a paper on the adsorption of a gas on a solid. In the above, u is the chemical potential and ε is the adsorption energy. The numerical values of K were derived from computers for various lattices with different values of the interaction parameters for nearest neighbors (c) and next nearest neighbors (c'), where c = exp(-w/kT) and w is the interaction energy, and the "order" of such lattices were plotted as the values of $\exp((\mu-\varepsilon)/kT)$ or $p/p_0=\exp((\mu-\varepsilon)/kT)$ versus θ or the degree of occupancy of the lattice. A method for approximating the lattice was accomplished mathematically by selecting basic figures such as the point ○, the bond ○—○, the triangle △, or the rhombus ▱. Other graphs were made which plotted the value of the pressure ratio vs. time. A model for the time was also selected from a previous publication (3) as L = (p/k)(1-exp(kt/p)) where p denotes the organism, k the rate concentration or exposure to chemical i, and L the toxicological measure. Results show that the lower the value of c (or the greater the antagonism or repulsion of cells or particles) the greater the chance of cancer. Also, the higher the chemical potential, the more the chance of cancer. Remedies are also indicated by changing the pressure or the diffusion of cells. The results were matched with experimental evidence from humans living near several chemical plants near the city of Pittsburgh (4).

The value of K given above has the chemical potential in it, and solving for this quantity we obtain K^Z $(\theta/1-\theta)$ where Z is the coordination number of cells in a tissue approximated as a lattice.

The lattice has been approximated as triangular and was broken up into basic figures mentioned above. The larger the basic figure the more complicated the algebra. The bond yields $K = (\beta-1 + 2\theta)/2\theta c$, where $\beta = [1+4c\theta(1-\theta)(c-1)]$ and the triangle as basic figure yields a quartic equation $K^4 - a_1 K^3 + a_2 K^2 - a_3 K + a_4 = 0$ where $a_1 = ((2-5\theta)c + 2 - 3\theta)/c(1-2\ )$,

$a_2 = (c+5)/c$, $a_3 = ((3 - 5\theta)c + 1-3\theta) / c(1-2\theta)$, and $a_4=1/c^2$. Every point then derived for the triangle basic figure subfigure is then found for the solution of the above quartic equation for given values of c and θ. The solution to the rhombus approximation is yet more formidable and requires a special computer program to approximate the answers.

One defines an order variable as order = $K^z\theta/(1-\theta)$ and using the model for time above, we obtain cancer = $K^Z(\theta/1-\theta)p/k(1-\exp(-kt/p))$. p has a different value for each organism and for Man it is unity. k has a value for each different environmental chemical. t is the time of exposure of a person to that chemical in years. The procedure used then is to select a basic figure, select a value for Z, select the proper value for k, and then use a sequence of values for θ and t. Such work was done on Quattro Pro on a PC. The value chosen for k was .05, the range of values for θ was from .0125 to .975, and the range of values for t was from 1 to 75 years.

The model was called ORDKIN (abreviated from order and kinetics). The data was taken by the graduate students at the University of Pittsburgh's Department of Public Health of some 50,000 residents in three zip number areas within dispersion distance of Neville Island which contains about a dozen industrial plants. In the graphs below b stands for bond, t stands for triangle and r stands for rhombus as the basic figure. Occupancy or theta (θ) or lattice occupation stands for the fraction of sites covered. Cancer and order are the expressions given above, u is the chemical potential, k in u/kT is the Boltzmann constant, and T is the temperature (Kelvin).

In Figure 90, the top two curves are for c = 0.36 (tp most) and 0.9, whereas the bottom curves are all for c = 2 or 3 and they all denote curves for cancer as in the equations and parameters listed above. Now it was of interest that for the values of c below unity, which denotes repulsion between particles, the chemical potential was higher than for those cases where c was above unity which indicates attraction between particles. It is also of interest that when the chemical potential is higher the system tends to be more unstable than when the chemical potential is lower. Indeed, when the chemical potential is at a minimum the system tends to be at equilibrium. These plots are versus occupation of the sites on a lattice which means θ = 0 when the occupation is zero and unity when it is full.

In order to test what is responsible for the separation of the chemical potential curves as shown in the graphs above, the order parameter was examined and two plots were made, one where the c values were >1 and one where the c values were <1, corresponding to regions of attraction and repulsion, respectively, and shown as Figures 91 and 92, respectively. Both graphs for c < 1 are shown as ln(order) plotted vs. age.

We have neglected some terms because actually $(u^G+e)/kT=P/P^*$ and $P^*=(2pmkT)^{3/2}kTj^G[\exp(-e/kT)/h^3$ where j^G is the internal partition function for the gas and ε is the adsorption energy with the other symbols having their usual meaning. These factors will be thought of as scaling factors in this work where individuals are construed to have approximately the same values and introduce some error.

Figures 93 and 94, respectively, compare the data for all observed cancer cases and breast cancer cases from the study conducted at and near Pittsburgh.

The graphs show that the worst fit for the data for all cases of cancer is the triangle basic figure with c = 0.1. The computer failed to obtain solutions for young victims in this case. The regression of all cancers and breast cancers show the linear nature of the regression in these cases. The zig-zag nature of the data from the field is clearly shown in these cases and it is possible that the linear curves are best in these cases.

The following table collates the value of c and their exponents.

	c	w/kT
> 1	3	1.10
> 1	2.77	1.02
>1	2	0.70
<1	.9	-0.105
<1	.36	-1.02
<1	.1	-2.30

from which we see that the best results are obtained with a small repulsive force between cells (c = 0.9). Another very important way to lower the chemical potential of cancer cells is to impose critical conditions on the system containing the cancer cells. Graphically this means that the curve for the order of the cancer cells be flat or parallel to the abcissa which can be the age of the people or the values of θ. This means the order would be constant in value for varying values of the age or θ. This curve or plateau must be both low and broad to be effective. It can be achieved by varying the temperature, the pressure, the concentration, or the constitution of the medium containing the cancer cells. This is similar to techniques in chemistry or chemical engineering where a foreign substance can bring about critical solution temperatures or cause the volatility to increase.

Results of the Study

1. If the malignant cells of the cancer can be related to the chemical potential, this would lead to many therapeutic methods to "cure" the can-

cer or prevent it from spreading. This is so because many physical processes can occur that are related to the chemical potential. It will be shown that the cancer cells have a higher chemical potential than the normal cells and it then becomes a task to lower the chemical potential. Some examples of ways to do this include changing concentrations. This can reduce the chemical potential according to the formula $\Delta u = kT\ln C/C_0$. Here the ratio of concentrations can change logarithmically. Another way the chemical potential can be lowered is by a change in pressure: $\Delta u = kT\ln(P/P_0)$. Yet another way for the reduction of the chemical potential is through electrochemical means so that $u_i = u_{i0} + z_i F\phi$ in the proper environment. In this equation u_{i0} is the standard chemical potential of the species, z_i is the charge on the species, F is the Faraday and ϕ is the potential.

2. In order to inhibit the growth of malignant cell and tissue, this study suggests that the carcinogenic tissue be washed or flushed with a liquid that can insert or replace molecules or ions with ones that do not interact with their neighbors as strongly as the original ones.
3. Also, the surrogate molecules or ions should have a radius more conducive to the blocking of deleterious interactions by changing the coordination number of the destructive carcinogenic molecules.
4. The signal communication or transduction between cells must be ameliorated for those that are beneficial and destroyed for those that are harmful. This can be expedited by studying the pathways, using the methods shown here, that are conducive to good health.
5. Critical regions must be avoided at all costs. The transition between states of matter resulting in the liquefaction of cells is a fatal phenomenon.
6. The results show that the effects noted in the literature as to the application of heat or the withdrawal of heat to tumors are in accord with the model given here and it would be well for medical and surgical specialists to study the ramifications of heat transfer to tumor size and how to lessen the pain involved in the process.
7. Photochemistry, photodynamics, and laser therapy are elucidated by the model and clinical observations are again predicted by this model.
8. A matrix technique is used to divide the molecular aspects from those of the interaction aspects and in so doing surrogate candidates can be found for these individual aspects to reduce or destroy carcinogenic tissue.
9. Biology does not render itself into simple order. It is governed by a systematic disorder and requires the mathematics of disorder or chaos to reflect reality. This model is strictly nonlinear.
10. Electomagnetic fields can be set up from cells or molecules aligned in the +-+-+- manner and can act to send signals or create a morphogenetic field.
11. The UNIFAC model for solutions has surface areas and volumes for chemical groups and interaction energies for many chemical groups that can be optimized to provide the best candidate molecular species to prevent or abate cancer.
12. Three computer programs exist to (a) provide kinetics for reaction involving chemical species involved in cancer (THERMOCHEMKIN), (b) provide thermodynamic functions for complex species involved in cancer (THERM) and (c) provide and optimize properties for chemical species involved in cancer (SYNPROPS).
13. The pathways between the subfigures of the basic figure become more numerous as the basic figure becomes larger and next-nearest neighbors appear and are taken into account. The calculated probabilities of these paths are close in value, yield logical values and provide signal possibilities that can be important to growth.
14. The refraction exaltation (difference between the experimental refraction and the calculated refraction) is due to a conjugated system of double bonds. Frequently such molecules are highly polyaromatic hydrocarbons, etc., and are carcinogenic.
15. Signals are transmitted by "antenna" as charges oscillate back and forth. According to field theory, these charges produce an electromagnetic wave. The wave reaches a receiving antenna and sets the charges in that antenna into oscillation, with results that are detected in the receiver. A paper presented before showed some ways that carcinogenic materials are deposited on a substrate of cells and are prepared for chemical reactions that form the basis for signal transmittal.
16. When the force between molecules is of a considerable magnitude and repulsive, then the probability for an occupied rhombus can become not only more sigmoidal, but also unstable, whether the cause is mathematical or real.
17. The expression for θ as a function of p/po, resembles a Fermi-Dirac distribution function and a plot shows a very sharp step function at body temperature from θ equals 0 to unity. This leads to a model from solid state physics where there are bands of energy within living creatures such as the valence band and the conduc-

tion band. When cancer cells can reach the conduction band, then metastasis is prevalent.

18. The model studied here concentrates on the order-disorder of the substrate. The molecules that then come in contact with a relatively stationary substrate can undergo chemical reactions with those in the substrate. It is the chemical reactions of such reactions, where the substrate is in the critical region, that can be one of the important causes of cancer.
19. Figure 74 elucidates the above. It utilizes the Michaelis-Menten equation as a model. This defines the quantitative relationship between the enzyme reaction rate and the substrate concentration [S] if bothVm and Km are known. Substitution of θ for [S] then shows that when the attractive forces in the substrate are large the reaction rate is much above that appropriate for the Langmuir isotherm, whereas when the forces are very repulsive, the reaction rate can fall much below.
20. In the chaotic region, the dynamics are very sensitive to initial conditions. The transition from the ordered to the chaotic regime constitutes a phase transition, which occurs as a variety of parameters are changed. The transition region, on the edge between order and chaos, is the complex region. Complex systems exhibit spontaneous order. Thus it is possible that adaptive evolution achieves the kind of complex systems which are able to adapt.

5.11 What Chemical Engineers Can Learn From Mother Nature

Economic pressures on the chemical process industries (CPI), particularly on R&D, are quite severe. The high cost of innovation must be reduced if the prosperity of the CPI is to endure. The primary function of the engineer is neither analysis nor design. It is creating new processes, products, concepts, and organizations. Conducting such creative activities can be accomplished by mimicking the evolutionary processes of nature.

Increasing the economy of evolutionary activity includes recent results of nonlinear dynamics and complexity theory, and provides some of the powerful innate organizing forces in the physical world of as yet unrealized potential. Evolution can be defined as an increase in functional efficiency manifesting itself as the spontaneous generation of useful information.

The nature of variation, selection, and heredity differ greatly among various evolutionary processes and these differences will have a major impact. If we were to benefit from biological examples, we must be able to identify generic characteristics of evolution dynamics and to evolve specific strategies effective for our purposes. Driving energies will vary. For many engineers the immediate source is money, but one cannot underestimate nonfinancial motivation.

In diversity, there are no well-defined species, only groups of closely related individuals. Also, diversity is a source of robustness for all organisms and ecosystems. This has relevance for all human organizations: hiring in one's own image is a dangerous procedure leading to inflexibility and limited capacity for dealing with changing circumstances.

Evolutionary processes do not depend entirely upon random events, but they are favored by the self-organizing nature of these systems.

In a research organization there is an optimum degree of interaction between individuals: too much isolation or too much hierarchical control from the top leads to stagnation, and too much interaction to chaos. Most creativity consists of rearranging known components in new ways. This is a generalization of the unit operations concept.

5.12 Design Synthesis Using Adaptive Search Techniques & Multi-Criteria Decision Analysis

Safety and real-time requirements of computer-based life-critical applications dramatically increase the complexity of the issues that need to be addressed during the design process. Typically quantitative analysis of such requirements are undertaken in an ad-hoc manner after the artifact has been produced. A more systematic approach which uses Adaptive Search and Multi-Criteria Decision Analysis techniques to provide analytical support during the design-decision process is described in this paper from the University of York.

5.13 The Path Probability Method

The Path Integral from Quantum Mechanics and Theoretical Physics is used to plan the best chemical groups for a given constrained stoichiometry. This then is utilized to ascertain whether there is a reaction scheme or mechanism to produce a molecular species that can appear in the program Envirochemkin and to find whether the reactions are feasible under a set of conditions (P, T. t. mode and mechanism) used in Envirochemkin. The program THERM is used to obtain thermodynamic functions for the molecule if they are not known or contained in the of the assigned chemical groups. The program used to find the set of groups is SYNPROPS and Envirochemkin file. It is to be noticed that the variable emphasis in this undertaking are chemical groups (of which there are 380 choices as in the program THERM) rather than chemical species as originally used in SYNPROPS in which there were 32 in the Linear Model and 33 in the Hierarchical Model, many of which were duplicated in both models.

Each of the groups in THERM has thermodynamic functions associated with it: namely, heat of formation at 298 K, entropy at 298 K, and heat capacities at constant pressure at 300, 400, 500, 600, 800, 1000, and 1500 K. Thus, the free energy can be found at any temperature (F = H-TS) and thus the free energy difference between the species and its precursor or descendent or that for any reaction between precursor and descendent can be obtained if the free energies of the precursor or descendent species are known as well. The data for 380 groups include that for free radicals, etc., so activated complexes can be included in the scheme and thus a tree can be drawn for the progress of the reactions from the initial reaction to the final reaction with the probability of occupancy of the different levels of the tree as well as the various participants in the progress of the reaction. This probability can be assumed to be proportional to the equilibrium values of the species which is proportional to the value of the quantities exp(- F/RT) which is obtained from the change in free energy of the reactions involved. The process can thus be used to arrive at the mechanism of the overall reaction and each of its constituent parts.

Another way to obtain mechanisms of reaction is from Rate Distortion Theory. Here a tree can be constructed to decode messages that are used in communication theory but will now be used to ascertain and depict needed mechanisms.

The path probability method of irreversible statistical mechanics has been applied to pollution prevention and waste minimization. The most probable path in time taken by a system is derived by maximizing the path probability after adding a space axis to equilibrium statistical mechanics. The path probability formulation is based on the Markoffian character of the process and this depends on the choice of variables used in describing the system. In a cooperative process all cooperating degrees of freedom must be taken into account. The cluster-variation method in which a finite size of a cluster is used to represent the whole system violates the Markoffian requirement of the process. The formulas for the most probable path can be interpreted based on a superposition approximation. This is a shortcut to the expressions for the most probable path without going through the path probability formulation and its maximization each time, and will greatly increase the maneuverability of the technique.

It is interesting to note that Feynman's space-time approach can quickly be used to write down results rigorously derived from quantum field theory (or second-quantized Dirac theory). Matrix elements derived using quantum field theory can be obtained much more quickly using the space-time approach of Feynman. Feynman's approach (based on the particle wave theory of Dirac) is simple and intuitive. It visualizes the formula correctly, which we derive rigorously from field theory. The Feynman graphs and rules have had a profound effect on a number of areas of physics including quantum electrodynamics, high energy (elementary particle) physics, nuclear many-body problems, superconductivity, hard-sphere Bose gases, polaron problems, etc. Although we do not use the technique here, there is a relation between the path probability method and Feynman's approach.

Consider a system of N atoms, each of which has two energy levels, g and e (for ground and excited). During a short time interval, t, states of atoms may change by exchanging energies with a heat bath of temperature, T. When we look at the system, its configuration may change in time as shown on the left where a system of two-level atoms changes in time. (e and g stand for the excited and the ground states, respectively). At the right is a configuration of an assembly of one-dimensional Ising model. x and o are a plus and minus spin, respectively.

t- t t+ t	k-1 k k+1
atom 1→e→g→g→e→g→e→	system 1 -x–o–o–x–o–x
atom 2→g→g→e→e→e→g→	system 2 -o–o–x–x–x–o-
————	————
————	————
atom N→g→g→e→g→e→e→	system N -o–x–o–o–x–x-

The correspondence between these cases suggests that the irreversible problem on the left can be treated in analogy with the equilibrium problem when the time axis is treated as the fourth space axis. Thus the probability function **P** to be written for the case on the left is expected to be constructed in analogy with the state probability function **P** or the free energy **F** for the case on the right.

Let us consider a tree, upside down with the root at the top. As we proceed from top to bottom we have choices as to which compounds form from the original compound.

C_0^0

a^c

C_1^1 C_1^2

^ ^

bd fg

$C_2^1 C_2^2$ C_2^3 C_2^4

Thus the first compound, which might be PAN or some other pollutant we wish to eliminate, is made to react with additives so that it forms the first tier of compounds, indicated by the subscript 1, and the two compounds in the first tier react with additives to form the four compounds in the second tier. The question is which is the most probable path that will be followed, a→d, a→e, c→f, etc?

To decide this, one may utilize the path probability method of Kikuchi together with some of my own publications on order-disorder theory. The probability that a given path is followed is proportional to the reaction rates that take place. In the reaction A + B = C + D, this can be approximated by knowing the structure of the molecules A and B and the activated complex A—B. A table is included in the recent book *Computer Generated Physical Properties* to approximate the range of such rate constants. Also needed is the energy of the reaction that can be obtained from the program THERM. This is usually printed out in a table and may be the enthalpy or the free energy of reaction under reaction conditions. The equation for the probability of reaction comes from the order-disorder theory, which is an equilibrium method because the kinetic theory reduces to equilibrium expressions as stated in Kikuchi's paper. Finally, we need the concentration (and conditions) of the pollutant and this is available either from Envirochemkin calculations or actual measurements in plant processes.

Thus, we have

P4 = (conc. of pollutant)(path probability)
(rate constant)(energy factor) =
Pollution Prevention Path Probability

The concentration of pollutant is derived from the Envirochemkin program or measurements, the rate constant comes from the structure of the reactants and activated complex and Table or the value of A_f/A_r from the output of the Thermrxn program, the (free) energy factor also comes from the subroutine Thermrxn of the THERM program and the path probability comes from the formalism developed here. Initially, the path probability can be set equal to unity and the three factors can be checked, the rate constant from the estimated rate table, the ratio of Arrhenius factors from THERM and the equilibrium constant from THERM. If the magnitude of these three constants are positive and the magnitude sufficiently large, then the proposed reaction, concentrations, conditions, and/or additional reactants can be tried in Envirochemkin for a final computation in a proper molecular environment. Thus one can control the environment and have clean production.

Order-disorder theory, also called cooperative phenomena or chaos, has been used to find critical conditions for systems with interactions between particles. This can be very helpful to find discontinuities in chemical potentials and phase equilibrium which can affect the reaction kinetics and equilibrium of systems that we are studying here.

5.14 The Method of Steepest Descents

The "method of steepest descents" can be used for the approximate evaluation of integrals in the complex plane. It is appropriate for the treatment of many integrals encountered in statistical mechanics.

Consider a function exp[Nf(z)] where f(z) is an analytic function of its argument and so the exponential is also an analytic function of z. We divide the f(z) into its real and imaginary parts:

$$f = u + iv$$

Because of the analytic character of f(z), its parts u and v must both satisfy Laplace's equation:

$$\partial^2 u/\partial x^2 + \partial^2 v/\partial y^2 = 0$$

$$\partial^2 u/\partial^2 x + \partial^2 v/\partial^2 y = 0$$

These equations show that u and v cannot, in the region where f is analytic, attain an absolute maximum or minimum value. Starting at any point in this region, one can follow a line of steepest increase of u indefinitely, either to • or to the boundary of the region; following a line of steepest descent one can go downward, either to • or to the boundary of the region. The surfaces representing the functions u(z) and v(z) have no peaks or bottoms, but they do have horizontal tangent planes. At any point where df/dz = 0, the rate of change of f, or of its parts u and v, in any direction is zero:

$$\partial u/\partial x = \partial u/\partial y = 0 \; ; \; \partial u/\partial x = \partial u/\partial y = 0$$

At such points both the u and v surfaces have horizontal tangent planes.

First consider the u-surface near such a point. It must have maximum curvature downward along one line through this point, and equal maximum curvature upward along a perpendicular line. The point itself is called a "col" or "saddle point." The directions of maximum curvature are lines of steepest descent and of steepest ascent, respectively.

The line of steepest ascent passes from the col along the crest of two "ridges" on the u-surface and the lines of steepest descent plunge into valleys separated (locally, at least) by these ridges. A u-surface may contain many valleys separated by many ridges, each of which can be crossed at a saddle point.

Since u and v are conjugate functions, the lines of steepest ascent or descent for u are contours of const v and conversely; otherwise the behavior of the v surface near a col is like that of a u-surface.

At a saddle point of f(z), exp (Nf(z)) is also stationary. Along the line of steepest descent, exp(Nf(z)) has maximum magnitude at the col, and its phase factor is stationary. Consider a line integral in the complex plane,

$$I = \int_A^B \exp(N(z))\, dz$$

between points A and B in valleys separated by a ridge, passing through a simply-connected region in which f(z) and exp(Nf(z)) are analytic. The value of the integral will be independent of the particular path between A and B. If one chooses a path through a col in the ridge between A and B, the integrand will have its maximum magnitude near the col, and the phase will be stationary, so that contributions to the integral from parts of the path near the col will not tend to cancel out. The major contribution to the integral along such a path will come from the neighborhood of the col — a region that is smaller the larger N is. To get an approximation to the value of the integral that will be asymptotically exact as $N \rightarrow \infty$, one needs only consider the contribution from the neighborhood of the col. One finds in this way

$$I + \mid 2\Pi \;/\; Nf''(z_0) \mid \exp(i\alpha)\exp(Nf(z_0))$$

where z_0 is the position of the col and α is the angle between the positive direction on the x-axis and the direction of the line of steepest descent at z_0.

5.15 Risk Reduction Engineering Laboratory/ Pollution Prevention Branch Research (RREL/PPBR)

The PPBR is responsible for projects that develop and demonstrate cleaner production technologies, cleaner products, and innovative approaches to reducing the generation of pollutants in all media. The PPBR is organized into two sections, Products and Assessments, and Process Engineering, and it is conducting projects in five major areas.

1. Cleaner production technologies
2. Tools to support pollution prevention
3. Cleaner products program
4. Pollution prevention assessments
5. Cooperative pollution prevention projects with other federal agencies

A summary of each project area is provided below based on descriptions published in the November, 1994 edition of PPBR's publication. An overview of the five major areas and their programs is presented below.

Summary of PPBR Program Areas

1. Cleaner Production Technologies
   a. Waste Reduction Innovative Technology Evaluation (WRITE)
   b. Support for RCRA Hard to Treat Wastes
   c. Support for the 33/50 Program
   d. Support for the Source Reduction Review Program (SRRP)
   e. Clean Technology Design and Development Projects
2. Tools to Support Pollution Prevention
   a. Life Cycle Assessment Development and Demonstrations
      (1) Life Cycle Assessment Demonstration: Carpeting
      (2) Development of Pollution Prevention Factors
      (3) Streamlined LCA Model Development and Demonstration
   b. Measurement Methodology Tools Development
      (1) A Measurement Methodology for Pollution Prevention Progress
      (2) Measurement Tools to Support Pollution Prevention
3. Cleaner Products Research Program
   a. Evaluating Potential for Safe Substitutes
   b. Clean Products/Source Reduction Case Studies
   c. Product and process design for Life-Cycle Risk Reduction and Environmental Impact Mitigation
4. Pollution Prevention Assessments and Support Program
   a. Small Generator Waste Minimization Assessments
   b. Industrial Assessment Centers Program
   c. Pollution Prevention for Public Agencies
   d. NATO/CCMS Project: Pollution Prevention Strategies for Sustainable Development
   e. Clean Technology Guides
5. Cooperative Pollution Prevention Projects with Other Federal Agencies

a. Waste Reduction Evaluations at Federal Sites (WREAFS) Program
b. Strategic Environmental Research and Development Program (SERDP)

5.16 The VHDL Process

Although hardware is concurrent, VHDL allows you to implement algorithms with a series of sequential statements that occur inside what is called a PROCESS. Understanding the operation of PROCESSes is critical to understanding how VHDL synthesizes synchronous designs. A PROCESS is a CONCURRENT statement used in an architecture which requires a WAIT statement or SENSITIVITY list.

A SENSITIVITY list is a list of variables which if a change in them occur, will trigger the associated PROCESS. Inside of the PROCESS statements execute sequentially. That is to say they execute in order, like a standard programming language does.

An example of a SENSITIVITY LIST PROCESS is given.

Any change in the my_set or my_reset will trigger the PROCESS to run. Let us reexamine our AND GATE example, except this one will use a process with the two inputs Ain and Bin being on a sensitivity list.

The operation is equivalent logically, but it wouldn't make much sense to do this since it adds a measure of complexity which is unneeded. The process sits there until either Ain or Bin change, then does whatever is inside of its begin and end process statements. The fact that all we have is the original AND expression does not discount the fact that we could put many other statements there.

Using a WAIT UNTIL statement is another way of implementing a PROCESS which is commonly used to simulate an edge triggered device. For SIGNALS which are changed in the PROCESS the changes are SCHEDULED. That is to say, the changes that occur within the PROCESS are reflected on the output of the associated SIGNAL on the next occurrence of the condition specified in a WAIT UNTIL statement. In the next piece of code we add an input clock line and use a WAIT UNTIL statement in the process as shown:

The code does not update COut until a rising edge is seen on the CLK line. This is essentially the following circuit

```
CLK———————————————————————|———|—
Aln———————|———\ | \/ |———COut
 Bln——— |———/ |—————|
```

The statements inside the PROCESS occur on the NEXT clock edge. This can be thought of as a LATCHED vs. COMBINATORIAL process. Combinatorial functions like ANS/OR/XOR occur essentially immediately. While logic is LATCHED, it does not become apparent until the CLOCK latches it out. And so in the statement A<=B which occurs outside a process, A essentially changes exactly as B does. However, when placed inside a process with a WAIT UNTIL statement, A will not reflect the change in B until the next WAIT UNTIL event (CLOCK) occurs.

Conclusions

It appears as if the successful work to determine analytically global solutions for pollution prevention and waste minimization, while simultaneously engaged in plant design or simulation, has begun. Here we are not concerned with heuristic methods but in designs that are necessary and sufficient.

This requires a new kind of engineer; one that is very adept in three subjects; chemical engineering, computer science, and mathematics. It requires yet another prerequisite: the engineer must be very creative.

There are not many engineers of this caliber today, but it is hoped that with proper training there will be more such engineers in the future.

It is to be emphasized that the mathematics required is not the same as that taught today but includes "less conventional" subjects or aspects of mathematics, such as discrete mathematics, etc.

This book has introduced many topics but has not gone into each of them very deeply. It was felt more important to expose the reader to more of the matter lightly so that his or her preferences would gel. This is true of the first book from this author as well: *Computer Generated Physical Properties.*

The book is divided into five sections. The first is called Pollution Prevention and Waste Minimization and it serves as an introduction. It reviews both computer process simulation as well as computer designed pollution prevention and waste minimization. It discusses the meaning and utilization of these methods at government agencies, industrial corporations, research centers, and countries of the world. It introduces the terminology "Clean Technology." It examines the effect of such methods on "the bottom line." It examines the effect of upsizing, novel chemical reactors, OHSA regulations, and risk on the design of clean technology, rather than the design of "dirty" technology with clean-up at a later time.

The second section entitled Mathematical Methods reviews many of the methods available to achieve an optimum using a computer. Such knowledge may be necessary to optimize cost, optimize yield, etc., in a chemical process while at the same time minimizing waste production. Some ideas are also introduced that can achieve or help to achieve the results such as Petri Nets, KBDS, Dependency-Directed Backtracking, and the Control Kit for O-Matrix. There is even a chapter on the construction of new types of computers.

The third section is called Computer Programs for Pollution Prevention and Waste Minimization. This actually considers computer programs of considerable assistance to computer simulations and models of pollution prevention and waste minimization. They include: Process Synthesis (Synphony), Mass Integration, LSENS, Chemkin, Multiobjective Optimization, Kintecus, the Simulation Science program, etc. Specialized programs such as BDK-Integrated Batch Development, Super Pro Designer, P2-Edge Software, CWRT Aqueous Stream Pollution Prevention Design Options Tool, and OLI Environmental Simulation Program (ESP) are also discussed. The concepts of Green Design and chemicals and materials from renewable resources are also examined.

The fourth section is Computer Programs for the Best Raw Materials and Products of Clean Processes. It shows the invaluable contributions of Cramer's papers to the SYNPROPS method of designing molecules with the most desirable physical and environmental properties available. It also describes Friedler et al.'s method for the design of molecules with desired properties by combinatorial analysis. It also examines the program THERM for its important contribution of thermodynamic functions to programs of Section three. It discusses the Pinch Technology, economics, Geographical Information Systems, health, HAZOP and other features that combine with the computer-assisted simulations.

The fifth section is called Pathways to Prevention. It has some theoretical considerations for the rest of the book. Examples include the Grand Partition Function, Cumulants, Generating Functions, the Path Probability Method, and the Method of Steepest Descent. It also combines Order and Kinetics to obtain the chemical potentials of cancer cells. It also studies the mechanisms and chemical reactions that play a part in pollution and pollution prevention.

End Notes

My thanks to Dr. L. T. Fan for sending me three items. One is the paper by R. W. H Sargent, "A Functional Approach to Process Synthesis and its Application to Distillation Systems", *Computers Chem. Eng.*, 22(1-2), 31-45, 1998. In it he shows that Douglas's hierarchical approach to process design, with successive refinement of models as required to resolve choices, can be embedded in a rigorous implicit enumeration procedure for finding the optimal design, within the accuracy implied by the final model. This is an advantage because the final design is verified by use of models as detailed and accurate as desired, while limiting computational effort by use of simpler models during development of the design. He also uses the representation of a process as a state-task network which contains a connected path from each feedstock to some product and conversely from each product to at least one feed; moreover each intermediate state and task must be on at least one such path. We can then devise an algorithm which generates all feasible state-task-networks. These can then be evaluated with an implicit enumeration procedure, at the same time refining models as required to resolve the choices.

Dr. Fan also sent me the latest flowsheet for the structure of SYNPHONY. It is shown as Figure 83.

He also brought to my attention the article "Unique Features of the Method for Process Synthesis Developed by F. Friedler, L.T. Fan, and Associates", which was discussed earlier.

Figure 61 shows paths followed in going from one occupied rhombus figure to another. It turns out that a direct product expression

$$Q = (x)^{g-h}(x')^{h}(y)^{r}(z)^{t}(y')^{s}.$$

Here x and x' are different sites on a geometrical figure and y, y', and z are interactions between different bodies on these sites. The exponents g-h, h, r, t, and s are the counts of the number of such sites and interactions that there are. Now I will multiply the above equation by $(u)^{w}(v)^{t}$. Here u and v are the reactor and the separator, etc. W and t are the number of reactors and separators, etc. present. It remains to find the expressions for (A): x, x', y, y', u, and v and also the values of the exponents (B): g-h, h, r, t, s, w, and z. This is done by the methods of Bumble and Honig and Hijmans and DeBoer for (A) by setting up 3 sets of equations: Equilibrium Equations, Consistency Equations, and Normalizing Equations from Statistical Mechanics. The $_{\text{valisues}}$ for the set B is then found by inserting the problem into SYPROPS and using the Optimization routine for Q with proper constraints. When done there will be an optimized chemical flowsheet.

Another way to proceed involves the Path Integral

$$M = \int_{a}^{b} \exp^{(i/h)\, S[b,a]}\, Dx(t)$$

and the entropy can be given analytically

$$S\{p_{is}^{(n)}\} = -N\sum_{m=a}^{n} y_n^{(m)} \sum_{is} L_{is}^{(m)} p_{is}^{(m)} \ln p_{is}^{(m)}$$

Also other techniques viewed were the random walk method, order-disorder methods, and the Wiener method.

Consider a flexible chain of fixed length constrained to lie on a square lattice. If one end is fixed at the origin, how many configurations of the chain will give the other end x coordinate c?

At each point n the chain may follow any of 4 paths. If it follows plus or minus y it contributes no new value to the x coordinate. However, plus or minus x paths will contribute plus or minus 1 to the x coordinate, so the generating function is

---

$$G(L, x) = (1/z+2+z)^{L} = (1+2z+z^{2})^{L}/z^{L} = (1+z)^{2L}/z^{L}$$

By the binomial theorem the coefficients can be seen to be

$$(2L)!/(L-D)!(L+D) \text{ where } D = pL$$

Then $g(L,x) = (2L)!/((1-p)L)!((1-p)L)!$

And utilizing $N! = (N/e)^N (2pN)^{1/2}$, we find

$$g(L, N) = 4^L/(pL)^{1/2}(1-p^2)^{1/2}(1-p)^{1-p}(1+p)^{1+p}$$

$$\ln g(L, x) = L[\ln 4-(1-p)\ln(1-p)-(1+p)\ln(1+p)]$$

Expanding ln(1+p) and ln(1-p) and neglecting higher terms we obtain

$$g(L,x) = 4^L \exp(-x^2/L)/(pL)^{1/2}$$

References

Section 1.1

Mah, R. S., *Chemical Process Structures and Information Flows*, Butterworths, London, 1989.

Section 1.2

Turton, R., Bailie, R. C., Whiting, W. B., Shaeiwitz, J. A., *Analysis, Synthesis and Design of Chemical Processes,* WVU Chemical Engineering, 1998.

Section 1.3

Kirk-Othmer Concise Encyclopedia of Chemical Technology, Kroschwitz, J. I., Ed., John-Wiley & Sons, by e-mail.

Section 1.4

Sheppard, L. M., Process Simulator Guide, Chem Proc. Simulation Guidebook, Chem. Eng. Dept., Louisiana Tech. U.

Section 1.5

Gopal, M. Ramdoss, P., El-Hawagi, M. M., Integrated Design of Reaction and Separation, Systems for Waste Minimization. AIChE, Annual Meeting, 1997.

Section 1.6

600R94128, A Review of Computer Process Simulation in Industrial Pollution Prevention, EPA.

Section 1.7

OECA, Office of Enforcement & Compliance, EPA Sector Notebooks Profile of the Inorganic Chemical Industry (1995) EPA/310-R-95-009.

Section 1.8

infochem Thermodynamic Models Ther...ransport Properties Phase Enveloper http://www.infochem.demon.co.uk/models.htm#bin

Section 1.9

Krieger, J. H., Chem. & Eng. News, 3/27/95. http://pubs.acs.org/cenear/950327/art08101.html

Section 1.10

Yang, Yihua, Huang, Yinlun, 1988 Annual Tech Program, AICHE.

Section 1.11

Del Mar, R., Fowler, K. M., Kuusinen, T., Discussion Draft, 6/30/97.

Section 1.12

Butner, Scott, Environment & Society Group, Battelle Seattle Research Center, http://www.seattle.battelle.org/p2online/p2design.htm

Section 1.13

Pollution Prevention Information Clearinghouse (PPIC), www.epa.gov/opptr/library//libppic.htm

Section 1.14

EPA, Development, Wash., D. C., Sept., 1998, www.epa.gov

Section 1.15

CCT Pollution Prevention, Dec. 19, 1995.

Section 1.16

Environmental Chemistry Process Laboratory, ecpl.chemistry.uch.gr/top.html

Section 1.17

Dasgupta, S., Lucas, R. E. B., Wheeler, D., Small Plants Pollution and Poverty:New Evidence from Brazil and Mexico, The World Bank Group, a Working Paper, 1998.

Section 1.18

Groenendijk, A. J., Plantwide Controllability and Flowsheet Structure of Complex Continuous Process Plants, OSPT ospt@ct.utwente.nl, 1996.

Section 1.19

Lobor, D. J., *J. Organizational Change Mgt.*, 11(1), 26-37, 1998, MCB University Press, 0953-4814.

Section 1.20

Satoh, Y., Soejima, T., Koga, J., Matsumoto, S., Homma, S., Sakamoto, M., Takansshi, Nammo, A., Computer Aided Process Flowsheet Design and Analysis System of Nuclear-

Fuel Reprocessing, *J. Nuclear Sci. Technol.*, 32(4), 357-368, 1995.

Section 1.21

Development of COMPAS, Computer-Aided Process Flowsheet Design and Analysis System, *J. Nuclear Sci. Technol.,* 32(4), 357-368, 1995.

Section 1.22

Design & Development of Computer-Based Clean Manufacturing : A Decision Tool for Industrial and Academic Use, Technology Reinvestment Project # 1051, NSF Grant #CIJ-9413104, 4/15/94-9/30/97, NJIT, MIT.

Section 1.23

Yi, J., Chah, S., Computer Aided Chemical Process Design for Pollution Prevention, Environmental Chemical Engineering Lab, School of Chemical Engineering, Seoul National University.

Section 1.24

MicrosoftExcel 7, Spreadsheet for Windows 95.

Section 1.25

P2TCP Pollution Prevention Tool for Continuous Processes.

Section 1.26

Clean Process Design Guidelines, es.epa.gov/ncerqa/cencitt/year5/process/process.html

Section 1.27

Singh, H., Zhu, X. X., Smith, R., Session 7 of AICHE Annual Meeting, 1998 Technical Program.

Section 1.28

Fritjof Capra's Foreword to Upsizing: The Road to Zero Emissions-More Jobs, More Income and No Pollution, ZERI Newsletter, Oct. 1998.

Section 1.29

ZERI Theory, zeri - org.

Section 1.30

Asher, W. J., SRI's Novel Chemical Reactor-PERMIX, 1998 huestis@mplvax.sri.com

Section 1.31

www.chemicalonline.com, Jegede, F., Process Simulation Widens the Appeal of Batch Chromatography, Chemical Online, N. Basta, Ed.

Section 1.32

http://www.epa.state/oh.us/opp/aboutopp.html

Section 1.33

Federal Register, Vol. 62, No. 120, Monday, June 23, 1997, Notices, pages 33868-33870, EPA, Notice of Availability of Waste Minimization Software and Documents.

Section 1.34

CLARIT web Image:530F95010 Env Fact Sheet http://www.epa.gov/cgi.bin/clariy.gov

Section 1.35

ASTDR Information Center/ATSDRIV@cdc.gov/ 188842-ATSDR or 1888-422-8737, 1999.

Section 1.36

OHSA Software/Expert Advisors, Occupational Safety & Health Admin., U.S. Dept. of Labor.

Section 1.37

EPA Headquarters-For Release Fri., Dec. 18,1998, EPA solicits grants for Public access to environmental monitoring, http://www.epa.gov/empact (application).

Section 1.38

Environmental Health Information Service, ehis.niehs.nih.gov/

Section 1.39

Analytical Chem. News and Features, A. C. S., Aug. 1, 528A-532A, 1998.

Section 1.40

Bumble, S., in *Clean Production,* Misra, K. B., Ed., Springer-Verlag, Berlin, 1996.

A. C. S., *Chemical Engineering In Medicine,* Advances In Chemistry 118, Washington, D.C., 1973.

Aho, A. V., Hopcroft, J. E., and Ullman, J. D., *The Design and Analysis of Computer Algorithms,* Addison-Wesley, Reading, PA, 1974.

Alfrey, T. Jr., and E. F. Gurney, *Dynamics of Viscoelastic Behavior in Rheology,* Vol.1, E. R. Eirich, Ed., Academic Press, New York, 1956.

Bumble, S., *Application of Order-Disorder Theory to Gas Adsorption,* Ph.D. Thesis, Purdue University, 1958.

Bumble, S. and J. M. Honig, *J. Chem. Phys.,* 33, 424, 1960.

Cheng, R. C. H., and G. Jones, Optimal Control of Systems with Markov Jump Disturbances, A Comparison of Exact and Approximate Solutions, in *Third International Mathematical Association Conference on Control Theory,* Marshall, J. E., W. D. Collins, C. J. Harris and D. H. Owens, Academic Press, London, 1981, 473.

Collins, W. D., Approximate Controllability of Multipass Systems Described by Linear Ordinary Differential Equations, in *Third International Mathematical Association Conference on Control Theory,* Marshall, J. E., W. D. Collins, C. J. Harris, and D. H. Owens, Academic Press, London, 1981, 685.

Frank, D., Control of Distributed Parameter Systems with Independent Linear and A Bilinear Modes, in *Third International Mathematical Conference on Control Theory,* Marshall, J. E. W. D. Collins, C. J. Harris and D. H. Owens, Academic Press, London, 1981, 827.

Gibson, J. E., *Nonlinear Automatic Control,* McGraw-Hill, New York, 1963.
The Toxicology Handbook, Principles Related to Hazardous Waste Site Investigations, ICAIR and PRC, for Office of Waste Programs, Enforcement, EPA.
Kauffman, S. A., *The Origins of Order,* Oxford University Press, London, 1993.
Karman, Y. V., M. A. Biot, *Mathematical Methods in Engineering,* McGraw-Hill, New York, 1940.
Lenard, R. X., Utilizing Low Grade Power Plant Waste Heat to Assist in Production of Commercial Quantities of Methane, page 671 of Vogt, W. G. and M. H. Mickle, Modeling and Simulation, Vol. 12, Part 2, Systems, Control and Computers, Proceedings of the Twelfth Pittsburgh Conference, April 30-May 1, 1981, School of Engineering, U. of Pittsburgh, Published and Distributed by the Instrument Soc. of America.
Lotka, A. J., *Elements of Mathematical Biology,* Dover Publications, New York, 1956.
Mah, R. S., *Chemical Process Structures and Information Flows,* Butterworths, London, 1989.
Moore, G. T., *Emerging Methods in Environmental Design and Planning,* M.I.T. Press, Cambridge, MA, 1968.
Nemhauser, G. L. and L. A. Wolsey, *Integer and Combinatorial Optimization,* John Wiley & Sons, New York.
Owens, D. H., *Multivariable and Optimal Systems,* Academic Press, London, 1981.
Papadimitriou, C. H. and K. Steiglitz, *Combinatorial Optimization: Algorithms and Complexity,* Prentice-Hall, Englewood Cliffs, NJ, 1982.
Pierre, D. A., *Optimization Theory with Applications,* Dover Publications, New York, 1969.
Poppinger, M. Optimization by Evolution on a Parallel Processor System, page 393 of Vogt, W. G., and M. H. Mickle, Modeling and Simulation, Vol. 12. Part 2, Systems and Computers, Proceedings of the Twelfth Pittsburgh Conference, April 30-May 1, 1981, School of Engineering, U. of Pittsburgh, Published and Distributed by the Instrument Soc. of America.
Reddick, H. W. and F. H. Miller, *Advanced Mathematics for Engineers,* 3rd ed., John Wiley & Sons, New York, 1938.
Reza, F. and S. Seely, *Modern Network Analysis,* McGraw-Hill, New York, 1959.
Rodiguin, N. M. and E. N. Rodiguina, *Consecutive Chemical Reactions, Mathematical Analysis and Development,* Van Nostrand Co., Princeton, NJ, 1964.
Saaty, Y. L., *Modern Nonlinear Equations,* Dover Publications, New York, 1981.
Saaty, T. L. and J. Bram, *Nonlinear Mathematics,* Dover Publications, New York, 1964.
Science Advisory Board to U.S. EPA, (A-101), 3AB-EC-90-021, Reducing Risk: Setting Priorities and Strategies for Environmental Protection, Sept. 1990.
Sethi, S. P., and G. C. Thompson, *Optimal Control Theory, Applications to Management Science,* Martinus Nijhoff Publishing Company, Boston, MA.
Soroka, W. W., *Analog Methods in Computation & Simulation,* McGraw-Hill, New York, 1940.
Thomas, R., Logical Versus Continuous Description of Systems Comprising Feedback Loops: The Relation Between Time Delays and Parameters in Chemical Applications of Topology and Graph Theory. A Collection of Papers from a Symposium at the University of Georgia, Athens, GA., 18-22 April, 1983, R. B. King, Ed., *Studies in Physical and Theoretical Chemistry,* Elsevier Publishers, Amsterdam, 28, 307-321, 1983.
Wist, A. O., J. A. McDowell and W. A. Ban, A Hybrid Computer System for Determination of Drug Dosage Regimens, Page 559 of Vogt, W. G. and M. H. Mickle, Modeling and Simulation,Vol. 12, Part2, Systems, Control, and Computers, Proceedings of the Twelfth Pittsburgh Conference, April 30-May 1, 1981, School of Engineering, University of Pittsburgh, Published and Distributed by the Instrument Society of America.
B. Crittendezi and S. Kolaczkowski, *Waste Minimisation Guide,* Institution of Chemical Engineers, London, 1994.
Waste minimisation: a route to profit and cleaner production. An interim report of the Aire and Calder project. Centre for Exploitation of Science and Technology, 1994.
R. A. Sheldon, *Chem.Tech.,* 24, 38, 1994.
K. G. Malle, in *Waste Minimisation: A Chemist's Approach,* K. Martin and T. W. Bastock, Eds., RSC, Cambridge, 1994, 35.
K. Smith, Chem. *Commun.,* 469, 1996.
M. Poliakoff and S. Howdle, *Chem. Br.,* 31, 118, 1995.
Encyclopedia of Chemical Technology Kirk-Othmer, 4th ed., vol 13, p 1048. Wiley, New York, 1995.
A. Mittelman and D. Lin, *Chem. Ind.,* September 1995, 694.
Chemistry of Waste Minimisation, J. H. Clark, Ed., Blackie Academic, Oxford, 1995.
J. F. Hayes and M. B. Mitchell, *Chem. Br.,* 29, 1037, 1993.
M. J. Braithwaite and C. L. Ketterman, *Chem. Br.,* 29, 1042, 1993.
I. G. Laing, in *Waste Minimisation: A Chemist's Approach,* K. Martin and T. W. Bastock, Eds., RSC, Cambridge, 1994, 93.
G. Steffan, Optimisation of classical processes and combination with modern reactions for the synthesis of fine chemicals. Chemspec '95, Essen.
Australia Centre for Cleaner Production, The Clean Advantage, p 1. February 1996.
P. T. Anastas and C. A. Farris (eds), Benign by design — alternative synthetic design for pollution prevention, ACS symposium series 577. ACS, Washington D.C., 1994.
K. Fischer and S. Hunig, *J. Org. Chem.,* 52, 564, 1987.
Design Expert, distributed by QD Consulting near Cambridge.

Section 1.41

Carmichael, H., info@hsl.gov.uk

Section 1.42

Hettige, H., Martin, P., Dingh, M., Wheeler, D., IPPS, The Industrial Pollution Projection System, Policy Research Working Paper WPS#1431, New Ideas in Pollution Regulation, 1994.

Section 2.1 through 2.2

Pierre, D. A., *Optimization Theory with Applications,* Dover Publications, New York, 1986.
Churchman, C. W., Ackoff, R. L., Arnoff, E. L. *Introduction to Operations Resarch,* John Wiley & Sons, New York, 1968.
Cooper., L., Bhat, U. N., LeBlanc, L. J., *Introduction to Operations Research Models,* W. B. Saunders, Philadelphia, PA, 1977.

Section 2.3

Pamadimitriou, C. H. and Steiglitz, K., *Combinatorial Optimization: Algorithms and Complexity,* Prentice-Hall, Englewood Cliffs, NJ, 1982.
Nemhauser, G. L., and Wolsey, L. A., *Integer and Combinatorial Optimization,* John Wiley & Sons, New York, 1989.

Section 2.4 through 2.17

Computer-Assisted Molecular Design (CAMD), http://http1.brunel.ac.uk:8080/depts/chem/ch2415/refer.htm

Pierre, D. A., *Optimization Theory with Applications,* Dover Publications, New York, 1986.

Goicoechea, A., Hansen, D. R., Duckstein, L., *Multiobjective Decision Analysis with Engineering and Business Applications,* John Wiley & Sons, New York, 1982.

Nemhauser, G. L., and Wolsey, L. A., *Integer and Combinatorial Optimization,* John Wiley & Sons, New York, 1989.

Owens, D. H., *Multivariable and Optimal Systems,* Academic Press, London, 1981.

Mah R. S., *Chemical Process Structures and Information Flows,* Butterworths, London, 1989.

Rashevsky, N., Mathematical Biophysics Physico-Mathematical Foundations of Biology, Vol. 2, Dover, NY, 1960

Rice U. and Rice's Computer Information Tech. Institute, the Center for Research on Parallel Computation (CRPC); Digital Equip. Corp. And the Keck Foundation, as part of Rice's W. M. Keck Center for Computational Discrete Optimization.

Section 2.18

http://www.cs.sandia.gov/opt/survey/intro.html

http://www.cs.sandia.gov/opt/survey/madr.html

Section 2.19

Glen, R. C., Payne, A. W. R., A Genetic Algorithm for the Automated Generation of Molecules Within Constraints, *J. Computer-Aided Molecular Design*, 181-202, 1995.

Section 2.20

Molecular Phylogenetic Studies, life.anu.edu.au/~weiller/wmg/wmg.html

Section 2.21

Design Synthesis Using Adaptive Search Techniques and Multi-Criteria Decision Analyais, www.cs.york.ac.uk/~mark/mndp/mndp.html

Section 2.22

Prof. J. Gasteiger, Research, Computer-Chemie-Centrum, Erlangen.

Dr. F. Friedler, Head of the Computer Science Dept., Vesprem, Hungary.

Section 2.23

Optimization of Chemical Processes for Waste Minimization and Pollution Prevention, pprc.pnl.gov/pprc/statefnd/gulfcoas/optimiz.html

Section 2.24

Multisimplex Electronic Newsletter. 12/8/97, webmaster @ multisimplex.com

http://www.multisimplex.com

Section 2.25

www.aps.org/meet/CENT99/baps/abs/G7755012.html

Section 2.26

http://www.daimi.au.dk/PetriNets/

Section 2.27

R. Srinnivasan, V. Venkata Subramanian, Laboratory for Intelligent Process Systems, School of Chemical Engineering, Purdue University, West Lafayette, Indiana, Dec. 8, 1996.

Section 2.28

Los Alamos Nonlinear Adaptive Computation, X windows, the X Division newsletter, Summer, 1993.

Section 2.29

R. Benares-Alcantara, J. M. P. King, G. H. Ballinger, Dept. Of Chem. Eng., U. of Edinburgh, Scotland, U.K., June, 1995.

Section 2.30

http://www.chem.eng.ed.ac.uk/ecosse/kbds/cp3/nodeb.html

Egide: A Design Support System for Conceptual Chem. Proc. Design.

Section 2.31

Sandia National Laboratories, Albuquerque, N, M,. Interactive Collaborative Environments, 1/23/95-6/24/98.

Section 2.32

Rapid Data: Control Kit, Harmonic Software, Inc.

Section 2.33

Radecki, P., (CenCITT), Baker, J., The Clean Process Advisory System: Building Pollution Prevention Into Design, CWRT, CenCITT, NCNS, Envirosense.

Section 2.34

Energy Systems Standards/Requirements Identification, http://www.bechteljacobs.com/pqa/compliance/~sproject/smrfg~16wm.htm

Section 2.35

es.epa.gov/ncerqa_abstracts/centerscenc...lean/barna.html

Section 2.36

Global Bytes, Chembytes, Chemistry in Britain, Dennis Rouvray.

Section 3.1

Minns, D., Zaks, D., Pollution Prevention Using Chemical Process Simulation, NRC-Institute for Chemical Processes and Environmental Technology Computer Modeling and Simulation, 1998.

Section 3.2

Hendrikson, C., Conway-Schempf, N., McMichael, F., Introduction to Green Design, Green Design Initiative, Carnegie Mellon University, Pittsburgh, PA.

Section 3.3

http://syssrvq.nrel.gov/st~it.html, NREL Research and Technology: Industrial Technologies.

Section 3.4

http://www.chemweek.com/marketplace/links/simsci.html

Sowa, C. J., Explore Waste Minimization via Process Simulation, CEP, No. (11), 40-42, 1994.

Section 3.5

EPA/NSF Partnership for Environmental Research, Technology for Sustainable Environment, Interagency Announcement of Opportunity, National Center for Environmental Research and Quality Assurance, ORD, US EPA, Opening Date, No. 18, 1997.

Section 3.6

Hypotech, Calgary, Canada.

Section 3.7

Varga, J. B., Fan, L. T., Risk Reduction Through Waste Minimizing Process Synthesis, 21st Annual RREL Research Symposium, Cincinnati, OH, 1995.

Friedler, F., Tarjan, K. Huang,Y. W., Combinatorial Algorithms for Process Synthesis, *Computers Chem. Eng.,* 16, S1-S548, 1992.

Friedler, F., Varga, J. B., Fan, L. T., Decision-Mapping for Design and Synthesis of Chemical Processes: Application to Reactor-Network Synthesis, AIChE Symposium Series No. 304 Volume 91, 1995.

Friedler, F., K. Tarjan, Y. W. Huang, L. T. Fan, Computer-Aided Waste Minimizing Design of a Chemical Process, Seventh Annual Conferenceon Hazardous Waste Research, Boulder, CO, June 1-2, 1992.

Kovacs, Z., F. Friedler, L. T. Fan, Algorithmic Generation of the Mathematical Model for Separation Synthesis, European Symposium on Computer Aided Process Engineering-3, Escape-3, 5-7 July 1993 Graz, Austria.

Friedler, F., Z. Kovacs, L. T. Fan, Unique Separation Networks for Improved Waste Elimination, Emerging Technologies for Hazardous Waste Management, A.C.S., Atlanta, GA, 1992.

Friedler, F., K. Tarjan, Y. W. Huang, L. T, Fan, Graph-Theoretic Approach to Process Synthesis: Polynomial Algorithm for Maximal Structure Generation, *Computers Chem. Eng.* 17(9), 929-942, 1993.

Hangos, K. M., F. Friedler, J. B. Varga, L. T. Fan, A Graph-Theoretic Approach to Integrated Process and Control System Synthesis, Presented at IFAC Workshop on Integration of Process Design and Control, Baltimore, MD, June 27-28, 1994.

Imreh, B., F. Friedler, L. T. Fan, An Algorithm for Improved Bounding Procedure in Solving Process Network Synthesis by a Branch-and-Bound Method, I, *Developments in Global Optimization,* M. Bomze et al., Eds., Kluwer Academic Publishers, Netherlands, 1997, 315-348.

Friedler, F., K. Tarjan, Y. W. Huang, L. T. Fan, Graph-Theoretic Approach to Process Synthesis: Axioms and Theorems, *Chem. Eng. Sci.,* 47(8), 1973-1988, 1992.

Varga, J. B., F. Friedler, L. T. Fan, Parallelization of the Accelerated Branch-and Bound Algorithm of Process Synthesis: Application in Total Flowsheet Synthesis, Acta Chimica Slovenica, 42/1/1995, pp. 15-20.

Friedler, F., K. Tarjan, Y. W. Huang, L. T. Fan, Combinatorial Foundation of Process Synthesis, 42nd Canadian Chemical Engineering Conference, Toronto, Ontario, Canada, Oct. 18-21, 1992.

Kovacs, Z., F. Friedler, L. T. Fan, Recycling in a Separation Process Structure, *AIChE J.,* 39(6), 1087, 1993.

Kovacs, Z., F. Friedler, L. T. Fan, Parametric Study of Separation Network Synthesis: Extreme Properties of Optimal Structures, *Computers Chem. Eng.,* 19, S107-S112, 1995.

Friedler, F., J. B.Varga, L. T. Fan, Decision_mapping: A Tool for Consistent and Complete Decisions in Process Synthesis, *Chem. Eng. Sci.,* 50(11), 1775-1768, 1995.

Personal transmission from Dr. L. T. Fan. An early Flowchart for APSCOT (Automatic Process Synthesis with Combinatorial Technique).

Friedler, F., K. Tarjan, Y. W. Huang, L. T. Fan, Graph-Theoretic Approach to Process Synthesis: Application to Waste Minimization, a preprint, July 21, 1990.

Section 3.8

T. J. Willman, President, EPCON, International Mathematical Software for Total Flowsheet Software, Flowsheet Synthesis, 1997 Vaaler Award Winner, The newsletter of EPCON International, Fall, 1997.

Section 3.9

Aspen Tech Aspen Custom Modeler, http://www.aspentech.com/pspsd/modeler.htm

Section 3.10

Hogg, T. draft paper for sixth Forsight Conference on Molecular Nanotechnology, http://Forsight.org/Conferences/MNT6/Papers/Hogg/ghindex.htm

Section 3.11

University of New Mexico, Biology 576: Landscape Ecology & Macroscopic Dynamics.

Self-Organizing Systems, E. H. Decker & B. T. Milne.

Section 3.12

El-Hawagi, M. M., Spriggs, H. D., Mass Integration: Now a Comprehensive Methodology for Integrated Process design, Version Edited, 1/15/97, *Chem. Eng. Prog.,* in press.

Section 3.13

1999 Spring Technical Program AICHE, Process Design for Pollution Prevention II, Wilson, S., Manousiouthakis, V., Minimum Utility Cost for Non-IdealMulticomponent Mass Exchange Networks, http://www.aiche/meetapp/programming/techprogram/sessions/T2008.htm

Section 3.14

El-Halagi, M., *Pollution Prevention Through Process Integration: Systematic Design Tools,* Academic Press, New York, 1997.

Section 3.15

Stowers, M. A., Lesniewski, T. K., Manousiouthakis, V., Pollution Prevention by Reactor Network, AICHE, Annual Meeting, 1998.

Section 3.16

A General Chemical Kinetics & Sensitivity Analysis Code for Gas-Phase Reactions, Radhakrishnan, K., Bittker, D. L., Lewis Research Center, PRCM Poster, http://www.osc.edu/pcrm/Marek.html

Section 3.17

Kee, R. J., Miller, J. A., Jefferson, T. H., Chemkin-A General Purpose, Problem-Independent, Transportable, Fortran Chemical Kinetics Code Passage, SAND80-8003.

Section 3.18

Bumble, S., Emerging Computer Simulation and Control of Plant Design and Retro-Design of Waste Minimization/Pollution Prevention in the Late Twentieth and Early Twenty First Centuries inEPA Region III Waste Minimization/Pollution Prevention Technical Conference for Hazardous Waste Generators, Philadelphia, PA, June 3-5, 1996.

Crittendezi and S. Kolaczkowski, *Waste Minimisation Guide,* Institution of Chemical Engineers, London, 1994.

Waste minimisation: a route to profit and cleaner production. An interim report of the Aire and Calder project. Centre for Exploitation of Science and Technology, 1994.

R. A. Sheldon, *Chem.Tech.,* 24, 38, 1994.

K. G. Malle, in *Waste Minimisation: A Chemist's Approach,* K. Martin and T. W. Bastock, Eds., RSC, Cambridge, 1994, 35.

K. Smith, Chem. *Commun.*, 469, 1996.

M. Poliakoff and S. Howdle, *Chem. Br.,* 31, 118, 1995.

Encyclopedia of Chemical Technology Kirk-Othmer, 4th ed., vol 13, p 1048. Wiley, New York, 1995.

A. Mittelman and D. Lin, *Chem. Ind.,* 694, September 1995.

Chemistry of Waste Minimisation, J. H. Clark, Ed., Blackie Academic, Oxford, 1995.

J. F. Hayes and M. B. Mitchell, *Chem. Br.,* 29, 1037, 1993.

M. J. Braithwaite and C. L. Ketterman, *Chem. Br.,* 29, 1042, 1993.

I. G. Laing, in *Waste Minimisation: A Chemist's Approach,* K. Martin and T. W. Bastock, Eds., RSC, Cambridge, 1994, 93.

G. Steffan, Optimisation of classical processes and combination with modern reactions for the synthesis of fine chemicals. Chemspec '95, Essen.

Australia Centre for Cleaner Production, The Clean Advantage, p 1. February 1996.

P. T. Anastas and C. A. Farris, Eds., Benign by design — alternative synthetic design for pollution prevention, ACS symposium series 577. ACS, Washington D.C., 1994.

K. Fischer and S. Hunig, *J. Org. Chem.,* 52, 564, 1987.

Design Expert, distributed by QD Consulting near Cambridge.

Section 3.19

Chang, C. T., Huang, J. R., Multiobjective Programming Approach to Waste Minimization in the Utility Systems of Chemical Processes, *Chem. Eng. Sci.,* 15(16), 3951-65, 1996.

Section 3.20

Friedler, F., Varga, J. B., Fan, L. T., Algorithmic Approach to the Integration of Total Flowsheet Synthesis and Waste Minimization, *AICHE Symp. Ser.*, 90(303), 88, 1995.

Varga, J. B., Friedler, L. T., Risk Reduction Through Waste Minimizing Process Synthesis, 21st Annual RREL Research Symposium, Cincinnati, 1995.

Section 3.21

Ianni, J., New Powerful Kinetics Program: KINTECUS (Dec, 20, 95).

http://www.cpma.u-psud.fr/ccl/244.htm

http://www.ioc.ac.ru/chemistry/soft/kintecus.html

http://www.coma.u-psud.fr/ccl/243.html

Section 3.22

Stategic Waste Minimization Initiative.

UDRI Software, Version 2.0, 4/3/97.

Section 3.23

Compute Software-Chemical Engineering-SUPERPRO, superset of BatchPro, EnviroPro and BioPro Designer, www.chempute.com

Section 3.24

What is available from P2 by Design.

Design for Environment (DFE) for DOE Home Page, Pollution Prevention by Design.

Section 3.25

CH2M Hill, The National Center for Clean Industrial and Treatment Technologies (CenCITT), http://www.cpas.mtu.edu/tools/tooo7.htm

Section 3.26

Webmaster@olisystems.com, 1997-1998.

Section 3.27

webmaster@olisystems.com, 1997-1998.

Section 3.28

Syngen Web Page, http://syngen2.chem.brandeis.edu/syngen.html

Section 3.29

Sohn, J., Reklaitis. G. V., Okes, M. R., Annual Meeting, AICHE, 1997.

Section 3.30

Life Cycle Analysis (LCA) and its role in Product and Process Development, 2(2), 13, 1993, J. A. Isaacs, IJECDM Abstract, J. P. Clark, Archive.

Section3.31

Computer Modeling of Environmental Systems, http://home.istar.ca/~ece/model.html

Section 3.32

Hopper, J. R. Yaws, C. L., Lamar U., Funder Cantacts Dobbs, R., Primary Funder, Gulf Coast Hazardous Substance Research Center.

Section 3.33

Glen, R. C., Payne, A. W. R., A Genetic Aldorithm for the Automated Generation of Molecules Within Constraints, *J. Computer-Aided Molecular Design,* 9, 181-202, 1995.

Section 3.34

White, W. B., S. M. Johnson, and C. B. Dantzig: Chemical Equilibrium in Complex Mixtures, *J. Chem. Phys.,* 28, 751-755, 1958.

Section 4.1

Cramer, R. D., J. Am. Chem. Soc., 102:6, pages 1837 and 1843, 3/12/80.

Cramer, R. D., Quant.Struct.-Act. Relat. 2, 7-12, 1983.

Section 4.2

Bumble, S., Emerging Computer Simulation and Control of Plant Design and Retro-Design of Waste Minimization/Pollution Prevention in the Late Twentieth and Early Twenty First Centuries in EPA Region III Waste Minimization/Pollution Prevention Technical Conference for Hazardous Waste Generators, Philadelphia, PA, June 3-5, 1996.

Section 4.3

Bumble, S., Emerging Computer Simulation and Control of Plant Design and Retro-Design of Waste Minimization/Pollution Prevention in the Late Twentieth and Early Twenty First Centuries in EPA Region III Waste Minimization/Pollution Prevention Technical Conference for Hazardous Waste Generators, Philadelphia, PA, June 3-5, 1996.

Section 4.4

EDF, Roe, D., Pease, W., Florini, K., Sibergeld, E., Summer, 1997, www.edf.org., EDF@edf.org

Section 4.5

Bumble, S., Emerging Computer Simulation and Control of Plant Design and Retro-Design of Waste Minimization/Pollution Prevention in the Late Twentieth and Early Twenty First Centuries in EPA Region III Waste Minimization/Pollution Prevention Technical Conference for Hazardous Waste Generators, Philadelphia, PA, June 3-5, 1996.

Section 4.6

Hart, Terence, Peptide Therapeutics, 1998.

Section 4.7

Byrne, Miriam, Imperial College for Environmental Technology, 1998.

Section 4.8

User's Guide, Borland Quattro Pro for Windows, Version 5.0, Houghton Mifflin, Scotts Valley, CA, 1991.

Section 4.9

http://http1.brunel.ac.uk:8080/depts/chem/ch2415/refer.htm

Section 4.10

Ebeling, H. O., Lyddor, L. G., Covington, K. K., Proceedings of the 77th GPA Annual Conference, Gas Association, Tulsa, OK, 1998.

Section 4.11

Texaco Chemical Company Plans to Reduce HAP Emissions Through Early Reduction Program By Vent Recovery System, Envirosense, Case Study: # 170, Texaco Chemical Co., Port Neches, TX.

Section 4.12

Friedler, F., Fan, L. T., Design of Molecules With Desired Properties by Combinatorial Analysis, 1997, Preprint.

Section 4.13

Friedler, F., Fan, L. T., Design of Molecules With Desired Properties by Combinatorial Analysis, 1997, Preprint.

Section 4.14

Globus, A., Lawton, J., Wipke, T., Automatic Molecular Design Using Evolutionary Techniques, draft paper for the Sixth Foresight Conference on Molecular Nanotechnology, final version submitted for publication in the special Conference issue on Nanotechnology. http://science.nas.nasa.gov/globus/home.html

Section 4.15

Friedler, F., Varga, J. B., Fan, L. T., Algorithmic Approach to the Integration of Total Flowsheet Synthesis and Waste Minimization, Pollution Prevention via Process and product modifications, AICHE Symposium Series, 90(303), 86.

Section 4.16

Testmart Project to Promote Faster, Cheaper, More Humane Lab Tests, Academic Environmental Experts Awarded

Joint Grant by Vira I. Heinz Endowment, Feb. 24, 1999, http://www.edf.org/pubs/NewsReleases/1998/Oct/b_cma.html

Section 4.17

Overcash, M., Dept. Chem Eng., overcash@che.ncsu.edu http://www.sfo.com/naer (volume 1, number 1).

Section 4.18

Cleaner Synthesis by Tim Lester, http://www.chemsoc.org/gateway/chembyte/cib/lester.htm

Section 4.19

Ritter, E. R., THERM User's Manual, Department of Chemical Engineering and Environmental Science, NJIT, 1980.

Section 4.20

Thurston, D., Product and Process Design Tradeoffs for Pollution Prevention, Pafific NW Pollution Prevention Research Center, 1996.

Section 4.21

Environmental Simulation Programs (ESP), http://www.olisystems.com/oliesp.htm

Section 4.22

Thurston, D., Product and Process Design Tradeoffs for Pollution Prevention, Pacific NW Pollution Prevention Research Center, 1996.

Section 4.23

Department of Energy Design for Environment (DfE) Publications.

Section 4.24

The Toxic Substances Control Act, http://es.epa.gov/oeca/ore/tped/tscatp.html

Section 4.26

Boyd, James, RFF 98-30.

Section 4.26

NRC's Institute for Chemical Process and Environmental Technology (ICPET), Dr. David Minns, www.icpet.nrc.ca/projects/simu.html

Section 4.27

Chemical Process Simulation for Waste Reduction, pprc.pnl.gov/pprc/rpd/fedfund/epa/epastd/chemproc.html

Section 4.28

SRI's Consulting Process Economics Program, http://process-economics.com

Section 4.29

American Process, http://apiweb.com/pinchtech.htm

Section 4.30

Reichhardt, T., Environmental G. I. S.: The World in a Computer., *Environmental Sci. Tech.,* Aug., 1996.

Section 4.31

Reible, D. D., *Fundamentals of Environmental Engineering,* Lewis Publishers, Boca Raton, FL, 1999.

Section 4.32

Mills, K., Griffith, C., Health: The Scorecard That Hit a Home Run, EDF (www.edf.org), 1999.

Section 4.33

www.plg-ec.com/riskman.htm, PLG, Inc. Risk Managementand Process Safety.

Section 4.34

Kletz, Trevor, Safer by Design, Chemistry in Britain, Jan. 1999. 64 Twining Brook Rd., Cheadle Hulme, Cheadle, Chesshire, U.K.

Section 4.35

Thurston, D. L., Carnahan, J. V., Hazardous Waste, Research and Information Center, 7/31/95.

Section 5.1

Fowler, R. H., *Statistical Mechanics,* Cambridge University Press, 1966.

Section 5.2

Carslaw, N., Jacobs, P., Piling, M., Atmospheric Research Group in the School of Chemistry at Leeds University, U.K.

Section 5.3

Blurock, E. S., Reaction: Modeling Complex Reaction Mechanisms, Methods of Computer Aided Synthesis, Johannes University, Research Institute for Symbolic Computation, 1995.

Section 5.4

http://www.c-f-c.com/supportdocs/cl2recycle.htm

Section 5.5

Hendrikson, J. B., Chem. Tech., Sept.98, (2819), 35-40, ACS, Teaching Alternative Synthesis. The Syngen Program, in Green Chem: Designing Chem. For the Enviromental FDAT.

Section 5.6

Anastas, P. T., Williamson, T. C., *Am. Chem. Soc.*, 214-231, Wash. D.C.

Section 5.7

Technology, The New York Times, Software Simulations Lead to Better Assembly Lines, Claudia H. Deutsch, 1999.

Section 5.8

Rice, S., Gray, P., *The Statistical Mechanics of Simple Liquids,* Interscience Publishers, New York, 1965.

Section 5.9

Rainville, E. D., *Special Functions,* The MacMillan Company, New York, 1960.

Section 5.10

J. Hijmans and J. de Boer, *Physica* 21, 471, 485, 499, 1955.

S. Bumble and J. M. Honig, *J. Chem Phys.* 33, 424, 1960.

S. Bumble, *Reducing Risk by Controlling the Environment in Clean Production,* K.B. Misra, Ed., Springer, New York, 1996.

E. O. Talbott, M. Arnowitt, N. Davis, K. P. McHugh, Cancer Incidence in the Neville Island Area: 1990-1994, Data from the Pennsylvania Cancer Registry, Graduate School of Public Health. University of Pittsburgh, Pittsburgh, PA and Clean Water Action, Pittsburgh, PA.

Kikuchi, R., *Phys. Rev.,* 81, 988, 1951.

Magnussen, Ramussen & Fredenslund, Copyright, A. C. S., used by permission, 1981.

Bumble, S., Emerging Computer Simulation and Control of Plant Design and RetroDesign of Waste Minimization/ Pollution Prevention in the Late Twentieth and Early Twenty-First Centuries, Proceedings Manual, EPA, Region III, Waste Minimization/Pollution Prevention Technical Conference, Philadelphia, PA, June 3-5, 1996.

Kauffman, S. A., *The Origins of Order,* Oxford, 1993.

Lehninger, A. L., *Biochemistry,* Worth Publishers, New York, 1972.

Ling, G. N., *A Physical Theory of the Living State: The Association-Induction Hypothesis,* Blaisdell Publishing Company, New York, 1962.

Section 5.11

What Chemical Engineers Can Learn From Mother Nature, CEP, AICHE, P. 67, 1998.

Section 5.12

Design Synthesis Using Adaptive Search Techniques and Multicriteria Decision Analysis.

Section 5.13

Feynman, R. P., Statistical Mechanics, A Set of Lectures, W. A. Benjamin, Reading, MA.

Kikuchi, R., The Path Probability Method, *Progress of Theoretical Physics,* Supplement No. 35, 1966.

Section 5.14

Fowler, R. H., *Statistical Mechanics,* Cambridge University Press, 1966.

Section 5.15

Risk Reduction Engineering Lab/P2 Research Branch (RREL/ PPRB), http://es.epa.gov/techinfo/research/cp11949.html

Section 5.16

http://www.synthworks.com/, http://www.x-tekcorp.com/index.htm, http://mikro.e-technik.uni-ulm.de/vhd/anl-engl.vhd/htm/

End Notes

Friedler, F., Tarjan, K., Huang, Y., Fan, L. T., Graph-Theoretic Approach to Process Synthesis: Axioms and Theorems, *Chem. Eng. Sci.,* 47, 1973-1988, 1992.

Friedler, F., Tarjan, K., Huang, Y. W., Fan, L. T., Graph-Theoretic Approach to Process Synthesis: Polynomial Algorithm for Maximum Structure Generation. *Comput. Chem. Eng.* 17, 929-942, 1993.

Friedler, F., Varga, J. B., Huang, Y. W., Fan, L. T., Decision Mapping: A Tool for Consistent and Complete Decisions in Process Synthesis. *Chem. Eng. Sci.,* 50, 1755-1768, 1995.

Friedler, F., Varga, J. B., Feher, E., Fan, L. T., Combinatorially Accelerated Branch-and Bound Method for Solving the MIP Model of Process Network Synthesis. In *The State of the Art in Global Optimization,* C. A. Floudas and M. Pardalos, Eds., Kluwer Academic Publishers, the Netherlands, 1996, 609-626.

Grossman, I. E., Sargent, R. W. H., Optimum Design of Heat Exchanger Networks, *Comput. Chem. Eng.* 2, 1-7, 1978.

Kondili, E., Pantelides, C. C., Sargent, R. W. H. A General Algorithm for Short-term scheduling of Batch Operations-I: MILP Formulation. *Comput Chem. Eng.,* 17, 211-227, 1993.

Safrit, B. T., Westerberg, A. W., Synthesis of Azeotropic Bacth Distillation Separation Systems. *Ind. Eng. Chem. Res.* 36, 1841-1854, 1997.

Sargent, R. W. H. A Functional Approach to Process Synthesis and its Application to Distillation Systems. A Report of the Center for Process Systems Engineering, Imperial College of Science, Technology and Medicine, 1994.

Sargent, R, W. H. A Functional Approach to Process Synthesis and Its Application to Distillation Systems. *Comput. Chem. Eng.,* 22, 31-45, 1988.

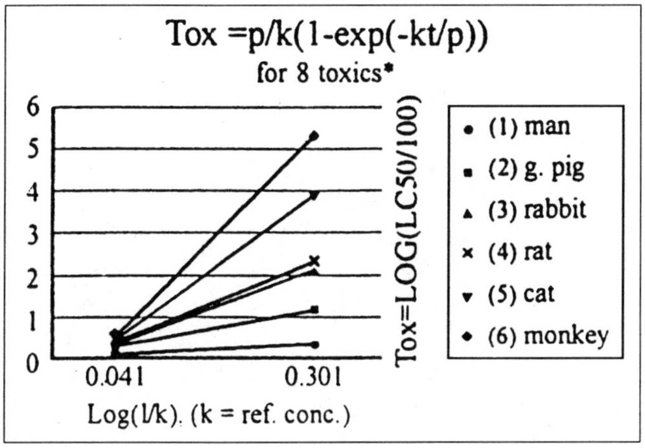

FIGURE 1 Toxicity vs. log (reference concentration).

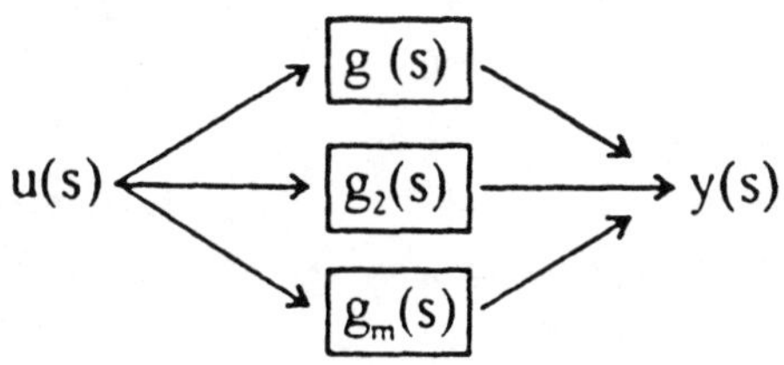

FIGURE 2 Parallel control.

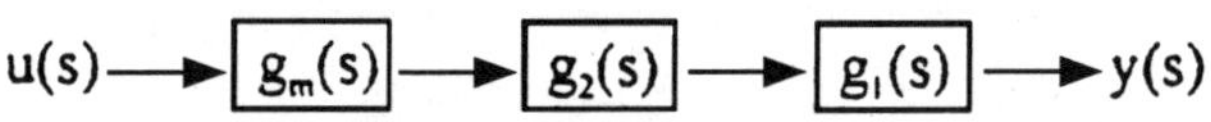

FIGURE 3 Series control.

FIGURE 4 Feedback control.

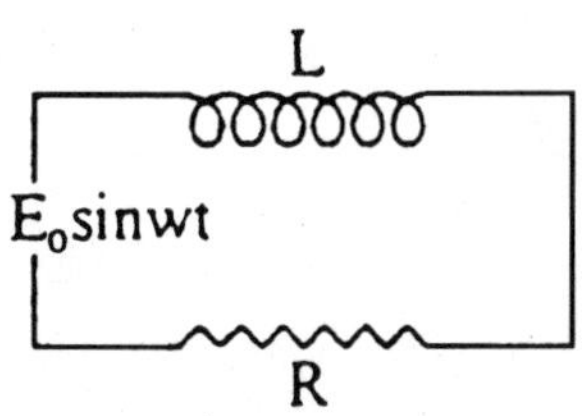

FIGURE 5 A simple series circuit.

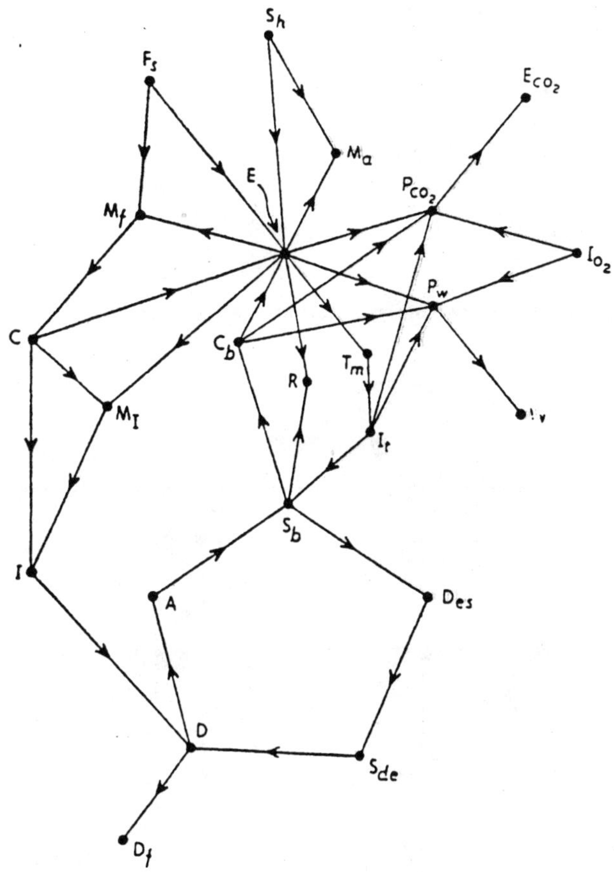

FIGURE 6 The Feeding Mechanism

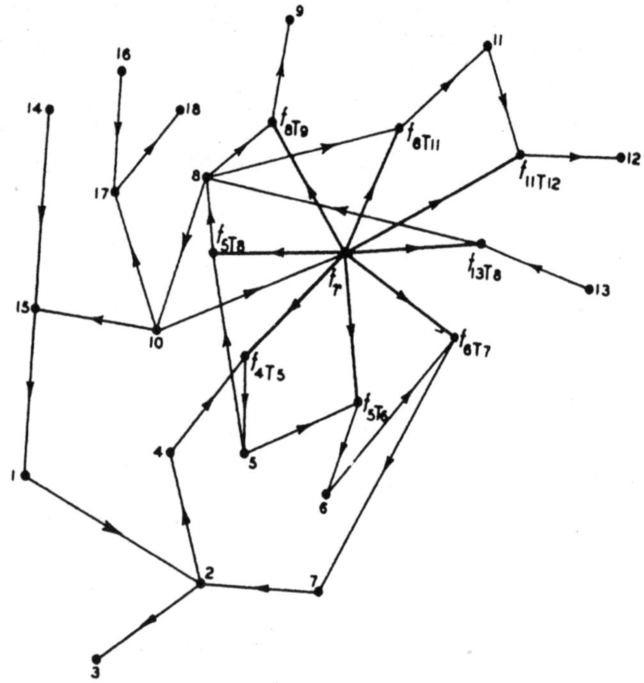

FIGURE 7 Organisms and graphs.

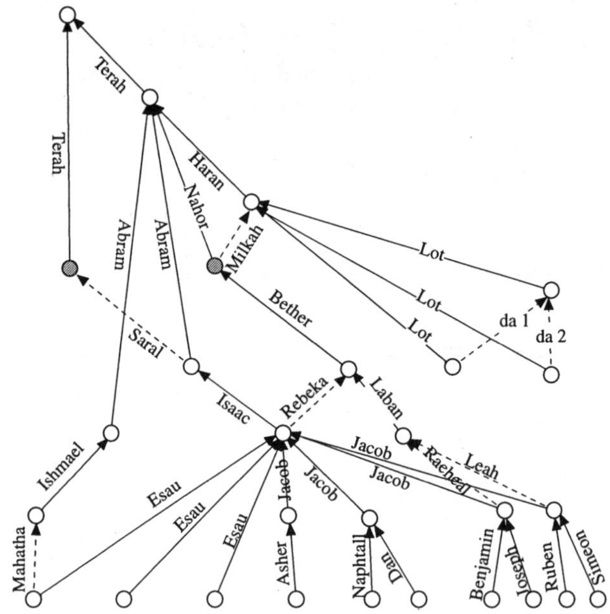

FIGURE 8 P-graph of Canaan geneology made by Papek program.

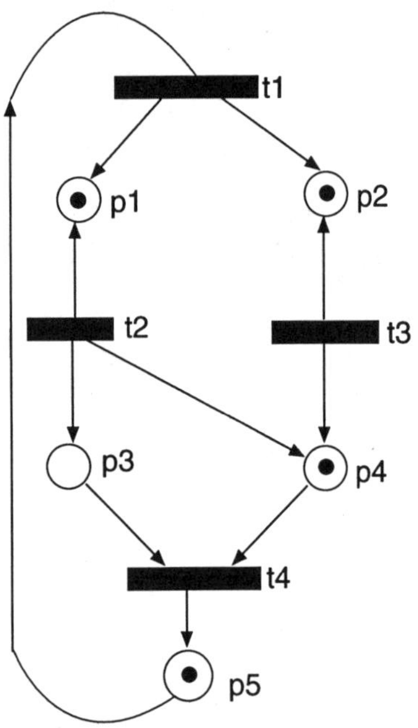

FIGURE 9 Example and matrix representation of Petri net.

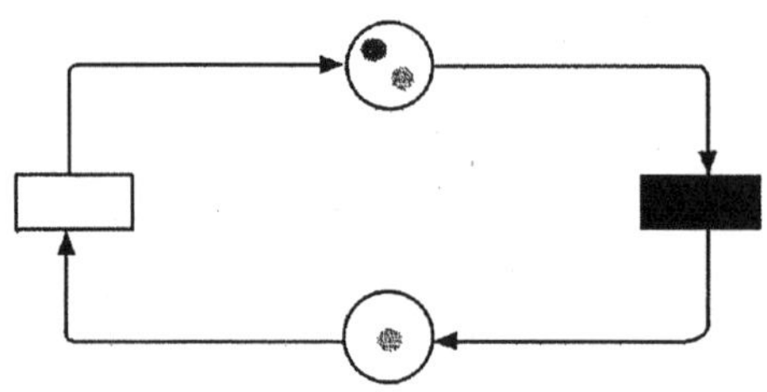

FIGURE 10 Petri nets.

$$
\begin{aligned}
H(s) &= \frac{100(s+1)(s+2)}{s(s+0.3)(s+0.7-j2)(s+0.7+j2)(s+15)} \\
&= \frac{100(s+1)(s+2)}{s(s+0.3)(s^2+1.4s+4.49)(s+15)} \\
&= \frac{(100s^2+300s+200)}{s(s^4+16.7s^3+30.41s^2+74.997s+20.205)}
\end{aligned}
$$

$$
\begin{aligned}
H(s) &= \frac{9.8985(s+1)\left(\frac{2}{2}+1\right)}{s\left(\frac{s}{0.3}+1\right)\left[\left(\frac{s}{2.119}\right)^2+0.6607\left(\frac{s}{2.119}\right)+1\right]\left(\frac{s}{15}+1\right)} \\
&= \frac{9.8985(0.5s^2+1.5s+1)}{s(0.0496s^4+0.8265s^3+1.5051s^2+3.7118s+1)}
\end{aligned}
$$

FIGURE 11 Ratio of s in two transfer functions.

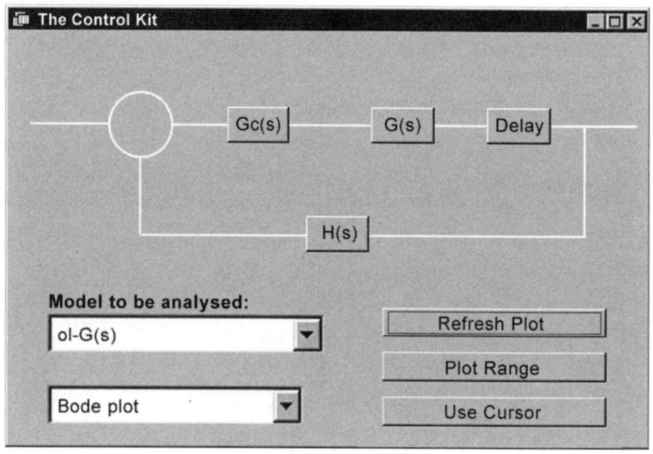

FIGURE 12 The Control Kit.

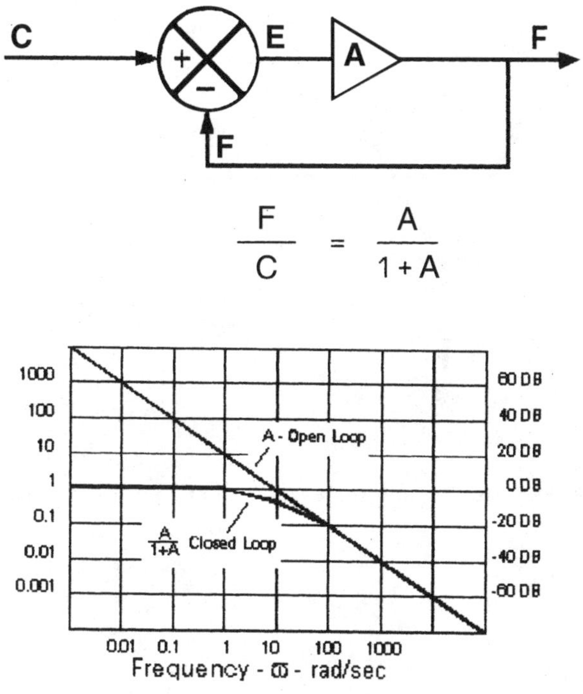

FIGURE 13 The bode diagram.

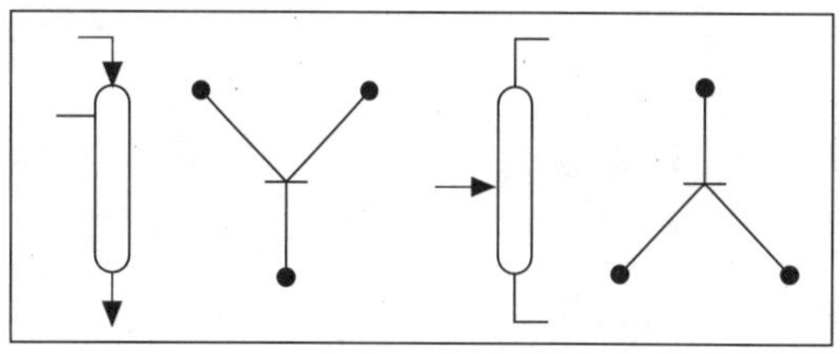

FIGURE 14 Conventional and P-graph representations of a reactor and a distillation column.

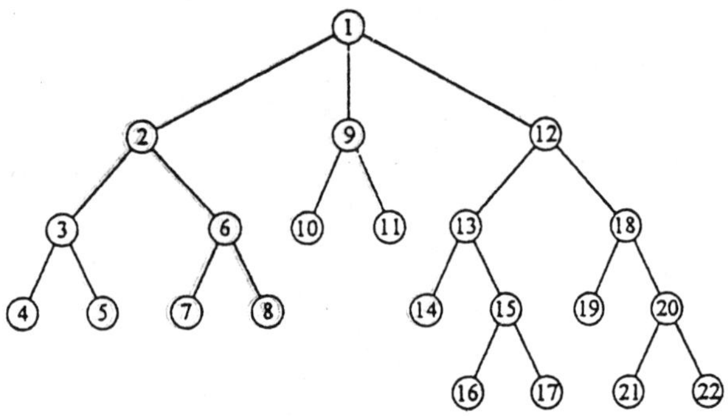

FIGURE 15 Tree for accelerated branch-and-bound search for optimal process structure with integrated in plant waste treatment (worst case).

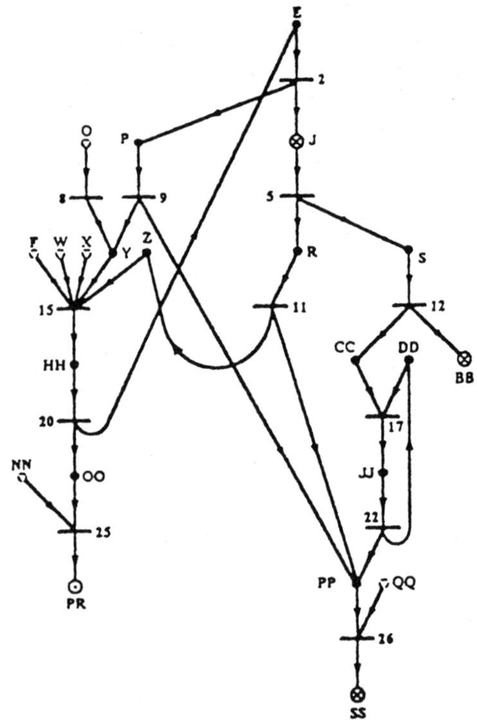

FIGURE 16 Optimally synthesized process integrating in-plant treatment.

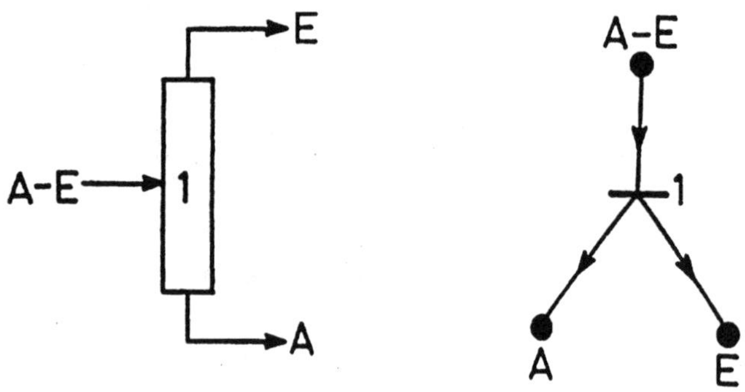

FIGURE 17 Conventional and P-graph representations of a separation process.

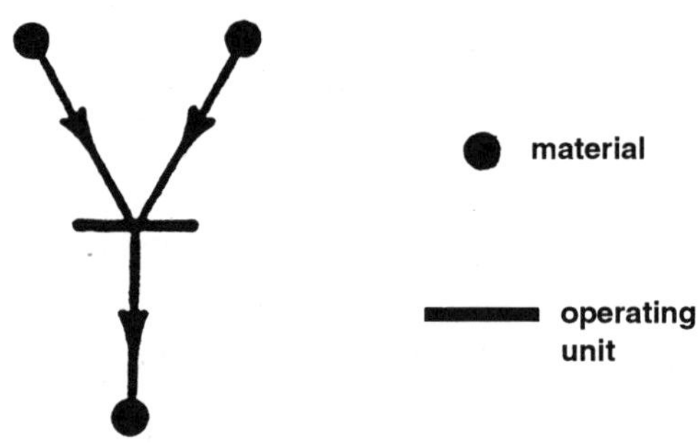

FIGURE 18 P-graph representation of a simple process.

Product: J
Raw Materials: A, B, E, F
Toxic waste: C
Reactions:

A + B —>	C + D	(1)
E + F —>	G + H	(2)
D + G —>	I + J	(3)

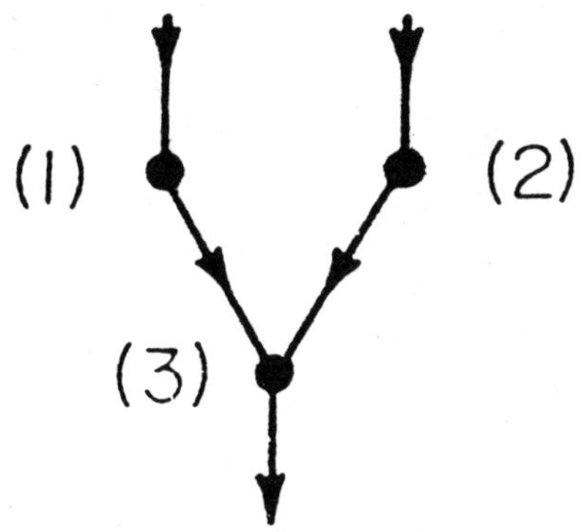

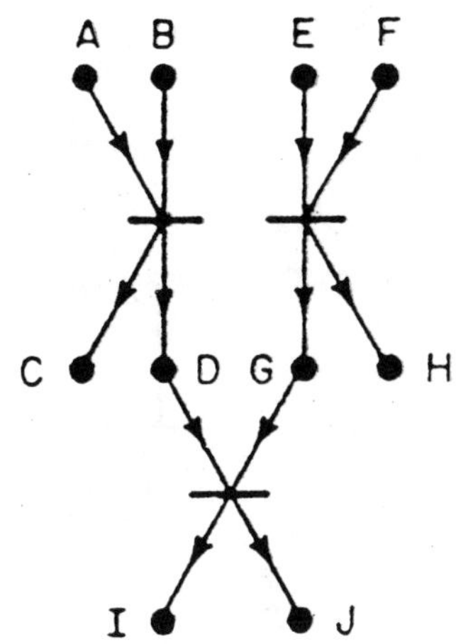

FIGURE 19 Representation of Separator: a) conventional, b) graph

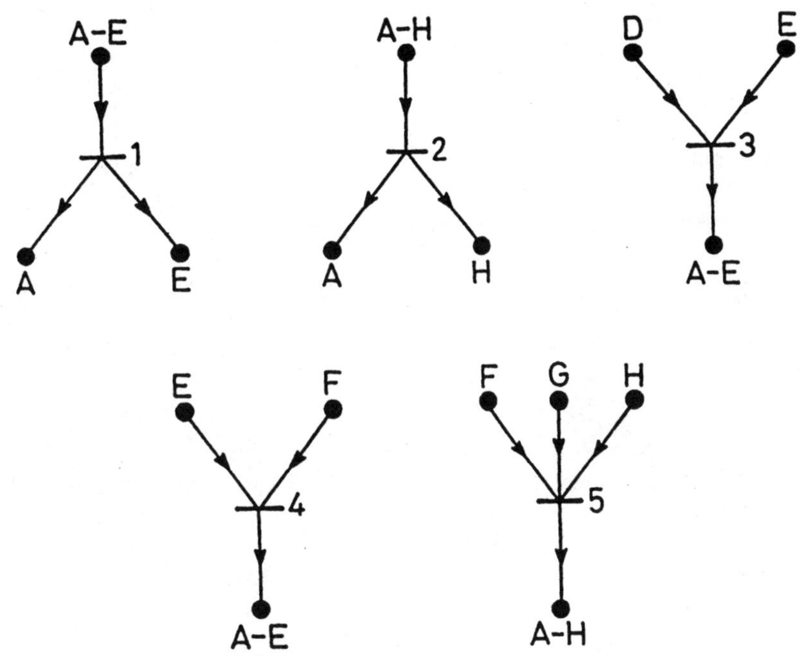

FIGURE 20 Graph representation of the operating units of the example.

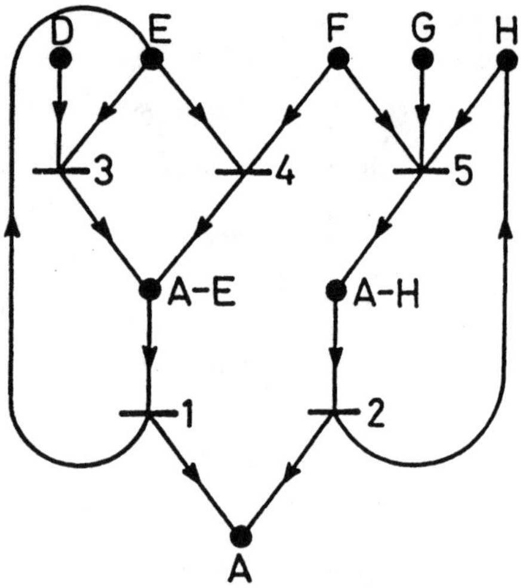

FIGURE 21 Maximal structure of the example.

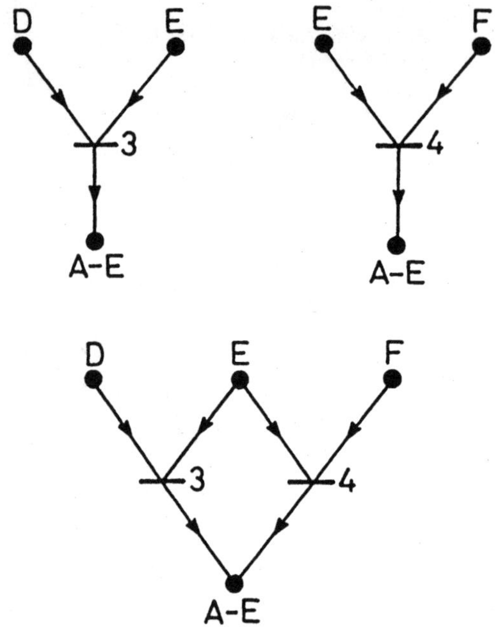

FIGURE 22 Three possible combinations of operating units producing material A-E for the example.

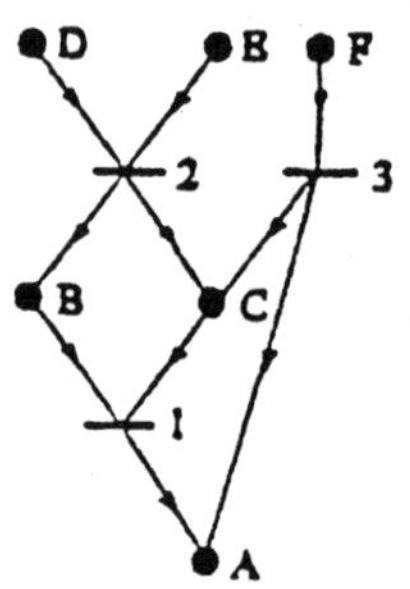

FIGURE 23 P-Graph where A, B, C, D, E, and F are the materials and 1, 2, and 3 are the operating units.

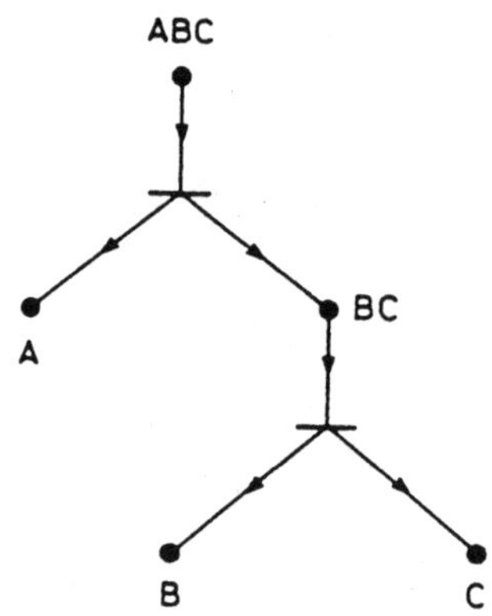

FIGURE 24 P-graph representation of a process structure involving sharp separation of mixture ABC into its three components.

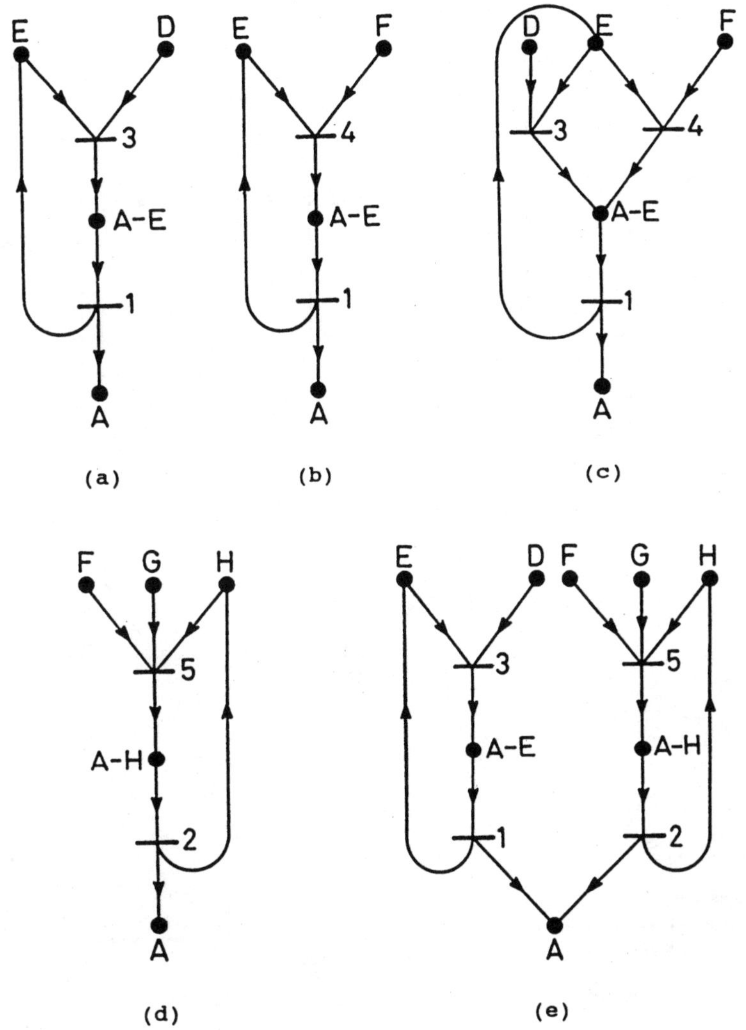

FIGURE 25 Feasible process structures for the example.

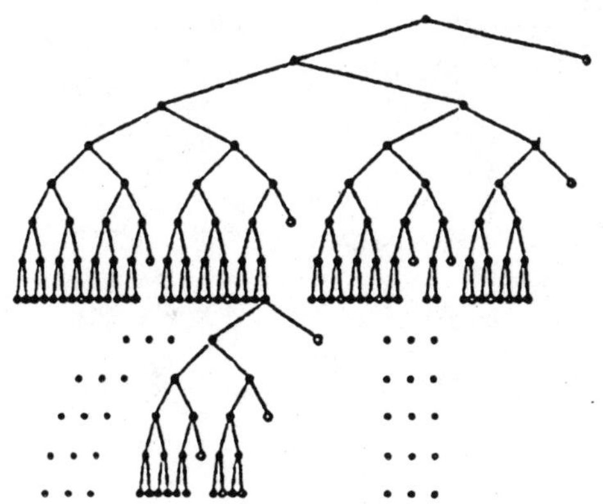

FIGURE 26 Enumeration tree for the basic branch and bound algorithm which generates 9991 subproblems in the worst case.

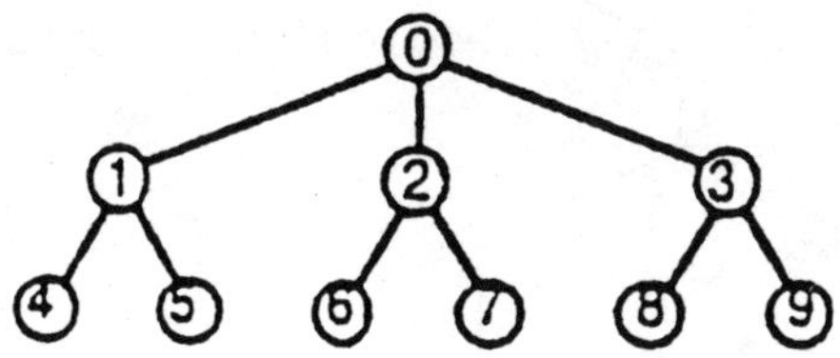

FIGURE 27 Enumeration tree for the accelerated branch and bound algorithm with rule a(1) which generates 10 subproblems in the worst case.

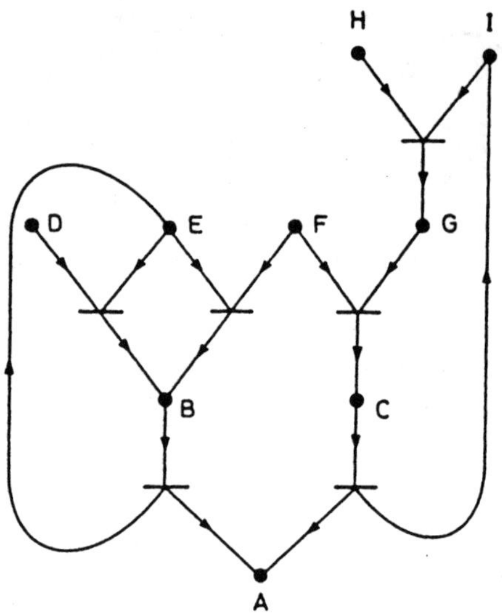

FIGURE 28 Maximal structure of synthesis problem (P_3, R_3, O_3)

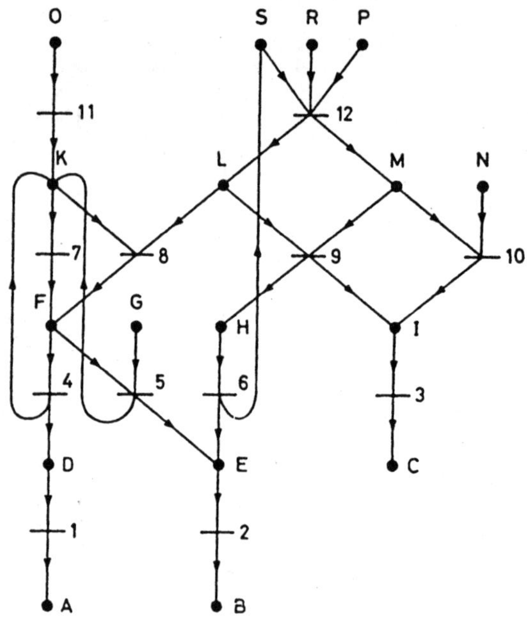

FIGURE 29 Maximal structure of synthesis problem (P_4, R_4, O_4).

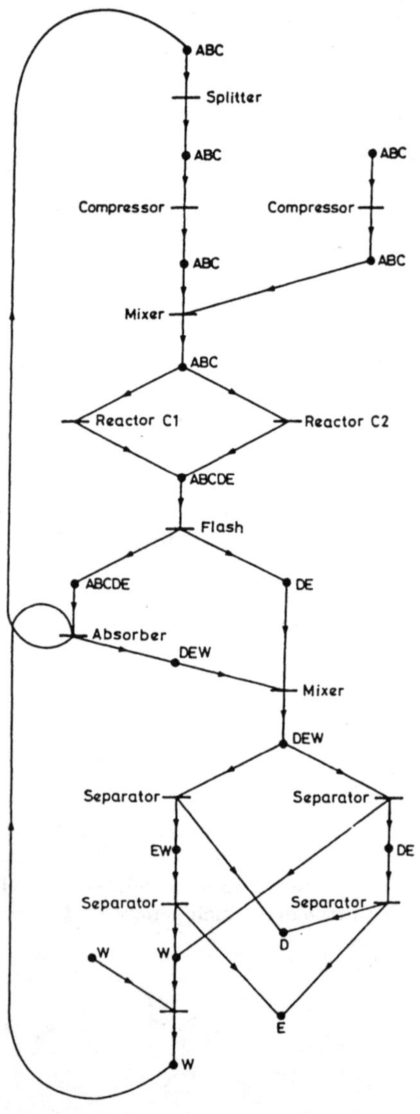

FIGURE 30 Maximal structure of the synthesis problem of Grossman (1985).

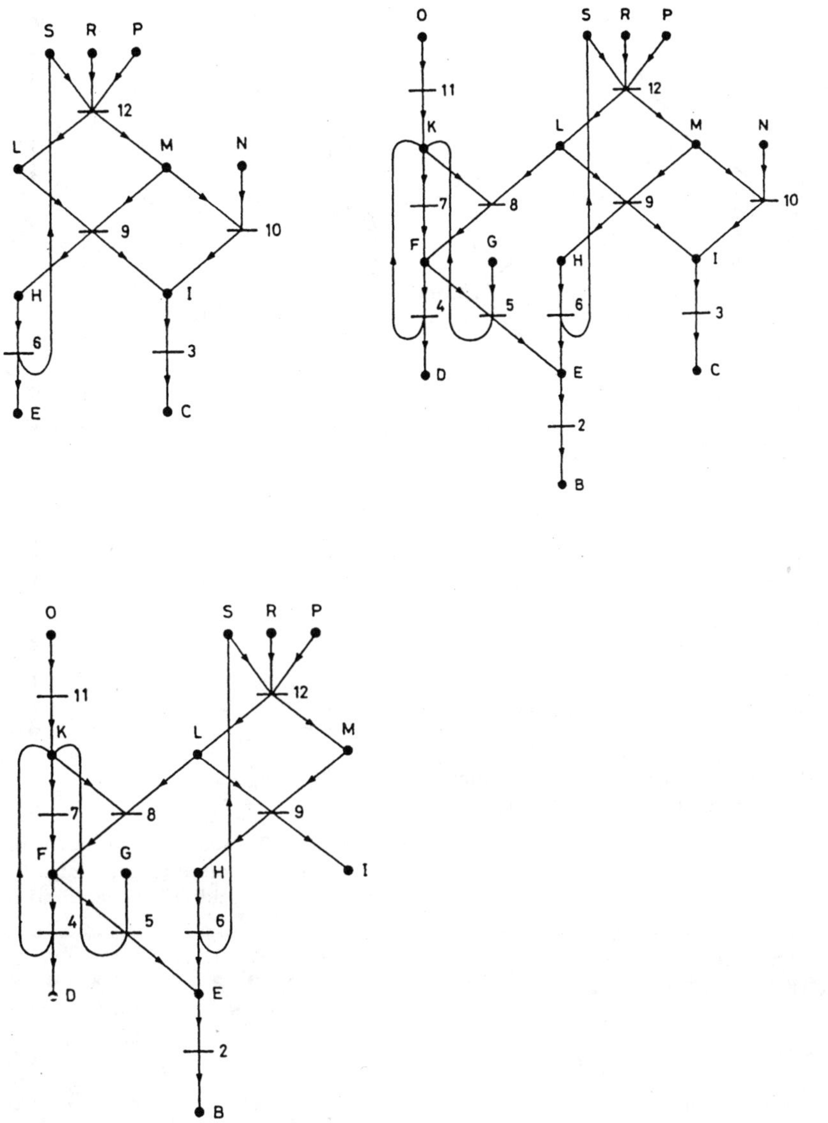

FIGURE 31 Maximal structures of 3 synthesis problems.

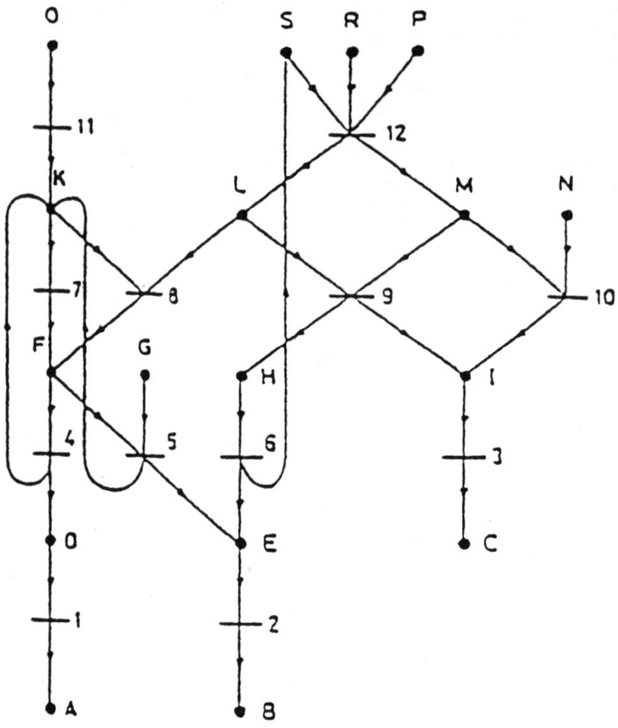

FIGURE 32 Maximal structure of the example for producing material A as the required product and producing material B or C as the potential product.

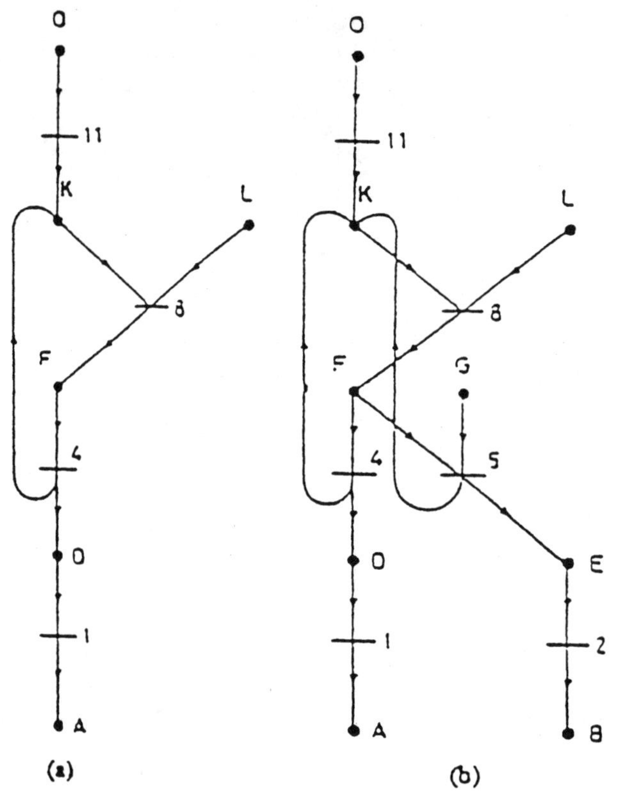

FIGURE 33 Solution-structures of the example: (a) without producing a potential product; and (b) producing potential product B in addition to required product A.

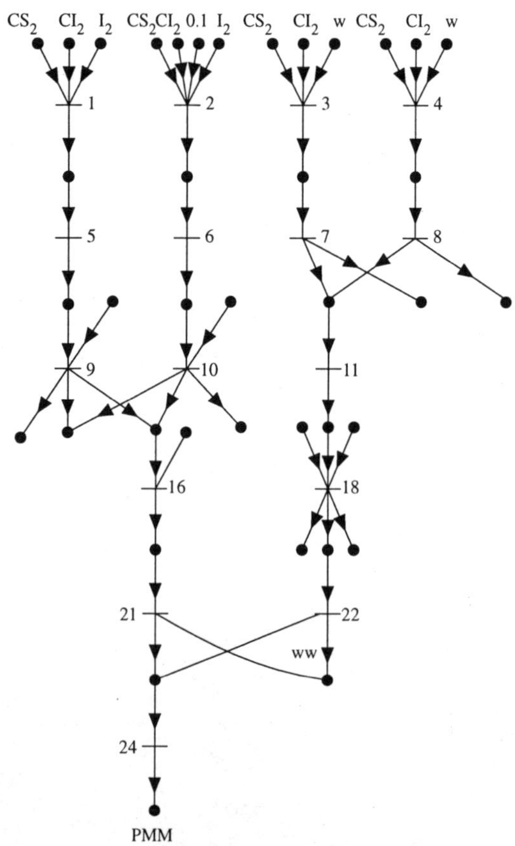

FIGURE 34 Maximal structure of the PMM production process without integrated in-plant waste treatment.

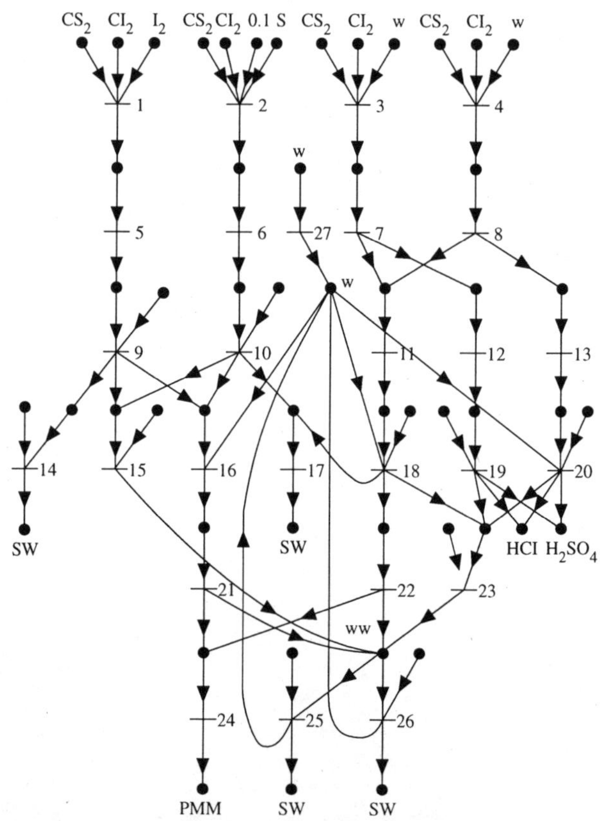

FIGURE 35 Maximal structure of the PMM production process with integrated in-plant waste treatment.

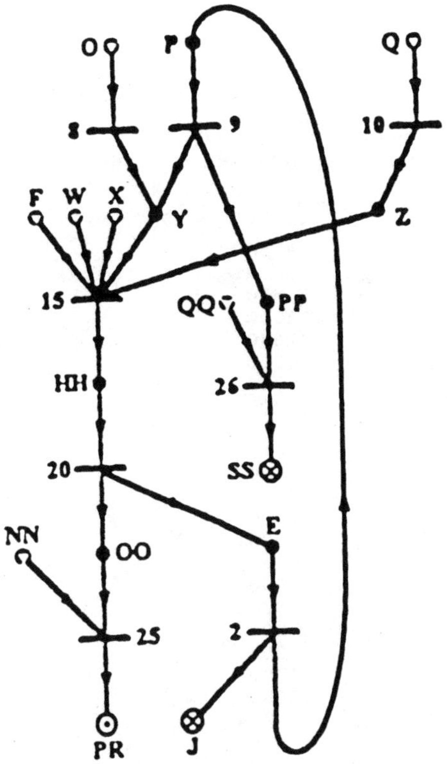

FIGURE 36 Structure of the optimally synthesized process integrating in-plant waste treatment but without consideration of risk.

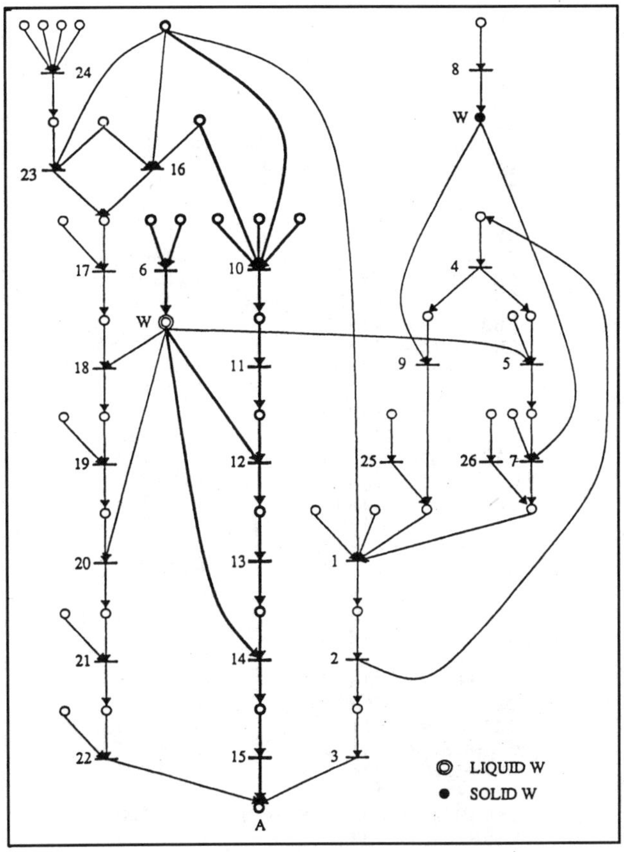

FIGURE 37 Maximal graph for the Folpet production with waste treatment as an integral part of the process.

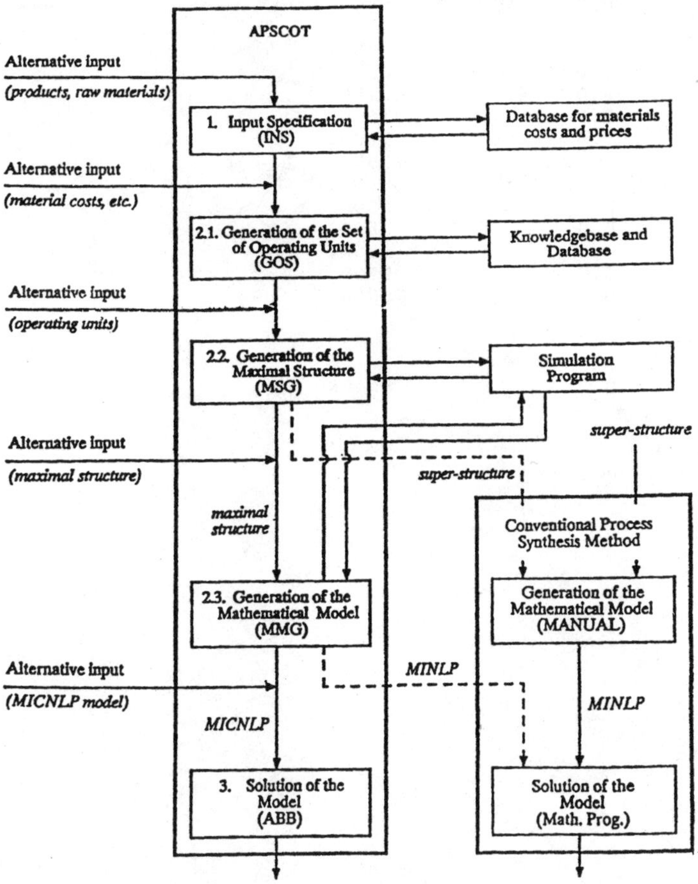

FIGURE 38 Flowchart for APSCOT (Automatic Process Synthesis with Combinatorial Technique).

```
ELEMENTS
C H S
END
SPECIES
CH4 H CH3 C2H6 H2 C2H5 C2H4 C2H3 C4H6 C2H2 C4H3 C4H2 C5H12 CH2 CH4S C2H6S C2H6S.
C4H10 C3H8 C3H6 H2S I*C3H7 N*C3H7 C3H5 C3H4 C4H8 BUTENE2 OCTENE* IC4H10
N*C4H9 S*C4H9 C4H7 P*C4H9 C6H6 C6H5 C12H10 C3H3 ETBENZENE OCTANE CUMENE
TOLUENE TMBEN* OXYLENE MXYLENE PXYLENE C6H12 C6H14 ISOOCTANE HEPTANE STYRENE
END
REACTIONS
CH4+M=CH3+H+M            2E17          0.0     88000.
CH4+H=CH3+H2             2.24E4        3.0      8750.
CH3+CH3=C2H6             1E13          0.0         0.
CH3+CH3=C2H5+H           7.94E14       0.0     26520.
CH3+CH3=C2H4+H2          1E16          0.0     32000.
CH3+M=CH2+H+M            1.95E16       0.0     91600.
C2H6+CH3=C2H5+CH4        5.49E-1       4.0      8280.
C2H6+H=C2H5+H2           5.37E2        3.5      5200.
C2H5+M=C2H4+H+M          2E15          0.0     30000.
C2H5+C2H5=C4H10          7.94E12       0.0         0.
C3H8=C2H5+CH3            1.7E16        0.0     85840.
C3H6=C2H3+CH3            6.31E15       0.0     85800.
C2H5+C2H3=C4H8           8.91E12       0.0         0.
C2H3+C2H3=C4H6           8.91E12       0.0         0.
C2H4+M=C2H2+H2+M         9.33E16       0.0     77200.
C2H4+M=C2H3+H+M          6.31E18       0.0    108720.
C2H4+C2H4=C2H3+C2H5      5.01E14       0.0     64700.
C2H3+M=C2H2+H+M          7.94E14       0.0     31500.
C2H4+H=C2H3+H2           1.51E7        2.0      6000.
C2H3+H=C2H2+H2           2E13          0.0       250.
C2H3+C2H4=C4H6+H         1E12          0.0      7300.
C2H2+C2H2=C4H3+H         1E13          0.0     45000.
CH2+CH2=C2H3+H           5.01E12       0.0         0.
CH2+CH2=C2H2+H2          3.16E13       0.0         0.
CH2+C2H3=CH3+C2H2        3.02E13       0.0         0.
C4H3+M=C4H2+H+M          1E16          0.0     60000.
I*C3H7+C3H8=N*C3H7+C3H8  3.16E10       0.0     12900.
C3H8+H=I*C3H7+H2         8.71E6        2.0      5000.
C3H8+H=N*C3H7+H2         5.62E7        2.0      7700.
C3H8+CH3=I*C3H7+CH4      1.1E15        0.0     25140.
C3H8+CH3=N*C3H7+CH4      1.1E15        0.0     25410.
C3H8+C2H3=I*C3H7+C2H4    2E11          0.0     10400.
C3H8+C2H3=N*C3H7+C2H4    2E11          0.0     10400.
C3H8+C2H5=I*C3H7+C2H6    2E11          0.0     10400.
C3H8+C2H5=N*C3H7+C2H6    2E11          0.0     10400.
C3H8+C3H5=I*C3H7+C3H6    7.94E11       0.0     16200.
C3H8+C3H5=N*C3H7+C3H6    7.94E11       0.0     16200.
N*C3H7=C2H4+CH3          9.55E13       0.0     31000.
N*C3H7=C3H6+H            1.26E14       0.0     37000.
I*C3H7=C2H4+CH3          2E10          0.0     29500.
I*C3H7=C3H6+H            6.31E13       0.0     36900.
C3H6=C3H5+H              1E13          0.0     78000.
C3H6+H=C3H5+H2           5.01E12       0.0      1500.
C3H6+CH3=C3H5+CH4        8.91E10       0.0      8500.
C3H6+C2H5=C3H5+C2H6      1E11          0.0      9200.
C3H5=C3H4+H              3.98E13       0.0     70000.
C3H5+CH3=C4H8            1.35E13       0.0         0.
C3H5+H=C3H4+H2           1E13          0.0         0.
C3H5+CH3=C3H4+CH4        1E12          0.0         0.
C4H10=N*C3H7+CH3         1E17          0.0     85400.
```

FIGURE 39 Reaction file for a refinery study of hydrocarbons using Chemkin.

```
C4H10+H=P*C4H9+H2            3.98E14     0.0      9700.
C4H10+H=S*C4H9+H2            2E14        0.0      8300
C4H10+CH3=P*C4H9+CH4         2.95E12     0.0      11700.
C4H10+CH3=S*C4H9+CH4         1.32E12     0.0      10100.
C4H10+C2H3=P*C4H9+C2H4       1E12        0.0      18000.
C4H10+C2H3=S*C4H9+C2H4       7.94E11     0.0      16800.
C4H10+C2H5=P*C4H9+C2H6       1E11        0.0      13400.
C4H10+C2H5=S*C4H9+C2H6       1E11        0.0      10400.
C4H10+C3H5=P*C4H9+C3H6       3.98E11     0.0      18800.
C4H10+C3H5=S*C4H9+C3H6       7.94E11     0.0      16800.
N*C4H9=C2H5+C2H4             1.58E12     0.0      28000.
S*C4H9=C3H6+CH3              2.82E13     0.0      31900.
N*C4H9=C4H8+H                1E13        0.0      36000.
S*C4H9=BUTENE2+H             2E13        0.0      39800.
C4H8+H=C4H7+H2               5.01E13     0.0      3900.
BUTENE2+H=C4H7+H2            5.01E13     0.0      3800.
C4H8+CH3=C4H7+CH4            1E11        0.0      7300.
BUTENE2+CH3=C4H7+CH4         1E11        0.0      8200.
C4H7=C4H6+H                  1.2E14      0.0      49300.
C4H7=C2H4+C2H3               1E11        0.0      37000.
C4H7+C4H7=C4H8+C4H6          1.58E12     0.0      0.
C4H7+C4H7=BUTENE2+C4H6       1.58E12     0.0      0.
C4H7+CH3=C4H6+CH4            7.94E12     0.0      0.
C6H6+C6H5=C12H10+H           4E11        0.0      4000.
C6H6=C6H5+H                  1E16        0.0      108000.
C6H6+H=C6H5+H2               2.5E14      0.0      16000.
C6H5=C4H3+C2H2               4.5E13      0.0      72530.
2C6H5=C12H10                 3.1E12      0.0      0.
C2H2+CH2=C3H3+H              3E13        0.0      0.
M+C3H4=C3H3+H                1.2E13      0.0      6600.
M+H2=2H                      2.2E14      0.0      96000.
I*C3H7+I*C3H7=C6H14          6.2E13      0.0      0.
S*C4H9+S*C4H9=OCTANE         3.16E12     0.0      0.
ETBENZENE=C6H6+C2H4          1E11        0.0      0.
CUMENE=C6H6+C3H6             1E11        0.0      0.
TOLUENE+H2=C6H6+CH4          1E11        0.0      0.
TMBEN*+H2=CH4+OXYLENE        1E11        0.0      0.
PXYLENE+H2=C6H6+C2H6         1E11        0.0      0.
OXYLENE+H2=C2H6+C6H6         1E11        0.0      0.
MXYLENE+H2=C2H6+C6H6         1E11        0.0      0.
I*C3H7+N*C4H9=HEPTANE        1E11        0.0      0.
IC4H10=C4H10                 1E11        0.0      0.
C2H5+N*C3H7=C5H12            1E11        0.0      0.
C2H5+I*C3H7=C5H12            1E11        0.0      0.
2CH4S=H2S+C2H6S              1E11        0.0      0.
CH4S+H2=H2S+CH4              1E11        0.0      0.
2CH4S=C2H6S2+H2              1E11        0.0      0.
C6H6+3H2=C6H12               1E14        0.0      0.
STYRENE+H2=ETBENZENE         1E11        0.0      0.
C2H6S+H2=H2S+C2H6            1E11        0.0      0.
END

C4H8+C4H10=ISOOCTANE         1E11        0.0      0.
IC4H10+C4H8=ISOOCTANE        1E11        0.0      0.
C2H4+C6H14=ISOOCTANE         1E11        0.0      0.
C3H6+C5H12=ISOOCTANE         1E11        0.0      0.
END
```

FIGURE 39 (continued) Reaction file for a refinery study of hydrocarbons using Chemkin.

C	A	B	C	D	E	F	G	H	I	J	K	L	M	N
271	Table IV. Influence of Chemical Groups on Physical and Biological Properties													
272														
273		c-c	c=c	arom c-c	c-h	c-f	c-cl	c-br-	c-i	-oh	-o-	-(c=o)-	-ch=o	COOH
274	mol wt	12.011	12.011	12.011	1.0079	22.0012	41.4585	79.904	129.907	17.0073	15.9994	28.0104	29.0183	45.0177
275	E rho g/cc	0.77088	0.92221	0.86191	1.04469	0.97668	1.07694	1.09606	1.2308	0.93436	0.56381	1.05755	1.16617	1.34829
276	LOG(E)	0.44399	0.65845	0.58626	0.84031	1.01692	1.08331	1.12519	1.31281	1.62385	0.66326	1.69075	0.9389	0.57877
277	CHI	56.7428	55.9138	55.903	54.7448	50.9537	52.4528	52.6309	53.2114	40.5835	48.5121	40.2792	56.2226	67.3651
278	-LOGKb	3.18452	9.87741	7.13092	13.6189	17.0432	18.7158	19.2756	24.2292	23.9952	10.4689	33.4726	13.1548	7.20745
279	DIPOLE M	0.53093	1.26178	0.965	1.67255	2.08538	2.24095	2.29465	2.81348	2.93033	1.435	3.98627	1.58459	0.80528
280	SUM INT	1	1	1	1	1	1	1	1	1	1	1	1	1
281	Tb-Tm	-29.9583	-0.28007	-14.1315	8.47485	18.3882	37.1615	44.4312	74.0207	41.5766	-19.0278	65.9648	18.6812	20.7144
282	MOL. WT	74.4389	95.0569	84.8099	99.6357	93.7709	112.49	117.782	141.349	84.4072	56.647	102.056	111.225	141.923
283	COUNT	0.5	1	2	0.5	0.5	0.5	0.5	0.5	0.5	1	1	0.5	0.5
284	logben/h2(	1.69579	2.08787	1.82243	2.22098	1.31452	1.85378	1.91984	2.56852	-1.68803	-0.27861	-0.5613	2.5932	5.72719
285	Tm	50.9968	93.9958	73.9957	93.8849	105.902	143.255	156.571	199.387	179.096	68.7713	186.008	111.772	93.8413
286	B rho g/cc	1.4249	1.65684	1.56903	1.91562	1.77672	1.87133	1.87518	2.05092	1.49447	1.04364	1.81652	2.06181	2.45407
287	B	-0.01386	0.07064	0.02624	0.04507	0.06432	0.15772	0.19286	0.28862	0.18168	0.00281	0.17555	0.09202	0.10663
288	C	-0.03557	-0.00497	-0.01357	-0.0012	0.06887	0.06919	0.07579	0.08459	0.3048	0.1154	0.28494	-0.01507	-0.22306
289	D	0.02886	0.01666	0.01966	-0.00897	0.00728	-0.00934	-0.01456	-0.03911	0.00602	0.0982	0.02406	-0.04621	-0.09652
290	E	0.0161	0.0126	0.0121	-0.01055	-0.0128	-0.00797	-0.0077	-0.0149	-0.00825	0.0342	-0.0141	-0.02335	-0.03715
291	F	0.01942	-8.2E-05	0.00912	-0.00794	-0.00968	-0.0115	-0.00969	-0.02059	0.01322	-0.00064	-0.04232	0.00544	0.0042
292	Tb	21.0385	93.7157	59.8642	102.36	124.29	180.416	201.002	273.408	220.673	49.7435	251.973	130.453	114.556
293	SOL PAR	7.67708	8.86673	8.42151	9.52256	10.2695	10.962	11.286	12.4375	13.5208	8.47188	13.4809	10.2206	8.72223
294	Tc	162.716	285.156	228.601	306.318	333.426	423.868	454.736	574.505	464.198	198.168	536.862	348.908	338.618
295	Pc	32.098	40.8459	37.8583	50.4576	50.8236	54.3897	55.3476	64.0027	58.0568	27.307	65.6732	55.9699	57.3895
296	log AC CC	1.04089	1.67597	1.40813	1.29215	2.36511	2.80899	3.07508	3.4687	6.35351	3.38011	5.66295	1.22844	-1.78726
297	log par co	1.75402	1.79991	1.73349	1.79208	1.22416	1.64721	1.76402	2.13379	-0.03235	0.15277	-0.287	2.35548	4.28878
298	Hv	7.49072	8.90619	8.27664	8.95017	9.78361	11.2261	11.8788	13.5243	14.0871	8.10912	12.9285	10,1757	8.95851
299	Surf Ten	-0.56839	2.89611	1.07571	1.84782	2.63731	6.46665	7.90742	11.8334	7.44902	0.1154	7.19761	3.77283	4.37203
300	log visc	-0.36792	-0.18204	-0.25343	-0.06784	0.06158	0.25717	0.36309	0.63406	0.72541	-0.45029	0.43481	0.23564	0.16201
301	log(et/h2o	1.86326	1.93357	1.86545	2.23638	1.38608	1.81318	1.90726	2.41376	-0.82304	-0.58672	-0.81028	3.02456	5.96288
302	logoil/h2o	1.46273	1.55436	1.43058	1.36454	0.57351	1.08516	1.17783	1.55365	-1.4269	-0.26142	-1.28461	1.75585	4.14396
303	logcyh/h2(	1.21785	1.49556	1.28154	1.63244	0.60312	1.06021	1.08031	1.61937	-2.79542	-0.9519	-1.62811	1.95785	5.31111
304	logchl/h2o	1.43078	2.13348	1.73322	2.24972	1.70001	2.28167	2.37258	3.14907	-0.45177	0.43515	1.08723	2.38902	4.56676
305	logcl4/h2o	1.10466	1.71065	1.35993	2.07234	1.27072	1.66764	1.65874	2.35848	-2.06945	-0.43067	-0.04018	2.17368	5.1317
306	vol.Bondi	52.2414	57.3726	54.0525	52.0123	52.7776	60.1124	62.811	68.92	56.4795	54.2781	56.182	53.9454	57.8316
307	molvolexp	96.564	103.075	98.3977	95.3736	96.0099	104.453	107.459	114.843	90.3365	100.472	96.5019	95.3761	105.261
308	C	1	1	1	0	0.25	0.5	0	0.25	0	0	1	1	1
309	H	0	0	0	1	0	0	0	0	1	0	0	1	1
310	O	0	0	0	0	0	0	0	0	1	1	1	1	2
311	N	0	0	0	0	0	0	0	0	0	0	0	0	0
312	F	0	0	0	0	1	0	0	0	0	0	0	0	0
313	I	0	0	0	0	0	0	0	1	0	0	0	0	0
314	D1	-0.03778	-0.07636	-0.05485	-0.2049	0.22572	0.05405	0.03828	0.02192	0.01916	-0.03087	0.03197	0.05899	0.08579
315	D12													
316	isothcom	14.5644	11.154	12.62	9.78677	9.46812	6.40969	5.28656	1.29423	4.74905	16.3083	4.85498	6.88397	5.72347
317	w	0.06364	0.2098	0.15205	0.21508	0.2547	0.3176	0.3395	0.38916	0.38835	0.14363	0.37905	0.25592	0.21911
318	Zc	0.28591	0.27422	0.27884	0.27379	0.27062	0.26559	0.26384	0.25987	0.25993	0.27951	0.26068	0.27053	0.27347
319	AIR	0.42993	1.61038	0.73032	1.542	0.3515	0.57717	0.43084	0.92583	0.00048	0.2154	0.13267	0.38375	17.4373
320	RES	43.8237	106.449	57.731	53.6809	13.5589	24.4176	19.7893	30.0491	0.07309	28.3542	4.22733	14.3431	272.345
321	drug1	4.64817	4.31163	4.5154	3.9633	4.52295	4.64888	4.84304	4.66377	7.79025	5.21448	5.3819	4.44739	2.54276
322	drug4	131.364	138.671	129.782	81.9764	82.9048	114.185	123.628	127.82	131.704	170.88	101.408	69.6323	39.034
323	drug6	4.58414	4.58851	4.63506	4.73405	4.87777	4.79335	4.77368	4.71338	5.73559	4.84155	5.49692	4.75014	4.01661
324	rho (B-E)	0.65403	0.73463	0.70712	0.87093	0.80004	0.79439	0.77912	0.82012	0.56011	0.47984	0.75897	0.89563	1.10578
325	coefficien	0	0	0	0	0	0	0	0	0	0	0	0	0
326	intercept	0	0	0	0	0	0	0	0	0	0	0	0	0
327	mol. wt.	12.011	12.011	12.011	1.0079	22.0012	41.4585	79.904	129.907	17.0073	15.9994	28.0104	29.0183	45.0177
328	# o	0	0	0	0	0	0	0	0	1	1	1	1	2
329	# n	0	0	0	0	0	0	0	0	0	0	0	0	0
330	# s	0	0	0	0	0	0	0	0	0	0	0	0	0
331	c-c	1	0	0	0	0	0	0	0	0	0	0	0	0
332	arom. c-c	0	0	1	0	0	0	0	0	0	0	0	0	0
333	c=c	0	1	0	0	0	0	0	0	0	0	0	0	0
334	c=-c	0	0	0	0	0	0	0	0	0	0	0	0	0
335	c-h	0	0	0	1	0	0	0	0	0	0	0	0	0
336	c-f	0	0	0	0	1	0	0	0	0	0	0	0	0
337	c-cl	0	0	0	0	0	1	0	0	0	0	0	0	0
338	c-br	0	0	0	0	0	0	1	0	0	0	0	0	0
339	c-I	0	0	0	0	0	0	0	1	0	0	0	0	0
340	oh	0	0	0	0	0	0	0	0	1	0	0	0	0
341	o	0	0	0	0	0	0	0	0	0	1	0	0	0
342	c=o	0	0	0	0	0	0	0	0	0	0	1	0	0
343	ch=o	0	0	0	0	0	0	0	0	0	0	0	1	0
344	cooh	0	0	0	0	0	0	0	0	0	0	0	0	1
345	coo	0	0	0	0	0	0	0	0	0	0	0	0	0
346	sh	0	0	0	0	0	0	0	0	0	0	0	0	0
347	nh2	0	0	0	0	0	0	0	0	0	0	0	0	0
348	nh	0	0	0	0	0	0	0	0	0	0	0	0	0
349	-n--	0	0	0	0	0	0	0	0	0	0	0	0	0
350	=n-	0	0	0	0	0	0	0	0	0	0	0	0	0
351	cn	0	0	0	0	0	0	0	0	0	0	0	0	0
352	conh2	0	0	0	0	0	0	0	0	0	0	0	0	0
353	no2	0	0	0	0	0	0	0	0	0	0	0	0	0
354	cycles(rin	0	0	0	0	0	0	0	0	0	0	0	0	0
355	tert C	0	0	0	0	0	0	0	0	0	0	0	0	0
356	quat C	0	0	0	0	0	0	0	0	0	0	0	0	0
357	x-c-x	0	0	0	0	0	0	0	0	0	0	0	0	0
358	x3-c-x3	0	0	0	0	0	0	0	0	0	0	0	0	0
359	c=c-x	0	0	0	0	0	0	0	0	0	0	0	0	0
360	arom. fusi	0	0	0	0	0	0	0	0	0	0	0	0	0
361	S	0	0	0	0	0	0	0	0	0	0	0	0	0
362	CL	0	0	0	0	0	1	0	0	0	0	0	0	0
363	BR	0	0	0	0	0	0	1	0	0	0	0	0	0
364	loggamma	1.03928	1.67471	1.40692	1.2932	2.36639	2.80979	3.07585	3.47019	6.35434	3.37669	5.66436	1.23077	-1.78354
365	m.refract	24.2084	27.9077	25.6052	24.5253	23.9329	29.2213	30.9168	35.2172	25.1484	23.8787	25.5072	25.9201	29.8311
366	thermcond	2.53225	3.24284	2.942	3.54392	4.00972	4.23693	4.33776	4.88437	5.05105	3.46235	5.86253	3.52808	2.73322
367	COST $/g	0.00083	0.00083	0.00083	0.00992	0.00148	0.0006	0.0005	0.0004	0.00118	0.00063	0.00071	0.00103	0.00089
368	MCL	0.01078	0.0195	0.01221	0.00866	0.00227	0.00439	0.0039	0.00545	2.7E-05	0.00464	0.00045	0.00348	0.06097
369	TAP	0.56182	2.135	0.94192	1.63304	0.38957	0.74839	0.59852	1.35062	0.00078	0.31256	0.1554	0.42892	17.15
370	COM	131.022	395.687	181.888	134.788	25.1601	58.6303	47.061	78.6633	0.05568	91.4732	6.70838	25.4871	774.718
371	drug2	-15.7779	-59.3603	-35.733	-68.5847	-59.0981	-141.613	-178.757	-271.029	-97.728	118.865	-7.51009	-180.662	-329.967
372	drug3	3.56135	3.43407	3.57515	3.49393	4.10953	4.12115	4.26396	4.22215	7.10206	4.16885	5.37459	3.94753	2.10573
373	drug5	10.9364	11.0146	10.935	10.7338	10.8347	11.2569	11.477	11.7948	11.4955	10.6717	10.6269	11.2001	11.5058

FIGURE 40 Influence of chemical groups on physical and biological properties.

C	O	P	Q	R	S	T	U	V	W	X	Y	Z	AA	AB	AC	AD	AE
273	COO	-nh2	C=-C	-nh-	-n--	=n-	-cn	-conh2	-n02	ring	tertiary c	quatern c	x-c-x	x3c-c-x3	c=c-xtrace	arom fusic	sh
274	44.0098	16.0225	12.011	15.0146	14.0067	14.0067	26.0177	44.0329	46.0055	0	0	0	0	0	0	0	33.0679
275	1.319753	0.806346	1.078636	0.424951	0.083951	0.637517	1.322077	1.508334	0.971963	1.851865	0.931546	0.877828	0.960975	0.989079	0.987407	0.775092	1.093439
276	0.612922	1.184604	1.059116	0.655936	0.094559	0.897336	1.968489	1.596884	1.980109	1.492173	0.692404	0.627448	0.592201	0.362052	0.633389	0.535726	1.277495
277	65.77323	43.86395	51.22721	45.15278	48.73826	47.432	40.91554	56.71173	40.89096	50.50811	54.36756	54.3505	57.29184	63.04533	57.00632	57.36921	49.66124
278	6.068194	19.50588	18.25117	8.95659	-5.30538	12.63161	39.33272	23.29671	36.73823	28.18528	10.92067	9.80859	7.921609	5.373728	8.363088	4.198144	22.58483
279	0.696814	2.443673	2.212933	1.317844	-0.25362	1.657966	4.593869	2.613377	4.288837	3.28165	1.39798	1.28304	1.03274	0.69294	1.08081	0.61806	2.687483
280	1	1	1	1	1	1	1	1	1	0	0	0	0	0	0	0	1
281	12.8291	16.85098	23.39769	-30.6715	-81.648	2.50761	97.83492	94.33866	116.6214	42.07402	-10.6521	-16.8239	-13.3316	-7.9869	-11.7529	-13.1948	51.6724
282	132.648	78.99083	104.7441	43.07013	8.640814	66.10748	129.7243	156.3916	109.733	135.9891	85.99152	81.35054	90.23216	103.6348	91.43656	78.57205	115.9993
283	1	0.5	1	1	1.5	1	0.5	0.5	0.5	0	0	0	0	0	0	0	1.5
284	5.045897	-1.10436	1.391469	-1.79053	-2.11455	-0.50367	0.2376	3.872028	0.286799	2.33148	1.82748	1.69916	2.49124	4.23156	2.43264	2.1212	1.259133
285	90.77313	157.1746	138.5635	91.54221	15.80922	99.73102	231.1387	175.6289	158.5911	180.5466	72.1422	65.3311	61.59157	40.7934	66.2445	40.9908	172.7541
286	2.374953	1.332039	1.863104	0.738287	0.152466	1.128953	2.190916	2.574387	1.817421	3.065345	1.743491	1.651755	1.809896	1.942361	1.845239	1.477646	1.844633
287	0.091166	0.146167	0.119637	0.040898	-0.06737	0.081429	0.261366	0.261218	0.217158	0.1045	-0.0074	-0.0199	-0.0099	-0.0077	-0.0047	-0.0031	0.210336
288	-0.19206	0.249211	0.091531	0.211514	0.130518	0.152218	0.288349	-0.01925	0.221326	0.0996	-0.0012	-0.0016	-0.057	-0.1762	-0.0496	-0.0657	0.135122
289	-0.09474	0.055281	-0.00464	0.131455	0.18003	0.07373	-0.02571	-0.14406	0.006722	-0.1034	0.0144	0.0257	-0.0011	-0.0063	-0.0074	0.0181	-0.00762
290	-0.0353	0.034099	0.002799	0.070749	0.093	0.0195	-0.0338	-0.1005	-0.0989	-0.0285	0.003	0.0095	-0.00397	-0.0142	-0.0058	-0.014	-0.00585
291	0.014939	-0.00354	-0.01308	0.011098	0.045839	0.012239	-0.05164	0.011435	-0.02684	-0.042	-0.0067	-0.0061	0.0019	-0.0051	0.0031	0.0241	-0.01453
292	103.6022	174.0256	161.9612	60.8707	-65.8388	102.2386	328.9736	269.9675	275.2125	222.6206	61.4901	48.50718	48.25994	32.8065	54.49156	27.79602	224.4266
293	8.883435	11.50663	10.83618	8.772861	5.944492	9.785168	15.04262	13.46987	14.16326	12.9627	8.713742	8.385682	8.281705	7.114736	8.507256	7.882378	12.00465
294	316.126	399.5599	399.9567	205.921	-14.7019	271.733	668.4895	562.9141	540.9111	537.0894	241.0615	219.2912	219.267	198.2504	229.6046	164.2224	490.7655
295	56.24577	44.91046	54.58701	19.79236	-4.9197	32.55076	80.49766	80.8648	66.78044	87.39527	44.1739	41.02477	43.19966	39.88033	44.92685	33.40237	57.89303
296	-1.34804	5.755165	2.996437	5.32707	4.253433	4.189816	5.773351	1.213893	4.48194	2.26618	1.206747	1.225153	0.401662	-1.3412	0.503344	0.477538	3.967612
297	4.038747	-0.11736	1.298699	-0.58692	-0.47055	0.368271	0.2444	3.621189	0.556727	1.61106	1.468493	1.362109	2.021701	2.940192	2.026834	2.119062	1.308304
298	9.143559	11.84001	10.67395	9.035392	6.468189	10.20837	14.76034	14.52403	14.38753	11.30403	7.823296	7.467259	7.58848	6.546604	7.82888	7.901625	12.56017
299	3.737808	5.992833	4.905107	1.676812	-2.76221	3.338592	10.716	10.70994	8.903494	4.2845	-0.3034	-0.8159	-0.4059	-0.3157	-0.1927	-0.1271	8.62377
300	0.20103	0.094559	0.092732	-0.50418	-0.92051	-0.00484	0.807692	1.390453	1.237696	0.266002	-0.32131	-0.41255	-0.28317	-0.37566	-0.23434	-0.10201	0.434531
301	5.54934	-1.16705	1.307907	-2.12485	-2.21047	-0.36199	0.071348	5.211034	0.890524	2.35857	1.72696	1.52797	2.54877	3.99303	2.56211	2.58119	1.224764
302	3.750252	-0.79516	0.822713	-0.95452	-0.70882	-0.39331	-0.85427	2.180766	-1.42417	1.13933	1.16444	1.11622	1.73166	2.89244	1.69607	1.51324	0.618447
303	4.611242	-2.03514	0.65238	-2.5581	-2.69637	-1.33871	-0.9243	2.920439	-1.0161	1.67729	1.31384	1.20525	2.02884	3.86909	1.95821	1.57949	0.338216
304	3.916323	0.15808	1.904513	-0.65932	-1.531	0.246249	1.898682	3.454733	1.540049	2.80962	1.87212	1.76017	2.19693	3.50313	2.14422	1.67561	2.022021
305	4.320988	-1.37499	1.232706	-2.23608	-3.04059	-0.87769	0.810261	3.411355	0.954327	2.58943	1.6904	1.55354	2.19502	4.0039	2.11437	1.58275	1.097881
306	55.85288	59.30068	56.2202	58.33788	56.67383	58.55646	59.21004	60.74906	60.37841	44.3634	49.32146	49.25096	49.85489	53.35508	49.55268	53.17379	62.88475
307	100.5097	97.96144	97.10794	101.3532	102.9272	103.6952	98.12158	103.685	112.8983	73.4336	92.3105	92.6725	93.89646	104.7791	92.6027	101.3713	106.0867
308	1	0	1	0	0	0	1	1	0	0	0	0	0	0	0	0	0
309	0	2	0	1	0	0	0	2	0	0	0	0	0	0	0	0	1
310	2	0	0	0	0	0	0	1	2	0	0	0	0	0	0	0	0
311	0	1	0	1	1	1	1	1	1	0	0	0	0	0	0	0	0
312	0	0	0	0	0	0	0	0	0	0	0	0	0	0	0	0	0
313	0	0	0	0	0	0	0	0	0	0	0	0	0	0	0	0	0
314	0.069144	0.146999	0.149894	-0.02757	-0.0151	-2.87085	0.01721	0.009997	0.011231	0.052255	-0.04326	-0.03369	-0.04909	-0.037	-0.05932	-0.13626	0.031988
315																	
316	6.09708	9.08648	7.383675	16.98557	24.11414	12.84632	-0.40964	-3.24726	3.01071	0.96928	12.49524	13.51764	12.37411	13.0457	11.82006	13.79838	4.564005
317	0.208922	0.330691	0.291651	0.17985	-0.55178	0.258512	0.415263	0.394953	0.443916	0.589039	0.241435	0.196348	0.197806	0.136037	0.220382	0.133054	0.636534
318	0.274286	0.264545	0.267668	0.276612	0.335143	0.270319	0.257779	0.259404	0.255487	0.243877	0.271685	0.275292	0.275176	0.280117	0.273369	0.280356	0.240077
319	3.612272	0.038136	0.836395	0.045561	0.009932	0.01582	0.203326	0.01019	0.000937	13.34969	2.218549	2.580087	2.108886	22.57195	1.607512	0.100079	0.239149
320	83.25222	7.445183	44.935	22.087	13.16714	2.445051	3.70429	0.100444	0.006363	208.6515	101.9182	135.5601	89.29924	460.6415	69.5148	5.674471	12.01479
321	3.229118	6.655875	4.784981	6.706092	7.12427	6.152652	5.204371	4.747794	5.074499	-34.6821	17.64802	44.7003	-41.9713	-120.019	-49.5765	-131.842	-153.237
322	43.12285	205.964	127.8082	261.825	291.723	163.0167	77.36653	-50.7736	-62.7655	4.065816	3.340209	3.265889	3.11314	1.813892	3.25602	3.619028	4.696108
323	4.201741	5.449069	5.037096	5.170799	4.809806	4.903628	5.557505	4.555205	4.658206	9.67129	10.42361	10.3655	10.65852	10.89198	10.67058	11.44347	11.40244
324	1.055201	0.525693	0.784468	0.313337	0.068515	0.491436	0.86884	1.066054	0.845458	1.21348	0.811945	0.773927	0.848921	0.953282	0.857832	0.702555	0.751194
325	0	0	0	0	0	0	0	0	0	0	0	0	0	0	0	0	0
326	0	0	0	0	0	0	0	0	0	0	0	0	0	0	0	0	0
327	44.0098	16.0225	12.011	15.0146	14.0067	14.0067	26.0177	44.0329	46.0055	0	0	0	0	0	0	0	33.0679
328	2	0	0	0	0	0	0	1	2	0	0	0	0	0	0	0	0
329	0	1	0	1	1	1	1	1	1	0	0	0	0	0	0	0	0
330	0	0	0	0	0	0	0	0	0	0	0	0	0	0	0	0	1
331	0	0	0	0	0	0	0	0	0	0	0	0	0	0	0	0	0
332	0	0	0	0	0	0	0	0	0	0	0	0	0	0	0	0	0
333	0	0	0	0	0	0	0	0	0	0	0	0	0	0	0	0	0
334	0	0	1	0	0	0	0	0	0	0	0	0	0	0	0	0	0
335	0	0	0	0	0	0	0	0	0	0	0	0	0	0	0	0	0
336	0	0	0	0	0	0	0	0	0	0	0	0	0	0	0	0	0
337	0	0	0	0	0	0	0	0	0	0	0	0	0	0	0	0	0
338	0	0	0	0	0	0	0	0	0	0	0	0	0	0	0	0	0
339	0	0	0	0	0	0	0	0	0	0	0	0	0	0	0	0	0
340	0	0	0	0	0	0	0	0	0	0	0	0	0	0	0	0	0
341	0	0	0	0	0	0	0	0	0	0	0	0	0	0	0	0	0
342	0	0	0	0	0	0	0	0	0	0	0	0	0	0	0	0	0
343	0	0	0	0	0	0	0	0	0	0	0	0	0	0	0	0	0
344	0	0	0	0	0	0	0	0	0	0	0	0	0	0	0	0	0
345	1	0	0	0	0	0	0	0	0	0	0	0	0	0	0	0	0
346	0	0	0	0	0	0	0	0	0	0	0	0	0	0	0	0	1
347	0	1	0	0	0	0	0	0	0	0	0	0	0	0	0	0	0
348	0	0	0	1	0	0	0	0	0	0	0	0	0	0	0	0	0
349	0	0	0	0	1	0	0	0	0	0	0	0	0	0	0	0	0
350	0	0	0	0	0	1	0	0	0	0	0	0	0	0	0	0	0
351	0	0	0	0	0	0	1	0	0	0	0	0	0	0	0	0	0
352	0	0	0	0	0	0	0	1	0	0	0	0	0	0	0	0	0
353	0	0	0	0	0	0	0	0	1	0	0	0	0	0	0	0	0
354	0	0	0	0	0	0	0	0	0	1	0	0	0	0	0	0	0
355	0	0	0	0	0	0	0	0	0	0	1	0	0	0	0	0	0
356	0	0	0	0	0	0	0	0	0	0	0	1	0	0	0	0	0
357	0	0	0	0	0	0	0	0	0	0	0	0	1	0	0	0	0
358	0	0	0	0	0	0	0	0	0	0	0	0	0	1	0	0	0
359	0	0	0	0	0	0	0	0	0	0	0	0	0	0	1	0	0
360	0	0	0	0	0	0	0	0	0	0	0	0	0	0	0	1	0
361	0	0	0	0	0	0	0	0	0	0	0	0	0	0	0	0	1
362	0	0	0	0	0	0	0	0	0	0	0	0	0	0	0	0	0
363	0	0	0	0	0	0	0	0	0	0	0	0	0	0	0	0	0
364	-1.34451	5.751755	2.996157	5.319995	4.244133	4.187866	5.776731	1.223943	4.49183	2.26903	1.206447	1.224203	0.402059	-1.33978	0.503924	0.478938	3.968198
365	28.3769	28.41878	27.6641	26.87328	24.54189	25.75538	28.22783	28.02546	21.73216	24.00456	22.96647	22.89774	23.28709	25.1828	23.20849	22.80997	30.80194
366	2.646942	4.47508	4.106527	3.433524	2.028218	3.83471	6.442334	4.774928	6.567492	4.719615	3.228784	3.116339	2.904463	2.600697	2.947877	2.737821	4.720956
367	0.000682	0.001872	0.000833	0.001332	0.000714	0.000714	0.000769	0.001136	0.000652	ERR	ERR	ERR	ERR	ERR	ERR	ERR	0.000907
368	0.023537	0.001501	0.007186	0.004596	0.004567	0.000644	0.000386	4.7E-05	1.3E-06	0.018998	0.014955	0.01935	0.015763	0.067463	0.012839	0.001732	0.002236
369	3.718476	0.067266	1.057422	0.093334	0.023758	0.025986	0.226721	0.010118	0.000881	10.60668	2.351911	2.801162	2.209469	21.48614	1.684623	0.118693	0.34138
370	188.7817	21.78315	135.4037	104.0392	62.1561	4.390136	5.125685	0.032971	0.001016	677.5463	321.015	477.472	253.7339	1594.534	186.1094	7.941331	26.76705
371	-313.795	47.9936	-56.2983	193.8081	246.6326	-8.59979	-122.114	-565.965	-343.679	3.681126	3.978137	3.996652	3.781645	2.331509	3.90764	4.528556	5.269632
372	2.683402	5.63247	4.206068	5.105538	4.878439	5.091195	5.490726	4.965908	5.767573	49.95195	98.50276	109.387	84.58779	59.07128	82.45671	71.72041	131.4678
373	11.52918	10.95202	10.81734	10.78824	11.04053	11.36215	10.85692	12.61822	12.23881	5.718561	4.775249	4.758098	4.587037	3.981852	4.647085	4.298454	4.972919

FIGURE 40 (continued) Influence of chemical groups on physical and biological properties.

B	A	B	C	D	E	F	G	H	I	J	K	L	M	N
1	NAME			B	C	D	E	F		tap	air	com	res	MCL
2	PCE		0	0.12942	-0.10019	-0.03781	-0.01863	-0.00022	b	0.681984	0.439809	2.273436	1.893833	2.084032
3	TCE		0	0.04489	-0.06222	-0.03297	-0.01174	-0.00249	c	14.21804	17.65265	17.73686	17.59624	14.7937
4	1,1-DCE		0	-0.04955	-0.08125	-0.0923	-0.00882	-0.00287	d	16.56847	14.10209	15.5473	21.39851	16.6342
5	CH3CCL3		0	0.04482	-0.06043	-0.01872	-0.01988	-0.00438	e	-71.6153	-58.6877	-54.5266	-102.557	-75.3489
6	CLCH2CH2CL		0	-0.02461	0.04194	-0.01907	-0.01085	-0.00735	f	58.15441	99.7817	105.4682	97.51875	80.87492
7	LINDANE		0	0.650492	0.227454	-0.00934	-0.01033	-0.00912		a				
8	CHLOROBENZENE		0	0.124999	0.023542	-0.02897	0.01476	-0.00966		4.59068	0	1	0	0
9	P-DICHLOROBENZ		0	0.224135	0.092169	-0.02532	0.017603	-0.01238		0	-0.18926	0	0	0
10	CCL4		0	0.097386	0.217174	-0.02268	-0.00025	-0.01634		0	0	-0.11241	0	0
11	CH2CLCHCLCH3		0	0.051662	0.003968	-0.00816	-0.01596	-0.00382		0	0	0	-5.08345	0
12	ENDRIN		0	1.144564	0.538338	-0.1744	0.003374	-0.07533		0	0	0	0	-4.06262
13	HEPTACHLOR		0	1.094874	0.435606	-0.15102	-0.00224	-0.05697						
14	HEXACHLOROBEN		0	0.620679	0.366677	-0.01069	0.028976	-0.02327						
15	C16H15CL3O2		0	1.087393	0.121172	0.156889	0.040672	-0.03669						
16	DICHLOROMETHAN		0	-0.12069	-0.03408	-0.03219	-0.01473	-0.00709						
17	C9H7CL3O3 SILVEX		0	0.5402	-0.02161	0.010827	0.017955	0.003468						
18	STYRENE		0	0.172774	-0.08803	-0.00505	0.019488	-0.00348						
19	1,2,4-TRICHLORO-		0	0.323271	0.160796	-0.02166	0.020446	-0.0151						
20	CLCH2CHCL2		0	0.074524	0.110571	-0.01542	-0.0081	-0.01008						
21	CHLORDANE		0	1.240548	0.120289	0.158875	0.040825	-0.03627						
22	C8H6CL2O3		0	0.36479	-0.05226	-0.00375	0.020127	0.002655						
23	HEPTACHLOREPO		0	1.013189	0.52041	-0.04063	0.035445	-0.03811						
24	BENZO(a)PYRENE		0	1.062891	-0.24077	-0.02347	-0.01537	0.068776						
25	VINYL CHLORIDE		0	-0.13409	-0.04328	-0.02438	-0.00192	-0.00515						
26	BENZENE		0	0.025863	-0.04509	-0.03263	0.011994	-0.00694						
27	CH2CLCHBRCH2BR		0	0.02346	0.087566	-0.01896	-0.01295	-0.0054						
28	CH2BRCH2BR		0	0.072707	0.058681	-0.03755	-0.01108	-0.0054						
29	O-DICHLOROBENZ		0	0.224135	0.092169	-0.02532	0.017603	-0.01238						
30	ENDRIN		0	1.144564	0.538338	-0.1744	0.003374	-0.07533						
31	ETHYL BENZENE		0	0.178412	-0.12104	-0.0108	0.001886	0.000137						
32	C5CL6		0	0.562942	0.331445	-0.00876	0.029889	-0.01989						
33	PENTACHLORPPHI		0	0.685816	0.604054	0.00065	0.028417	0.000613						
34	***TCDD DI***	***IOXIN***	0	0.875294	0.378511	0.125929	0.044079	0.008024						
35	TOLUENE		0	0.102138	-0.08306	-0.02171	0.006901	-0.0034						
36	TOXAPHENE		0	1.211823	0.306311	-0.01911	-0.0189	-0.02493						
37	XYLENES		0	0.178412	-0.12104	-0.0108	0.001886	0.000137						
38	BENZOBFLUORANT		0	1.065991	-0.17507	-0.04157	-0.00137	0.044676						

FIGURE 41 Structural parameters and structure to property parameter used in SYNPROPS.

A	P	Q	R	S	T	U	V	W	X	Y	Z	AA	AB	AC	AD	AE
49	TABLE IV	MOLE FRACTION S		LUTE X IN												
50	WATER	0	0.001	0.005	0.01	0.05	0.1	0.2	0.3	0.4	0.5	0.6	0.7	0.8	0.9	1
51	E rho g/cc	0.7782	0.778491	0.779654	0.781108	0.792739	0.807278	0.836356	0.865434	0.894512	0.92359	0.952668	0.981745	1.010823	1.039901	1.068979
52	LOG(E)	0.7782	0.98443	1.809351	2.840502	11.08971	21.40122	42.02425	62.64727	83.27029	103.8933	124.5163	145.1394	165.7624	186.3854	207.0084
53	CHI	-12.97	-8.78591	7.950465	28.87093	196.2347	405.4393	823.8486	1242.258	1660.667	2079.077	2497.486	2915.895	3334.304	3752.714	4171.123
54	-LOGKb	-19.405	-13.6317	9.46132	38.32764	269.2582	557.9214	1135.248	1712.574	2289.901	2867.227	3444.553	4021.88	4599.206	5176.533	5753.859
55	DIPOLE M	1.87	2.412477	4.582386	7.294772	28.99386	56.11772	110.3654	164.6132	218.8609	273.1086	327.3563	381.604	435.8517	490.0995	544.3472
56	SUM INT	66.06841	66.00634	65.75807	65.44773	62.96499	59.86157	53.65473	47.44789	41.24105	35.03421	28.82736	22.62052	16.41368	10.20684	4
57	Tb-Tm	100	185.7526	528.7631	957.5261	4387.631	8675.261	17250.52	25825.78	34401.05	42976.31	51551.57	60126.83	68702.09	77277.35	85852.61
58	MOL. WT	18.016	72.14869	288.6794	559.3429	2724.65	5431.285	10844.55	16257.82	21671.09	27084.36	32497.63	37910.9	43324.16	48737.43	54150.7
59	COUNT	455.9075	455.4536	453.638	451.3684	433.2121	410.5168	365.126	319.7353	274.3445	228.9538	183.563	138.1723	92.7815	47.39075	2
60	logben/h2	0	1.592112	7.960558	15.92112	79.60558	159.2112	318.4223	477.6335	636.8446	796.0558	955.267	1114.478	1273.689	1432.9	1592.112
61	Tm	0	106.6588	533.294	1066.588	5332.94	10665.88	21331.76	31997.64	42663.52	53329.4	63995.28	74661.16	85327.04	95992.92	106658.8
62	rho g/cc	0.3	0.301283	0.306413	0.312826	0.364131	0.428262	0.556524	0.684785	0.813047	0.941309	1.069571	1.197833	1.326094	1.454356	1.582618
63	B	1.803739	2.163167	3.600878	5.398018	19.77513	37.74653	73.68931	109.6321	145.5749	181.5177	217.4605	253.4032	289.346	325.2888	361.2316
64	C	0	-1.3E-06	-6.4E-06	-1.3E-05	-6.4E-05	-0.00013	-0.00026	-0.00039	-0.00051	-0.00064	-0.00077	-0.0009	-0.00103	-0.00116	-0.00128
65	D	7.593541	7.585904	7.555354	7.517166	7.211667	6.829794	6.066047	5.302299	4.538552	3.774805	3.011058	2.247311	1.483564	0.719816	-0.04393
66	E	2.425988	2.423519	2.413644	2.4013	2.302549	2.179109	1.93223	1.685351	1.438472	1.191593	0.944714	0.697835	0.450956	0.204077	-0.0428
67	F	-1.60276	-1.60119	-1.59491	-1.58707	-1.52429	-1.44583	-1.2889	-1.13197	-0.97503	-0.8181	-0.66117	-0.50424	-0.34731	-0.19038	-0.03345
68	Tb	100	292.4114	1062.057	2024.114	9720.571	19341.14	38582.28	57823.42	77064.57	96305.71	115546.8	134788	154029.1	173270.3	192511.4
69	SOL PAR	9.4	11.28665	18.83326	28.26652	103.7326	198.0652	386.7304	575.3956	764.0608	952.726	1141.391	1330.056	1518.722	1707.387	1896.052
70	Tc	374.1	652.3711	1765.455	3156.811	14287.65	28201.21	56028.31	83855.42	111682.5	139509.6	167336.7	195163.8	222990.9	250818	278645.2
71	**Pc**	218.3	215.5837	204.7187	191.1373	82.4867	-53.3266	-324.953	-596.58	-868.206	-1139.83	-1411.46	-1683.09	-1954.71	-2226.34	-2497.97
72	log AC CC	0	1.839201	9.196005	18.39201	91.96005	183.9201	367.8402	551.7603	735.6804	919.6005	1103.521	1287.441	1471.361	1655.281	1839.201
73	log par co	0	1.320204	6.601022	13.20204	66.01022	132.0204	264.0409	396.0613	528.0817	660.1022	792.1226	924.1431	1056.163	1188.184	1320.204
74	Hv	-80.4637	-74.9851	-53.0705	-25.6773	193.4684	467.4004	1015.265	1563.129	2110.993	2658.857	3206.721	3754.585	4302.449	4850.313	5398.178
75	Surf Ten	71.97	86.70853	145.6626	219.3553	808.8963	1545.823	3019.675	4493.528	5967.38	7441.233	8915.085	10388.94	11862.79	13336.64	14810.5
76	LOG visc	-1.3516	-0.46886	3.062119	7.475838	42.78559	86.92278	175.1972	263.4716	351.7459	440.0203	528.2947	616.5691	704.8435	793.1179	881.3922
77	log(et/h2o	0	1.194728	5.973639	11.94728	59.73639	119.4728	238.9456	358.4184	477.8912	597.3639	716.8367	836.3095	955.7823	1075.255	1194.728
78	logoil/h2o	0	1.26547	6.327349	12.6547	63.27349	126.547	253.094	379.641	506.188	632.7349	759.2819	885.8289	1012.376	1138.923	1265.47
79	logcyh/h2	0	1.121958	5.609792	11.21958	56.09792	112.1958	224.3917	336.5875	448.7833	560.9792	673.175	785.3708	897.5667	1009.763	1121.958
80	logchl/h2o	0	1.881192	9.40596	18.81192	94.0596	188.1192	376.2384	564.3576	752.4768	940.596	1128.715	1316.834	1504.954	1693.073	1881.192
81	logcl4/h2o	0	1.195153	5.975767	11.95153	59.75767	119.5153	239.0307	358.546	478.0614	597.5767	717.092	836.6074	956.1227	1075.638	1195.153
82	D1	0.0002	0.0002	0.000199	0.000198	0.000191	0.000182	0.000163	0.000145	8.7E-05	4.3E-05	1.7E-05	5.2E-06	1E-06	1E-07	1.6E-05
83	molvolexp	0.0002	50.65666	253.2825	506.5648	2532.823	5065.646	10131.29	15196.94	20262.58	25328.23	30393.87	35459.52	40525.16	45590.81	50656.46
84	isothcom	0.0005	-8.15546	-40.7793	-81.5591	-407.797	-815.595	-1631.19	-2446.79	-3262.38	-4077.98	-4893.57	-5709.17	-6524.77	-7340.36	-8155.96
85	volbondi	1	35.2149	172.0745	343.149	1711.745	3422.49	6843.98	10265.47	13686.96	17108.45	20529.94	23951.43	27372.92	30794.41	34215.9
86	m.refract.	11.2	30.31395	106.7697	202.3395	966.8973	1922.595	3833.989	5745.384	7656.778	9568.173	11479.57	13390.96	15302.36	17213.75	19125.15
87	thermcon	4.14	5.292311	9.901554	15.66311	61.75554	119.3711	234.6021	349.8332	465.0643	580.2954	695.5264	810.7575	925.9886	1041.22	1156.451
88	RHO(B-E)	1.0355	1.034978	1.032891	1.030281	1.009407	0.983314	0.931128	0.878942	0.826755	0.774569	0.722383	0.670197	0.618011	0.565825	0.513639
89	Zc	0.23	ERR	ERR	ERR	ERR	ERR	ERR	ERR	ERR	ERR	ERR	ERR	ERR	ERR	ERR

FIGURE 42 Properties of aqueous solutions.

	A	B	C	D	E	F	G	H	I	J	K	L	M	N	O	P	Q
72														B/A^.5	0.011302		
73														B	-39.8587	A^(1/2)	-3526.7
74							HIERARCHICAL MODEL							rho (B-E)	-152.734	mol wt	596.829
75	coefficient	0												coefficiem	0	E rho g/cc	141.742
76	intercept	1	-50.647	-50.647	-5.605	-5.605	0.663	0.663	3.12	3.12	2.8	2.8		intercept	1	LOG(E)	-764.54
77	mol wt	151.0887	0.2251	34.01007	0.0294	4.442009	-0.0669	-10.1078	-0.0051	-0.77055	-0.014	-2.11524		mol. wt.	151.0887	CHI	4347.02
78	O atoms	6.046346	0.82	4.958003	-0.32	-1.93483	-0.31	-1.87437	-0.39	-2.35807	-0.28	-1.69298		# o	6.046346	-LOGKb	-60913.6
79	N atoms	0	1.17	0	-1.31	0	0.55	0	-0.53	0	-0.3	0		# n	0	DIPOLE M	-6789.97
80	S atoms	0	2.31	0	8.1	0	8.67	0	4.73	0	0.58	0		# s	0	SUM INT	12
81	c-c bonds	0	-4.09	0	-3.91	0	3.69	0	1.67	0	2.11	0		c-c	0	Tb-Tm	-129431
82	arom. c-c	0	-0.08	0	-1.71	0	2.77	0	1.27	0	1.08	0		arom. c-c	0	MOL. WT	-128597
83	c=c bonds	0	4.36	0	-0.85	0	2.47	0	1.32	0	0.16	0		c=c	0	COUNT	12.56689
84	c≡-bonds	0	9.26	0	8.8	0	0.34	0	0.34	0	-1.14	0	#GRP/	c≡-c	0	logben/h2	-6002.83
85	c-h bonds	0	4.28	0	-0.15	0	-0.83	0	-1.05	0	-0.78	0	CPD	c-h	0	Tm	-135174
86	c-f bonds	0	1.48	0	6.24	0	2.2	0	-1.17	0	-0.66	0		c-f	0	B rho g/cc	-10.9916
87	c-cl bonds	0	6.44	0	5.7	0	1.84	0	-0.59	0	-0.57	0		c-cl	0	B	-0.13914
88	c-br bonds	0	1.3	0	5.23	0	3.89	0	-0.37	0	0.15	0		c-br	0	C	0.61
89	c-i bonds	3.512414	-0.38	-1.33472	4.64	16.2976	4.78	16.78934	-0.84	-2.95043	-0.24	-0.84298		c-I	3.512414	D	0.3292
90	-oh grps	0	13.52	0	30.3	0	2.05	0	-0.35	0	1.84	0		oh	0	E	0.045
91	-o- grps	3.482321	-4.14	-14.4168	11.39	39.66364	11.2	39.002	3.89	13.54623	0.44	1.532221		o	3.482321	F	333
92	-(c=o)- gr	0.122784	10.43	1.280635	27.99	3.436718	4.59	0.563578	0.88	0.10805	-3.56	-0.43711		c=o	0.122784	Tb	-264605
93	-ch=o grp	2.44124	1.85	4.516295	-2.04	-4.98013	-2.37	-5.78574	-1.8	-4.39423	1.23	3.002726		ch=o	2.44124	SOL PAR	-2553.99
94	cooh	0	-1.11	0	-22.99	0	-6.02	0	-2.71	0	1.61	0		cooh	0	Tc	-602688
95	-coo- grps	0	-2.43	0	-19.86	0	-5.91	0	-2.53	0	2.67	0		coo	0	**Pc**	-76590.7
96	-sh grps	0	11.28	0	4.44	0	-7.22	0	-5.15	0	-0.41	0		sh	0	log AC CC	2471.08
97	-NH2	0	9.84	0	25.76	0	6.05	0	4.02	0	0.17	0		nh2	0	log par co	1689.49
98	-nh- grps	0	-0.46	0	22.02	0	13.6	0	7.68	0	1.62	0		nh	0	Hv	4837.254
99	-n- - grps	0	-11.06	0	13.95	0	18.39	0	9.9	0	5.08	0		-n--	0	Surf Ten	-5.70454
100	=n- grps	0	3.82	0	16.12	0	7.76	0	2.55	0	1.72	0		=n-	0	log visc	1684.505
101	-cn grps	0	19.11	0	29.38	0	-1.38	0	-2.72	0	-4.5	0		cn	0	log(et/h2o	324.1547
102	-conh2 gr	0	14.22	0	-1.59	0	-11.7	0	-8.91	0	2.34	0		conh2	0	logoil/h2o	-1006.41
103	-no2 grps	0	8.55	0	22.73	0	3.82	0	-8.35	0	-1.18	0		no2	0	logcyh/h2	-6337.1
104	cycles(ring	1	10.45	10.45	9.96	9.96	-10.34	-10.34	-2.85	-2.85	-4.2	-4.2		cycles(ring	1	logchl/h2o	-8663.59
105	tertiary C	1	-0.74	-0.74	-0.12	-0.12	1.44	1.44	0.3	0.3	-0.67	-0.67		tert C	1	logcl4/h2o	-10997
106	quatern c	1	-1.99	-1.99	-0.16	-0.16	2.57	2.57	0.95	0.95	-0.61	-0.61		quat C	1	vol.Bondi	11699.49
107	x-c-x trace	0	-0.99	0	-5.7	0	-0.11	0	-0.44	0	0.19	0		x-c-x	0	v exp	-907.256
108	X3-C-C-X3	0	-0.77	0	-17.62	0	-0.63	0	-1.42	0	-0.51	0		x3-c-x3	0	C	4.320231
109	c=c-x trac	0	-0.47	0	-4.96	0	-0.74	0	-0.58	0	0.31	0		c=c-x	0	H	2.44124
110	arom fusio	2E-06	-0.31	-6.2E-07	-6.57	-1.3E-05	1.81	3.6E-06	-1.4	-2.8E-06	2.41	4.8E-06		arom. fusi	2E-06	O	6.046346
111		1	B=	-0.13914	C=	0.61	D=	0.3292	E=	0.04701	F=	-0.03233		S	0	N	0
112														CL	0	F	0
113														BR	0	I	*3.512414*
114														loggamma	2471.076	D1	8.3E-06
115														m.refract	-3995.09	D12	
116														thermcond	-4822.93	isothcom	7353.613
117														COST $/g	0.00201	w	ERR
118														MCL	0	Zc	ERR
119														TAP	0	AIR	0
120														COM	0	RES	0
121														drug1	13475.72	drug2	-685477
122														drug4	98165.33	drug3	8984.079
123														drug6	-135.681	drug5	6560.406

$$P_{ij} = a_i + b_i B_j + c_i C_j + d_i D_j + e_i E_j + f_i F_j \quad \textbf{for physical properties}$$

$$-LN(P_{ij}) = a_i + b_i B_j + c_i C_j + d_i D_j + e_i E_j + f_i F_j \quad \textbf{for toxic properties}$$

$$P = \exp(-a_i).\ \exp(-b_i B_j).\ \exp(-c_i C_j).\ \exp(-d_i D_j).\ \exp(-d_i D_j).\ \exp(-e_i E_j).\exp(-f_i F_j)$$

FIGURE 43 SYNPROPS spreadsheet of hierarchical model.

	A	B	C	D	E	F	G	H	I	J	K	L	M	N	O	P	Q
73	TABLE I - GROUP PARAMETERS / B, C, D, E, F					LINEAR	MODEL							B	0.021788	A^(1/2)	1.019675
74	-H	1	0.066	0.066	0.018	0.018	-0.027	-0.027	-0.019	-0.019	-0.019	-0.019		-H	1	mol wt	1.0079
75	-CH3	0	0.142	0	-0.02	0	-0.016	0	-0.023	0	-0.015	0		-CH3	0	E rho g/cc	1.035896
76	-CH2-	0	0.076	0	-0.038	0	0.011	0	-0.004	0	0.003	0		-CH2-	0	LOG(E)	0.335615
77	>CH-	0	0.003	0	-0.058	0	0.053	0	0.018	0	0.015	0		>CH-	0	CHI	52.4805
78	>C<	0	-0.075	0	-0.076	0	0.091	0	0.043	0	0.034	0		>C<	0	-LOGKb	-0.56798
79	-CH=CH-	0	0.147	0	-0.043	0	0.028	0	0.01	0	0.003	0		-CH=CH-	0	DIPOLE M	0.20565
80	-CH=CH2	0	0.212	0	-0.025	0	0	0	-0.009	0	-0.015	0		-CH=CH2	0	SUM INT	1
81	>C=CH2	0	0.147	0	-0.043	0	0.028	0	0.01	0	0.003	0		>C=CH2	0	Tb-Tm	-128.725
82	-C=-CH	0	0.171	0	0.074	0	0.027	0	0.002	0	-0.012	0		-C=-CH	0	MOL. WT	26.4226
83	-C6H5	0	0.467	0	-0.007	0	0.012	0	0.007	0	-0.017	0		-C6H5	0	COUNT	0
84	=CH-	0	0.088	0	0.002	0	-0.007	0	0.001	0	-0.003	0	#GRP/	=CH-	0	logben/h2	0.3314
85	AROMAT	0		0		0		0		0		0	CPD	AROM	0	Tm	-50.8455
86	-NAPHTH	0	0.766	0	0.018	0	-0.026	0	0.024	0	-0.028	0		-NAPTH	0	B rho g/cc	3.666394
87	-CYCLOH	0	0.489	0	-0.148	0	0.004	0	-0.029	0	-0.009	0		-CYCLOH	0	B	-0.4405
88	-F	0	0.078	0	0.088	0	0.009	0	-0.019	0	-0.02	0		-F	0	C	-0.038
89	-CL	0	0.165	0	0.087	0	-0.024	0	-0.012	0	-0.021	0		-CL	0	D	-0.02
90	-BR	0	0.213	0	0.095	0	-0.033	0	-0.008	0	-0.02	0		-BR	0	E	0.012
91	-I	0	0.302	0	0.103	0	-0.056	0	-0.01	0	-0.031	0		-I	0	F	0.009
92	-CF3	0	0.15	0	0.017	0	0.035	0	-0.037	0	-0.013	0		CF3	0	Tb	-179.57
93	-CCL3	0	0.41	0	0.015	0	-0.009	0	-0.017	0	-0.017	0		-CCL3	0	SOL PAR	6.28669
94	-OH	0	0.202	0	0.324	0	-0.012	0	-0.015	0	0.003	0		-OH	0	Tc	-102.365
95	-O-	0	0.044	0	0.155	0	0.061	0	0.019	0	-0.022	0		-O-	0	Pc	49.46995
96	-C=0	0	0.135	0	0.246	0	0.061	0	0.023	0	-0.021	0		-C=O	0	log AC CC	-1.43319
97	-CH=O-	0	0.219	0	0.244	0	0.01	0	-0.014	0	-0.027	0		-CH=O	0	log par co	0.444145
98	-COO-	0	0.167	0	0.17	0	0.062	0	0.015	0	-0.027	0		-COO-	0	Hv	1.39767
99	-COOH	0	0.323	0	0.342	0	-0.011	0	-0.017	0	0.008	0		-COOH	0	Surf Ten	-3.9E+13
100	-NH2	0	0.167	0	0.269	0	0.037	0	0.027	0	-0.014	0		-NH2	0	log visc	-1.38556
101	-NH-	0	0.082	0	0.251	0	0.095	0	0.056	0	-0.01	0		-NH-	0	log(et/h2o	0.98335
102	-N-	0	-0.006	0	0.189	0	0.125	0	0.069	0	0.014	0		-N-	0	logoil/h2o	0.34285
103	-CN	0	0.241	0	0.269	0	-0.007	0	-0.023	0	-0.041	0		-CN	0	logcyh/h2(	0.47705
104	-N=	0	0.102	0	0.183	0	0.031	0	-0.011	0	-0.02	0		-N=	0	logchl/h2o	-0.343
105	PYRIDINE	0		0		0		0		0		0		PYRIDINE	0	logcl4/h2o	0.27635
106	-NO2	0	0.238	0	0.241	0	-0.012	0	-0.027	0	-0.037	0		-NO2	0	vol.Bondi	7.2067
107	-CONH2	0	0.444	0	0.499	0	-0.019	0	-0.039	0	-0.012	0		-CONH2	0	v exp	25.507
108	-S-	0	0.136	0	0.13	0	0.028	0	0.032	0	-0.02	0		-S-	0	C	0
109	-SH	0	0.231	0	0.155	0	-0.026	0	-0.011	0	-0.013	0		-SH	0	H	1
110	MOLECUL	1	-0.5065	-0.5065	-0.056	-0.056	0.007	0.007	0.031	0.031	0.028	0.028		MOLECUL	1	O	0
111		0	B =	-0.4405	C =	-0.038	D =	-0.02	E =	0.012	F =	0.009		S	0	N	0
112		0												CL	0	F	0
113														BR	0	I	0
114														loggamma	-1.43439	D1	-0.01003
115														m.refract	1.272815	LC50mg/L	1.064864
116														thermcond	1.262905	isothcom	21.1437
117														COST $/g	ERR	w	-1.68893
118														MCL	45.83906	Zc	0.426114
119														TAP	3.133201	AIR	6.073325
120														COM	1583.394	RES	424.8742
121														drug1	3.694379	drug2	479.5124
122														drug4	19.56298	drug3	3.339547
123														drug6	6.089324	drug5	7.383173

```
A:D83:   +B83*C83

A:D111:  (@SUM(D74..D110))

A:Q87:   (-0.15*B84+6.24*B85+5.7*B86+5.23*B87+4.64*B88+30.3*B89+11.39*B90+27.99*B91-2.04*B92-22.99*B93-19.86*B
         94+4.44*B95+25.76*B96+22.02*B97+13.95*B98+16.12*B99+29.38*B100-1.59*B101+22.73*B102+9.96*B103-0.12*
         B104+8.8*B83-0.85*B82-5.605*B75+0.0294*B76-0.32*B77-3.91*B80-1.71*B81-17.62*B107-1.31*B78-0.16*B105-5.7
         *B106-4.96*B108-6.57*B109+8.1*B79)*0.01

A:Q93:   +248.6+770.4*Q86+343.1*Q87-927*Q88-68*Q89-1810*Q90

A:Q119:  @EXP(-(-4.06262+2.084032*Q86+14.7937*Q87+16.6342*Q88-75.3489*Q89+80.87492*Q90))

A:O119:  @EXP(-(-5.08345+1.893833*Q86+17.59624*Q87+21.39851*Q88-102.557*Q89+97.51875*Q90))
```

FIGURE 44 SYNPROPS spreadsheet of linear model.

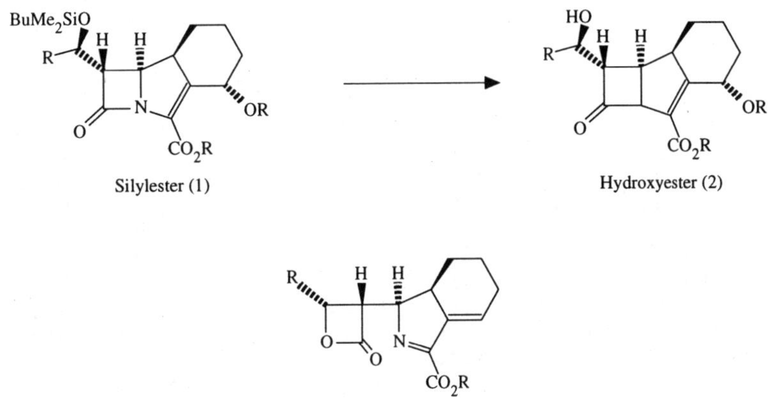

FIGURE 45 Synthesis and table from cleaner synthesis.

```
SPECIES
CY13PD
Thermo estimation for molecule
CY13PD                                                          C5H6
UNITS:KJ
GROUPS    4
  Gr # -   GROUP ID  - Quantity
     1 - CD/CD/H         -  2
     2 - CD/C/H          -  2
     3 - C/CD2/H2        -  1
     4 - CY/C5/DE        -  1
  Hf      S    Cp 300     400     500      600      800     1000     1500
  135.56  268.54  73.81 102.22 125.56  144.35  172.76  192.67  221.84
          CPINF =   257.72
SYMMETRY       2
 CREATION DATE:  4/11/89
 ENDSPECIES
```

```
SPECIES
OH
Thermo Data Entered:
OH                                                              OH
UNITS:KJ
GROUPS    0
NOCALC
  Hf      S    Cp 300     400      500      600      800     1000     1500
   39.75  183.59  29.96  29.71   29.46    29.71    29.96    30.71    32.93
          CPINF =   33.25
SYMMETRY       1
 ENDSPECIES
```

FIGURE 46 Thermo estimations for molecules in THERM.

UNITS:KCAL			4/12/89	dstbt w/ THERM					
80	Hf	S	Cp: 300	400	500	600	800	1000	1500
C/C/H3,	-10.20	30.41	6.19	7.84	9.40	10.79	13.02	14.77	17.58
C/CB/H3,	-10.20	30.41	6.19	7.84	9.40	10.79	13.02	14.77	17.58
C/CD/H3,	-10.20	30.41	6.19	7.84	9.40	10.79	13.02	14.77	17.58
C/CT/H3,	-10.20	30.41	6.19	7.84	9.40	10.79	13.02	14.77	17.58
C/C2/H2,	-4.93	9.42	5.50	6.95	8.25	9.35	11.07	12.34	14.20
C/C3/H,	-1.90	-12.07	4.54	6.00	7.17	8.05	9.31	10.05	11.18
C/C4,	0.50	-35.10	4.37	6.13	7.36	8.12	8.77	8.76	8.12
CD/H2,	6.26	27.61	5.10	6.36	7.51	8.50	10.07	11.27	13.19
CD/C/H,	8.59	7.97	4.16	5.03	5.81	6.50	7.65	8.45	9.62
CD/C2,	10.34	-12.70	4.10	4.61	4.99	5.26	5.80	6.08	6.36
CD/CD/H,	6.78	6.38	4.46	5.79	6.75	7.42	8.35	9.11	10.09
CD/C/CD,	8.88	-14.60	4.40	5.37	5.93	6.18	6.50	6.62	6.72
CD/CB/H,	6.78	6.38	4.46	5.79	6.75	7.42	8.35	9.11	10.09
CD/C/CB,	8.64	-14.60	4.40	5.37	5.93	6.18	6.50	6.62	6.72
CD/CB/CD,	7.18	-16.50	4.70	6.13	6.87	7.10	7.20	7.16	7.06
CD/CD/CT,	6.94	-16.50	4.70	6.13	6.87	7.10	7.20	7.16	7.06
CD/CB/CT,	6.70	-16.50	4.70	6.13	6.87	7.10	7.20	7.16	7.06
CD/CD2,	4.60	-16.50	4.70	6.13	6.87	7.10	7.20	7.16	7.06
CD/CT2,	6.46	-16.50	4.70	6.13	6.87	7.10	7.20	7.16	7.06
CD/CB2,	8.00	-16.50	4.70	6.13	6.87	7.10	7.20	7.16	7.06
CD/CT/H,	6.78	8.03	4.46	5.79	6.75	7.42	8.35	9.11	10.09
CD/C/CT,	8.53	-14.60	4.40	5.37	5.93	6.18	6.50	6.62	6.72
C/C/CD/H2,	-4.76	9.80	5.12	6.86	8.32	9.49	11.22	12.48	14.36
C/CD2/H2,	-4.29	10.20	4.70	6.80	8.40	9.60	11.30	12.60	14.40
C/CB/CD/H2,	-5.29	10.20	4.70	6.80	8.40	9.60	11.30	12.60	14.40
C/C/CT/H2,	-4.73	10.30	4.95	6.56	7.93	9.08	10.86	12.19	14.20
C/C/CB/H2,	-4.86	9.34	5.84	7.61	8.98	10.01	11.49	12.54	13.76
C/C2/CD/H,	-1.48	-11.69	4.16	5.91	7.34	8.19	9.46	10.19	11.28
C/C2/CT/H,	-1.72	-11.19	3.99	5.61	6.85	7.78	9.10	9.90	11.12
C/C2/CB/H,	-0.98	-12.15	4.88	6.66	7.90	8.75	9.73	10.25	10.68
C/C/CB/CT/H,	-1.55	-11.65	4.33	6.27	7.58	8.48	9.52	10.1	10.63
C/C/CB/CD/H,	-1.56	-11.77	4.50	6.57	8.07	8.89	9.88	10.39	10.79
C/C/CD/CT/H,	-2.36	-11.19	3.61	5.52	7.02	7.92	9.25	9.9	11.23
C/C/CD2/H,	-2.31	-11.31	3.78	5.85	7.51	8.30	9.61	10.33	11.39
C/C/CT2/H,	-2.54	-10.69	3.44	5.22	6.53	7.51	8.89	9.75	11.07
C/C/CB2/H,	-1.06	-12.23	5.22	7.32	8.63	8.45	10.15	10.45	10.19
C/CB/CD2/H,	-1.39	-11.39	4.12	6.51	8.24	9.00	10.03	10.53	10.89
C/CD3/H,	-2.14	-10.93	3.4	5.76	7.68	8.44	9.76	10.47	11.5
C/CB3/H,	-0.34	-12.31	5.56	7.98	9.36	10.15	10.57	10.65	9.7
C/C3/CD,	1.68	-34.72	3.99	6.04	7.43	8.26	8.92	8.96	8.23
C/C3/CB,	2.81	-35.18	4.37	6.79	8.09	8.78	9.19	8.96	7.63
C/C3/CT,	2.81	-35.18	4.37	6.79	8.09	8.78	9.19	8.96	7.63
C/C2/CD2,	1.61	-34.34	3.61	5.98	7.51	8.37	9.00	9.02	8.34
C/C/CD3,	2.54	-33.96	3.32	5.86	7.57	8.54	9.22	9.36	8.45
C/C2/CB/CD,	2.99	-34.8	3.99	6.70	8.16	8.92	9.34	9.16	7.74
C/C2/CD/CT,	2.99	-34.8	3.99	6.70	8.16	8.92	9.34	9.16	7.74
C/C/CB2/CD,	5.1	-34.88	3.99	7.36	8.89	9.58	9.76	9.16	7.25
C/C/CB/CD/CT,	5.1	-34.88	3.99	7.36	8.89	9.58	9.76	9.16	7.25
C/C/CD/CT2,	5.1	-34.88	3.99	7.36	8.89	9.58	9.76	9.16	7.25
C/C2/CB2,	1.16	-35.26	3.57	5.98	7.51	8.37	9.00	9.02	8.34
C/C2/CB/CT,	1.16	-35.26	3.57	5.98	7.51	8.37	9.00	9.02	8.34
C/C2/CT2,	1.16	-35.26	3.57	5.98	7.51	8.37	9.00	9.02	8.34
C/C/CB3,	6.23	-35.34	4.37	8.11	9.55	10.10	10.03	9.36	6.65
C/C/CB2/CT,	6.43	-35.34	4.37	8.11	9.55	10.10	10.03	9.36	6.65
C/C/CB/CT2,	6.23	-35.34	4.37	8.11	9.55	10.10	10.03	9.36	6.65
C/C/CT3,	6.23	-35.34	4.37	8.11	9.55	10.10	10.03	9.36	6.65
C/CB2/CD2,	5.48	-34.50	3.61	7.30	8.97	9.69	9.84	9.42	7.36
C/CB/CD2/CT,	5.48	-34.50	3.61	7.30	8.97	9.69	9.84	9.42	7.36

FIGURE 47 Table of Therm values for groups in THERM.

```
C6H5NO NITROSO-PHENYL or NITROSO-BENZENE  CALCULATED BY BOZZELLI USING GROUP
ESTIMATES. EXTRAPOLATED TO 6000 K USING WILHOIT'S POLYNOMIALS. HF298=48.0 KCAL
REF=Choo, Golden & Benson, Int. J. Chem Kinet 7,(1975),713. Max Lst Sq Error
Cp @ 1500 K 0.36%.
C6H5NO                 T 7/95C   6H   5N   1O   1G   298.150  5000.000  F 107.11184 1
 0.15129273E+02 0.20169394E-01-0.79009702E-05 0.14240839E-08-0.96649392E-13    2
 0.17118219E+05-0.56899184E+02 0.20849489E+01 0.34489960E-01 0.27707248E-04    3
-0.65460444E-07 0.29821820E-10 0.21870293E+05 0.15898367E+02 0.24154400E+05    4

C6H5O  PHENOXY RADICAL  SIGMA=2  STATWT=2  IA=14.8396  IB=30.4574  IC=45.297
NU=3087,3063,3027,1603,1501,1261,1168,1025,999,823,526,958,817,409,973,881,751,
686,503,225,3070,3049,1610,1472,1343,1277,1150,1070,619,403  REF=BURCAT,
ZELEZNIK & MCBRIDE HF298=47.698 KJ  MAX LST SQ ERROR CP @ 1300 K 0.81 % .
PHENOXY RAD            L12/84C  6.H  5.O  1.   0.G   300.000  5000.000  B  93.10490 1
 0.13833984E 02 0.17618403E-01-0.60696257E-05 0.91988173E-09-0.50449181E-13    2
-0.69212549E 03-0.50392990E 02-0.18219433E 01 0.48122510E-01-0.46792302E-05    3
-0.34018594E-07 0.18649637E-10 0.42429180E 04 0.33526199E 02 0.57367379E 04    4

C6H5OO  PEROXYPHENYL  SIGMA=1  STATWT=2   IA=16.0594   IB=51.5184   IC=67.5778
NU=3061,3032,3024,3014,3004,1526,1515,1461,1425,1303,1246,1223,1124,1115,1106,
1032,977,936,924,917,860,788,759,712,629,593,587,457,427,382.5,257.4,216.5,59.5
HF298=39.59 KCAL  REF=C. Melius Database AA3V  Max Lst Sq Error Cp @ 200 K 0.65%
C6H5OO                 T03/97C  6.H  5.O  2.   0.G   200.000  6000.000  B 109.10450 1
 1.61783950E+01 1.80959380E-02-6.61459065E-06 1.08059157E-09-6.52339007E-14    2
 1.28261324E+04-5.89741433E+01 1.99359550E-01 4.70697558E-02 8.34324919E-06    3
-5.63540961E-08 2.94168315E-11 1.77945712E+04 2.67267648E+01 1.99223478E+04    4

C6H5OH PHENOL  IA=14.854 IB=32.045 IC=46.8942 NU=3087,3063,3027,1603,1501,
1261,1168,1025,999,823,526,958,817,409,973,881,751,686,503,225,3070,3049,
1610,1472,1343,1277,1150,1070,619,403,3656,1176. HF298= -96.4 KJ.
INTERNAL ROTOR SYMNO=2,V=3468 CAL,IB=1.2236E-40
REF=BURCAT,ZELEZNIK & MCBRIDE  MAX LST SQ ERROR CP/R @ 1300K .75%
C6H5OH                 L 4/84C   6H   6O   1    0G   300.000  5000.000  B  94.11280 1
 0.14912073E 02 0.18378135E-01-0.61983128E-05 0.91983221E-09-0.49209565E-13    2
-0.18375199E 05-0.55924103E 02-0.16956539E 01 0.52271299E-01-0.72024050E-05    3
-0.35859603E-07 0.20449073E-10-0.13284121E 05 0.32542160E 02-0.11594207E 05    4

BENZENE Liquid, DATA TAKEN FROM TRC 4/83 TABLES. HF298(L)=49.08 KJ.
C6H6(L)                P10/86C  6.H  6.   0.   0.L   278.680   500.000  C  78.11364 1
 0.00000000E+00 0.00000000E+00 0.00000000E+00 0.00000000E+00 0.00000000E+00    2
 0.00000000E+00 0.00000000E+00 6.36690229E+01-6.00534398E-01 2.66792810E-03    3
-5.06308828E-06 3.63955562E-09-1.67085472E+03-2.43891797E+02 5.90293355E+03    4

C6H6  BENZENE  IA=IB=14.8396  IC=29.6792  SIGMA=12  NU=3062,992,1326,673,3068,
1010,995,703,1310,1150,849(2),3063(2),1486(2),1038(2),3047(2),1596(2),1178(2),
606(2),975(2),410(2)  REF=SHIMANOUCHI  MAX LST SQ ERROR CP @ 1300 K 0.87 % .
HF298=82.8 KJ REF=BURCAT, ZELEZNIK & MCBRIDE
BENZENE                L12/84C  6.H  6.   0.   0.G   300.000  5000.000  B  78.11364 1
 0.11815166E 02 0.19169778E-01-0.65425238E-05 0.98228425E-09-0.53280361E-13    2
 0.40707441E 04-0.43973511E 02-0.32181215E 01 0.47168836E-01-0.21254918E-05    3
-0.34879005E-07 0.18425386E-10 0.89017773E 04 0.36999313E 02 0.99586128E 04    4

C6H6  FULVENE (5-METHYLENE-1,3-CYCLOPENTADIENE)  SIGMA=2  IA=10.1825  IB=21.7319
IC=31.9144  NU=3059,3053,3051,3036,3028,2980,1677,1604,1524,1423,1341,1318,1227,
1081,1078,976,956,946,943,932,863,785,782,770,692,644,606,483,328,199.8
HF298=56.60 KCAL  REF=C. MELIUS DATABASE A70D  Max Lst Sq Error Cp @ 200 K 0.95%
C6H6  FULVENE          T03/97C  6.H  6.   0.   0.G   200.000  6000.000  B  78.11364 1
 1.19233607E+01 1.98993861E-02-7.21223888E-06 1.17141499E-09-7.04278845E-14    2
 2.27199368E+04-4.13488172E+01 1.25853571E-01 3.04056534E-02 4.01806332E-05    3
-8.27651456E-08 3.77645005E-11 2.68838408E+04 2.44628931E+01 2.84820633E+04    4

C6H7 1,4-CYCLO-RADICAL STATWT=2 SIGMA=2   IA=15.707  IB=16.0895  IC=31.2965
NU=3017,3002,2998,2984,2983,2799.5,2796,1469.5,1456,1435,1405,1379,1332,1258,
1160,1117,1112,1058.5,953,916,909,898,881.5,876,806,710,657,589.5,552,528,473,
335.5,159  HF298=47.942 KCAL REF= BAC/MP4 calculations of C. Melius in 1988.
Private communication    Max Lst sq Error Cp @ 200 K 0.9%.
C6H7 1,4 CYCLO     T 6/93C   6H   7    0    0G   200.000  6000.000  B  79.12158 1
```

FIGURE 48 NASA format for thermodynamic value used in Chemkin.

E	W	X	Y	Z	AA	AB
53	Variable C	Starting	Final	-0.00601	Increment	Decremen
54	E:B74..B7	0.205983	0.204412	-0.09892	0	0
55	E:B75..B7	0.231215	0.229862	0.017624	0	0
56	E:B76..B7	0.298318	0.295906	0.002207	0	0
57	E:B77..B7	0.379374	0.375771	0.10404	0	0
58	E:B78..B7	1.917173	1.912297	0	0	0
59	E:B79..B7	0.231438	0.333	0	0	0
60	E:B80..B8	0.43615	0.333	0	0	0
61	E:B81..B8	0.315743	0.333	0	0	0
62	E:B82..B8	0.23717	0.23602	0	0	0
63	E:B83..B8	0	0	0	0	0
64	E:B84..B8	0.251219	0.249277	0	0	0
65	E:B85..B8	0.267886	0.265246	0	0	0
66	E:B86..B8	0	0.001419	0	0	0
67	E:B87..B8	0.070914	0.073125	0	0	0
68	E:B88..B8	0.216802	0.21543	0	0	0
69	E:B89..B8	0.165472	0.164918	0	0	0
70	E:B90..B9	0.152152	0.152009	0	0	0
71	E:B91..B9	0.161103	0.161495	0	0	0
72	E:B92..B9	0.23536	0.234206	0	0	0
73	E:B93..B9	0.217888	0.218441	0	0	0
74	E:B94..B9	0.224692	0.22527	0	0	0
75	E:B95..B9	0.235279	0.233432	0	0	0
76	E:B96..B9	0.206551	0.205795	0	0	0
77	E:B97..B9	0.155899	0.156387	0	0	0
78	E:B98..B9	0.206878	0.206032	0	0	0
79	E:B99..B9	0	0.001559	0	0	0
80	E:B100..B	0.195312	0.194991	0	0	0
81	E:B101..B	0.292286	0.290415	0	0	0
82	E:B102..B	0.409739	0.406573	0	0	0
83	E:B103..B	0.11473	0.115834	0	0	0
84	E:B104..B	0.219987	0.219082	0	0	0
85	E:B105..B	0.267886	0.265246	0	0	0
86	E:B106..B	0.122624	0.123575	0	0	0
87	E:B107..B	0	0.011189	0	0	0
88	E:B108..B	0.200765	0.199575	0	0	0
89	E:B109..B	0.156013	0.156213	0	0	0
90	E:B107..B	0	0			
91	E:B108..B	0	0			

E	W	X	Y	Z	AA	AB
92	Cell	Value	Constrain	Binding?	Slack	Dual Valu
93	E:B74..B7	0.204412	>=0	No	0.204412	0
94	E:B75..B7	0.229862	>=0	No	0.229862	0
95	E:B76..B7	0.295906	>=0	No	0.295906	0
96	E:B77..B7	0.375771	>=0	No	0.375771	0
97	E:B78..B7	1.912297	>=0	No	1.912297	0
98	E:B79..B7	0.333	>=0	No	0.333	0
99	E:B80..B8	0.333	>=0	No	0.333	0
100	E:B81..B8	0.333	>=0	No	0.333	0
101	E:B82..B8	0.23602	>=0	No	0.23602	0
102	E:B83..B8	0	>=0	Yes	0	0
103	E:B84..B8	0.249277	>=0	No	0.249277	0
104	E:B85..B8	0.265246	>=0	No	0.265246	0
105	E:B86..B8	0.001419	>=0	No	0.001419	0
106	E:B87..B8	0.073125	>=0	No	0.073125	0
107	E:B88..B8	0.21543	>=0	No	0.21543	0
108	E:B89..B8	0.164918	>=0	No	0.164918	0
109	E:B90..B9	0.152009	>=0	No	0.152009	0
110	E:B91..B9	0.161495	>=0	No	0.161495	0
111	E:B92..B9	0.234206	>=0	No	0.234206	0
112	E:B93..B9	0.218441	>=0	No	0.218441	0
113	E:B94..B9	0.22527	>=0	No	0.22527	0
114	E:B95..B9	0.233432	>=0	No	0.233432	0
115	E:B96..B9	0.205795	>=0	No	0.205795	0
116	E:B97..B9	0.156387	>=0	No	0.156387	0
117	E:B98..B9	0.206032	>=0	No	0.206032	0
118	E:B99..B9	0.001559	>=0	No	0.001559	0
119	E:B100..B	0.194991	>=0	No	0.194991	0
120	E:B101..B	0.290415	>=0	No	0.290415	0
121	E:B102..B	0.406573	>=0	No	0.406573	0
122	E:B103..B	0.115834	>=0	No	0.115834	0
123	E:B104..B	0.219082	>=0	No	0.219082	0
124	E:B105..B	0.265246	>=0	No	0.265246	0
125	E:B106..B	0.123575	>=0	No	0.123575	0
126	E:B107..B	0.011189	>=0	No	0.011189	0
127	E:B108..B	0.199575	>=0	No	0.199575	0
128	E:B109..B	0.156213	>=0	No	0.156213	0
129	E:Q93..Q	8.17735	=8.17735	Yes	0	0
130	E:Q92..Q	127.5493	=127.549	Yes	0	0
131	E:Q119..C	0.000166	<=0.1	No	0.099834	0
132	E:B83..B8	0	=0	Yes	0	0
133	E:A111..A	9	=9	Yes	0	0
134	E:B79..B7	0.333	=0.333	Yes	0	0
135	E:B80..B8	0.333	=0.333	Yes	0	0
136	E:B81..B8	0.333	=0.333	Yes	0	0
137	E:B75..B7	1	=1	Yes	0	0
138	E:B76..B7	1	=1	Yes	0	0
139	E:B94..B9	1	<=1	Yes	0	-0.19865
140	E:B75..B7	1	=1	Yes	0	0
141	E:B83..B8	0	>=1000	Yes	0	0
142	E:B84..B8	0	>=1000	Yes	0	0
143	E:B85..B8	0	>=1000	Yes	0	0
144	E:B86..B8	0	>=1000	Yes	0	0
145	E:B87..B8	0	>=1000	Yes	0	0
146	E:B88..B8	0	>=1000	Yes	0	0
147	E:B89..B8	0	>=1000	Yes	0	0
148	E:B90..B9	0	>=1000	Yes	0	0
149	E:B91..B9	0	>=1000	Yes	0	0
150	E:B92..B9	0	>=1000	Yes	0	0
151	E:B93..B9	0	>=1000	Yes	0	0
152	E:B94..B9	0	>=1000	Yes	0	0
153	E:B95..B9	0	>=1000	Yes	0	0
154	E:B96..B9	0	>=1000	Yes	0	0
155	E:B97..B9	0	>=1000	Yes	0	0
156	E:B98..B9	0	>=1000	Yes	0	0
157	E:B99..B9	0	>=1000	Yes	0	0
158	E:B100..B	0	>=1000	Yes	0	0
159	E:B101..B	0	>=1000	Yes	0	0
160	E:B102..B	0	>=1000	Yes	0	0
161	E:B103..B	0	>=1000	Yes	0	0
162	E:B104..B	0	>=1000	Yes	0	0
163	E:B105..B	0	>=1000	Yes	0	0
164	E:B106..B	0	>=1000	Yes	0	0
165	E:B107..B	0	>=1000	Yes	0	0
166	E:B108..B	0	>=1000	Yes	0	0
167	E:B109..B	0	>=1000	Yes	0	0

E	S	T	U
75	Answer Report		
76	Variable C	Starting	Final
77	E:B74..E:	0.940345	0.920668
78	E:B75..E:	0.869702	0.826722
79	E:B76..E:	0.929356	0.906054
80		0.974882	0.966597
81	E:B78..E:	1.026686	1.035491
82	E:B79..E:	0.77708	0.703549
83	E:B80..E:	0.720565	0.628392
84	E:B81..E:	0.808477	0.745303
85	E:B82..E:	0.569859	0.427975
86	E:B83..E:	0.248038	0
87	E:B84..E:	0.868132	0.824635
88	E:B85..E:	1	1
89	E:B86..E:	0	0
90	E:B87..E:	0.503925	0.340292
91	E:B88..E:	0.755102	0.674321
92	E:B89..E:	0.660911	0.549061
93	E:B90..E:	0.580848	0.442589
94	E:B91..E:	0.382847	0.179104
95	E:B92..E:	0.798579	0.731609
96	E:B93..E:	0.359358	0.148076
97	E:B94..E:	0.197657	0
98	E:B95..E:	0.562009	0.417537
99	E:B96..E:	0.270016	0.029228
100	E:B97..E:	0.279435	0.041754
101	E:B98..E:	0.350079	0.135699
102	E:B99..E:	0	0
103	E:B100..E	0.739403	0.653445
104	E:B101..E	0.240188	0
105	E:B102..E	0.408163	0.212944
106	E:B103..E	0.246468	0
107	E:B104..E	0.521193	0.363257
108	E:B105..E	1	1
109	E:B106..E	0.450549	0.269311
110	E:B107..E	0	0
111	E:B108..E	0.488226	0.319415
112	E:B109..E	0.452119	0.271399
113	E:B107..E	0	0
114	E:B108..E	0.631638	0.58312

E	S	T	U	V
115	Cell	Value	Constrain	Binding?
116	E:B74..E:	0	>=0	Yes
117	E:B75..E:	0	>=0	Yes
118	E:B76..E:	0	>=0	Yes
119	E:B77..E:	0	>=0	Yes
120	E:B78..E:	0	>=0	Yes
121	E:B79..E:	0	>=0	Yes
122	E:B80..E:	0	>=0	Yes
123	E:B81..E:	0	>=0	Yes
124	E:B82..E:	0	>=0	Yes
125	E:B83..E:	0	>=0	Yes
126	E:B84..E:	0	>=0	Yes
127	E:B85..E:	0	>=0	Yes
128	E:B86..E:	0	>=0	Yes
129	E:B87..E:	0	>=0	Yes
130	E:B88..E:	0	>=0	Yes
131	E:B89..E:	0	>=0	Yes
132	E:B90..E:	0	>=0	Yes
133	E:B91..E:	0	>=0	Yes
134	E:B92..E:	0	>=0	Yes
135	E:B93..E:	0	>=0	Yes
136	E:B94..E:	0	>=0	Yes
137	E:B95..E:	0	>=0	Yes
138	E:B96..E:	0	>=0	Yes
139	E:B97..E:	0	>=0	Yes
140	E:B98..E:	0	>=0	Yes
141	E:B99..E:	0	>=0	Yes
142	E:B100..E	0	>=0	Yes
143	E:B101..E	0	>=0	Yes
144	E:B102..E	0	>=0	Yes
145	E:B103..E	0	>=0	Yes
146	E:B104..E	0	>=0	Yes
147	E:B105..E	0	>=0	Yes
148	E:B106..E	0	>=0	Yes
149	E:B107..E	0	>=0	Yes
150	E:B108..E	0	>=0	Yes
151	E:B109..E	0	>=0	Yes
152	E:Q87..E:	0	=0.05	Yes
153	E:Q89..E:	0	=0.05	Yes
154	E:Q88..E:	0	=0.01	Yes
155	E:Q90..E:	0	=0.01	Yes
156	E:Q89..E:	0	=0.05	Yes
157	E:Q88..E:	0	=0.01	Yes
158	E:Q90..E:	0	=0.01	Yes

FIGURE 49 Iteration history for a Run in SYNPROPS.

- **SYNGEN** is a unique program for automatic generation of the shortest, most economic organic synthesis routes for a given target compound.
- **SYNGEN** is based on Professor Hendrickson's *Half-Reaction Theory.* It does not require a reaction database.
- **SYNGEN** is easy to use. After input of a target structure, the program automatically generates all the shortest routes. Then, if you press

  Step button, **synthesis routes are ordered by reaction steps,** with the shortest one first.

  You can then press

  Next button to see the next shortest route, or press

  Prev button to see the previous route,

  or simply type in

  10 in the **Goto:** space if you want to see the 10th route.

  If you press

  Cost button, **synthesis routes are ordered by overall cost,** witht the cheapest one first.

  If you want to see those routes belonging to bond set **4**, for example, you just need to type in 4 in the **Bondset:** space.

FIGURE 50 SYNGEN.

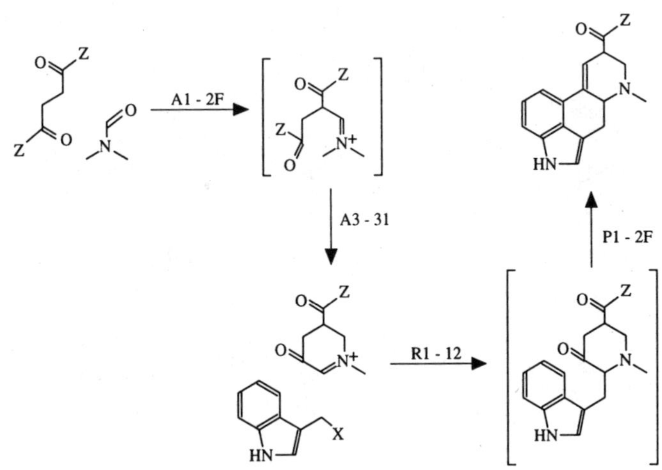

FIGURE 51 Building a synthesis for an estrone skeleton.

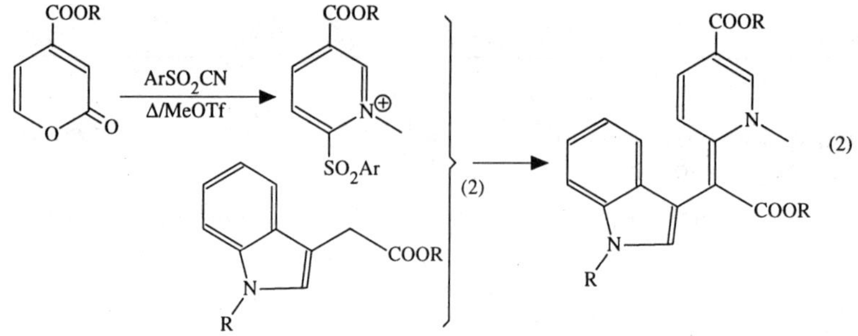

FIGURE 52 Any carbon in a structure can have four general kinds of bonds.

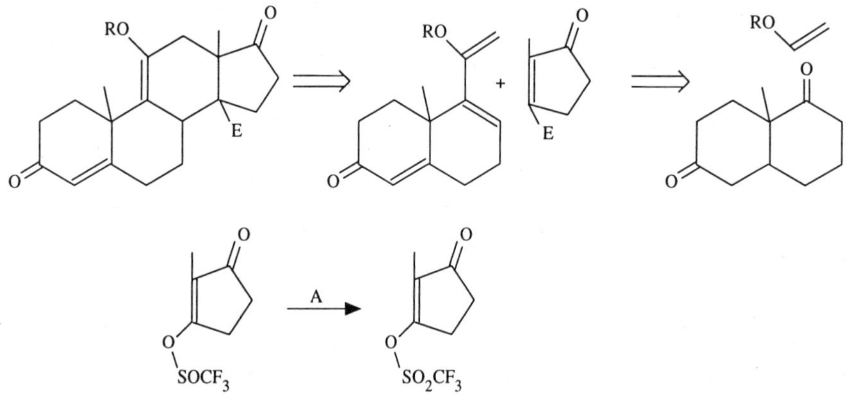

FIGURE 53 SYNGEN synthesis of cortical steroid.

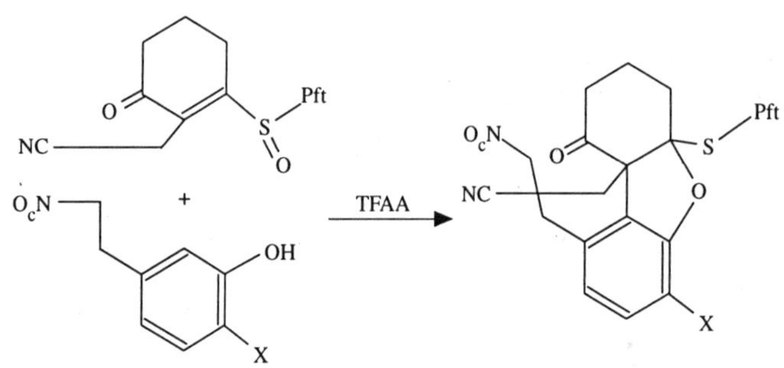

FIGURE 54 Pericyclic reaction to join simple starting materials for quick assembly of morphinan skeleton.

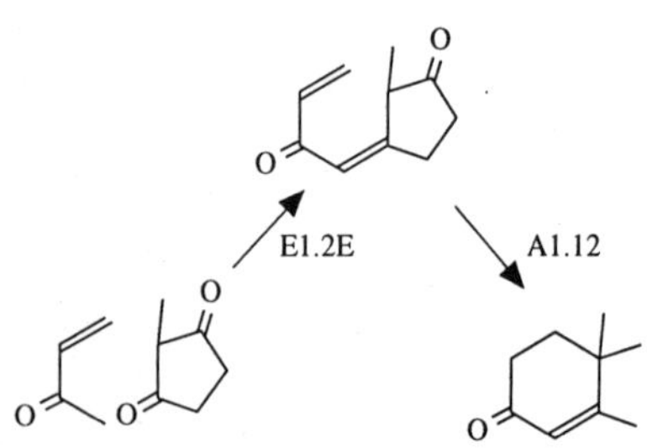

FIGURE 55 Sample SYNGEN output screen from another bondset.

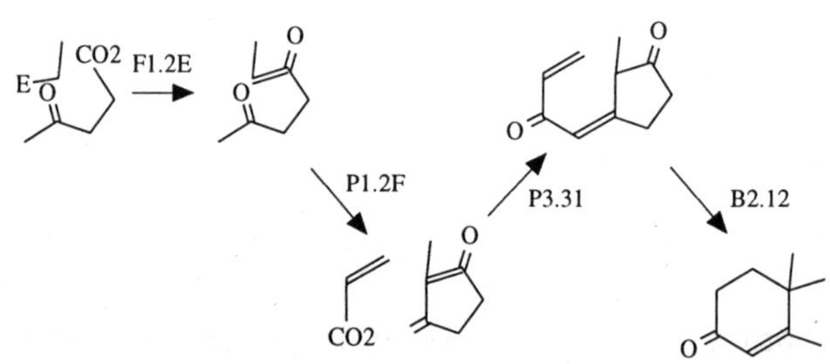

Figure 56 Second sample SYNGEN output screen.

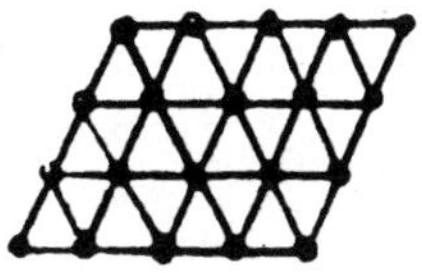

FIGURE 57 The triangular lattice.

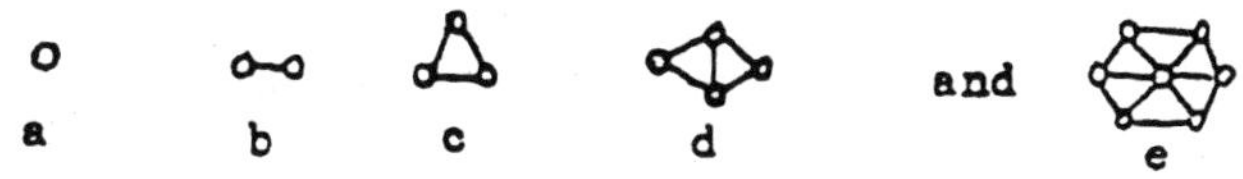

FIGURE 58 Essential overlap figures.

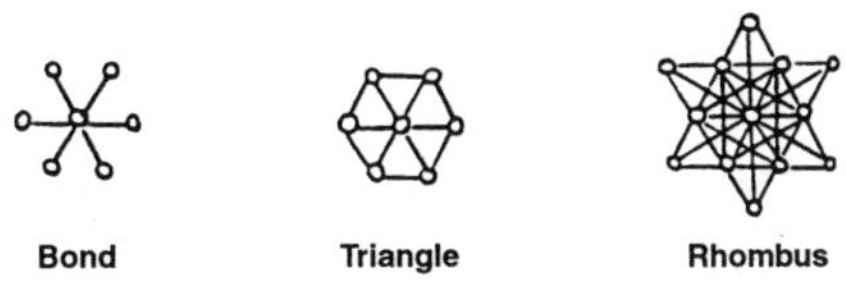

FIGURE 59 Effect of considering larger basic figures.

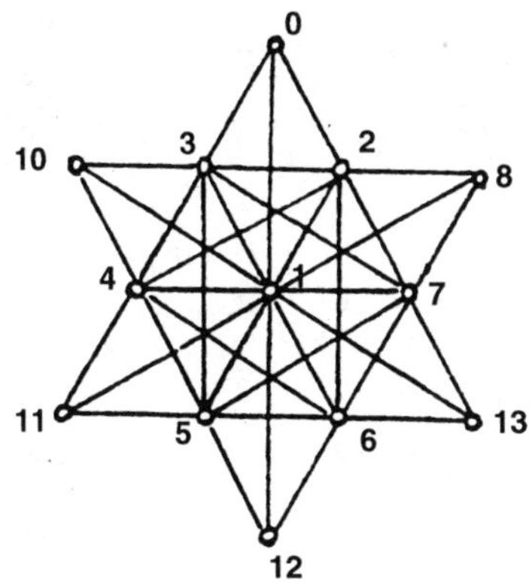

FIGURE 60 The rhombus approximation.

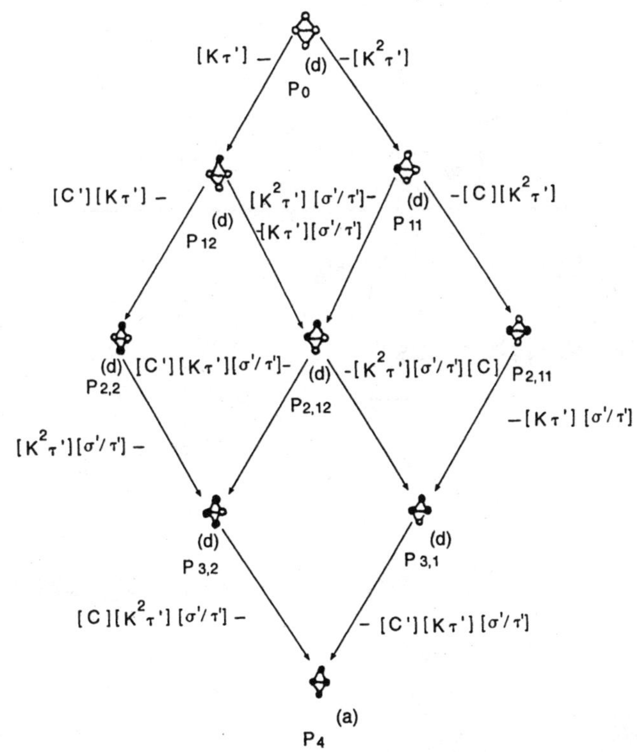

FIGURE 61 The successive filling of rhombus sites.

Entry	m	g,ηs	Figure	Distribution Numbers $P_{g,\eta_s}^{(m)}$	Multiplicity $\lambda g,\eta s^{(m)}$	Energy
1	a	0		$p0^{(a)}$	1	0
2	a	1		$p1^{(a)}$	1	ε
3	b	0		$p0^{(b)}$	1	0
4	b	1		$p1^{(a)}$	2	ε
5	b	2		$p2^{(b)}$	1	2 ε + w
6	c	0		$p0^{(c)}$	1	0
7	c	1		$p1^{(c)}$	3	ε
8	c	2		$p2^{(c)}$	3	2 ε + w
9	c	3		$p3^{(c)}$	1	2 ε + 3w
10	d	0		$P0^{(d)}$	1	0
11	d	11		$p11^{(d)}$	2	ε
12	d	12		$p12^{(d)}$	2	ε
13	d	2,11		$p2,11^{(d)}$	1	2 ε + w
14	d	2,12		$p2,12^{(d)}$	4	2 ε + w
15	d	2,2		$p2,2^{(d)}$	1	2 ε + w′
16	d	3,1		$p3,11^{(d)}$	2	3 ε + 3w
17	d	3,2		$p3,2^{(d)}$	2	3 ε + 2w + w′
18	d	4		$p4^{(d)}$	1	4 e + 5w + w′

FIGURE 62 Distribution numbers for a plane triangular lattice.

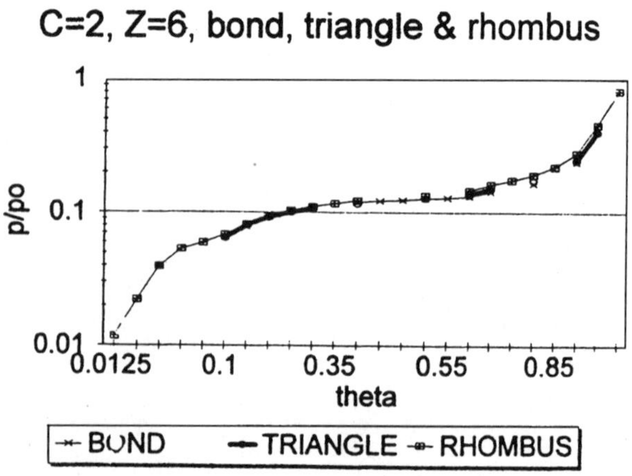

FIGURE 63 Order and complexity.

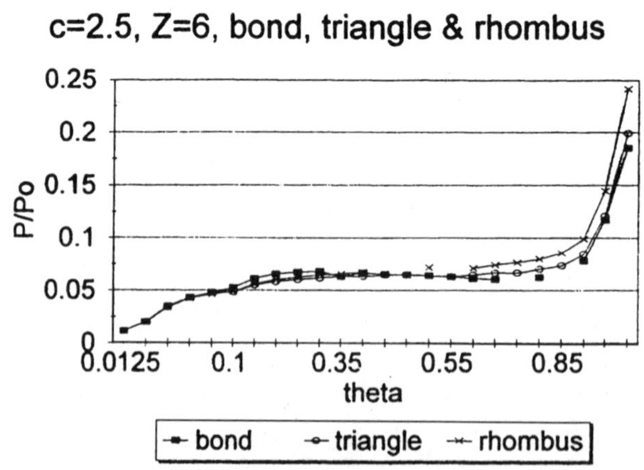

FIGURE 64 Order-disorder, c = 2.5.

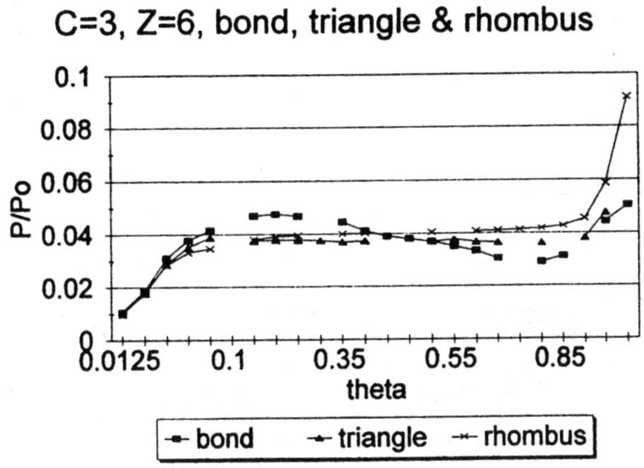

FIGURE 65 Order-disorder, c = 3.

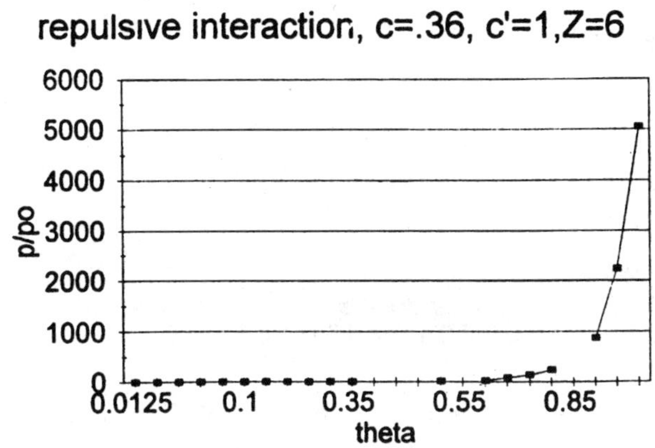

FIGURE 66 p/p0 for rhombus.

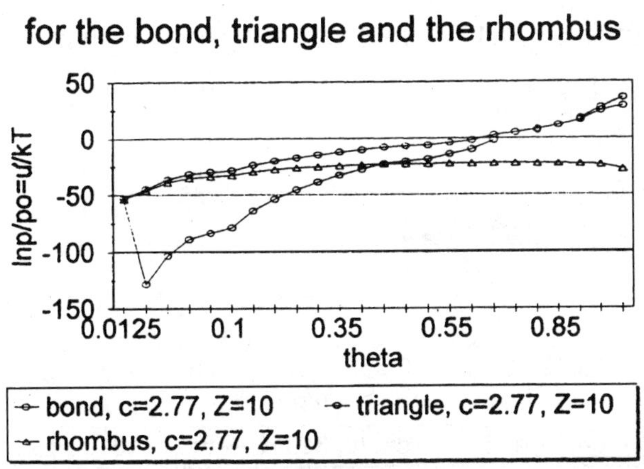

FIGURE 67 u/kT vs. occupancy.

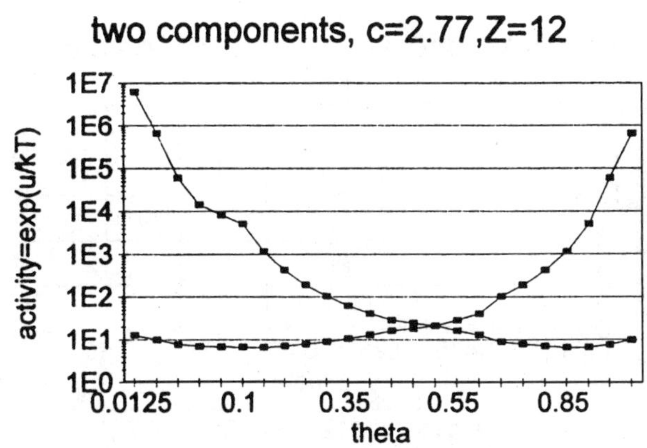

FIGURE 68 Activity vs. theta.

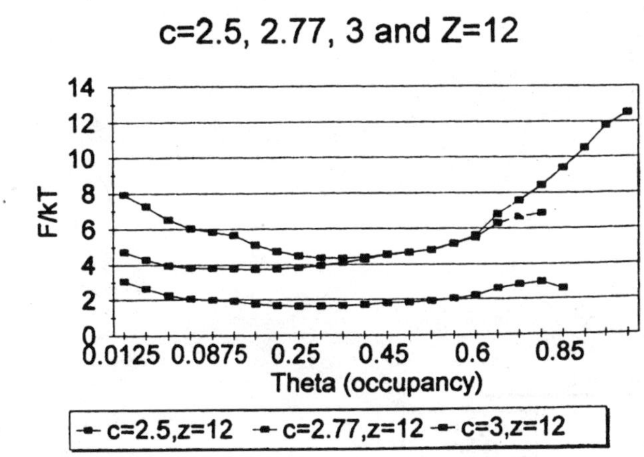

FIGURE 69 F/kT: Bond figure.

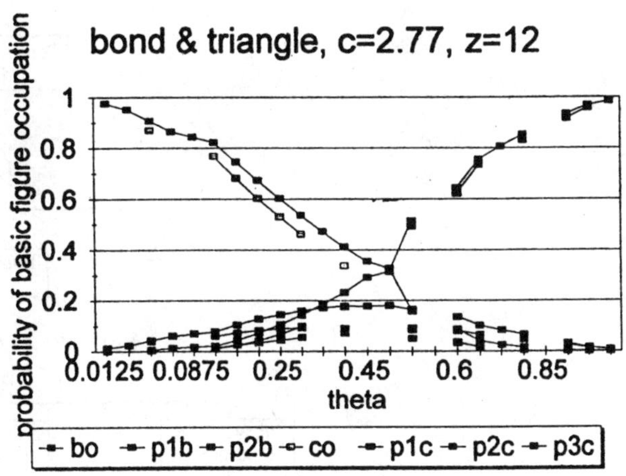

FIGURE 70 Probability vs. theta, c = 2.77.

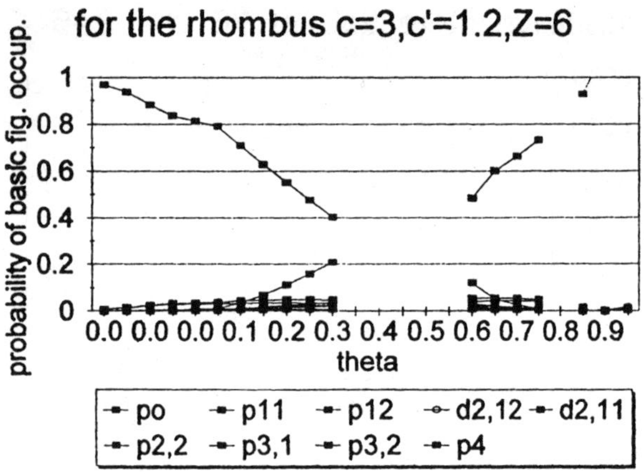

FIGURE 71 Probability vs. theta, c = 3.

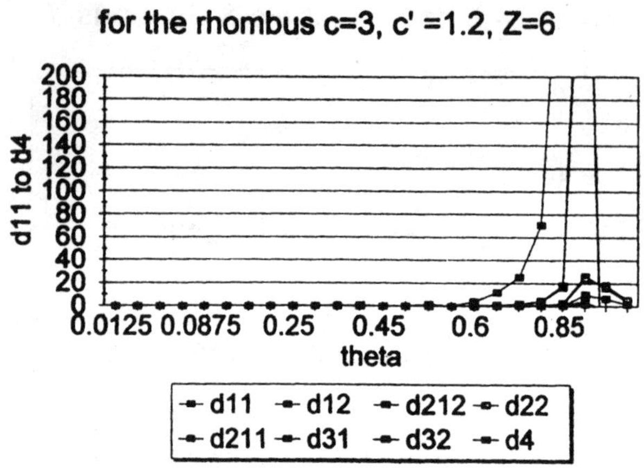

FIGURE 72 d vs. theta.

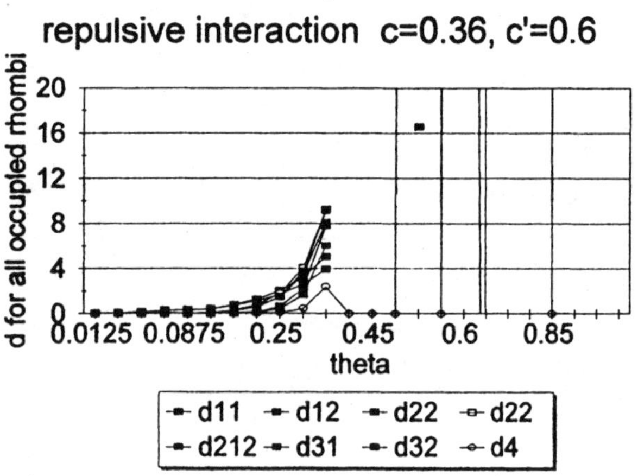

FIGURE 73 d for rhombus.

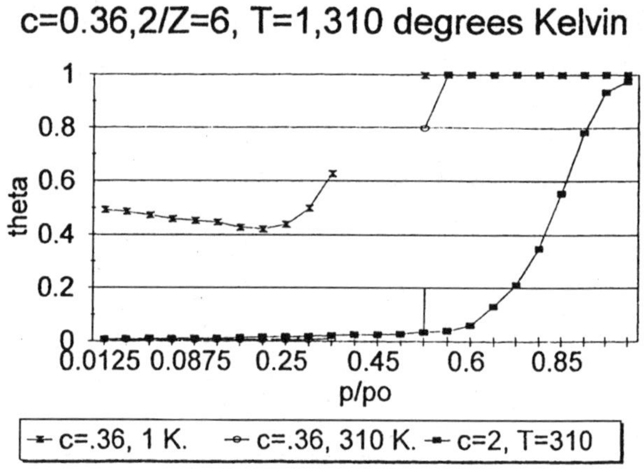

FIGURE 74 Metastasis/rhombus.

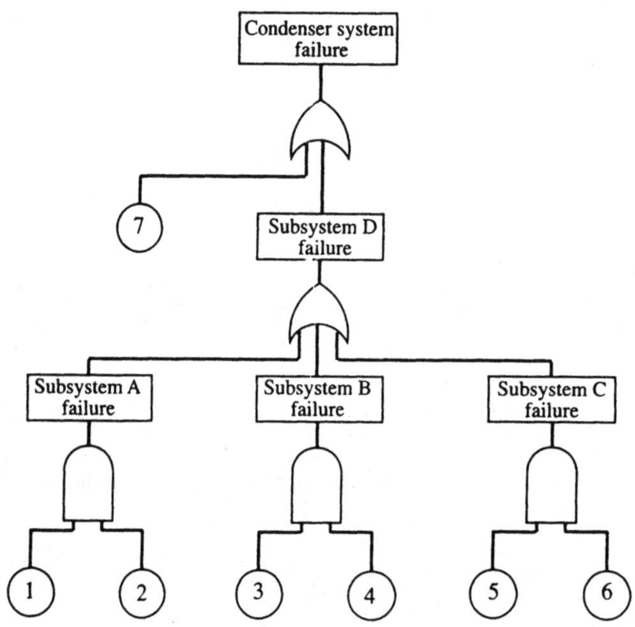

FIGURE 75 A fault tree network.

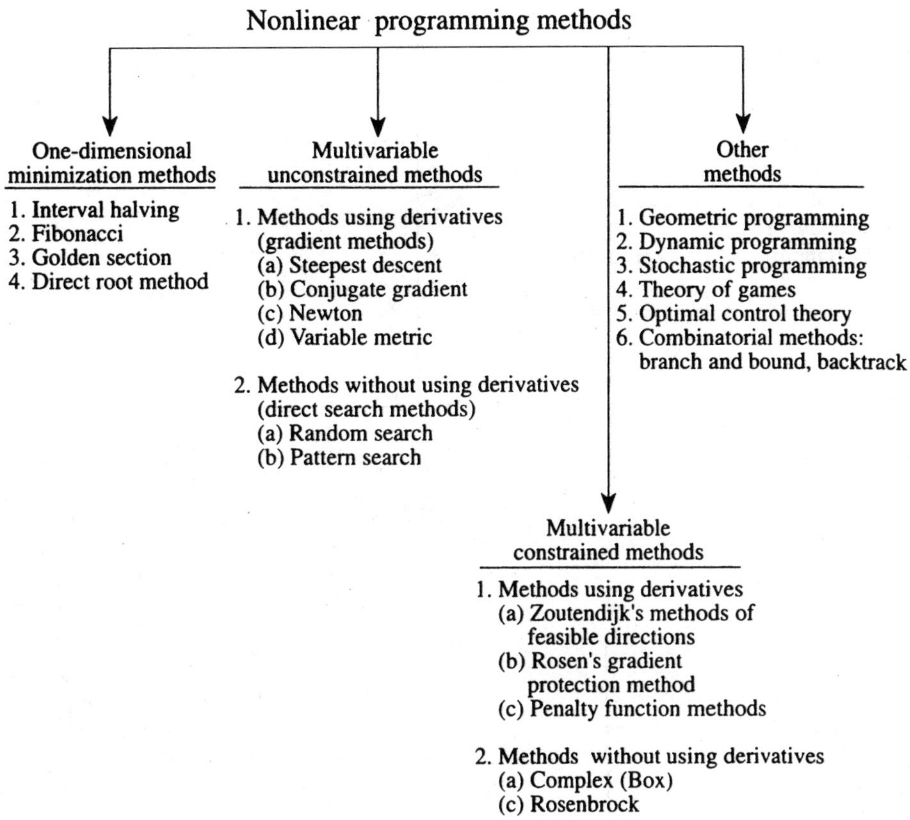

FIGURE 76 Selected nonlinear programming methods.

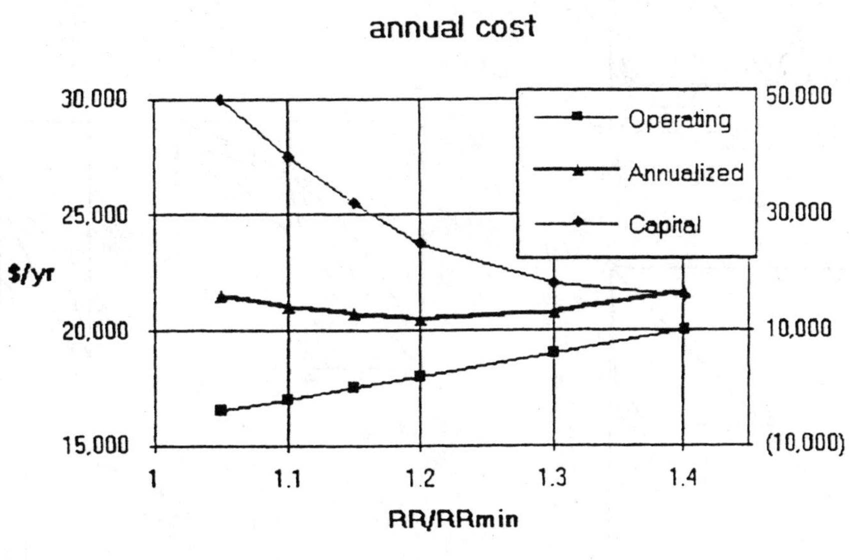

FIGURE 77 Trade-off between capital and operating cost for a distillation column.

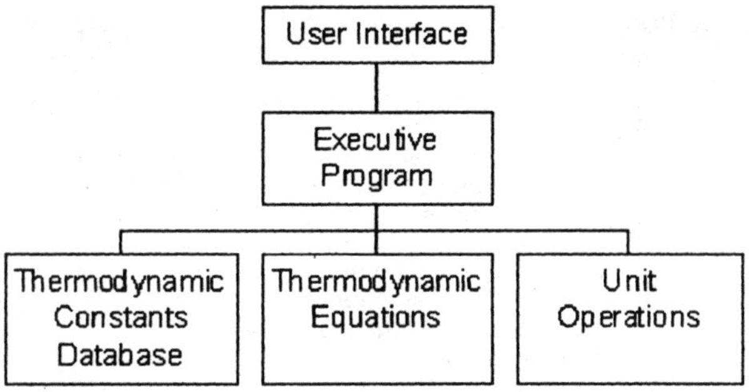

FIGURE 78 Structure of process simulators.

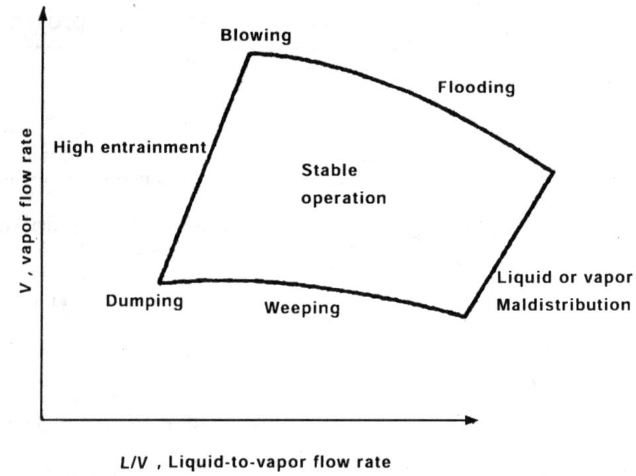

FIGURE 80 Tray malfunctions as a function of loading.

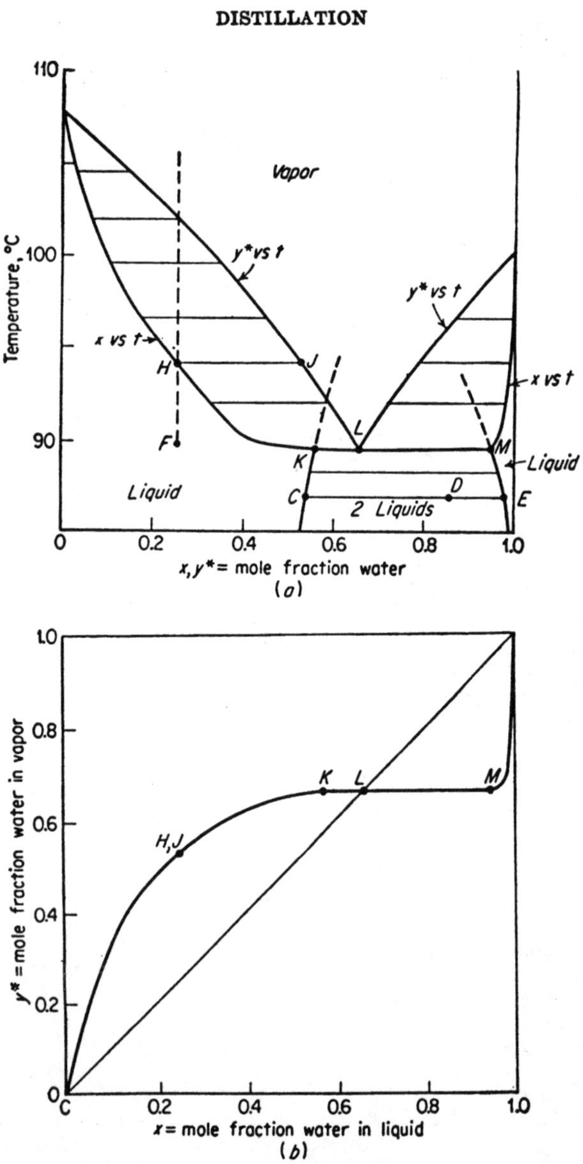

FIGURE 79 Acetone-Formamide and chloroform-methanol equilibrium diagrams showing non-ideal behavior.

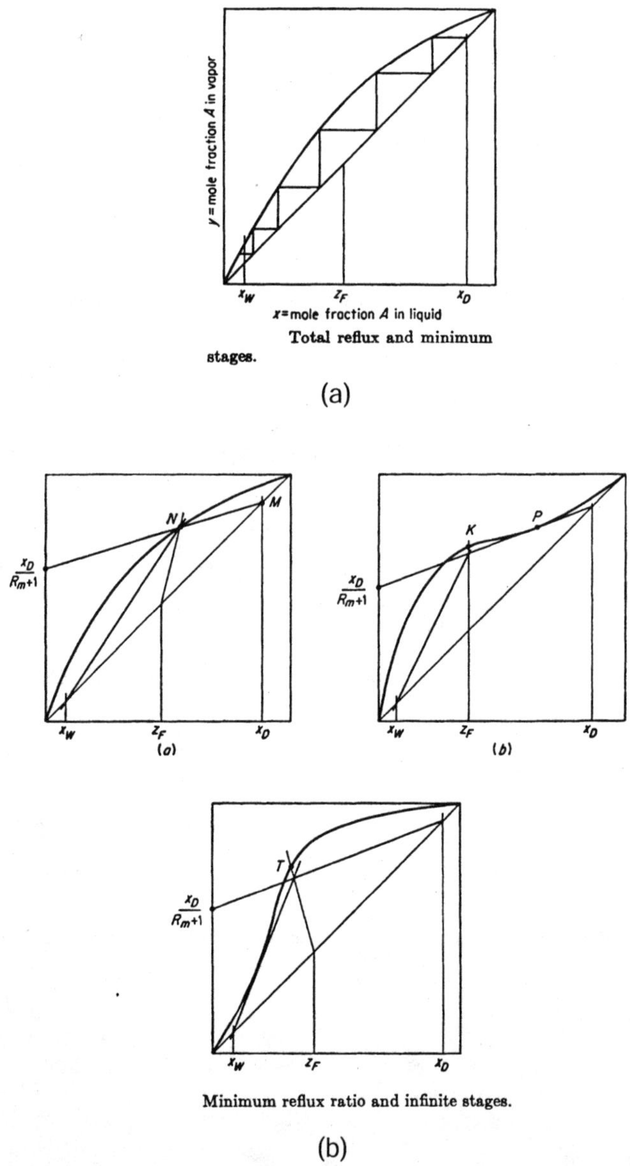

FIGURE 81 McCabe-Thiele for (a) minimum stages and (b) minimum reflux.

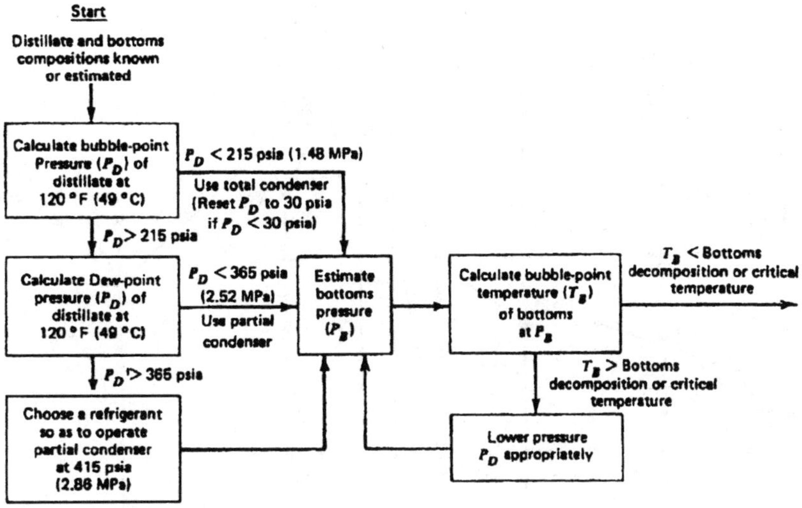

FIGURE 82 Algorithm for establishing distillation column pressure and type condenser.

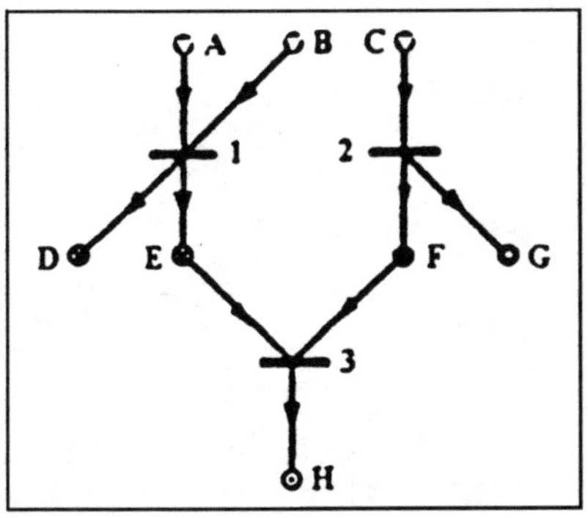

FIGURE 83 P-Graph of the process manufacturing required product H and also yielding potential product G and disposable material D from raw materials A, B, and C.

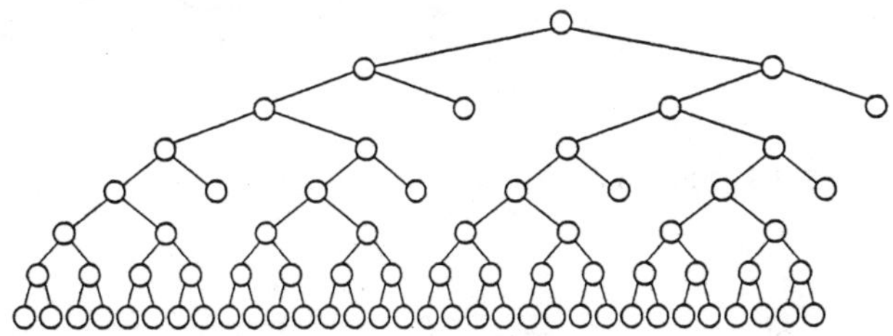

FIGURE 84 Enumeration tree for the conventional branch-and-bound algorithm.

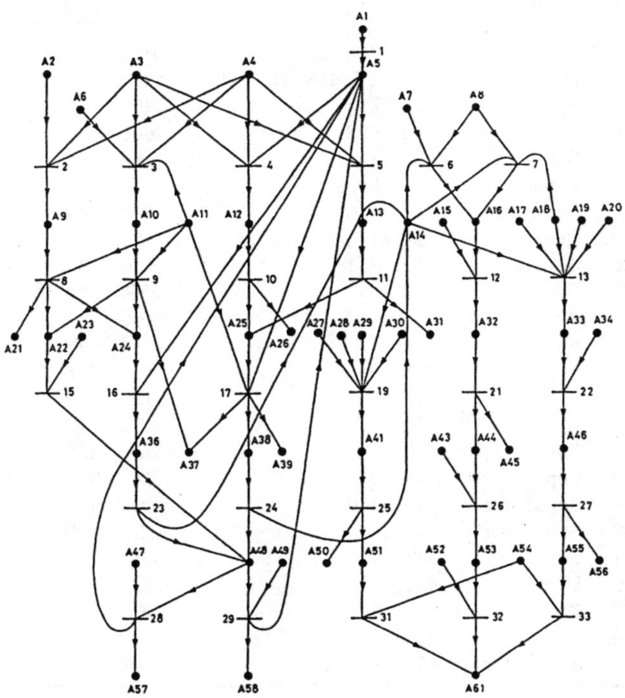

FIGURE 85 Maximal structure of example generated by algorithm MSG.

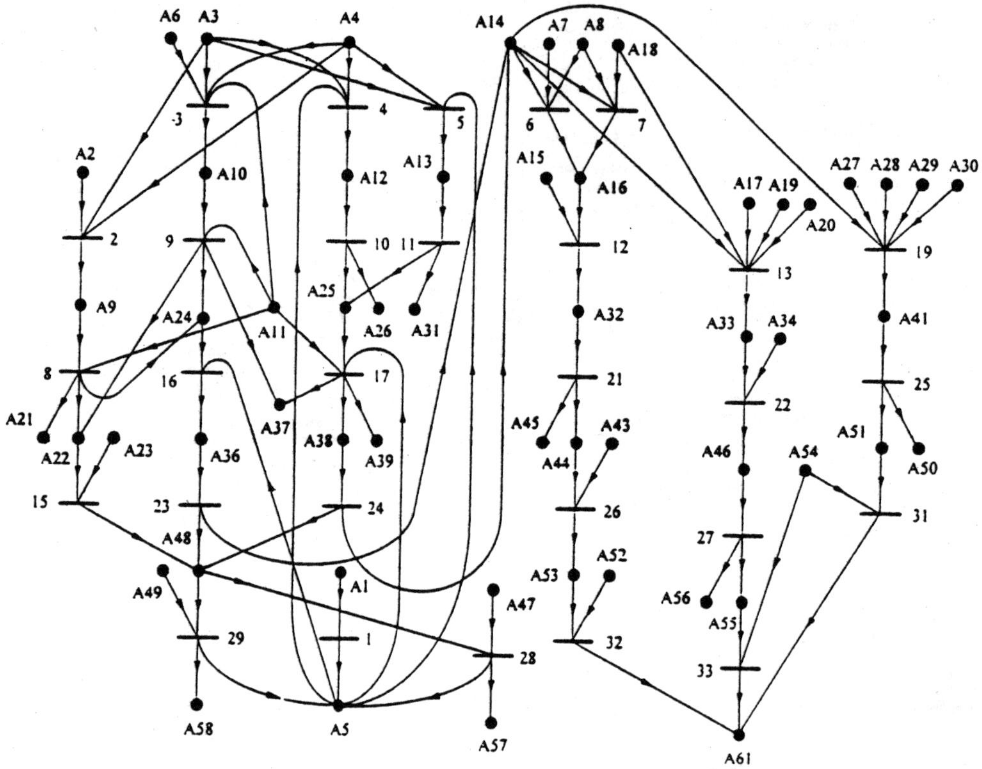

FIGURE 86 Maximal structure of example.

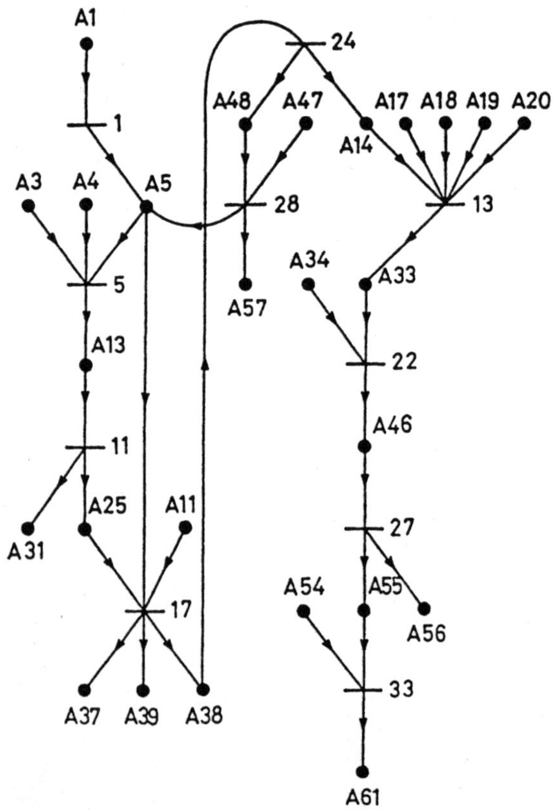

FIGURE 87 Solution-structure of example.

No.	Type	Inputs	Outputs
1.	Feeder	*A*1	*A*5
2.	Reactor	*A*2, *A*3, *A*4	*A*9
3.	Reactor	*A*3, *A*4, *A*6, *A*11	*A*10
4.	Reactor	*A*3, *A*4, *A*5	*A*12
5.	Reactor	*A*3, *A*4, *A*5	*A*13
6.	Reactor	*A*7, *A*8, *A*14	*A*16
7.	Reactor	*A*8, *A*14, *A*18	*A*16
8.	Separator	*A*9, *A*11	*A*21,*A*22,*A*24
9.	Separator	*A*10, *A*11	*A*22,*A*24,*A*37
10.	Separator	*A*12	*A*25, *A*26
11.	Separator	*A*13	*A*25, *A*31
12.	Dissolver	*A*15, *A*16	*A*32
13.	Reactor	*A*14,*A*17,*A*18,*A*19,*A*20	*A*33
14.	Reactor	*A*6, *A*21	*A*35
15.	Washer	*A*22, *A*23	*A*48
16.	Washer	*A*5, *A*24	*A*36
17.	Separator	*A*5, *A*11, *A*25	*A*37,*A*38,*A*39
18.	Separator	*A*11, *A*26	*A*40, *A*42
19.	Reactor	*A*14,*A*27,*A*28,*A*29,*A*30	*A*41
20.	Separator	*A*11, *A*31	*A*40, *A*42
21.	Centrifuge	*A*32	*A*44, *A*45
22.	Washer	*A*33, *A*34	*A*46
23.	Separator	*A*36	*A*14, *A*48
24.	Separator	*A*38	*A*14, *A*48
25.	Filter	*A*41	*A*50, *A*51
26.	Washer	*A*43, *A*44	*A*53
27.	Filter	*A*46	*A*55, *A*56
28.	Separator	*A*47, *A*48	*A*5, *A*57
29.	Separator	*A*48, *A*49	*A*5, *A*58
30.	Separator	*A*50	*A*59, *A*60
31.	Dryer	*A*51, *A*54	*A*61
32.	Dyer	*A*52, *A*53	*A*61
33.	Dryer	*A*54, *A*55	*A*61
34.	Distillation	*A*59	*A*62, *A*63
35.	Separator	*A*60	*A*64, *A*65

FIGURE 88 Operating units of example.

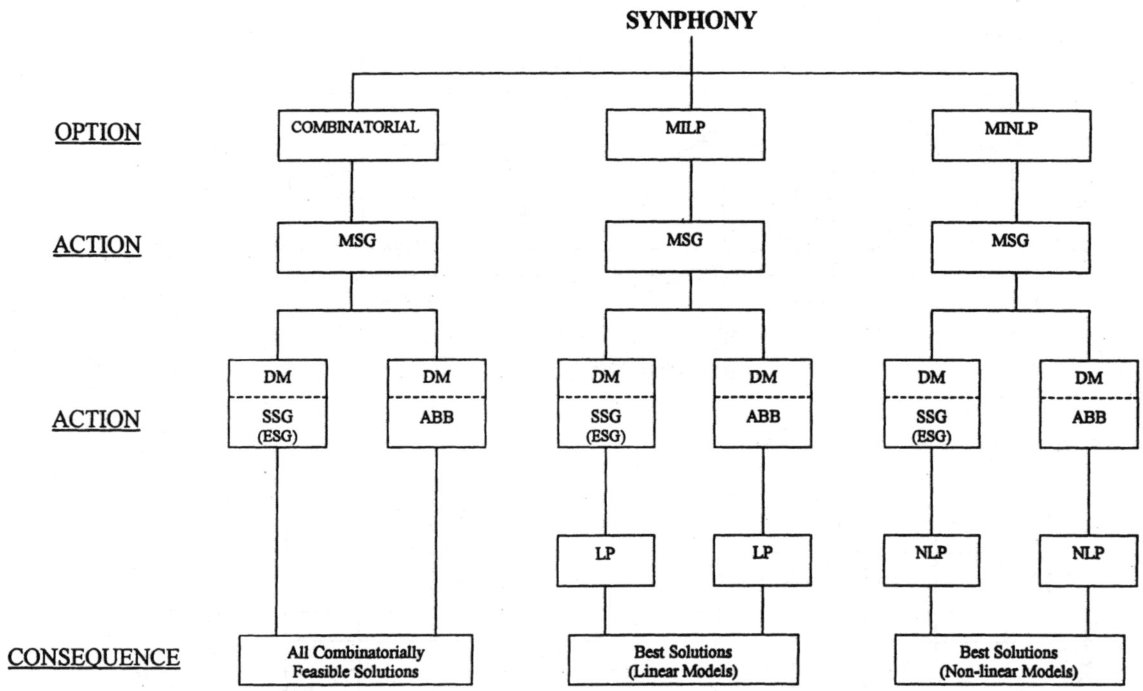

FIGURE 89 Structure of SYNPHONY.

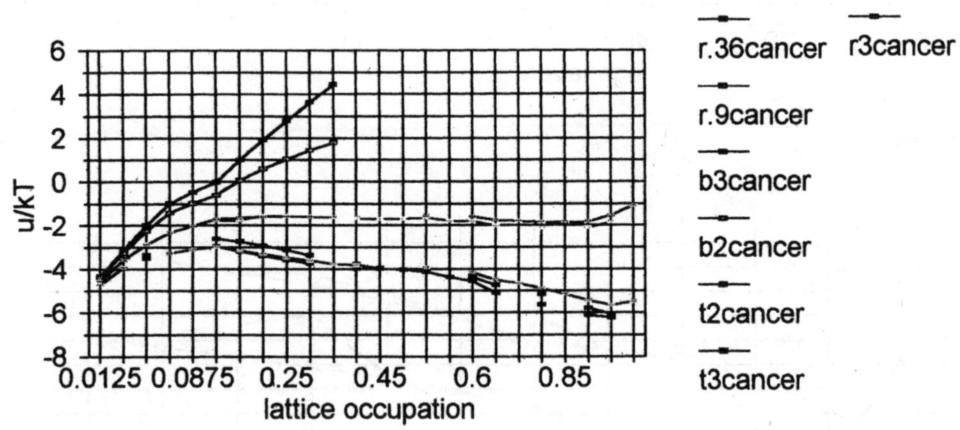

FIGURE 90 Cancer probability or u/kT.

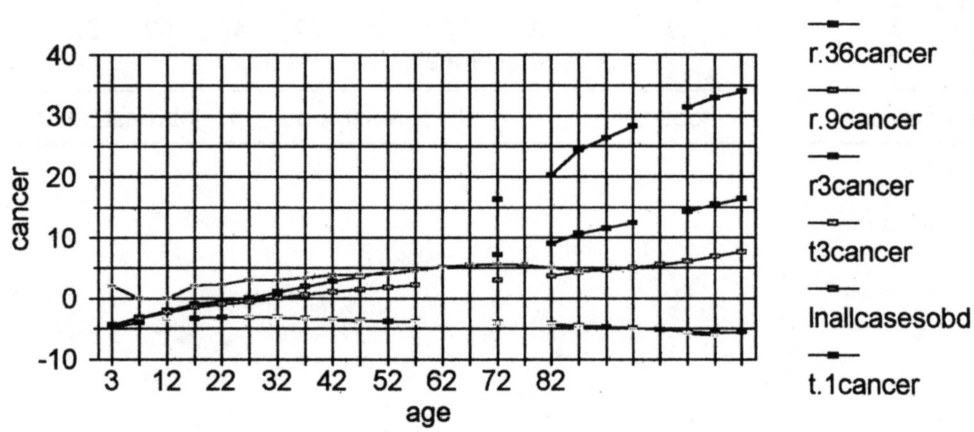

FIGURE 91 Cancer Ordkin-Function.

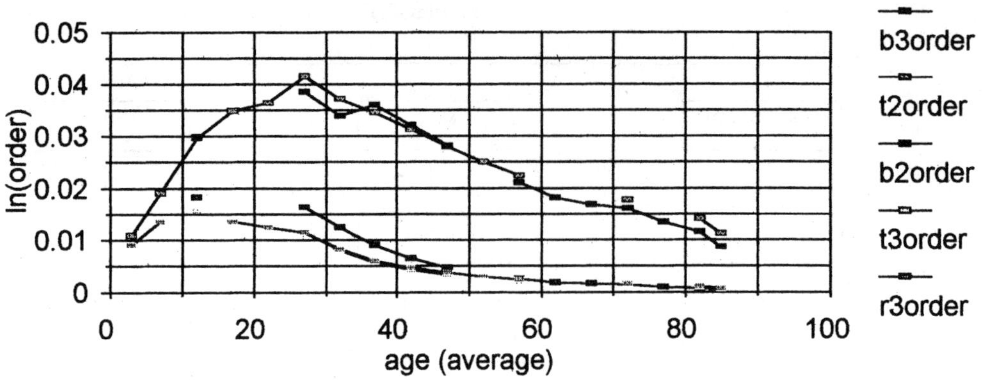

FIGURE 92 Order vs. age for attractive forces.

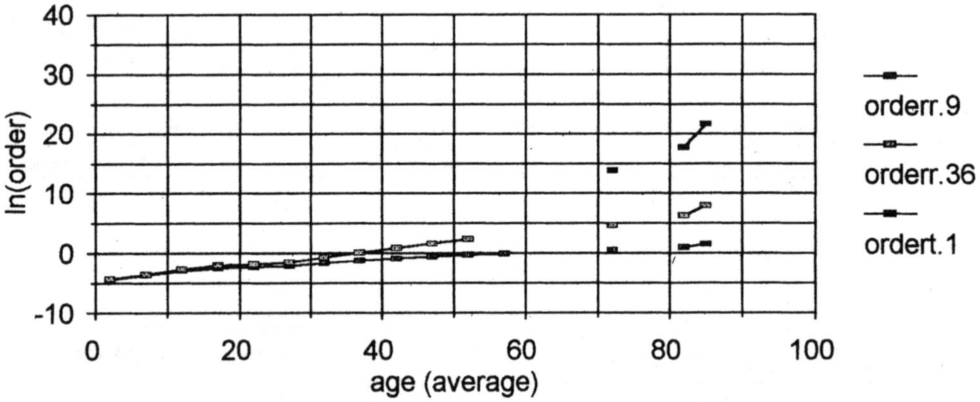

FIGURE 93 Order vs. age.

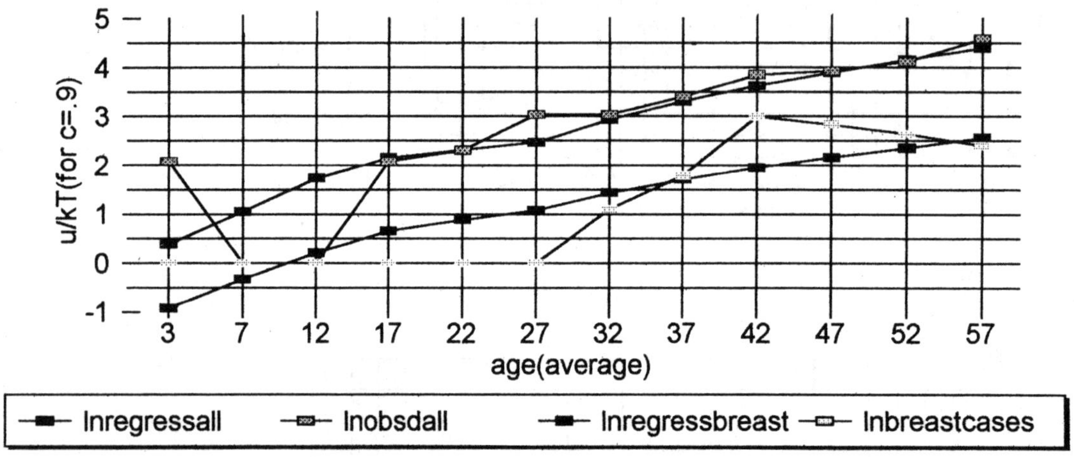

FIGURE 94 Regression of cancers.

Index

D

E

L

M

N

T

The Ecology of the Fish Pond Ecosystem

Developments in Hydrobiology 72

Series editor

H. J. Dumont

The Ecology of the Fish Pond Ecosystem

with special reference to Africa

by

Guy Delincé

Institute of Animal Ecology,
University of Ghent, Belgium

KLUWER ACADEMIC PUBLISHERS
DORDRECHT / BOSTON / LONDON

Library of Congress Cataloging-in-Publication Data

Delincé, Guy.
The ecology of the fish pond ecosystem : with special reference to Africa / Guy Delincé.
p. cm. -- (Developments in hydrobiology ; 72)
Includes bibliographical references and index.
ISBN 0-7923-1628-2
1. Fish-culture--Africa. 2. Fish ponds--Africa--Fertilization.
3. Fresh water--Fertilization. 4. Pond ecology--Africa. I. Title.
II. Series.
SH121.D45 1992
639.3'11--dc20 92-54

ISBN 0-7923-1628-2

Published by Kluwer Academic Publishers,
P.O. Box 17, 3300 AA Dordrecht, The Netherlands.

Kluwer Academic Publishers incorporates
the publishing programmes of
D. Reidel, Martinus Nijhoff, Dr W. Junk and MTP Press.

Sold and distributed in the U.S.A. and Canada
by Kluwer Academic Publishers,
101 Philip Drive, Norwell, MA 02061, U.S.A.

In all other countries, sold and distributed
by Kluwer Academic Publishers Group,
P.O. Box 322, 3300 AH Dordrecht, The Netherlands.

Printed on acid-free paper

Printed in the Netherlands

Contents

Preface

This exhaustive review on the potential of fertilization as a means to increase fish production in ponds under African tropical conditions is much overdue. Pond fish farming has been introduced to Africa over the past forty years. Although it is generally agreed that this continent has a considerable potential for freshwater aquaculture, past development efforts have met with variable but generally limited success. Several constraints have recently been identified while suggesting future research topics and ranking their respective priority. From the technical point of view, pond fertilization has been recognized as the main limiting factor for the development of tilapia farming.

In fact, even outside Africa, most of the present aquaculture production depends at least partly on the use of fertilizers. Over 70 % of the world production of farmed finfish and shrimp is realized with extensive to semi-intensive pond cultural systems, where natural productivity is enhanced through the controlled additions of fertilizers.

Most of the basic scientific research on pond fertilizations was carried out in temperate latitudes, mainly in Central and Eastern Europe, and more recently in the Southern United States and Israel. Not only should much of this past research be open to criticism on the design of experiments, but its results should only be applied in tropical Africa with the greatest caution, if at all. It is a fact that our understanding - and therefore our possible control - of the intermediate stages through which fertilizers are converted into fish food and fish flesh is still far from complete, particularly in African ponds.

The principles of pond fertilization are similar to those applied to agriculture. Fish production is increasing by the addition of certain essential substances present in limited concentrations and by the stimulating effect of these substances on the production of natural live fish food. Such effects relies on the improved operation of several interlinked food chains based on highly complex chemical and biological interactions taking place simultaneously in the bottom soil and the overlying water of the pond. This review is unique in bringing together and synthesizing the available knowledge on all factors involved and their multiple interactions. The importance of the mud surface as the "chemical laboratory" of the pond is duly stressed. Water quality, nutrient cycling and natural productivity are fully discussed. Such a complete illustration of the complexity of pond fertilization pathways towards fish production should hopefully promote further research on these particular subjects, in Africa as well as in other tropical regions.

The application of fertilizers to tropical waters, although showing great promise, is still in its infancy. For example no standard procedure has yet been devised for the on-farm determination of fertilizer requirements. Therefore the process is still largely empirical and depends on the fish farmer's judgement. In Far Eastern Asia, the latter is often based on the evolution, by trial and error over several centuries, of productive farming methods well suited to local conditions. In Africa, however, where the people have no background of practical experience and where both the farming conditions and the fish are different, such judgement can only be based on locally-acquired knowledge. Undoubtedly, this is one of the main reasons, which has prompted the author to prepare this multidisciplinary review. It should be a standard reference text to enable African scientists to become increasingly interested, it is hoped, in removing the numerous questions marks still attached to pond fertilization.

Dr. André G. Coche
Former Senior Fishery Resources Officer (Aquaculture)
Food and Aquaculture Organization of the U.N.
Rome, Italy

Acknowledgements

Words cannot express my profound gratitude to Prof. Dr. H. J. Dumont for the many ways in which he assisted me in bringing this manuscript to the press. I am grateful to him for his critical reading and commenting of the manuscript, and for the many suggestions for improvements, which helped me to shape the manuscript into its final form. I also thank him for his constant availability, whenever his assistance was required, and his patience with me.

My thanks go to Dr. A. G. Coche, former Senior Fishery Resource Officer (Aquaculture), FAO, for writing the foreword and for his useful comments and advice on the manuscript.

I also wish to gratefully acknowledge the material provided by Prof. Dr. Tilzer on primary productivity.

This handbook could only be completed thanks to the logistical assistance of a number of people, especially Mr. D. Claeys, my brother, Jean-Marie, Mr. K. Roche, Mr. E. Bruyneel, Dr. F. Fiers, Ms. N. Moortgat. Mr. R. S. Coppola, from the Fishery Information, Data and Statistics Service, FAO, computed the statistical analysis of the pond production data. I gratefully acknowledge his help.

Most of the information was gathered at various public libraries in Brussels, in particular the Library of the Royal Institute for Natural Sciences, the Library of the Institute for Hygiene and Epidemiology, the Centre for Research and Information on Tropics (CERDAT), and the library of the Laboratory for Ecology, University of Ghent. I wish to thank them all for making publications available to me.

I also thank Kluwer Academic Publishers for agreeing to publish this volume in the Developments in Hydrobiology series.

Lastly I wish to thank my wife, Martine, for her cheerful support throughout the work.

This volume is dedicated to my parents.

Introduction

Production in African fish farming relies largely on the natural fertility of ponds. Use of artificial or composite feed in fish culture in Africa is, for a number of economic and social reasons, limited. This lack of artificial fish feed has long been known, and research programmes on the use of fertilizers in fish farming were begun by Fisheries Administrations during the fifties and sixties in almost all African countries. Fish production relying on natural food produced within ponds, enhanced by manuring with livestock dung and organic compost, may achieve high yields. Therefore, there is a need to develop a technology based on management of the natural fertility of ponds. The specific conditions leading to such high productions, as well as the means to achieve them, have to be properly defined.

At the FAO Conference on Warmwater Aquaculture in Rome in 1966, one of the objectives had already been to review the status of knowledge on pond fertilization and the role of soil in pond productivity. The Conference concluded that more research was needed on the destiny of fertilizers within ponds, the processes leading to nitrogen-fixation in ponds, the influence of fertilization on the abundance of desirable algae, the need for lime, the effects of purity of phosphorus fertilizers on phosphorus mobilization, and the aquatic flora and fauna within ponds. Unfortunately, during the Symposium on Aquaculture in Africa in Accra, Ghana, in 1975, the emphasis on pond fertilization was weak, and, in the specific recommendations of the meeting, no statement on problems related to pond fertility was issued.

Previous approaches to fertilization needs in ponds considered differences between ponds to be so wide as to render general recommendations for the application of fertilizers impossible. The need for fertilizers had to be defined experimentally by comparing yields obtained with increasing amounts of such fertilizers as lime, and nitrogen and phosphorus compounds in various amounts and combinations, and assessing the cost on production. These trials did not take soil properties into account, nor did they consider the physical, chemical and biological variables of the water. The studies were conducted under various stocking conditions, and results achieved were not conclusive on fertilizer needs, and did not lead to a general procedure for assessing the need for fertilizers in pond culture.

To achieve a good fish yield in pond production based only on natural food, one needs a good understanding of the processes of plankton production and the relevant limnetic conditions. Water quality is still an abstract concept for fish farmers, as the criteria behind it have never been adequately described. Some attempts had been made to define a biogenic capacity to assess fish biomass, such as the Léger-Huet index proposed in 1949 and 1964, but never really refined or used. Plankton production is dependent on the availability of nutrients, which is determined by the cycling of nutrients within the environment. Nutrient cycling is the result of all processes consuming and producing nutrients. Nutrient production is achieved by recycling and regeneration processes, which are either physical, chemical or biological. The concentration of nutrients in water is the result of all processes mobilizing and immobilizing nutrients in soil and water, including the activities of aquatic organisms.

Maintaining optimum water quality means keeping the water at an optimum for the physiological requirements of fish and for plankton production.

The next step is to stimulate natural productivity, increasing the production of plankton, so as to ensure higher yields of fish. Intensification of farming requires a good understanding of the ecological processes of natural productivity, so as to ensure control of production factors.

Nutrients are the basic elements for the elaboration of organic matter by autotrophic production; understanding the nutrient cycling in ponds is therefore fundamental. The establishment of nutrient budgets in ponds will lead to the modelling of nutrient cycling and eventually to the setting up of indicators which will enable the calculation of fertilizer needs.

At the end of the seventies, Prof. C. E. Boyd, of the University of Auburn, Alabama, USA, set up a procedure for the establishment of the lime requirements of fish ponds. At that time, there was some hope that procedures for the requirements of nitrogen and phosphorus in ponds would follow. But, unfortunately, until now, no further step had been made. This contribution aims at bringing together the latest information on the subject of soil-water interactions, water chemistry and natural productivity, hoping to stimulate new ideas and foster anew research on the fertilizer needs of ponds.

*

* *

The ecology of aquatic environments has made much progress during the last decades. But, information gathered in limnology does not seem to be used in aquaculture, the two disciplines seemingly appearing isolated. There is, however, much to gain in bringing them together: the information gathered in limnology can help to improve production in ponds, while ponds can be used as research sites for limnological research. The pond is an environment which is very convenient for study, because it is of small size and many of its factors are easily controlled. An exchange of information between the hydrobiological domain and aquacultural research would therefore be mutually beneficial.

The objective here is to bring together, in one book, information which concerns all aspects of natural productivity in ponds. First, there is a need to summarize the latest information on nutrient cycling and to outline how this is related to production factors. The aquatic environment is partitioned into several compartments, which are often separately investigated without setting the information together in the overall context of the aquatic environment. Between those working on the dynamics of the sediments, the specialists of water chemistry, the plankton specialists, and the fisheries biologists, exchange of information is poor. It seems interesting to combine in one book the information and ideas of all these specialities, even when taking the risk of being considered too superficial by each of these specialists.

*

* *

The book is divided into three parts. The first is devoted to the sediments, with emphasis on the exchange of nutrients and substances across the sediment-water interface, and the transformations occurring within the sediments.

The second part is devoted to water chemistry, to the cycling of nutrients within the water column, and to effects of fertilizers on water quality.

The third part is devoted to the analysis of productivity, to the factors involved in plankton production, and to the interrelation of nutrients with aquatic organisms, which compose the aquatic community. The specific factors involved in fish production are described and discussed in the light of existing information. The maximum fish yields obtained in the ponds of some African farms are shown and compared with maximum yields gained outside the region.

Chapter 1.

Pond soil

1. Soil characteristics

Importance of soil for water productivity

The soil lining pond bottoms is the border zone between the terrestrial and the aquatic environment. It is an area of exchange between both, which links them together and makes them interdependent. Soil is more dense owing to its solid state and is rich in nutrients, unlike water which is liquid and with a more diluted nutrient content. The upper part of the sediment is microbially active compared to the water column and to lower level sediments. According to Doremus & Clesceri (1982), one centimeter of surface sediment in Lake George, New York, U.S.A., is approximately two times more productive than 18 m of water column and four orders of magnitude more productive than sediments at 5 or more metres depth.

The interpenetration of the two environments (Fig. 1) results in a transformation of the soil into a semi-liquid mud, soaked with water, which is chemically very active. Some nutrients are released from it after transformation, but mud also acts as a sink for chemicals, which are immobilized and retained by it. The water supplies the soil with energy in the form of organic matter. Solar radiation, the primary source of energy for living organisms, usually does not reach down to the pond bottom. The biological activity of the mud is therefore dependent on supply of energy-yielding organic matter from the overlying water, in combination with the physical, chemical and biological properties of the soil itself.

Life in the pond soil depends on specific conditions, related essentially to the capability of the environment to transform the organic compounds and recover some of their energy content. This capability relies on the presence of electron acceptors, needed for the redox reactions of decomposition. Hence, the specific nature of the soil is of crucial importance.

Pond bottom constitution

The pond bottom is a composite environment made up of solid matter, interstices, water and living organisms. It becomes more abiotic with depth. The water surrounding the soil particles plays an important role in the aquatic ecosystem and is discussed below (p. 7).

The solid matter is a conglomerate of mineral and organic particles. The mineral particles originate from the underlying soil and rocks or are brought in by water entering the pond (washload). They sediment onto the pond bottom in the course of time. These are the sediments *sensu stricto*. Mineral constituents of the soil appear as particles of sand, loam, silt or clay and range over a variety of mineralogical compounds, such as silica, clay, aluminium and iron oxides, carbonates, sulphides, etc. Their importance is dependant on their chemical reactivity. Some of these constituents, such as quartz sand (crystalline silica), are chemically inert, while others are reactive, such as colloidal clay. A soil is characterized by its major constituents.

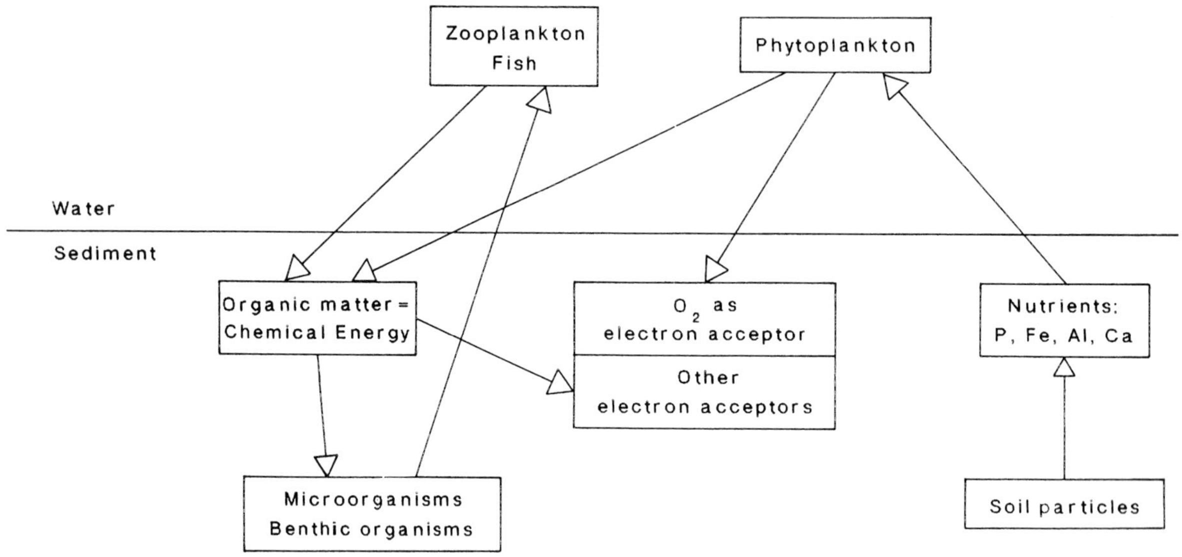

Fig. 1. Exchanges between water and soil. The aquatic environment provides the soil with (1) organic matter, which supplies energy and electron acceptors for life in the soil and (2) oxygen. The soil supplies nutrients of geochemical origin and some food to benthic organisms and microorganisms consumed by benthophagous organisms.

Morphological features of a soil

The morphological feature taking into account the size of soil particles is the texture of the soil or its granulometric composition, i. e. the relative frequency of particles of different sizes. Each soil has a specific and characteristic size distribution. The sizes of the mineral particles determine whether they are sand, silt, loam or clay. Different classifications exist. They vary in the limits of the size ranges of each class. According to the USDA (U. S. Department of Agriculture) classification, soil types are as follows (Coche, 1986):

type of soil	particles sizes
stones or boulders	: above 50 mm
gravel	: from 2 to 50 mm
sand	: from 0.05 to 2 mm
silt	: from 0.002 to 0.05 mm
clay	: below 0.002 mm

Particles smaller than 0.1 μm are colloids.

Other properties, such as structure, colour and mottling, specific weight or density, porosity, permeability, compressibility or plasticity, moist consistence, and shearing stress, are also used to characterize a soil. These properties are extensively discussed in Coche (1986).

The clay and organic matter content of a soil play key roles in the interaction between soil and water.

Some clay properties

Clay particles are small, of plate- and flake-like, tabular or even spheroidal shape, and many of their properties stem from their small size. The latter entails an elevated surface to volume ratio, defined by the specific surface or surface area, i.e. the external surface area of one unit weight of particles. A colloidal clay has a specific surface varying between 10 to 1000 $m^2\ g^{-1}$, compared to that of silt or sand, which is 1 and 0.1 $m^2\ g^{-1}$, respectively (Brady, 1984). Some other properties result from these particle size characteristics: for instance, water retention by capillarity, and adsorption of nutrients. Dried clay forms soil lumps very hard to break. Moist clay soil can easily be squeezed into a ball or a roll. The water retention capacity of clay soils is high as indicated by the high shrink-swell potential.

Clay minerals are alumino-silicates (feldspars) or iron or aluminium silicates (micas), made up of thin sheets of tetrahedral silica or octahedral alumina, arranged like the pages of a book. The minerals are classified according to the number of silica and aluminium/magnesium sheets contained in the crystal structure. Common clay minerals are kaolinite, smectite, illite, montmorillonite. Clay minerals are characterized by the variety of their chemical composition, wherein silicon and aluminium atoms may be substituted by other atoms of similar size. Montmorillonite is a mixed silicate of aluminium, sodium, iron and magnesium. Illite is a mixed silicate of potassium and aluminium. Clays found in tropical areas are rich in iron and aluminium oxides (Brady, 1984).

Fine clays are colloids and their chemical properties are related to surface chemistry, in connection with surface phenomena and exchange sites. Their importance and role in soil are preeminent.

Organic matter in soil

Organic particles are either dead or living organic matter. Sediments are often considered as abiotic material, with no account being taken of the microorganisms or benthic macroinvertebrates which live there. Yet, large populations of bacteria are associated with detritus in soils (Hart, 1982). They are not easy to isolate from their substrate, or to identify and enumerate. These microorganisms are a dynamic element of sediments, as they transform dead organic matter into living material. The dead organic particles are the remains of organisms living in the water column of the pond and sedimented on the pond floor after death and of organisms which live in the mud itself. Their dimensions vary over a large range of sizes, from coarse material almost the original size of the organism from which they originated, to transformed and aggregated matter. They are often characterized by a diffuse outline; in this case they are often called aggregates. The variety of compounds is great owing to the diversity of compounds available in living organisms and of patterns of decomposition. The ultimate form of organic material is humus, a group of organic compounds, in a dissipated (colloidal) state, coating particles of silica and clay and difficult to isolate. Its presence is revealed by the dark-brown colouration of a soil. Organic matter in colloidal form plays an important role in the exchange capacity of soils. It can be likened to a polyelectrolyte which adsorbs and releases solutes.

A soil is characterized by its proportion of organic matter versus mineral content (percentage of organic matter). It must contain more than 80 % of organic matter to be qualified as organic. Organic soils are common in marshes, wetlands, floodplains, bogs, swamp areas, and estuary zones, where they are subjected to continuous or long periods of inundations, which retard the oxidation of organic matter. Organic soil has a dark brown or black colour.

Stratification of flooded soil

The composition of pond bottom sediment also varies with depth. On the surface and in the upper layers, the influence of the overlying water column is strong, but decreases strongly at greater depths. The organic matter content of a soil is mainly generated by the water column. This content increases with depth, indicating accumulation over time (Twinch & Breen, 1982).

Three zones can be distinguished in flooded soil. Their characteristics are summerized in Table 1:

- a surficial oxidized microzone, with recently deposited, loose, uncompacted material, unstructured flocculent and fluffy flakes, approximately 1 cm thick (between 1 to 20 mm), with a patchy and intermittent distribution, easily disturbed by turbulent mixing (Twinch & Breen, 1982; Doremus & Cleresci, 1982; Sweerts *et al.*, 1986);
- a superficial layer (-1.5 to -5 cm stratum), more consolidated, still aerobic or oxidized, of variable thickness. (Levitan (1987) observed an oxidized capping extending at least 10 cm deep in Lake Mitchell, U.S.A.);
- a deeper layer (beyond -5 cm), anaerobic or anoxic, with reducing conditions, less microbially active.

Table 1. Stratification of soil - Characteristics of the strata.

	Physical state	Chemical state		Biological status		
				Microbiots	Benthic algae	Zoobenthos
Superficial microzone	loose	oxidized	aerobic	+++	+++	+++
Superficial layer	slightly compacted	oxidized	aerobic	+	+	++
Deeper layer	compacted	reduced	anaerobic	-	-	+

	Type of transformation within sediments	EXCHANGES WITH OTHER LAYERS AND WATER:	
		supplies	exports
Superficial microzone	Organic	- sedimentation of organic and mineral particles - adsorption - absorption	- resuspension - predation - transfer to superficial layer - diffusion
Superficial layer	Organic/chemical	- transfer from the microzone - upward migration	- resuspension - leaching
Deeper layer	Organic/chemical/ physical	- accumulation of leached nutrients	- gaseous ebullition

The three strata play an analogous role to those of a typical soil: the upper layer contributes most to biological activity. It accumulates organic material sedimenting from the water column and colonies of bacteria, benthic algae, zoobenthos, and a vast spectrum of decomposers, with a high heterotrophic potential (also shown by its high uptake potential). The layer below is a sediment layer, made up of partially transformed material, of poor nutritional quality, while the deepest layer is an original soil layer, where leached nutrients, such as calcium, magnesium, and phosphate accumulate. These nutrients may be lost from the environment if sequestered but otherwise may diffuse upwards by means of the interstitial water and be returned to the aquatic cycle.

The penetration of oxygen into the soil reaches a certain depth where it forms the aerobic/anaerobic borderline. This coincides with a change in chemical conditions, switching from an oxidizing to a reducing environment. This border is called the redox-cline or the oxic/anoxic boundary (or oxycline) and may not be sharply delineated, if the redox environment changes gradually. These two frontiers are of importance for life in the soil.

The borderline between the pond bottom and the water column is called the sediment-water interface and is more a concept than a distinct physical barrier. It is the exchange area between soil and water, allowing the transport of solutes and nutrients in and out of the sediments.

Pore water or interstitial water

Due to impoundment, the soil of pond bottoms becomes waterlogged. The pores between soil particles fill up with water. The total volume of pore water contained in sediments can be significant when compared with the amount of water in a pond, but depends largely on the underlying soil properties (its permeability) and the proximity of the water table. In Lake Södra Bergundasjön, Sweden, the water contained in the upper layers of soil represents up to 5 % of the total water volume (Enell & Löfgren, 1988). The major part of this water is a mobile liquid medium, which surrounds the soil particles and connects deeper layers of sediment with the aquatic environment through a complex network of tiny channels.

This water is named interstitial or pore water and contains solutes, which are transferable to the overlying water (Fig. 2). The composition of interstitial water is controlled by complex interactions between the ground-water recharge system (diluting the concentration of some of the more concentrated solutes released inside the sediments and carrying some solutes along with it), mineralogical dissolution and precipitation (due to the occurrence of interreactive solutes leading to authigenic mineral formation), biological activity (sequestering and release of some specific solutes) and physical interactions between mud and water (adsorption and desorption of solutes) (Enell & Löfgren, 1988).

Interstitial water as a transition compartment

Water in soil is predominantly interstitial or pore water. Only a small portion of the water is bound to solid material. The greatest part is in the chemically active form of a solvent.

Interstitial water is a transition compartment for solutes between soil and water. As pore water surrounds the soil particles, nutrients are released from the soil, or are removed from water. The composition and concentration of solutes in interstitial water are the result of concentration gradients and solubility kinetics of the solutes they encounter, together with the movement of the pore water. Changes in composition appear with depth (Nembrini *et al.*, 1982). The concentration gradients and mobility of water in soil ensure the transport of

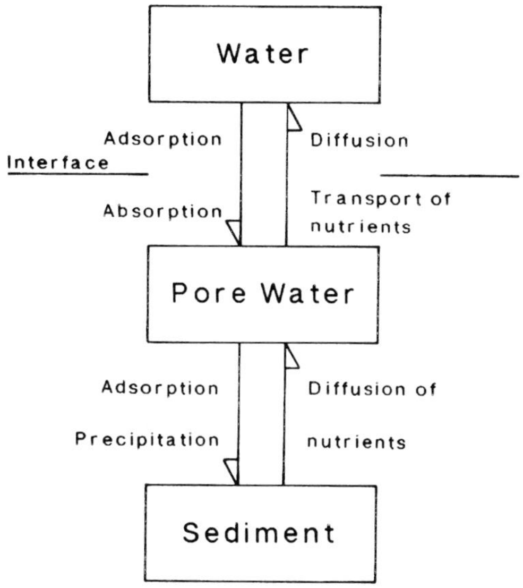

Fig. 2. Exchanges between water, pore water and sediment.

nutrients across the sediment-water interface. Exchange of nutrients across the interface is of considerable importance for the trophic status and productivity of the pond.

Movement in the groundwater influences the retention capacities of a soil. In laboratory experiments, Van Liere & Mur (1982) observed an increase of nutrient concentration (in their specific case, essentially phosphorus release) in the pore water, whenever there was an upward or downward movement of groundwater. During downward movement of groundwater, release of phosphorus occurred in an opposite direction to water movement, whereas diffusion and upward migration were working in the same direction.

The high water content (over 90 %) of sediment prompted Davison (1982) to conclude that sediments sometimes behave as liquids with a high concentration of particles. In the upper layers of the sediment, biological, chemical and physical activities are high, concentration gradients appear and the solute and nutrient content of the pore water changes in the course of time, whereas in the lower layers concentrations stabilize.

Mechanisms for transport across the interface

Transport processes are grouped according to the direction of the fluxes (Table 2). Sedimentation processes are downward transport mechanisms and encompass the sedimentation of mineral particulate matter suspended in water entering the pond as well as the sedimentation of living organisms after death. Sedimentation of plankton is a phenomenon of quite some magnitude in rich productive ponds. The sedimentation of a high load of poor nutritional value may contribute significantly to the silting-up of a pond. All these processes result in the growth of a sediment column in the pond and are macroscopic.

With regard to upward fluxes, the most important process is the transport of particulate matter and solutes owing to turbulence, bioturbation (turbulence caused by organisms) and other reworkings of sediment, while other upward migrations are essentially molecular diffusion phenomena, caused by concentration gradients or by changes in the flow of pore water. Upward flow as a macroscopic phenomenon is an erratic and random process, while the diffusion of solutes and nutrients occurs steadily.

When attempting to assess the importance of exchange between sediment and water, one should take into consideration the nature of the material exchanged. Sediments contain various types of material and their individual reactions to transport vary. For example, inert material, such as soil particles of sand, loam or silt, and fibrous organic material, such as remains of rice bran, will behave differently, according to the size of the particles and the cohesion forces (soil consistency and compaction) which hold them together.

Movements of particles with an electrical charge such as clay and organic particulate matter will depend on electrical forces (electrical potential) and fields and their electrophoretic mobility. In the absence of an electrical potential, they carry a cloud of ions of opposite charge. In the absence of an electrical field, the transport of ions, much smaller than colloidal electrolytes, depends on the intrinsic movement of the pore water (movements in the groundwater, modification of the level of the water table), and also on diffusion, the latter being dependent for each ion considered on the concentration gradient (chemical potential) generated and the ionic mobility of the ion species. Ions transported in the pore water, moreover, will react with other ion species they meet to form new compounds.

Table 2. Transport processes between sediment and water

TRANSPORT ACROSS THE INTERFACE (U = upward; D = downward)		
Physical transport	Chemical transport	Biological transport
Molecular diffusion (U, D)	Electrophoretic migration (U, D)	Bioturbation (U)
Gaseous ebullition (U)		Predation (U)
Turbulences: effect of wind and waves (U)		
Sedimentation (D)		

Concentration of some ions will lead to formation of stable or metastable mineral compounds.

Finally, the redox boundary separates the oxic from the anoxic region, and forms a chemical barrier, implying changes to the ions and compounds crossing it. Although this boundary is not commonly present at the sediment-water interface, its presence in deeper layers affects global transport across the interface. A typical example of ion species converted while crossing this redox boundary are iron and manganese. Oxidized forms do not have the same solubility as reduced ions. The ferrous (Fe^{2+}) and manganese ions (Mn^{2+}) are soluble and mobile in their reduced form, but are insoluble and precipitate when oxidized (Fe^{3+}, Mn^{3+}) (see p. 45). Once precipitated they form particles and are transported as particulate matter. The speeds at which manganese and iron are reduced are quite different. Manganese is reduced 50 times more slowly than iron (Davison, 1982).

Importance of soil for the productivity of water

1. Soil is rich in solutes and nutrients and exchanges them with the water across the soil-water interface.

2. Nutrients and solutes released by the soil and the rocks accumulate in the soil and can be supplied to the water column.

3. Organic matter sedimenting from the water column is recycled and returned to the water column in the form of living organisms or nutrients.

4. The recycling role of a soil depends on the chemical reactivity of the soil, essentially surface chemistry and low-energy bonds of ions of mineral, mainly clay, and organic colloids,and on the activity of decomposers.

5. Chemical transformations inside the soil are dependent on the prevailing oxidizing or reducing conditions, which change with depth.

6. Exchange of nutrients between water and sediment occurs in the interstitial or pore water, which transports dissolved solutes. The sediment-water interface plays a key role in the exchange.

2. Soil properties

Soil properties related to nutrient cycling are of direct interest to the fish farmers, as they influence the productivity of a pond. Soil activities are related to the quantity and quality of transformable material and to conditions conducive to reaction.

Clay and organic matter are amongst the most active components of soils. Nutrients and solutes are adsorbed and retained on the surface of colloidal particles and are released when conditions change. Besides exchange properties, a variety of chemical and biological transformations occur in a soil. These transformations are subject to the chemical conditions of the soil, which are either reducing or oxidizing, acidic or alkaline.

Organic or peaty soils have properties different from mineral soils; a notable difference is their retention capacity for water (and nutrients). A waterlogged mineral soil retains about 2 to 4 times its weight in water, while an organic soil has the capacity to hold 12 to 20 times its weight (Brady, 1984).

2.1. Colloids and their adsorption properties

In the past, the study of soil colloids was approached analytically. The main soil components suspected to be responsible for the colloidal properties were extensively studied *in vitro* and their adsorption capacities deduced from textural analysis. Soil fractions which exhibit colloidal properties are mainly clay particles and organic compounds, and so colloids were classified as mineral or organic. However, electron microscopic observations revealed that organic compounds and metallic oxides coat clay or silica particles and clay is likely to act only as a substrate (Hart, 1982).

Adsorption capacities of soil are further complicated by interference from biological absorption by soil microorganisms. Biological absorption and excretion of nutrients by microorganisms takes place rapidly and can cause confusion (Björk-Ramberg, 1985).

Thus, the analytical point of view has been substituted by a more functional approach, wherein total sorption capacities (all types of physical or chemical adsorption combined) are described using sorption isotherms and where the specific sorption of each species, for example phosphate or a heavy metal (in the case of pollution), is considered independently.

2.1.1. Mineral colloids

Even though clay does not play the role in nutrient sorption formerly ascribed to it, it still merits consideration.

Colloidal clay has particle dimensions ranging from 0.5 to 0.2 μm. The particles are negatively charged and the charges are localized on the surfaces, which, in the case of clays, are external and internal, between the crystal units or the silica and alumina layers. Free ions in solution in the pore water are adsorbed onto the external and internal surfaces of the colloid. Ions adsorbed onto internal surfaces are not easily removable. Water molecules, because of their polar character, penetrate the crystalline layers of some clays, such as smectite or montmorillonite, making the clay swell. The fixation sites on the external surface of the particles are surrounded by a swarm of electronic charges and those sites are essential for the exchange capacity of clays.

Electric charges on clays

From a chemical point of view, clays are alumino-silicates. Negative charges originate from atomic substitution inside the silicate tetrahedrons or the alumina octahedrons (Fig. 3). Atomic isomorphous substitution of silicon or aluminium atoms by heterovalent atoms leads to the appearance of negative non-neutralized charges. In silicate tetrahedrons, the silicon atom can be replaced by an aluminium or iron atom and in the alumina octahedrons, the aluminium atom by an iron, magnesium or zinc atom. These charges are permanent and do not depend on the pH of the environment. When such replacements occur, oxygen and hydroxyl ions, previously neutralized by the silicon or the aluminium, are so no longer, hence the appearance of negative charges.

Another cause of the appearance of negative charges on clay is the ionization of the hydroxyl groups at the broken edges of the clay lattice. These charges are pH-dependent and appear within certain pH ranges:

	$Clay{-}Al{-}OH_2^+$	$Clay{-}Al{-}OH$	$Clay{-}Al{-}O^-$
pH:	below 3	around 3	above 3

The pH at which clay has no net charge, called the point of zero charge, differs from clay to clay. Soils which are not highly acidic are mainly cation exchangers. In slightly to highly acidic soil, hydrogen ions (H^+) are tightly held and can be exchanged only with acidic cations, such as Al^{3+} or Fe^{3+}. At pH 6 and more, the hydrogen cation is exchanged against basic cations, such as Ca^{2+}, Mg^{2+}, K^+ and Na^+, or the hydroxyl group, which is more mobile than most other anions, is replaced by SO_4^{2-}, $H_2PO_4^{2-}$ or other dissolved anions (ligand exchange). The number of exchange sites depends on the type of clay.

Other types of charges

Metallic oxides, especially alumina, Al_2O_3, FeOOH (amorphous), and MnO_2, originating from iron and aluminium oxide clays, common in tropical soils, play an important role in adsorption mechanisms. In particulate material, iron and manganese hydroxides or oxides occur as surficial armouring on material and finely dispersed particles. They are present in amorphous, microcrystalline and more 'aged' crystalline forms (Förstner, 1982). In water, the metal oxides are covered with surface hydroxyl groups and adsorption of ions, cations or anions,

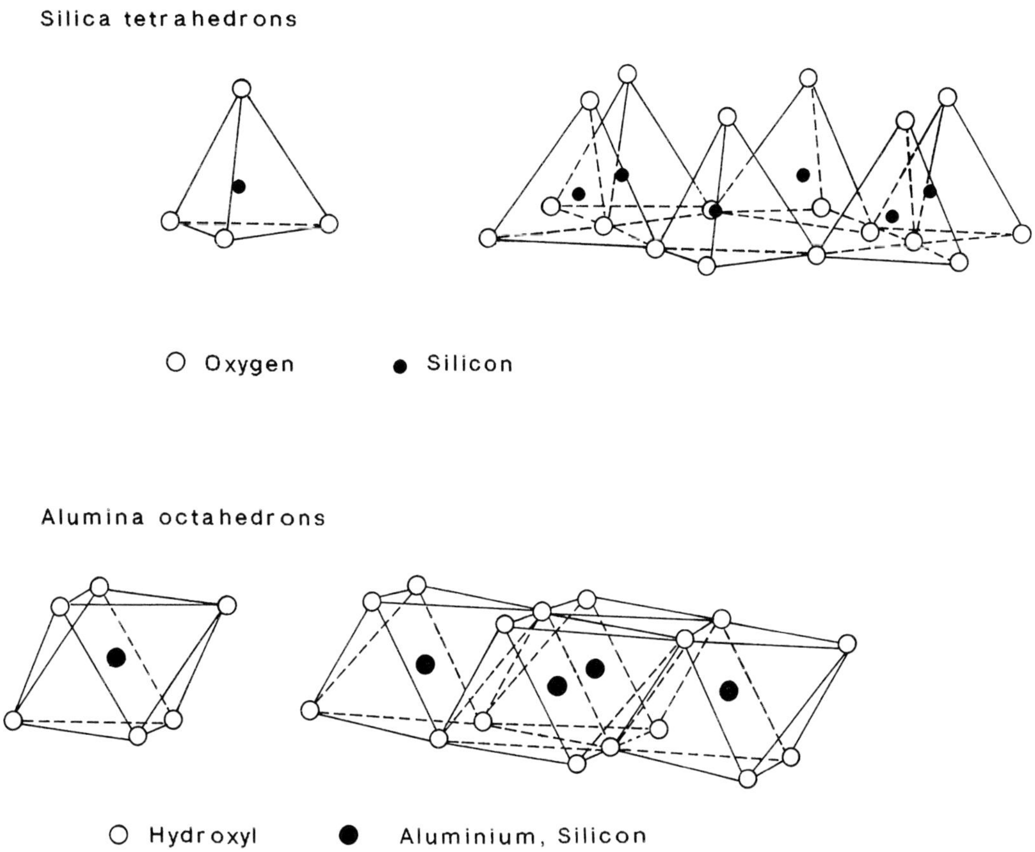

Fig. 3. Silica tetrahedrons and alumina octahedrons from the clay mineral lattice.

occurs at the water-oxide interface (Hart, 1982). In lake Léman, Switzerland, iron (III) forms polynuclear complexes which are highly reactive and have high sorption capacities (Nembrini *et al.*, 1982). The behaviour of aluminium or iron hydrous oxide is as follows (Holtan *et al.*, 1982):

$$\begin{matrix} OH_2^+ \\ \backslash\ / \\ Al \\ /\ \backslash \\ OH \end{matrix} \quad \underset{+H^+}{\overset{-H^+}{\rightleftharpoons}} \quad \begin{matrix} OH \\ \backslash\ / \\ Al \\ /\ \backslash \\ OH \end{matrix} \quad \underset{+H^+}{\overset{-H^+}{\rightleftharpoons}} \quad \begin{matrix} OH^- \\ \backslash\ / \\ Al \\ /\ \backslash \\ OH \end{matrix}$$

The charge (positive, negative or neutral) of the oxide depends on pH. The surface hydroxyl groups are the reactive sites and are binding ions. Exchange of cations or ligands occurs depending upon pH and redox potential, upon the formation constants of the complexes, and upon the concentration of the reacting species, the ionic strength, and competing ions (Hart, 1982). The strength of binding sites depends also on the sorption densities, sites not all being equal in their binding capacities. At higher sorption densities, the number of strongest binding sites decreases. Freshly precipitated hydrous iron and manganese oxides adsorb more than aged hydroxides (Nembrini *et al.*, 1982; Förstner, 1982).

Metallic oxides are also active in adsorbing organic matter. Adsorption on metal oxides is explained as surface complexing or ion exchange with other cations (Förstner, 1982); in Lake Neusiedler, Austria, organically complexed Fe^{3+} and Al^{3+} are the most likely sites for adsorption on organic matter surfaces (Gunatilaka, 1982). Various compounds, such as amino acids as well as natural fulvic and humic material, appear to be adsorbed to oxides. In Lough Neagh, Northern Irland, a eutrophic lake, adsorption of hydrous iron oxide on organic matter suggests involvement of ligand exchange with surface $FeOH^+$ or FeOH groups (Tipping *et al.*, 1982). In the presence of calcium and magnesium ions, competition between cations and the oxide for negative fixation sites on the humics increases with pH. The oxide adsorbs more humic material, in reducing the fixation sites of the humics (Hart, 1982).

Soil carbonates also show some adsorption properties. In the Neusiedlersee, Austria, freshly precipitated calcium carbonate has a high affinity for binding phosphate by coprecipitation (Gunatilaka, 1982). Phosphate anions are sorbed in a monolayer at the surface of the calcite crystals, replacing sorbed water molecules, bicarbonates or hydroxyl ions (Holtan *et al.*, 1988). The sorption mechanisms are summarized in Table 3.

2.1.2. Organic colloids

Solid organic matter in soil is present as various compounds, depending on its origin and transformation. There is no easy way to classify this organic matter, because of the variability of the compounds. Some are well defined, such as amino acids and organic acids, while most are uncharacterized and belong to the so-called natural fulvic and humic material. Fish farming is often sited in marshland areas where the acid water is rich in humic material, revealed by the dark colour of the water. Sometimes organic material is grouped according to molecular weight.

Organic material coats silica and clay particles and forms a biofilm. Organic surfaces originate in various ways:

- remains of substrate-dwelling microorganisms such as bacteria, algae and fungi
- condensation of organic compounds of low molecular weight forming organic aggregates
- sorption of organic compounds onto clay or metallic oxides.

According to Hart (1982), surfaces of solid matter in aquatic systems are more homogenous than might be expected from the finding that the particles all have a negative charge and that their electrophoretic mobility varies only slightly, indicating similar surfaces; the negative charge is the result of the presence of an organic film covering most of the particles.

The binding groups of organic material involve several chemical forms, but are predominantly carboxylic and phenolic groups, and in particular salicylic acid (Hart, 1982). Humic matter is normally negatively charged and binds preferably cations. However, in association with certain iron and aluminium cation species, it is able to form complexes which also bind anions. This mechanism of sorption is quite important as it supports the

Table 3. Different sorption mechanisms occurring in soil

SORPTION MECHANISMS		
Physical sorption	Adsorption (reversible)	- Clay - Metallic (iron and aluminium) oxides and hydroxides - Carbonate (calcite) - Organic matter (alone or with metals): humic and fulvic material
Chemical sorption	Chemisorption (partly reversible) Precipitation (non reversible)	- Clay - Carbonate and calcium salts
Biological sorption	Absorption, followed by excretion	- Microorganisms

fixation of phosphate in sediments. Organic ligands or compounds can block and reduce the number of sorption sites on mineral surfaces. The organic matter acts in two ways: in sorbing ions and in blocking sites for fixation.

Interaction between metallic oxides and organic compounds is complex, as iron and aluminium oxides adsorb organic matter and vice-versa. The mechanism is not well understood.

Binding between organic matter and soluble organic compounds or ligands also occurs. This leads to condensation of organic compounds. The nature of these reactions often makes it difficult to draw the line between physical and chemical sorption.

2.1.3. Ions sorbed by colloids

A colloidal particle is a polyelectrolyte micelle, carrying electric charges at its surface, and surrounded by adsorbed ions. The micelle as a whole is negatively charged, though the surface displays fixation sites for positive and negative ions. Charges associated with soil particles are simple cations and anions as well as complex organic and inorganic charge complexes.

The strength of the retention force of the cations to a micelle depends on the hydrated ionic radii and on the valency of the cation species, cations with short radii and those with higher valency being more tightly sorbed, and on the ionic strength of the pore water solution surrounding the particles. The composition and the ionic strength of the pore water change considerably locally and with depth. This explains why calcium and magnesium concentrations in Lake Léman, Switzerland, increase with depth, while nitrates and sulphates decrease (Nembrini *et al.*, 1982).

Interaction between different species of cations is of importance for the sorption properties of soils, though the mode and range of interactions are not well understood. For instance, magnesium and calcium ions significantly affect the amounts of organic matter sorbed to a solid surface, increasing the degree of sorption of fulvic material to goethite (Hart, 1982); iron, aluminium and manganese are essential for the availability of nutrients (Nembrini *et al.*, 1982). Competitive reactions between solid surfaces, colloids and pore water solutions are not well understood.

Colloids in suspension are surrounded by water molecules, and monovalent hydrated ions, which are not strongly held, but increase the electronegativity of the colloidal particles. This permits individual particles to repel each other and stabilizes them. This state is termed dispersion as opposed to flocculation, which is a reduction of electronegativity accompanied by an increase of the size of the colloidal particles and settling rate. Flocculation is enhanced by divalent and trivalent ions. These ions are more tightly adsorbed by the colloids and enhance the attractive forces between particles. Aluminium, hydrogen and calcium ions favour flocculation (Brady, 1984). Flocculation of mineral particles and organic solids modifies the real or effective particle size. In Wilton Creek, Ontario, U.S.A., this phenomenon is seasonal and related to biological activity in sediments (Ongley *et al.*, 1982).

A distinction should be made between acidic, and alkaline ions: hydrogen, aluminium and iron are the acidic, and magnesium, calcium, potassium and ammonium alkaline cations. Acidification of soil is the result of two different processes: a release of acidity caused by decomposition of organic matter by microorganisms (nitrification and sulphate reduction); and leaching of basic cations from the upper layers of the soil towards deeper horizons. The acidic or alkaline character of a soil will be discussed later (p. 15).

2.1.4. Ionic exchange capacity

Cation exchange capacity

The capacity of a soil to sorb and release ions is of paramount importance for the fertility of the overlying water. As soil particles are negatively charged, ions withheld by the colloids are primarily cations. The amount of cations sorbed to a soil can be determined and is termed its cation exchange capacity (CEC). It represents the quantity of cations adsorbed to the soil and is expressed as milliequivalents of cations per 100 g of dry soil or as centimoles of positive charge per kg of soil (cmol kg^{-1}), according to the International system unit. The conversion is: 1 meq $(100\ g)^{-1}$ is equal to

1 cmol kg^{-1}. Exchangeable cations are Al^{3+}, H^+, Fe^{3+}, Ca^{2+}, Mg^{2+}, K^+, Na^+, NH_4^+.

The CEC of soil is related to its clay content. Some types of clays have higher CEC than other: illite has higher exchange capacities than kaolinite, owing to mean particle size or to surface area per gram of particles. The CEC of a few clays is listed below (expressed as meq (100 g of dry $soil^{-1}$)):

Type of clay mineral	meq (100 g of dry $soil^{-1}$)
Kaolinite	3-10
Illite	15-40
Montmorillonite	80-125
Vermiculite	85
Smectite	118

The CEC also depends on the organic content of a soil; the greater its proportion of organic matter, the higher its CEC. In soils rich in humus, CEC's in excess of 180 meq (100 g of dry $soil^{-1}$) are not uncommon. Samples of soil taken from deeper parts of fish ponds at Auburn, Alabama, U.S.A., have higher CEC's than from shallower parts (Boyd, 1976b). This is explained by a higher organic content in deeper parts of the pond compared to the shallower zones, where the mineralization processes are faster. In sediments below cages installed in two Polish lakes (Szczytno Małe and Łętowo Lakes) used for trout culture, CEC's were some 20 to 30 meq (100 g of dry $soil)^{-1}$) higher than in the adjacent sediments. Similarly, the organic carbon content of the soil underlying the cages was some 0.5 to 9.5 % higher than in littoral sediments, reaching a concentration of 12 to 17 % (Korzeniewski & Moczulska, 1985).

Changes in acidity or alkalinity of sediments affect the CEC. In acidic conditions, hydrogen ions are more tightly sorbed and not exchanged against other cations as easily as in alkaline conditions. Furthermore, many pH-dependant charges are no longer available and the amount of exchange sites is reduced. Acidification of a sediment can considerably reduce its amounts of exchangeable cations, principally alkaline ones such as calcium and magnesium; along with these, sulphate or chloride anions are also washed out. This process is called leaching. The relative proportion of acid cations such as aluminium increases and the sediment develops an acid reaction. Restoration of alkaline conditions is achieved through liming.

The fraction of CEC (expressed as percentage) occupied by basic cations is called base saturation and the fraction of acidic cations is the base unsaturation. Base saturation, pH of the soil and total hardness of pond water are closely related (Boyd, 1979).

Anion exchange

Most nutrients essential for the productivity of a pond, such as carbonates, nitrates, and phosphates, are anions. Their availability in sediments as exchangeable anions is important. Other common anions present in soil are chloride, sulphate, sulphide and hydroxide ions. In some tropical sediments, colloids carrying positive charges are present. They are mostly clays rich in hydrous aluminium or iron oxides, of reddish or yellowish colour. Minerals found in such soils are gibbsite, $Al_2O_3.3H_2O$, goethite, α-FeOOH, and hematite, α-Fe_2O_3. In most soils, anion exchange is more complex, because sorption of anions is associated with cations such as Fe^{2+}, Al^{3+} and Ca^{2+} which are themselves exchangeable (Holtan *et al.*, 1988). Measurement of the anion exchange capacity is not commonly carried out, the availability of specific anions being estimated with the help of sorption isotherms.

Sorption isotherms

Sorption isotherms express the relation between the quantity of ion sorbed and the concentration of this ion in solution. This relation was first developed to account for the sorption of gases on solids. The Langmuir isotherm assumes that the adsorbed ions cover the surface until full saturation of the exchange sites occurs, after which there is no further adsorption. This assumes that there are no interactions between adsorbed ions, that interchange of ions can proceed as readily as on a bare site, and that there is no variation in sorption energy with the degree of surface saturation. The Freundlich isotherm introduces a correction factor into the relation; when the exchange sites are almost saturated, the exchange constant is reduced and when the exchange sites are free the exchange constant is increased.

The experimental data fit these isotherms relatively well, provided the concentration ranges are not too wide.

Importance of the colloidal properties of a soil

1. Soil colloids bind solutes and nutrients by low-energy bounds and form a store of exchangeable nutrients.

2. Soil colloids are usually classified as mineral or organic. Mineral colloids consist mainly of clays and metallic oxides, while organic colloids are composed of a group of largely unidentified compounds, of which fulvic and humic material are the most important.

3. Some observations by electron microscopy give evidence that most soil particles are coated by an organic film, which carries negative charges able to bind cations. This gives a certain homogeneity to the surface of solid matter.

4. The sorption of cations occurs at the negative fixation sites, while the sorption of anions occurs in association with iron, aluminium and calcium ions. Cations are separated into acid and alkaline species.

5. The amount of cations sorbed to a soil is measured by the CEC, the cation exchange capacity. The proportion of alkaline cations sorbed is termed the base saturation and the proportion of acid cations sorbed is the base unsaturation.

6. The amount of anions sorbed is determined by isotherm sorption curves.

7. Investigations on the sorption properties of the soil are hampered by interference from biological absorption by microorganisms.

2.2. The acid or alkaline reaction of soil

Many soil properties depend on its acidic or alkaline character, as this determines its oxidizing or reducing properties. Furthermore, in an acidic soil, the fixation sites on colloidal particles are saturated with ions not suitable for productivity, while the ions essential for productivity are unavailable. A soil is said to be acidic if its pH is below 7, and alkaline if above 7.

A compound is defined as an acid if it yields dissociated hydrogen ions in solution. A base yields hydroxyl ions in solution and per extension a solution is defined as acidic when the concentration of hydrogen ions is higher than that of hydroxyl ions. pH is thus a measure of the concentration of hydrogen ions, expressed as the negative logarithm of the hydrogen ion concentration.

The pH of a soil is determined not only by the presence of hydrogen ions but by all other acidic ions, mostly aluminium and iron cations, or by the hydroxyl and alkaline ions, such as calcium, magnesium, potassium, and sodium.

Soil pH is a measure of its acidity at a given moment, but as a soil is a dynamic system, changes occur continuously as reactions proceed. It is therefore important to determine the factors responsible for modifying soil pH. In some cases, pH is able to resist changes and is said to be buffered. The buffer systems of soils are of much interest.

Many factors affect soil pH: texture, with the contribution of soil minerals generated by weathering; organic matter content, which has is own acidic or alkaline reaction and which, through decay processes, acidifies the soil; biological activities of microorganisms, releasing carbonic acid, and whose densities are correlated to organic matter content; pore water concentration and the supply of alkaline and acidic ions to the soil; supply of oxygen to the soil which favours oxidizing or, in its absence, reducing reactions; soil drainage, which washes away alkaline ions, such as calcium, magnesium, sulphate and nitrate, which are not tightly sorbed. Organic soils have pH with high buffering capacities and shifting the acidic pH of these peaty soils towards values well suited to production is not easy.

In short, all processes generating protons acidify a soil, while all proton-consuming processes have an alkalizing effect.

Acidifying role of aluminium and iron

Aluminium and iron are said to be acidic cations. In soil with pH below 4.5, aluminium is in the soluble form of Al^{3+} ions; in pH between 4.5 to 9, aluminium is precipitated as $Al(OH)_3$; above 9, it appears as an aluminate ion, $Al(OH)_4^-$. In water, all aluminium species are subjected to hydration. The Al^{3+} ion, for example, is surrounded by water molecules and forms:

$$Al^{3+} + 6H_2O \longrightarrow [Al(H_2O)_6]^{3+}$$

The aluminium ion in soil of pH above 4.5 reacts with water to form two soluble species of hydrous ions. They can be written as follows, leaving out the hydrating water molecules and considering only the anhydrous ions:

$$Al^{3+} + H_2O \rightleftharpoons AlOH^{2+} + H^+$$

$$AlOH^{2+} + H_2O \rightleftharpoons Al(OH)_2^+ + H^+$$

$$Al(OH)_2^+ + H_2O \rightleftharpoons Al(OH)_3 + H^+$$

At each step, hydrogen ions are released into solution adding, to the acidification. These ions formed are partly adsorbed and partly free in solution, and their relative concentrations depend on the equilibrium constants and reaction rates. Moreover, these ions are able to polymerize and form chelates.

In the presence of silica, which is common in soils, the aluminium hydrous oxides react to form:

$$H_4SiO_4 + AlOH^{2+} \longrightarrow H_3SiO_4\text{-}AlOH^+ + H^+$$

which is an alumino-silicate ion, and

$$H_4SiO_4 + Al(OH)_2^+ \longrightarrow H_3SiO_4\text{-}Al(OH)_2 + H^+$$

which is the base molecule of kaolinite.

In acidic soil, hydrogen ions are not the only exchangeable ions responsible for acidity, as aluminium Al^{3+} ions are also involved. When these exchangeable aluminium ions are removed from the soil, they are replaced after a short time by some other aluminium ions which spontaneously appear at the same locations. They are probably extracted from the clay lattice to occupy positions at the exchange sites (Ségalen, 1973).

Iron reacts in a similar way, forming hydrous oxides. It is present as Fe (II) and Fe (III) ions, depending on the redox potential of the soil. Ferrous ions are soluble and relatively free from complexation, while ferric ions are insoluble (Davison, 1982) and in Lake Erie, U.S.A. and L. Windermere, U.K., tend to form colloidal complexes with organics (Förstner, 1982). Ferrous ions begin precipitating as ferrous hydroxide, $Fe(OH)_2$ at pH 6.0, and the reaction is completed at pH 7.8, while precipitation of ferric ions as $Fe(OH)_3$ occurs at a pH between 2.3 and 3.4. Free ferric ions are not likely to be found in soil, as acidity rarely drops so low. The transformation of Fe^{3+} and Fe^{2+} ions into hydrous oxides gives rise to intermediary ions similar to the aluminium-oxide reaction, which increase acidity, but to a lesser extent.

Moreover, iron easily forms chelates and other complexes with organic compounds. Complex aggregates between clay minerals, ferric ions and humic compounds can be formed. Included in these aggregates, iron loses its property as an acidic ion.

With increasing pH, the influence of aluminium and iron diminishes and they are transformed into precipitated hydrous oxides. The exchange sites they occupied are freed and made accessible to other ions, especially alkaline ones.

Active and potential acidity

Active acidity measures the amount of acidic ions in soil, present and active at the time of record. It is an intensity measure of the acidity at a given time. This acidity depends on a potential, an exchange or a reserve acidity, which is the total amount of acidic ions in reserve or adsorbed to soil particles and which will eventually be released. The potential acidity is related to CEC, base unsaturation and buffering capacity of the soil. Correction of the acidity of a pond mud must take into consideration the total, i.e. active and potential acidity.

Acidification and alkalinization factors

Acidification is generally attributed to three factors: high concentrations of free acidic ions, leaching of alkaline ions, and release of acidic and hydrogen ions by microorganisms, through respiration, nitrification (see p. 24) and sulphide oxidation (Table 4). Decaying organic matter releases organic and inorganic acids, carbonic acid and nitrate ions. Acidic ions displace alkaline ions from the adsorbed sites, which are subjected to leaching.

Alkalinization is caused by an increase of alkaline and hydroxyl ions in the mud. It results from a neutralization of acidic ions or a proton-consuming process. Ageing of sediments causes a neutralization of the soil. Soils subjected to high evaporation in arid regions display an alkaline pH. Proton-consuming processes are: solubilization of limestone by carbonic acid freeing alkaline ions, denitrification, and sulphate reduction.

Nutrient availability and pH

The increase of aluminium, iron, and manganese ion concentrations with decreasing pH entails a decrease in the availability of phosphate, which reacts with the former to form insoluble compounds such as $FePO_4$ (Golterman, 1982). Furthermore, metallic ions are toxic at high concentrations. In alkaline pH, the concentration of calcium increases and with phosphate forms calcium apatite, $Ca_5(PO_2)_6(OH)$, which is insoluble. Iron and manganese are precipitated at high pH, thus becoming unavailable. The requirements of the microorganisms for those oligo-elements are no longer met, and deficiencies arise.

An optimum pH is found within a median zone where the solubility, and thus availability, of the different nutrients is at a maximum. At a suitable pH, microbial decomposition is stimulated and nutrient regeneration optimized. pH acts on the composition of the benthic fauna as well (see p. 50).

2.3. Redox potential

Redox potentials indicate the type of oxidations likely to take place in sediments. The redox or oxidation-reduction potential of a solution is the proportion of oxidized to reduced substances. It is called E_h and measured by means of a hydrogen electrode (Boyd, 1979). Reducing conditions are indicated by negative potentials and oxidizing conditions by positive values. The redox potential of natural water shows values around 0.5 V (Boyd, 1979).

Values of redox potential are standardized to pH 7, using the following formula:

$$E_7 = E_m + 244 - 57\ (7 - pH)$$

where E_m is the value of the redox potential

Table 4. Acidification and alkalinization factors occurring in soils.

ACIDIFICATION FACTORS	ALKALINIZATION FACTORS
Leaching - soil drainage	Evaporation
Organic decomposition (release of carbon dioxide)	Ageing of sediment
Production or release of acidic ions: Al^{3+}, Fe^{2+}, SO_4^{2-}, NO_3^-, HCO_3-	Production of alkaline or hydroxyl ions (solubilization of limestone): Ca^{2+}, Mg^{2+}
Nitrification	Denitrification
Sulphide oxidation	Sulphate reduction

recorded, all values of potentials being expressed in mV (Smayda, 1990). Each change of one pH unit is accompanied by a change of E_h of 57 mV, 58 mV or 59 mV according to Smayda (1990), Pearsall & Mortimer (1939) and Boyd (1979), respectively. For pH values above and below 7, the redox potential is respectively increased or decreased by 57 mV per unit of pH.

As indicated earlier, most transformations in sediments are related to the content of organic matter which settles from the overlying water. Non-photosynthetic organisms meet their energy requirements by transforming reduced organic matter, recovering part of the energy of the high energy bonds. The most common oxidizing agent or electron acceptor is the oxygen molecule, but a number of other electron acceptors is available. The organic loading of sediment and mineralization appear to be responsible for changes in redox potential depth profile (Rippey & Jewson, 1982; Lijklema & Hieltjes, 1982). As the load increases, the change in redox potential at different depths indicates a shift from an oxic to an early anoxic environment below the sediment surface, which later again becomes oxic (Rippey & Jewson, 1982). The free energy released by each of these reactions decreases as reducing conditions increase or redox potential decreases.

The sequence of microbially mediated redox processes is as follows:

- O_2 reduction to O^{2-}
- denitrification, NO_3^- to NO_2^- and NH_4^+
- MnO_2 reduction to Mn(II)
- FeOOH reduction to Fe(II), $FeCO_3$ for example
- SO_4^{2-} reduction to S^{2-}
- methane fermentation or methanogenesis, CO_2 or acetate to CH_4.

The redox potentials at which these reactions take place are listed below (Seitzinger, 1988).

Table 5. Redox potentials associated with the principal electron acceptors.

Electron acceptors	Redox potential (mV)
O_2	720-740
NO_3^-	710
MnO_2	470
FeO(OH)	60
SO_4^{2-}	-200
CO_2	-250

Oxygen concentrations required for nitrate-nitrite and nitrite-ammonium reductions to proceed are 4 mg l^{-1} and 0.4 mg l^{-1}, respectively. For the ferric-ferrous reduction, the oxygen concentration is 0.1 mg l^{-1}, and nil for the sulphate-sulphide reduction.

Efficient - in terms of energy-yield - transformation of the organic matter will depend on the prevailing redox conditions or on the supply of oxygen to the sediments. Redox conditions are also important for the mobilization-immobilization of phosphate in sediments. Ferric iron precipitates as hydrous ferric oxide, which adsorbs phosphate, whereas ferrous iron sets phosphate free.

Importance of pH and redox potential properties

1. The acid or alkaline character of a soil indicates the type of prevailing ions (alkaline or acidic) in solution and measured by the pH.

2. Two types of acidity occur: active and potential. Active acidity is the actual value recorded, whereas potential acidity is an indication of the total amount of acidic ions adsorbed to the soil and likely to be released in the course of time. It corresponds to the acid buffer capacity of the soil.

3. Aluminium and iron play an important role as acidic ions in soils.

4. The pH of the sediment acts essentially on the availability of nutrients, especially phosphate.

5. The pH results from a series of processes proceeding simultaneously in the soil. Microbiological activity in the soil has acidifying and alkalinizing effects on soils.

6. The redox potential indicates the oxidizing or reducing character of the soil.

7. The bulk of reducing reactions which occur in sediments results from activity of heterotrophic organisms degrading organic matter.

8. The amount of energy yielded by transformation depends on the availability of electron acceptors.

9. Redox potential plays an important role in the immobilization or mobilization of phosphate in sediments.

3. Mineral and organic components of a soil

A soil is a huge store-house for nutrients, provided by the parental soil or rock, by water, by decomposition of organic matter. The cycling and availability of nutrients in mud makes the latter of interest to fish farmers.

Soils are a mixture of chemical substances, some in large quantities, some in minute amounts. A soil can be characterized by its carrier substances, which are the major constituents onto which are attached or into which are incorporated trace elements (Jacquet *et al.* 1982), and major nutrients. The most important carriers are clays, metallic oxides, silicates, limestone and dolomite.

Some of the major elements of soil, such as phosphorus, calcium and magnesium are closely associated with living organic matter, while others, such as aluminium, iron and potassium, are linked to inorganic matter, especially clays (Ongley *et al.*, 1982).

Models, particularly mass-balance models, are often used to study interactions between water, sediments, and cycling of a particular element, such as phosphorus. They help to test assumptions underlying our knowledge, bring about an awareness of the complexity of the system, and identify areas in which more knowledge is required. Assessment of parameters needs to be made on a set of observations, while validation necessitates an independent data set. Only then can predictions and extrapolations be made (Lijklema & Hieltjes, 1982).

These models divide the environment into a series of boxes, simulating the major compartments of a water body. The fluxes between the different boxes are formulated mathematically. Such models vary in complexity and in resolution and represent a more or less comprehensive description of the real environment. A discussion of the use of a hierarchy of models is found in Lam *et al.* (1982). The ability of models to cope with reality is better when considering only physical and chemical phenomena than when including complex biological situations. Examples of failures in the use of models as a result of incomplete biological information are given by Fryer (1987).

3.1. Mineral component of soil

With regard to the mineral component of soil, nutrients are of most relevance for pond productivity. However, one must consider not only elements such as calcium, phosphorus and nitrogen, but also others such as sulphur and iron, which play a more significant role than their importance in living organisms might suggest.

3.1.1. Calcium

Calcium is present in soils as calcareous sediments, e.g. calcium carbonate or calcite, the mineral form. This element is found in a great variety of common soil minerals such as apatite or gypsum, and also in feldspars, zeolites, amphiboles, garnets, epidotes, pyroxenes, and clay minerals such as illites, smectite, vermiculites, etc., and in a solid form.

The adsorption of dissolved calcium ions onto soil colloids strengthens the base saturation or base buffer capacity of the soil. As free ions, they contribute to the concentration of alkaline ions in the pore water. Dissolution of calcium carbonate releases calcium and carbonate ions. These latter ions participate in the carbonate-bicarbonate system buffering the pH against sudden changes. The bulk of carbonate ions in this alkalinity system does not have a geochemical origin, but originates mainly from organic matter and decaying processes. In these ways, calcium participates in the regulation of soil pH.

The availability of phosphate is reduced in the presence of high concentrations of calcium, as apatite is formed, which is insoluble. Calcium acts on the flocculation of dissolved organic matter.

Calcium and magnesium influence sorption of other cations, especially trace metals, such as manganese, zinc, copper, on metallic oxide (Al_2O_3 and FeOOH) surfaces. This is so because the former are normally present at concentrations many orders of magnitude higher than the trace metals and thus occupy most surface binding sites, even though they form less stable surface complexes. Calcium and magnesium have a significant effect on the amount of natural organics sorbed onto the solid surface of the oxide. Twice as much natural fulvic material was sorbed to goethite when calcium and magnesium were present than when absent (Hart, 1982). Furthermore, calcium promotes flocculation of colloids.

Lastly, calcium is also biologically important as it is required by organisms for their physiological needs, such as in enzyme regulation.

Calcium is subject to leaching and therefore its concentration in soil increases with depth. Magnesium follows a similar trend (Nembrini *et al.*, 1982).

Control of soil pH is effected through liming, a common practice in fish ponds. The objective is to secure a lasting increase of the pH of the pond water; this subject will be discussed on p. 78.

Importance of calcium in soil

1. Calcium is present in numerous minerals in soils, mostly in crystalline and occluded forms, limestone being the most common form.
2. When dissolved, calcium carbonate generates calcium ions, which, being alkaline, strengthen the base saturation of the soil, and carbonate ions, which participate in the buffer capacity. Calcium thus has an important role in soil pH regulation.
3. Calcium acts on the flocculation of organic matter.
4. Many organisms require calcium for their physiological needs.
5. Calcium ions are subject to leaching.

3.1.2. Nitrogen

Nitrogen compounds present in pond muds

Usually, both inorganic and organic nitrogen compounds are present in soils. Organic nitrogen is associated with decaying organic matter, while inorganic nitrogen is essentially present as ammonium, nitrate and nitrite ions. These ions are soluble and occur free in pore water or adsorbed on colloids. The quantities of ions adsorbed depend on availability of fixation sites for sorption on the colloids and on the equilibrium between ions adsorbed and in solution. Ammonium at high concentrations and at alkaline pH forms free ammonia; however, a high pH is not common in soils. Ammonium and nitrate ions in solution never account for more than 1 to 2 % of total nitrogen, with the main proportion of nitrogen in an immobilized state.

Adsorption of ammonium to soil is not homogenous and is related to the organic matter content of the sediment: areas rich in organic matter tend to accumulate nitrogen. The amounts of ammonia adsorbed onto soil particles may be significant: sediments in fish ponds in three different fish breeding regions in Israel (Beith Shean valley,

Jordan valley and the coastal plain) contained from 300 to 900 μg N g^{-1} dry sediment (Fig. 4) (Diab & Shilo, 1986).

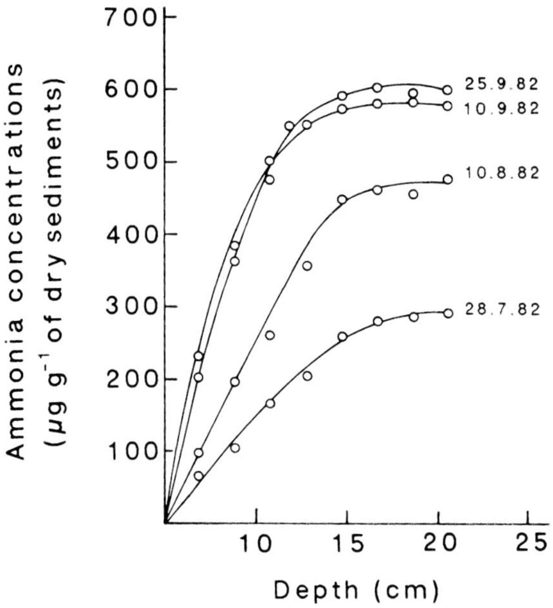

Fig. 4. Changes in ammonia concentration in the sediment profile of a fish pond in the course of a production period (adapted from Diab & Shilo, 1986).

Origin of nitrogen in sediments

Almost all nitrogen in pond mud is of biotic origin, stemming from the decay of organic matter produced in the pond. A small portion of organic matter may be of external origin, falling into the pondwater or imported by water fed into the pond. Depending on the state of decomposition, this nitrogen is in an organic or inorganic form. Only a small portion of the organic matter sedimenting on the pond bottom is mineralized every year.

While humic material is considered as organic matter, it is refractory, of poor nutritional value for decomposers, and supplies little nitrogen to the sediments.

Transfer of nitrogen across the interface

Most of the nitrogen ion species are well soluble and easily transferred from sediment to water. The means of transport are either diffusion across concentration gradients, or sediment perturbations, due to stirring of sediments and bioturbation. Blue-green algae make use of this upward transport of nitrogen compounds at the interface. When concentrations of nitrogen ions were low in the water, these algae were observed in shallow brackish water fish ponds in Israel to make daily downward migrations to the interface, where they took advantage of the diffusion of nitrogen compounds from the sediments (Van Rijn *et al.*, 1987).

Conditions at the interface

Pond sediments are predominantly anaerobic when a pond is impounded and in production, and aerobic when the pond is harvested and dried out.

In ponds in the Beith Shean and Jordan valleys, Israel, used for tilapia and carp rearing, Diab & Shilo (1986) showed an accumulation of ammonium in the sediments to a depth of 15 cm (Fig. 4). Klapwijk & Snodgrass (1982) observed that ammonium concentrations in pore water of Hamilton Harbour, Lake Ontario, Canada, subject to large ammonia and carbon loadings from a waste-water treatment plant and from steel industries, were some 2 to 100 times higher than in the overlying water (0.5-5.0 mg l^{-1} and 10-50 mg l^{-1} ammonia for water and sediment, respectively). This is in concordance with observations of Schindler *et al.* (1987) in Lake 227, of the Experimental Lake Area, Ontario, Canada (Fig. 5). Ammonium concentrations of sediments are dependent on sedimentation of organic matter from the water and on the ammonification potential of the sediments. In Lake Gunillajaure, northern Sweden, Björk-Ramberg (1984) showed a significant correlation between organic matter content and nitrogen content. At the surface (0 to -3cm), ammonium concentrations were lower owing to diffusion across the interface, and to microbial nitrification and denitrification.

Denitrification in sediments

Denitrification is the reduction of nitrate to gaseous molecular nitrogen or even nitrogen oxide, N_2O (Fig. 6). The concentration of the latter is always

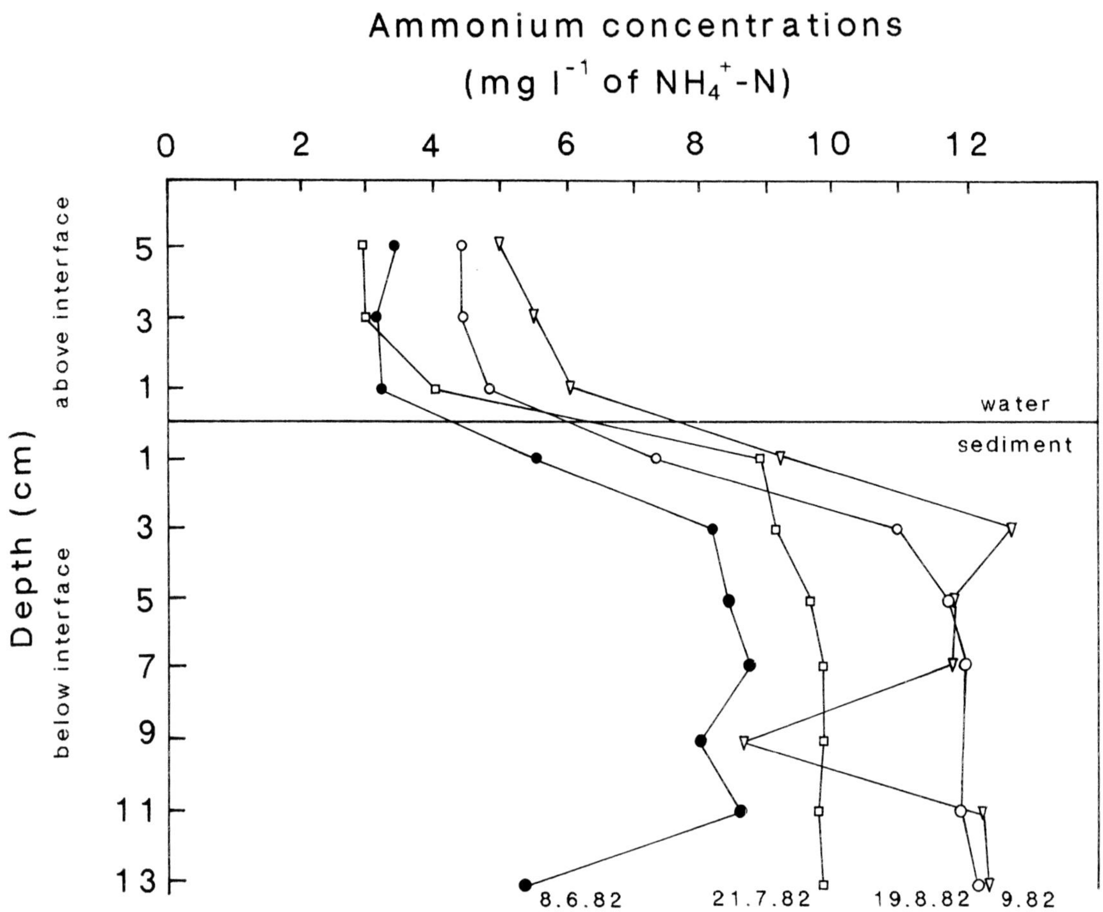

Fig. 5. Seasonal changes in ammonia concentration in the pore water of a sediment profile and in the overlying water of a Canadian lake. The samples were taken in the sediment under 9 m of water (modified from Schindler *et al.*, 1987).

low (Seitzinger, 1988). In this reduction, which is microbially mediated, nitrate is used as a final electron acceptor in the absence of oxygen. Nitrate ions migrate through aerobic sediments to the reaction site where the reduction takes place. Diffusion alone cannot account for the flux observed: bioturbation is partly responsible for the transport of nitrate, ammonium and molecular nitrogen (Klapwijk & Snodgrass, 1982). Nitrate concentration also decreases with depth (Nembrini *et al.*, 1982).

Nitrate for denitrification is essentially provided by sediments and not by water, and the sediments do not act as a sink for nitrate from the water (Klapwijk & Snodgrass, 1982; Gardner *et al.*, 1987; Seitzinger, 1988). During mineralization of organic matter in sediments, a major portion of the nitrogen is denitrified, but considering the external supply of nitrogen to aquatic ecosystems, the amount denitrified does not represent more than one quarter to one third of the supply (Klapwijk & Snodgrass, 1982; Seitzinger, 1988).

The activities of denitrifiers in a sediment vary according to the nature of the sediment and with depth: in loamy sediments, activities are five times higher and extend deeper than in sandy ones. Redox potentials required must not be over +300 mV and maximum activity of the nitrate reductase is around +210 mV; dissolved oxygen concentrations must not be in excess of 0.06 mmol dm^{-3} (1.92 mg l^{-1}) (Cooke & White, 1987a). Nitrate in soil is provided by nitrification of ammonium resulting from ammonifi-

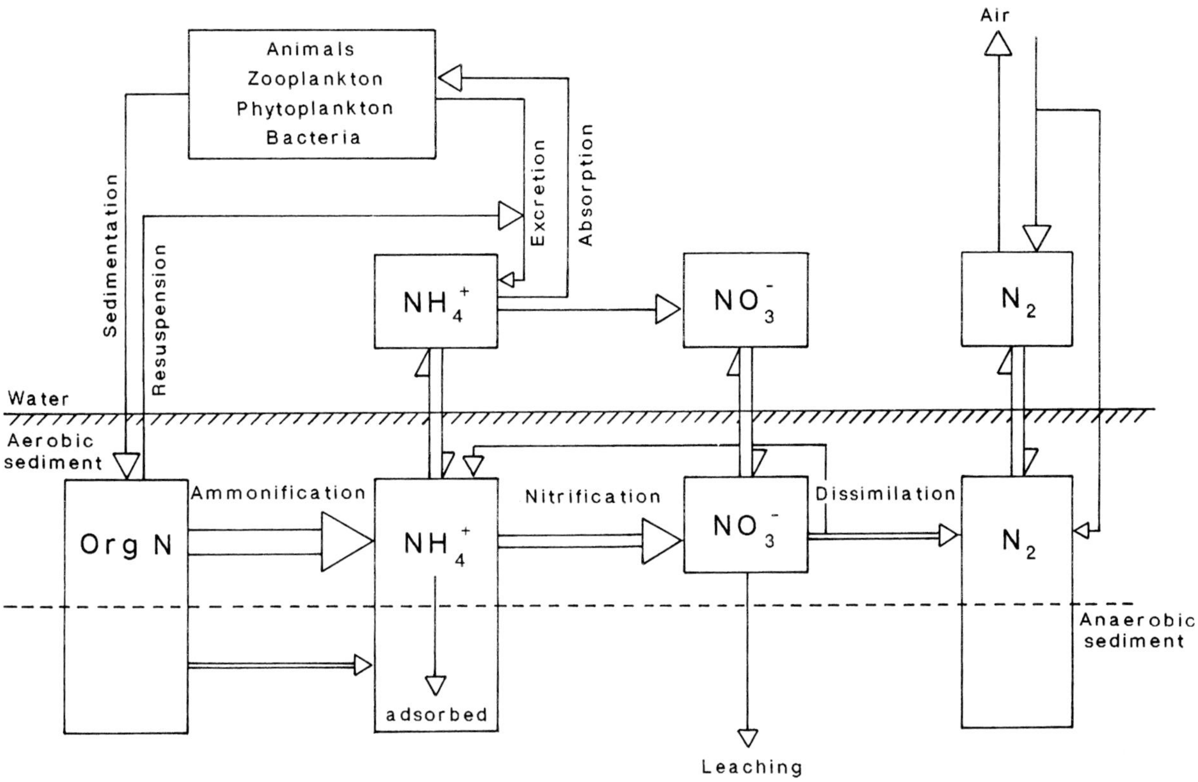

Fig. 6. Transformation of the nitrogen compounds within the sediment of an impounded pond. The nitrogen-fixation process is not shown: the molecular nitrogen is fixed into organic nitrogen.

cation of organic matter. Because of denitrification, ammonium concentration in deeper layers is depleted (Cooke & White, 1987b).

Nitrogen fixation

Nitrogen fixation in sediments has been intensively investigated by Blauw (1988). Molecular nitrogen is used by various microorganisms and blue-green algae to cover their needs of organic nitrogen (Fig. 6). Cellular enzymes controlling the process are nitrogenases and the activity of such enzymes in sediments is correlated to the availability of organic substrate, namely carbohydrates, organic alcohols and acids, and to the maximum rate of glucose absorption by the heterotrophic population. Nitrogen fixation in sediments is essentially an activity of heterotrophic organisms. Such fixation in anaerobic conditions is less efficient than in aerobic conditions and therefore more substrate is needed: *Clostridium pasteurianum* needs about 10 mg nitrogen per g of sugar, while *Azotobacter chroococcum* in microaerobic conditions is able to incorporate 46.5 mg of nitrogen per g of sugar consumed. The abundance of microorganisms in the rhizosphere of macrophytes where excretions from the roots are available is explained in the same way, these excretions providing the energy supply for nitrogen fixation. Such organic excretions are the limiting factor for nitrogen fixation, which is thus linked to the carbon cycle.

Nitrogenase synthesis is activated when interstitial ammonium concentrations drop below a minimum level. Ammonium represses the synthesis and inhibits the activity of the aerobic fixation enzyme, while the synthesis of the anaerobic fixation enzyme is repressed, but its activity is not inhibited. The intracellular concentration of ammonium has a greater effect on the regulation of the synthesis of nitrogenase than the concentration of ammonium in pore water. However, sediments with high ammonium contents still show some nitrogen fixation. This apparent contradiction could result from the huge sorption capacities for ammonium of sediments, but this does not seem a sound explanation. Probably the areas where nitrogen fixation takes place are protected from interstitial ammonium, and are located on microsites in a microenvironment not in direct contact with interstitial water. Protection or barriers consist of quaternary structures of macromolecules, attached to particular fractions or could be bacterial mucus or excretion products. Both bacteria and benthic organisms secrete mucus, within sediments and along burrow walls, which bind sediment particles (Ongley *et al.*, 1982). The barriers constituted by soil particles contribute to the maintenance and renewal of microgradients. Epipelic diatoms are also known to produce extended matrices covering sediments. These consist of mucopolysaccharides, which consolidate the sediments (Paterson, 1989).

Nitrogenase activity is highest at the mud-water interface, where organic matter is in abundant supply and of good nutritional quality; the organic matter content of sediments is reduced by half in the first cm of depth (Blauw, 1988).

Nitrification

Nitrification is the microbial transformation of ammonium into nitrate. It takes place in the upper 2 to 3 mm of the sediment, i.e. the aerobic layer (Fig. 6). This microzone or biofilm is thin, sometimes even less than a fraction of a millimetre. The thickness of the layer is dependent on the biological activity of the sediment, the dissolved oxygen concentration and the depth of overlying water. The shallower the water, the greater the quantity of oxygen reaching the sediment.

Denitrification depends on oxygen transport and biological and chemical oxygen demand. A good supply of oxygen is needed to allow nitrification to proceed. If oxygen consumption exceeds supply, anaerobic conditions appear; the oxygen demand of the sediment can no longer be met and denitrification commences (Bouldin *et al.*, 1974). Denitrification activities occur just below nitrification sites, which indicates a coupling of the two processes (Cooke & White, 1987a).

Nitrification stops when dissolved oxygen concentrations drop below 2.24 to 0.96 mg l^{-1} (Cooke & White, 1987a). The penetration of oxygen into the sediment results in inactivation of the denitrifying enzymes and decreases the amount of mineralizable organic carbon. In deeper layers, denitrifiers, which are facultative anaerobes, are the main consumers of organic carbon and the main mineralization agents (Cooke & White, 1987a). The activity of nitrifying bacteria is maximum above pH 6.5, decreasing with pH.

Ammonification

Organic matter settling from the water column, such as phytoplankton, remains of fish feeds, and fish excreta, decays (Fig. 6). The transformation of organic nitrogen into ammonium is called ammonification. As a consequence, the sediment of a fish pond steadily accumulates ammonium in the upper 10 to 20 cm layer, where sorption onto soil particles takes place till full saturation of sorption sites. Very little ammonium is in solution in the pore water as long as the sediments are not saturated with ammonium (the adsorption properties of sediment for ammonium are high). At the same time, the concentration of N-NH_3 in the water generally remains below 0.1 mg l^{-1}. Sediments act as a sink for the ammonium in the water. Concentrations of nitrate or nitrite in the pond mud are virtually nil.

Modifications occurring during the drying of pond bottoms

In a harvested pond, left to dry, sediments are exposed to the air and become oxidized. Organic matter accumulated during the production period is mineralized (Fig. 7). Liming of the pond bottom speeds up the process (Legendre *et al.*, 1987). Ammonium trapped during the production period is transformed into gaseous (volatile) ammonia if soil pH increases above 9.5; otherwise, it is nitrified and transformed into nitrate. Bacterial activity is dependent on water content of the sediment and on oxygenation. Nitrate in sediments is then used by denitrifiers or leaches to deeper horizons. Concentrations of nitrate in surface layers decrease rapidly

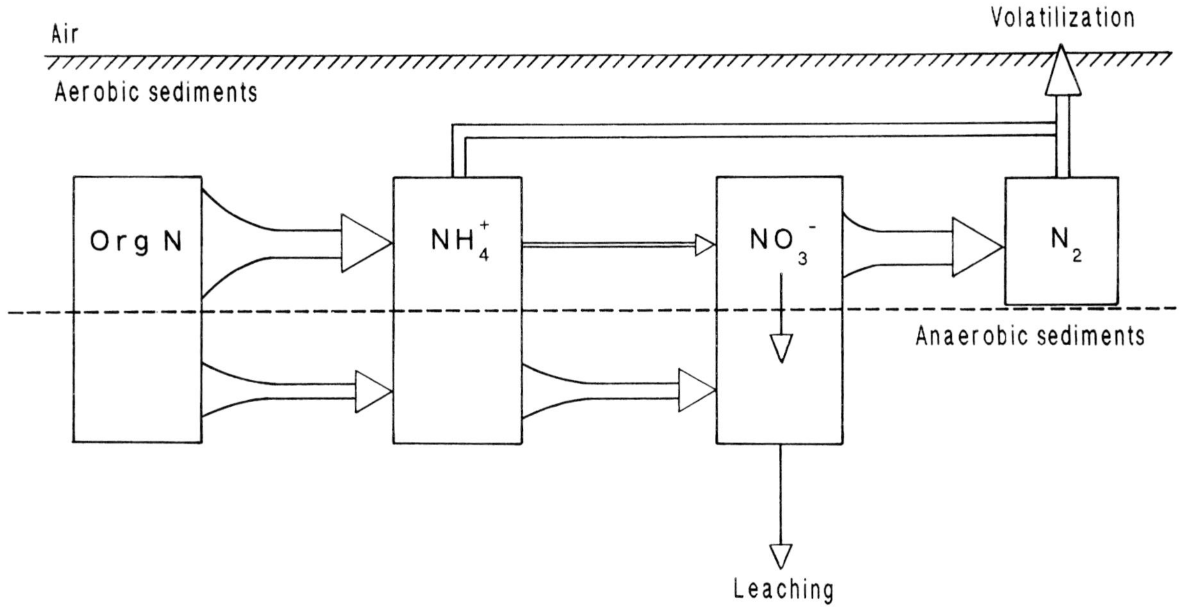

Fig. 7. Transformation of nitrogen compounds within the air-exposed sediment of a pond, during the period of drying.

and are soon virtually insignificant (Fig. 8).

Ammonium, nitrate and nitrite production, as well as nitrification, denitrification and ammonification activities are essentially concentrated in the upper layers of a sediment and decrease with depth (Diab & Shilo, 1986: 75, Fig. 5a). Nitrate increases with depth due to leaching (Diab & Shilo, 1986: 76, fig. 5b). Denitrification rates in deeper layers are low. Rewetting of sediments during the drying phase increases microbial activity and accelerates mineralization process after impoundment.

In soil exposed to air, some nitrogen fixation by bacteria and blue-green algae may occur, but this is not necessarily so if the drying period is short.

Changes occurring after refilling

In the refilled pond, nitrates and nitrites in the soil disappear quickly, owing to diffusion to the water column or to denitrification. Ammonification activities resume and ammonium accumulates in sediment and water. Immediately upon refilling, the ammonium present in the sediments diffuses to the water and high concentrations (up to 3.5 mg l^{-1} of N-NH_4^+, in the case reported) of ammonium and nitrite temporarily appear (Diab & Shilo, 1986). Nitrate is assimilated and its concentration in the water decreases. Ammonification, nitrification and denitrification are microbial processes, sited in the upper layers of the sediments, and occurring in such a sequence that ammonium concentration always remains low, being transformed into nitrate that is assimilated or transformed into molecular nitrogen. Volatilization of gaseous ammonia from sediments occurs only at alkaline pH or after transformation of ammonia into gaseous nitrogen by denitrification. The first process is merely a chemical phenomenon, while the second is microbial.

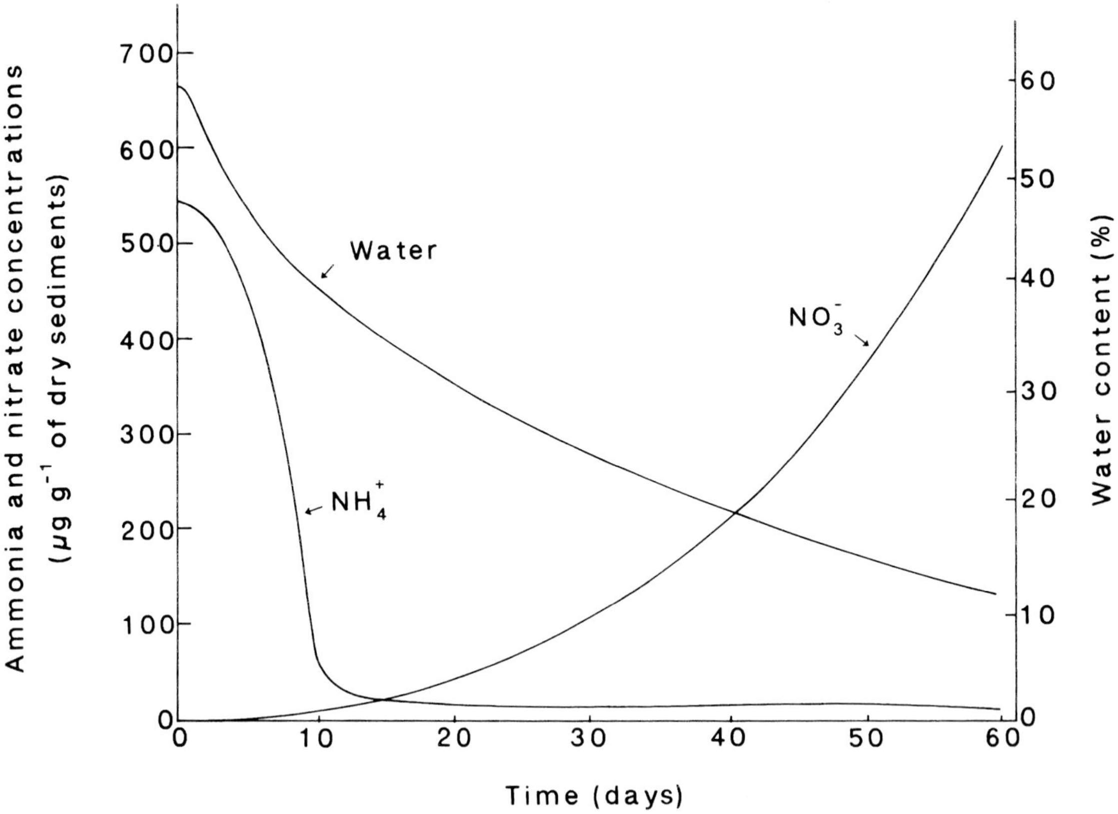

Fig. 8. Changes in ammonia and nitrate concentration and water content of the sediment of a fish pond with time. The values are the mean of four samples along a 10 cm core, and water content is expressed as percent of wet sediment weight (adapted from Diab & Shilo, 1986).

Importance of nitrogen in the sediments

1. Nitrogen is present in soil as the inorganic ions ammonium, nitrate, and nitrite, and as organic compounds. The total amount of nitrogen sorbed onto soil may be important, as ammonium is well soluble and easily sorbed onto colloids.

2. Most nitrogen is of organic origin and supplied by the organic loading of the pond.

3. The cycling of nitrogen in sediment is essentially microbial, different sets of microorganisms being responsible for the transformation of various nitrogen species.

4. Ammonification is the transformation of organic nitrogen into ammonium; nitrification is the transformation of ammonium into nitrate; denitrification is the transformation of nitrate into molecular nitrogen; nitrogen fixation is the transformation of molecular nitrogen into organic nitrogen.

5. During pond production, the most important process is ammonification, followed by nitrification in the aerobic strata of the sediment. The denitrification process is energy-yielding,

using nitrate as an electron acceptor in place of oxygen, and takes place in anaerobic portions of the sediment, where leached nitrate ions occur.

6. In drying pond soil, ammonification leads to nitrification.

7. Nitrogen ion species are easily transferred across the sediment-water interface.

3.1.3. Phosphorus

3.1.3.1. Introduction

Concentration of phosphorus in sediments

Great disparities between the phosphorus content of water and sediments exist. Phosphorus concentration in water is always low, while in sediments it can be found in large amounts. Hepher (1966) found up to 12 g m^{-2} of total phosphorus (P_{tot}) in the mud of a freshwater fish pond at Dor, in Israel. Schindler *et al.* (1987), studying lake sediments (Lake 227, Experimental Lake Area, Ontario, Canada), observed that amounts of phosphorus in solution in the upper layers of the sediments were much higher than in the water near the sediment-water interface (Fig. 9). Sediments are said to act as a sink for phosphorus.

Complexity of phosphorus dynamics in sediments

Phosphorus accumulates in sediments, but can also

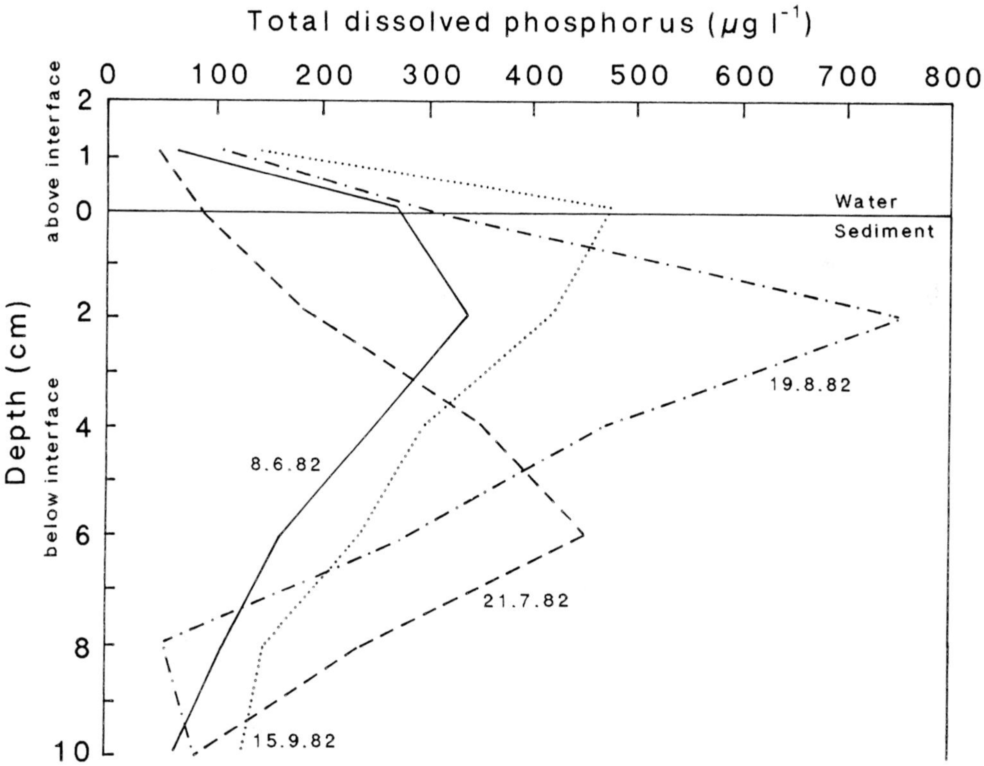

Fig. 9. Seasonal changes in total dissolved phosphorus concentration in the sediment profile and overlying water of a Canadian lake. The samples were taken in the sediment under 9 m of water (modified from Schindler *et al.*, 1987).

be released to the water. The mechanisms of sediment phosphorus release form a complex and dynamic system of which only some aspects are understood (Marsden, 1989). The processes responsible for phosphorus immobilization and mobilization differ in nature:

(1) physical and chemical sediment properties determine its phosphorus sorption capacities, and consequently, by way of sorption equilibria, the concentration of interstitial phosphorus;
(2) the sedimentation rate of organic matter, its mineralization rate, microbial assimilation and excretion rates, and to a lesser extent authigenic mineral formation, determine phosphorus loading to sediment and turn-over rates;
(3) disturbances of sediment causing reworking and resuspension of sediment is responsible for transport of sediment and phosphorus fluxes.

It is therefore impossible to assess, based only on total phosphorus of a sediment, the amounts which will be released in any period of time. There is also no simple procedure to establish the portion of phosphorus requirements, for fish production needs, which can be met by the sediments or the amount of phosphorus fertilizer which will be immobilized in pond mud.

Although phosphorus in sediments has been extensively studied and many limnologists have devoted their life to research on this subject, many aspects still remain poorly understood. A summary of present knowledge is given hereafter.

3.1.3.2. Different forms of phosphorus in sediments

Sediment-bound and interstitial phosphorus

Phosphorus present in pore water is termed interstitial phosphorus; that sorbed to particles is particulate phosphorus. Distinction between liquid and particulate phases is arbitrary (Logan, 1982) and is made by filtration through a membrane filter of 0.45 μm pore size. The filtrable material is defined as dissolved and the retained material as particulate. The filtration procedure neglects colloidal material less than 0.45 μm in size (Stumm & Morgan, 1981), the arbitrary size limit for colloids being 0.3 μm (Broberg & Persson, 1988). Dissolved phosphorus is mainly present as orthophosphate and is called soluble reactive phosphorus (SRP). (For the analytical determination of orthophosphate, see p. 109). The concentration of dissolved phosphorus is often overestimated, due to occurrence of particulate phosphorus in the filtrate (Logan, 1982).

Importance of interstitial phosphorus

Phosphorus in pore water is essentially soluble, and the most reactive species of phosphorus of sediments (Fig. 10). The pore water is an intermediary compartment between soil particles and the water column and is the transient cell for the phosphate carried in by the pond water to be sorbed onto particles or to be carried away after desorption. It is a communication network for solutes, connecting soil particles together. Because of higher concentrations than in pond water, transformations and chemical reactions between ions and solutes take place in the pore water.

The portion of soluble phosphorus is small (estimated to be 1 % (Boström *et al.*, 1988a)) compared to particulate phosphorus and is about 5 to 20 times more concentrated than in pond water. Concentration of interstitial phosphorus reflects the trophic state of the water body and the level of accumulation of phosphorus in the sediments. Solubilization of phosphorus in pore water depends on the solubility coefficients of the phosphorus compounds immobilized onto soil particles and chemical concentration gradients between soil particles and pore water. When the sediment concentration of P_{tot} exceeds 2 mg l^{-1}, phosphorus is not considered limiting (Böstrom *et al.*, 1982).

The concentration of interstitial phosphorus changes during the year. Highest concentrations are recorded when sediments are oxygenated. Interstitial phosphorus concentrations are sensitive to changes in the environment (Enell & Löfgren, 1988). Most changes occurring in soil are the result of microbial activities, reacting to changes in the availability (load) of organic matter, in temperature, in availability of electron acceptors, etc., which all modify the dynamics of phosphorus release.

Orthophosphate is the dominant form of phosphorus in the upper layers of sediments, while in deeper layers, condensed forms occur.

Data on concentrations of interstitial phosphorus

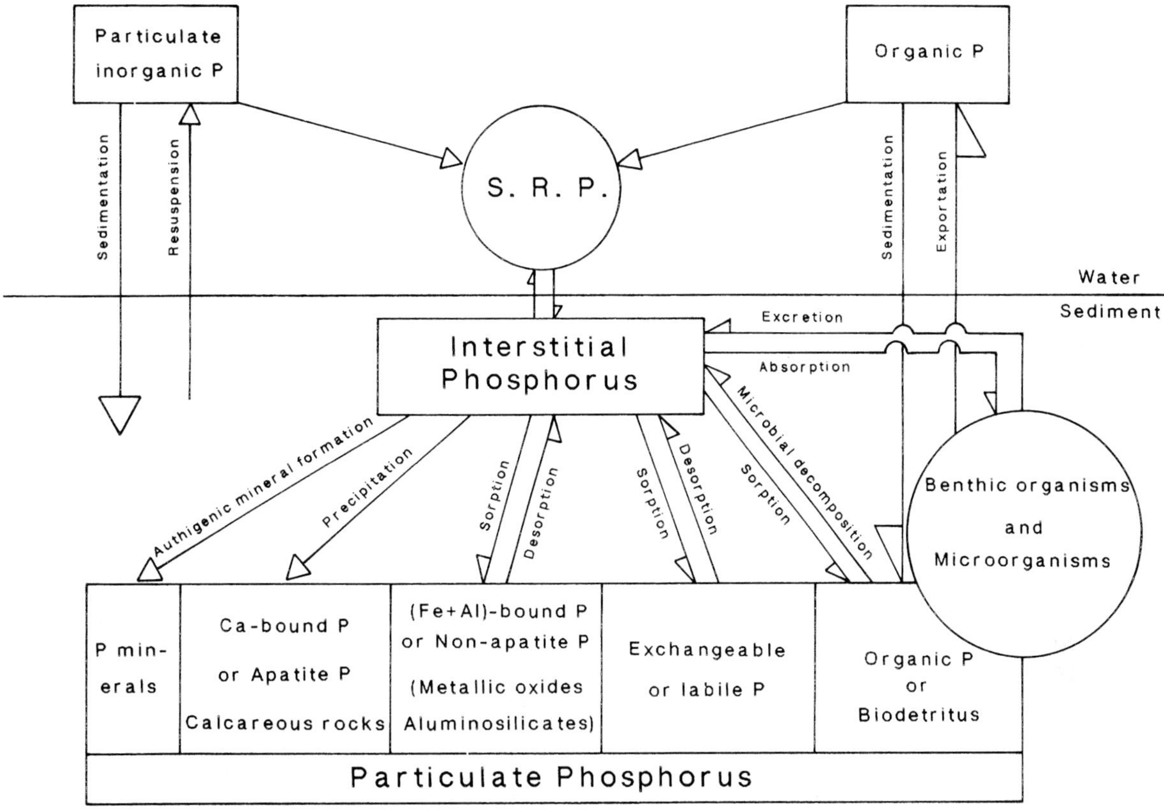

Fig. 10. Importance of interstitial phosphorus. Relations of interstitial phosphorus with soluble reactive phosphorus (S.R.P.) in water and with particulate sedimentary phosphorus. The overlap of organic phosphorus and benthic organism compartments indicates the difficulty in separating biotic organic phosphorus from non-living organic phosphorus.

are often more important than data on concentrations of total phosphorus in soil, but are more difficult to obtain. A review of methods for sampling interstitial water is given in Enell & Löfgren (1988).

Particulate phosphorus

Phosphorus associated with particles or sediment-bound phosphorus is a mixture of various chemical compounds, ranging from mineral, non-reactive (crystalline or amorphous) forms to labile or complex organic compounds (Fig. 10). Distinction is made between adsorbed and precipitated, and between mineral and organic phosphorus. Labile compounds include adsorbed phosphate, easily hydrolyzable or dissolved. Sequential phosphorus extraction schemes are numerous and their results difficult to compare (Pettersson *et al.*, 1988). Such procedures attempt to quantify the different forms of phosphorus present and, through fractionation, partition particulate phosphorus into major groups, which generally comprise (Logan, 1982; van Eck, 1982; Pettersson *et al.*, 1988; sequential extraction procedure according to Hieltjes & Lijklema, 1980):

- (Fe + Al) bound P (24 h extraction with 0.5 N HCl)
- calcium bound P (apatite) (17 h extraction with 0.1 N NaOH)

- reductant soluble P or exchangeable P (2 h extraction repeated twice with 1 M NH_4Cl)
- organic P, including labile and resistant organic P (digestion with $HNO_3/HClO_4$).

Comparisons between results concerning different phosphorus fractions obtained with different extraction methods are difficult and one should be cautious in interpreting such data.

Mineral phosphorus is usually separated into forms sorbed to apatite and non-apatite minerals. Apatite phosphorus is associated with calcium, while non-apatite phosphorus is essentially associated with iron or, to a lesser extent, aluminium. Acid-extracted phosphorus encompasses most of the phosphate bound to aluminium and iron; the most important forms of phosphorus in sediments are amorphous iron-hydroxy complexes with an atomic ratio of Fe to P between 5 and 10 (van Eck, 1982; Lijklema & Hieltjes, 1982: 230, Fig. 4), while Al-bound phosphorus plays only a minor role (van Eck, 1982). The atomic ratio of iron to phosphorus of the acid-extract is between 1.5 and 1.8. Besides the other compounds extracted (exchangeable-P, carbonate-P and labile organic-P followed by clay and other compounds), tinctite, $Fe_3(PO_4)_2(OH)_3.3H_2O$, or cacoxenite, $Fe_4(PO_4)3_2(OH)_3.12H_2O$, two stable Fe(III) phosphates under oxidizing conditions, are assumed to be present (van Eck, 1982). The abundance of iron in aerobic sediments is important in governing phosphorus sorption. The oxidation of ferrous iron should immobilize phosphorus only when the atomic Fe/P ratio in the interstitial water is above 1.8 (Marsden, 1989). A great number of phosphorus minerals - more than 200 have been described - are also likely to be found in sediments, as a result of diagenetic processes. (Diagenesis refers to abiotic and biogenic changes taking place within the sediment during and after burial).

In the humic lake Tjeukemeer, Netherlands, organic phosphorus compounds have high molecular weights, belonging to the fulvic or humic compound groups. Fulvic acids are aromatic polycondensates, with phenolic and carboxylate groups, which give them chelating properties. (The aromatic nature of the fulvic material could be an artefact). In contrast to iron, humic acids are not able to bind inorganic phosphorus directly to form a colloidal phase (De Haan & De Boer, 1986).

A distinction exists between the various fractions of phosphorus characterized in sediments and phosphorus available for production, designated as bioavailable. The fractions from which phosphorus is taken by microorganisms and autotrophs to satisfy their needs are not well defined and will be discussed on p. 42.

The concentrations of phosphorus in sediments are generally expressed in weight per unit weight of dry sediment, though it might be better to use weight per unit surface or per unit volume of sediment, specifying the depth of sediment considered (Holtan *et al.*, 1988).

Sorbed or immobilized sediment phosphorus

Immobilization of phosphorus in soil is the result of sorption (see Table 3, p. 12). Sorption is removal from solution to a solid phase and involves binding at fixation sites on the surface of colloids. Distinction is made between physical sorption and chemisorption. Physical sorption is due to forces between the solid surface and the adsorbed molecules or ions. These forces are similar to van der Waals forces between molecules and are undirected and relatively non-specific. The energies involved are of the order of a few hundred to a few thousand calories per mole. Physical adsorption is generally reversible. In contrast, chemisorption is the result of strong binding forces, comparable to those leading to the formation of chemical compounds; such an adsorption may be regarded as the formation of a surface compound and is seldom reversible. The energies of chemisorption range from 10^4 to 10^5 calories per mole (Morris, 1956; Holtan *et al.*, 1988).

Adsorption in sediments was first studied on clay minerals and on metallic oxides and hydroxides. Acid muds with high concentrations of hydrous iron (III) oxides, together with small quantities of alumina, silica and organic matter, show high sorption, with peaks at pH 2 to 5 (Hepher, 1958). Sorption capacities for phosphate of non-calcareous sediments are much higher than calcareous sediments. Crystalline iron and aluminium hydrous oxides sorb five to ten times more phosphorus than crystalline alumino-silicates or calcium carbonates. In contrast, short-range order components, such as iron oxide gel, sorb ten to one hundred times more phosphorus than their crystalline counterparts, i.e.

up to almost one thousand times more than crystalline alumino-silicates and calcium carbonate (Holtan *et al.*, 1988).

In alkaline soils with high concentrations of carbonates, phosphates are adsorbed to the surface of calcite crystals, where they displace water molecules, or carbonate or hydroxyl ions, or co-precipitate with calcium carbonate, or form calcium phosphate.

Organic matter can act in a twofold, antagonistic way. It sorbs phosphate directly onto fixation sites with positive charges or indirectly with the help of intermediary ions, but can also block sorption sites on oxides, preventing the sorption of phosphates. Organic anions prevent crystallization of aluminium oxide (Holtan *et al.*, 1988). Organic colloids, such as humic material, do not adsorb phosphates, as they carry negative surface charges, but in the presence of iron, aluminium and calcium, they are able to adsorb great amounts of phosphates. Humic acids have a high affinity for chelating iron, which is decreased in the presence of calcium. In a pH range of 2 to 5, phosphates react to form ferri-phosphate-humate complexes (Hepher, 1958 & 1966). These complexes are highly refractory, once formed.

Decomposition of organic matter releases carbon dioxide and phosphate, which raises the concentration of free phosphate and increases the concentration of phosphate sorbed. An increase in acidity causes a drop in pH, which modifies the composition of the pore water and decreases the sorption of phosphate on calcium carbonate, which in turn affects phosphate sorption.

Microorganisms involved in decomposition excrete enzymes responsible for hydrolysis of organic matter and these extracellular enzymes and chelating agents modify the mobilization of phosphorus.

Precipitation of phosphorus or authigenic mineral formation

Formation of stable or metastable minerals is a geochemical process taking place in soils, but still poorly understood since identification of mineral species in the solid phase is difficult (Nembrini *et al.*, 1982). Phosphates in soil react with cations to form insoluble salts. Ionic activity product calculations for different mineral phases in the pore fluid are used to estimate conditions under which precipitations may occur, though direct identification is the only proof of existence of a mineral. Supersaturation of the pore fluid with respect to the solubility products constant has been observed for a variety of minerals, including vivianite, apatite, and others (Nembrini *et al.*, 1982; Enell & Löfgren, 1988).

Examples of some minerals formed are as follows: with calcium, phosphate forms apatite; with aluminium, phosphate forms variscite and wavellite; with iron, vivianite, $Fe_3(PO_4)_2.8H_2O$, and strengite are formed. The number of phosphate minerals identified in soils reaches almost 200 (Psenner *et al.*, 1984), which amply shows the complexity of this topic. Chemical formulae of these minerals may be found in Pettersson *et al.* (1988).

3.1.3.3. Phosphorus loading to sediments

Phosphorus accumulation in sediments

As a result of sorption onto sediments, phosphorus accumulates (loading) and is trapped in the soil. Identification and quantification of the mechanisms responsible for immobilization and mobilization of phosphorus are important. Mobilization is also called desorption, but according to Golterman (1988), the latter term is unfortunate and is a *contradictio in terminis*, as it is in reality an adjustment of sorption equilibria.

Origin of phosphorus in sediments

Mechanisms of transfer of phosphorus from water to sediments are as follows (Boström *et al.*, 1988a):

- sedimentation of detrital phosphorus and particles in suspension, brought in by the water feeding the pond
- adsorption or precipitation of phosphorus with inorganic compounds, such as coprecipitation with iron, manganese or calcium, and adsorption on clays, carbonates, or oxyhydroxides, and sedimentation of these aggregates
- sedimentation with allochthonous or autochthonous organic material
- assimilation of phosphorus in the water column by plankton or other biota and sedimentation, as well as sedimentation of all dead organisms

- direct adsorption by sediments.

The largest portion of phosphate in soil originates from decomposition of organic matter (Fitzgerald, 1970; Sinke & Cappenberg, 1988) and from the fertilization of the pond. The contribution of phytoplankton to organic loading of sediment depends on natural productivity of the pond, and grazing upon phytoplankton. Decomposition and release of phosphates from organic matter is a function of temperature and activity of microorganisms (Sinke & Cappenberg, 1988). When considering immobilization of phosphorus in sediments, the first approach was to study the physical and chemical sorption processes occurring in soils, as if the latter were an abiotic environment. However, it soon became clear that biotic aspects of phosphorus cycling were essential for availability of phosphorus; these aspects involve microbial decomposition of organic matter sedimenting to the sediments. Compartmentalization of phosphorus between microbial, non-living organic matter and labile sorbed phosphorus is difficult, because of the rapid turnover of such compounds.

Loading models and kinetics of sorption

Continuous loading of phosphorus to the pond bottom leads to a saturation of its sediment. Sedimented phosphorus is transformed diagenetically and is sorbed. Sorption processes are saturation processes, which were first described with the help of models. But to be useful, these models should allow prediction of sorption as well as release of phosphorus from sediments.

Equilibrium models calibrated at various phosphate concentrations have been used, describing accumulationaccording to a saturation curve, levelling off at a maximum internal load (sigmoid curve). Phosphorus sorption onto soil is best described by sorption isotherms, of which those of Langmuir and Freundlich are most popular. They are based on equilibrium or non-equilibrium kinetics, and partition sorbed (or labile) and dissolved phosphorus. Most models do not, however, satisfactorily describe desorption (Logan, 1982). Quantitative comparisons of adsorption isotherms reported in the literature are often difficult, because experimental designs are not standardized (Enell & Löfgren, 1982).

When using adsorption isotherms to describe phosphate sorption, it often appears that using a single exponential function to describe phosphorus uptake kinetics is inadequate. Analysis of the kinetics leads to a separation into two, three or more distinct phases, each described by an exponential function, the observed kinetic being the combined effects of several reactions with different rates. Graphical resolution of uptake curves into distinct phases does not, however, constitute proof of the existence of an equivalent number of reactions (Twinch & Breen, 1982).

Diphasic Langmuir plots have often been reported to indicate dual phosphate adsorption mechanisms, separated into a fast and a slow uptake phase (Twinch & Breen, 1982; Logan, 1982; Gunatilaka, 1982; Froehlich, 1988). The significance of each phase remains obscure (Twinch & Breen, 1982). According to some authors, sorption proceeds at various binding energies at distinct sites of higher and lower energy (Gunatilaka, 1982).

A more realistic explanation suggests a biotic-abiotic partitioning of phosphorus uptake, which has been investigated with the help of sterilized versus unsterilized sediments (Doremus & Clesceri, 1982).

Sterilized sediments show much slower uptake kinetics. Abiotic sorption is determined by the relative presence of different carrier substances, i.e. metal oxides, calcium, carbonates, found in sediments. Sorption of phosphate to them will depend on their amounts and their binding forces.

The most mobile sediment phosphorus pool is linked to biodetritus and to lower bonding energies. Absorption of phosphate by microorganisms is rapid, followed after a short time by the excretion of phosphate, which accounts for the observed desorption rate. The rapid uptake matched by a corresponding efflux is indicative of a dynamic steady-state system. Turnover time of phosphate within the biodetritic compartment, i.e. the abiotic organic material and the microorganism population, is short, and the biodetritus acts as a transitory compartment between organic and inorganic phosphorus or in a buffer capacity for phosphate (Doremus & Clesceri, 1982; Twinch & Breen, 1982). This buffer capacity is expressed as the quantity of phosphate adsorbed per unit increase of phosphate in solution (Twinch & Breen, 1982). Inorganic phosphorus in excess of the phosphate sorption and buffering capacities of the biodetritus is adsorbed onto ferric oxides and other sorbent compounds (Doremus & Clesceri, 1982).

Phosphate sorption properties vary with depth and with changes from aerobic to anaerobic conditions. In oxidized sediments, microbial activity is at a maximum and available phosphorus is important, while in reduced sediments, microbial activity is less important and less efficient, an iron oxide sequestering mechanism is lacking and non-sorbed phosphate accumulates (Doremus & Clesceri, 1982).

Surface sediments have higher sorption properties than deeper layers; the actively exchanging layer in Midmar Dam, South Africa, consists almost exclusively of the upper 2 cm of sediment (Twinch & Breen, 1982). Langmuir adsorption isotherm curves show that phosphorus sorption and bonding energy constants are maximum in the uppermost sediment layers (the loose surface material) being some 20 % and 60 % respectively lower than those of sediments from the -3 to -5 cm strata (the original flooded terrestrial soil, which is more mineral).

Phosphorus sorption in sediments - the Einsele-Mortimer paradigm

Phosphorus is immobilized in sediments because surface sediments act as a barrier to exchange between water and sediment. Immobilization of phosphorus in sediments has been described by a now classical model, based on interaction between iron and phosphorus under aerobic and anaerobic conditions, or oxidized and reduced conditions (Einsele, 1936; Mortimer, 1941 & 1942) (Fig. 11). A few millimeters below the sediment-water interface, an oxidized microzone of amorphous or poorly crystallized ferric oxyhydroxides forms which has high sorption properties for phosphate and constitutes a protective oxidized surface (Hepher, 1958; Callender, 1982). These iron oxides sequester the phosphate and considerably reduce its concentration in pore water.

In deeper strata, iron is reduced owing to anoxic conditions to ferrous ions, which do not have sorbing properties, and release phosphate (Fig. 12). When anoxic conditions, i.e. redox conditions decreasing from 450 to 200 mV and dissolved oxygen concentration decreasing from 4 to 0.1 mg l^{-1}, extend to the sediment surface, phosphate is released into water. Optimum pH and redox conditions are shown in Fig. 13. The concentration of ferrous iron is negatively correlated with redox potential of sediments: oxic sediments sorb some 25 times more phosphorus than anoxic sediments (Callender, 1982).

Reduction of iron partly results from the use of iron as an electron acceptor in the absence of other electron acceptors (Boström *et al.*, 1988a). Redox conditions are mainly modified by the activity of microorganisms; the destruction of the protective oxidized microzone and shift to anoxic conditions is

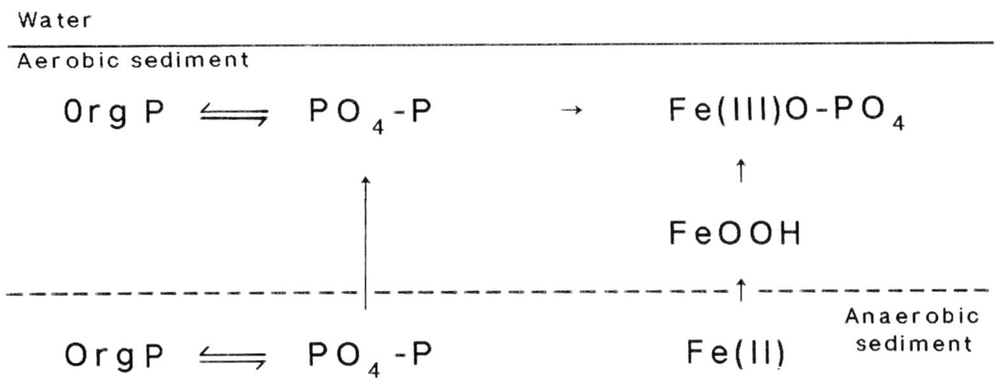

Fig. 11. Sorption of phosphate to ferric oxyhydroxide in aerobic sediment and oxidation of ferrous ions migrating from the anaerobic zone of the sediment.

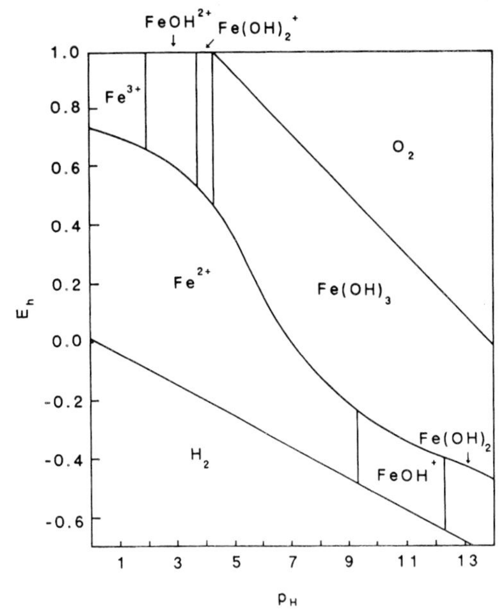

Fig. 12. Different iron ion species present at various pH and redox conditions (after Segalen, 1964).

the result of an organic loading in excess of the processing capabilities of microorganisms and dissolved oxygen supply to the sediments. It results in a change in the phosphate flux from sediment to water, the concentration gradient existing at the interface promoting the exchange.

The Einsele-Mortimer paradigm for phosphate sorption in the sediments has created the widespread opinion that phosphorus cycling is entirely linked to iron chemistry (Boström *et al.*, 1988). Though an important process, the contribution to phosphate sorption of other carrier substances should not be ignored.

Factors affecting phosphorus immobilization

Internal phosphorus loading or immobilization of phosphorus is dependent on sediment characteristics, such as pH, porosity, compaction, texture and sorption properties (ion exchange capacities), on

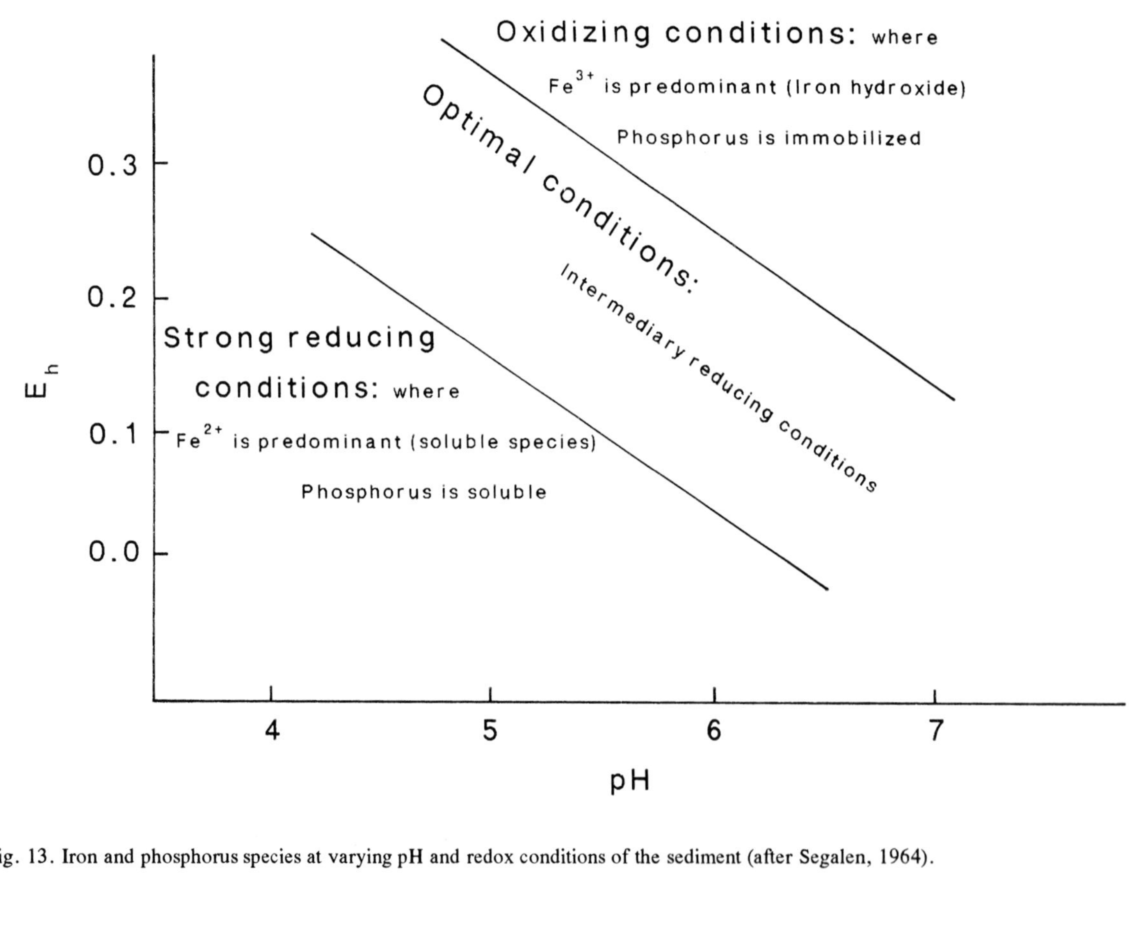

Fig. 13. Iron and phosphorus species at varying pH and redox conditions of the sediment (after Segalen, 1964).

concentrations of dissolved ions, and on the presence of carrier substances. Ostrofsky (1987) established, on data collected from 66 lakes (of different trophic levels) of eastern North America, significant correlations between iron and labile phosphorus (NH_4Cl-extractable), calcium-bound phosphorus (HCl-extractable) and iron plus aluminium-bound phosphorus (NaOH-extractable), total inorganic content, and total phosphorus in sediments. His results suggest that sediments with high sedimentary iron levels have higher concentrations of labile and iron and aluminium-bound phosphorus (Fig. 14).

Immobilization or trapping of phosphorus in sediments is essentially the result of microbiotic (Gächter *et al.*, 1988) and biogeochemical processes. Factors affecting phosphorus mobilization operate differently according to the type of process and must be discussed in connection with controlling environmental factors. Net phosphorus release is also typically seasonal, linked to microbial activities (Marsden, 1989).

A raise in temperature increases microbial activity and, as such, has an indirect effect on mobilization. Mineralization processes and metabolism of microorganisms, including all processes related with assimilation of phosphorus, excretion of phosphate from cells and transformation of organic detrital matter, are stimulated. Temperature also modifies sorption equilibria on sorbent substrates. Increase in microbial activity induces changes in the chemical environment of sediments: release of acidity and drop in pH, consumption of oxygen, release of reduced electron acceptors and drop in redox potential. The increase of microbial activity due to a raise in temperature may be locally restricted to patches where nutritionally rich organic matter is

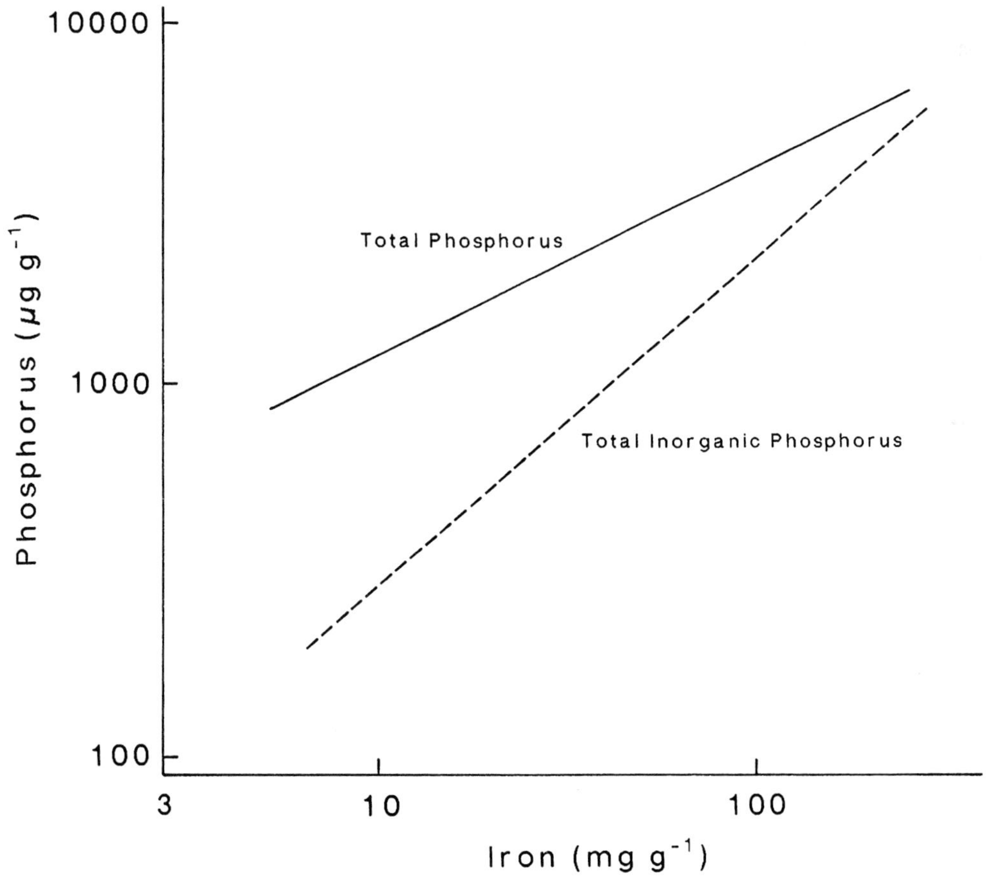

Fig. 14. Relations between the amount of total and total inorganic phosphorus and the amount of iron within the sediment (modified from Ostrofsky, 1987).

available. At those loci, anaerobic micropatches can appear which pierce the oxidized microzone and allow diffusion of phosphate.

An increase in pH reduces the sorption potential of iron and aluminium ions, primarily by competitive ligand substitution in which hydroxyl ions replace orthophosphate (Boström *et al.*, 1988a; Marsden, 1989). Changes in pH are a consequence of modifications in carbonic acid or dissolved oxygen concentrations. Microbial activity has specific pH requirements or pH ranges at which activity is maximum. The spectrum of a bacterial population changes according to prevailing pH conditions of mud. Moreover, pH changes affect sorption potential on pH-dependent fixation sites on colloids. Liming of the pond bottom increases pH, but also adds calcium carbonate to the sediment. Phosphate freed from its sorption to ferric hydrous oxides by a pH raise will coprecipitate with or be adsorbed onto calcium carbonate to form apatite.

A reduction in redox potential means an increase in reducing conditions and a reduction in dissolved oxygen concentration. Under reducing conditions, ferric iron is transformed into ferrous form, which is soluble. Sorption potentials for phosphate of the ferrous iron are greatly reduced in comparison to ferric compounds and interstitial phosphate concentrations increase (Holtan *et al.*, 1988; Søndergaard, 1989). Progressive changes from aerobic to anaerobic conditions induce modification of the spectrum of microorganisms capable of living in mud and making use of alternative electron acceptors. In ponds with high biomass and/or organic load, the water at night is not supplied with oxygen, and consumption of the latter leads to its depletion. The sediments, in which heterotrophic activity goes on, deplete their oxygen reserves and anaerobic conditions set in, which allows nightly release of phosphate. A diel diphasic phosphate release and immobilization may hence be present.

Efficiency of energy transfer in transformation of energy-yielding organic compounds is reduced and the diversity of microorganisms is lowered. Absorption of phosphate by microorganisms also differs according to aerobic and anaerobic conditions. Rapid uptake of phosphate, in excess of needs, by microorganisms is stored in polyphosphate granules; when conditions turn anaerobic, polyphosphates are used and phosphate, then in excess in the cell, diffuses out by increased osmotic pressure (Boström *et al.*, 1988a).

Among the geochemical factors responsible for immobilization of phosphorus, the abundance of iron in sediment is important. The oxidation of iron to the ferric form immobilizes phosphorus in aerobic sediments only when the atomic ratio of Fe/P in the interstitial water is greater than 1.8. Mechanisms that maintain low interstitial iron concentration (such as precipitation with sulphide ions) may be important for leakage of phosphorus from sediment to water (Marsden, 1989). Other mechanisms which reduce phosphorus sorption during its transport across aerobic sediments include: organic acids which act as chelating agents sorbing calcium, iron, manganese and aluminium; high concentrations of soluble silica, which compete with phosphate for adsorption onto iron (Marsden, 1989).

Authigenic mineral formation may change the availability of phosphate and other ions, removing them definitively from pore water. Geochemical transformations occurring in sediments have been studied intensively to assess whether there are any equilibrium reactions which can 'buffer' high pore water concentration by precipitation to a solid phase. Environments suitable for formation of these minerals are mostly found in highly reduced sediments. Moreover, kinetics of precipitation reactions are slow, while most sorption reactions involving active biological exchange, as occur in the uppermost layers, are rapid (Enell & Löfgren, 1988). The former are not likely to be a major importance in surficial strata.

Influence of mineralization on the availability of phosphate

The transformation of organic matter in sediments results in a transfer of detrital phosphorus, via living biomass, to the pool of interstitial or labile phosphorus. The cycling of phosphorus is not easy to quantify, owing to the rapidity at which it is transferred from one compartment of the soil to the other.

The effects of mobilization processes on phosphorus depend on (Boström *et al.*, 1988a):

- the state of decomposition of the organic matter sedimenting on the sediments and the prevailing type of transformation

- the initial phosphorus content of the organic matter and the C:N ratio of settling organic matter
- the concentration of interstitial phosphorus and the saturation percentage of sediments with regard to phosphorus
- the sedimentation rate of organic matter, and whether steady or irregular, as in the case of an algal crash (see p. 148)
- environmental conditions, such as temperature, redox potential, and dissolved oxygen concentration.

Microbial activities modify the concentrations of sedimentary phosphorus. Microorganisms provide for their needs by drawing phosphorus from organic matter. This involves a transfer of phosphorus from the biodetrital to the biotic compartment and is realized with the help of hydrolytic enzymes. During this extracellular hydrolysis, some phosphorus is released to the environment. The amount of phosphorus released depends on the nutritional value of organic matter. Nutrients are absorbed by microorganisms in a given proportion; nutrient in excess of this ratio is wasted by the organisms and released to the environment.

Decomposition of substrata of poor nutritional value results in a net immobilization of nutrients (Doremus & Clesceri, 1982), while decomposition of substrata of better value results in nutrient regeneration (Boström *et al.*, 1988a). In lake George, New York, U.S.A., sedimented organic matter was often of poor nutritional quality, so that decomposers were nutrient-limited (Doremus & Clesceri, 1982).

The intensification of decomposition activities mobilizes various fractions of phosphorus between compartments and increases the amount of phosphorus available.

Effect of nitrate on mineralization

The presence of nitrate affects mineralization and mobilization of phosphorus by either promoting or inhibiting its release.

Release of phosphorus is inhibited when the concentration of nitrate is high: the presence of nitrate buffers the redox potential, preventing the use of iron as an electron acceptor.

On the other hand, high nitrate concentration increases the mineralization rate of phosphorus as a result of increased denitrification, which promotes more rapid transfer of phosphorus from degradable organic matter to the mobilized phosphorus pool (Boström *et al.*, 1988a).

Changes in phosphorus loading with depth

In an oligotrophic lake in northern Sweden, Björk-Ramberg (1984) noticed that the amounts of sediment phosphorus correlated positively with sediment depth and negatively with organic matter content. Phosphorus concentrations in surficial sediments were some 40 % lower than in deeper strata. Sedimentary phosphorus increases with depth, when expressed on a volumetric basis (weight per unit volume of sediment), but decreases when expressed as weight per dry weight of sediment. This is most likely explained by an increase in compaction of sediments, and a decrease in water content and of mineralization rates with depth. The concentration of phosphorus in surficial sediments increases as a result of:

- an increase in concentration of recently sedimented material, due to an increase in phosphorus loading
- a decrease in mineralization rate due to an increase in volume of settling matter (increased sedimentation rate), and/or a decrease in activities of microorganisms as a consequence of reduced oxygen supply
- an upward diffusion along a concentration gradient of sorbed phosphorus accumulated in deeper horizons, where oxygen depletion and redox potential reduce sorption capacity. The phosphate migrating to the surface layer is sorbed, due to changes in redox potential and pH (Holtan *et al.*, 1988).

On the other hand, reworking of the sediments by benthic macroorganisms opposes the establishment of this phosphorus gradient. Sediments from deeper strata are mixed with sediments from upper layers and thus brought in contact with the aerated water of the pond.

Buffering mechanisms of sediment phosphorus

Various buffering mechanisms of sediment phosphorus are as follows and are summarized in Table 6:

- the first or rapid step of the sorption process, in which a large pool of dissolved phosphate is retained by low-energy bonds and set aside as labile phosphorus, which is easily released into solution, should conditions change to promote desorption (Froehlich, 1988). This phosphorus reflects a mobile sediment phosphorus pool, with increased phosphate desorption potential (Twinch & Breen, 1982)
- the combined phosphate immobilization ability of the microbial biomass and its associated organic matter acts as a phosphate buffer for surface sediments (Doremus & Clesceri, 1982)
- the temporary immobilization of phosphate due to rapid uptake of available phosphate by microorganisms followed by rapid excretion resulting in a mobile pool. Gross phosphorus transport in and out of bacterial cells far exceeds net transfers measured (Boström *et al.*, 1988a)
- the storage of phosphate in microorganism cells under aerobic conditions, for use and release when conditions shift to anaerobic (Boström *et al.*, 1988a)
- lastly, the authigenically formed phosphorus minerals in sediments have been examined as a possible phosphate buffering system for pore-water phosphorus, but they can hardly buffer changes which occur at a much faster rate than do these chemical precipitations (Enell & Löfgren, 1988).

The various mechanisms which keep phosphate available, despite its low solubility, indicate that phosphorus is a 'strategic' element for life in the sediment.

3.1.3.4. Phosphorus release from sediments

Phosphorus release and sediment characteristics

Transfer of phosphorus from sediments to the overlying water is prevented by the presence of the oxic/anoxic barrier at the interface.

Phosphorus dynamics, sorption and desorption, depend on sediment characteristics. With regard to desorption, three types of sediments exist: sediments which do not release phosphorus under any circumstances; sediments which release phosphorus at a given rate, not influenced by the addition of nitrate or acetate, the release rate increasing independently up to a certain level; sediments which release phosphorus when acetate is added. The addition of acetate stimulates bacterial activity, while the addition of nitrate stabilizes the redox potential and prevents release of phosphorus. The differences in behaviour are linked to the varying content in sediments of apatite-P, non-apatite-P and labile phosphorus (Boström & Pettersson, 1982).

Phosphorus sorbed to aluminium and iron is released when conditions of pH and redox potential of the sediments and phosphorus concentration in pore water change; phosphorus sorbed to calcium is released according to the solubility product constant of the calcium/phosphate compound formed; organic phosphorus releases phosphorus under the action of enzymes. Microbial activity releases acid fermenta-

Table 6. Principal buffering mechanisms of phosphorus in sediments

BUFFERING MECHANISMS OF SEDIMENTARY PHOSPHORUS	
PROCESS	TEMPORARILY IMMOBILIZED FORM
Adsorption/desorption on mineral particles	Labile phosphorus
Adsorption/desorption on organic particles	Biodetritus and humic compounds
Absorption/excretion (Phosphorus exchange)	Microorganisms
Absorption for stock constitution	Microorganisms

tion products which lower the pH of pore water. The consequent dissolution of carbonate minerals allows mobilization of calcium, magnesium and associated phosphorus. Increase of pH in pond water does not affect pore water pH, as the latter is buffered by the acidic products of bacterial metabolism (Marsden, 1989).

Phosphorus release and trophic level of a water body

Phosphorus immobilization in sediments is dependent on the saturation of fixation sites, while exchange of nutrients is dependent on the trophic level of the water body. Soil profiles in the pond mud give indications of the amount and type of accumulation occurring, and of the depth of oxygen penetration.

Eutrophication is the result of nutrient enrichment, especially phosphorus, which has accumulated on the pond bottom beyond the buffer capacity of the sediments. This excess loading entails a shift from aerobic to anaerobic metabolism in sediments, which further reduces the retention capacities of the latter.

Factors in sediments promoting desorption of phosphorus

Desorption is a readjustment of sorption equilibria and the factors which influence the sorption equilibria of phosphorus can be separated into those promoting removal of phosphorus from sorption sites and dissolution of phosphorus, and those promoting immobilization or sorption of phosphorus (Table 7). Amongst the factors promoting mobilization, different effects may be distinguished: those which modify the physical and chemical conditions required for sorption, and those which increase the concentration of soluble phosphorus, leading to saturation of sorption sites and thereby increase the concentration of soluble phosphorus.

Physical and chemical factors promoting sorption include temperature, pH, redox potential, dissolved oxygen concentration, and iron content of the sediments. They were discussed extensively above.

Pore water as a source of phosphorus

Interstitial phosphorus concentrations reflect the amounts of phosphorus sorbed onto sediments and the type of sorption. Loosely sorbed labile phosphorus and dissolved phosphorus are in a steady equilibrium. Soluble phosphorus is the phosphorus species most easily transferred out of sediments. The internal loading with regards to phosphorus refers to the amount of phosphorus immobilized by sediments, accrued from the pore water phosphorus. Saturation of sorption sites entails a saturation of pore water which brings about a concentration gradient appearing at the sediment-water interface and a directional diffusion flux from sediment to water.

Orthophosphate is the main form of phosphorus mobilized from sediments, but soluble forms of organic phosphorus, characterized by molecular weights of around 20,000 to 30,000, also ooze out of sediments enriched with organic matter (Enell & Löfgren, 1988). Physical mixing of sediment with water, as a result of turbulence acting on the surficial layers of sediments, promotes exchange of pore water with pond water and cause resuspension of sedimented particles (Gunatilaka, 1982). These aspects of phosphorus transport will now be discussed.

Aerobic and anaerobic transfers

Phosphorus release from sediment to water depends on prevailing conditions in sediments and in water: distinction is made essentially between anaerobic versus aerobic sediments and aerobic versus anaerobic water. Water is considered aerobic when dissolved oxygen concentration is higher than 2 mg l^{-1} (Marsden, 1989). Shifts between oxic and anoxic conditions may happen overnight when the supply of oxygen to the sediment is interrupted or following an algal crash. It results in an alteration of the patterns of phosphorus release and immobilization.

The sediment-water interface forms a physical and chemical barrier. Crossing this barrier involves a change of phase: the sediments are solids (or semi-solids), while water is liquid. Moreover, a chemical borderline is set by the redox-cline or oxycline, which is also often situated near the interface. The energy required to cross these borders comes from the establishment of concentration gradients, which promotes the diffusion of interstitial solutes which do not possess electrochemical charges. Molecular diffusion and gaseous ebullition, especially of methane,

Table 7. Factors affecting phosphorus release from sediments

FACTORS AFFECTING RELEASE:	promoting	impeding
Nature of the sediment	non-apatite minerals, (rich in aluminium and iron)	apatite minerals or calcareous rocks
Redox potential	negative redox values	positive redox values
pH	light acidic pH and slightly reducing conditions	high pH and presence of alkaline salts
Phosphorus loading or trophic level of the water body	saturation of the sediment with phosphorus	low total phosphorus concentration
Microbial activity	high microbial activity	low microbial activity
Factors affecting microbial activity	high temperature enhancing microbial activity	low temperature
Primary production, dissolved oxygen production	good oxygen supply to sediment ensuring continuous aerobic conditions in sediment	low oxygen supply to sediment and onset of anaerobic conditions in sediment
Supply of organic material to the sediment	high sedimentation rate of organic material of good nutritional value	low sedimentation rate

take place that way. Another way to cross the barrier is by the destruction of the protective oxidized surface. This microzone is destroyed by the onset of anaerobic conditions. Lastly, any turbulence, causing resuspension of sediments, insures transport of sediment material and pore water which mixes with pond water and permits release of any soluble phosphorus (Gunatilaka, 1982).

Importance of phosphorus release rates

Diffusion coefficients of sedimentary phosphorus across the interface of sediments are in the range of 1 10^{-6} cm^2 s^{-1} or 1 μmole PO_4^{-3} m^{-2} d^{-1} (Pomeroy *et al.*, 1965) or 0.6 10^{-6} to 2.2 10^{-6} cm^{-2} s^{-1} (Marsden, 1989). This transport is quite slow, but is considerably modified by ancillary processes, thereby being increased by several orders of magnitude. Release rates of phosphorus in aerobic sediments and in turbulent waters range from 1 to 2 mg m^{-2} d^{-1} to 40 mg m^{-2} d^{-1} and sometimes up to 550 mg m^{-2} d^{-1} (Marsden, 1989).

In laboratory conditions, recorded phosphorus effluxes from sediments, over defined time periods, were in the range of -7.7 to 1330 μmole PO_4^{-3} m^{-2} d^{-1}. They were not correlated to phosphorus concentrations, but to the NH_4 and CH_4 concentrations of pore water, which strongly suggests that phosphate flux is linked to mineralization processes, of which methane and ammonia are by-products (Sinke *et al.*, 1990).

Diffusion of phosphate is promoted by movement of groundwater; sediments release phosphate, when the movement of water is either upward or downward, even though in the latter case, the movement of water is in the opposite direction (Van Liere & Mur, 1982). Upward migration of phosphate might be hampered by adsorption and chemical precipitation.

Phosphorus is also transported by benthic organisms. Sorbed to biodetritus, it is temporarily immobilized; consumption of biodetritus by detritivorous organisms results in its exportation from the sediments. Part of the phosphorus ingested by these organisms is assimilated, the other part is excreted. This biodetritic phosphorus is therefore more mobile than when sorbed to colloids.

Gaseous ebullition

Methane, hydrogen sulphide, carbon dioxide, and molecular nitrogen are gases produced by microbial metabolism. When gas production and pressure conditions allow bubble formation, the gases rise and, in their upward movement, transport soil particles and dissolved substances. Spotte (1979) observed that about 90 % of the reactive phosphorus is removed from aquarium water in a time lapse of 24 hours by a normal aerator. Turbulence and disturbance induced by wind and benthic organisms cause the resuspension of sediment. Ions and solutes adsorbed to the disturbed material are desorbed and released to the water. On the other hand, soil particles with a high sorption capacity can, during their migration phase through water, sorb important amounts of dissolved phosphorus and increase phosphate removal from water (Boström *et al.*, 1988a).

Changes caused by resuspension of soil particles

Perturbations, such as turbulent mixing, bioturbation, and reworking of sediments caused by activities of benthic organisms, affect sediment characteristics and exchanges with water. With regard to animal activities, the depth to which sediments are affected varies. Turbulent mixing affects only the loose surficial strata. The water layer affected by sediment turbulences is sometimes called the lutocline, the interphase between the 'clean' water and the sediment-water mixture.

Effect of benthic organisms on sediment-water exchange

The activities of benthic organisms on and in the sediments increase the penetration of water into sediments and thereby the exchange between water and mud (phosphorus fluxes can be increased by a factor 5 to 10). The specific effects of organisms on sediments is termed bioturbation and involve transporting activities, including advective transport of both solutes and particulate matter (Andersson *et al.*, 1988).

Benthic activities include irrigation of the sediments by the burrow holes of benthic macroinvertebrates, and ventilation of the burrow holes by water currents caused by ciliary and muscular actions. These water movements are important in gas exchange, and in the removal of metabolites, egested wastes and detritus. Interstitial water is pumped in and out and mixes with the water in the burrow holes.

Benthic organisms (the composition of the benthic fauna is described in the paragraph on organic matter decomposition, p. 50) ingest and transport sediments (Rippey & Jewson, 1982). The transport of sediments from the surface to deeper strata removes organic rich (or easily degradable) sediment of high nutritional value from the surface, where decomposition proceeds in oxic conditions and is submitted to intense microbial activity, to deeper strata, where conditions are reducing and microbial activity diminished. The penetration of dissolved oxygen deeper into sediments (Rippey & Jewson, 1982) entails a higher sorption of phosphate onto the sediments; the release rate of phosphorus by sediments is decreased. The transport of sediment in the opposite (upward) direction has an opposite effect on phosphorus release, but the organic material carried from below to the surface is generally refractory (Kamp-Nielsen *et al.*, 1982). The effect of benthic organisms on the sediments varies according to their mode of life and their activities. Detritivorous organisms recycle phosphorus trapped in the biodetrital complex (Doremus & Clesceri, 1982). Their transport of particulate material is considerable.

In shallow ponds, the mixture of sediment and water by bioturbation and other disturbances allows contact between the sediments and the trophogenic layers of the water, which promotes exchange of nutrients, especially of phosphate, and permits a greater mobilization of sedimentary phosphorus (Boström *et al.*, 1988a) (see also p. 179).

The mass balance models of phosphorus in sediment and water generally do not take into account bioturbation and other sediment disturbances.

Removal of phosphorus from sediments by epipelic algae

Epipelic algae provide for their needs by removal of phosphorus from sediment. Superficial sediments in the euphotic zone are submitted to diel formation and destruction of an oxidized microzone. During photosynthetic activity, the surficial sediments are oxidized due to the release of oxygen, a by-product of photosynthesis. Phosphate is prevented from diffusion by the presence of this oxidized microzone. At night, this microzone is destroyed and phosphorus is taken up by the algae. On sediment samples from Lawrence Lake, an oligotrophic lake in Michigan, U.S.A., phosphorus efflux clearly showed these two distinct phases (Carlton & Wetzel, 1988). Advantage and disadvantage of periphyton and benthic algae over planktonic algae with regard to phosphorus uptake is discussed later (p. 106).

Transfer and phosphorus regeneration

Transfer of nutrients, especially phosphorus, from sediments to overlying water is often understood or depicted as a nutrient regeneration, benefitting the organisms of the water column. The importance of this regeneration in the supply of nutrients to the phytoplankton can be assessed by comparing the benthic fluxes to the uptake of nutrients in the water. This process contributes some nutrients, especially phosphorus, required for primary production. The regeneration of phosphorus is mainly effected by release of phosphorus sorbed to iron and by mineralization of organic matter. Interstitial phosphorus, assessed on the basis of C/N ratios, originates, to a major extent (50 %), from organic matter decomposition. The other part is supplied by the dissolution of iron(III)-phosphate compounds.

This regenerated phosphorus can diffuse, be sorbed, form complexes with organic or inorganic ions present, or precipitate as mineral. Significant correlations between methane, a final product of anaerobic organic decomposition, the dissolved reactive phosphorus, ammonia, calcium ions and iron existed in the eutrophic Loosdrecht lakes, Netherlands, and these concentrations increased with depth (Sinke *et al.*, 1990). This shows that available phosphate is supplied by decomposition.

Assessment of bioavailability of phosphorus

The amounts of total phosphorus found in sediments are often important. But this information is of little use to the fish farmer who wishes to know whether the stocks of phosphorus in the sediments can be mobilized and made available to primary production in his pond. As discussed earlier, phosphorus bound to different carriers can be determined analytically by sequential extraction, but, unfortunately, there is no concordance between the different fractions of phosphorus determined chemically and the bioavailable phosphorus, i.e. the amount available for primary production (Twinch & Breen, 1982; Klapwijk *et al.*, 1982; Logan, 1982). Bioavailability can be defined as the sum of phosphorus directly available, i.e. soluble reactive phosphorus and labile phosphorus, accrued from the phosphorus yielded by natural processes, such as desorption, dissolution, autolysis and hydrolysis. The specific rate of mobilization of each fraction of phosphorus depends on the total amount immediately available and on the processes by which phosphorus is released from its bonds. Distinction must be made between actual and potential or maximum availability.

Several methods have been used to assess available phosphorus. The first approach was to search for a concordance between chemical extractions and amounts of available phosphorus, but the correspondence was poor and unsatisfactory. Therefore, autotrophs (bacterioplankton and algae) were cultured on the sediments, directly or diluted with water, to determine how much, and from which fractions, the organisms were capable of extracting phosphorus. These bio-assays are time-consuming and tedious and are therefore not common. Also, they are biased by the fact that they do not exactly conform to reality, insofar that the algae commonly used for these assays are not the same as those in the pond; algal species have specific requirements for growth, and the response of a natural population might be drastically different from that of any given species. Another approach to bioavailability of phosphorus is through factors responsible for mobilization of phosphorus under natural conditions (Boström *et al.*, 1988b).

In the case of algal culture in the presence of sediments, the amount of phosphorus removed from the sediment is determined. The culture medium does not contain any phosphorus other than the sediment, and the algae have been phosphorus-starved before the beginning of the experiment. Algae in general use are *Scenedesmus* and *Selenastrum*. A quantitative budgeting of particulate and interstitial phosphorus indicates the amount of phosphorus removed by the algae. The phosphorus taken up by algae consists mainly of non-apatite inorganic phosphorus, i.e. phosphorus sorbed to aluminium and iron (Young & DePinto, 1982; van Eck, 1982) and clay (Klapwijk *et al.*, 1982). Little of the phosphorus linked to carbonates appears to be usable by the algae (Klapwijk *et al.*, 1982; van Eck, 1982). Phosphorus associated to apatite does not support any algal growth. Organic phosphorus is mostly biologically inert, though it may be hydrolysed by phosphatases. This is in contradiction with observations; in many situations, phosphates are supplied by organic matter decomposition. This apparent contradiction can be explained by the intermediary role of microorganisms in the transformation of organic matter and the regeneration of nutrients to the ultimate benefit of autotrophs (Boström *et al.*, 1988b).

Phosphorus compounds taken up by algae are orthophosphate and certain organic compounds, such as phosphate monoesters with an organic part of low molecular weight, and more complex molecules, such as complex organic phosphates. They cannot, however, make use of ATP or DNA. Portions of organic phosphorus of high or low molecular weight (HMWP and LMWP) are only used in an erratic way (Boström *et al.*, 1988b).

Effect of pond fertilization on pond bottoms

Fertilization of ponds will be discussed later, but the effects on the sediments can be outlined here. Phosphates in water will be sorbed by soil colloids and saturate sorption sites in the soil. Pond soils which are saturated with phosphorus are able to support higher phytoplankton populations for a long time. In Auburn, Alabama, U.S.A., Boyd & Musig (1981) observed that decrease in phosphorus concentrations in pond water during the first seven days after impoundment is negatively correlated to the initial phosphorus concentration of the soil. Large amounts of phosphorus immobilized in soil decrease the ability of sediment to remove inorganic phosphorus from the water and make each addition of phosphorus through fertilization more available for production.

Importance of phosphorus in sediments

1. Total phosphorus concentration in sediments is often high. This is the consequence of an internal loading process.

2. Sedimentary phosphorus is separated into interstitial or particulate. Interstitial phosphorus is, for the most part, the soluble reactive phosphorus (SRP), while particulate phosphorus is divided into different fractions, namely labile phosphorus, iron+aluminium associated phosphorus, calcium associated phosphorus and organic or biodetritic phosphorus. The relative proportion of each of these fractions can be determined by sequential extraction and chemical characterization.

3. Diagenetic transformation of phosphorus in the soil includes the mineralization of organic phosphorus, the sorption of phosphorus and eventually the authigenic mineral formation. Each of these transformations proceeds on its own time scale and specific kinetics; the former two are relatively rapid, while the latter occurs more slowly.

4. Most sediment phosphorus has an organic origin and the internal loading processes are of paramount importance for the saturation of sediments with phosphorus.

5. The most important sorption process in soil is the bonding of phosphate to ferric oxyhydroxides in aerobic sediments. The phosphate is released from its bond when the sediments become anaerobic and the iron is reduced to its ferrous form. All sorption processes result in a removal of soluble phosphorus, to a solid form, and are characterized by their reaction kinetics and

irreversibility or reversibility.

6. The sediment-water interface forms a barrier made of an oxidized microzone, preventing release of free phosphate from reduced sediments below.

7. The immobilization capacities of soil are linked to the sediment characteristics, to environmental conditions, and to the saturation of sediment with phosphorus. Sediments rich in iron have better 'labile' sorption abilities than sediments rich in calcium, which precipitate insoluble phosphate compounds. Environmental conditions which promote mineralization of organic phosphorus, without deterioration of the required living conditions (oxygen supply or availability of other relevant nutrients, for instance) for microorganisms, will lead partly to immobilization of phosphorus in biotic forms, which can be then exported out of the sediments by direct consumption by macro-detritivorous organisms. The phosphorus in excess of microbial needs will be immobilized by physical or chemical sorption. If environmental conditions deteriorate to anaeroby, the pace of organic decomposition will slow down and all accumulated phosphate in labile sorbed forms will be, little by little, released and diffuse to the overlying water. The sequence of diagenetic transformation is dependent on soluble phosphorus fluxes and kinetics of each sorption process, the reaction material and prevailing conditions.

8. The release rate of phosphorus by diffusion from sediments is generally low, but is increased by several orders of magnitude by ancillary phenomena, such as gaseous ebullition, and resuspension of sediments by wind induced turbulence and bioturbation. Bioturbation is the reworking of sediments due to activities of benthic organisms.

9. The portion of phosphorus in sediments available for primary production is called bioavailable phosphorus. This bioavailable fraction can be assessed with bio-assay methods and concords roughly with the portion of iron-aluminium associated phosphorus.

3.1.4. Sulphur

Importance and forms of sulphur present in sediments

Sulphur is an important element for living organisms, but is rarely in short supply. Sulphur is present in organic matter and released by decomposition. Sulphate is used as an alternative electron acceptor and plays a role in oxidative processes.

Organic matter decomposition generates sulphide compounds, which are transformed, for instance, into sulphate and hydrogen sulphide. In organic compounds, sulphur is bound either to carbon or to sulphate esters.

Sulphur is present in different oxidation stages, as thiol, sulphide or thioester (-2), disulphide (-1), sulphoxide (0), sulphone (+2), sulphonic acid (+4), and sulphate esters or sulphate (+6). All these compounds are intermediary products of decomposition. Not much is known about the different steps of the transformation processes of sulphur compounds in the sediments (Urban & Baker, 1989; Kelly & Rudd, 1989); sulphide ions eventually form iron sulphide and pyrite (Callender, 1982; Schuurkes & Kok, 1988).

Sulphate and organic matter decomposition

Microbial organic matter decomposition in sediments makes use of a sequence of available electron acceptors, as discussed earlier (*vide* p. 18). Sulphate is resorted to when sediments are fully anoxic, redox potential has reached a negative value of 60 to 100 mV (Boström *et al.*, 1988a) or -200 mV (Seitzinger, 1988), and other acceptors have been exhausted. Sulphate is partly supplied from the water column (Nembrini *et al.*, 1982). Sulphide ions, resulting from the reduction of sulphate, react with iron (II) to form iron sulphide (Nembrini *et al.*, 1982). Reaction is also possible with iron oxyhydroxide resulting in the same ferrous sulphide. Iron sulphide reacts further with elementary sulphur to form pyrite. Sulphate reduction is a proton-consuming reaction and therefore increases alkalinity in soils.

Sulphate oxidation and alkalinity increase

The loss of sulphate is accompanied by an equivalent loss of protons, which represents an alkalinity gain, but not a net production of new alkalinity. It is a transfer, from sediment to water, of acid neutralizing capacity. Ferrous ions in the sediments accumulate in the pore water of anoxic sediments and diffuse upward till they meet with oxygen, where they precipitate as ferric oxide. This oxyhydroxide gel is quite common and occurs in high concentrations at the sediment-water interface.

In sediment rich in carbonate, iron precipitates to form ferrous carbonate and consumes alkalinity. The iron oxyhydroxide can be reduced and constitutes a potential source for acid neutralization when coupled to a process of removal of soluble ferrous iron and irreversible fixation in a solid compound. This process is called the acid neutralizing capacity of iron. Iron oxyhydroxide is therefore assimilated into a solid-phase alkalinity (Anderson & Schiff, 1987).

Microorganisms responsible for decomposition

Bacteria responsible for sulphate reduction in free sulphide are not sensitive to pH changes and easily support a pH varying between 6 and 4 (Schuurkes & Kok, 1988). Mazumder & Dickman (1989) observe that purple and green sulphur bacteria require light and an absence of oxygen. They are found in the upper part of the hypolimnion or in the metalimnion.

3.1.5. Iron

As shown in the paragraphs on phosphorus and sulphur, iron holds a key position in the cycling of those nutrients. Iron is specially active when soluble and therefore only the portion of iron which is soluble and easily desorbed is to be considered. At high concentrations, iron (II) is toxic for most organisms. Iron activity is often discussed in the context of chelators.

Iron and manganese concentrations vary with sediment depth, as a result of cycling and transport. Their oxidation stage, as well as the position of the redox barrier, are important for their transport. This redox barrier is however not a sharp border, but more a discontinuous, sometimes diffuse line. When the soluble ferrous ion migrates and meets with oxic conditions, it is oxidized and precipitates. However, ferrous ions can also be included and occluded in bonds that resist oxidation, such as $FeOH^+$ or $FeCl^+$. The compounds most commonly found in hydromorphous soil are $Fe(OH)_3$, with a solubility product coefficient of 10^{-35}, $Fe(HCO_3)_2$ and $FeSO_4$. Iron carbonate and bicarbonate are used as energy sources by some microorganisms.

Iron (II) ion transport is achieved by various diffusion processes. The vertical concentration profile in the sediment gives an indication of the specific type of supply and transport. The two most common profiles are either concave or convex, and reflect different rates of transport in the sediments (Davison, 1982). Physical modifications at the interface affect transport and removal of products. Oxidation and transport kinetics of different ions vary; manganese ions are oxidized 50 times more slowly, but are more rapidly transported than iron (Davison, 1982).

The dissolution of iron in soil requires a drop in pH or in redox potential. Iron reduction is under microbial control.

Importance of sulphur and iron

1. Sulphur and iron are not often specifically considered because their cycles are not well understood. Sulphur forms many intermediary products. Soluble iron has great chelating capacities and forms part of many complexes. Soluble iron is toxic at high concentrations.

2. Sulphate and iron are both electron acceptors; iron oxyhydroxide is preferentially used to sulphate. Sulphate consumes protons when oxidized and, in combination with iron hydroxide, produces alkalinity in the soil. This is termed the acid neutralizing capacity of iron.

3. Sulphate and iron oxidation are microbial processes.

3.2. Organic matter and humus

3.2.1. Importance of organic matter

Origin of organic matter

Pond organisms which die in the course of the production period sediment to the bottom and become part of the organic loading of the sediments. Other contributions come from dead biomass of benthic organisms and unconsumed artificial feed. Depending on the location of the pond and the course of its feeder canal, some allochthonous organic matter can find its way into the pond and eventually to the sediments. This material is mostly leaves and plant debris (Cuffney *et al.*, 1990). The amount of sedimented organic matter is a function of primary production (Cappenberg *et al.*, 1982), and the sedimentation rate changes seasonally (Rippey & Jewson, 1982).

The superficial layers of the sediments are rich in organic matter, with a content decreasing with depth, according to the rate of mineralization (Rippey & Jewson, 1982). Organic matter is mainly made of organic carbon; the decomposition of organic matter is part of the carbon cycle.

The flocculent layer as the centre of organic matter decomposition

Organic matter settles on the pond bottom and forms a flocculent layer, a stratum of loose, uncompacted fluffy material, easily resuspended by water turbulence. This material is rapidly colonized by a microbial population (epipelic community) with a high respiration rate and a high heterotrophic, and sometimes even photosynthetic, activity, composed of bacteria, benthic algae, fungi and zoobenthos (Sweerts *et al.*, 1986). This flocculent layer is the centre of detritus decomposition in the aquatic environment (Doremus & Clesceri, 1982; Adams *et al.*, 1982; Schroeder, 1987).

3.2.2. Decomposition and nutrient regeneration

Decomposition

Decomposition is a generic term for a group of processes, which transform organic material into living matter and nutrients. It can also be viewed as a nutrient regeneration process, benefitting the water column. Processes differ according to the type of organic material, prevailing conditions and the type of microorganisms present. The aim of the microorganism is to benefit from the energy and nutrients contained in organic matter.

Dying and sinking phytoplankton cells settle to the pond bottom and their decomposition starts immediately after death, by autolysis during the sedimentation of the cell. Leafy and plant material needs shredding before it can be attacked by microorganisms; this is efficiently done by benthic macroinvertebrates (Cuffney *et al.*, 1990).

Various decomposition strategies can be chosen according to prevailing environmental conditions. They are either oxic or anoxic, oxidizing or reducing. The spectrum of organisms found in sediments is great and still, to a large extent, unexplored and unknown. The most active decomposition occurs at the sediment-water interface.

Organic matter is not a material of defined value; its nutritional value decreases with time, labile compounds being quickly hydrolysed, and with depth, as fresh sediments bury older ones. The ultimate forms are humic and fulvic compounds of no nutritional value.

Microorganism population growth is induced by the availability of organic material, and the size of the population is a function of the presence of material and suitable environmental conditions.

Decomposition is divided into three main steps, in which the end products of one process are inputs to the next process (Fig 15). This has been termed the trickle-down theory (Seitzinger, 1988). The first step is a depolymerization of organic polymers, e.g. polysaccharides, lipids, DNA, RNA, etc. into monomers, i.e. monosaccharides, amino acids, fatty acids, and nucleotides. The second step is a fermentation, releasing carbon dioxide and other fermentation gases, and yielding acetate, alcohols and methylamines. The last step is the terminal metabolism or the consumption of fermentation end products by terminal consumers, releasing methane,

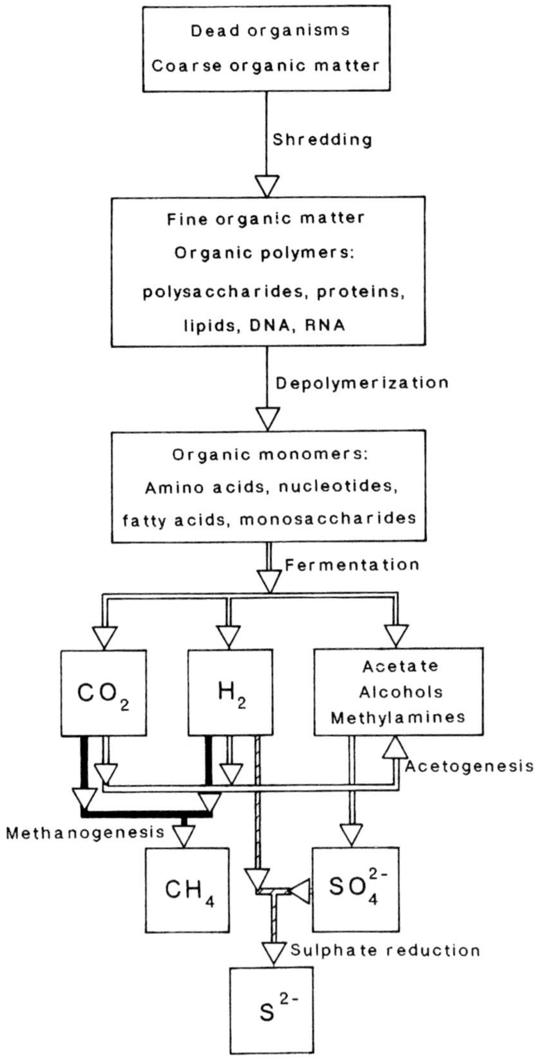

Fig. 15. Main steps of the decomposition process in sediment, including the shredding, depolymerization and fermentation of organic matter. Fermentation includes different alternative processes, depending on prevailing conditions and on the presence of electron acceptors (modified from Seitzinger, 1988).

carbon dioxide and reduced electron acceptors, by processes such as methanogenesis, acetogenesis, and sulphate reduction (Seitzinger, 1988). Each process requires specific conditions, with regard to, for example, pH, redox potential, or availability of some electron acceptor, and of course, availability of organic matter. The amount of energy released is a function of the type of reaction taking place. The aerobic condition of the sediment is largely determined by the input of organic matter. Planktonic production is the most important factor influencing sediment oxygen demand (Marsden, 1989). The amount of energy released by the transformation of organic matter has been established as follows (Seitzinger, 1988):

Electron acceptor	Energy released (in kJ mol^{-1} CH_2O)
O_2	475
NO_3^-	448
MnO_2	349
FeOOH	114
SO_4^{2-}	77
CO_2	58

Oxidation of organic carbon needs electron acceptors and this can be a limiting factor. Two groups of electron acceptors can be distinguished: those in soluble or diffused form, such as oxygen, sulphate and nitrate, and those in particulate form, such as iron and manganese oxides (Kelly *et al.*, 1988). In marine water, sulphate reduction is a more important process than in freshwater, where methanogenesis takes over, as salinity and sulphate concentration decrease. Energetic yield for the metabolization of organic matter through sulphate reduction is higher than by methanogenesis (Capone & Kiene, 1988).

Composition of the benthic microbial fauna

The benthic fauna is by and large made of detritivorous organisms, which live on the organic matter present. This organic matter is first ingested by the benthic macrofauna, which reduces the size of the organic material and makes it accessible to microbial attack. Along with the ingested material, the macrofauna consumes microorganisms which have colonized the material. The microorganisms living in the sediments are unfortunately not yet satisfactorily described and appear important in density and diversity. They have been classified in groups according to their basic metabolism. Often, they are difficult to separate from organic material; the term biodetritus encompasses the organic matter along with the microbial biomass living in it (Johnson *et al.*, 1989).

Bacterial populations of fish pond waters and sediments have not been inventorized and there is a dearth of information on them. A study on the bacteria present in the water and sediment of a carp pond in Japan identified 489 different strains, which were capable of decomposing substrates as different as casein, tributyrine, starch, cellulose and chitin. The first three substrata are also those most common in fish pond sediments and in unconsumed fish feed (Sugita *et al.*, 1987).

Aerobic decomposition and sediment oxygen demand

At the interface and in the surficial layer of sediments, conditions are aerobic: oxygen is provided by the overlying water. Decomposition processes are most active in the flocculent layer, where microbial respiration coincides with a demand for electron acceptors; the oxygen demand there is greatest, with the highest rate of oxygen depletion occurring at this interface (Adams *et al.*, 1982, on sediments of Lake Erie). Total oxygen requirements are termed the sediment oxygen demand (SOD), and indicate the importance of decomposition processes happening in the sediment (Jurkowska, 1987). Primary production is significantly related to annual sediment oxygen uptake. Nutrient-rich waters with high levels of primary production have anaerobic sediments; their rate of oxygen supply is not sufficient to satisfy the oxygen demand of mineralization (Rippey & Jewson, 1982; Marsden, 1989).

The oxygen exchange across the sediment-water interface depends on the renewal of the water in deeper portions of the pond. The transfer kinetics of oxygen have been studied by Hall *et al.* (1989). Oxygen supply to the sediments by diffusion processes has been estimated to reach a maximum of 4 g m^{-2} d^{-1} of oxygen. At Orissa, India, the oxygen demand of sediments in fresh water fish ponds was found to vary between 0.39 to 3.39 g m^{-2} d^{-1}, which represents some 5.1 to 29.5 % of primary production and 5 to 30 % of community respiration (Oláh *et al.*, 1987). This is in line with reported undisturbed sediment oxygen uptake rates in fish ponds in Israel peaking at 3 to 4 g m^{-2} d^{-1} (Schroeder, 1987). This represents about 2 g m^{-2} d^{-1} of organic carbon processed aerobically. Based on cellulose digestion in anaerobic conditions, the organic carbon processed would have reached 3 to 8 g m^{-2} d^{-1} (Schroeder, 1987), which is in the previously recorded range of oxygen consumption per molecule of methane transformed of 1:1.2-1.8 (Adams *et al.*, 1982). The relationship between sediment oxygen uptake rate and oxygen gradients in pore water argues against a molecular or eddy diffusion process operating alone (Rippey & Jewson, 1982). In sediments where the activity of benthos is high, vertical transport is best described by advection due to irrigation. Bioturbation has been found to increase the oxygen transfer rates by a factor 1.5 to 10, according to the nature of the sediments (Schroeder, 1987).

Sediment oxygen demand is divided into two components, biological and chemical. The biological SOD corresponds to heterotrophic respiration, i.e. the respiration of microorganisms and benthic macroorganisms. The chemical SOD pertains to the action of all reducing substances or electron-donors, such as iron (II), manganese (II) or sulphide, which rapidly oxidize in the presence of oxygen.

Sediment oxygen uptake is measured by oxygen profiles in the sediment; the oxygen concentration in sediments is considered to result from the difference between oxygen supply via diffusion across the interface, and oxygen consumption within the sediment. It is interesting to quantify the proportion of SOD due to the upward or benthic flux of dissolved reduced electron acceptors, such as Fe^{2+}, Mn^{2+}, NH_4^+ and CH_4, which are mobile and accumulate beneath the surface of sediments with low redox potential or with high decomposition activity. In the sediment of Lake Erie, the transport of these reduced constituents represented a sizeable oxygen demand, ranging from less than 1 % for ferrous iron and manganous manganese to as high as 26 % for ammonia and 30 % for methane (Adams *et al.*, 1982). Their transport depended on their concentration gradients. The reductant-oxygen ratio (in moles) for these substances is as follows (Adams *et al.*, 1982): for Fe^{2+}, 1:0.25; for Mn^{2+}, 1:0.5; for NH_4^+, 1:2 and for CH_4, 1:1.5.

Variation of dissolved oxygen concentration in the water at the interface causes changes in thickness of the aerobic layer, called the biofilm. The rate of decomposition is related to biofilm thickness. The diurnal exchange rate of the sediments versus the oxygen concentration and temperature in the water show a hysteresis effect due to the amplification of the discrepancy between exchanges under transient and under steady state conditions. In transient conditions, when oxygen supply increases, accumulated anaerobic end-products are oxidized, owing to the increase of the aerobic layer (Jeppesen, 1982). The diel oxygen fluctuations in the water due to phasic photosynthetic activity of autotrophs indicate that transient conditions are more realistic than steady state.

The importance of sediment oxygen demand increases with pond shallowness. Renewal of the water at the interface to increase benthic fluxes and to supply fresh dissolved oxygen may be of prime importance in fostering regeneration of nutrients in a pond. Practical means of implementing this water circulation will be discussed later (e.g. p. 200).

Anaerobic decomposition

Organic carbon mineralization in anaerobic conditions leads to methane as end product and, in aerobic conditions, to carbon dioxide (Cappenberg *et al.*, 1982). The various mineralization processes in water, and on and in the sediments, and the transfers across the interface constitute the carbon cycle. Carbon cycling in aquatic environments is the means by which the autotrophic populations are supplied with the carbon needed for production. Our knowledge of the various anaerobic breakdown pathways and turnover rates of carbon, and between its input and output in sediments is inadequate (Cappenberg *et al.*, 1982).

Nutrient enrichment experiments have shown that carbon can be limiting the growth of microorganisms. Growth of microorganisms attached to refractory organic matter, such as lignin, is not limited by nitrogen or phosphorus, as enrichment with these nutrients has been found not to enhance degradation. The limiting factor is the quality and the quantity of the organic carbon present. Carbon is limiting when the C:N ratio falls below 10:1 (Peters *et al.*, 1987). In sediments, the C:N ratio increases with depth (from 11:1 to 14:1) indicating that the proportion of organic carbon to organic nitrogen compounds increases too (Cappenberg *et al.*, 1982).

In waters with a high primary productivity, most organic matter comes from algal cell walls, which settle from the photogenic zone to the sediment (Cappenberg *et al.*, 1982; Rippey & Jewson, 1982). The decomposition is engineered by anaerobic bacteria and leads to the production of acetate, as an intermediary metabolite (Fig. 15). Acetate is further broken down either by anaerobic, sulphur-reducing, acetate-oxidizing bacteria in the upper layers of the sediment or by acetate-fermenting methanogenic bacteria in the lower layers. In the overlying water, acetate is broken down under aerobic conditions. The most active sedimentary transformation occurs in the lower sediments (Cappenberg *et al.*, 1982).

Acetate diffuses from the lower sediments towards the surface and high concentrations indicate an acetate point source localized at an intermediary depth. Acetate sinks exist both in the deeper layer (due to methanogenic activities) and at the sediment interface (due to aerobic, metabolic activities). Part of the acetate in sediment diffuses downwards, where it accumulates (Cappenberg *et al.*, 1982).

Mineralization rate is the combination of various simultaneous processes and their kinetics: production of acetate, methanogenic transformation, oxidative acetate transformation, and downward migration. To this should be added all other transformations discussed earlier, involving transformation of nitrogen and phosphorus. All these processes are drastically modified by bioturbation and physical and chemical disturbances of the sediments, such as gaseous ebullition and wind-induced turbulence.

Acetate is not the only intermediary product in organic carbon decomposition: lactate and formate are produced as well. Lactate, which is produced on a smaller scale, is a more voluminous molecule and therefore diffuses more slowly. Formate is unstable and transforms easily into hydrogen and carbon dioxide, which serve as precursors for the reactions of the hydrogen/carbon dioxide fermenting methane-producing bacteria in the upper sediments. These last intermediates, which are less oxidized than acetate, play a less important role (Cappenberg *et al.*, 1982).

The breakdown of complex organic matter into low-molecular weight intermediates such as acetate is a rate-limiting step. The ultimate transformation of acetate to carbon dioxide or methane has a rapid turnover rate. In organic matter decomposition, the major part of the input is returned to the overlying water by gaseous diffusion and complementary processes of gas transport (Cappenberg *et al.*, 1982).

The anaerobic decomposition processes lead to an accumulation of reduced electron acceptors, which accumulate in deeper horizons and migrate upwards to be oxidized. These aspects have been discussed earlier (p. 9).

Decomposition rate and nutritional value of organic matter

Sedimented organic matter is often of poor nutritional quality and detritus decomposers may be nutrient-limited. Nutrient requirements of surface sediment decomposers can be satisfied in two ways: diffusion from lower sediments or removal from the overlying water column. In the sediment, pore water is the intermediary compartment, concentrating and conveying soluble nutrients (Doremus & Clesceri, 1982).

Decomposition first involves an enzymatic attack, which detaches the most labile radicals from the organic compounds. The first to be removed are the nitrogen radicals, by ammonification. These transformations are reflected in the evolution of the C:N ratio, which increases with depth.

Mineralization is confined to the surficial layers of sediments, where acid-soluble and hydrolyzable compounds are metabolized, with removal of carbon and nitrogen atoms. The C:N ratio reaches its maximum value in the lower portion of this layer. The bacteria active here are anaerobic, belonging to the group of sulphate-reducers in marine and brackish waters, or to the acetate-fermenters or acetate-oxidizers in freshwater. They are almost always present, though their activities can be latent, depending on the volume of substrate present.

In deeper strata of the sediments of a tropical lagoon (Biétri bay, in the urban zone of Abidjan, Côte d'Ivoire), decomposition was slowed down by geochemical complexation processes and by immobilization of organic matter. Organic matter is sequestered in clay-humate complexes within fine mineralogical particles. Humic acid undergoes polymerization, in which compounds of molecular weight below 100,000 make up 70 % of the total. In anoxic sediment, the mineralization processes by heterotrophic organisms diminish in favour of humification processes and immobilization of organic detritus (Guiral, 1984). This humic material is composed of organic complexes, largely uncharacterized chemically, in a reduced and dispersed state, so that they are not easily separated from the mineral and the other undecomposed organic material present. Humic matter is a gel-like product, black or dark-brown in colour, synthesized by microorganisms. Being a colloid, this matter is able to sorb solutes.

In eutrophic waters (Ebrié Lagoon, Côte d'Ivoire), the resuspension of sediment in well oxygenated water reactivates heterotrophic bacteria, which drastically increases the oxygen consumption in the overlying water. This may have serious consequences for life in a pond (Carmouze & Caumette, 1985).

Effects of decomposition processes on sediments

The decomposition of organic matter in sediments modifies the conditions of the latter, inducing a decrease in pH and an increase in the concentration of exchangeable aluminium, which affects phosphate sorption and benthic fluxes (Twinch & Breen, 1982). The rate of decomposition of organic matter is seemingly reduced by acid conditions, although studies in acidified Swedish lakes do not support the hypothesis of decreased microbial activity in acidified freshwater sediments (Hasselrot & Hultberg, 1984).

The surficial flocculent layer prompts changes in the redox potential profile beneath it, which are essentially a shift from aerobic to anaerobic conditions.

Effects of benthic organisms on decomposition and benthic fluxes

Based on a survey of Polish lakes, Kajak (1988) stated that the abundance of the benthic fauna reflects the trophic state of ponds. The presence and diversity of the fauna is linked to the importance of aerobic decomposition and to pH. The number of taxa of benthic macroinvertebrates present in sediments is related to sediment pH; below 5.5, molluscs, crustaceans and larvae of mayflies are absent and the number of taxa of caddis-fly larvae (Trichoptera) and chironomids is low. This may be linked to the non-availability of calcium ions and high concentrations of free aluminium ions, which are toxic for these organisms (Friday, 1987).

The macrofauna belongs to few main groups only: tubificids (Oligochaeta Tubificidae, e.g. *Pomatothrix*), chironomids (Diptera Nematocera, e.g. larvae of *Chironomus* and *Chaoborus*), amphipods, and burrowing bivalves. Other organisms found include copepods, ostracods, nematodes, rotifers and protozoans (Andersson *et al.*, 1988). Species assemblages of benthic fauna are incompletely known, due to a lack of identification keys.

Oligochaetes of the tubificid family feed at shallow depths (3 to 6 cm below the surface) and deposit their faeces on or near the surface of the sediment (Andersson *et al.*, 1988). In lake Esrom, Denmark, their daily ingestion rate was estimated at four times their biomass (Kamp-Nielsen *et al.*, 1982). Their transport of particulate material is therefore considerable. The renewal of water inside their burrow holes, required for gaseous exchange, removal of metabolites and egested wastes (ventilatory movements), is provided by undulatory muscular contractions of the posterior end. Oligochaeta do not construct permanent galleries.

Tubificids promote exchanges at the water-sediment interface, in particular the diffusion of ferrous iron, which is transported to the surface, where it is subjected to oxidation and forms clouds of colloidal ferric oxide at the interface.

Ammonium concentration in the upper layers of a sediment can reach saturation levels. This is the result of intense ammonification by microorganisms, as a consequence of benthic macrofaunal activities, such as ingestion, egestion and mixing of the sediment with excreta.

Nitrate supply to sediment causes an increase in the disappearance of carbon, which indicates an increase in respiration or heterotrophic activity, and an intensification of microbial denitrification. Tubificids are responsible for anaerobic decomposition of organic matter, as denitrifying microorganisms are present in the intestine of these worms as well as in sediments.

Tubificids affect phosphorus cycling by increasing the redox potential and causing stronger binding of phosphate, depending on sediment type (Andersson *et al.*, 1988). They increase the mineralization rate of organic matter, promote exchange by bioturbation, especially the benthic flux of nitrogen compounds (Fukuhara & Sakamoto, 1988, in Lake Suwa, a shallow eutrophic lake in Japan), stimulate the growth of microbial biomass and of their predators, and add to the organic load of the sediment by their excreta (Kurihara, 1989, in ricefields in Japan).

Chironomids are surface filter or surface deposit feeders, which transport sediment from the interface to subsurface layers. They live in straight or U-shaped tubes. Water is pumped inside these tubes by body movements (Andersson *et al.*, 1988).

Chironomids are important with respect to biomass, production and recycling of organic matter. They influence water quality, as they drastically modify oxygen circulation above sediments, as well as oxygen penetration to sediments. By their burrowing activities, they greatly enhance the transport of dissolved substances from sediment to water. In Lake Balaton, Hungary, the presence of chironomids accelerates organic matter transformation, and organic material is also made more accessible to bacterial attack (Dévai, 1990).

Amphipods feed at the sediment-water interface, while bivalves burrow inside sediments (Andersson *et al.*, 1988).

The various effects of bioturbation can be summarized as follows: bioturbation transports sediment downward, i.e. buries organic rich sediment and prevents rapid degradation, and carries sediment upward, i.e. reexposes it to aerobic decomposition. At the same time, burrow holes facilitate penetration of aerobic water, electron acceptors and bacterial decomposers, which leads to a more efficient sediment mineralization. Other changes include redox potential increase and modifications in phosphate sorption. By ingesting material, deposit feeders increase the rate of particulate matter mineralization, though most detrital material ingested is not assimilated because of its low nutritional value (Andersson *et al.*, 1988). The shredding of fibrous organic material by macroinvertebrates makes it available to further bacterial attack (Cuffney *et al.*, 1990). However, for macrobenthic fauna, aerobic conditions are required. Under prolonged anoxic conditions, most organisms cannot survive. Under low dissolved oxygen conditions, the presence of chironomids and tubificids increases the dissolved oxygen flux.

Exchanges with the water column

The transformation of organic material is essentially a process involving the recovery of nutrients and energy for decomposer metabolism. Hence, it transforms dead organic material into living organic material, along with release of some chemical by-products. Regeneration of organic matter benefits the biotic communities of the water column primarily via consumption of benthic organisms by pelagic organisms. Besides this, and under specific conditions, excess soluble nutrients are released by the sediments (benthic fluxes). As discussed above, the most active stratum involved in decomposition is the flocculent layer, which is an entangled mixture of microorganisms and organic matter. In lake Valencia, Venezuela, this flocculent layer is intensively browsed upon by tilapias (*Oreochromis mossambica*) (Bowen, 1981). In Awassa Lake, Ethiopia, 10 cm sized tilapias (*O. niloticus*) were found to feed actively on benthic epipelic diatoms, belonging to the genera *Synedra*, *Cymbella*, *Neidium*, and *Gomphonema*. The food of tilapia of 25 to 30 cm size (standard length) has been found to consist of chironomid larvae and adult insects. On increasing in age and size, these fish change to a more herbivorous diet, though their regime may still be diversified. They consume phytoplankton, loose benthic detritus rich in bacteria, epiphytic diatoms, and periphyton (Tudorancea *et al.*, 1988).

The chemical by-products are essentially reduced electron acceptors, which migrate downwards and/or upwards and constitute benthic fluxes. They confirm the importance of the mineralization processes.

Importance of organic matter and its decomposition in the sediment

1. Organic matter settles from the water column to the sediment; this sedimentation rate is related to the primary productivity of the pond. The organic matter forms the flocculent layer on the sediments, which is a biodetrital layer, very active in decomposition and made of an entanglement of bacteria, fungi, algae and detritus.

2. Decomposition processes are varied, and depend on environmental conditions. The main two groups of decomposition processes are aerobic and anaerobic.

3. The sediment oxygen demand is an indicator of the oxygen consumption of the sediment, i.e. community respiration, especially by decomposers (microorganisms and others), and chemical oxidation by the reductants. The SOD is a measure of the importance of aerobic decomposition occurring in the sediment.

4. The rate of organic matter decomposition depends on the nutritional value of the organic matter, i.e. the nutrient content, expressed by the C:N:P ratio, and the energetic value.

5. The presence of benthic macroorganisms completely modifies the decomposition processes, in particular the sedimentation stratigraphy, the penetration of oxygen inside the sediments, and the benthic fluxes.

6. Regeneration to the water column of nutrients resulting from decomposition occurs via consumption of benthos by pelagic organisms and benthic fluxes.

Chapter 2.

Water and productivity

1. Water characteristics

Water is the culture environment for fish or other aquatic organisms. It is the physical support, in which they carry out their life functions, such as feeding and excretion. Water properties can be divided into two main categories: physical characteristics, like density and temperature, and chemical characteristics, the concentration of solutes and nutrients.

The water column is bordered below and at the sides by the pond sediment, and above by the atmosphere. Exchanges between water and sediment have been discussed in the previous chapter; exchanges with the air will now be discussed.

One of the main problems facing aquatic organisms is the low concentration of oxygen in water as compared to the aerial environment. The chemical composition of water is also much more variable than that of air and values of acidity or alkalinity can vary drastically, deeply modifying the living conditions of the organisms. The difference in density of water as compared to that of air has also influenced the shape and locomotion of aquatic organisms.

Overall water characteristics are summarized in the concept water quality, but, unfortunately, there is no universally accepted definition of this concept (see Couillard & Lefevre, 1985, for 20 different indices used). Because the specific requirements of cultivated species vary over such wide ranges, a general definition of suitable ranges of different water characteristics, or water quality criteria applicable to fish farming, is not possible.

2. Physical characteristics of water

2.1. Temperature, density and viscosity

The sun is the primary energy source on earth. The sunlight brings solar energy to ponds, which is absorbed, warming up the water mass. The low thermal conductivity of water restricts the heating to the upper water layers. Evaporation of water and diffusion of gases in solution in the water cause heat losses.

Air and water temperature depend on latitude, altitude and mean daily number of sunshine hours. Water temperature is also affected by specific characteristics, such as turbidity, wind force, plant cover, and microclimatic conditions of the immediate vicinity of the pond. Variations in water temperature at the surface closely follow changes in air temperature.

Temperature variations in shallow ponds are wider and affect the water mass to a greater extent than in deep ponds. In water bodies less than 1 m in depth, water temperature can reach 40° C or more under direct sunlight. In deeper ponds, heating is restricted to the upper layers and water stratifies thermally. Temperatures of 30 to 35° C are tolerated by fish, but above these temperatures, aquatic life is threatened. During periods of high temperatures, fish seek refuge in deep, cool water.

Increase in water temperature results in evaporation, which may entail considerable losses of water from a pond, depending on degree of humidity of air and wind velocity. Edwards *et al.* (1981) recorded evaporation rates of 25 $l\ m^{-2}\ d^{-1}$, in a humid tropical climate (Bangkok, Thailand).

Turbid water increases more rapidly in temperature than non-turbid water, regardless of whether turbidity is mineral or organic. Under the same sunshine conditions, the increase in temperature is more rapid in water rich in plankton than in water poor in plankton.

Temperature is a factor of great importance for aquatic ecosystems, as it affects the organisms as well as the chemical and physical characteristics of water. For example, the metabolism of cold-blooded organisms or poikilotherms is dependent on temperature, as are the solubility of dissolved oxygen, and the density and viscosity of water. There is, therefore, a need to keep pond water within some range, avoiding temperatures which are either too high or too low, and in agreement with the temperature requirements of cultured organisms. Excessively high water temperatures can be controlled by water renewal; water temperature in feeder canals is generally lower because it is not standing and poor in plankton.

Water density and viscosity decrease with temperature, while the former increases with salinity. These water properties are not commonly taken into consideration or measured, but do affect stratification (see p. 130) and motility of the aquatic organisms and seston.

2.2. Water movements and stratification

Water movements are either horizontal or vertical convective movements. Water movements connected with transport are not discussed here. They are treated in details by Coche & Van der Wal (1981).

Most ponds are impounded before stocking and no major water renewal occurs, except for occasional additions to restore a correct water level.

Effects of wind and waves

Waves are produced by wind. Their height is proportional to wind speed and fetch. Waves move a thin, sometimes even thick, layer of surface water, and homogenize water temperature. As referred to earlier (see p. 41), wind-induced turbulences even affect the flocculent layer on the pond bottom. In Lake George, Uganda, wave induced water movements cause resuspension of the sediment layer at the water-sediment interface to a depth of 5 to 14 cm (Viner & Smith, 1973). Wave action also inflicts long-term damage on dykes, situated perpendicular to the predominant wind direction.

Wind blowing across a water surface induces, besides waves, circulatory currents known as Langmuir cells. These are horizontal tubes of rotating water, their axes aligned approximately parallel to wind direction. Adjacent tubes rotate in opposite directions. Tubes of varying diameter can be present at the same time. Both wind and waves must be present to generate these roll vortices, but even a light wind over small-amplitude waves can set them up. Cells have diameters ranging from a few centimeters to several meters: at a wind speed of 5 m s^{-1}, a typical cell might have a diameter of 10 m and a surface speed of 1.5 cm s^{-1} (Kirk, 1986).

Wind action in tropical countries is responsible for a daily destruction of thermal stratification of ponds and shallow water bodies. This mixing of water affects all production parameters of a pond, and specifically photosynthetic activities (Kirk, 1986; Barbosa & Tundisi, 1989, in a shallow tropical Brazilian lake).

Convection currents

Vertical convection currents occur in all deep water bodies and appear at the change of season, when daily temperature regime changes. Water masses at higher temperature, and therefore lighter, move upward and displace other water masses, causing a seasonal overturn of the entire water mass. The superficial water temperature cools down as a consequence of a drop in air temperature, leading to the creation of warmer water layers at greater depth. This overturn mixes water thoroughly, carrying upwards nutrients released from the sediment by benthic flux and supplying deeper layers with surface water rich in dissolved oxygen. Upward flux of nutrients during overturn may be partly prevented by sorption, complexation or trapping reactions occurring during transport, as observed in Biétri Bay, in the urban zone of Abidjan, Côte d'Ivoire. As a matter of fact, organic colloids and humic matter readily sorb nutrients and solutes which come in their vicinity (Guiral *et al.*, 1989). Positive and negative responses to mixing exist for diatoms and blue-green algae respectively, though a simple relationship between mixing and gross estimates of the heterogenous phytoplankton populations is unlikely. Mixing may result in the dispersion of diatom populations to diatom-bare water (Talling, 1969).

Different types of stratification of ponds and other water bodies

Thermal stratification has been abundantly described in limnology manuals. The upper zone, in which wind homogenizes the water, is called the epilimnion. This stratum rests on a water mass in which water temperature decreases rapidly with depth. This is called the thermocline or metalimnion. The deepest water mass, termed hypolimnion, is defined by a stable temperature. Thermal stratification of water bodies has been studied extensively in temperate climates.

In tropical waters, thermal stratification exists as well, but to a minor extent. There is no seasonal overturn, as temperatures change little over the months. In shallow lakes, such as Lake Chad and Lake George, with mean depths 4 m and 4.5 m, respectively, thermal stratification is destroyed by wind action. These lakes are polymictic. In some cases, especially deeper lakes, overturn is only partial and does not affect the whole water mass; these lakes are called meromictic, such as Lake Malawi (Talling, 1969). In other lakes, such as Lake Upemba (1500 ha and 2 m mean depth), on the Lualaba river, Zaire, overturn occurs daily. These lakes are polymictic and holomictic.

A fish pond resembles a shallow lake, and daily thermal stratification occurs (Boyd, 1979; Chang & Ouyang, 1988). During daytime, sunlight heats up the water and heat is accumulated in the upper 1 m layer. Beneath this superficial layer, water temperature of the intermediary strata changes drastically over a narrow depth, located between 1 m and 1.75 m. Below this, water temperature remains stable (Fig. 16).

This thermal stratification is labile and may change at any moment, subject to water movements or eddies, caused by wind or other disturbances, such as inflow of water from a feeder canal, or heavy rain (Boyd, 1979). Chang & Ouyang (1988) observed a steady stratification during daytime and distinguished in the tropical fish pond (in the Pearl River Delta, Guangdong Province, China) the three main thermal zones defined above. In the late evening or at night (between 8 to 10 pm), a convective

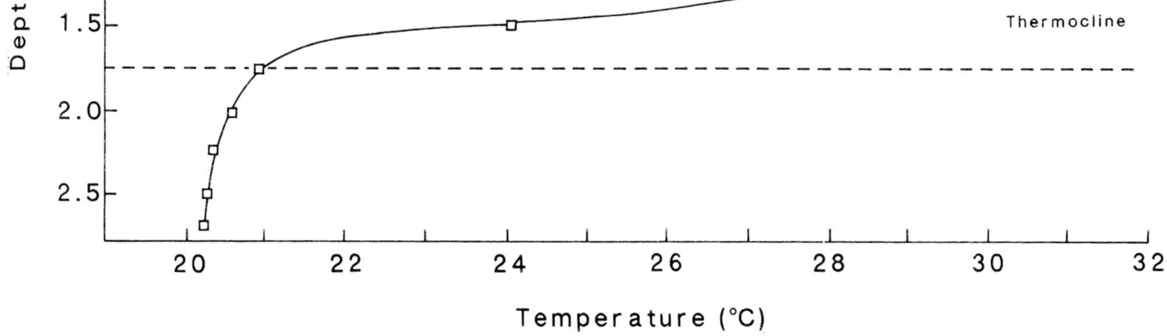

Fig. 16. Thermal stratification in a fish pond. Changes of water temperature with depth, showing the presence of a thermocline (after Boyd, 1979).

water movement caused destruction of the daytime stratification, mixing the upper 0.7 to 1.0 m water layer. The cooling off of the superficial layers was the result of a breeze (less than 100 cm s^{-1}), which is common in the late evening in tropical countries.

In shallow ponds in Auburn, Alabama, U.S.A., overturn homogenized the whole water mass, while deeper ponds remained stratified over the warmer months of the production season (Boyd, 1979). The accumulation of heat is greater in more turbid ponds, with stratification being more marked. Disappearance of plankton in ponds may cause disappearance of thermal stratification.

2.3. Colour, transparency and turbidity

Water colour

The colour of an object results from absorption of wavelengths in visible light. Absorption is the sum of partial absorptions, by water itself, by particles in suspension (seston), by dissolved matter in the water, and by phytoplankton.

A distinction between soluble and particulate matter is drawn arbitrarily by filtration through a membrane pore filter of 0.45 μm pore size. Dissolved matter ranges over a wide variety of organic and inorganic material, such as pigments released by plankton, humic matter, mineral salts, and nutrients. Large quantities of colloidal iron in suspension provide water with a brown opalescent colour. Particles in suspension, termed seston, are either non-living or living. The living ones are, in fact, the plankton. Sometimes, particles in suspension are brought by water fed to the pond, especially at flood times. An example is fine clay, which gives a milky appearance to the water and takes long to settle. In the Niger, water loaded with montmorillonite-kaolinite clay is called white flood, and these particles appear to be coated with bacteria and algae (John, 1986).

The abundance of phytoplankton and zooplankton in suspension is mainly responsible for the colour of pond water. Green, blue-greenish or brown-greenish water indicates a good plankton population. Turbidity caused by clay is brown, beige, or milky in appearance. Turbidity modifies the penetration of light and directly affects photosynthesis and primary productivity. Seston is also responsible for light scattering in water. Such scattering is stronger in the range of short waves than in that of long waves.

Light propagation in water and attenuation of light by plankton

The solar rays reaching the water surface are partly reflected, and partly refracted, according to their angle of incidence. The closer to the perpendicular the rays fall, the deeper they penetrate; if rays fall outside a certain angle, they are totally reflected. When the water surface is calm and smooth, the light penetration is better than when it is agitated.

The light penetrating the water covers a spectrum of wavelengths, including the visible spectrum. Each wave is absorbed and scattered according to its wavelength and specific absorption coefficient (Kirk, 1986). About 53 % of the incident light is absorbed within the first meter of water (Yentsch, 1980; Kirk, 1986: 105, Fig. 6.1). Fong & Ju (1987),in shallow water ponds located in Tainan county, Taiwan, established a relation between incident light intensity and depth:

$$\text{Intensity (in log of lux)} = 5.517 - 0.052 \times \text{depth (in cm)},$$

and between the extinction coefficient and turbidity:

$$K = 0.0084 + 0.0020\ T,$$

where turbidity is expressed in mg.l^{-1}.

The zone receiving more than 1 % of the incident light is called the euphotic zone. There phytoplankton can develop. Beneath this zone, respiration exceeds photosynthesis.

Phytoplankton algae possess photosynthetic pigments adapted to the light available in water (Yentsch, 1980). The various pigments cover almost the full width of the spectrum (Fig. 17), so as to make best use of the light available at any depth. In the water, the vertical zonation of algae is determined by the specific pigments they possess.

Water colour is rarely measured. Transparency or turbidity is more commonly recorded. Transparent water indicates an absence of plankton, while turbid water signifies presence of plankton and/or suspended particles.

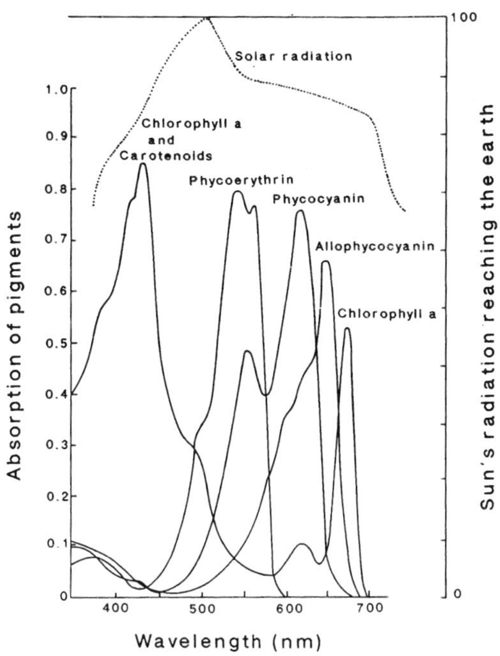

Fig. 17. Absorption spectra of different photosynthetic pigments and their relation to solar radiation reaching the earth (from Stockner & Antia, 1986) (with kind permission of Prof. Stockner on behalf of authors & Can. J. Fish. Aq. Sci.).

Water turbidity

Turbidity results from the presence of seston and from the resuspension of sediment (see p. 180). Most abiotic matter is colloid or non-colloidal material, which eventually settles downwards, as soon as water movements permit. The abiotic particles in suspension can be separated into mineral and organic material. Mineral particles consist of silt and clay, brought in by feeder canals, while organic material originates from the pond or from feed or manures added.

Movement and maintainability of suspended particles

The movements of particles in suspension are dependent on water movements, caused by waves, roll vortices, convective currents, upwellings, and currents generated by living organisms, such as filtrating zooplankton. Such particles are further affected by bacterial mineralization, which changes their shape and density, and by the ionic composition of the water, which causes flocculation or dispersion of colloids (see p. 13) (Weilenmann *et al.*, 1989). They are nevertheless subject to gravity.

Sedimentation of particles in suspension depends upon their size and electrostatic charge. Small sized particles take more time to settle. Most particles carry surface charges, usually negative, which originate from carboxylic or phenolic radicals, in the case of organic material, or from hydroxyl radicals, in the case of mineral particles. The ability to remain in suspension depends on surface chemistry, mutual electrostatic repulsion, or formation of aggregates.

The surface charge of a particle attracts a cloud of ions of opposite charge, and determines the dimension of the adsorbed cloud and the possibility of particles to come close to each other. Water conductivity, i.e. the total amount of electrolytes in solution, reflects the size of the ionic cloud. Electrophoretic mobility measures the size of the ionic cloud, or the stability of the particles, owing to electrostatic repulsion. The mobility of particles is altered by changes in conductivity, dissolved organic matter, and pH (Gerritsen & Bradley, 1987). Increase of cation concentration, such as calcium, in the water entails coagulation of the aggregates, and causes sedimentation. Increase of dissolved organic matter enhances dispersion and prevents sedimentation, and can be considered as a eutrophication factor (Weilenmann *et al.*, 1989). Particles coated with a film of humic matter are more stable than those without. The presence of blue-green and green algae, and of bacteria, confers a greater homeostasis to pH gradients and prevents flocculation (Gerritsen & Bradley, 1987).

The mobility of particles in water is particularly important for filtering organisms, as it affects the suspension of phytoplankton, bacterial adherence to substrates, and algal flocculation (Gerritsen & Bradley, 1987).

The role of coagulation in particles sedimentation

After collision, particles can remain together and form aggregates. The stability of such aggregates is a function of the concentration equilibria of ions in solution, and of surface properties of the particles.

The aggregation process takes place in two distinct steps: transport of the particles, and their attachment. Transport of particles is subject to movement of fluids (brownian diffusion, fluid shear) and to differential sedimentation, a consequence of differences in size and densities. The tendency of particles to form aggregates is essentially a chemical process (see p. 112). Aggregation increases the size of particles and modifies sedimentation rate. Even zooplankton grazing can be considered a coagulation process. However, it reduces sedimentation rate, rather than accelerating it (see p. 164).

Effect of turbidity on production

Organic particles in suspension are important for life in a pond, especially for decomposers and filtering organisms living in water, with a heavy organic load, such as that of ponds receiving organic manures.

Mineral particles reduce light penetration and, thus, photosynthetic rate. Mineral particles also interfere directly with life. In experiments in limnocorrals in a small lake in North Carolina, Cuker (1987) observed that clay in suspension reduced net community productivity, chlorophyll a concentration and algal cell numbers. The composition of the community changed to the benefit of flagellates and at the expense of blue-green algae.

Coarse clay particles (less than 2 μm), at high but not at low concentrations (50 to 100 mg l^{-1} and 10 mg l^{-1}, respectively), and fine particles (less than 1 μm), at both high and low concentrations, have been found to differently affect various species of cladocerans, by reducing their growth, especially under borderline conditions of food availability. Rotifers in the same experimental conditions, were not affected (Kirk & Gilbert, 1990). As we shall explain later (p. 162), low turbidity reduces the degree of vertical migration of cladocerans and copepods, become they become less vulnerable to predators in conditions of reduced visibility (Cuker, 1987).

Among effects on the benthos, one should note that large amounts of sedimenting material tend to bury life at the sediment-water interface.

Suspended material also affects fish. It deposits on the gill filaments and in the opercular cavity, forming a film of fine particles, which impede respiration and cause death by suffocation. Eggs, larvae and fry are much affected. High turbidity, such as 200,000 mg l^{-1}, causes death of fish. In natural conditions, turbidity seldom exceeds 200,000 mg l^{-1}, with values around 2,000 mg l^{-1} common in muddy waters (Boyd, 1979). Tilapia do not fare well in turbidities above 1300 mg l^{-1}. Many species living in bogs and shallow waters became adapted to turbid conditions by reduction of eye size and by mucus secretion. Mucus helps to precipitate particles in suspension, surrounding the fish with an envelope of clarified water. Some aquatic plants also clarify water.

Turbidity measurement

Methods for estimating turbidity are: measurement of the amount of settleable matter, or of total solid matter, Secchi disk measurement, and nephelometric measurement. The first two are analytical and quantitative, while the other two are qualitative.

Procedures for measuring amounts of settleable matter and total solid matter are detailed in Boyd (1979).

The Secchi disk, first used by Angelo Secchi in 1866 during a scientific expedition, is painted black and white on alternate quadrants, weighted and fixed to a graduated rope. The disk is immersed in water and the depth at which it disappears from sight is noted as P_1. The disk is lowered further into water and then slowly hauled up again; the depth at which the disk reappears is noted as P_2. The Secchi disk measure is the mean of the two measures:

$$P = \frac{1}{2} (P_1 + P_2)$$

The disappearance depth is a visual measure of water clarity (Preisendorfer, 1986). The Secchi disk measurement should be made between 10 am and 2 pm on clear days and with a non-overcast sky. The observer should stand perpendicular to the disk and not wear sunglasses. The sun should preferably be at his back (Boyd, 1979).

The depth of the euphotic zone corresponds roughly to 2.5 times Secchi disk visibility (Fig. 18). Minimum visibility requires about 15 % of total daylight and the vertical extinction coefficient of incident light is about 1.7/P to 1.9/P.

Almazan & Boyd (1978) established a correlation between Secchi disk visibility and phytoplankton concentration, and between the former and organic particulate matter concentration in ponds (Boyd, 1979: 278, Fig. 13.8 & 13.9). Measurements made by different observers do not vary in a significant

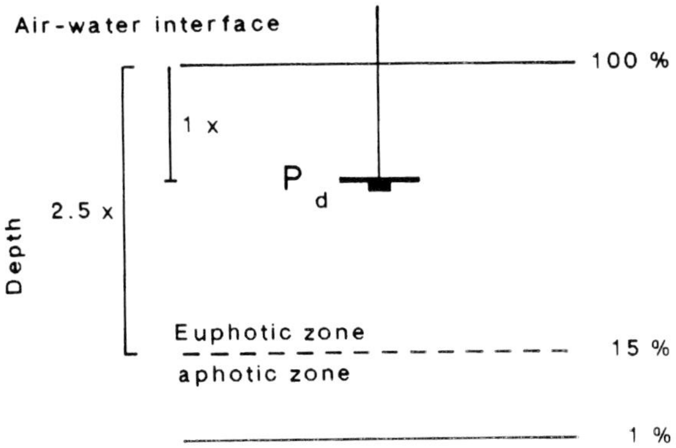

Fig. 18. Absorbance of light in water with depth and Secchi disk visibility according to light absorbance.

way, which indicates that the measures are objective. A detailed study of the conditions for use of the Secchi disk was made by Preisendorfer (1986).

The use of the Secchi disk to assess the abundance of plankton in ponds is a current practice. An acceptable plankton productivity in a pond corresponds to a Secchi disk visibility of 0.4 to 0.8 m (Boyd, 1979), and coincides with an organic particulate matter density of 7.5 to 14 mg l^{-1}.

The nephelometer is an optical instrument used to measure turbidity in comparison to standard solutions. Absorption in the water, of which the turbidity is to be measured, is compared to absorption of a solution of known turbidity. The measure is expressed in nephelometric turbidity units, NTU. This instrument is an attachment to the spectrophotometer, but can be used for field work. Nephelometric measurements give a good indication of absorption of light in water. Sigler *et al.* (1984) established an equation for transforming nephelometric units into amounts of suspended material, expressed in mg l^{-1}:

$$NTU = 10 + 0.178\ (mg\ l^{-1})\ \text{with}\ r^2 = 0.764.$$

Turbidity control

Methods to control turbidity include the use of lime (see p. 79 & p. 112) and alum, $Al_2(SO_4)_3.14H_20$, at the rate of 20 mg l^{-1}. Alum has an acidifying effect on water, 1 mg l^{-1} of which destroys 0.5 mg l^{-1} of alkalinity and produces 0.44 mg l^{-1} of carbon dioxide. Alum is not toxic for fish at these concentrations. The acidifying effect of alum is neutralized by the use of lime, 0.37 mg l^{-1} of which, in the form of $Ca(OH)_2$, neutralizes 1 mg l^{-1} of alum (Boyd, 1979).

Importance of physical characteristics of water

1. Optimum water temperature is essential for fish farming, as temperature should remain in a range suitable to the requirements of cultivated organisms (temperature preferendum); water temperature affects a series of other variables of which dissolved oxygen concentration is certainly the most important, but which includes also, for instance, the activities of decomposer microorganisms.

2. Water temperature is governed by thermal exchange with air and by direct heating by insolation. The temperature exchange within the water mass is low and thermal stratification in fish ponds does occur. However, thermal stratification is destroyed daily.

3. Control should be exerted on water temperature, when the latter exceeds values which may directly affect fish and cause their death. An influx of cooler water helps to reduce high temperature.

4. Turbidity of water is caused by mineral and/or organic particles in suspension. Mineral turbidity is composed of clay and silt particles, and is detrimental to living organisms, while organic turbidity is made up of plankton organisms, though organic colloids can also be involved, especially in ponds supplied with liquid manure. Colour of the water characterizes the type of seston load of the water.

5. Water turbidity impedes light transmission in water and reduces photosynthetic rate.

6. Control of abiotic turbidity is done by precipitating particles, using flocculating agents, such as lime.

3. Chemical variables of water

Factors determining water chemistry

The chemical composition of water depends on the ions and the gases in solution. Three main sources provide water with these substances: exchange with sediments, exchange with overlying air, and cycling of elements within the water column. To this should be added the supply of feed and fertilizers to ponds during production time. *Sensu stricto*, only ions participate in the chemical composition of the water, as the properties of gases in water are strictly part of its physical properties. However, gases and ions are often interrelated, like for example, the transformation of gaseous carbon dioxide into carbonate and bicarbonate ions.

The amounts of gases dissolved in water, such as oxygen, carbon dioxide, and molecular nitrogen, depend on atmospheric pressure and on specific solubility of each gas. Solubility increases with a decrease in temperature. Compared to the composition of atmospheric air, the gas composition of water is fundamentally different: oxygen concentrations are much lower, carbon dioxide concentrations are much higher, and ammonia, hydrogen sulphide and methane, almost non-existent in the atmosphere, are present in varying amounts, according to specific anaerobic decomposition processes going on. The concentrations of these gases depend on production and exchange processes prevailing in the water, on agitation and movement of the water, and on temperature. Moreover, their concentrations vary with depth. Water is usually aerobic in the upper layers, but may be anaerobic at greater depths.

Water is a solvent. Owing to its polar nature, the water molecule attracts ions, surrounding itself with an envelope of molecules. The ions are said to be hydrated. Ionization of salts in water gives a certain mobility to the ions, which can be set in motion under the influence of electrochemical forces. The electrolytic mobility of an ion depends on its charge, its diameter and on electrochemical potential. The concentration of ions in solution depends on the solubility coefficient and the reaction equilibria with the solid compound, generating these ions.

3.1. Dissolved oxygen

Importance of dissolved oxygen

Dissolved oxygen is one of the key factors of life. Oxygen is much more important to aquatic than to terrestrial life, as oxygen has a low solubility in water and is often a limiting factor for life in water. Some species are able to breathe atmospheric oxygen, by surfacing and gulping air, and do not rely entirely on oxygen dissolved in water. However, oxygen is also needed for all oxidation, nitrification and decomposition processes. Oxygen concentration in water is controlled by four factors: photosynthesis, respiration, exchanges at the air-water interface, and supply of water to the pond (Krom *et al.*, 1989b; Erez *et al.*, 1990).

Oxygen transfer at the air-water interface and within water

The quantity of oxygen dissolving in water depends on atmospheric pressure and on the partial pressure of oxygen in contact with the water surface. Between the two media, an equilibrium is established, responsible for exchange of oxygen between air and water, according to the degree of saturation of the water. At the peak of production in the early afternoon, oxygen saturation can reach 250 %. In these cases, the exchange is towards release of oxygen into the air. Oxygen is then not the only gas diffusing out of the water; other gaseous products, such as nitrogen are also carried along. If the water is below saturation levels, oxygen dissolves in water.

Transfer at the interface is proportional to the ratio of saturation level to actual concentration, to the surface area of contact, to the volume of water, to the period of contact, and to the transfer coefficient. The transfer coefficient depends on the degree of agitation and turbulence and ranges between 0.01 to 0.03 g O_2 m^{-2} h^{-1} in calm water, to about 0.75 g $m^{-2}.h^{-1}$ in rough water (Boyd, 1979). These numbers have been determined under laboratory conditions and give an indication of the range of values in controlled conditions, but are an underestimation of the real importance of the phenomenon in nature.

The diffusion of oxygen to deeper strata within the water column decreases with the logarithm of depth. It, therefore, rapidly becomes an insignificant process as depth increases. In Lake George, the surface water was strongly supersaturated (216 %), but an oxygen depletion (49 % of saturation) was observed at a depth of only 1.8 m (Talling & Talling, 1965). A change in atmospheric pressure modifies the saturation level, i.e. the maximum concentrations of oxygen soluble in water. These changes are, however, modest, being around 7 % for a variation of the atmospheric pressure equivalent to 50 mm of mercury at 25 to 30° C. Atmospheric pressure decreases with altitude.

Solubility of oxygen in water

The maximum amounts of oxygen soluble in water (the saturation level) depend on temperature and/or salinity. With increasing temperature or salinity, the solubility of oxygen decreases (Table 8).

Oxygen sources and consumption in water

Only a small part of the oxygen dissolved in water originates from air, the major part coming from photosynthetic activities. In Israel, Milstein *et al.* (1989) have shown that water quality of reservoirs used for fish farming is primarily the result of algal activities, then decomposition processes, and, lastly, wind action.

Oxygen concentrations in water between the surface and the euphotic limit are mainly the result of photosynthetic activities, and cannot be explained solely by diffusion.

Heterotrophic respiration is one of the principal sources of oxygen consumption in a pond. Heterotrophs include all animals contained in the pond, from aerobic microorganisms and zooplankton to fish. At night, the action of the autotrophs has to be added. Decomposition of organic matter also consumes much oxygen, though this is actually a part of the heterotrophic respiration (being consumption of oxygen by microorganisms).

Dynamics of change in dissolved oxygen in ponds

The dissolved oxygen dynamics in a pond are still largely unknown, particularly the vertical dynamics, i.e. interactions between the supersaturation at the surface and oxygen depletion in deeper water.

The photosynthetic rate varies in the course of day, and with it, the rate of oxygen release to the water. In the epilimnion, algae produce oxygen, while in the hypolimnion, they consume oxygen. In the metalimnion, consumption and production are balanced. The total production of O_2 per unit surface area can be considered fairly stable. Surface production is compensated by a reduction of production with depth, or a reduction of compensation depth, i.e. the depth at which production and consumption of O_2 are balanced (Chang & Ouyang, 1988).

In the epilimnion of ponds in Guangdong Province, China, the ranges of diurnal variations of dissolved oxygen were important and reached up to 10 mg l^{-1}. The 20 to 30 uppermost centimeters were frequently supersaturated with oxygen, which consequently involved a loss, by diffusion, of oxygen from water to the atmosphere. In hypereutrophic ponds, the amounts of nutrients available daily are sizeable and favour the growth of blue-green and green algae. Blue-green algae possess positive buoyancy properties; they rise to the surface, to form thick algal mats. This upward migration is promoted by the supersaturation of gases in the water and gaseous ebullition. Along with them, they pull to the surface plankton, microorganisms, and detritus. At the surface, oxygen produced by photosynthesis does not dissolve in water, but the major part diffuses into the air. Their presence at the surface not only reduces the impact of the wind on the water surface, but also prevents exchanges with the water beneath; in particular, the diffusion of gases is inhibited.

In the hypolimnion, dissolved oxygen concentrations usually do not exceed 2 mg l^{-1} and diurnal changes are not greater than 1 mg l^{-1}. The lowest oxygen concentrations are recorded at mid-day, when solar radiation is most intense and thermal

Table 8. Dissolved oxygen concentration in relation to temperature and salinity (after APHA, 1976)

Temperature (° C)	Dissolved Oxygen (in mg l^{-1})			
	Chloride concentration in water (g l^{-1})			
	0	5	10	15
15	10.2	9.7	9.1	8.6
16	10.0	9.5	9.0	8.5
17	9.7	9.3	8.8	8.3
18	9.5	9.1	8.6	8.2
19	9.4	8.9	8.5	8.0
20	9.2	8.7	8.3	7.9
21	9.0	8.6	8.1	7.7
22	8.8	8.4	8.0	7.6
23	8.7	8.3	7.9	7.4
24	8.5	8.1	7.7	7.3
25	8.4	8.0	7.6	7.2
26	8.2	7.8	7.4	7.0
27	8.1	7.7	7.3	6.9
29	7.9	7.5	7.1	6.8
30	7.6	7.3	6.9	6.5
31	7.5			
32	7.4			
33	7.3			
34	7.2			
35	7.1			

stratification is at a maximum. During daytime, the oxygen supplies are reduced, but consumption never stops. At night, vertical circulation of the water reinjects new quantities of dissolved oxygen into these layers and homogenizes the dissolved oxygen concentrations of the epilimnion and hypolimnion. The oxygen consumption rates are dependent on the amount of oxidable organic matter present. A relationship between oxygen consumption and dissolved oxygen has been established (Chang & Ouyang, 1988):

$$\text{dissolved oxygen consumption (in mg } l^{-1} h^{-1}) = 0.149 \times [OD]^{0.668}$$

Oxygen consumption rate is dependent on temperature and pH, being maximum at a pH of around 8. In the case of a regularly resuspended sediment, consumption has been found to double (Chang & Ouyang, 1988).

Supersaturation of the upper water layers should help to improve the conditions of the hypolimnion, if circulation of the pond water occurs during the hottest hours of day (Chang & Ouyang, 1988).

Factors controlling dissolved oxygen

The photosynthetic activity of algae changes in the course of the day. Factors modifying photosynthetic activity also affect dissolved oxygen concentration. They are, among others, temperature, light, cloud cover (Fig. 19), turbidity, nutrient concentration, and abundance of autotrophs. Photosynthetic rate decreases with depth, though the vertical distribution of phytoplankton is more complex than a simple decrease of abundance with depth (see p. 146).

In the presence of a phytoplankton bloom, daily oxygen fluctuations are high: the phytoplankton releases much oxygen during the photosynthetically active period, but, at night, consumes more oxygen than was released during daytime. In ponds with an abundant benthic flora, the pattern of dissolved oxygen stratification is reversed. The oxygen produced by benthic algae is released at the sediment level, and diffuses upwards towards the surface.

The dissolved oxygen regime in ponds is heavily dependent on the total organic matter (dissolved, particulate, on the pond bottom) in the process of being decomposed. Decomposition, aerobic or anaerobic, is a result of the availability of dissolved oxygen and organic load. Almost 90 % of the recycling of organic matter in deep lakes, such as Lake Titicaca, occurs in the euphotic zone, supplying the major part of the inorganic carbon for photosynthesis. The euphotic zone is characterized

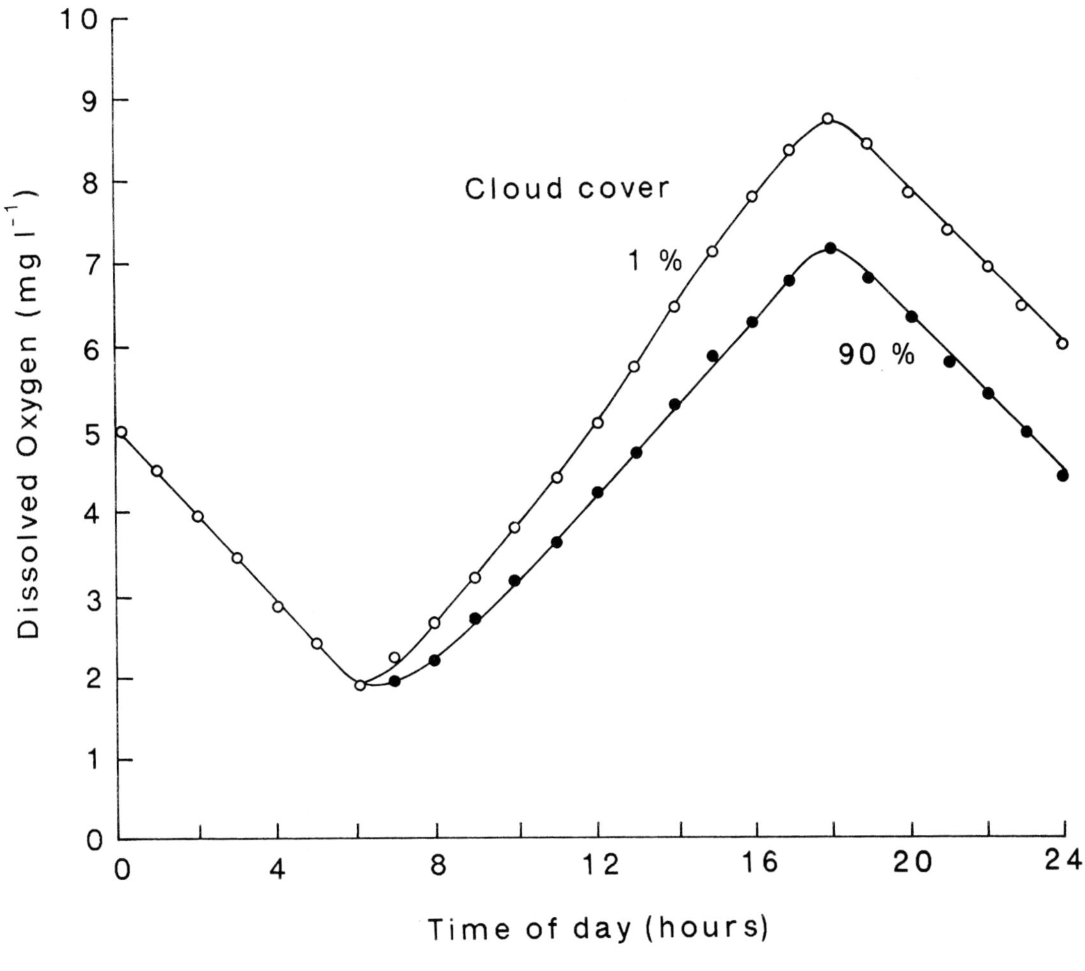

Fig. 19. Diel changes of dissolved oxygen according to a cloud cover of 1 and 90 % (from Piedrahita *et al.*, 1984) (with kind permission of Prof. Piedrahita & J. World Maric. Soc.).

by intense heterotrophic activity. The metabolism of the water body is tightly correlated to oxygen supply (Carmouze *et al.*, 1984).

In catfish production ponds in the United States, Cole & Boyd (1986) found a negative correlation between the fish feeding rate (amount of feed supplied to the pond per hectare per day) and dissolved oxygen concentration at dawn. The most important factors which influence aerobic bacterial activity are temperature, pH of the substrate, and the nature of the organic material. One criterion used to assess the nutritional value of organic matter is the C:N ratio. The closer this ratio in organic matter is to the ratio of the bacteria themselves, the more nutritive is the substrate, the faster organic matter is decomposed and the higher the rate of oxygen consumption (see p. 198 for bacterial oxygen consumption in ponds). Fibrous or structural material, such as lignin and chitin, takes a long time to decompose (Boyd, 1979).

The dissolved oxygen concentration (DO) varies in the course of the day, being at minimum in the morning, slightly before dawn, and at maximum in the early afternoon. The minimum DO concentration in the morning is the result of the night respiration of all heterotrophs and autotrophs combined. By early afternoon, photosynthesis has attained its optimum rate, the release of oxygen is highest and the water reaches its highest DO level. Oxygen decrease during the night can render the water anoxic, causing problems for the fish. Supersaturation of the water by oxygen may also be a problem, as it can provoke gaseous embolies to fry and fingerlings. The influence of low DO concentrations has a greater impact when other water quality factors are unfavourable, and the fish is already stressed. There is thus a need to reduce daily oxygen fluctuations, i.e. the difference between the daily maximum and minimum, and to contain the nightly drop in DO.

Boyd and associates (Romaire *et al.*, 1978; Romaire & Boyd, 1979; Boyd *et al.*, 1978) tried to predict nocturnal decreases in DO. Of the night consumption of oxygen, 75 and 85 % was found to be accountable to the plankton, in non-fertilized and fertilized ponds respectively, with the rest attributed to fish and benthic respiration (Romaire *et al.*, 1978). Oxygen production is directly and significantly correlated to the intensity of solar radiation, concentration of chlorophyll a, and to percentage O_2 saturation at dawn. Similarly, a model has been developed, combining all factors, allowing prediction of the DO concentration at dawn, using chemical oxygen demand, COD, and temperature, these two factors being correlated. COD is estimated from measurements of turbidity or Secchi disk visibility (Boyd *et al.*, 1978). The most important factors affecting dissolved oxygen concentrations at dawn are biotic or algal turbidity, planktonic respiration, inorganic nutrient depletion, and algal biomass (Smith & Piedrahita, 1988).

Measurement of dissolved oxygen

The measurement of dissolved oxygen is usually done by a modified Winkler method or by oxygen electrodes. Dissolved oxygen is expressed as mg l^{-1}, ml l^{-1}, or ppm. The factor to convert mg l^{-1} into ml l^{-1} is 1.43 mg ml^{-1} at normal pressure and a temperature of 0° C.

Dissolved oxygen concentration is sometimes expressed as percentage saturation, i.e. the ratio between the amount of oxygen per litre and the maximum amount soluble at the prevailing temperature.

Example: the saturation value at 35° C is 6.98 mg l^{-1}, the recorded oxygen concentration at 35° C reads 3.5 mg l^{-1}
the percentage saturation is:

$$\frac{3.5 \times 100}{6.98} = 50.1\ \%$$

Methods for controlling dissolved oxygen

Interventions of the fish farmer to prevent oxygen depletion are of two types: ensure aeration of the pond, or control phytoplankton production. Solutions proposed to control phytoplankton production include the use of various chemicals, such as simazine, copper sulphate or potassium permanganate. Planktivorous fish (biological control) can also be used. Among the species used are the silver carp (*Hypophthalmichthys molitrix*), the bighead carp (*Aristichthys nobilis*), and different species of tilapia (*Oreochromis aureus*, *O. niloticus*, and their hybrids). Plankton consumption by fish is discussed on p. 184.

Artificial aeration of ponds is discussed on p. 199.

Sudden death of a great amount of plankton, such as the dying off of a plankton bloom, results in a dramatic decrease in dissolved oxygen in the water. It is therefore of great importance to keep a vigilant eye on phytoplankton density and, if necessary, to reduce it (preventing mass destruction of phytoplankton). A discussion on factors promoting the appearance of phytoplankton blooms is given later. Filamentous algae at excessive densities can be removed with the help of a seine (Heinen *et al.*, 1988). Boyd *et al.* (1975) observed the dying off of blue-green algae of the genus *Anabaena*, induced by a change in light intensity; this resulted in mats of algae floating at the surface, constituting a light screen. Any such observation should lead to the removal of this floating material. Another method is to replace part of the pond water, thereby diluting the plankton (see p. 142).

Simazine is a herbicide toxic to phytoplankton, but apparently not toxic to fish. A treatment of 0.5 mg l^{-1} or 0.25 mg l^{-1}, in repeated applications, kills off blue-green algae. Simazine is commercialized under the trade name Aquazine. Tucker & Boyd (1978) used a treatment comprising three applications: initially 0.8 mg l^{-1}, followed by two doses, of 0.25 mg l^{-1}, after 3 and 1 months. This gave a good control of phytoplankton, but was accompanied by a decrease in dissolved oxygen, unfavourable for production. Copper sulphate ($CuSO_4$) is much more toxic for fish and should be used at a dose of 2 mg l^{-1} or 0.84 kg ha^{-1}. However, at this concentration, Tucker & Boyd (1978) did not observe an effect on *Microcystis* blooms.

Potassium permanganate is a strong oxidizing agent and is used in fish ponds to reduce biological oxygen demand, by oxidizing surplus organic matter. Tucker & Boyd (1977) studied the effects of applications of 2, 4, and 8 mg l^{-1} in ponds and observed that the product decreased biological oxygen demand, reduced phytoplankton density and photosynthetic rate, but did not affect dissolved oxygen concentration. *In vitro*, permanganate is bactericidal, but this property has not been confirmed in pond use. To increase its bactericidal effect, it has been proposed to treat ponds with a dose of 4 mg l^{-1}, in addition to the amount determined by the so-called '15-minute $KMnO_4$ demand'. The quantity of permanganate reduced after 15 minutes by organic matter present in the fish pond is determined and, to this amount, 4 mg l^{-1} is added.

Tucker (1984) discussed the oxidation processes which go on during the 15 minute period. He concluded that it is essentially the soluble organic matter which is oxidized, and this represents only a variable part of total organic matter. He recommended the extension of this period to 180 minutes, to achieve a more thorough destruction of excess organic matter. Tucker (1989) proposed multipling the 15-minute demand by 2.5. The concentrations of potassium permanganate used in these ranges are not harmful to American catfish.

The use of these products should be restricted and a better approach would be to investigate ways of combatting the causes of algal proliferation, such as by decreasing nutrient supply, in the form of fertilizers or feed, or by controlling algal biomass mechanically. Toxic products may indeed eventually reappear in fish (Boyd, 1979; Smith, 1988).

Importance of dissolved oxygen in ponds

1. Dissolved oxygen is essential for pond life, being required by all organisms for heterotrophic respiration. All aerobic chemical oxidation processes require oxygen as well. Oxygen find its way into a pond either by diffusion from the air or by photosynthesis. It is this latter process which provides the bulk of oxygen to the water mass.

2. The dynamics of photosynthesis, involving changes in the course of the day and with depth, are the major determinant of the spatial and temporal partitions in oxygen concentration within the pond, imposing concentration gradients and a vertical stratification. This stratification is destroyed by homogenization processes, such as periodic overturns of the water, and wind and wave action.

3. Control of dissolved oxygen dynamics in a pond, entails, firstly, management of the phytoplankton mass and the photosynthetic rate, and, secondly, making proper use of the overturn process, with the aim of facilitating the transport of oxygen to deeper levels, where dissolved oxygen can easily become depleted.

3.2. Conductivity and salinity

Definition

The conductivity of water is its capacity to transport an electrical current. The transport is ensured by the ions present. The measure gives an indication of the total amount of ionizable salts in solution and is expressed in $\mu S\ cm^{-1}$. Earlier, one spoke of resistivity, which is the reverse of conductivity.

Distilled water has a conductivity of 0.5 to 2 $\mu S\ cm^{-1}$, which increases to 4 $\mu S\ cm^{-1}$, when stored. During conservation, carbon dioxide and ammonia from the atmosphere dissolve in water. The conductivity of fresh water varies between 50 to 1500 $\mu S\ cm^{-1}$ (Boyd, 1979), but some polluted waters reach 10,000 $\mu S\ cm^{-1}$. Seawater has a conductivity around 35,000 $\mu S\ cm^{-1}$. The degree of mineralization or total load of salts of a water is in direct relation with its conductivity.

Factors responsible for mineral load of water

The total amount of mineral salt in water is the result of several processes (Golterman & Kouwe, 1980):

- the solubilization of salt, in particular, calcium bicarbonate, originating from soil and rocks, with which the water comes in contact during percolation and run-off
- dilution by rainfall
- precipitation due to interaction with other salts or to evaporation of water.

The last two processes lead to a dominance of sodium and chloride ions (Golterman & Kouwe, 1980). African waters are largely dominated by rock erosion, concentration by evaporation, and precipitation of calcium carbonate. Calcium carbonate begins precipitating when the saturation concentration of calcite is reached, which corresponds to an alkalinity of approximately 2.5 meq l^{-1} of solid matter in solution. When dissolved matter reaches beyond 1 mg l^{-1}, most calcium ions are removed from solution. The evaporation-crystallization process leads to a substantial loss of carbon at each step of concentration (Kilham, 1990).

Dominant ions

To characterize a water, its ionic composition needs to be analysed. The amounts of nutrients present give an indication of the potential production of the water. The chemical environment is important, though it does not give sufficient information to assess the production capacities of an environment. It is not the total amount of nutrients present in any form, but their availabilities, seasonal fluctuations and their minimum concentrations that matter. Also, nutrients appearing in limiting concentrations have a direct impact on life in a pond.

Temperate freshwaters are characterized by an abundance of calcium, magnesium and sodium, and bicarbonate, sulphate and chloride. The ionic composition of African lake waters is distinctive, being characterized by high contents of sodium bicarbonate and carbonate, and a wide range of salinities (Talling & Talling, 1965; Welcomme, 1972). In some cases, magnesium is present in large amounts (Golterman & Kouwe, 1980). Among the minor constituents, a high proportion of silica and low concentration of nitrate are features probably widespread in tropical lakes (Talling & Talling, 1965).

Types of water based on conductivity

According to conductivity, three main groups are distinguished: waters with conductivities below 600 $\mu S\ cm^{-1}$; between 600 and 6000 $\mu S\ cm^{-1}$; and beyond 6000 $\mu S\ cm^{-1}$ (Talling & Talling, 1965).

Waters with low conductivities are found in marshland areas. They are usually of dark colour and have high concentrations of organic matter, in particular humic matter. Waters with intermediate conductivities contain mainly sodium, chloride and bicarbonate ions. Waters with high conductivities result from high evaporation rates and salt concentrations (Talling & Talling, 1965).

With increasing conductivity, the influence of major ions changes: at low conductivities, increase of conductivity is accompanied by an increase in calcium ion concentration; in waters of intermediary conductivities, bicarbonate and carbonate ions play a more important role; at high conductivities, calcium is replaced by sodium (Talling & Talling, 1965). The predominance, in African lakes, of carbonate and bicarbonate accounts for the correlation between conductivity and alkalinity (Talling & Talling, 1965).

Interactions between ions occur, such as between calcium, carbonate and phosphate ions. Calcium carbonate is dissolved by carbon dioxide. If carbon dioxide is removed from solution, calcium precipitates. In some cases, supersaturation of calcium and carbonate beyond the solubility constant of calcite ($K_{sp} = 1.10^{-8}$) occurs (Golterman & Kouwe, 1980). Such supersaturation of carbonates may be related with organic load, organic compounds sorbing calcium ions, and preventing precipitation.

Conductivity is used to assess the expected yield of fisheries reservoirs, with the use of the morphoedaphic index (MEI), establishing a relationship between conductivity (in $\mu S\ cm^{-1}$) and mean depth (in meters), and between MEI and yield ($kg\ ha^{-1}$) for African water bodies (Welcomme, 1983):

$$MEI = \frac{\text{conductivity}}{\text{mean depth}} \quad \text{and} \quad Y = 14.3136\ MEI^{0.4681}$$

Modes of representation of ionic composition

The chemical composition of ions in solution can be represented by the Kufferath ionic diagram. To draw such a diagram, four concurrent axes, equidistant by the same angle, are drawn. Each axis is graduated in percentages, representing calcium, sodium, potassium, magnesium, bicarbonate, carbonate, chloride and sulphate. The concentration of each ion is calculated in milliequivalents and expressed as a percentage of the sum of the ions of each sign. Each percentage calculated is placed on scale on one half-axis; the cations are arranged on one side, opposite to the anions, and the different points joined. Sometimes, only three axes are drawn, sodium and potassium, and carbonate and bicarbonate, being grouped together (Fig. 20).

Salinity and chlorinity: definition

Salinity is the sum of all solid substances in solution in 1 kg of water, when all carbonate ions have been converted to oxide ions, all bromide and iodide ions replaced by chloride ions, and all organic matter oxidized (Spotte, 1979). Salinity is expressed in $g\ kg^{-1}$. Sea water normally has a salinity of 35 $g\ kg^{-1}$. Salinity is difficult to measure by direct chemical methods and is often measured refractometrically (salinometer). It is also possible to measure chlorinity (S) by titration, and establish salinity with the help of the relation:

$$S\ (g\ kg^{-1}) = 1.80655 \times Cl\ (g\ kg^{-1}).$$

Chlorinity measures the concentration of halide ions, expressed in $g\ kg^{-1}$. Chlorosity is the same measure, expressed in $g\ l^{-1}$.

Water with a salinity below 0.5 $g\ kg^{-1}$ is considered fresh; salinities between 0.5 and 30 $g\ kg^{-1}$ indicate brackish water; when salinity is beyond 30 $g\ kg^{-1}$, the water is either seawater, or hypersaline.

3.3. Alkaline cations

(1) Calcium and magnesium

Calcium and magnesium are the most abundant ions in temperate freshwater. However, in African waters, sodium and magnesium are often dominant. According to Weninger (1985), in humid tropical areas, waters are commonly dominated by calcium and bicarbonate ions and are alkaline. In areas where soil erosion is important, owing to heavy seasonal rains, dominant ions are magnesium and bicarbonate. In waters rich in humics, calcium is retained by a strong ion exchange capacity system. In tropical rainforests, ammonium and iron are present in high concentrations.

The ratio of monovalent to divalent ions (M/D), which is equivalent to the ratio of sodium and potassium to calcium and magnesium, has, for a long time, been considered useful for predicting the phytoplankton assemblage in a given water body

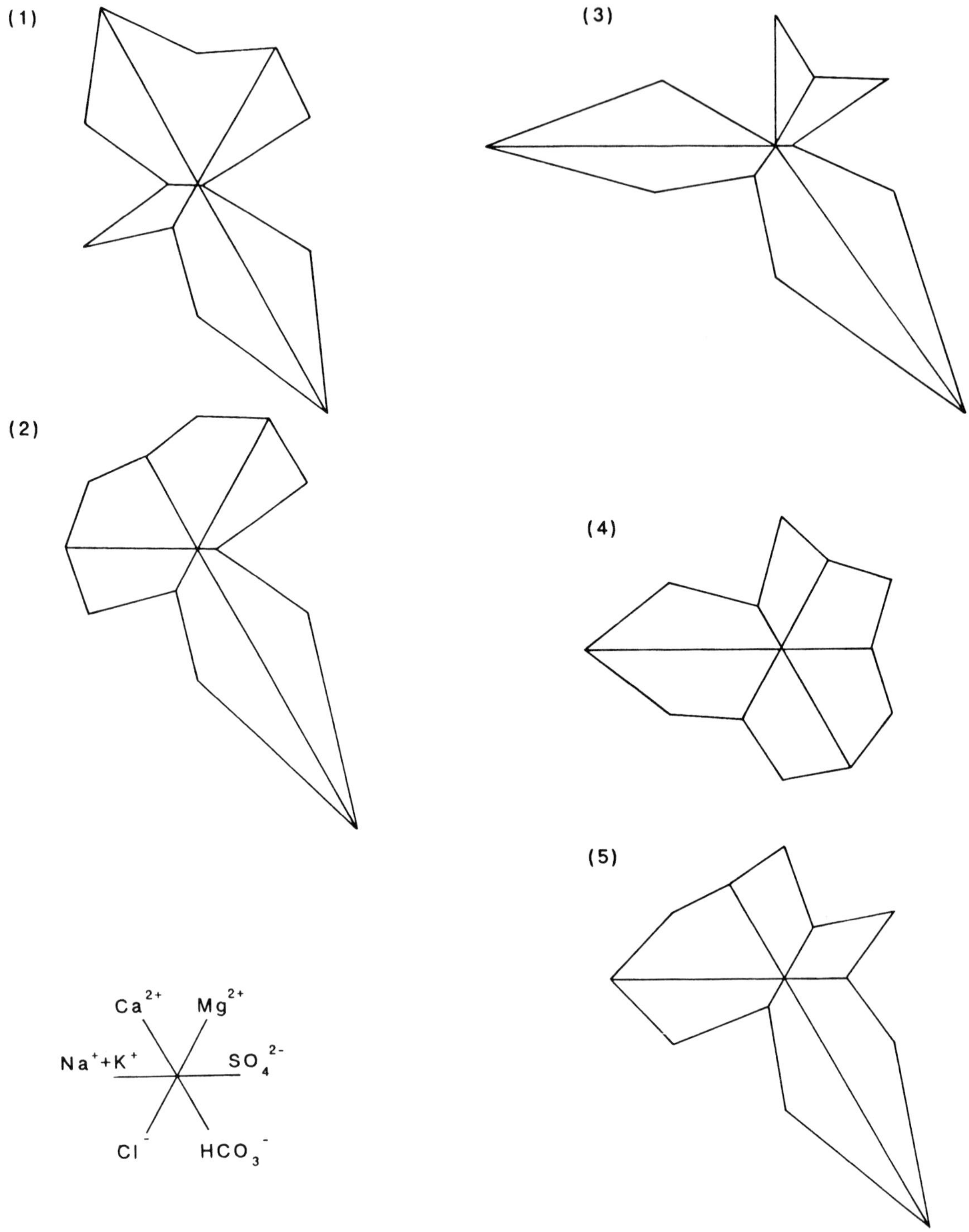

Fig. 20. Ionic diagrams for African waters belonging to the four main types of water, drawn using data of Weninger (1985): (1) Calcium-bicarbonate water, from Madagascar, groundwater (85, Table 2); (2) Magnesium-bicarbonate waters, from Reunion, Cascades (86, Table 3); (3) Sodium-rich waters, from Madagascar, spring pool (88, Table 4,); (4) and (5) Water enriched with humic acids, ammonia and iron, from Kinshasa, lake and S. E. Nigeria, streams (89, Table 5)

(Pearsall, 1924). In a study of 99 Scottish freshwater lochs, Shoesmith & Brook (1983) showed that a M/D ratio below 1 was favorable for diatoms and blue-green algae, and a M/D ratio above 1 suitable for desmids.

This ratio alone cannot explain the dynamics of phytoplankton; ions, such as nitrate, phosphate and silicate, must be considered individually. The importance of the M/D ratio is difficult to disentangle from the usual correlation between increasing M/D ratio values and decreasing amounts of phosphate, nitrate and ammonium salts (Lund, 1965). Correlations between a phytoplankton ratio, using the proportion of the number of species of blue-green algae, Chlorococcales, centric diatoms and euglenids to desmid species numbers, and the concentrations of calcium and magnesium have been established, illustrating the role of these ions in the biology of these species. This ratio also gives information on the trophic state of the water body. Values below 1 indicate oligotrophy, values from 0.0 to 0.3, dystrophy, and values above 1, eutrophic conditions (Shoesmith & Brook, 1983). However, there seems to be little evidence for limitation of production by a direct lack of calcium. Calcium usually occurs in excess of requirements, even in soft waters (Lund, 1965). The importance of calcium is largely in relation to pH and the carbonate-bicarbonate buffer system, and to microbial processes.

The ratio attempts to distinguish between calciphobe and calciphile algae, but there seem to be relatively few planktonic forms, whose distribution is restricted by calcium. In some examples commonly quoted, such as desmids, other factors may be equally important, such as pH or alkalinity (Talling & Talling, 1965). In lakes with different calcium contents, phytoplankton assemblages can be grossly the same. Only a few algae seem to have strict requirements (Lund, 1965).

Calcium and magnesium concentrations are interdependent with those of carbonate, through the carbonate-bicarbonate buffer system and chemical reactions. Biological processes, such as photosynthesis, respiration, and decomposition, which modify the concentrations of carbon dioxide, act on the availability of these cations. Moreover, these cations are involved in many physiological reactions, alone or with phosphates, such as the formation of bones and shells. Calcium is quite abundant in macrophytes and algae, being some 2.5 and 0.5 mg g^{-1} in dry weight, respectively.

(2) Sodium and potassium

Sodium is generally predominant in waters of high salinity, and its concentration increases fairly regularly with conductivity and total ionic concentration (Talling & Talling, 1965).

Potassium is rarely considered important for the ecology of algae (Lund, 1965). This ion is apparently always present in smaller amounts than sodium, the difference tending to increase with salinity.

Importance of conductivity and alkaline ions

1. Conductivity is a measure of the amount of electrolytes dissolved in water, or the total ionic charge of the water.

2. Dominant ions in African lake waters are the cations sodium and magnesium, and the anions carbonate and bicarbonate. In comparison to temperate waters, the predominance of calcium is decreased; this may involve several other profound changes in water quality.

3.4. Alkalinity, pH, hardness and carbon dioxide

3.4.1. pH

Definition of pH and pH regulation

pH is a measure of concentration of hydrogen ions in water and, as such, determines its acid, alkaline or neutral character. Any water is a mixture of acids and bases and is the result of acid-base titration processes. The compounds in solution come from geochemical and biological sources. The most important acid in freshwater is carbonic acid, originating from carbon dioxide dissolving from the air into the water, and from carbonaceous rocks. Respiration and photosynthesis releases and takes up, respectively, carbon dioxide, thus modifying the pH of the water. Some microbiological processes, such as denitrification and sulphate reduction, generate acidity. Finally, the pH of water is strongly influenced by soil pH.

Equilibria exist between the concentrations of carbon dioxide, bicarbonate and carbonate. The reactions between the different ion species are rapid and act as a pH buffer system. There are, however, other buffer systems, though not of the same importance.

pH is an intensity factor, while the acid neutralizing capacity (ANC) and the base neutralizing capacity (BNC) are capacity factors. The buffer system is an implicit function of both intensity and capacity (Stumm & Morgan, 1981).

Optimum pH values

pH is a general measure, its value being determined by a number of processes occurring in ponds. Little is known about the actual way in which pH affects organisms. pH values below 4 and above 11 are lethal for most fish. Generally, pH values ranging between 6.5 and 8 are considered as optimum.

Productivity studies in Canadian acid lakes have shown that the relationship between primary productivity and concentrations of total phosphorus are not modified by pH; biological production stays high if the nutrient supplies remain steady (Schindler *et al.*, 1985: in a Canadian Shield lake after a multiyear experimental acidification; Kerekes *et al.*, 1990: in twenty lakes across Nova Scotia, Canada). Productivity limitation due to acidification can result from inhibition by elevated trace metal concentrations, particularly inorganic monomeric aluminium, and/or increased phosphorus limitation as a consequence of adsorption and/or co-precipitation of phosphate with aluminium (Mulholland *et al.*, 1986).

3.4.2. Alkalinity and hardness

Definition of alkalinity

Alkalinity, or total alkalinity, is defined as the net negative charge of all ions which react with the hydrogen ion. This definition corresponds with that of the acid neutralizing capacity (ANC), which measures the net deficit in protons, by titration of all bases with a strong acid to a preselected reference point (Stumm & Morgan, 1981). It is the sum of all anions, but mainly carbonate, bicarbonate, and hydroxyl ions. The latter are less important:

$$Alk = [HCO_3^-] + [CO_3^{2-}] + [OH^-] + [A^-] - [H^+],$$

where A stands for the proteolytic anions.

Alkalinity is expressed in meq l^{-1} or mg l^{-1} equivalent $CaCO_3$. The factor of conversion is as follows: 1 meq l^{-1} of alkalinity is equal to 50 mg l^{-1} equivalent $CaCO_3$. The concentration of anions determines the buffer capacity of water.

The base neutralizing capacity (BNC) is the equivalent sum of all acids that can be titrated with a strong base to an equivalent point. Conceptually, it is a measure of the concentration of the species containing protons in excess minus the concentration of the species containing protons in deficiency of the proton reference level. It is a measure of the net excess of protons over the reference level of protons. The reference level is defined by the composition of a pure solution, where the proton concentration balances the anion concentration (Stumm & Morgan, 1981).

Processes modifying alkalinity

Equilibria between the different carbonate ions, and

their impacts on alkalinity, have been summarized by Spotte (1979) as follows (Fig. 21):

- when carbon dioxide is added or removed from water, alkalinity does not change, as carbon dioxide does not bear any electrical charge
- when bicarbonate ions are added or removed, alkalinity increases or decreases according to 1 meq l^{-1} per meq l^{-1} of bicarbonate
- addition or subtraction of carbonate entails a change of 2 meq l^{-1} of alkalinity per meq l^{-1} of CO_3^{2-}
- addition of a strong base or acid causes a modification in alkalinity according to the amount of meq l^{-1} added, without, however, changing the total concentration in carbon dioxide.

In his reasoning, Spotte does not take into account equilibria between the amounts of carbon dioxide and carbonic acid in solution. The relations between these equilibria and alkalinity are discussed in Stumm & Morgan (1981).

Various biological and chemical processes directly affect alkalinity. Photosynthesis, denitrification, sulphate reduction, and dissolution of lime or other calcareous rocks increase or produce alkalinity (consume protons), while respiration, nitrification, and sulphide oxidation decrease or consume alkalinity (release protons) (Stumm & Morgan, 1981; Cook *et al.*, 1986). The relationship between alkalinity and pH is illustrated in Fig. 22.

Alkalinity can be exchanged between water and sediments. The loss of alkalinity equivalents in the

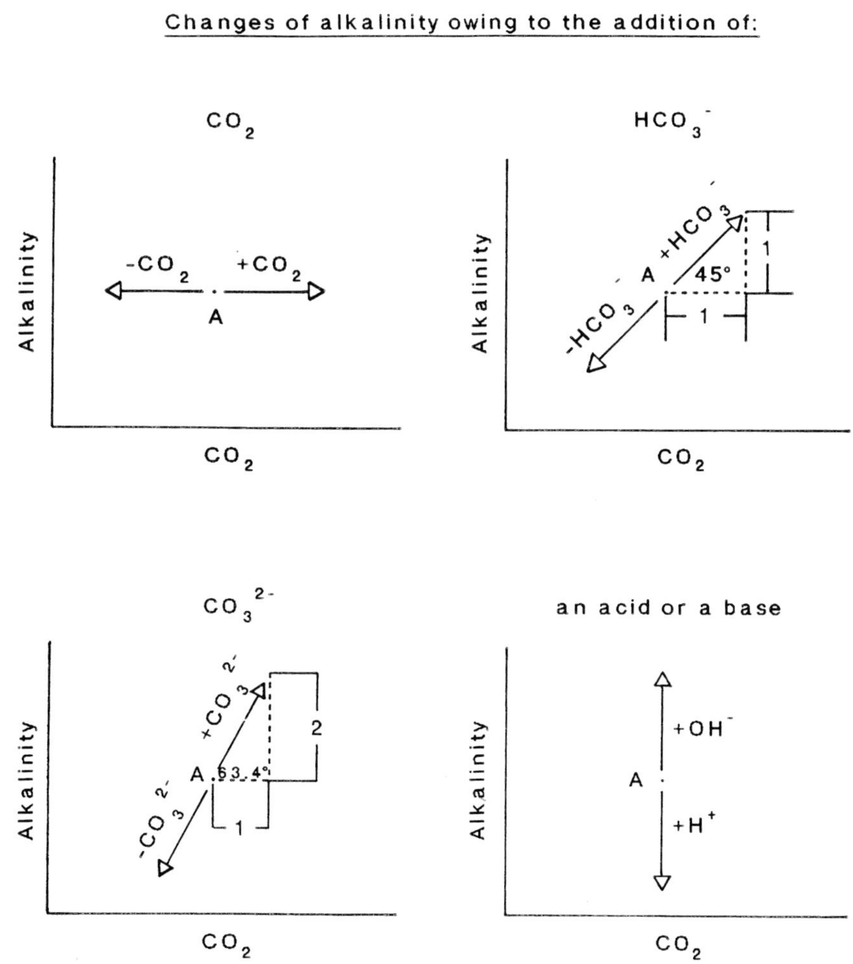

Fig. 21. Impact of different alkaline ions on alkalinity (from Spotte, 1979) (with kind permission of Dr. Spotte & Wiley).

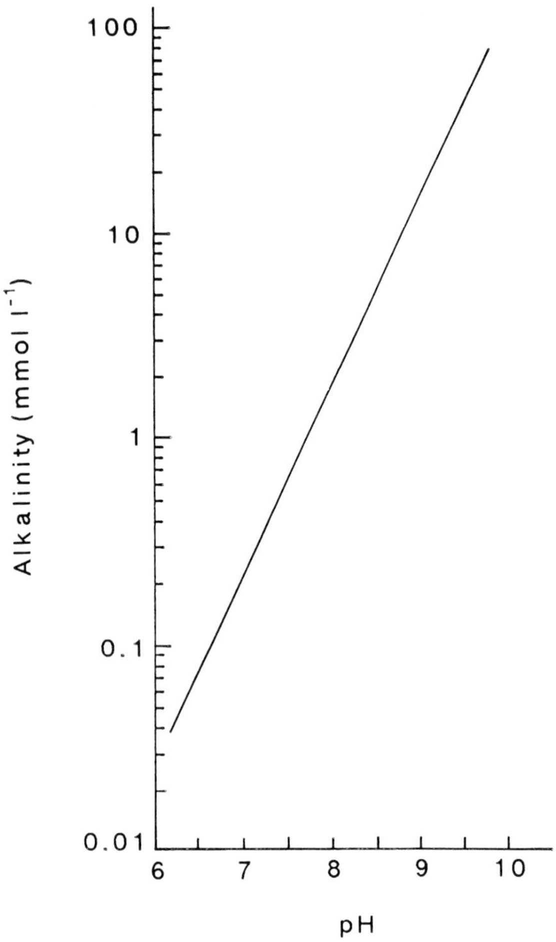

Fig. 22. Relation between alkalinity and pH (after Golterman & Kouwe, 1980, with kind permission of Dr. Golterman on behalf of the authors).

water column can be compensated by supply from the sediments, and vice-versa (see p. 45).

Total alkalinity of natural waters varies between 5 to several hundred mg l^{-1}. Water is considered to be hard when total alkalinity is above 40 mg l^{-1} of $CaCO_3$, while waters with values below this are said to be soft (Boyd, 1979). Hard water is usually more productive than soft water.

Other definitions of alkalinity have been used in the past: carbonate alkalinity, phenolphthalein alkalinity, and carbon dioxide alkalinity.

Definition of hardness

Hardness is a measure of the amount of calcium and magnesium ions associated with carbonate. Total hardness is expressed in mg l^{-1} equivalent $CaCO_3$ and measures the total concentration of calcium and magnesium. As for alkalinity, other definitions have been used in the past, such as temporary hardness and permanent hardness.

3.4.3. Carbon dioxide

Importance of carbon dioxide

Carbon dioxide takes part in the carbon cycling of the ecosystem, forming an input to organic carbon production and other processes related with primary productivity. Moreover, it appears as an output or byproduct of respiration and organic matter decomposition. Dissolution of carbonaceous salts in water is also a source of carbon dioxide (Fig. 23). Decomposition of organic matter in water will be discussed later (p. 115, and also p. 100).

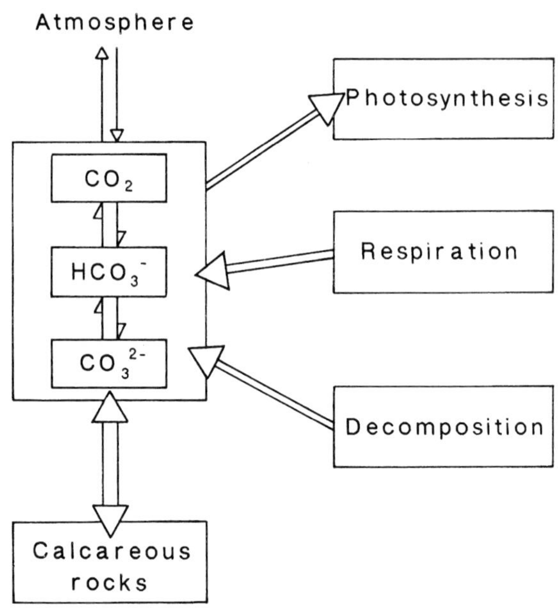

Fig. 23. Relationship between the different carbonate ions within the aquatic environment *sensu lato*.

Carbon dioxide forms ion species which play an important role in controlling pH and restricting its fluctuation range in water. Carbon dioxide dissociates into bicarbonate and carbonate ions; their respective concentrations are maintained in steady equilibrium, so that a change imposed on the concentration of one species is spread and 'absorbed' by the different ions, owing to equilibrium interactions, which buffer changes in concentration.

Carbon dioxide penetrates into water across the soil-water and water-air interfaces. The exchange with the sediments involves carbon dioxide of geochemical (limestone from the soil or added by liming) and biological origin (microbial and benthic activities). The different carbon species in equilibria in water are as follows:

$$CO_{2\,(g)} \rightleftharpoons CO_{2\,(aq)} + H_2O \rightleftharpoons H_2CO_3 \rightleftharpoons H^+ + HCO_3^- \rightleftharpoons 2H^+ + CO_3^{2-}$$

Carbon dioxide is well soluble in water. Its solubility is modified by temperature and salinity; a rise in either decreases solubility (Table 9). The higher the pH, the more the equilibrium shifts to the right.

Most carbon dioxide is continuously recycled in the water column: removal of carbon dioxide during daytime by autotrophs, and release of carbon dioxide by all the organisms at night. Community respiration depends on the biotic load of a pond. If total biomass is high, the decomposer population is significant, with microbial respiration representing a great part of the community respiration. The carbon dioxide which dissolves in water makes a small contribution, (some 0.2 g m^{-2} d^{-1}, less than 5 %) to the dioxide demand of primary productivity in productive ponds (Schroeder, 1987). The carbon dioxide content of the air is about 0.03 %.

Carbon dioxide affects fish respiration. High concentrations of carbon dioxide interfere with the capacity of hemoglobin to fix and transport oxygen and even cause release of oxygen already fixed. This phenomenon is called the Bohr effect, and it slows down the metabolism of fish. Fish tolerate, without too many problems, concentrations of carbon dioxide up to 60 mg l^{-1}, provided that dissolved oxygen concentrations are high. It is recommended that the concentration of carbon dioxide remains below 70 mg l^{-1}; however, in hard waters, concentrations as low as 10 to 15 mg l^{-1} cause nephrocalcinosis.

Table 9. Solubility of carbon dioxide in relation to temperature and salinity (according to Spotte, 1979).

Temperature	Solubility of CO_2 (10^{-4} mol l^{-1} atm^{-1})		
(° C)	Concentrations in chlorides (mg.l^{-1})		
	0	15	20
16	440	385	367
18	413	364	347
20	389	344	328
22	367	325	310
24	347	308	294
36	328	291	278

Carbon dioxide and buffer capacity

The buffer capacity of water is its ability to maintain pH within a narrow range of values, resisting changes in pH upon addition or production of acids or bases. Water is characterized by an acid neutralizing capacity and a base neutralizing capacity

(Stumm & Morgan, 1981). Among the ions in freshwater, carbonate ions play a key role in the buffer capacity of water.

Carbon dioxide is the most important input to photosynthesis. The consumption of carbon dioxide (primary production) in a productive pond in Israel is around 10 g m^{-2} d^{-1} of carbon, of which only about 5 % is supplied from the air (Schroeder, 1987). This means that 95 % of the carbon dioxide originates from autochthonous production, i.e. respiration (about 71 %) and decay processes. Respiration is a continuous process, while photosynthesis is restricted to the daytime. Therefore, the removal of carbon dioxide occurs roughly in half the time needed to supply it. The stocks of carbon dioxide are not limited to the amounts needed for photosynthesis: they are always much larger and, in the above cited example of Schroeder, amounted to three days' gross primary productivity demand. The unused carbon dioxide must be prevented from acidifying the water to such an extent as to cause damage to living organisms. The changes in concentration of such large amounts of carbon dioxide associated with primary productivity and respiration can induce strong fluctuations of pH if the buffer system controlling such variations is weak. This is the case in waters with low or high alkalinities. In waters of high alkalinity and high pH, carbonate is the dominant ion and little carbon dioxide is made available by the buffer system, thus reducing photosynthesis (Boyd, 1972). pH variations are subject to photosynthetic activities during the day and to respiration at night; pH increases during the day and decreases at night to reach a minimum at dawn.

The ionization constants, K_1 and K_2, of carbonic acid H_2CO_3 are, respectively, $10^{-6.35}$ and $10^{-10.33}$. With these values, the concentrations of carbonate ions can be calculated (Fig. 24). Carbonate is only present in measurable amounts when pH is above 8.34, when all carbon dioxide is ionized. The two ionization constants decrease with temperature and salinity.

Carbon dioxide in water reacts with carbonate salts to form bicarbonate:

$$CaCO_{3\,(solid)} + CO_2 + H_2O \rightleftharpoons Ca^{2+} + 2HCO_3^-$$

Changes in the carbon dioxide concentration induce changes in bicarbonate ion concentration. If carbon dioxide is removed from solution, the bicarbonate concentration decreases; this happens when the concentration of carbonate and carbonic acid increases, which implies an increase in pH, as carbonate ions are more alkaline than bicarbonate. In the case of carbon dioxide addition, the reactions are reversed and pH decreases. The combination of the two ionization constants enables the calculation of pH:

$$[H^+] = \sqrt{K_1 \cdot K_2 \frac{[CO_2 tot]}{[CO_3^{2}-]}}$$

pH is directly proportional to the concentration quotient of carbon dioxide to carbonate. It is, therefore, the bicarbonate ions which maintain the balance in the concentrations of carbon dioxide and carbonate and buffer changes in pH.

Carbon dioxide and photosynthesis

Carbon dioxide and light are required for photosynthesis. Their availability may limit the photosynthetic rate. The photosynthetic rate varies at any given light intensity with carbon dioxide concentration, approximately in accordance with the Michaelis-Menten equation:

$$P = \frac{P_m\,[CO_2]}{K_m(CO_2) + [CO_2]}$$

where P is the photosynthetic rate,
P_m is the maximum photosynthetic rate,
K_m is the substrate concentration which gives half the maximum rate.

Observed curves of photosynthetic rate vs. CO_2 concentration are approximately fitting this relation; *in situ* dioxide concentrations may indicate to which extent dioxide availability limits photosynthesis in natural waters. The K_m values for the first intracellular enzymatic reaction of CO_2 (carbon dioxide reduction and combination with ribulose 1,5-diphosphate, in the presence of ribulose carboxylase (Lehninger, 1965)) is about 20 μM in the isolated state. This indicates the maximum rate at which CO_2 can be processed within the cell. Values of K_m determined for freshwater phytoplankton range between 20 to 170 μM (determined at pH 5.5 and 560 μE PAR m^{-2} s^{-1} are for *Pediastrum boryanum*,

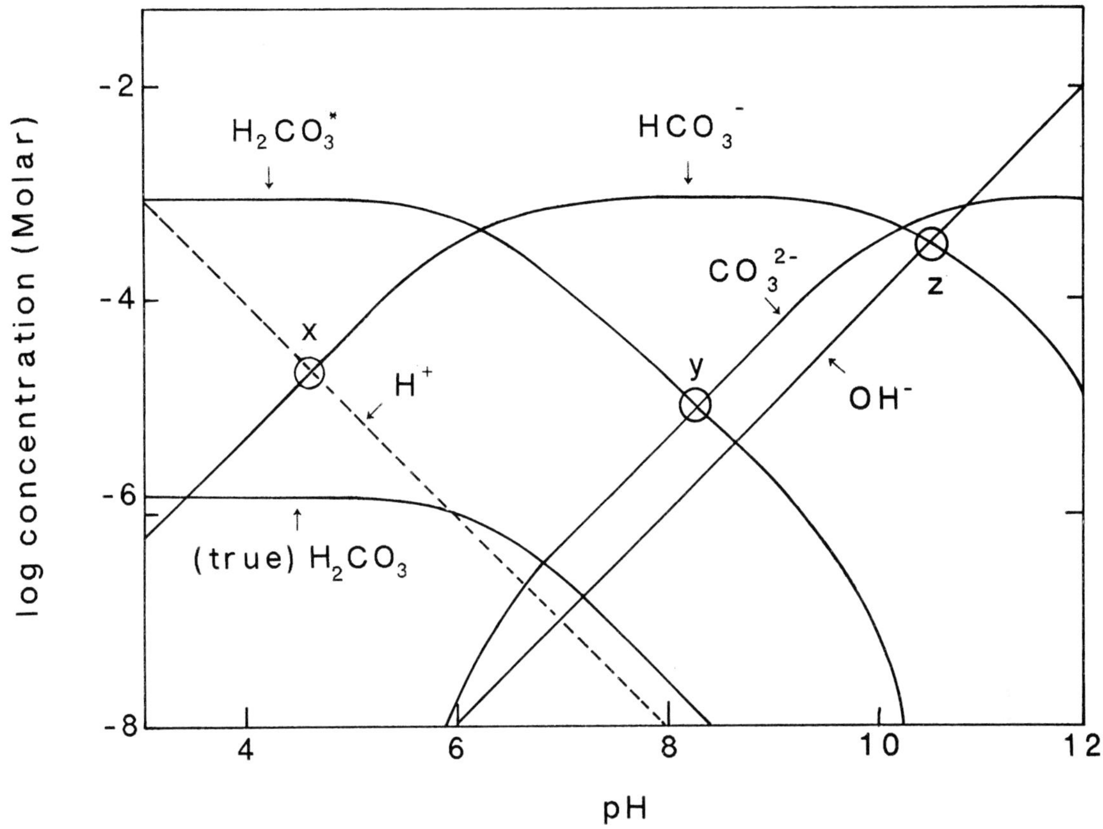

Fig. 24. Distribution of solute species in aqueous fresh water carbonate systems at 25 °C. Logarithmic equilibrium diagram for fresh water. $[H_2CO_3^*] = [H_2CO_3] + [CO_{2\,aq}]$, but here equivalent to $CO_{2\,aq}$, as $CO_{2\,aq}$ is much more abundant than H_2CO_3. H_2CO_3 is a much stronger acid than $CO_{2\,aq}$. The equivalence points marked x, y, z, correspond to pure solutions (molar concentrations) of $H_2CO_3^*$, $NaHCO_3$ and Na_2CO_3 respectively (from Stumm & Morgan (1981), with kind permission of Prof. Stumm & Wiley).

(Chlorococcales), 40 μM; for *Cosmarium botrytis*, a desmid, 170μM; for *Anabaena cylindrica*, a blue-green alga, 60μM. A full noon summer sunlight is about 2000 μE PAR m^{-2} s^{-1}) (Kirk, 1986).

Fresh water in equilibrium with the atmosphere at 15° C contains dissolved dioxide at about 14 μM. It could therefore be expected that aquatic plant photosynthesis in natural waters is undersaturated with respect to dioxide concentration, and should respond to an increase in dioxide concentration with an increase in photosynthetic rate. This does not only apply at saturating light intensity, but even at low light intensities. Most inland water contain more inorganic carbon in the form of bicarbonate ion, HCO_3^-, than dioxide, and increases with pH (Kirk, 1986).

It would clearly be an avantage for aquatic plants to use bicarbonate as a carbon source for photosynthesis. In fact, many do, but not all. (The interconversion reaction is catalysed by a chloroplast enzyme, the carbonic anhydrase; OH^- accumulates and is excreted to counterbalance the uptake of bicarbonate). Most higher plants can utilize bicarbonate, but amongst freshwater phytoplankton this ability varies widely. A possible reason why bicarbonate is less effectively used is that energy must be used for its active transport into the plant, whereas dioxide diffuses in freely. Carbon is not obtained from bicarbonate, except at high pH, above 9. The ability of blue-green algae such as *Anabaena*, *Microcystis*, *Spirulina* to continue photosynthesizing in water of high pH, and effectively zero free carbon dioxide could be one of the main factors contributing to their frequent domination of eutrophic lakes in late summer, and their dominance of highly alkaline waters such as the African soda lakes (Kirk, 1986).

The 14 μM concentration applies only to waters in equilibrium with the atmosphere. But any productive water is likely not to be in equilibrium so far as carbon dioxide is concerned. Carbon dioxide concentration varies with time and depth in accordance with processes of consumption and production.

Photosynthesis decreases the concentration of free dioxide, by increasing pH. Sinking rate of diatoms increases when carbon dioxide becomes limiting (Kirk, 1986).

Influence of alkaline ions on buffer capacity

In waters with a low proportion of calcium, sodium and potassium ions, i.e. soft waters, production of carbon dioxide leads to the formation of sodium and potassium carbonate, which are much more soluble than calcium carbonate, and the increase of bicarbonate ions entails a decrease in pH. In the presence of calcium ions, the precipitation of $CaCO_3$ buffers the pH. Waters with low calcium and high sodium concentrations are much more frequent in Africa (see p. 66).

The importance of bicarbonate ion concentration in the buffer system is expressed by the Henderson-Hasselbach equation:

$$pH = p_{K1} + \log \frac{[HCO_3^-]}{[CO_{2\,tot}]}$$

p_{K1} is the negative logarithm of the value of the ionization constant, $-\log 10^{-6.35} = 6.35$.

The effects of the ionization of carbon dioxide to form carbonate species are modified by other ions, such as sodium, magnesium, and sulphate (Table 10 & Fig. 25). Cations are able to bond with carbonate and sulphate ions, but the bicarbonate ions are always the most important.

Table 10. Reactions which take part in the bicarbonate-carbonate buffer system (according to Spotte, 1979).

Type of reaction								p_K
Ca^{2+}	+	H_2CO_3	$\rightleftharpoons$	$CaHCO_3^+$	+	H^+	(1)	5.450
Mg^{2+}	+	H_2CO_3	$\rightleftharpoons$	$MgHCO_3^+$	+	H^+	(2)	5.502
H_2CO_3			$\rightleftharpoons$	HCO_3^-	+	H^+	(3)	6.124
Na^+	+	HCO_3^+	$\rightleftharpoons$	$NaHCO_3$	+	H^+	(4)	6.745
Mg^{2+}	+	HCO_3^+	$\rightleftharpoons$	$MgCO_3$	+	H^+	(5)	7.665
Ca^{2+}	+	HCO_3^+	$\rightleftharpoons$	$CaCO_3$	+	H^+	(6)	7.913
Na^+	+	HCO_3^+	$\rightleftharpoons$	HCO_3^-	+	H^+	(7)	9.186
HCO_3^+			$\rightleftharpoons$	CO_3^{2-}	+	H^+	(8)	9.776
H_2O			$\rightleftharpoons$	OH^-	+	H^+	(9)	13.791

In the case of a dying algal bloom, the bicarbonate-carbonate buffer system reacts in a surprising way: the production of ammonium and carbon dioxide, originating from decomposition of the algae, in water with a weak buffer system, leads to a drastic drop in pH. In alkaline waters, with a strong buffer system, a large fraction of the ammonium ions forms gaseous ammonia, which is toxic and can cause fish kills (Krom *et al.*, 1989b) (see also p. 148).

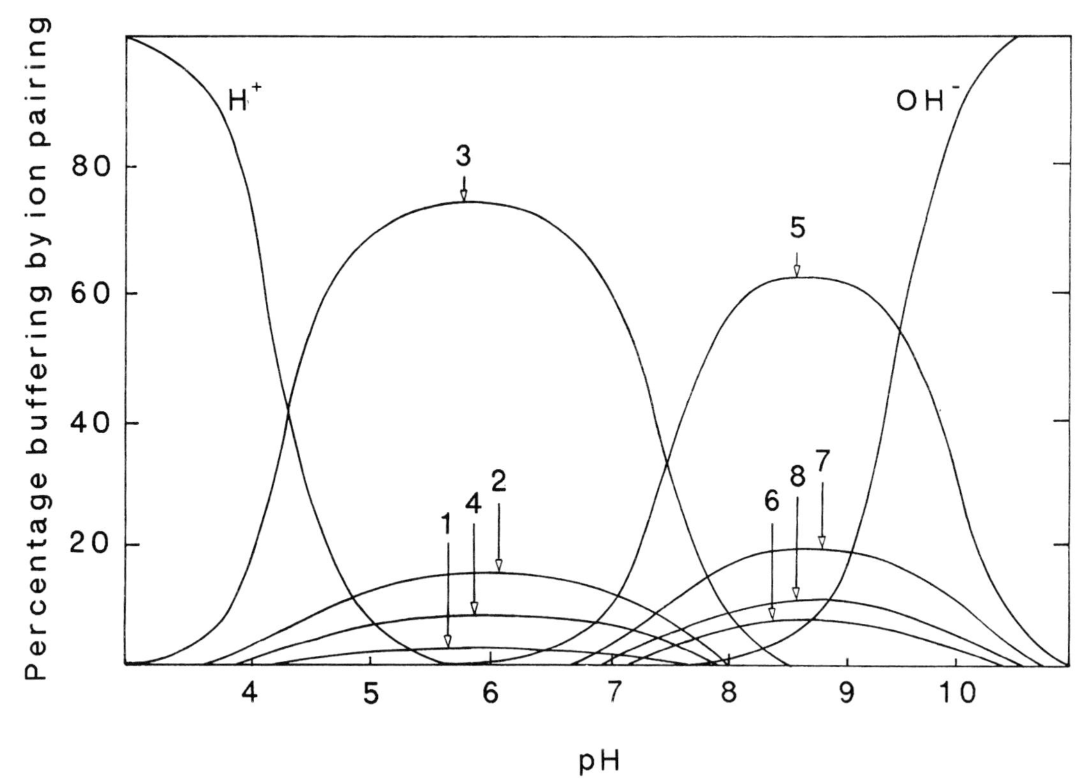

Fig. 25. Contribution of the different ions in solution in water to the buffering capacity of water, expressed as a percentage of buffering by ion pairing, according to the pH (from Spotte, 1979) (with kind permission of Dr. Spotte & Wiley).

Importance of alkalinity, hardness, pH and carbon dioxide

1. pH is a measure of the acidity, alkalinity or neutrality of water. Alkalinity measures the concentration of all species capable of reacting with hydrogen ions, especially carbonate and bicarbonate ions. Hardness is a measure of the amount of calcium and magnesium ions in water.

2. Carbon dioxide is essential for life in a pond as it provides the inorganic carbon needed for photosynthesis. Carbon dioxide dissociates in water to form carbonate and bicarbonate, and participates in the carbonate-bicarbonate buffer system, which regulates pH.

3. Most of the carbon in the water is continuously recycled but some exchanges with air and with sediments exist as well.

3.5. Liming

3.5.1. pH control of water by liming

pH variations in water result from the interaction of various abiotic and biotic processes, as discussed above. Biological processes are responsible for the most rapid modifications, as well as for cyclic variations. In ponds with high plankton biomass and productivity, pH rises slowly from dawn to early in the afternoon. These are the two most important values for a fish farmer to record, as they represent the daily minimum and maximum. If the range between these two values is wide, buffer capacity is weak. If the values are close to each other, buffer capacity is strong. Most fish do not tolerate large pH changes. pH values between 6.5 to 8 are considered optimum.

In areas most commonly available for fish farming, such as marshlands and swampy areas, waters are generally acidic. Natural waters with alkaline pH are less frequently found.

The most common aquacultural practice to control pH is liming, which consists of spreading lime on the pond soil or onto the water.

3.5.2. Effects of liming

Liming is a management practice, aimed at improving the abiotic and biotic conditions of the environment. The effects of lime on soil and water are as follows (Huet & Timmermans, 1970):

a) increase of sediment pH
Lime neutralizes the acidity in soil and water, by adding alkalinity. Proton and hydroxyl ion concentrations decrease and increase respectively, as $CaCO_3$ reacts with protons to form bicarbonate and calcium ions. Laboratory studies of sediment taken at 10-11 m depth in lake Hovvatn, in southern Norway showed effects of lime up to a depth of 6 cm (Smayda, 1990). Thus, modification of soil pH affects water pH. The reaction between lime and acid water is first order and is controlled by the concentration of protons on the surface of the lime particles, and not by mass transport of protons to these particles (Compton *et al.*, 1989).

The calcium ions of lime react with and neutralize the free acid ions of actual acidity, and acid ions sorbed (potential acidity). The decrease in acidity also reduces the solubility of toxic iron (II), aluminium and manganese (II) ions, which precipitate.

The neutralizing action of lime is more important in soil than in water. Lime, even when used in fine particles, eventually settles down to the pond sediment. Here, the concentration of particles is denser, carbon dioxide is more abundant, the actual and potential acidity is higher, and reaction kinetics are more rapid.

b) decrease in redox potential of sediments
The redox potential and pH control the transfer of ions across the water-sediment interface. In a fish pond, sediment is usually oxidized on the surface and reduced underneath. Liming significantly decreases the redox potential, E_7, of the sediment to a depth of 5 to 6 cm, with recorded changes in the order of 100 to - 66 mV. Oxidized conditions persisted at the surface in a Norwegian lake, but, below this, conditions were highly reductant (Smayda, 1990). This pronounced reduction is perhaps due to a significant mineralization within the sediment, which makes use of all electron acceptors available to oxidize organic matter.

c) increase of the base unsaturation of soil, hardness and alkalinity of water, and buffer capacities
Lime reacts with the free carbon dioxide present in water, thus competing with phytoplankton. As a result, the photosynthetic rate decreases (Fig. 26). This transformed carbon dioxide is, however, not lost from the environment; it increases the alkalinity and the buffer capacity of the water.

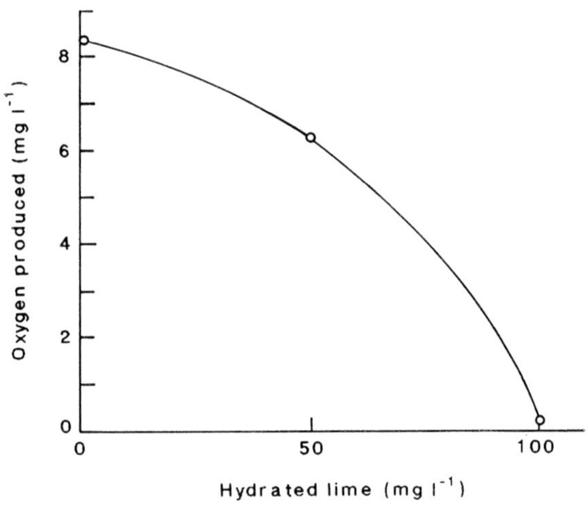

Fig. 26. Effect of different doses of hydrated lime on oxygen production (by-product of photosynthesis) in fish ponds (average value of three ponds), owing to the fixation of carbon dioxide by hydrated lime.

The calcium and magnesium ions from lime are adsorbed onto colloids, thereby increasing the concentration of alkaline ions and the base reserve and base saturation of the sediment. This addition of calcium and magnesium increases the water hardness.

d) acceleration of the decomposition rate of organic matter
Decomposition rates increase with liming. Turnover times of free glucose in surface sediments become

shorter, indicating great, persisting increases in microbial activity after neutralization. Easily degradable organic compounds, which accumulate in sediments under acidic conditions, are rapidly decomposed by invertebrates and microorganisms (Hasselrot & Hultberg, 1984).

Lime increases the decomposition rate of organic matter (Fig. 27), especially cellulosic material in sediment, and is responsible for the crumb structure of dried pond mud. This decomposition reduces the risk of propagation of bacterial and parasitic diseases, and decreases the sediment oxygen demand. Acid products resulting from fermenting processes of decomposition are neutralized by lime, thereby dissolving carbonate salts and allowing the mobilization of calcium, magnesium and associated phosphates (Marsden, 1989). This nutrient mobilization stimulates bacterial activities.

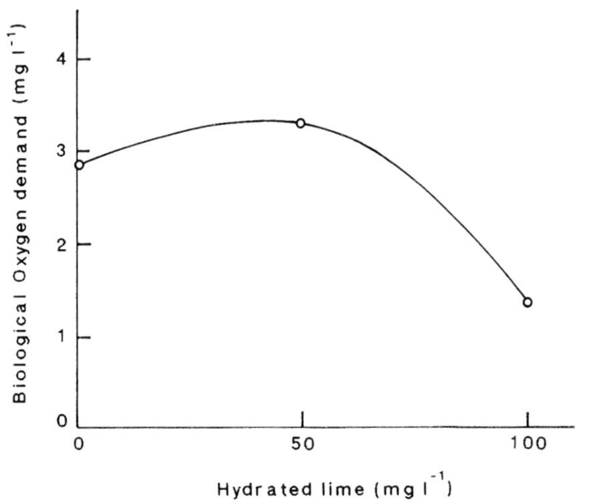

Fig. 27. Effect of different doses of hydrated lime on the biological oxygen demand in fish ponds (average value of four ponds).

Decomposition of organic matter in water is also stimulated by liming, probably as a result of flocculation of dissolved organic matter by calcium ions. Further, dissolved organic matter is adsorbed onto calcium carbonate and precipitates (Saunders, 1980). In three small lakes in Michigan, the precipitation of dissolved organic matter after liming considerably increased light penetration, resulting in the stimulation of green photosynthetic sulphur bacteria in deep, anoxic waters; bacterial blooms resulted. The stimulation of organic matter decomposition had the additional effect of reducing hypolimnetic oxygen levels, with the consequence of hydrogen sulphide entering the photic zone (Leavitt *et al.*, 1989). Though the presence of sulphur bacteria is unlikely in fish ponds, better penetration of light in ponds where, for example, liquid manure is used, can be achieved with lime.

e) precipitation of suspended organic matter

In waters with high concentrations of humic matter, lime reacts with it to form stable aluminium-humate complexes at the surface of the limestone, thus inactivating the lime (Hasselrot & Hultberg, 1984). But, in low concentrations, humic matter, which gives a brownish colour to water and affects light penetration and productivity, precipitates. Precipitation of calcite is prevented by the presence of organic matter. Polyphenols, derivatives of phosphonic acid, polyphosphate, fulvic acids, magnesium and inorganic phosphate, at high concentrations, also hinder the precipitation of calcite, while, at low concentrations, they participate in adsorption and coprecipitation (Stabel, 1986).

Calcite coprecipitates with some algae, by surface nucleation occurring between solutes and cell walls, or by physiological reactions. A possible explanation of this phenomenon is as follows: the photosynthetic assimilation of bicarbonate and release of hydroxyl ions increase the pH of the microenvironment of the cell, so that supersaturation concentrations of calcite are reached, which lead to crystal formation on the cell wall and to precipitation. Algae can also act as surface catalysts for precipitation (Stabel, 1986).

f) acceleration of nitrification processes

Nitrifying bacteria react strongly to hydrogen ion concentration. Although a wide range of pH optima have been reported, there is almost general agreement that, as pH moves into the acid range, nitrification rates decline (Kaiser & Wheaton, 1983). In the light of present information, pH should be above 6.5. Thus, the alkalinizing effect of liming stimulates nitrification.

g) detoxifying and sterilizing effect

Liming promotes aerobic decomposition, limits fermentation and, thus, prevents accumulation of intermediary toxic products. Quick lime and calcium cyanamide kill parasites and pathogens. Spread on soil, they also eradicate aquatic macrophytes with superficial roots.

h) increase of the availability of phosphate and molybdate ions

In water or soil of acid pH, the acid ions aluminium and iron, react with phosphate to form precipitates; in alkaline conditions, phosphate combines with calcium to form apatite. The highest availabilities of phosphate, as well as molybdate, are at intermediate pH. Lime in excess has a deleterious effect, not only on the availability of phosphate, but also on that of iron, manganese, copper, zinc and boron.

i) increase in planktonic and benthic species

In acid waters, most studies revealed a reduced species richness, biomass, and productivity (Almer *et al.*, 1974; Hendrey & Wright, 1976; Kwiatkowski & Roff, 1976; Yan, 1979; Bleiwas *et al.*, 1984), though others observed a shift in phytoplankton groups composing the community, without a decline in biomass or productivity (Schindler, 1985). Liming induces a shift in community structure, with the appearance of species characterized by faster growth and greater nutrient requirements than most acid tolerant species, suggesting a subsequent increase in primary productivity. With regard to the zooplankton communities of acid waters, cladocerans are lacking, while rotifers and copepods are reduced in numbers. Liming of Swedish lakes resulted in increased rotifer density and species number, due likely to increased availability of food for filtering species (Hasselrot & Hultberg, 1984). Changes in the benthic fauna also occurred.

3.5.3. Types of lime used and their equivalences

Different types of lime are on the market and it is useful to know the differences between the products and to establish equivalences (Brady, 1984):

- agricultural lime, crushed limestone, marl, and oyster shells are all calcium carbonate compounds, mixed with varying amounts of magnesium carbonate. They consist, on average, of 75 to 95 % $CaCO_3$, and are almost insoluble in water, dissolving slowly under the influence of carbon dioxide. Dissolution depends on the size of the particles; fine particles of less than 1 mm in size are most efficient and dissolution extends over 1 to 2 months. Amongst the above-cited products, agricultural lime is the most easy to use.
- Quick-lime is actually calcium oxide, CaO, obtained by the burning of lime. It also contains varying amount of magnesium oxide. The action of quick lime is fast, caustic and toxic and, when the base saturation is low, the induced pH changes are great. It is used to kill off parasites and disinfect ponds, tanks, etc. It contains about 85 to 95 % CaO.
- Pulverized, hydrated or slaked lime, $Ca(OH)_2$, is obtained by mixing quick lime with water. The product is caustic and toxic for fish. Slaked lime hardens in air, due to the presence of other oxides as impurities. It contains about 95 to 96 % $Ca(OH)_2$.
- Calcium cyanamide, $Ca(CN)_2$, is toxic, with a residual effect lasting 2 to 3 weeks and is in powdered or pelletized form. The cyanide radical contains nitrogen, which is released after chemical transformation and acts as a nitrogen fertilizer.

Lime is assessed on its chemical composition and the equivalences of the different products are expressed according to three different, but similar, criteria (Brady, 1984):

- the calcium oxide equivalence or the conventional amount of oxide. The different concentrations of calcium and magnesium compounds are converted to their calcium oxide equivalents. Specific conversion factors are used for each compound and the results of the conversions are added to give the total equivalent amount in calcium oxide. The conversion factors are established using the molecular weights of the calcium or magnesium compounds. For example, the conversion factors of calcium and magnesium carbonate are 0.65 and 0.56, respectively
- the neutralization power or calcium carbonate equivalence. The principle is the same as for the calcium oxide equivalence, the calcium and magnesium compounds present (the chemical composition is usually listed by the manufacturer on the bag) being converted into calcium carbonate equivalents. Factors of conversion are established by using the molecular weight of the compounds and are, for example, 1.786 and 1.186, respectively, for calcium oxide and magnesium carbonate
- the percentage of calcium and magnesium, expressed as a percentage in weight of each of these elements.

The calculation of chemical equivalences should not mask the fact that differences between different lime products exist. Each product has its specific reaction and none are totally equivalent.

3.5.4. Liming methods

If the pond has been drained, lime is cast on the bottom and dykes, while, after impoundment, it is sprinkled on the water surface or fed into the pond by the feeder canal. Each of these practices has specific aspects. Liming of soil before impoundment accelerates decomposition processes in the sediment, kills off parasites and eradicates weeds. To obtain this effect, it is better to use quick lime or calcium cyanamide. Huet & Timmermans (1970) recommended a dose of 1000 to 1500 kg ha^{-1} for quick lime or 1000 kg ha^{-1} for calcium cyanamide. The total dose is divided into several applications separated by 8 to 15 days, preferably during periods without rain. The impoundment and stocking of the ponds should be given a supplementary 10 to 15 days lead time. In liming pond water, care should be taken to prevent the pH from becoming too alkaline, especially from reaching values beyond 9.5. Liming of the water fed into the pond is most efficiently carried out by the rotary drum system (Zurbuch, 1984).

When handling lime, protective clothing, especially gloves and goggles, should be worn, and it is recommended to rub some protective ointment on exposed skin.

3.5.5. Identification of the need for lime

The procedure for establishing the need for lime in a fish pond has been extensively described by Boyd (1979). In summary, the hardness of water, which gives a reflection of soil pH, and the base unsaturation of the soil, are measured. Base unsaturation is estimated by adding a soil sample to a buffer solution at pH 8, and measuring the resulting decrease in pH. If the hardness of the water is above 20 mg l^{-1}, no liming is necessary. Also, if hardness is low, but alkalinity high, liming would not be useful. Lime requirements are read from a chart (Table 11), based on initial sediment pH and the decrease in pH of a buffer solution, caused by the soil sample (Boyd, 1976c).

Table 11. Lime requirements, expressed in kg ha^{-1} of calcium carbonate (neutralization power = 100), to increase the total hardness or the alkalinity of the pond water to 20 mg l^{-1} or more (Boyd, 1976c)

Initial pH of the soil	Amount of calcium carbonate (in kg ha^{-1}) needed in connection with the pH decrease of the buffer solution									
	7.9	7.8	7.7	7.6	7.5	7.4	7.3	7.2	7.1	7.0
5.7	91	182	272	363	454	544	635	726	817	908
5.6	126	252	378	504	630	756	882	1008	1134	1260
5.5	202	404	604	806	1008	1210	1411	1612	1814	2016
5.4	290	580	869	1160	1449	1738	2029	2318	2608	2898
5.3	340	680	1021	1360	1701	2042	2381	2722	3062	3402
5.2	391	782	1172	1562	1953	2344	2734	3124	3515	3906
5.1	441	882	1323	1765	2205	2646	3087	3528	3969	4410
5.0	504	1008	1512	2016	2520	3024	3528	4032	4536	5040
4.9	656	1310	1966	2620	3276	3932	4586	5280	5980	6552
4.8	672	1344	2016	2688	3360	4032	4704	5390	6048	6720
4.7	706	1412	2116	2822	3528	4234	4904	5644	6350	7056

The number of samples to be taken to determine lime requirements depends on pond size. Soil properties vary according to its depth. The sediments in the deeper portion of the pond, having higher CEC, are usually more anaerobic, have a more acid pH, and, therefore, require more lime. According to Boyd (1976b), the number of samples should be 12 in ponds of less than 1 hectare, and 25 in ponds between 1 to 4 hectares. The effect of liming lasts for a period of about 10 impoundments and drainages (Boyd & Cuenco, 1980). After each draining of the pond, a dose one fourth to one fifth of the initial dose should be supplemented.

3.5.6. Other methods of neutralization

Neutralization of sediments with sodium bicarbonate

Sodium bicarbonate (soda) has been injected into sediments enriched with organic matter, to saturate them with sodium. The impact of these sodium ions on the neutralization of sediments was compared to that of calcium, used in liming. Sodium ions were exchanged with hydrogen ions, involving consumption of alkalinity, as the protons reacted with bicarbonate to form carbonic acid. Phosphorus was released from the sediments, stimulating algal growth. This gave rise to the idea of using soda, by adding solutions of alkaline sodium, to control adsorption and exchange processes in acid sediments and create a long-lasting neutralization capacity. Sediments act as reversible cation exchangers. Lime is inactivated by humic matter and calcium-humate complexes have a restricted reversibility. This signifies that some of the calcium involved in these complexes is lost to further exchanges; this is not the case with sodium (Lindmark, 1982). In some other experiments, in two small lakes in eastern Norway, soda was deposited onto lake sediments, but here results were unsatisfactory. Because of low levels of water turbulence and mixing, the soda and neutralized water were trapped in a stable layer, formed by differences in density (Rosseland & Skogheim, 1984). This subject is clearly in need of further studies.

Use of nitrogen fertilizers

In waters with high pH and low hardness, one solution to reduce the pH is the use of ammonium fertilizers. Ammonium fertilizers set protons free during the process of nitrification:

$$NH_4^+ + ½\, O_2 \longrightarrow NO_2^- + 2H^+ + H_2O \quad (1)$$

$$NO_2^- + ½\, O_2 \longrightarrow NO_3^- \quad (2)$$

The drop in pH has, however, no extended effect. The presence of nitrate induces plankton blooms, which bring about wide pH fluctuations. Moreover, all ammonium ions are not nitrified and those which subside are transformed into ammonia. This ammonia, at high pH values, forms toxic gaseous ammonia.

Other products

Alum, an aluminium sulphate, has the immediate effect of diminishing pH (1 mg of alum decreases the alkalinity by 0.5 mg l^{-1} and produces 0.44 mg l^{-1} of carbon dioxide), but without alleviating the causes for the high pH. It is therefore only a temporary measure against critically high pH.

Gypsum, $CaSO_4.2H_2O$, is a substance which contributes calcium ions to water and increases hardness in a proportion of 2.15 mg l^{-1} per mg l^{-1} of total hardness. It is, however, not comparable to a liming product.

Importance of liming

1. Liming is a common practice in aquaculture, aimed at keeping soil pH in an optimum range. The beneficial effects of liming include a long-lasting increase of water and sediment pH, an increase of organic matter decomposition rates in the sediments, precipitation of dissolved organic matter in the water column, and a sterilizing effect on the sediment.

2. A few products are available for liming, but agricultural lime is the most suitable for routine use.

3. Other non-calcareous products have also been tested, some hopes arising from the use of soda, sodium bicarbonate, but a proper methodology has yet to be outlined.

4. Nutrients in water

Relation between nutrients, water fertility and fertilizers

Nutrients represent the natural fertility of the water, on which primary productivity and, ultimately, fish production depend. This natural production is achieved with minimum operational costs, as no feed or fertilizer are added to the pond. In many cases, it is financially rewarding to take maximum advantage of natural productivity, especially plankton production. The increase of production beyond this base level requires knowledge of nutrient requirements and their cycles. Unfortunately, there is as yet no standard procedure for the use of fertilizers, and recommendations found in the aquacultural literature are empirical.

Nutrients are supplied to ponds in the form of inorganic or organic fertilizers. Inorganic fertilizers are characterized by their composition of primary nutrients, i.e. nitrogen, phosphorus and potassium (N:P:K), in mineral form and in finite amounts. Organic fertilizers are either animal excrements or products of organic decomposition, such as compost. Compared to inorganic fertilizers, the composition of organic fertilizers is never well defined, the respective amounts of nitrogen and phosphorus varying according to the type of material used to feed the animal, in the case of excrement, or used to make the compost, and to the time and conditions of stocking. Organic fertilizers contain, in addition, decaying organic material. This material is essentially an energy reservoir and a 'mine' of organic carbon. Potassium is seldom considered, as it is usually accepted that the needs for this element are satisfied by the amounts naturally contained in water. Therefore, potassium fertilizer is usually not added; it is the N:P proportion that matters.

The amount of autotrophic producers is a function of the concentrations of the various nutrients. This relationship has often been considered as the sole determinant of the dynamics of aquatic systems and, on that basis, mathematical models have been constructed. However, persisting presence of algae in great numbers, despite conditions of low phosphorus and nitrogen concentrations, were observed in White Lough, a hard water lake, Northern Irland, and implied that biological factors alter, in a significant way, the relationship between algae and phosphorus (Foy & Fitzsimons, 1987). Actual concentrations of nutrients have sometimes no direct relationship with the physiological state of the autotrophs, while past concentrations do. An improved knowledge of the ecology of algal species and their interactions is also needed to better understand the factors involved in production.

Recycling and regeneration of nutrients

In a steady dynamic system, inputs or supplies of nutrients have to be balanced by outputs. If the supplies of nutrients from outside are less than the rate of utilization, nutrients have to be regenerated or mobilized from the pool of non-available nutrients or storage compartment, by recycling. The importance of the recycling process depends not only on the deficit of nutrients or the proportion of available to non-available forms, but also on production and utilization rates of regenerated nutrients, described in terms of instantaneous changes in available forms (Howard-Williams, 1985).

The regeneration rate is the rate at which a nutrient re-enters the system and is re-utilized; this rate decreases as the proportional amount of the available form increases and vice-versa. Each nutrient has a maximum loading capacity for its available forms and if the nutrient concentration is close to the maximum load, recycling and regeneration rates are low. In running waters, the continuous unidirectional water flux carries away nutrients and their utilization is only possible during a short time. This gave rise to the concept of nutrient spiralling, defined as the distance a nutrient will travel before being taken up; this will depend on the upstream supply rate and the utilization or exchange rate (Howard-Williams, 1985). Similarly, in a production system, there is exportation of nutrients in the form of fish at harvest; the supply flux of nutrients must be regulated with regards to utilization rates.

4.1. Rules for the use of fertilizers

Some general rules for the assessment of fertilizer needs should be followed based on their modes of action.

The limiting factor(s) of natural productivity should first be determined. If they are physical factors, such as light or temperature, which are difficult to modify drastically, there is not much one can do. However, other physical and chemical factors can, by proper management, be adjusted to suit production needs.

Empirical determination of fertilizer needs, by comparative trials, requires a careful choice of ponds. Firstly, the experimental ponds have to be of intermediate size, not less than one fifth of the area of the target ponds. They should have been previously used for fish farming for at least three years, have the same fertilization history, and their natural productivity should be known. The stocking rate of fish should be the same as in the target fish ponds. The feeding or non-feeding of the fish should be considered. If the actual contribution of fertilization to production is to be assessed, no feeding should be carried out. But if feeding and fertilization are supplementary, a comprehensive sequence of the conditions should be tested: no fertilizer and no feed; fertilizers only; fertilizers and feed. Concerning general pond conditions, the pond should be weeded, and pH of the water should be neutral or within the optimum pH range and the hardness and alkalinity within the normal range (above 20 mg l^{-1}). If liming is required, fertilization should not take place prior to a period of 8 to 15 days after liming. Lime precipitates with phosphate and therefore liming and fertilization should be applied separately.

Before discussing the use of inorganic fertilizers in fish farming in Africa, it is important to mention that intensive fish farming never depends exclusively on inorganic fertilizers; their use is supplementary to artificial feeding. Inorganic fertilizers, because of their fixed chemical composition, are a useful experimental tool to determine the nitrogen or phosphorus needs in specific ponds, and serve especially in the first steps of the empirical determination of fertilizer needs.

Fertilizers stimulate primary productivity and therefore some questions need to be asked beforehand, such as how much plankton is required, what role this plankton should play, and what proportion of natural feed to artificial feed is desirable. The production of natural feed is also justified by the fact that most artificial feeds in common use (in Africa) are not nutritionally balanced, so that natural feed can provide missing nutrients, such as essential amino acids, vitamins and minerals.

The cost of inorganic fertilizers argues strongly against their exclusive use, without any other feed or manure, in intensive culture. Any use of inorganic fertilizer should therefore take production costs into account. This criterium alone will indicate whether use of inorganic fertilizer makes sense.

Organic fertilizers should be used in small amounts, and preferably at short intervals. The pond has no way to store, in the long term, the nutrients (at least nitrogen) supplied; a consequence of this is a rapid decline of plankton biomasses, as soon as fertilization stops (Shearer *et al.*, 1987). In ponds with sediments of low exchange capacities, it is especially important to apply fertilizers frequently, but in small amounts.

Importance of nutrients for the production of natural feed

1. Nutrients are the primary inputs for autotrophic production, along with light. Knowledge of the nutrient needs is essential for achieving high primary production and, eventually, high fish production.

2. No standard procedure for the determination of fertilizer needs has yet been established, partly because nutrient cycling in ponds is not well understood. Mathematical modelling would require quantitative knowledge of the various processes involved.

3. Inorganic and organic fertilizers are available for use in ponds. Nutrients are present in defined amounts in inorganic fertilizers. The latter are rather costly. Organic fertilizers are cheap, and provide, besides nitrogen and phosphorus, organic carbon, which is an energy source. However, the supply of nitrogen and phosphorus, from organic fertilisers, is usually not properly controlled as the concentrations of these nutrients are variable. Potassium is usually not required in ponds.

4. The determination of fertilizer needs requires identification of the factors which limit production.

4.2. Nitrogen and nitrogen compounds

Nitrogen, along with carbon, oxygen and hydrogen, is one of the four major elements, of which organic molecules are composed, and is therefore quantitatively important in living matter. Nitrogen needs of animals are met principally by proteins, while plants use mineral nitrogen compounds. Nitrogen assimilation occurs simultaneously with that of other nutrients, in accordance with a given proportion between nutrients. The carbon-nitrogen-phosphorus ratio (C:N:P) most commonly used with reference to algal needs is the Redfield ratio of 106:16:1, though other ratios are mentioned, such as a C:N:P of 100:12:1 (Blažka *et al.*, 1980). This proportion is not fixed, as the levels of nitrogen and phosphorus in algal cells change constantly, through assimilation and other metabolic activities.

4.2.1. Nitrogen cycling in water

Nitrogen compounds in water

Nitrogen is found in water as gaseous molecular nitrogen, (N_2), NO_3^-, NO_2^-, NH_4^+, gaseous ammonia, NH_3; and organic compounds: urea, amino acids, and polypeptides. The nitrogen compounds found in natural waters are of various origins:

- nitrate ions come from run-off water, especially from fertilized agricultural soils, and from nitrogen fixation in soils. They are not firmly sorbed onto soil particles and are easily washed away with heavy rains. They are present in sediments and polluted water as well
- nitrite ions, NO_2^-, are produced by bacterial oxidation of ammonium or by reduction of nitrate. They are only present in water with low dissolved oxygen
- ammonium concentrations are usually low, but may fluctuate widely in time and space. Ammonium is produced by decomposition of organic matter or animal excrements, by deamination (McCarthy, 1980). High concentrations of ammonium ions inhibit phytoplankton development (Pańczakowa & Szyszka, 1986)
- organic nitrogen substances include urea, released by animal excretion, free amino acids and peptides, which are of importance for microorganisms, and a number of unidentified organic compounds.

The transformation cycle of the different compounds is shown in Fig. 28.

Nitrogen fixation in water

Nitrogen enters the cycle as atmospheric molecular nitrogen, which dissolves in water and is fixed by nitrogen-fixing bacteria and blue-green algae. In freshwater, blue-green algae are responsible for most of the planktonic nitrogen fixation. In mesotrophic and eutrophic waters, benthic nitrogen fixation is done mainly by heterotrophic bacteria, while, in oligotrophic waters, blue-green algae are important. In oligotrophic and mesotrophic waters, nitrogen fixation does not represent an important nitrogen supply, whereas in eutrophic waters, where phosphorus is not limiting, nitrogen fixation contributes significantly to primary production (Howarth *et al.*, 1988a).

Nitrogen fixation is significant when the N:P supply ratio is below the Redfield ratio of 16:1 (Horwarth *et al.*, 1988b). Nitrogen fixation is carried out in the algal heterocysts and is proportional to their density (McFarland & Toetz, 1988). In some African lakes, like Lake George and L. Tanganika, nitrogen fixation accounts for about half of total nitrogen input (Hecky *et al.*, 1991). The fixation rate of atmospheric nitrogen in Lake George, Uganda, (mean depth of 2.5 m) was 0.44 to 1.69 mg m^{-2} h^{-1} (Horne & Viner, 1971) and 11 to

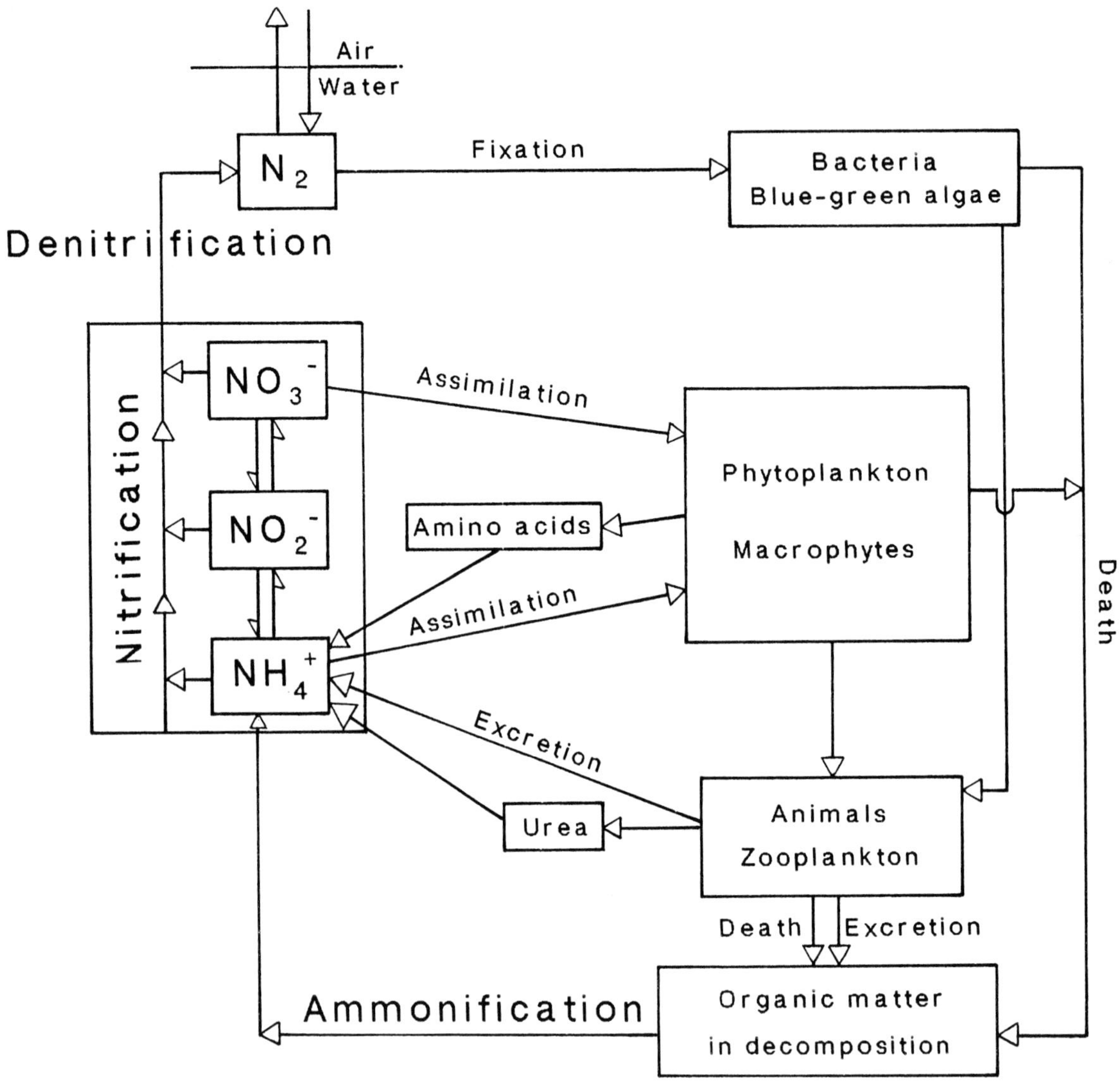

Fig. 28. Nitrogen cycling within ponds.

58 mg m^{-2} d^{-1} of N (Ganf & Horne, 1975). Nitrogen fixation was maximum in Lake George at the surface and at a depth of about 40 cm and minimum at depths of 15 cm and 150 cm (Horne & Viner, 1971). Nitrogen fixation in Hungarian fish ponds amounted to 4.7 mg m^{-2} d^{-1} or 0.33 mg m^{-2} h^{-1} of N, and was temporarily inhibited by the use, in the pond, of nitrogen fertilizers (El Samra & Oláh, 1979).

Factors inhibiting nitrogen fixation

Considerable N-fixation occurs in the dark. Oxygen inhibits the nitrogenase, as photorespiration and nitrogenation compete for reductants (Ganf & Horne, 1975). The activity of nitrogenase depends more on the previous history of the cell than on prevailing nutrient concentrations (Wynne & Berman, 1990).

As nitrogen fixation is inhibited by oxygen, the

occurrence of the process in water has to be explained. According to Paerl (1985), calm conditions in the water allow the creation of anoxic microzones around cells, conducive to fixation. These microzones are also conducive to the extracellular reduction of microelements, such as molybdenum and iron.

Ammonium and ammonia (NH_{4+} and NH_3) are strong inhibitors of nitrogen fixation (McFarland & Toetz, 1988), their effect being less in water than in sediment (Howarth *et al.*, 1988b).

Favourable conditions for nitrogen fixation include calm waters of low mixing depths and long stratification periods, light intensity, and, probably, high temperatures (Horwarth *et al.*, 1988b).

Ammonification

Ammonification is a transformation of organic matter (Fig. 29), yielding ammonia. It transforms into the ammonium ion in the presence of water or is released from water in gaseous form, when the solubility of ammonia is exceeded.

Ammonification is performed by heterotrophic bacteria, whose density is strongly dependent on organic substrate available (Barat & Jana, 1987). Their activity is more important in the presence of light and when dissolved oxygen is abundant (Jana *et al.*, 1983). High oxygen concentrations are related to active metabolism of phytoplankton, which find, because of the activities of ammonifying bacteria, high concentrations of available nutrients; this relationship benefits both the bacteria and phytoplankton. Harrison (1978) established a correlation between ammonification and ammonium and carbon dioxide assimilation, suggesting a coupling between nitrogen recycling and photosynthesis. High concentrations of calcium, magnesium and phosphate ions also have a positive effect on ammonification.

Drying up of the pond bottom between production periods reduces, by ammonification, the amounts of organic matter in the sediment and, as a consequence, the amount of ammonium in solution after the dying off of planktonic blooms, in subsequent productions. Thus, Krom *et al.* (1989a) observed that ammonium concentrations in water were higher

Denitrification: $NO_3^- \rightarrow NO_2^- \rightarrow NH_3$; $NO_2^- \rightarrow N_2O \rightarrow N_2$; $NO_2^- \rightarrow N_2$

Ammonification: $Org\ N \rightarrow HN_4^+$

Nitrification: $NH_4^+ \rightarrow NO_3^-$

Fig. 29. Chemical transformations involved in the main processes of the nitrogen cycle.

after the dying off of algal blooms in ponds which had not been dried during the production interphase as compared to those which had. This would mean that the sediments of undrained ponds were almost or fully saturated with respect to ammonium, and could not adsorb ammonium from the water.

Nitrification

Ammonium ions are absorbed directly by plants or phytoplankton, or first transformed, by nitrification, into nitrite and, later, nitrate. The oxidation of ammonium is done by chemoautotrophic bacteria, to which this reaction provides energy, as shown in equations (1) and (2) (p. 82).

The complete oxidation of 1 mg of $N\text{-}NH_4^+$ requires 4.6 mg of oxygen. The bacteria responsible for reaction (1) belong to the genus *Nitrosomonas*. This reaction produces 13.9 to 20.1 kJ per mole of ammonium oxidized. The bacteria of the genus *Nitrobacter* perform reaction (2), in which 3.68 to

5.0 kJ per mole of nitrite are liberated. Optimum pH are 7.8 for reaction (1) and 6.6 to 6.7 for reaction (2), at temperatures of 25 to 35° C (Kaiser & Wheaton, 1983).

This process is largely dependent on the availability of a substrate, to which bacteria can attach. The process is inhibited by high light intensity (Lipschultz *et al.*, 1985), but presence of oxygen is a prerequisite for this photoinhibition, which is maximum at 420 nm, i.e. blue light. The substrate or analogous compounds (ammonium, methylamine or methanol, and nitrite or citrate) partially protect the enzymes. The temporary accumulation of nitrite in the water column is sometimes due to excessive light (Diab & Shilo, 1988) or to a paucity of phytoplankton (Krom *et al.*, 1989b).

Gaseous ammonia and nitrous acid also inhibit nitrification. At high concentration, ammonia inhibits the oxidation of ammonium and nitrite. Nitrification releases protons and is therefore a source of acidification of the water, consuming alkalinity. This can be controlled by use of lime (Hunt & Boyd, 1981).

Denitrification or dissimilation

In the absence of oxygen, some bacteria use nitrate ions instead of oxygen as electron acceptors for respiration, releasing molecular gaseous nitrogen as a byproduct (Jørgensen & Sørensen, 1985). This process is called dissimilation or nitrate respiration. The products released are nitrite, ammonium, nitrogen oxide or molecular nitrogen. The gaseous products tend to escape from the water and are lost to the environment. When the end-product is nitrogen oxide or molecular nitrogen, the process is called denitrification. The number of bacteria capable of denitrifying are numerous and not all are obligate anaerobes. They can also make use of alternative energy metabolisms (Kaspar, 1985).

Dissimilation is inhibited in the presence of oxygen, which affects the synthesis of nitrate and nitrite reductases, the enzymes which perform this transformation.

Assimilation of nitrogen by plants and phytoplankton

Nitrogen ions are physiologically required by autotrophs, especially for amino acid synthesis. Among the different nitrogen species, ammonium is preferred to other forms, i.e. nitrate, nitrite and urea, because the uptake of these other forms requires a supplementary energy expenditure to transform them into ammonium, NH_4^+, for example, 18.5 kJ mole^{-1} at 25° C for nitrate (Thompson *et al.*, 1989). The uptake of ammonium increases with light intensity and is maximum at a light intensity which inhibits nitrification (Lipschultz *et al.*, 1985). Nevertheless, ammonium uptake also takes place in the dark, and can reach 70 % of the diurnal maximum.

Nitrate or nitrite absorption requires the presence of nitrate or nitrite reductase, yielding ammonium as an end product. The enzymes depend on the intracellular substrate pools which reach concentrations much in excess of those in the medium. Intracellular nitrite never appears in detectable amounts, which suggests the existence of small pools, compared to those of intracellular nitrate and ammonium (McCarthy, 1980). Nitrite reductase activity is inhibited in the presence of nitrate. Enzyme activities vary in the course of the day and absorption is a process which happens within minutes (Krom *et al.*, 1989a). Nitrate reductase is more active during daytime, with a maximum activity at mid-day, while nitrite reductase activity increases during nighttime, to reach a maximum which peaks at around five times the diurnal minimum. Glutamate dehydrogenase is also more active at night; this enzyme regulates the incorporation of the amine radical (amination) in the α-ketoglutarate. Assimilation is maximum during the first half of the day, and minimum during the first half of the night (McCarthy, 1980).

When the concentrations of NH_4^+ in the water are above 5 μg-at l^{-1} of nitrogen, phytoplankton absorbs ammonium in preference to nitrate, even when nitrate concentrations are high (McCarthy, 1980: 213, Fig. 5.4). Ammonium accounts for 92 % of the nitrogen uptake of the cells, with nitrate contributing not more than 7 %, when ammonium (N-NH_4^+) ion concentrations are above 1 μg-at l^{-1} (McCarthy, 1980). The importance of ammonium uptake is therefore 4 to 30 times greater than that of nitrate; ammonium recycling is effected in 0.3 to 1.23 days.

Nitrogen uptake affects water quality. The uptake of mineral nutrients, such as nitrogen and phosphorus ions, which are metabolites, cleanses the water. At the same time, plants and phytoplankton release amino acids and vitamins into the water, improving its quality.

Relative importance of the various processes

Budgeting the amount of nitrogen recycled in a pond is difficult: measurements of the amounts involved in the various processes, such as nitrogen fixation, denitrification and ammonification, are cumbersome and time-consuming and need to be considered simultaneously to give a real picture of the nitrogen fluxes.

A considerable amount of nitrogen is supplied by the organic loading to the pond. This material is continuously recycled; for some compounds, this occurs rapidly as they leach from the excreta and are almost immediately taken up again, while others go through the decomposition process, being gradually released. Some of the material released, such as the faecal pellets of the zooplankton, for example, are small quanta suspended in the water column, and thus not easily discriminated.

Bacteria and zooplankton make use of large amounts of amino acids for their energetic needs, and release much ammonia into water. This ammonia supply is difficult to assess (Golterman & Kouwe, 1980). The turnover time of ammonia is short, varying between some tens to some fractions of hours, and is important in sustaining production.

Nutrient enrichment experiments on a natural seawater plankton assemblage gave following results. In conditions of low concentrations, most ammonia uptake was bacterial. Increase in dissolved nitrogen concentration benefitted phytoplankton. Its increased uptake resulted in increased primary production. This trend was similar as regards phosphate, where the uptake could also be partitioned between bacteria and phytoplankton, according to availability. Availability of a nutrient determines the outcome of exploitative competition between bacteria and phytoplankton, and between species of phytoplankton. As more of a limiting resource is directed into larger cells under a patchy nutrient regime, these cells should do better when nutrient supplies are in pulses, than when present at a low, homogeneous concentration (Suttle *et al.*, 1990). Availability does not imply persisting concentrations, but includes short-lived high concentrations or pulses of nutrient supply. Pulsed supplies permit a higher diversification of the assemblage, where species with different requirements can coexist.

An example of a nitrogen budget in a fish pond in Israel is given by Schroeder (1987). The different inputs of nitrogen into the water, and outputs and losses to the sediment and to fish were quantified. The principal sources of nitrogen in the pond were phytoplankton respiration (670 mg m^{-2} d^{-1} of N) (51.6 %), followed by manure and fertilizers added to the pond (29.7 %), and fish metabolism (15.6 %). Oxic microbial processes at the sediment surface and zooplankton metabolism were insignificant, bioturbation processes could not be quantitatively assessed, and nitrogen fixation was not mentioned. The total quantity of nitrogen supplied to the water column was estimated to be 1.3 g m^{-2} d^{-1} of N. The gross primary productivity removed nitrogen from the water at a rate of 1.31 g m^{-2} d^{-1} of N, representing 89.7 to 96.3 % of the daily withdrawals. Of this algal production, 38.9 % sedimented to the bottom, out of which 70 % accumulated in the sediments and 20 % was returned to the water by oxic decomposition and benthic fluxes. Of the nitrogen added by manuring and fertilization (386 mg m^{-2} d^{-1}, i.e. 30 kg N ha^{-1} wk^{-1}), 18 % was incorporated into fish, and 52 % was lost through seepage, diffusive flux to the atmosphere, and denitrification.

Ammonia toxicity for aquatic organisms

Ammonia concentrations together with pH need to be monitored, especially when ammonia-containing fertilizers are added to the water. With increasing pH, the ammonium ion-ammonia equilibrium tends to shift towards ammonia:

$$NH_4^+ + OH^- \rightleftharpoons NH_3 + H_2O$$

The dissociation constant of the reaction is $10^{-4.74}$ at 25° C. The proportion of gaseous ammonia at different pH and temperature is shown in Table 12 and Fig. 30.

Gaseous ammonia is toxic to fish; the fish farmer should make sure that concentrations do not reach lethal levels. There is, however, disagreement between authors on the lethal dose.

The maximum acceptable ammonia concentration is 0.1 mg l^{-1} of N-NH_3; the conversion factor for N-NH_3 to NH_3 is 0.8235 (Meade, 1985). Ammonia toxicity causes osmoregulatory imbalance, kidney

Table 12. Percentage of ammonia in solution at temperatures between 16 and 30° C and pH values between 6 and 10 (according to Emerson *et al.*, 1975)

Temperature (° C)	pH								
	6.0	6.5	7.0	7.5	8.0	8.5	9.0	9.5	10.0
16	0.0295	0.093	0.294	0.925	2.87	8.54	22.8	48.3	74.7
17	0.0318	0.101	0.317	0.996	3.08	9.14	24.1	50.0	76.1
18	0.0343	0.108	0.342	1.07	3.31	9.78	25.5	52.0	77.4
19	0.0369	0.117	0.368	1.15	3.56	10.5	27.0	53.9	78.7
20	0.0397	0.125	0.396	1.24	3.82	11.2	28.4	55.7	79.9
21	0.0427	0.135	0.425	1.33	4.10	11.9	29.9	57.5	81.0
22	0.0459	0.145	0.457	1.43	4.39	12.7	31.5	59.2	82.1
23	0.0493	0.156	0.491	1.54	4.70	13.5	33.0	60.9	83.2
24	0.0530	0.167	0.527	1.65	5.03	14.4	34.6	62.6	84.1
25	0.0569	0.180	0.566	1.77	5.38	15.3	36.3	64.3	85.1
26	0.0610	0.193	0.607	1.89	5.75	16.2	37.9	65.9	85.9
27	0.0654	0.207	0.651	2.03	6.15	17.2	39.6	67.4	86.8
28	0.0701	0.221	0.697	2.17	6.56	18.2	41.2	68.9	87.5
29	0.0752	0.237	0.747	2.32	7.00	19.2	42.9	70.4	88.3
30	0.0805	0.254	0.799	2.48	7.46	20.3	44.6	71.8	89.0

failure, suppressed excretion of endogenous ammonia, resulting in neurological and cytological failure, and damage to the gill epithelium, which leads to suffocation. Establishment of safe concentrations of ammonia is based on gill histology, ammonia poisoning giving rise to hyperplasia of the gill epithelium, separation of the epithelium from the underlying membranes, and hypertrophy and fusion of gill lamellae, leading to suffocation (Meade, 1985).

As ammonia molecules are small and do not carry electrical charges, they are transported by passive diffusion and penetrate cells by crossing the cell membranes. They have to be eliminated by active excretion. A correlation between ammonia concentration in water and blood has been demonstrated. Ammonia is transformed into ammonium and exchanged for sodium ions. The ionic concentration of the water directly affects the excretion process; low concentrations of sodium, calcium and chloride ions increase ammonia toxicity. Any factor affecting the osmotic balance influences fish tolerance to ammonia.

Various factors affect ammonia toxicity, including urea, amine and amine oxide derivatives, creatine, creatinine, uric acid, carbon dioxide, and dissolved oxygen concentrations. Carbon dioxide is excreted by the gills and causes some alkalinization of the gill lamellae, when carbon dioxide forms bicarbonate on contact with water. These lamellae are thus exposed to high, though locally restricted, concentrations of ammonia, resulting from high pH.

Fish have specific tolerances, changing with the period in their life history, and with past history. A fish, previously exposed to low concentrations of ammonia, better supports high concentrations after-

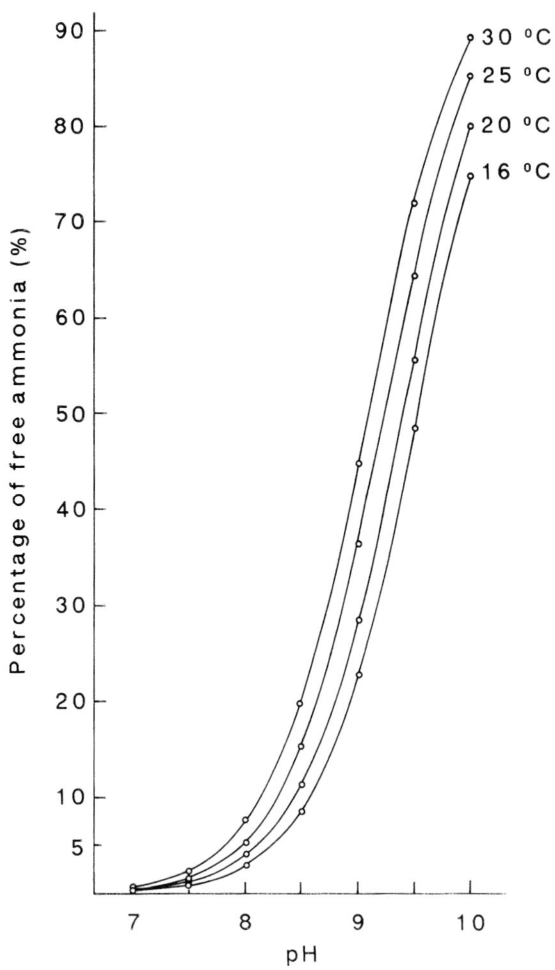

Fig. 30. Proportion of gaseous ammonia, expressed as a percentage of free ammonia, resulting from the equilibrium reaction between the ammonium ion and ammonia at different temperatures and pH's (based on data of Emerson *et al.*, 1975).

wards. Varying concentrations are more harmful than steady ones. Stress reduces resistance of the fish.

Gaseous ammonia is seldom measured, although suitable electrodes exist. Usually, the concentration of total ammonium, gaseous and soluble, is titrated and, with the help of Table 12, the concentrations of gaseous ammonia and ionized ammonium are established, in compliance with pH and water temperature.

Amounts of ammonia excreted by fish

The main ammonia sources in ponds are fish excretion and decomposition of excess unconsumed feed. Attempts to assess the amounts produced per pond have been made using different methods: consumption of feed containing nitrogen, and respiration (Meade, 1985):

- ammonia production (in kg d^{-1}): W_t x R_F x N_D x N_M x N_E
  where W_t = fish weight
  R_F = feeding rate as percent of body weight per day
  N_D = percent protein-nitrogen in the feed
  N_M = percent protein metabolized
  N_E = percent excreted nitrogen excreted as ammonia
- the ammonia production : 0.053 x the oxygen consumption of the fish (in kg d^{-1} of O_2).

In experiments on nitrogen loading and ammonification rates in cement tanks (at the facilities of University of Kalyani, West Bengal) stocked with carnivorous and forage fish, Barat & Jana (1987) observed that daily nitrogen excretion rates of carnivorous fish, *Heteropneustes fossilis* and *Clarias batrachus*, were more important than those of carp, (common and Indian carp, *Labeo rohita*, *Cirrhinus mrigala*, *Catla catla*).

A good approximation is 20 to 80 g N-NH_3(tot.) kg^{-1} d^{-1} of feed. Nitrogen excretion by fish varies in the course of the day, according to the nature of feed consumed, and the period of feeding and duration of digestion. Some 72 to 79 % of the nitrogen of the consumed feed is contained in excreta (Edwards, 1980).

Ammonia is more of a problem when fish are cultured in artificial environments, such as tanks. In African ponds, during periods of active photosynthesis, high alkaline pH values (readings close to 11) are not uncommon, and ammonia toxicity could be a menace.

Nitrogen and nitrogen compounds in water: salient points

1. Nitrogen is present in large amounts in organic matter, and, therefore, the needs for nitrogen are high. Requirements are also linked to those of carbon and phosphorus, in a relatively fixed proportion.

2. Five inorganic nitrogen compounds, i.e., ammonium, nitrate, nitrite, molecular nitrogen and ammonia, and a multitude of organic compounds are present in water. Their respective concentrations change continuously.

3. Nitrogen enters water by nitrogen fixation, mainly by blue-green algae; this process can contribute significantly to the nitrogen budget and primary productivity, even in eutrophic waters. Oxygen inhibits nitrogen fixation, but the process proceeds, nevertheless, in anoxic microzones surrounding the cells.

4. The major nitrogen source in a pond is the ammonification process, which removes ammonium from organic molecules. This bacterial process requires oxygen and light. There is some speculation concerning a coupling between ammonification and photosynthesis. Ammonium released to the water by decaying organic matter is eventually sorbed onto the sediment.

5. Ammonium is subject to nitrification, i.e. transformation into nitrate. This is also a bacterially mediated process, yielding energy, of which light is an inhibitor.

6. Nitrate is used by some bacteria as an electron acceptor, in place of oxygen, being transformed into molecular nitrogen, nitrogen oxide or ammonium. The process is termed denitrification, dissimilation or nitrate respiration.

7. Ammonium is the nitrogen species preferentially absorbed by autotrophs, but, in the absence of ammonium, nitrate or nitrite are taken up as well. The assimilation of nitrate or nitrite requires the intervention of nitrate or nitrite reductase, the activities of which seem to be highest under conditions in which the presence of nitrate and nitrite compounds is most likely. Algal nitrogen uptake competes with that of bacteria.

8. The nitrogen budget of a pond is difficult to establish, as nitrogen compound cycling is rapid and the many nitrogen sources are difficult to measure, e.g. the contribution of zooplankton excretion. A nitrogen budget in a fish pond has shown that primary productivity daily consumed almost all nitrogen dissolved in the water, but half of this nitrogen was returned daily through respiration. The addition of fertilizers compensates for the phytoplankton lost due to settling to the sediments.

9. Ammonium transforms into ammonia at alkaline pH, and is toxic to fish.

10. The amounts of ammonium excreted by fish can be estimated from their oxygen consumption or from calculations based on nitrogen content in feed.

4.2.2. Nitrogen fertilizers

Nitrogen fertilizers available

Different inorganic fertilizers are available (Table 13).

Acidity released by fertilizers

Fertilizers containing ammonium and urea have an acid reaction in water: the transformation of the ammonium ions into nitrate releases protons (as discussed earlier p. 82). Hunt & Boyd (1981) have calculated the decrease in alkalinity produced by the principal nitrogen fertilizers and the amount of lime necessary to neutralize the acidity produced. Each milliequivalent of ammonium liberates 2 meq of hydrogen ions, which react with 2 meq of calcium carbonate; knowing the molecular weight of the different fertilizers, the theoretical amounts of lime necessary can be calculated (Table 14).

Table 13. Chemical composition and nitrogen content (in percent of weight) of the most common nitrogen fertilizers (Brady, 1984)

Name of the fertilizer	Chemical composition	Percentage of nitrogen
Sodium nitrate	$NaNO_3$	16 %
Ammonium sulphate	$(NH_4)_2SO_4$	21 %
Ammonium nitrate	NH_4NO_3 or NH_4NO_3 + dolomite	33 - 34 %
Urea	$CO(NH_2)$	42 - 45 %
Calcium cyanamide	$Ca(CN)_2$	22 %
Liquid anhydrous ammonia	NH_3	82 %
Ammonia	NH_4OH dil. or NH_4NO_3 in NH_4OH or urea in NH_4OH	20 - 25 % 27 - 53 %
Ammonia phosphate	$NH_4H_2PO_4$	11 % (48 % in P_2O_5)
Diammonium phosphate	$(NH_4)_2HPO_4$	21 % (53 % in P_2O_5)

Table 14. Amounts of agricultural lime (in kg of calcium carbonate) per 100 kg of nitrogen fertilizer, to neutralize the fertilizer acidifying effect

Type of fertilizer	Amount of agricultural lime (kg 100 kg^{-1} of fertilizer)
NH_4NO_3	125
$(NH_4)_2SO_4$	152
$(NH_4)_2PO_4$	114
$NH_4H_2PO_4$	87
$(NH_4)_2CO$	167

Acid release by the fertilizers appears only after some time. This is explained as follows. The population of nitrifying bacteria is directly dependent on the amount of substrate available, responding to the supply of fertilizer by a slow transformation rate at first, but increasing after some time.

Destiny of fertilizers in water

The addition of nitrogen fertilizer to a pond is followed by short-lived peaks in ammonium and nitrate. The disappearance of inorganic nitrogen added to water could result from biological immobilization (nitrogen uptake and incorporation into organic compounds), denitrification of nitrate, transformation of ammonium into ammonia and subsequent gaseous escape of the latter, or nitrification of ammonia, followed by denitrification (Bouldin *et al.*, 1974). The relative importance of each process should be assessed (see p. 89 for results of Schroeder, 1987). Amounts of nitrogen sorbed onto and accumulating in the sediment are relatively small. Gaseous escape of ammonia is restricted to periods of high pH and does not appear to be significant.

Relation between nitrogen needs and needs for other nutrients

Nitrogen addition is required when the amounts of available inorganic nitrogen are limiting. After addition, nitrogen is no longer the limiting factor, but some other nutrient or environmental factor takes over. Nutrient needs must always be considered interconnected (see p. 138 & p. 140). Nitrogen and phosphorus are generally more limiting than carbon, but light can also become limiting, for instance in cases of self-shading, when production is very high (see p. 132 & p.135).

Nitrogen and phosphorus needs in a pond are assessed on the basis of concentrations of dissolved nitrogen species and orthophosphate in water. Ideally, the stocks accumulated in sediments should also be taken into consideration. The optimum concentrations of dissolved nitrogen species and phosphate, above which no fertilization is required, have not been clearly outlined. Boyd (1979) considered that 2.0 mg l^{-1} of total inorganic nitrogen and 0.5 mg l^{-1} of dissolved orthophosphate are satisfactory. He (Boyd, 1981b) measured average concentrations of inorganic nitrogen of 200 $\mu g\ l^{-1}$ in productive ponds (sunfish ponds in Auburn, Alabama, U.S.A.), where Secchi disk visibility was 30 cm, and chlorophyll a concentrations reached up to 200 $\mu g\ l^{-1}$, though, on average, values were around 33 to 63 $\mu g\ l^{-1}$. In concrete ponds (in Bangkok, Thailand) where phytoplankton abundance was around 120 mg l^{-1}, Edwards *et al.* (1981) recorded values of 11 to 13 mg l^{-1} of total nitrogen, of which 33 to 68 % was organic, between 0.08 and 0.56 mg l^{-1} as $N\text{-}NO_3^-$, and 2.62 to 8.82 mg l^{-1} as $N\text{-}NH_4^+$. A concentration of 0.1 mg l^{-1} of inorganic nitrogen in ponds (in Auburn, Alabama) is considered low (Boyd & Sowles, 1978). In African lakes, records on nitrogen and phosphate concentrations indicate that nitrate concentrations are usually very low, below 30 $\mu g\ l^{-1}$ of $N\text{-}NO_3^-$ in surface waters, and often are not readily detected. The distribution of ammonia is practically unknown. Concentrations of phosphate and total phosphorus are, on the other hand, generally high, between 150 and 1000 $\mu g\ l^{-1}$ (Talling & Talling, 1965).

Fertilization frequencies

Fertilizers should preferably be applied at low doses and frequently, for example, every two weeks. Fertilization should also take into consideration the feeding regime of the fish. According to the present state of our knowledge, it is impossible to forecast the specific plankton assemblage, and the relative density of algae, which will appear sebsequent to fertilization. Thus, polyculture is more suitable than monoculture, because a variety of species allow a more effective use of the natural feed produced.

Controversy about the need for nitrogen fertilizers

Some authors (Boyd, 1979; Miller, 1976; Murat & Boyd, 1987; Yusoff & McNabb, 1987) consider that nitrogen fertilizers are not necessary, as nitrogen needs can be met by recycling and regeneration of organic matter, or by nitrogen fixation by blue-green algae and bacterioplankton. Other authors (Talling & Talling, 1965; Horne & Viner, 1971; Wurtsbaugh *et al.*, 1985) state that tropical waters are chronically deficient in nitrogen, resulting in severe limitation of primary productivity, the presence of blue-green algae confirming this fact, as they provide missing nitrogen to the water. Observations from African waters confirm low nitrogen concentrations. However, concentrations do not indicate actual fluxes. Also, it is difficult to make general statements on all water types; proper experimentation on the effect of nitrogen fertilizers on production will give a more reliable answer than a long debate. Yusoff & McNabb (1989) discussed this controversy in the light of experiments with varying nitrogen-phosphorus ratios (see p. 154).

Doses of nitrogen fertilizers applied to ponds

Data on dose sizes of inorganic nitrogen fertilizers in the literature vary from 0 to 15 kg ha^{-1} (2 weeks)$^{-1}$ of N. In his classical experiments in fish ponds at Malacca, Indonesia, Hickling (1962) showed that nitrogen fertilizers, in combination with phosphate fertilizers, did not improve the production of the ponds.

Boyd (1976a) tested doses of 0, 1.125 and 4.5 kg ha^{-1} (2 weeks)$^{-1}$ of N, combined with phosphate fertilizers, in ponds in Alabama, U.S.A. Phosphate fertilizers, applied at a rate of 4.5 kg ha^{-1} (2 weeks)$^{-1}$ of P_2O_5 and of 1.125 kg ha^{-1} (2 weeks)$^{-1}$ of K_2O, did not give a significant increase in production. However, the dose of 1.125 kg ha^{-1} (2 weeks)$^{-1}$ of N gave a yield of 947 kg ha^{-1} $yield^{-1}$ of *O. aureus*, as compared to 651 kg ha^{-1} $yield^{-1}$ in the control. This represents an increase of 17.5 kg of fish per kg of nitrogen.

In other experiments at the same location, Boyd (1981b) tested five combinations of nitrogen and phosphate fertilizers, using application rates varying between 0 to 9 kg ha^{-1} (2 weeks)$^{-1}$ of N, i.e. 1, 1.2, 1.6 and 9 kg ha^{-1} (2 weeks)$^{-1}$ of N. He observed that the concentrations of dissolved inorganic nitrogen in the water remained virtually unchanged, at around 155 $\mu g\ l^{-1}$ of total N. The effects of nitrogen fertilizers on fish production are indeed difficult to assess.

Msiska (1983) in ponds at Chikwawa, southern Malawi, tried various combinations of fertilizers at doses of 1.23 kg ha^{-1} (2 weeks)$^{-1}$ of N and/or 2.6 kg ha^{-1} (2 weeks)$^{-1}$ of P_2O_5, but did not obtain a significant increase in fish production. Hepher (1962) recommended the use of 14 kg ha^{-1} (2 weeks)$^{-1}$ of N for fish ponds in Israel, while Miller (1976) reports that Vincke recommends 8.5 to 17 kg ha^{-1} (2 weeks)$^{-1}$ of N, for ponds in Madagascar.

Recommendations for use of nitrogen fertilizers

Fish production depending on natural production necessitates a primary production of about 4 to 6 g m^{-2} d^{-1} of carbon, which requires about 800 mg m^{-2} d^{-1} of nitrogen. About half of the nitrogen is returned to the environment by phytoplankton respiration (Schroeder, 1987). Fish and phytoplankton excretion supply some more nitrogen, which is recycled in the water column. Depending on the nature of the sediment and previous fertilizations, the sediment supplies, by recycling and regeneration, the rest of the required nitrogen to the water column. Losses by denitrification and diffusion are low and a supply of 3 to 7 % of the daily needs may be satisfactory to maintain the natural production of phytoplankton. This would correspond to some 3.4 to 8.4 kg ha^{-1} (2 weeks)$^{-1}$ of N. This also grossly corresponds to the amount eventually reaching the fish in Schroeder's experiments (see p. 89).

To conclude, in the light of actual knowledge, it is suggested to use doses of 5 to 10 kg ha^{-1} (2 weeks)$^{-1}$ of N, along with phosphate fertilizers (see p. 114 for the rates), and to monitor, on a regular basis, the inorganic nitrogen level in solution and the density of phytoplankton, or, in the absence of quantitative measurements, to assess the abundance of phytoplankton. Fertilizer application should increase proportionally to fish load, the amount applied increasing in the course of the production period. A regular adjustment of doses with time is justified by the absence of a storage system for nitrogen in the environment.

Nitrogen fertilizer application

1. An assortment of nitrogen fertilizers is available on the market; nitrogen fertilizers containing ammonium or urea have an acidifying effect on the water.

2. The partitioning of nitrogen added by nitrogen fertilizers to water between the different 'N-users' is difficult to quantify: dissolved nitrogen concentrations are not related to amounts added.

3. Fertilizers should preferably be applied frequently at low doses rather than less frequently at high doses. The rates should take into consideration the increase of fish biomass and be increased accordingly.

4. Nitrogen fertilization is more necessary in African waters than in temperate waters, as nitrogen appears to be more limiting in the former region.

5. The application rates recommended are 5 to 10 kg N ha^{-1} (2 $weeks)^{-1}$.

4.3. Phosphorus and phosphorus compounds

Of the three major elements, phosphorus is considered most frequently limiting, because its concentrations in water are often low. At the cellular level, phosphorus is contained in ATP and ADP, in RNA and DNA, and in numerous enzymes. In animals, phosphorus combined with calcium forms bone, teeth, and scales. The phosphorus needs of animals are met through their food.

For primary productivity, phosphorus is an essential nutrient, its importance having already been appraised by Justus von Liebig in 1855. The intracellular concentrations of phosphorus condition the photosynthetic rate (Senft, 1978). The amounts of available phosphorus are often limiting primary productivity (Hepher, 1958 & 1966, based on his experiments on fish ponds in Israel), and the abundance, the species assemblage and the succession of phytoplankton are a function of phosphorus availability. Over long periods, the amounts of dissolved and available phosphorus determine phytoplankton biomass present in an environment (Straškraba, 1980). According to Schindler (1987), a large portion of the variability in the overall production of fresh waters is explained by external phosphorus loading. In excess, phosphorus causes eutrophication.

4.3.1. Phosphorus cycling

Phosphorus species present in water

As for this element in the sediments, distinction is made between particulate and dissolved phosphorus. Dissolved phosphorus is either reactive (SRP) or non-reactive. Dissolved phosphorus is separated from particulate phosphorus by filtration through a filter of 0.45 μm pore size. Total phosphorus is the sum of all phosphorus forms in water, dissolved and particulate, available and occluded.

The various phosphorus forms in water and the phosphorus stocks of the sediments are (at least partly) in steady equilibrium. Particulate phosphorus makes up the major part, and its relations with soluble phosphorus are quite important (Nalewajko & Lean, 1980).

Sources of particulate phosphorus

Particulate phosphorus is supplied from four sources: organic matter produced by living organisms in water (organic phosphorus), sediments (geochemical phosphorus), from chemical reaction, precipitation or sorption within the water column, and, finally, from feeds and fertilizers. Such phosphorus can be separated into inorganic or organic, and colloidal or non-colloidal particles (Broberg & Persson, 1988):

- plants, bacteria and animals cells, either alive or

dead, and constituting coarse organic particles
- organic detritus; excreta of animals, which constitute fine organic particles
- organic aggregates generated by the flocculation of organic macromolecules, in the form of true organic colloids
- inorganic/organic coprecipitates, complexes or chelates, produced by the inclusion of phosphorus in organic aggregates by metal-phosphorus binding, forming mixed inorganic-organic colloids
- weathering products, such as primary and secondary minerals, and resuspended fine non-colloidal sediment particles
- Inorganic phosphorus precipitated or sorbed onto mineral particles, such as clay, forming true mineral colloids.

Composition of dissolved phosphorus

Dissolved phosphorus mainly consists of orthophosphate and inorganic polyphosphate (DIP, dissolved inorganic phosphorus), and the dissolved organic pool (DOP), part of which probably are nucleotides or polynucleotides (Broberg & Persson, 1988). Polyphosphates and metaphosphates are condensed compounds, with chain and ring structures, produced by biological activity. The soluble organic phosphorus fraction contains phosphate esters and organic acids. Part of these compounds are hydrolyzed by chemical or biological reactions.

The various inorganic phosphorus species are not well soluble and their solubility is pH-dependent. Their successive dissociations are as follows:

$$H_3PO_4 \rightleftharpoons H^+ + H_2PO_4^- \qquad K_1 = 10^{-2.13}$$

$$H_2PO_4^- \rightleftharpoons H^+ + HPO_4^- \qquad K_2 = 10^{-7.21}$$

$$HPO_4^- \rightleftharpoons H^+ + PO_4^- \qquad K_3 = 10^{-12.36}$$

H_2PO_4- and HPO_4^- concentrations are maximum at pH 4.67 and 9.78, respectively.

Composition of dissolved organic phosphorus

A relatively large part of dissolved phosphorus is organic, especially in waters with high biotic loading. The major sources of the DOP are known, but not how much they contribute to the pool. It is also unknown whether DOP is released during active cell metabolism or after cell death. Part of the DOP pool is readily available to biota and rapidly turns over, while part is refractory, probably consisting of nucleotides or polynucleotides (Broberg & Persson, 1988).

The dissolved organic phosphorus pool is comprised of a great number of unidentified compounds, which have been grouped into high (HMWP), intermediate (IMWP or XP), and low molecular weight (LMWP) fractions. This grouping is based on the results of gel chromatography analysis (Broberg & Persson, 1988).

The HMWP fraction consists mainly of colloids, of great ecological importance as carriers and binding surfaces for dissolved organic and inorganic phosphorus. Colloidal phosphorus is biologically available, at least to some extent, though part of it is resistant to enzymatic hydrolysis. It appears to come from percolation through the soil profile. The compounds of this fraction have not been identified, but seem to be made up of a great number of molecules of varying composition and nature, including humic acid-metal-phosphate complexes, organic matter-iron(or aluminium)-phosphate complexes, and ferri-hydroxide-phosphate peptized by organic matter.

Evidence for inorganic phosphorus saturation of high molecular weight humic acids (organic material-Fe-P-Fe-organic material) has been found. This HMWP fraction sorbs orthophosphate only in the presence of microorganisms, which suggests some reversible isotopic exchange with sorbed orthophosphate on iron and/or aluminium oxide surfaces. The nature of these high molecular weight organic compounds probably varies considerably between water bodies. Part of these compounds decompose when exposed to low dosages of UV radiation, which confirms their organic nature. But decomposition by UV radiation does not exclude the presence of metal(calcium or iron)-phosphate polymers, associated or not with organic material. The HMWP fraction is more important in oligotrophic waters, while in eutrophic waters, IMWP and LMWP are more important. Most of the organic phosphorus from the HMWP fraction is retained in the fulvic acid fraction, as opposed to the humic acid fraction.

Phosphorus is, however, not necessarily associated

with organic material. Colloidal iron and phosphorus occur in a ferri-hydroxide-phosphorus complex, which may be peptized by organic carbon. The inorganic nature of this colloidal fraction is further suggested by its acid lability. HMWP compounds may be important regulators of dissolved and total phosphorus contents in the water, being involved in sorption/desorption, precipitation/dissolution and flocculation/dispersion processes (Broberg & Persson, 1988).

The IMWP fraction has not yet been characterized and is probably formed by incorporation of phosphate into seston (Jansson, 1988), by biological activity, leading to HMWP formation. It apparently holds a central position in phosphorus cycling, as it is excreted from seston before being bound to the HMWP fraction, involving a release of orthophosphate. Exposed to UV, IMWP releases phosphate, but enzymes cannot cause such a release. Thus, it reacts as an organic anion complex, of low molecular weight, protected from enzymatic degradation. The possibility of it consisting of phosphomonoesters seems to be ruled out because of its resistance to enzymatic attack. It is more likely that this fraction consists of polyphosphates, as they resist enzymatic decomposition, but not UV radiation. Some have speculated that it could be made up of inositol hexaphosphates (Boström *et al.*, 1988b), which are components of cell walls. These products are protected from hydrolysis by associated lipids and proteins (Broberg & Persson, 1988).

The LMWP fraction is composed of orthophosphates and organic phosphorus compounds of low molecular weight. Several IMWP and LMWP compounds are rapidly degraded by natural enzymes and are, therefore, only temporarily present in water (Broberg & Persson, 1988). The presence of organic phosphorus compounds is not easily detected, as organic compounds release phosphate in acid phase, during the characterization process of inorganic phosphate and the released phosphates artificially increase the dissolved phosphate concentration of the inorganic phosphorus pool.

Among the more easily identifiable organic compounds are excretion and decomposition products, such as phospholipids, phosphomononucleosides, sugar phosphates, and aminophosphoric acids (Chrost *et al.*, 1984).

Phosphorus concentration in water

Phosphorus concentration in natural waters vary seasonally, but, on average, levels fluctuate around 5 to 30 $\mu g\ l^{-1}$ of soluble phosphorus, after filtration through a 0.45 μm filter. Orthophosphate is always present in water, though often in low concentration; the presence of organic compounds is more erratic and probably depends on the biotic load of the pond. Auer *et al.* (1986) have classified waters according to their phosphate concentrations (Table 15).

Oligotrophy is defined by concentrations below a critical level, while at or above the eutrophic level, plankton growth is insensible to any further increase of phosphorus concentration in the water.

In fertilized fish ponds (Auburn, Alabama), concentrations of 20 $\mu g\ l^{-1}$ of $P\text{-}PO_4^{3-}$ and 0.17 mg l^{-1} of total phosphorus are normal (Boyd, 1979). In studies of water eutrophication, maximum phosphorus concentrations and loadings have been assessed and criteria have been established. In temperate waters, phytoplankton production can only be reduced, in the long term, if phosphorus loading is reduced to less than 0.5 to 2.0 g $m^{-2}\ yr^{-1}$ and phosphorus concentrations to less than 80 to 150 $\mu g\ l^{-1}$ (Jeppesen *et al.*, 1990a), or 100 to 200 $\mu g\ l^{-1}$ of total phosphorus or 10 $\mu g\ l^{-1}$ of dissolved phosphorus (Marsden, 1989). These values indicate concentrations in eutrophic conditions, which coincide with those of maximum productivity.

Transformations of phosphorus in water

The phosphorus cycle in the natural environment is complex (Fig. 31), mainly due to the high number of organic and inorganic phosphorus compounds in water (Psenner *et al.*, 1984), and to the difficulty of quantifying phosphorus fluxes between compartments. In this respect, it is more exact to speak in terms of kinetics, than in terms of average concentrations (Shapiro, 1988). The turnover time of some compounds is often short. The reincorporation of phosphorus into sedimentary rocks is calculated in millions of years; this is worrying in view of the rate at which phosphorus compounds are disposed of in sewage, leading to enrichment of natural waters and sediments (Holtan *et al.*, 1988).

The two processes responsible for the decrease of total phosphorus concentrations in water, are the

Table 15. Total phosphorus and dissolved phosphorus concentrations in waters of different trophic levels (according to Auer et al., 1986)

Trophic level	P_{diss} (in $\mu g\ l^{-1}$ of P)	P_{tot} (in $\mu g\ l^{-1}$ of P)
ultraoligotrophic	< 0.5	< 4.3
oligotrophic	0.5 - 1.2	4.3 - 11.5
mesotrophic	1.2 - 8.0	11.5 - 37.5
eutrophic	> 8.0	> 37.5

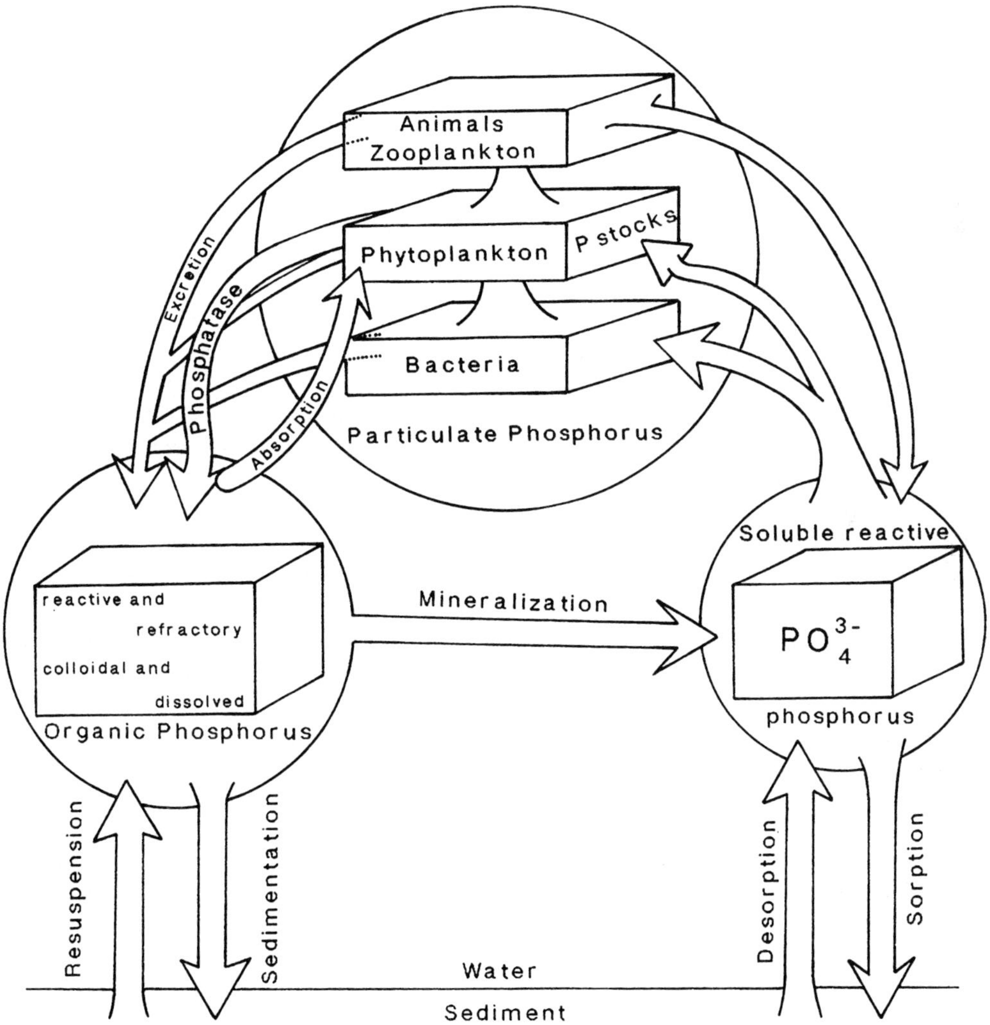

Fig. 31. Phosphorus cycle in water. Transfers of phosphorus between the particulate, the organic and the dissolved phosphorus compartments.

sedimentation rate and the renewal rate of the water (Marsden, 1989).

The parts of the phosphorus cycle which attract most interest are phosphorus assimilation, decomposition of organic phosphorus, phosphorus excretion by organisms, and chemical adsorption.

Time needed for various transformations

In studying the cycling of phosphorus in ponds, the various transformation and transport processes should be quantified (Lijklema & Hieltjes, 1982). The time scale of each process is important.

- Orthophosphate uptake by algae and its incorporation into the foodweb, its active excretion, and its release from dead organisms by decomposition are processes which proceed rapidly, in a matter of hours or days. The incorporation kinetics of phosphorus into biomass are quick. The concentration of dissolved reactive phosphorus and organic phosphorus is directly proportional to the photosynthetic rate and to the excretion rate of zooplankton in water (Golachowska, 1986)
- Sedimentation of organic particulate phosphorus and phosphorus saturation of sediments are slow processes. Algal cells sink and disintegrate at a pace which depends on the volume, the surface to volume ratio, and the density of the cells. Phosphate diffuses out of dead cells during lysis. Part of the phosphorus is quickly immobilized by bacteria, while the rest is taken up again by algae. The sedimentation rate of algae and bacteria is enhanced in the presence of concentrations of dissolved calcium in the water. Self-flocculating aggregates of calcium carbonate, along with algae and bacterial cells and dissolved phosphate, form and sink. In the presence of blue-green algal blooms, all dissolved phosphorus was observed to be removed due to coprecipitation of algae with phosphate and calcium carbonate, while in the absence of the calcium salt, the concentration of dissolved reactive phosphorus remained unchanged (Boström *et al.*, 1988a)
- Supplies and losses to the environment due to fertilization and drainage of the pond, respectively
- Exchanges across the sediment-water interface and release of phosphorus by the sediment. This involves the dynamics of phosphorus release by the sediments, extensively discussed in the first chapter
- The sediment mineralization rate includes the production of dissolved phosphate and leads to anaerobic conditions at the sediment surface. This rate varies greatly
- Adsorption/desorption processes in the water column govern the balance between the different forms of phosphorus, according to adsorption isotherms
- Sediment dilution is the accumulation and burial of sediments in the course of time (Lijklema & Hieltjes, 1982).

Importance of phosphorus mineralization

Organic phosphorus mineralization recycles large amounts, estimated to be 55 to 85 % of the phosphorus required by primary productivity. Microbial decomposition is the main process, accounting for some 70 % of phosphorus recycling. The activity and population size of the microorganisms increase in proportion to the availability of substrate. The remaining part of the decomposition involves physical and chemical processes, such as phosphorus release mediated by UV radiation. Some organic phosphorus is refractory to phosphate release. But, with changing conditions, release of phosphorus becomes possible. The ferri-humate complexes release phosphate under the influence of UV radiation, owing to the photo-reduction of ferric ions, but not under enzymatic attack.

The organic phosphorus pool includes the biotic particulate phosphorus, i.e. phosphorus bound to living organisms in water, which is of course not available. The particulate phosphorus most quickly recycled is that contained in zooplankton and fish excreta, rich in phosphorus and constituting the most important phosphorus source for the water column (Andersson *et al.*, 1988). Actively feeding zooplankton generates plumes with inorganic phosphorus concentrations several orders of magnitude higher than ambient levels, and epilimnetic zooplankton can supply ten times more inorganic phosphate to the surface layer of a natural water body than all outside sources (Tarapchak & Nalewajko, 1986a). Zooplankton therefore holds a primary role in phosphorus regeneration. According to Bartell & Kitchell

(1978), the supply of organic soluble phosphorus by zooplankton in eutrophic waters reaches maximum values of 35 to 125 mg m^{-3} d^{-1} of total phosphorus. Zooplankton faecal material varies in form and density according to taxonomic group, calanoid copepods producing faecal pellets 30 to 300 μm in equivalent spherical radius, which sink rapidly, while cladocerans produce flocculent material or a slurry, more easily recycled (Kitchell & Carpenter, 1987; Andersson *et al.*, 1988). Part of these excreta is soluble in water, while part is refractory. Zooplankton is also responsible for vertical transport of phosphorus, due to their vertical migration behaviour (see p. 162). Healthy phytoplankton cells release soluble and refractory excretion products, while lysing phytoplankton cells promptly release their phosphate content (De Pinto *et al.*, 1986).

Without phosphorus mineralization, the phosphorus stocks in the water column would be rapidly exhausted, i.e. in between 0.2 to 20 days, photosynthesis would stop. Golterman & Kouwe (1980) calculated the recycling rate using the life expectancy of algal populations. The maximum oxygen production of algae is 25 mg (mg chl a)$^{-1}$ h^{-1}, which is equivalent to 9.4 mg (mg chl a)$^{-1}$ h^{-1} of carbon. The chlorophyll a to cellular carbon ratio is 1:35. This means that carbon fixation is 3.22 mg (mg of C_{cell})$^{-1}$ d^{-1}. Algal respiration is assessed at 1 mg (mg chl a)$^{-1}$ h^{-1} of oxygen, or 4 % of the photosynthesis, or 0.38 mg (mg chl a)$^{-1}$ h^{-1} of carbon. The cellular carbon to produced oxygen quotient, expressed in mg, is 9.4:25. The loss of cellular carbon on the death of algae is,

$$\frac{0.34 \times 24}{35} \times \frac{\text{respiration depth}}{\text{euphotic zone depth}}$$

A depth quotient equal to 2 means that there is a loss of 0.52 mg (mg of C_{cell})$^{-1}$ d^{-1} of carbon. The net budget of photosynthesis is 3.22 - 0.52 = 2.7 mg (mg of C_{cell})$^{-1}$ d^{-1} of carbon. This coincides with the maximum oxygen production. In the whole water column, production is calculated to be about 2.6 times that of maximum production. For the euphotic zone, the net photosynthetic budget is therefore 1 mg (mg of C_{cell})$^{-1}$ d^{-1} of carbon. This is also the average growth rate of the algae. Considering a maximum production of 25 mg (mg chl a)$^{-1}$ h^{-1} and a respiration rate of 4 % of the oxygen production, average life expectancy of algae calculated for a period of 1 year is 1 day, which implies that, on average, daily decomposition rate is equal to mean algal biomass. If oxygen production is less, life expectancy of algae increases. This continuous mineralization or metabolic phosphorus cycling is therefore an important contribution to primary productivity, (De Pinto *et al.*, 1986). Phosphorus losses to the sediments have been discussed on p. 32.

Phosphate uptake and influences on phosphorus availability

Phosphorus loading to the water is regulated by calcium concentration and pH, which set the limits to phosphate solubility in water. In hard waters, calcium controls phosphate solubility, while, in acid waters, phosphate solubility is under the control of iron. The range of the solubility product of apatite (10^{-50} to 10^{-56}) sets the limits within which soluble phosphate is likely to be found (Golterman, 1982).

Dissolved organic matter modifies the concentration of dissolved phosphate in water. Inorganic phosphate added to water becomes attached to compounds of high molecular weight (above 100,000 and between 10,000 and 20,000). The degree of this 'attachment' is greater in waters of high humic or iron concentrations. A balance between the fractions of high molecular weight and soluble phosphate exists and the equilibrium is reached in two phases: one quick and one slow. Using Langmuir adsorption isotherms on water samples from the Neusiedlersee, Austria (Gunatilaka, 1982), and from lakes of different trophies (eutrophic, acid bog lake and freshwater estuarine marsh) of Ohio, Georgia and South Carolina (Heath, 1986), two distinct phases could be shown: a first phase which lasts around 1 minute, but is almost completed within 10 seconds, probably corresponding to a rapid adsorption on calcium carbonate, and a second phase lasting for 1 to 10 minutes, corresponding to fixation onto low energy sites on seston. In studies carried out in three acidic lakes of southern Finland (Evo district), Jones *et al.* (1988) noticed that the high molecular weight phosphorus was not reactive (or only partly) to orthophosphate reagents, and exposure to solar light did not release phosphate (Jones *et al.*, 1988). The influence of light on phosphorus release from the humic lake Tjeukemeer, Netherlands, was only significant at acid pH (De Haan & de Boer, 1986).

In water from acidic lakes of southern Finland, humic acids, unlike iron, were not able to directly sorb phosphate to form colloids. However, ferri-humic complexes sorbed orthophosphate and organophosphate ions. Adding iron (III), at a concentration of 1 mg l^{-1}, to water stimulated the movement of soluble phosphate, provided humic matter was not present at high concentrations. In the absence of humic matter, even a concentration as low as 0.1 mg l^{-1}, precipitated all added phosphate (Jones *et al.*, 1988).

In acid waters enriched with humic acids as that of the Tjeukemeer, ferric iron did not show cationic properties and the formation of colloidal iron oxyhydroxide (FeOOH) was postponed, allowing most dissolved phosphate to remain in solution (De Haan & de Boer, 1986). In aerobic water from an acidic lakes in northcentral Oklahoma, the interaction between colloidal organic matter and iron was found to decrease the amount of phosphate adsorption onto iron-organic complexes, thus freeing inorganic phosphorus for phytoplanktonic utilization. Binding of phosphorus occurred when iron was present in the ferric state; ferrous ion-DHM complexes could not bind iron (Franko, 1986).

In alkaline waters, from non-humic lakes, with low concentrations of humate, iron oxyhydroxide dominates. Most of the particulate phosphorus is sorbed onto humic matter, to form humic-iron-phosphate complexes (DHM-Fe-PO_4^{3-}); this reduces the total amount of available phosphate (De Haan & de Boer, 1986).

Fulvic acids are aromatic polycondensates with a molecular weight around 1000. They carry phenolic and carboxylic radicals, which provide them with chelating properties. They are polyelectrolytes, on which some sites, for steric reasons, are not available for ion sorption. They compete for the sorption of metal ions. Inorganic phosphate is sorbed onto inorganic colloids, ionic and colloidal iron, and fulvic acids (De Haan & de Boer, 1986).

Iron sequestration on phosphate by humic matter (in combination with iron) reduces the availability of phosphate. These adsorbed ions remain chemically active, but biologically non-available (Franko, 1986). In acid lakes of southern Finland, as equilibrium between the dissolved phosphorus and these fractions was established, addition to or removal from water of phosphate resulted in additional sorption or release according to the equilibrium. The total amounts dissolved and available remained unchanged (Jones *et al.*, 1988). It is a form of short-term storage, which stimulates the biological incorporation of phosphorus, but on which the addition of nitrogen and iron does not have much effect. The dissolved humic matter complexes formed a chemical regulation system for available phosphates, analogous to a phosphate buffer (Jones, 1990).

Reduction in the size of the dissolved phosphate pool affected the cycling and uptake of phosphorus, phosphatase activities, and carbon uptake. Organisms with low uptake constants, as well as those with alternative strategies for meeting phosphorus requirements, were favoured (Jones *et al.*, 1988).

Bacterial mineralization and role of phosphatase

Phosphomonoester and organic phosphorus compounds release phosphate due to the hydrolytic activity of phosphatases. These enzymes contribute to the decomposition of organic matter, by removing orthophosphate from the organic moiety. They also participate in algal phosphorus uptake. The reaction is a hydrolysis of esters and anhydrides of phosphoric acid, and the enzymes are phosphomonoesterases and phosphomonoesterhydrolysases. These enzymes are also involved in transport across cell membranes and synthesis of new organic phosphate, i.e. transphosphorylation, not discussed here. Phosphatases carry a metal ion as a co-factor, and it has been observed that the addition of divalent cations enhances their activity (Jansson *et al.*, 1988).

Phosphatase sources

Two types of phosphatases are present in water: particulate and dissolved. The particulate phosphatases are associated with phytoplankton and bacterial cells, while the dissolved ones are found in the filtrate after filtration (through a 0.45 μm filter). The latter represent 31 to 87 % of the total amounts in water. Their occurrence is as yet unexplained. They probably originate from bacteria, phytoplankton, or zooplankton (Currie *et al.*, 1986).

Enzymes produced by bacteria have been little studied. Of interest are the amount of bacterial excretions and whether they are passive or active. These dissolved enzymes are able to hydrolyse external substrata, but there is little information on the importance and role of such extracellular excretions.

Algae of all major groups produce intracellular and extracellular phosphatases. The role of these extracellular enzymes in phosphate regeneration, to the benefit of the algae, and secretion during periods of dissolved phosphate deficiency are mainly based on indirect evidence (Jansson *et al.*, 1988).

Zooplankton also produces phosphatases, but as digestive enzymes, and not as components of their food. Apparently, the portion attributed to zooplankton secretion makes up a large part (up to 50 % in Lake Hymenjaure, Sweden (Jansson, 1976)), of the dissolved amounts present in the water. In most cases, the fraction due to zooplankton secretion or excretion cannot singled out.

Dissolved phosphatases do not appear to differ from other enzymes and could originate from the lysis of dying or dead cells. The process leading to their presence in the water, outside of cells, is not yet fully understood, with the importance of active secretion not yet established. Dissolved enzymes are active over periods of up to several days, with a progressive reduction of activity. The inactivation mechanisms are not fully understood either (Jansson *et al.*, 1988).

Differences between alkaline and acid phosphatases

Alkaline and acid phosphatases are separated on the basis of their pH optimum, which is between 9 and 10, and 4 and 6, respectively. Their activity extends, however, beyond these pH zones. The two types are present in algae and bacteria, as internal as well as external enzymes. The alkaline enzymes have been most studied, because they are more commonly found. Differences between the phosphatase types include their location in the cell and their mode of synthesis. Acid phosphatase is more abundant within the cell than outside; the synthesis of the alkaline enzyme is repressed by orthophosphate, while this is not the case for the acid enzyme. This would suggest that acid phosphatases are constitutive, while alkaline enzymes are induced, and dependent on, the cellular supply of phosphate. Several studies have shown that the internal cellular pool of phosphate regulates the synthesis of repressible phosphatases and that the synthesis is derepressed when the internal pool gets exhausted or falls below a certain critical level (Jansson *et al.*, 1988). Dissolved acid phosphatases have been discovered in acid lakes, nearby Stockholm (Sweden), where they appear to substitute the role of alkaline phosphatases present in neutral or alkaline water. They seemed inhibited or repressed by the presence of inorganic extracellular phosphate (Olsson, 1990).

Aluminium and iron have affinities for phosphomonoesters and thereby reduce the action of alkaline phosphatase, via competitive inhibition between the metals and the enzyme for the substrate. In acid waters, this results in a higher phosphatase activity, attributed to the production of greater levels of phosphatase (sometimes approximately ten times higher) by the biota, thus enabling the enzymes to outcompete aluminium for the substrate (Mulholland *et al.*, 1986; Jansson *et al.*, 1988).

This information makes the distinction between the roles of alkaline and acid phosphatases still more unclear and raises the question whether the differences between the two groups of enzymes are not methodological rather than ecological (Olsson, 1990).

Measurement of enzyme activity

Enzyme activity depends upon the concentration of the substrate (Boavida & Heath, 1988) and is expressed by the Michaelis-Menten constant, which establishes the concentration at which the reaction proceeds at half its maximum speed. Low Michaelis-Menten constants indicate high affinities or reactions which proceed at low substrate concentrations, while high constants indicate the reverse. Phosphate uptake has been considered either to follow Michaelis-Menten kinetics (Nalewajko & Lean, 1980), but also to deviate from them (Brown *et al.*, 1978).

According to uptake characteristics, some algae survive in conditions of phosphate concentrations insufficient for others. Phytoplankton is capable of adapting to low concentrations of substrate. Phosphorus starved algae are able to produce phosphatases with higher affinity than those produced by algae in phosphorus-abundant conditions.

Relation between phosphorus deficiency and phosphatase concentrations

Enzyme activity has be used as an indicator of the degree of possible phosphorus deficiency in water (Chrost *et al.*, 1984: 16, Fig. 8). This is only possible if a relationship between the potential activity of alkaline phosphatase (APA) and the algal biomass can be established; this is expressed by the specific APA, which is the APA per particulate phosphorus content (nmol PO_4^{3-} hr^{-1} μg P_{part}^{-1}). The derepression of alkaline phosphatase is associated with low cellular total phosphorus or polyphosphate concentrations, in the range of 6 to 7 μg (mg of C)$^{-1}$ of phosphorus or 3 to 3.5 μg (mg of particulate organic matter)$^{-1}$ of phosphorus for *Chlamydomonas reinhardtii* in chemostat culture (Olsen *et al.*, 1988) or 1.4 μg P mg POC^{-1} for an assemblage of green algae and diatoms from Green Bay, Lake Michigan, U.S.A. (Auer *et al.*, 1986). The characteristic deficiency threshold is encountered when the stocks in reserve fall below 4 times the minimum cellular value (Fig. 32).

Gage & Gorham (1985), studying nine Minnesota lakes in Clearwater and Hubbard Counties, of varying (meso-eutrophy to eutrophy) trophic levels,

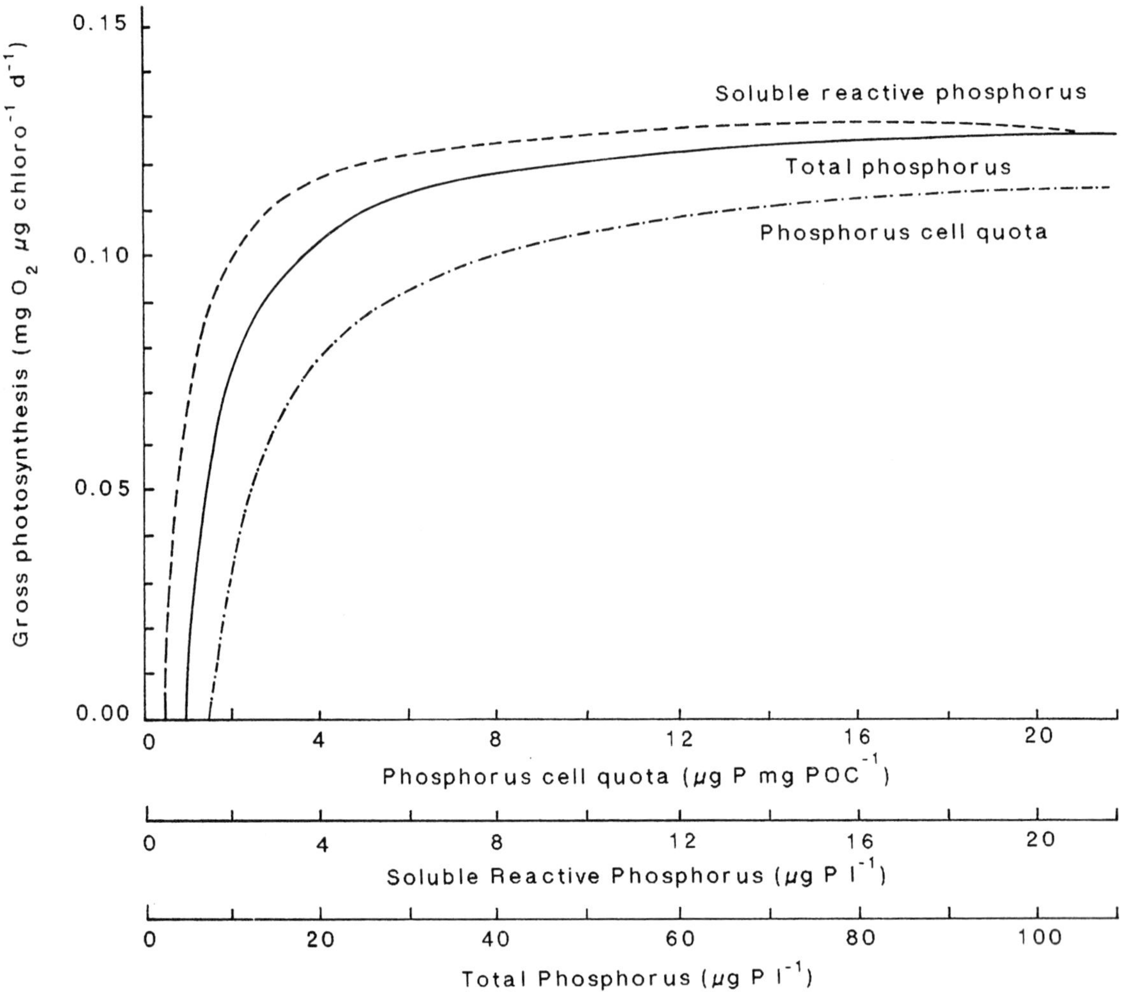

Fig. 32. Gross photosynthesis per unit chlorophyll as a function of (a) soluble reactive phosphorus concentration, (b) total phosphorus concentration and (c) phosphorus cell quota, calculated as phosphorus per unit of particulate organic phosphorus (POC) (modified from Auer *et al.*, 1986).

established phases of alkaline phosphatase activity related to cellular phosphorus concentrations, considering that situations of severe starvation, of rationing, and of excess or surplus phosphorus correspond to equivalent situations of phosphatase activity. A deficiency situation exists when the phosphorus concentration is less than 5 μg (mg of particulate organic matter)$^{-1}$, rationing between 8 to 10 μg (mg of POM)$^{-1}$ of P, and excess above 10 μg (mg of POM)$^{-1}$ of P. The three phases of enzyme activity are: high enzyme activity, medium level activity, and constitutive activity. These activities are in a ratio of 1:2:5, the high level activity being 5 times the activity of the base level (24 nmol PO_4^{3-} (mg POM)$^{-1}$ h^{-1}).

Heath (1986) found that release rates of phosphate from DOP accounted for less than 1 % of the phosphate uptake by seston in five different fresh-water systems (a eutrophic kettle lake, an acidic dystrophic bog lake, an estuarine marsh, in Ohio, and two reservoirs in Georgia and South Carolina) during summer. Abundance of phosphatase during summer phytoplankton blooms suggested that enzyme presence was related with P limitation, but comparisons of calculated phosphate release from DOP (based on Michaelis-Menten kinetics) and measurements of phosphate uptake lead the conclusion that DOP was not a major source of phosphate in summer. Previous observations had shown an increase in phosphatase activity in relation with P limitation and a depletion of internal P stores, specifically in the presence of the blue-green alga, *Aphanizomenon flos-aquae* in eutrophic East Twin Lake, Ohio. Heath (1986) concluded that DOP compounds are of negligible nutritional benefit to phytoplankton.

Although Lake Memphremagog, Quebec is considered highly P-limited, phytoplankton grows at relatively high rates, but does not synthetize large quantities of alkaline phosphatase. Algae obtain their phosphate from other sources than the low dissolved concentrations (Smith & Kalff, 1981).

In four trophically distinct lakes of north-central Oklahoma, Franko (1986) showed high plankton-associated alkaline phosphatase activity related to low ionic orthophosphate concentrations (suggesting P-deficiency) in late summer. It was not, however, the only DOP regenerative mechanism. Photodependent release from UV-sensitive complexes was also operating.

Alkaline phosphatase activity was related to the severity of phosphorus deficiency in 13 lakes in the region of Quebec, Vermont and New York, and correlated positively with the rate constant of orthophosphate uptake (a parameter close to phosphate turnover rate) of phytoplankton (particles above 3.0 μm) (Currie *et al.*, 1986). Other studies on relationships between APA and phosphorus in the ecosystem are listed by Jansson *et al.* (1988).

The relation between low inorganic phosphate concentrations and high APA has lead to the idea that the synthesis of alkaline phosphatase is induced by environmental factors, though the exact mechanism of control of induction and inhibition is obscure (Jansson *et al.*, 1988).

From observations on APA and orthophosphate uptake in several lakes of different trophy, Currie *et al.* (1986) concluded that algae are responsible for more of the APA than for orthophosphate uptake. Bacterioplankton has a higher uptake rate of inorganic phosphate than algae and, therefore, can outcompete them for inorganic phosphate. How exactly algae meet their requirements is not understood. One explanation is that they use organic phosphorus (Currie & Kalff, 1984b), or they might have adapted to low bulk inorganic P concentrations in nature by synthesizing exceptionally high-affinity transport systems (Tarapchak & Herche, 1986). High-affinity systems do not preclude that algae may rely on other sources, however (Tarapchak & Nalewajko, 1986).

In the relationship between dissolved inorganic phosphate deficiencies and extracellular phosphatase production, it is necessary to establish the origin of the phosphatase in solution, and to separate the portion which is not actively secreted by algae from that which is passively released. As long as there is no certainty concerning the amounts and the induction mechanisms of enzyme actively secreted by phytoplankton, the hypothesis of phosphatase secretion by phytoplankton to hydrolyse DOP in view of alleviating the deficiency of ambient inorganic phosphorus, cannot be contented.

The dissolved organic phosphorus concentrations in water are often very low and short-lived, so that the activity of phosphatase on its substrate is not easy to show (Jansson *et al.*, 1988).

Phosphorus uptake by autotrophs

The phosphorus species taken up by bacteria and algae is orthophosphate, PO_4^{3-} (Nalewajko & Lean, 1980). This ion does not change its valence, being neither reduced nor oxidized inside the cell, and is directly dependant on the available phosphorus concentration. Phosphorus is mainly excreted at the same oxidation stage, i.e. orthophosphate, but some is also in organic form, as exudations of living or dead, lysed algal cells, or contained in animal excrements.

Biologically available phosphorus consists of orthophosphate in solution, dissolved reactive phosphorus (SRP), some dissolved non-reactive phosphorus, part of the organic phosphorus pool which, after enzymatic hydrolysis yields inorganic phosphorus, and labile phosphorus, which is likely to yield also some inorganic phosphorus by desorption.

Phosphate uptake by macrophytes, phytoplankton, periphytic and benthic algae, and bacteria is a process of active transport. This transport is mediated by electronic or non-electronic ion pumps, which may be sensitive to iron and/or humates (Franko, 1986; Tarapchak & Nalewajko, 1986a). The energy for such transport comes from photosynthesis or respiration. Phosphate is essential for cellular metabolism and photosynthesis (Fig. 32) and, below a critical cellular phosphorus level, growth (Auer *et al.*, 1986) and photosynthesis stop (Senft, 1978).

Phosphate uptake by phytoplankton is stimulated by light, but high phosphorus concentration is the prime prerequisite for such stimulation (Nalewajko *et al.*, 1986). Uptake is modified by the presence of iron, and humic and fulvic matter in water, which have inhibitory or stimulating influences on the activity of dissolved and particulate alkaline phosphatases, and which form complexes and chelates with phosphates (Franko, 1986).

Phosphate assimilation by phytoplankton depends upon cell size. Smaller cells remove proportionally more phosphate from water than larger cells, owing to the higher surface:volume ratio of the former. (Organism size is considered by Smith & Kalff (1985) as a major determinant of community structure, more than taxon-specific specialization.) In taking up large amounts of phosphate, small cells constitute a pool of stocked phosphate, partly due to the rapid turnover time between the different absorption and excretion phases (Broberg & Persson, 1988).

Periphytic and benthic algae (*Scenedesmus sp.*, *Chlorococcum sp. Oedegonium sp.*, *Tribonema sp.*, *Anabaena sp.*, *Ankistrodesmus sp.*, *Oscillatoria sp.*) collected from a small eutrophic lake (Sövdeborgssjön) in southern Sweden were able, in laboratory experiments, to take up phosphorus from water, but also from sediment or from substrata on which they live (see p. 42) (Hansson, 1989). This gives them an advantage over planktonic algae. The inherent disadvantage of fixation onto a substratum is that light requirements cannot be met to the same extent, so that this factor often becomes limiting. In conditions of limiting dissolved nutrient concentrations, phytoplankton biomass is low and light penetrates deep into the water and, thus, periphyton and benthic algae are favoured, while, in conditions of high nutrient concentrations (eutrophic conditions), plankton algae have the advantage (Hansson, 1988).

Processes of phosphate uptake by algae and bacteria

Phosphate absorption by bacteria is the result of two kinetically distinct systems, one of high and one of low affinity. The high affinity system is activated when the internal reserves of phosphorus are low. The two systems function simultaneously and replenish the inorganic phosphorus pool of the cell. When the pool is full, only the low affinity system is active. The low affinity system supplies the cell with phosphorus at a rate equal to the rate at which phosphorus is metabolized in the cell.

The internal inorganic phosphorus pool has a limited capacity and all the phosphorus which enters the cell is stored in the pool, before being used by the cellular metabolism. The high affinity system is repressible, while the low affinity system is constitutive. Their respective energy supplies are different, the low affinity system being fuelled by protons, while the high affinity system functions with ATP.

Bacteria excrete phosphorus at a high rate. The absorption/excretion cycle is rapid, even when the stocks are full, but excretion is reduced when the cell is deprived of phosphorus (Jansson, 1988).

Knowledge on absorption of phosphate by algae is

less complete. The two-phase absorption scheme also exists, but is not so common. The uptake rate of a phosphorus starved cell is 10 to 100 times greater than that of a non-starved cell. The uptake system is controlled by the internal stocks of phosphate, especially polyphosphates. The internal polyphosphate pool is supplied at times when the phosphorus concentrations in the water are high; this pool can support several generations of growth in the complete absence of an external phosphorus supply. It reflects past phosphorus concentrations in the water. A depleted pool indicates a disruption in supply and low ambient phosphorus concentrations in the past. The energy required for phosphate uptake by algae comes from ATP through photophosphorylation or oxidative phosphorylation. The electronic transport chain is the first energy source.

The positive influence of light on phosphorus assimilation shows the importance of photophosphorylation (Jansson, 1988). Phosphate uptake by algae is 1.8 times higher in light than in the dark. It is maximum at times of highest illumination and inhibited at dawn and dusk (Nalewajko *et al.*, 1986); nevertheless, phosphate uptake also occurs in the dark.

Uptake kinetics of phosphate by algae, and the changes over time in phosphate concentrations of water, influence the specific composition of the planktonic community. Concentrations of phosphate in water can be steady or fluctuating, varying with pulses. This also applies for other nutrients. Competition for dissolved phosphate appears to favour smaller cells, either for reasons of interspecific differences in phosphate uptake or by the constitution of a variable assemblage of species along a resource and nutrient gradient (Kilham, 1986). Picoplankton (0.5 to 2 μm in size) contributes an important part of the total number of phytoplankton species, and, though low in biomass, participates notably in primary productivity. Carbon fixation by picoplankton has been estimated to be 20 % of total carbon fixation (Tarapchak & Herche, 1986).

Phosphate competition between phytoplankton and bacteria

Bacteria and bacterioplankton appear to be the largest consumers of dissolved inorganic phosphorus in water. Changes in dissolved phosphate concentrations are directly related to glucose assimilation in the water and not to bicarbonate absorption or chlorophyll concentrations. Hayes & Phillips (1958) observed that inactivation of bacteria by antibiotics caused the fixation of phosphates by sediments, while, normally, utilization by bacteria kept the phosphate in solution, available to the plankton.

Phytoplankton competes with bacterioplankton for absorption of inorganic phosphate (Currie & Kalff, 1984a; Tarapchak & Nalewajko, 1986b; Smith & Kalff, 1985). Investigations in six different lakes of the North German Baltic area showed that the greatest portion of orthophosphate is absorbed by particles less than 1 μm in size (Rai & Jakobsen, 1990). Competition for phosphorus acts as a phosphorus regulator (Currie & Kalff, 1984a & b).

Phosphorus consumption by bacteria in Lake Nesjøvatn, central Norway, was about four times more important than that by phytoplankton. The P:C ratios of phytoplankton and bacteria were different, being 2.5 to 55 mg P (mg of carbon)$^{-1}$ and 34 to 360 mg P (mg C)$^{-1}$, respectively (Vadstein *et al.*, 1988). Bacteria appear to have an important advantage over phytoplankton (Currie & Kalff, 1984a; Currie *et al.*, 1986), except during conditions of high phosphate concentration. The phosphate uptake system of bacteria is saturated at concentrations lower than those required to saturate the algal uptake system. The proportion used by each of the groups is difficult to quantify, as the use of labelled ^{32}P tracer measures gross absorption of phosphates, and the inorganic phosphorus pool in water cannot be measured precisely (Tarapchak & Nalewajko, 1986). Soluble regulating molecules secreted by some groups of algae affect phosphate absorption by phytoplankton.

Bacterial and algal synergy

The existence of an algae-bacteria synergy has been suggested: bacterial growth is restricted by the supply of energy, in particular, by dissolved, reduced organic carbon. Organic carbon released by growing phytoplankton is therefore a major carbon source for bacteria. It can provide up to 50 % of bacterial carbon requirements. Bacteria, in return, regulate phosphorus supply to phytoplankton (Jansson, 1988).

As phosphate uptake by bacteria is faster and

characterized by a higher affinity, competition for phosphate uptake should be at the advantage of bacteria. However, when bacteria have no direct metabolic use for the phosphorus absorbed, their absorption rate is slowed down and accompanied by a steady excretion. These bacterial excretion products are the phosphorus source for algae, either as organic phosphorus, from which algae remove the phosphate, or, more likely, as directly available inorganic phosphate. Orthophosphate excretion is approximately 100 times faster than excretion of organic compounds (Jansson, 1988). The orthophosphate flux across the bacterial cell membrane is quick, while the absorption/excretion cycle is much slower. The same transport mechanisms are used for the intake and the outflow of phosphorus from the cell. This process is, however, not restricted to bacteria, and has been observed in some algae (Jansson, 1988). It is called phosphate exchange and explains the following phenomena:

- bacteria regulate the net phosphorus uptake. Even when saturated with phosphorus, bacteria steadily import phosphate into the cell, and thereafter reject it. The relationship between absorption and excretion is adjusted to the rate at which intracellular phosphorus is metabolized
- bacteria remove phosphate from the water, but release it rapidly, which contributes to keeping it available for algae; the bacteria thus form some kind of transitory compartment (Jansson, 1988). Phytoplankton develops at high growth rates, in spite of low phosphate concentrations in water. This is explained by one or more of these assumptions
- algae possess a high affinity absorption system (Tarapchak & Nalewajko, 1986a) or
- their requirements for phosphate are met by sources other than the dissolved inorganic phosphorus pool (Currie & Kalff, 1984b; Tarapchak & Nalewajko, 1986a) or
- algae build up stocks during periods when phosphate concentrations in water are high and live on these when supplies become scarce.

Intracellular phosphate stocks in algae

The formation of phosphate stocks is an adaptation of algae to provide for their needs in times of shortage, as pointed out by Ketchum & Redfield in 1949. Intracellular stocks consist of phosphates and polyphosphates, in vacuoles and in the cytoplasm. They reach maximum and minimum values of 30 to 100 mg g^{-1} and 1 mg g^{-1} of dry weight, respectively (Broberg & Persson, 1988). According to Hepher (1958), stocks can reach up to 25 times the minimum cellular phosphate concentration. This has been called luxury consumption of phosphate. At temporary deficiency of phosphate in water, the organisms resort to the use of these stocks. Cellular multiplication is proportional to cellular phosphorus content. In nature, excess intracellular phosphorus seldom exists, as the biomass increases to adjust to increasing phosphorus availabilities, reaching a balance between internal and external phosphorus concentrations. Normal phosphorus amounts in algae are about 2 to 10 mg P g^{-1} of dry weight; in comparison, bacterial phosphorus reaches amounts 5 to 10 times that of algae (Broberg & Persson, 1988).

Importance of phosphate effluxes

In phosphorus cycling, phosphorus effluxes and kinetics of cellular phosphorus also have to be taken into account. The amounts of phosphate rejected are significant, so that the net assimilation of phosphorus appears quite low. Few quantified data exist on these effluxes. The phosphorus metabolism in the cell is important; in certain cases, absorbed orthophosphate is excreted after a lapse of only 1 to 8 minutes, although it is usually longer (Hayes & Phillips, 1958; Nalewajko & Lean, 1980).

Amounts of phosphorus excreted by fish

With their excrements, fish release sizeable amounts of phosphate to the water. Excretion is dependent upon the quantity and quality of ingested food, and on environmental and metabolic factors, such as stress or blood pH. Phosphate absorption in the gut, after the ingested feed has been submitted to enzymatic catalysis, is effected in a passive, but restricted, way. The major part is egested with the excrement, and some more is contained in the urine. The rejected phosphorus content sometimes represents between 61 to 87 % of the feed phosphorus (Edward, 1980; Sternik, 1983, for common carp).

Fish feed contains around 1 g P (100 g of feed)$^{-1}$,

on a dry weight basis, for trout pellets and zooplankton, and some 3 g P (100 g feed)$^{-1}$ for forage fish. When the fish is healthy and feeding conditions are good, a large proportion of the phosphorus in the faeces is orthophosphate, but when fish is starving, most of the phosphorus is organic. For a carp of 100 g, Sternik (1983), found an average daily excretion of around 78 mg P_{tot} kg^{-1} d^{-1} and 4.4 mg P-PO_4^{3-} kg^{-1} d^{-1}, at 25° C in laboratory conditions. This is an impressive amount for a normal stocked fish population, i.e. for a fish loading of 50 kg m^{-3}, it represents a phosphorus supply of 200 μg l^{-1} d^{-1} of P-PO_4^{3-}.

Measurement of phosphorus in water

Phosphorus is measured after transformation into orthophosphate (Boyd & Tucker, 1980). Orthophosphate is analyzed by a method involving the reaction of orthophosphate ions with molybdate ions to form phosphomolybdate, subsequently reduced to a soluble blue complex, detected spectrophotometrically. This analytical method has a tendency to overestimate phosphate concentrations and does not achieve the required precision of 0.01 μg l^{-1} (Shapiro, 1988; Broberg & Pettersson, 1988).

Phosphorus and phosphorus compounds in water: salient points

1. Phosphorus is the most frequent limiting nutrient to primary production in water. Concentrations of dissolved phosphorus are usually low. Phosphorus deficiency affects algal metabolism, growth and photosynthetic activity. In excess, phosphorus causes eutrophication.

2. Phosphorus is present in water as dissolved orthophosphate, organic phosphorus, and particulate phosphorus. Some degree of equilibrium exists between the various phosphorus species. Sorption of inorganic dissolved phosphate onto organic colloidal particles or onto mixed inorganic-organic colloidal aggregates is an important process. Phosphorus is supplied to ponds by organic matter, as a byproduct of primary productivity, animal excreta (zooplankton and fish), and uningested fish feed and fertilizer. Phosphorus is removed from water by sedimentation of seston and drainage of ponds (water renewal).

3. The dissolved organic phosphorus pool contains a group of organic compounds, which has an important influence on dissolved phosphorus availability. Three broad groups have been distinguished, namely, high (HMWP), intermediate (IMWP), and low molecular phosphorus (LMWP) fractions. The HMWP fraction seems to be a decomposition-refractory complex of organic (humate or fulvate) and iron (metal) oxides, having high sorption capacities for inorganic and organic phosphate, sometimes saturated with phosphates, and acting as a regulator of dissolved phosphate. The IMWP or XP fraction results from decaying processes (perhaps, algal cell walls fragments). This fraction releases phosphate under UV radiations, but is refractory to enzymatic action. The LMWP fraction is made of organic compounds of low molecular weight. The IMWP and LMWP fractions are probably decomposition products, only temporarily present in water.

4. Phosphorus cycling in a pond is difficult to quantify, as the turn-over time of many compounds is short and hard to detect, especially phosphate uptake, phosphate excretion by zooplankton and fish, and sorption processes. Other processes, such as sedimentation of phytoplankton to the pond bottom, are much slower, and greatly modified by external and water quality factors. The most important processes responsible for phosphorus changes in water are organic matter decomposition, and phosphate uptake by bacteria and algae.

5. Organic phosphorus mineralization is mostly a microbially mediated process. It provides almost three-quarters of the phosphorus supply. Some physical and chemical decomposition also occurs: UV decomposition or photoreduction. Phosphate sorption and desorption processes by dissolved organic compounds are numerous and as yet unquantified. Most phosphorus release from organic compounds is due to the activity of phosphatases, but some release from humic-iron-phosphate complexes is induced by UV irradiation. Dissolved humic substances act as a kind of buffer for dissolved phosphate.

6. Particulate and dissolved phosphatases are present in water; particulate phosphatase is contained in algae and bacteria, while dissolved phosphatase is released into water by algae, bacteria and zooplankton. The most likely role of these phosphatases is hydrolysis of phosphate from organic compounds, at times of dissolved inorganic phosphate deficiency.

7. Phosphate is taken up by algae and bacteria and competition can occur between these two groups for limited amounts of dissolved phosphate. The advantage in this competition is with the bacteria, which have an uptake system with higher affinity. However, bacteria excrete large amounts of phosphate, so that bacterial phosphate uptake seems to act more as a temporary storage. Bacteria and algae may live in synergy, involving phosphate and organic carbon exchange. Algae form intracellular stocks of phosphate, the process being called luxury consumption; beyond these stocks, any surplus of phosphate in water induces algal multiplication. Stocks of intracellular phosphate can be passed through several generations of algae.

4.3.2. Phosphorus fertilizers

Phosphorus fertilizers available

Phosphate deposits are of three broad apatite groups: deposits of igneous and metamorphic origin, sedimentary phosphorites, and guano and related deposits. Guano is formed on land by sea birds and was the earliest source of phosphate fertilizers. Phosphate rocks are mined for fertilizers and industrial chemicals, such as detergents, toothpaste and pharmaceuticals (Holtan *et al.*, 1988).

When apatite is treated by sulphuric acid, the following reaction occurs:

$$[Ca_3(PO_4)_2]_3.CaF_2 + 7H_2SO_4 \longrightarrow 3Ca(H_2PO_4)_2 + 7CaSO_4 + 2HF$$

A list of phosphorus fertilizers is given in Table 16.

Fertilizers containing a mixture of nitrogen, phosphorus and potassium are characterized by their respective concentrations of each of these elements. A value of N:P:K of 8:16:16 indicates 8 % total nitrogen, 16 % P_2O_5 (or 7 % phosphorus) and 16 % K_2O (or 13 % soluble K).

Destiny of phosphorus fertilizers in ponds

Phosphorus fertilizers added to ponds in Israel caused a rapid increase in dissolved orthophosphate concentration in the water, which disappeared after a short time (Hepher, 1958). Disappearance of phosphorus is the result of absorption by bacteria, macrophytes and phytoplankton, which occurs in a short space of time, sometimes in few minutes, and adsorption by organic colloids in solution or in suspension in water, as well as adsorption by sediment or interaction (chelation or complexation) with ions in the water, as observed by Hayes & Phillips (1958) in eight lakes of different trophy in Nova Scotia, Canada. The main part of added phosphorus to the water of Lake Gunillajaure in northern Sweden was used by the seston, and a small portion sorbed onto sediments (Björk-Ramberg, 1984).

In experiments conducted in ponds at the Auburn University Fisheries Research Unit, Alabama, U.S.A., some portion of the phosphorus supplied by the fertilizers accumulated in pond sediments, which, although acting primarily as a sink for dissolved phosphorus, were also a phosphorus source. Draining the fish pond at harvest had generally a negative effect on pond fertility (Hollerman & Boyd, 1985). In fish ponds in production and undrained for three years, total phosphorus concentration in water was significantly higher than that of ponds drained annually. The phytoplankton biomass of the former was comparatively higher than that of annually-drained ponds, owing to phosphorus mobilization from the sediments (Hollerman & Boyd, 1986). In an undrained pond compared to normally drained ponds of a commercial farm in Upper Galilee, Israel, sediments lost their capacity to remove phosphorus from water (Eren *et al.*, 1977), which resulted in higher average phosphorus concentrations in water to the benefit of primary production.

Table 16. Chemical composition and phosphorus content of the principal phosphorus fertilizers (according to Brady, 1974).

Fertilizer name	Chemical formula	Percentage	
		in P_2O_5	in P
Superphosphate(*)	$Ca(H_2PO_4)_2$ $CaHPO_4$ $CaSO_4$	16 - 50	7 - 22
Triple superphosphate (*)	$Ca(H_2PO_4)_2$	44 - 47	17 - 21
Ammonium superphosphate	$NH_4H_2PO_4$ $CaHPO_4$ $(NH_4)_2SO_4$	16 - 18	7 - 8
Ammonium phosphate	$NH_4H_2PO_4$	48	21
Ammonium polyphosphate	$(NH_4)_4P_2O_7$ and others	56 - 60	
Diammonium phosphate	$(NH_4)_2HPO_4$	46 - 53	20 - 23
Basic slag	$(CaO)_5P_2O_7.SiO_2$	15 - 25	7 - 11
Steamed bones	$Ca_3(PO_4)_2$	23 - 30	10 - 13
Calcium metaphosphate	$Ca(PO_4)_2$	62 - 63	27 - 28
Phosphoric acid	H_2PO_4	54	24
Superphosphoric acid	$H_3PO_4 + H_4P_2O_7$	76	33

(*) 85 % soluble in water

Interaction of fertilizer with other water characteristics

A high calcium ion concentration induces precipitation of phosphate, thereby decreasing the concentration of dissolved orthophosphate. An alkaline pH also promotes precipitation. If calcium concentration in water is 20 mg l^{-1}, the maximum concentrations of dissolved orthophosphate are 10 mg l^{-1} at pH 8 and 0.01 mg l^{-1} at pH 10 (Boyd, 1979). Alkalinity and pH have a direct effect on the orthophosphate concentration in water. A decrease of carbon dioxide, resulting from photosynthesis or gaseous diffusion, increases pH and causes phosphate precipitation (Hepher, 1966).

The use of phosphorus fertilizers reduces water turbidity caused by clay particles (mineral turbidity). This reduction in mineral turbidity is due to coprecipitation of clay particles, calcium carbonate and algae. An algal polysaccharide inducing flocculation is released during photosynthesis. This algal sensitivity to particles in suspension, shown in limnocorral experiments in a small lake in North Carolina, U.S.A., is specific to certain algal species. *Anabaena* is more sensitive than *Chlamydomonas* or *Chlorella*. *Euglena* does not flocculate at all. Flagellates do not flocculate, and escape by swimming away (Cuker, 1987).

Phosphorus fertilizer doses applied

The use of phosphorus fertilizers is commonly advocated, but doses recommended vary between 2.5 and 20 kg P_2O_5 (eq.) ha^{-1} (2 $wks)^{-1}$. In Indonesia, Hickling (1962) tested effects of three different doses, namely 22.4 kg, 44.8 kg and 67.2 kg P_2O_5 (eq.) ha^{-1} $yield^{-1}$, on the production of hybrid tilapia during a period of 6 months (6 replications), in monoculture, and in polyculture with grass carps and *Puntius*. Though results were variable, he obtained an increase in production, in comparison with the control pond, of 11.2 kg, 9.67 kg and 10.03 kg of fish per kg of P_2O_5 (eq.) added, for the three doses, respectively. There was a direct proportionality between increase in production and increase in phosphorus fertilizer dose, in a proportion of about 10 kg per kg of equivalent P_2O_5 added.

Boyd (1976a & 1981b) at the experimental fish farm of Auburn University, Alabama, U.S.A., used phosphorus fertilizer applications at rates varying between 2.2 and 9 kg P_2O_5 (eq.) ha^{-1} (2 $wks)^{-1}$, in fish ponds stocked with tilapia, *O. aureus*, or with *Lepomis*. At the same time, doses of 1.2 kg, 16 kg, and 9 kg ha^{-1} (2 $wks)^{-1}$ of N were applied. The average concentration of dissolved orthophosphate in the water increased with fertilizer dose, being 27 μg, 43 μg, and 82 μg l^{-1} of P-PO_4^{3-} for application rates of 2.2 kg, 4.5 kg, and 9 kg P_2O_5 (eq.) ha^{-1} (2 $wks)^{-1}$, respectively. The differences in production between various fertilizer treatments were not large, being relatively good with 9 kg P_2O_5 (eq.) ha^{-1} (2 $wks)^{-1}$ alone or in association with nitrogen fertilizer (considered by Boyd as conditions of superfertilization), and with 4.5 kg P_2O_5 (eq.) ha^{-1} (2 $wks)^{-1}$ with 1.6 kg ha^{-1} (2 $weeks)^{-1}$ of N.

In Israel, Hepher (1962) recommended application of 12 to 16 kg P_2O_5 (eq.) ha^{-1} (2 $wks)^{-1}$, along with 14 kg ha^{-1} (2 $wks)^{-1}$ of N. These doses generated an abundance of algae equivalent to 109 mg of chlorophyll a m^{-3}, a carbon fixation rate of 1.45 to 4.47 g C m^{-2} d^{-1}, and a net algal production of 19 to 32 g m^{-2} d^{-1}. Daily carp production amounted to 0.43 g m^{-2} d^{-1}. A doubling of fertilizer doses, maintaining the application doses, but increasing the application frequency (every week instead of every other week), increased algal biomass to 134 mg of chlorophyll a m^{-3}. The great abundance of plankton caused self-shading of the 20 uppermost centimetres of water and a sizeable reduction of primary productivity. Eren *et al.* (1977) recommended, for Israeli fish ponds, 9 kg P_2O_5 (eq.) ha^{-1} (2 $wks)^{-1}$, three times during the first weeks of production, followed by the same dose once a month thereafter.

In Malawi, Msiska (1983) tested, with success, an application of 2.6 kg P_2O_5 (eq.) ha^{-1} (2 $wks)^{-1}$, while Miller (1976) reported trials by Vincke in Madagascar using 10 to 20 kg P_2O_5 (eq.) ha^{-1} (2 $wks)^{-1}$, along with 8.5 to 17 kg ha^{-1} (2 $wks)^{-1}$ of N. In Côte d'Ivoire, trials with triple superphosphate, at a dose of 25 to 30 kg P_2O_5 (eq.) ha^{-1} mth^{-1}, gave good results, and the dose was recommended.

The Fisheries Department in Malaysia recommended to fish farmers doses of 12 to 15 kg P_2O_5 (eq.) ha^{-1} mth^{-1} (Yusoff & McNabb, 1989). Experiments made in ponds at the Agriculture University of Malaysia at Serdang, to test the efficiency of these doses, alone or with urea at a rate of 16.6 kg ha^{-1} wk^{-1} of N, in comparison with the use of organic fertilizers, gave net productions of up to 1700 kg ha^{-1}, using a combination of four species of carps, at a low stocking rate (0.5 per m^{-2}). The natural productivity was 0.26 g of C m^{-2} h^{-1} and the chlorophyll concentration in water was 109.2 mg m^{-3}. They concluded, in the light of ratios between dissolved inorganic nitrogen and phosphorus in water, that phosphorus was the limiting factor for production (Yusoff & McNabb, 1989).

Application mode of phosphorus fertilizers

Most phosphorus fertilizers are in pellet form, and are applied as such or after solubilization in water.

Sometimes, they are kept inside the sack in which they were packed; this is perforated and immersed in the pond. The fertilizer slowly dissolves and oozes out of the sack. Pellets dumped into the pond sink and get buried in or dissolve on contact with the sediment. Dissolving phosphate becomes sorbed by the mud and is lost to the water. Dissolution prior to the application of solid fertilizer to the water column or hanging of a container containing the solid fertilizer in the mid-water is recommended, as dissolved fertilizers are directly available to plankton (Boyd, 1981a).

In ponds at Auburn, Alabama, U.S.A., Boyd *et al.* (1981) compared three phosphorus fertilizers, triple superphosphate, diammonium phosphate and liquid ammonium polyphosphate, at doses of 9 kg P_2O_5 (eq.) ha^{-1} $(2\ wks)^{-1}$, along with 3.5 kg ha^{-1} $(2\ wks)^{-1}$ of N. The assessment involved monitoring of orthophosphate and chlorophyll a concentrations. After 10 to 15 days, dissolved orthophosphate and total phosphorus reached the same levels in all ponds, being 5 to 30 $\mu g\ l^{-1}$ of $P\text{-}PO_4^{3-}$ and 0.05 to 0.17 mg l^{-1} of total P. Chlorophyll a concentrations were higher in ponds treated with liquid polyphosphate and this was confirmed even with doses two thirds or one third of the original dose (Fig. 33).

If dissolved prior to application, the effect of tri-

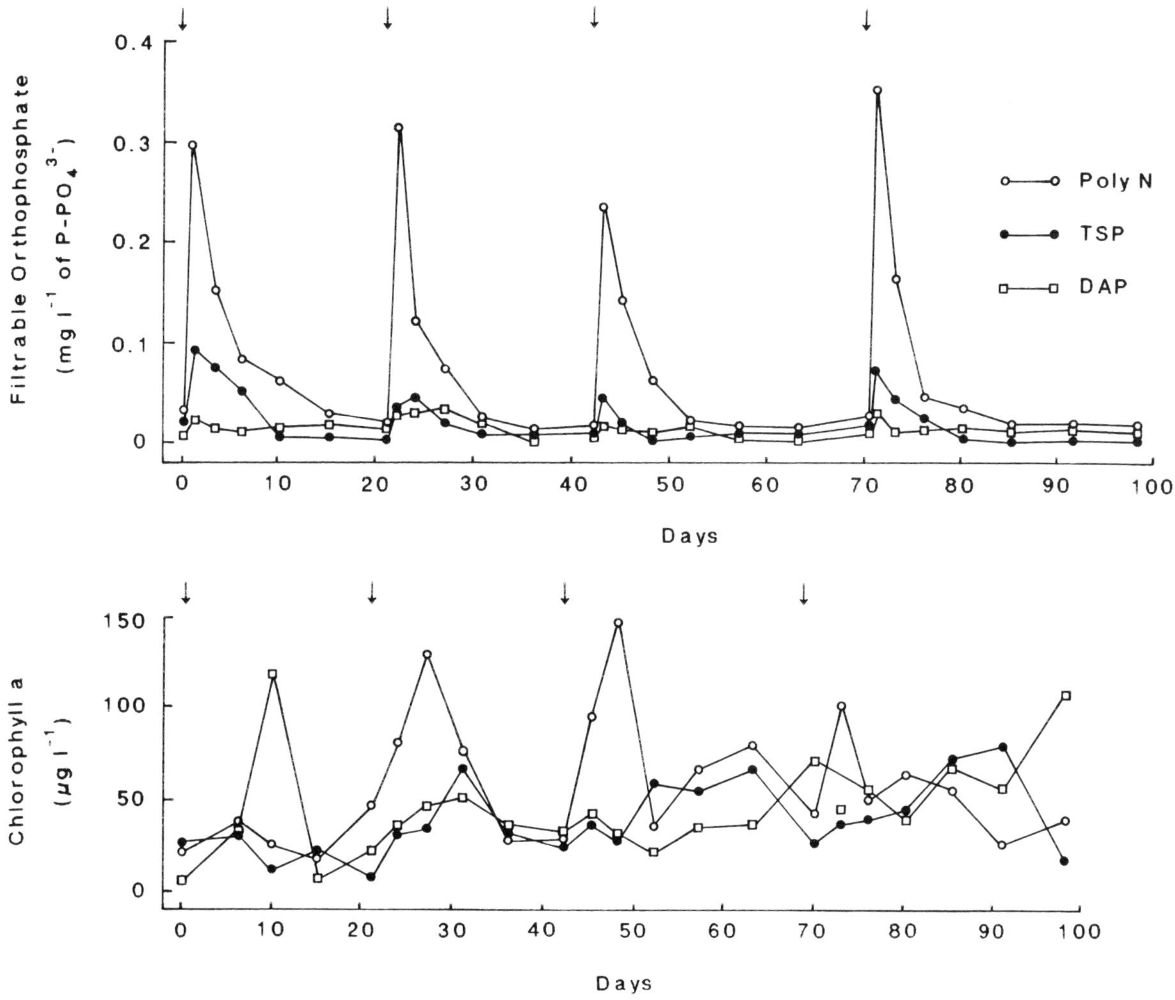

Fig. 33. Comparative effect of three phosphorus fertilizers, i.e. diammonium phosphate (DAP), triple superphosphate (TSP) and liquid ammonium-phosphate fertilizer (Poly N) on filtrable orthophosphate and chlorophyll a concentrations in pond water. The arrows (↓) indicate the date of fertilizer applications (from Boyd *et al.*, 1981) (with kind permission of Prof. Boyd on behalf of the authors & Aquaculture).

ple superphosphate (TSP) was similar to that of liquid polyphosphate. When applying liquid fertilizers, Boyd (1981c) recommends that an atomizer or sprayer be used or that the water is churned to insure a good mixing of fertilizer with water. Fertilizer can also be poured in the eddies of the screw of an outboard engine. Fertilizer has a higher density than water, and, if not thoroughly mixed with the latter, will sink.

Experimentation in ponds to determine the best application rates for nitrogen and phosphorus fertilizers should conform to the following scheme: production trials in stocked ponds in which only fertilizers are applied must run concurrently to trials in which fertilizer is supplemented with the usual feed at the usual feeding rate. The trials need to be repeated at low and high stocking densities, to compare with the growth of fish obtained with natural feed at different densities. The analysis should take into consideration production data, feed conversion rates, and growth rates during different periods (between sampling dates) of production.

Recommended application rates for phosphorus fertilizers

The dynamics of phosphorus transformation within aquatic systems is yet to be fully understood, and the different fluxes and exchanges between the sediment and the water column, between particulate and dissolved compartments, and between the different groups of organisms remain unquantified. It is therefore difficult to predict the amounts of phosphorus, provided by a fertilizer, which really benefit phytoplankton production, and to assess the amounts made available through internal loading of the pond. The amounts of phosphorus required for phytoplankton production cannot be calculated on the basis of an analytical budgeting of the phosphorus cycling in the pond. Some insight can be gained from the data of limnologists studying eutrophication problems and the ways and means of reducing internal phosphorus load. According to this information, the upper limit of external phosphorus loading (in temperate lakes) is considered to be 0.5 to 2 g P m^{-2} yr^{-1} (Jeppesen *et al.*, 1990a). This is equivalent to a fertilizer supply of 7.5 kg P ha^{-1} (2 $wks)^{-1}$. Considering that frequent renewal of water at each harvest represents an important loss of phosphorus, and that, at each harvest, phosphorus is exported in the form of fish, fertilizer application rates in the range of 5 to 10 kg P ha^{-1} (2 $wks)^{-1}$ seem reasonable.

These fertilization rates are equivalent to doses of 8.75 to 17.5 kg P_2O_5 (eq.) ha^{-1} (2 $wks)^{-1}$, and are supplementary to regular feeding of the fish and use of nitrogen fertilizers at 5 to 10 kg ha^{-1} (2 $wks)^{-1}$ of N. This is equivalent to 12 to 24 kg of urea (or 23 to 47.5 kg of ammonium sulphate) and 17.5 to 35 kg of triple superphosphate per hectare per 2 weeks. In comparison to nitrogen fertilizer, doses of phosphorus fertilizer need to be adjusted less to production. Increasing doses during the production period are less justified, the objective being to maintain surplus phosphorus and saturated intracellular stores.

Phosphorus fertilizer application

1. Many inorganic phosphorus fertilizers are available on the market. Phosphorus fertilizers added to the water interact with other water features, particularly with calcium carbonate in alkaline water, and with iron and aluminium in acid water.

2. Phosphorus fertilizers are preferably applied in liquid form, as solid pellets sink to the pond bottom and are lost from the water column.

3. Recommended application rates for phosphorus fertilizers are 8.75 to 17.5 kg P_2O_5 (eq.) ha^{-1} (2 $wks)^{-1}$.

4.4. Manure or organic fertilizers

Differences between inorganic fertilizers and manure

A number of organic fertilizers, such as livestock dung, and plant or vegetal detritus, have long been used for fertilizing ponds. Yields obtained with organic fertilizers are usually higher than those obtained with inorganic compounds. In ponds, highest productions are reached in conditions of intense continuous manuring of pond water, in association with animal production. The cost of manure is lower than that of inorganic fertilizers. Their use can, however, be problematic, for several reasons. For example, manure in excess can have adverse effects on production. Also, manure may pass diseases on to humans.

Organic fertilizers contain nitrogen and phosphorus, but also decaying organic matter. Nutrients benefit phytoplankton (and phytobenthos), while the organic matter promotes the development of bacteria in suspension in the water and in the sediment. The heterotrophic production cycle is thus stimulated leading to an increase in zooplankton and zoobenthos. Manure also supplies organic colloids, which increase the sorption capacities of soil and seston.

While the microbial decomposition cycle is boosted with increased supplies of organic matter, the oxygen consumption of the water increases. The supplies of nutrients by manure lead to an increase in primary production, which, in turn, provide more oxygen to the decomposer community. Manure has therefore a dual effect on autotrophic and heterotrophic production cycles. The two cycles are deeply intricated. The phytoplankton population supplies organic matter for decomposition, as most phytoplankton is not consumed by predators. Decomposers recycle and regenerate nutrients to the benefit of phytoplankton and are themselves consumed by zooplankton and fish. Zooplankton is consumed by fish.

Fish fed on plankton and produced in manured ponds have lipid contents (6.2 %) less than those fed pellets of high protein content (15 % lipids) or those fed cereals (20 % lipids) (Moav *et al.*, 1977).

Decomposition processes in water

Manure is a loose, fibrous material, made of particles of various sizes. Dumped into water, it takes time to sediment. Nutrients and colloids ooze out, and organic matter dissolves from the particles in suspension. All these organic flocs are colonized by microbial fauna.

Decomposition processes take place either in water or in sediments. The importance of recycling of nutrients in water is often underestimated, partly because there is controversy concerning methods for measuring bacterial productivity (Moriarty, 1990; Costa-Pierce, 1990). Costa-Pierce & Craven (1987) concluded from studies on microbial production in aquaculture ponds in Kahaku, Hawaii, using the incorporation of adenine into ATP, RNA and DNA, that more than 50 % of heterotrophic production is microbial. In Lake Titicaca, Peru/Bolivia, Carmouze *et al.* (1984) observed that almost 90 % of the organic matter is recycled within the euphotic zone whence heterotrophic activities are particularly intense.

Manure provides a substrate for microbial development, and microbial population size and activities are tightly correlated to the available substrate. The substrate is characterized by its C:N or C:N:P ratios; ratios of 17:1 or 20:1:0.2 ensure a good microbial growth. In an aerobic environment, 20 to 50 % of the available carbon in organic matter is used by bacteria, while in anaerobic water, the efficiency of carbon utilization is reduced. Microbial production is essential for fish production, as the former can contribute to almost half of the latter. Microbial production is consumed by zooplankton, especially ciliates (microzooplankton), and by detritivorous fish.

In investigations on Canadian shield lakes in the Experimental Lake Area (ELA), decomposition of organic carbon in the hypolimnion of three lakes of different trophy (artificially eutrophied and artificially acidified) proceeded aerobically or anaerobically, according to dissolved oxygen concentrations. In the absence of oxygen, concentrations of available electron acceptors and redox potential determined the reduction process, which included, for instance, sulphate and nitrate reduction (Kelly *et al.*, 1988).

The decomposition process on sediments is also determined by oxygen supply. Sediment oxygen demand (almost equivalent to the biological oxygen demand, BOD) increases with microbial activity. Aerobic decomposition in sediment is more limited by oxygen than that in water. It is useful to remember the importance of oxygen demand for the decomposition of organic fertilizer, when estimating the maximum manure load to be used in a pond (Schroeder, 1980). Decomposition in water and on/in sediment is a function of bacterial population densities.

Liquefied dung supplies organic colloids to the water. It remains in suspension and serves as food and substrate for the bacterial population of decomposers, which live in the water column. As such, fertilization benefits the water column more than the sediments. The surfaces exposed to microbial attack are much greater than those of lumps of solid dung. Oxygen availability is greater in water than in sediments. The released nutrients are directly available to the phytoplankton, and do not have to diffuse from sediments to the euphotic zone. With the use of liquid cow dung, good productions have been achieved in Israel (Moav *et al.*, 1977).

In lake Aydat, in Central France, two major types of bacteria are involved in the degradation of suspended organic matter, i.e. free living organisms, of small size (0.2 to 1.2 μm), and fixed organisms, of greater size (1.2 to 63 μm) (Marvalin *et al.*, 1989). A great proportion of the dissolved organic matter is polymeric material, of high molecular weight, which can only be assimilated by bacteria after decomposition by extracellular enzymes.

Exoenzymatic hydrolysis is the first and probably the limiting step of organic matter decomposition. In turbid waters, exoenzymes are closely associated with particles in suspension and remain active even when adsorbed. The activity of exoenzymes increases in the presence of high concentrations of suspended material. Exoenzymes most frequently considered are alkaline phosphatases (see p. 102), aminopeptidases, and endopeptidases. Aminopeptidases decompose proteins; they appear, for instance, quickly after the death of an algal bloom, but their activity does not persist (Boon, 1990).

The microfauna which develops on organic matter is consumed by ciliates, flagellates, and rotifers. The recovery of energy from organic matter to the benefit of the foodweb is called the microbial food loop and is discussed on p. 170.

Exoenzymes release nutrients, which are used by algae, thus fostering primary productivity. An increase of algal biomass is not necessarily accompanied by an increase in grazing. The algal cells disappear after some time and are returned to the decomposer organisms.

Types of fish cultured in manure-fertilized ponds

Fertilization with manure can be applied to ponds where fish with various diets are cultured in polyculture, making use of the different feed resources of the pond. Primary productivity is used by phytoplanktivorous fish, secondary productivity by zooplanktivorous fish, and detritus by detritivorous fish. Some fish, especially tilapias, have wide feeding regimes, capable of using a broad range of food items (see p. 184). For example, *O. mossambicus*, generally considered a phytoplanktivorous fish, can browse on the flocculent layer at the surface of the sediment, as in Lake Valencia, Venezuela (Bowen, 1981). This flocculent layer consists of organic matter undergoing decomposition, colonized with decomposers and microzooplankton. *O. niloticus* from Lake George, L. Albert, and the middle Niger Delta was observed feeding on bottom deposits on some occasions (Trewawas, 1983). Different emphasis on possible food items is reflected in the relative coarseness or fineness of the pharyngeal dentition and, to some extent, of the teeth on the jaws (Trewawas, 1983).

Detritivorous or benthophagous fish ingest organic mud, digesting bacteria which develop within, and egesting, in their excrement, all what cannot be digested. These excrements are immediately recolonized by microorganisms.

About 90 % of the phytoplankton produced in ponds consists of nanoplankton (cell sizes between 3 and 20 μm), unavailable to fish, because of their small size. This assertion is, however, refuted by some researchers, who state that tilapias are able to agglutinate, using mucus secreted by the oral and pharyngal epithelium, the small phytoplankton cells brought into the buccal and branchial cavities. In Lake Kivu, nanoplankton and the planktonic bacterium *Spirillum* are indeed used as food by *O. niloticus* during most of the year (Trewawas, 1983).

Fish feeding on phytoplankton can foster some taxa, such as the blue-green alga *Microcystis*. Fish have direct and indirect effects on phytoplankton assemblages in ponds. This aspect will be discussed below (p. 176).

Nutritional value of manure

Some fish, especially tilapia, feed directly on fresh excrements. They can be observed, for instance, below pigsties or at the entrance of feeder canals bringing sewage from stables, feeding on excreta floating in the water. This lead to the idea of using dung directly in the preparation of artificial feed pellets. Analysis of cow dung and chicken manure indicated that this contains an amount of metabolizable energy, in the order of 2510 to 3347 kJ kg^{-1} and 3766 to 5021 kJ kg^{-1} and 10 to 15 % and up to 25 % gross N-proteins, respectively. Msiska (1981) found chicken manure to consist of 14.2 $\pm$ 4.0 % gross protein.

Trials with pellets containing 30 % protein, made out of dung, as feed has been tested with a variety of fish (common carp, tilapia, channel catfish, goldfish) and has not given satisfactory results (Schroeder, 1980). Other trials in outdoor tanks in northern Israel, where part of the ingredients were replaced by cow dung, gave results which varied with the species cultivated. The replacement of 50 % of the pellet ingredients by cow dung gave better results in tilapia (*O. aureus*) production, but not in *Clarias gariepinus*. These differences appeared only when feed was given in daily amounts corresponding to 5 % of the actual biomass; at 2.5 %, no difference appeared. The most likely explanation for this increase in tilapia production is that the dung contained in the feed had a stimulatory effect on primary productivity, which benefitted the tilapia culture, but was not used by the catfish (Degani *et al.*, 1984).

The organic nitrogen portion of manure, previously considered to be gross protein and measured by Kjeldahl analysis, is actually mainly uric acid and non-protein nitrogen. These substances have no direct nutritional value for fish. More precise analysis has indicated that cow dung contains less than 1 % pepsin-digestible proteins, more than 65 % total fibres, less than 1 % starch or carbohydrates, and a caloric level below 3347 kJ kg^{-1} of metabolizable energy. This makes it a feed of low nutritional value. In comparison, a feed of medium level quality contains about 25 % proteins, 65 % starch and 12552 kJ kg^{-1} of metabolizable energy (Schroeder, 1987).

Detritus and dung directly ingested by tilapia are mainly stripped of their bacterial colonies and digested. This was confirmed by observations in aquaria at the University of Sterling, UK; *O. niloticus* (average weight of about 3 g) fed actively on free-living non-attached bacteria, which were introduced in the water at concentrations of 10^5 to 10^6 bacteria ml^{-1}. Bacteria were found in the fish stomachs (Beveridge *et al.*, 1989). See also the observations of Bowen (1981), already mentioned (p. 52).

Disadvantages of the use of manure

A disadvantage of manure is the presence of excess amounts of organic fertilizer, causing a dissolved oxygen deficit, in particular during the hottest hours of the day and in the early morning (shortly before dawn). This is a consequence of high microbial activity. Another threat in the use of manure is the release of ammonia, especially when pH is alkaline. As discussed earlier, ammonia is toxic. The use of dung in ponds increases the risk of transmission to humans of parasitic diseases with fish as intermediate and humans as final hosts. However, no such disease has yet been detected.

Mode of use of fertilizers - precautions for their use

Precautions to be taken when using organic fertilizers are related to the disadvantages listed above. Manure should only be spread in water in small amounts, several times a week. The spreading should be as homogeneous as possible over the whole surface. In experiments conducted at the Wuxi Municipal Fish Farm, in Jiangsu province, China, ponds of large size (7333 m^2) are more difficult to manage than ponds of smaller size (between 667 and 3800 m^2), and give lower production values. The effects of manure were more drastic in ponds of small size; the fluctuations of dissolved oxygen were wider and were causing problems to fish (a variety of Chinese carp) more quickly (Zhang *et al.*, 1987).

One of the first factors affected by high decomposition rates is DO. The value of dissolved oxygen at or a little before dawn should be regularly checked. A low or critical DO value in the early morning indicates excessive heterotrophic respiration. The presence of fish gasping for oxygen at the surface is typical of oxygen depletion. A decrease of luminosity, owing to an increase of cloud cover or rain, decreases photosynthetic rate (Fig. 20) and, consequently, oxygen release. The spreading of manure should be suspended as soon as indications of a dissolved oxygen deficit are perceived and physical and chemical features of the water should never be allowed to fluctuate widely.

Manures containing much ammonia should be used in extra-small amounts at a time, in combination with close monitoring of pH. The presence of ammonia in water in concentrations around 30 to 100 mg l^{-1} of $N\text{-}NH_4^+$ or the presence of hydrogen sulphide indicate a threatening anaerobic situation.

Manure in an advanced stage of decomposition fertilizes better than less decomposed organic matter, because the nutrients diffuse more easily to the water and because the ultimate decomposition can take place during the sinking of the organic matter towards the pond bottom. This allows enrichment with nutrients of the whole water column and a decrease in oxygen demand of the sediment; the microorganisms digesting the organic matter remove the oxygen needed for their activities from the water and not from the sediment. Dissolved oxygen removed from water is more easily replaced than that taken from sediments.

In laboratory experiments using soil collected from a brackish water fish pond, in West Bengal, India, manure decomposed less rapidly in brackish water than in fresh water, the rate of decomposition decreasing with increasing salinity. The amount of nitrogen released into the water, in the form of nitrate and ammonium, decreased and the amount of organic carbon found in the soil after a given period was higher at increased salinity (Chattopadhyay & Mandal, 1980).

Manures are usually added to the water. Another method is to spread dung or compost on the bottom of dried ponds before impoundment, or even to work it into soil by shallow ploughing. The exchange capacities of the soil are increased by incorporation of organic matter; the water retention capacity of the soil is greater and infiltrations are reduced (Chakroff, 1976).

Organic fertilization requires proper pond management and constant monitoring of production conditions. Ponds best suited for organic fertilization should be shallow, ideally with a mean depth of 1 m. A judicious choice of species and of proper stocking densities allows the best use of different food resources produced and the maintenance of a balance between the various natural production processes within the pond.

Types of organic manure used and their composition

Organic fertilizers used are mainly solid or liquid excreta of livestock animals (cattle, pigs, sheep, ducks, chicken), slaughter house wastes (rumen contents, blood, etc.), compost, and domestic or industrial (from agricultural industries) sewage water. Fish culture may be combined with animal production: livestock is held in stables nearby or even over the pond, so that excreta and feed surplus find their way directly to the pond.

The chemical composition of animal excreta is variable, depending not only upon the nature of feed ingested, but also on other ingredients added to form the final manure product, such as urine, straw litter, and/or excess feed. Moreover, they are characterized by a great variability in water and nutrient content. Their nutrient to dry matter content ratio is almost 20 to 30 times lower than that of inorganic fertilizers. A ton of organic fertilizer is equivalent to about 45 kg of N:P:K of 10:5:10 (Delmendo, 1980). The nitrogen/phosphorus ratio is not balanced, phosphorus being less abundant. Organic matter contributes to the formation of humus. An organic fertilizer not properly stored or mixed with soil quickly looses its quality. Nitrogen is leached or lost by gaseous diffusion.

In Tables 17 and 18, average compositions of excreta of livestock animals and amounts of excrement produced daily are indicated.

Comparative value of manures

At the Ginosar Farm, Israel, comparative trials of dry and liquid cow dung and chicken manure, at a rate of 5 kg of manure (dry weight) per day, with applications five times a day, to carp ponds in which fish feed was in pellet form, demonstrated the superiority of chicken manure over cow dung. As

Table 17. Average composition, in humidity, nitrogen, phosphorus, and potassium, and faeces/urine ratios, of excreta of common livestock animals (modified from Little & Muir, 1987)

Type of animal	Faeces to urine ratio	Humidity	Nitrogen (in %)	Phosphorus (in % of P_2O_5)	Potassium (in % of $K^2 0$)
Dairy beef	75:25	85	4.0	2.7	3.4
Feedlot beef	75:25	85	7.8	2.5	3.6
Chicken	100	65	6.0	7.9	3.7
Pig	53:47	85	4.7	4.3	2.7
Sheep	67:33	70	3.9	2.4	3.5
Horse	75:25	75	2.3	1.3	1.4

Table 18. Amounts of excrements, expressed in fresh weight and dry weight per animal, and per 100 kg of live weight, per day, produced by common livestock animals (modified from Little & Muir, 1987)

Animal	Fresh excreta per animal and per day (kg)	Dry weight per animal and day (kg)	Fresh excreta per 100 kg live weight and day (kg)	Dry weight per 100 kg live weight and day (kg)
Beef	30.0	4.5	6.0	0.9
Chicken	0.7	0.27	35.0	13.4
Ducks	1.0	0.38	33.3	12.7
Pigs	8.0	1.2	14.5	2.17
Sheep	2.1	0.71	7.0	2.4
Horse	24.0	8.2	6.3	2.15

compared to the control, chicken manure increased fish production by 21 %, while this value was 11 % for cow dung. The feed conversion rate of pellets was decreased by 12 and 6 %, respectively, for chicken manure and cow dung, as compared to the control. These manures, in liquid or dry form, at the rates used, were well mineralized, as shown from a comparison between the dissolved organic matter, organic carbon, total nitrogen, and ammonium concentrations in sediment of fertilized ponds versus unfertilized ponds. This indicates that organic fertilizers did not accumulate unduly on the pond sediment (Rappaport & Sarig, 1978).

In ponds in Camayagua, Honduras, other comparative trials between manures and inorganic fertilizers, in production ponds of *O. niloticus* indicated that better productions are achieved with chicken manure than with cow dung or inorganic fertilizers, at equivalent fertilization rates (Green *et al.*, 1989). In Ibadan, Nigeria, trials using chicken sweepings, containing 30 to 40 % manure, and spread 6 days a week on ponds as feed and fertilizer, yielded a production of 6.75 tons ha^{-1} yr^{-1} of *Clarias isheriensis* (Faybenro & Sydenham, 1988).

Effects of manure on water quality

Manures affect water quality in different ways, but mainly with regard to nutrient content. Ammonium and ammonia concentrations vary but sometimes reach high values: 0.16 to 0.78 mg l^{-1} and 0.22 to 1.24 mg l^{-1} of $N\text{-}NH_3$ in ponds near Kinmundy, Illinois, U.S.A., fertilized with the excreta production of 39 to 66 pig ha^{-1} (Buck *et al.*, 1978); and 0.56 to 4.05 mg l^{-1} and 4.25 to 9.28 mg l^{-1} of $N\text{-}NH_4^+$ in concrete tanks in Bangkok, Thailand, fertilized with sewage water (Edwards *et al.*, 1981). Such values do not seem to affect the environment adversely. Nitrate values, however, remained low: less than 0.01 mg l^{-1} $N\text{-}NO_3^-$ (Buck *et al.*, 1978); and 0.15 to 0.83 mg l^{-1} and 0.01 to 0.03 mg l^{-1} $N\text{-}NO_3^-$ (Edward *et al.*, 1981).

Plankton concentration and primary productivity obtained with manure

The levels of phytoplankton production reached in ponds subjected to organic manuring vary widely. Buck *et al.* (1978), using 39 to 66 pigs ha^{-1}, which is equivalent to about 40 to 70 kg of excrements (dry weight) per day, to fertilize ponds in Illinois, recorded a primary productivity of 15,000 organisms ml^{-1}, equivalent to 125 μg l^{-1} of chlorophyll a. In Malawi, Msiska & Cantrell (1985), conducted experiments in cement tanks fertilized with chicken manure. They obtained variable chlorophyll a concentrations reaching up to 80 μg l-1, with an average of 50 μg l^{-1}.

In Israel, at Dor Research Station, Noriega-Curtis (1979), fertilizing with liquid cow dung at a rate of 67 to 70 kg (dry matter) d^{-1} $ha^{-1,}$ recorded a planktonic population, collected on a plankton screen of 37 μm, of 2.1 to 10.6 μg l^{-1}, and a primary production of 3.6 to 5 g C m^{-2} d^{-1}. In Bangkok, Edwards *et al.* (1981), using sewage water for fertilization, reported phytoplankton concentrations between 36.5 and 175 mg l^{-1}, and primary productivity between 1.62 and 6.67 g C m^{-2} d^{-1}. They considered a phytoplankton production of 70 mg l^{-1} sufficient to sustain optimum growth in *O. niloticus*.

Considering the variability in size of the plankton production obtained, it appears that primary production parameters are not easy to forecast. Neither plankton concentrations achieved, nor the plankton assemblage which develops, can even approximately be predicted (see p. 157). The different phytoplankton species present in an assemblage do not have the same nutritional value and are not grazed upon in the same proportions. The problem is to know which species should be promoted and how to promote them. These two questions have not yet been answered, mainly because information on the selectivity of fish feeding upon natural food is scarce and the ecophysiology of tropical plankton (and also epiphytic and epilithic fauna and flora) is still largely unknown. The plankton assemblage changes with time, with species appearing and disappearing. In cement tanks in Bangkok, Edwards *et al.* (1981) observed that euglenoid algae, *Euglena*, and *Cryptomonas*, which concentrate at the surface, induced a decrease of dissolved oxygen in deeper water. A relationship between decreased fish production and dissolved oxygen depletion has not been clearly established, but is surmised. The sedimentation rate of these algae is sizeable, which would suggest that they are not consumed to a great degree. The dynamics of species changes within plankton communities has still to be explored. This whole subject will be discussed in details in various sections of chapter 3.

Yields obtained with manuring

The values of fish production obtained vary widely and depend upon stocking density, which needs to be balanced with the volume of natural food available. In ponds of the Aquaculture Research Centre of Texas A & M University, U.S.A., Burns & Stickney (1980) observed productions of *O. aureus* in the range of 2.5 to 16 kg ha^{-1} d^{-1}, in conjunction with manuring with poultry droppings from 500, 1000, 2000, and 4000 birds per ha. However, as stocking rates were low, at 0.16 g m^{-2}, fish yields were also low. In Israel, Noriega-Curtis (1979) recorded fish productions of 21 kg ha^{-1} d^{-1}: the stocking density was low at 8000 ha^{-1}, consisting of 4000 carp, 3000 tilapia and 1000 silver carp per hectare. Common carp does not readily consume phytoplankton and, thus, does not exploit properly all natural food resources present.

In Israel, at the Dor Research Station, Moav *et al.* (1977) reported fish productions of 32.6 kg ha^{-1} d^{-1}, with 4 species in polyculture: common carp, tilapia (*O. aureus* or hybrids), silver carp, and grass carp. The ponds were manured with liquid cow dung. The conversion rate of fertilizer into fish was calculated to be 2.7, based on the dry matter weight of the liquid fertilizer. Comparison of the results of these trials with results from ponds cultured simultaneously and in analogous conditions, where fish were fed a spectrum of artificial feeds, such as protein-rich pellets, showed that liquid manure fertilization resulted in yields similar to feeding with protein-rich pellets.

In Bangkok, Tilapia (*O. niloticus*) productions, resulting from manuring with sewage water, of about 50 kg ha^{-1} d^{-1} (44 to 51.5 kg ha^{-1} d^{-1}) were found by Edwards *et al.* (1981). The organic matter present in the culture medium was assessed through COD, chemical oxygen demand, measurements. Production in tanks during a period of 12 weeks reached values, which, when recalculated and extrapolated, were equivalent to 20 tons ha^{-1} yr^{-1}. Only proper management, accompanied by strict monitoring of water quality, and an optimum fish density, allow the achievement of such high production rates; sometimes, even such monitoring and control cannot prevent unexpected drops in production (see p. 190).

Unexplained water quality deterioration in the culture medium has been observed in intensive conditions. The reasons for this may be suboptimum dissolved oxygen levels, accumulation of growth inhibiting factors, a decrease in natural food organisms, and a decrease in nutritional value of primary producers (Moav *et al.*, 1977).

As indicated earlier, the rate of diffusion of nutrients from sediments to water is slow. A means to accelerate this diffusion would be to promote convective water movement. Water from deeper layers would bring nutrients to the surface, while surface water would carry dissolved oxygen downward, thus making use of the supersaturation of oxygen in the surface waters during periods of high photosynthesis. Such a device for convective water circulation may considerably improve pond conditions (see p. 199).

Number of livestock heads needed for integrated fish farming

To combine pond fish culture with animal production, one needs to know how many heads of livestock animals should be used to fertilize a pond, as the number of animals determines the amount of droppings which will find their way into the pond. This number varies with intensity of production, and with size and weight of the animals. For pigs, the numbers recommended vary between 30 to 85 pigs ha^{-1}, and, for ducks, between 1000 and 3500 ha^{-1} (Buck *et al.*, 1978; Delmendo, 1980; Stickney *et al.*, 1979; Little & Muir, 1987). Pigsties and duck pens can be built directly over ponds, on stakes, or nearby. In the latter case, the pens need to be cleaned regularly with water and the excreta and excess feed washed into the pond. When the pens are over the water, the floor should be a lattice work, such that detritus falls directly into the pond.

Pigs are raised for 5 to 6 months, until they reach a commercial weight (60 to 100 kg). During this period, they require suitable and balanced feed, and proper health care. Chicken and duck are raised for 1 to 3 months, till they reach their market weight.

Maximum amounts of manure to be used

Organic fertilizers are more efficient in aerobic than in anaerobic pond conditions. It is useful to determine the maximum amount of organic matter that a pond can digest. At ideal temperatures, i.e. above 18° C, maximum sediment oxygen consumption has been estimated to be 3.5 g O_2 m^{-2} d^{-1}. Each gram of oxygen consumed is able to oxidize 0.9 g of organic matter, which is equivalent to the decomposition of about 32 kg ha^{-1} d^{-1} of organic matter. If the bacterial population of the water (of 1 m depth) is equivalent to 1 square meter of bottom sediment, the total amount of digested organic matter is around 70 kg ha^{-1} d^{-1} (dry weight), or about 100 kg ha^{-1} d^{-1} of manure. An equal amount of organic matter is synthesized each day by the bacterial population. In practice, up to 150 kg ha^{-1} d^{-1} (dry weight) of manure have been dumped, without problems, into ponds with high fish biomasses. In stocked ponds, manuring is increased concurrently with fish growth, the daily applied fertilization rate (in dry weight of organic matter) being 2 to 4 % of the standing fish biomass (Schroeder, 1980).

Schroeder (1987) studied the carbon and nitrogen cycling in manured fish ponds in Israel in an effort to understand the reasons why fish production seemed to be limited to a maximum of 30 to 35 kg ha^{-1} d^{-1}. In ponds stocked with silver carp or grass carp, hybrid tilapia and common carp, he observed, using stomach content analysis, that only the silver carp directly consumed live algae, while the other species fed on detritus, generated by algal sedimentation and decomposition by microorganisms, and algal consumption by invertebrates and transformation by microbial processes. These microbial processes, and the passage through the invertebrate pathway, were estimated to represent 3 and 10-fold losses of organic carbon, respectively. The primary production of the pond supplied, directly or indirectly, with the manure, about 50 % of the carbon required for fish production.

Organic fertilizers, role and use

1. A great variety of organic fertilizers are available, their advantages being that they are cheap and efficient. Organic fertilizers or manures supply not only nitrogen and phosphorus to the pond, but also organic carbon and organic colloids, which promote organic matter decomposition within the water column. The heterotrophic production cycle is thus stimulated.

2. Decomposition processes taking place in the water are still largely unexplored and represent a bottleneck to organic fertilizer use in ponds. Microbial productivity has been only grossly assessed and the different production processes affected by organic fertilizer additions to ponds have not been quantitatively differentiated.

3. The quality of fish raised in organically fertilized ponds is higher than that of fish raised on artificial feed.

4. The maximum recommended application rates for organic fertilizers are 100 to 150 kg ha^{-1} d^{-1} (dry weight) in intensive production, or an equivalent amount of livestock excreta.

4.5. Other nutrients involved in productivity

4.5.1. Silicon

Importance of silicon

Silicon is a non-metallic element present in the valves and cell walls of diatoms, representing 10 to 30 % of their dry weight. Diatoms are an important phytoplankton group, also with epiphytic and epilithic forms, with regard to primary productivity and natural food for some fish species. As diatoms are the organisms which make most use of silicon, their biology is closely linked to the cycling of this element.

Species and concentrations of silicon in water

Silicon is present in water as orthosilicic acid, $Si(OH)_4$, and the hydrated form of silica, SiO_2. The transformation of dissolved silicic acid into amorphous silica and back by organisms forms the core of the biological silicon cycle (Paasche, 1980). Silica is also present in colloidal form, which is a highly polymerized form of silica, and biogenic silicon, i.e. amorphous silica, included in diatoms and Chrysophyceae. In interstitial water of Lake Ontario, L. Erie and L. Superior, Canada, aluminium and iron (III) oxyhydroxides and biogenic silicon formed a cryptocrystalline complex of ferro-aluminium silicate, $(Al_{1-x}.Fe_{1-x}\ (SiO_2)OH)_{6(1-x)}$ (Nriagu, 1978).

The concentration of silicon in freshwater lakes of

temperate regions, in different parts of the world, is around 1 to 5 mg l^{-1} (Paasche, 1980: 261, Table 7.1), or less than 10 mg l^{-1} (Lund, 1965), but up to 100 mg l^{-1} in saline African lakes of class III conductivity (Talling & Talling, 1965: 445, Fig. 6). In ponds, values range from 1 to 80 mg l^{-1} (Boyd, 1979). Concentrations in seawater are low, silicon being present almost exclusively in living or dead diatoms, or as biogenic silicon.

Silicate comes mainly from rock and soil erosion and run-off water. In rocks, the average ratio of phosphorus, nitrogen and silicon atoms is 35:2:10 000. The Si:P ratio of soils, which should be around 285:1, is in the range of 110:1, according to field determinations (Golterman & Kouwe, 1980). The solubilization rate of silicate is 2 to 3 times less than that of phosphate. Besides run-off water, sediment-water exchanges, and dissolution of diatom valves supply silicon to the water. Dissolution of diatom shells takes place in the sediments and is accelerated by the presence of salt (Paasche, 1980). High pH and alkalinity promote dissolution of silicon compounds (Talling & Talling, 1965). Lake overturn and vertical mixing of water return significant amounts of silica to the photosynthetically active upper zone, thus boosting diatom growth (Talling, 1969). In Lake Titicaca, Peru/Bolivia, sedimentation of diatom valves proceeds fairly quickly and this silicon is largely lost to production (Carmouze *et al.*, 1984).

Concentrations less than 25 $\mu g\ l^{-1}$ of Si-$Si(OH)_4$ are considered silicon depletion; concentrations of 0.5 to 0.8 mg l^{-1} are needed to sustain a diatom population. Decreases in silica concentration in water are linked to increases in diatom biomass, examplified in phytoplankton species succession in temperate lakes described in the PEG-model (see p. 158).

Silicon assimilation by algae

Silicon is assimilated by algae as orthosilicate; the colloidal form is unavailable to them (Lund, 1965). Formation of new valves (silicification) in diatoms is restricted to the period of cell division which lasts around 1 to 2 hours (Paasche, 1980). All silicon is, however, not contained in the valves, part of it being involved in algal cellular functions, such as regulation of protein synthesis, and hydrocarbon and chlorophyll formation. This metabolic silicon does not exceed 10 % of total silicon (Paasche, 1980). Absorption kinetics of silicon are directly dependent on silicification rate, as cells can store only insignificant amounts of silicon beyond that used for cell wall formation (Paasche, 1980). Factors affecting silicon absorption, as well as concentrations limiting to diatom growth, are not fully understood. Silicification rate of cells is dependent on temperature, and light; high illumination reduces absorption (Lund, 1965).

Decrease in size of diatom cells may be related to decreases of silicate concentrations in water. When silicate concentrations are low, the limiting effect of phosphate on growth is more pronounced.

Influence of silicon concentrations on plankton composition

Varying proportions of silicium, phosphorus and nitrogen in water promote different planktonic groups. According to Kilham *et al.* (1986), dominance of diatom species in different East African lakes is the result of specific requirements for silicium, phosphorus and light: *Synedra*, *Nitzschia*, and *Stephanodiscus* have increasing phosphorus and decreasing silicium requirements; *Melosira distans* has high light but low phosphorus requirements. The opposite is the case for *M. nyassensis*. Paleolimnological studies in lakes extending from Lake Abhé, Ethiopia to L. Cheski, Zambia, have shown the dominance of *Stephanodiscus astraea*, the eutrophic character of their water, and the importance of internal nutrient regeneration processes (Kilham & Kilham, 1990).

In Lake Michigan, U.S.A., in the presence of Si:P ratios of over 100 and high N:P ratios, diatoms are abundant; when the Si:P ratio is below 5, green algae dominate, while for low N:P ratios, blue-green algae are more abundant. Diatoms have high silicium requirements, but lower phosphorus requirements than green algae (Kilham, 1986).

Measurement of silicon

Silicon concentrations are determined colorimetrically by transformation of silicon ions to molybdosilicic acid, reduced to a blue heteropolyacid, and measured spectrophotometrically. Phosphate in the sample reacts to form molybdophosphoric acid, which is destroyed by oxalic acid. The separation of biogenic silica from quartz and silicate minerals is important, but not always easy. Several methods exist, mainly aiming at making use of the greater solubility of diatom silica as compared to mineral particles, or attempting to convert diatoms cell counts into cell surface areas and thus into silica amounts (Paasche, 1980).

4.5.2. Sulphur

Sulphur is one of the six major elements of living matter. It appears in the amino acids cystine, cysteine, and methionine, and is, therefore, present in all proteins. It is also part of some co-enzymes.

Sulphur is abundant in water and soil and makes up some 7.8 % of the solutes in sea water. The sulphate ion, SO_4^{2-}, is soluble and therefore the most common form. Sulphate is absorbed by plants, thus entering the cellular metabolism. In the cell, sulphate is reduced to sulphide and is present as the -SH radical in molecules. This radical is chemically active and forms disulphide bonds between chains of amino acids and participates in the quaternary structure and active sites of macromolecules. Incorporated in organic molecules, sulphur is transferred to animals. Excess sulphur is rejected in urine, in the form of sulphate.

Another sulphur source is decomposition of organic matter of plant or animal origin, including excreta. The transformation of these products by bacterial action, in particular by the thiobacilles, the chlorothiobacilles and the rhodothiobacilles, or by chemical action produces sulphide, hydrogen sulphide and elementary sulphur in reducing conditions, and sulphate in oxidizing conditions:

$$2H_2S + O_2 \longrightarrow 2H_2O + 2S + \text{Energy}$$

$$2S + 2H_2O + O_2 \longrightarrow 2\,SO_4^{2-} + 4H^+ + \text{Energy}$$

In coastal environments, such as mangroves, the presence of sulphur drastically modifies the chemical and biological characters of the water. The accumulation of organic matter in mangrove soil causes an enrichment in sulphide, which profoundly affects sediment acidity once sulphides are transformed into sulphate.

Sulphur deficiency causes a sudden increase in the synthesis of polyphosphates in bacteria. Their role is related to the regulation, at the intracellular level, of ATP, ADP, phosphonucleotides and orthophosphate; a deficiency in sulphur halts synthesis of nucleic acids.

4.5.3. Iron

Importance of iron

Iron is an oligoelement, abundant in rocks and sediment. It plays an important role in living matter, is involved in a number of redox reactions, in the synthesis of chlorophyll, and occurs in proteins and enzymes, such as myoglobin, hemoglobin, and cytochromes. Iron, in dissolved or available forms, is needed at concentrations between toxicity and deficiency limits (Huntsman & Sunda, 1980).

Iron species in water

Iron is usually present in oxidized form, iron (III) hydroxide, in natural waters. Iron oxide is sometimes combined to organic matter, such as humic substances, and to phosphorus, silicium, calcium, and trace metals. In natural waters, particles containing ferric oxide are of different origins. In some organic complexes, iron atoms are coordinated to functional organic groups and may be stabilized in ferrous form (Tipping *et al.*, 1982).

In Eastwhite Water, a small productive lake in Cumbria, U.K., iron concentrations in water samples taken at different periods showed an important increase of iron (III) in water at overturn, which can be explained by the ascendant movement of sedimentary iron (II) or iron from anoxic water layers to be oxidized in aerobic water.

Part of the iron in suspension in water, not sedimented by centrifugation, was in the form of small particles stabilized against flocculation by adsorption onto organic matter. Adsorbed humic substances impart negative charges. Iron (III) oxide produced by ferrous iron oxidation at or nearby the oxic-anoxic border during stratification was associated with humic substances, which were also negatively charged and underwent slow flocculation.

The degree of flocculation increased with increasing ion concentration, especially of multivalent ions, of opposite charge, such as magnesium or calcium. Flocculation depended also on total ionic charge and the amount of material in suspension. It involved associations with particles similar to those containing iron. Flocculated iron was mainly amorphous, consisting exclusively of iron or iron associated with clay minerals or particulate organic matter (peptized form). Calcium, in large amounts (equimolar amounts), silicium, phosphorus and sulphur were also consistently present. Iron complexes with organic compounds also occurred; they consisted of monomeric iron (III) atoms coordinated to humic functional groups, sufficiently stable not to undergo hydrolysis. A group of largely undefined iron compounds was found in the supernatant after centrifugation (Tipping *et al.*, 1982).

The importance of chelators

Most ferrous compounds are poorly soluble. However, owing to the presence of chelators and various complexing compounds, considerable amounts of iron are kept available in the water. EDTA is one of the artificial chelators most used in laboratories, while citric acid is used in bio-assays (Huntsman & Sunda, 1980).

In natural waters, humic and fulvic acids, and phenolic, quinone, enolic, carbonyl, amino and carboxylic groups have chelating and complexing properties. They are derived from soil and decaying plant material. Humic matter sometimes reaches concentrations of up to 70 mg l^{-1}, binding metals on the carboxylic and phenolic radicals, which are primary binding sites. Fulvic acids are less effective than humics in stimulating algal growth. As a matter of fact, humic acids are characterized by their capacity to bind iron and there are strong correlations between humate concentrations and amounts of dissolved iron. Iron may exist in peptized form rather than in true solution. Adsorption of other ions occurs, with a preference for divalent cations, such as copper, cobalt, and nickel, which are three to four times more abundant than the trivalent cations such as aluminium and iron (Huntsman & Sunda, 1980).

Humic substances stimulate algal growth, probably because they hold not only essential oligo-elements, but also common nutrients, such as phosphate and silicate. Iron is mainly linked to hydroxy and carbonate groups, which predominate in fresh water and are pH and alkalinity dependent. Chelate formation is a common binding mode, with low kinetics, but is not fully understood. It allows the temporary fixation of ions, which are transported through water, the pH of which would normally cause precipitation, and released when the environment permits. The rate at which they are released corresponds to the uptake rate of the organisms present. This system maintains suitable concentrations of these ions for organisms, preventing toxic concentrations from being reached. Chelators thus act as buffers for metals. The total amount of chelating agent should, however, not be excessive, leading to over-chelation (Huntsman & Sunda, 1980).

Iron concentration in water

Ferrous ions should be absent from aerobic water. They are, however, present, because a certain number of organic compounds, such as humic and tannic acids, and high chlorinity reduce the oxidation rate of ferrous iron. The chelating capacity decreases with salinity. Dissolved iron concentrations in natural waters are higher than those predicted by the solubility constant of iron hydroxide. In freshwater, Fe^{2+} and $FeOH^{+}$ are predominant, while, in sea water, $FeCl^{+}$ ions are more common. In freshwater, Fe^{2+} can compete successfully with calcium and magnesium ions for organic chelating agents, forming moderately stable Fe (II) complexes. In sea water, high concentrations of ions of the alkaline earth group outcompete ferrous ions for available chelators. Mercury, silver and cadmium ions are preferably complexed to chloride ions (Huntsman & Sunda, 1980).

Generally iron concentrations in natural waters are variable, depending upon rainfall, the amount of material in suspension, erosion, and human contamination (pollution) (Huntsman & Sunda, 1980). Normal average concentrations in temperate oxygenated water are in the order of 10^{-7} mol l^{-1}. Considering the solubility constants of iron compounds, the concentrations should be no more than 10^{-15} mol l^{-1}. The difference is due to the presence of chelators (Huntsman & Sunda, 1980).

Iron is always more abundant in anoxic sediment, in reduced form, and is recycled during mixing or overturn events, when sediments are resuspended.

Effect of iron on fish and algae

Excessive concentrations of ferric hydroxide in water cause suffocation of fish. Iron hydroxide precipitates in alkaline water, at a pH of 7.8. Ammonia excretion alkalinizes water, on and near the gills, which results in precipitation of iron hydroxide on the gill, leading to suffocation.

Phytoplankton is able to live in water containing variable amounts of iron and trace metals. Some species have evolved mechanisms enabling them to make use of varying metal availabilities. Some are able to bind metals outside the cell, by producing extracellular chelating agents, in particular hydroxamic acids. These hydroxamate compounds virtually outcompete all other ligands. They are produced by many species of bacteria and fungi, but the only phytoplankton group to do so seems to be the blue-green algae. To be effective, these hydroxamates have to outcompete other ligands, such as humates and fulvates. Some eucaryotic algae have developed a different strategy; insoluble iron (III) is reduced to the soluble ferrous form, by the mediation of extracellular products at the external cell surface. The general importance of this reduction mechanism in iron uptake has, so far, not been determined (Huntsman & Sunda, 1980). The high affinity process of blue-greens for impounding available iron in water influences competition with other algal groups. During blooms of blue-green algae, other plankton algae cannot thrive. The growth of blue-greens results in a high nitrogen fixation rate and a high iron demand. This induces production and excretion of hydroxamate siderophores or siderochromes by *Anabaena flos-aquae*, which has chelating properties for iron, outcompeting other species in culture (Murphy *et al.*, 1976; McKnight *et al.*, 1980). By controlling water quality, these algae insure their own dominance (Hart, 1982; Smayda, 1980).

Iron determination in water

Iron in water is determined spectrophotometrically. The amounts of dissolved iron measured are often higher than predicted, due to the occurrence of particulate iron. Some dissolved iron hydroxide is sometimes pooled with the particulate fraction. The separation of particulate and dissolved iron is thus not easy to perform. The rapid development of bacteria and phytoplankton on the walls of aquaria is due to the adsorption of iron onto surfaces (Huntsman & Sunda, 1980).

Importance of silicon, sulphur and iron in water

1. Silicon is supplied to the water by the sediments and run-off water. The silicon concentration in water is usually not limiting for algae, but may determine the structure of the plankton assemblage and its succession, as some algae, like the diatoms, have strict and high silicon requirements.

2. Sulphur is supplied to water by organic matter decomposition; microbial sulphur cycling is related to organic matter decomposition.

3. Iron is supplied by the sediments and is an important oligo-element in many organic molecules. Iron in oxidized form is insoluble in water, being present as iron oxyhydroxide. Iron ions are also present in organic complexes, such as humates. The Fe-DHM complexes are binding sites for phosphate and other nutrients. These iron complexes maintain iron (and other trace metals) in water at concentrations which range between deficiency and toxicity levels. Iron oxide in water harms fish, as it may precipitate on the gills causing suffocation. Algae have developed mechanisms to guarantee their iron supply: blue-green algae excrete hydroxamates, which have a high chelating affinity; other algae are able to reduce iron at the cell surface, before absorption.

Chapter 3.

Productivity and fish production

1. Primary productivity of water

1.1. Natural productivity and fish production

Types of aquatic organisms present in water

The aquatic environment supports various communities of living organisms. They constitute the biotic load of a pond. Natural productivity is the capacity to increase this biotic load over time. Assessment of this productivity is required in order to determine how much an environment can generate or produce using its natural resources.

Living communities are made up of different groups: benthos, plankton, periphyton, and fish. The names of the first three designate the place where they occur in the environment: in or on the sediment, within the water, and fixed on a substratum. Organisms can also be grouped according to their role in the ecosystem: producers, consumers, decomposers. Primary production is the rate of formation of organic matter due to photosynthesis, by all photosynthetically active organisms. Secondary production refers to increases in biomass of zooplankton, zoobenthos and fish, while decomposers recycle dead organic matter, using the energy and nutrients contained therein. These separations are somewhat arbitrary, as will be seen later.

Plankton designates the community of pelagic organisms, composed of various groups, which are in suspension in water and hence restricted mobility, often less than that of the water which carries them. Plankton is grossly divided in phytoplankton or photosynthetic organisms, and zooplankton or heterotrophic organisms. The smallest algae composing the phytoplankton belong to the group of blue-green algae and are also called bacterioplankton. Plankton may also be classified according to the size group to which they belong.

Seston is the particulate matter in suspension in the water column, comprised of approximately two thirds detritus and one third living organisms.

Periphyton consists of different types of communities, firmly attached to or moving freely on substrata of various kinds, such as leaves of aquatic plants or stones, forming dense felts of brown and green scums and composed of living cells thriving on dead cells underneath. It can be divided in phytoperiphytonand zooperiphyton. Phytoperiphyton is made of groups of algae (diatoms and green algae), thick mats of algal filaments or gelatinous lumps of algae (greens, such as desmids and *Spirogyra*, and blue-greens); algae which move freely on substrata, especially sediments, such as green algae and diatoms (Westlake, 1980). Zooperiphyton is composed of ciliates, rotifers, crustaceans, Bryozoa, Hydrozoa, sponges, Oligochaeta (Naididae). The periphyton attached to aquatic plants is sometimes called aufwuchs. The species composition of periphyton of tropical water is still largely undocumented.

Benthos designates the organisms on and in sediments and can, again, be divided in phytobenthos and zoobenthos. Algae of the phytobenthos are found on the surface of sediments, as, for instance, epipelic diatoms and blue-green algae. Organisms belonging to the zoobenthos have been discussed earlier (see p. 50). Zoobenthos is grouped according to size range, in microbenthos (bacteria, ciliates), meiobenthos (rotifers, nematods, cladocerans, ostracods, copepods) and macrobenthos (Turbellaria, molluscs, Oligochaeta, insect larvae, decapods).

Little information exists to date on the relative importance of periphyton and benthos and their roles in fish ponds. Periphyton develops in littoral zones of lakes, at shallow depths, where nutrient availability is important, and they are readily available food for invertebrates and fish. The response of phytoperiphyton to nutrient enrichment is usually poor. Larger individuals increase in proportion with increasing trophy, but the response to eutrophication is variable. Periphyton communities are dependent upon substratum quality, slope and fetch (Cattaneo, 1987). The development of benthic algae in shallow brackish water ponds in the Philippines is promoted to serve as food for the milkfish (lab-lab). In the following discussion, it should be remembered that, though the information is centered on pelagic algae, the contribution of periphyton and benthic algae must be included in primary production. However, their relative contributions remain to be quantified.

The last group of the living community focusing our interest is the fish. Fish species consist of primary herbivores or phytoplanktivores, and secondary consumers, zooplanktivores or predators, and their growth and production is dependent on natural food available to them (but see p. 179).

According to Noriega-Curtis (1979), fish production cannot be predicted solely on the basis of primary productivity of ponds fertilized with manure, because autotrophic production is not the only energy source for the fish. Protein biosynthesis through the microbial decomposition cycle (or the heterotrophic production cycle) is certainly as important as that of phytoplankton production. Apportionment of each of the production cycles to fish production will need to be assessed by future work.

Spectrum of algae found in ponds

The phytoplankton community consists of a large number of taxa. According to cell size, the following groups can be distinguished: picoplankton (from 0.02 to 0.2 μm), nanoplankton (from 0.2 to 2.0 μm), and phytoplankton (from 2.0 to 200 μm) (Sieburth *et al.*, 1978). Organisms over 20 μm in size are called netplankton. Picoplankton and nanoplankton, often ignored in plankton studies, can contribute significantly to primary productivity, often in excess of what their biomass would suggest (Hardy *et al.*, 1986; Stockner & Antia, 1986; Stockner, 1988). In experiments with bags suspended in lake Brooker, Florida, U.S.A., and which were filled with lake water ten times concentrated, Scheda & Cowell (1988) showed that they were efficiently consumed by filtrating zooplankton and played a key role in energy transfer towards higher trophic levels. Increases of herbivorous (phytoplanktivorous) rotifers affected nanoplankton biomass.

The most diverse algal group in tropical African water bodies are the green algae or Chlorophyceae, which make up between 50 to 85 % of the total number of species found. Amongst the green algae, the desmids are the most important algal order. Ponds and pools recently filled with rainwater are invaded by green algae, of the order Chlorococcales, belonging to the genera *Scenedesmus*, *Ankistrodesmus*, *Crucigenia*, and *Chlorella*, and also brown algae or Cryptophyceae, Chrysophyceae (of the genus *Ochromonas*) and Euglenophyceae. Green algae form dense population blooms in aquaria left in open air. If left undisturbed, such tanks will eventually fill with blue-green algae or Cyanophyceae. Zygnematalean algae, another order of the green algae, are filamentous, and can almost completely fill temporary pools where they form loosely entangled, gelatinous masses. *Spirogyra*, along with the cyanophyte *Lyngbya borgertii*, is reported to form entangled masses in pools in the Ivory Coast. In fertilized ponds, the phytoplankton is almost entirely made up of euglenoids. Green, flocculent, or felty growths of algae covering submerged stems and leaves are commonly slimy and consist of various zygnematalean algae, belonging to the genera *Spirogyra*, *Zygnema*, *Mougeotia*, and occasionally entangled with *Oedogonium*, which is usually a truly benthic form. Diatoms are opportunistic, and rapidly colonize newly submerged surfaces. They are the dominant algal class colonizing wooden surfaces. They become associated with surfaces in large numbers when forming blooms. They are a major component of epiphytic and epilithic colonies, along with blue-greens (John,

1986).

Species numbers of blue-green algae or Cyanophyceae (also termed Cyanobacteria) and of Euglenophyceae represent each ca. 5 to 15 % of the total number of algae species. Bacillariophyceae or diatoms are quite diversified and make up 15 to 25 % of species diversity. Chrysophyceae, Xanthophyceae, and Dinoflagellates represent only 1 to 2 % of the total number of species (John, 1986). Diatoms require lower light intensities than green algae (Chlorophyceae) for their development. Blue-green algae can withstand higher light intensities and higher temperatures. Almost half of the species are widely distributed, not specific to Africa (Iltis, 1980).

Algae associated with sediments are called epipelic and consist of assemblages of diatoms and filamentous blue-green algae. These algae possess a vertical migration rhythm similar to that reported in temperate regions. Upward migration occurs just before dawn (John, 1986).

Effects of phytoplankton on the environment

Phytoplankton development benefits ponds for the following reasons (Erez *et al.*, 1990):

- a dense phytoplankton bloom can produce ten times more oxygen than it consumes. This oxygen production is important in compensating for respiration and is supplied to the pond without added energy expenditure
- phytoplankton represents an efficient means of removing excess nutrients or toxic micronutrients, especially ammonium, from the water
- phytoplankton acts as a screen to light penetration in the water and prevents development of benthic algae. The development of scums of blue-green algae on the bottom promotes production within the sediment of a reductant surficial layer, rich in hydrogen sulphide, which is highly toxic. These algal scums remove nutrients and postpone the return to normal growth for algae after an algal crash
- phytoplankton is good food for planktivorous fish. Exploiting primary productivity is a cheap way to produce fish
- the development of phytoplankton prevents the development of macrophytes, which is unfavourable for fish culture.

Relations between phytoplankton and planktivorous fish

In ponds relying on primary productivity for fish nutrition, daily phytoplankton production, as well as ways to increase the latter, should be known. The increase in supply of available phytoplankton is used to calculate the amount by which the stocking rate of fish can be increased, or the amount by which the fish load in a pond has increased, due to fish growth. Phytoplankton availability has to be adjusted to fish consumption to achieve high production. Consumption of phytoplankton by fish and effects on pond primary production will be discussed later (p. 184 & p. 179).

Common features for lakes and ponds

An abundant limnological literature on nutrient cycles and on the dynamics of living communities exists. Application of information collected in natural water bodies to fish ponds must be done cautiously, taking into consideration the common features, but also the differences between the two types of environments.

Morphometric differences are linked to the respective sizes of the water bodies; natural lakes are always some order of magnitude larger than ponds, the area of which never exceeds few hectares. The water mass of lakes is relatively stable against changes in physical and chemical variables. Ponds are usually not more than 3 m deep, while lakes with maximum depths of 3 m are considered shallow. This entails large differences in the exchanges between sediment and water. Lakes are permanent water bodies, with an external water supply, from a river or at least run-off water, which also supplies nutrients. In a fish pond, the water supply is limited to pond impoundment and does not supply a significant amount of nutrients.

Despite such differences of scale, and the greater stability of lakes, lacustrine information is important in understanding the functioning of ponds.

Production in tropical and temperate regions

According to Nilssen (1984), production processes in tropical and temperate regions are fundamentally similar. Kilham & Kilham (1990) consider that, in tropical aquatic environments, the biological control of elementary cycles is predominant over physical processes and morphometric parameters, unlike in temperate lakes. The consequence is that nutrient regeneration is more important and nutrient recycling in the tropics faster. Endogenous processes of a lake dominate the external loading. Lakes show less long-term stratification and the frequent overturn of water allows resuspension of sediments, which promotes mobilization of sequestered nutrients and a better oxygenation of the water. The reduction of water density makes mixing easier, and to a greater depth. In a study of 39 natural lakes in Cameroon, Kling (1988) observed that light penetration is deeper and water temperature fluctuations are more narrow than those of temperate lakes. The regeneration rate of phosphorus is, in the long term, more important than its immobilization. In a great number of lakes, more than 85 % of phosphorus is recycled from zooplankton excretion, and phytoplankton and detritus decomposition, the external supply not exceeding 5 %. Temperature also acts to increase enzymatic reactions. Silica is supplied internally from the sediments by solubilization (Kilham & Kilham, 1990). More profound differences exists in the taxonomic composition of tropical and temperate zooplankton (p. 165).

Natural productivity: what the concept covers

The natural productivity is the capacity of the living community to increase its biotic load over time. The community of living organisms can be separated according to their position in the food web as producers, consumers and decomposers. This separation is based on their mode of harnessing the energy needed for their metabolic activities. Three main processes emerge: photosynthesis, the use of solar energy to elaborate organic compounds; grazing and predation, or consumption of living organisms containing high-energy compounds; decomposition or consumption of dead organic matter, again to make use of high-energy compounds. These three processes are complementary and interdependent.

1.2. Determinants of primary productivity

Primary productivity characteristics

Primary productivity is the production of organic matter or carbon fixation by photosynthesis, using inorganic nutrients and light as an energy source.

To measure primary productivity is to assess the rate of photosynthesis, i.e. the rate at which chemical energy and organic matter is added to the system. The rate of production of organic matter is governed by the rate of input of light into the water column and by the capacity of the organisms present to use it for gross photosynthesis. Gross photosynthesis is the transformation rate of radiant energy to chemical energy, accompanied by equivalent conversions of carbon dioxide and water into organic compounds, with release of oxygen. However, while performing photosynthetic activities, the plant uses energy, converting some of the organic compounds back into carbon dioxide. Respiration performed in the light is called photorespiration, while respiration which occurs at night is called dark respiration. Photorespiration and dark respiration are different; thus, net photosynthesis cannot be correctly and precisely measured. This is why the term photo-assimilation was coined. This term designates the difference between the rate at which organic carbon becomes assimilated into the plant and the dark respiration rate, stressing that this is not the true net photosynthetic rate (Westlake, 1980) (see p. 132). For blue-green algae, it has been shown that photorespiration may be twenty times greater than dark respiration (Westlake, 1980). Net photosynthesis is a function of the availability of nutrients, and, eventually, of the increase in phytoplankton biomass. In the presence of high biomass, photosynthesis decreases owing to self-shading.

Nutrients needed for photosynthesis are supplied by the environment or from outside. Carbon is supplied by carbon dioxide and is primarily used to produce glucose, the basic energy molecule. Nitrogen and phosphorus are mainly supplied by decomposition of organic compounds, partly also by the atmosphere, through nitrogen-fixing organisms, and by geochemical phosphorus released by the sediment. Supplies of nitrogen and phosphorus contained in the feed and fertilizers should also be included. The major share of nutrients available to primary production comes from recycling and regenerative processes. The nutrient concentration in water may limit the specific rate of gross photosynthesis directly, but except in situations where carbon is limiting, this is not common (Westlake, 1980) (see also p. 155). Very rarely, situation arise (in conditions of extremely dense algae blooms, and hard water with high pH (Tilzer, 1989)) where the supply of CO_2 severely restricts the extent of phytoplankton photosynthesis (Wetzel, 1983). In an oligotrophic lake in Pennsylvania, U.S.A., in conditions of intense production, the growth of periphyton (86.1 % of which were Chlorophytes) was limited by the availability of dissolved inorganic carbon and was enhanced by a supplement of organic and inorganic carbon and nitrogen (Fairchild & Everett, 1988).

The photosynthetic capacity

Photosynthetic capacity is defined as the maximum specific rate of photoassimilation reached at high irradiances or in light-saturating conditions, expressed per unit of chlorophyll a, in mg O_2 (mg chl a)$^{-1}$ h^{-1}, or sometimes in dry or wet weight (Westlake, 1980). This capacity is dependent on temperature, on nutrient availability and on the amount of photosynthetic organisms present in the water, i.e. standing crop (Fig. 34). For phytoplankton, photosynthetic capacity varies in the range of 0.5 to 38 mg O_2 (mg chl a)$^{-1}$ h^{-1}, which corresponds to a population renewal (or a life expectancy for the algae) of 180 to 3 hours. For macrophytes, it is estimated to be ca. 2 mg O_2 (mg chl a)$^{-1}$ h^{-1} (Westlake, 1980).

Primary productivity is always a function of light; production versus light intensity reveals a production optimum. This production optimum can be divided by a measure of biomass (chlorophyll, biovolume or particulate carbon) to give a light-independent measure of the carbon turnover in phytoplankton (Heyman & Lundgren, 1988).

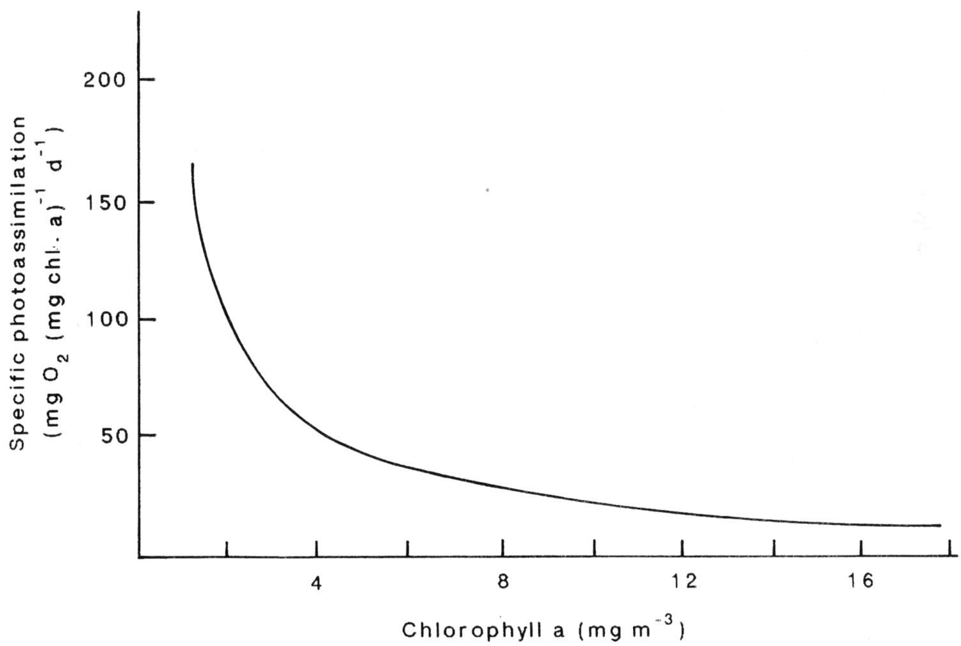

Fig. 34. Relationship between mean daily specific photoassimilation and mean algal biomass concentration (euphotic zone), expressed as chlorophyll a concentration, in a Czech reservoir (after Javornický, 1980).

Methods to measure photosynthetic rate

To measure photosynthesis, one of two conventional methods is usually adopted: oxygen production in dark and light bottles, the oxygen method of Gaardner & Gran (1927), or carbon-14 fixation (Steemann Nielsen, 1952; Harris, 1980). Both methods have serious limitations. With the spiked carbon method, no reliable estimate of respiration is possible (Westlake, 1980). Some of the fixed carbon is directly lost to respiration and part of this released carbon dioxide is used anew in photosynthesis.

The increase of oxygen in a bottle is not a true reflection of the oxygen released by photosynthesis, as some amount is consumed by respiration (community respiration) of seston (the heterotrophic and autotrophic organisms present in water) (Stack, 1973). The respiration rate in the light, the photorespiration, and in the dark, the dark respiration, are obviously not equal. Nighttime carbon losses usually comprise 30 to 50 % of the previous day's uptake and are far in excess of daytime respiration (Tilzer, 1990). Therefore, the assumption that the respiration rate in the dark bottle provides a basis for the calculation of net photosynthesis in the light bottle is challenged (Westlake, 1980).

Carbon fixation calculated by the dark and light bottle method gives a value for carbon fixation about twice that found by the carbon-14 method. The following rule is therefore adopted: results obtained using the carbon-14 method with phytoplankton are multiplied by three to obtain the oxygen output of photoassimilation. With higher plants, result obtained by the carbon-14 method are multiplied by 2.5 (Westlake, 1980).

In a given environment, oxygen consumption increases in direct proportion to the size of the bacterial population. The respiratory activities of bacteria can constitute up to 30 % of the total oxygen consumption. The daily bacterial consumption rate is estimated to be 0.1 10^{-9} mg O_2 per cell. The average rate of specific community respiration is assessed to be 1 mg O_2 (mg chl a)$^{-1}$ h^{-1}, with a normal range between 0.1 to 4.5; up to 10 mg O_2 (mg chl a)$^{-1}$ h^{-1} have been recorded, during periods when high bacterial activity was suspected (Westlake, 1980).

Factors determining photosynthetic rate

Factors determining photosynthetic rate are besides light, nutrient concentrations and volume of producers present, i.e. the biomass of phytoplankton, periphyton and macrophytes, including their species composition. In addition, the presence of inhibitors of photosynthesis, and the density of herbivores should be taken into account. Producer biomass is assessed on the basis of chlorophyll measurements. Factors affecting cell growth, such as temperature, will be discussed on p. 137.

Maximum densities of chlorophyll a

The maximum chlorophyll content is around 2 g of chlorophyll a m^{-2} for macrophytes and 500 mg (range: 180-620 mg) of chl a m^{-2} for phytoplankton, which is equivalent to 200 and 50 mg m^2 of dry matter for the macrophytes and phytoplankton, respectively. In Lake George, Uganda, phytoplankton density has been found to reach 400 mg of chl a m^{-2} (Horne & Viner, 1971).

In chemostat culture, Agustí *et al.* (1987) showed that the maximum density of algae in a population is a function of the -1.27 power of cellular volume. In a study involving 11 species of freshwater algae, followed in batch culture, the maximum density varied as the -0.79 power of cellular volume, independently of environmental conditions. Physiological constraints and self-shading are probably responsible for this value (Agustí *et al.*, 1987; Agustí & Kalff, 1989).

Phytoplankton density and photosynthetic rate

With increasing phytoplankton density, i.e. increasing biomass, photosynthetic rate decreases, owing to alteration of pigment content or modifications of photosynthetic enzymes. The daily mean specific photoassimilation, expressed in terms of algal volume, declines rapidly as cell volume increases. This suggests a deterioration of certain factors in the environment of the planktonic algae, leading to a decrease in chlorophyll a content; chlorophyll synthesis lags behind cell growth, which leads to a decrease of specific photoassimilation in terms of carbon volume, while, in terms of chlorophyll, it remains more or less constant. Two main factors are involved: self-shading in shallow waters by an abundance of plankton, and nutrient depletion (Westlake, 1980).

1.2.1. Influence of light

Autotrophic and heterotrophic organisms

The energy required for cellular metabolism is supplied to animal organisms by oxidation of energy-rich organic compounds, while plant-like organisms use solar light, transform it into chemical energy and elaborate organic compounds. Animals are heterotrophic organisms, while macrophytes and algae are autotrophs. Autotrophic organisms introduce energy into the system, which is captured with the help of photosynthetic pigments. They are therefore the base of production. Solar energy is also the cheapest source available.

Among heterotrophs, different subgroups are distinguished: chemoheterotrophs are organisms growing without solar light, while photoheterotrophs use light as an energy source, but organic compounds for their nutrient requirements. Mixotrophic organisms can adjust their metabolism to prevailing conditions. A certain number of phytoplanktonic organisms are mixotrophic, especially amongst the blue-green algae of the genus *Oscillatoria* (Feuillade & Feuillade, 1990). Observations on two lakes in Quebec, Canada revealed that some mixotrophic Chrysophyceae of the genus *Dinobryon* depend more on ingested bacteria than on photosynthesis for their metabolism (Bird & Kalff, 1987).

Light absorption by water and stratification according to light intensity

Part of the solar rays falling on the water surface is reflected, while rays which penetrate the water mass are refracted. The photosynthetically active radiation (PAR) at zero depth, I'_o, represents about 40 % of the incident radiation, I_o, and is composed of the visible spectrum, between 400 and 700 nm (Westlake, 1980). It is progressively absorbed with depth. Absorption depends on wavelength, on turbidity, and on colour.

Photosynthesis is nil when irradiation drops below 1 % of incident radiation. This is the limit of the euphotic zone, or the extinction depth. The optical depth defined by the subaquatic irradiance relative to the surface irradiance (Westlake, 1980), creates a vertical stratification of the water column. The vertical attenuation coefficient varies with wavelength, is different for each wave-band (Westlake, 1980) and is dependent on the biotic and non-living particles in the water.

Photo-assimilation rates as a function of depth present a vertical stratification. Three zones can be distinguished: a lower one where light is limiting and photosynthetic rate is low, corresponding to about 15 to 1 % of the incident light; a middle one where light is saturating and photosynthetic rate is at a maximum, corresponding to 60 to 15 % of the incident light; and an upper or inhibition zone, of low photosynthetic activity, where high irradiances lead to photoautoxidation of the photosynthetic pigments (Fig. 35) (Westlake, 1980).

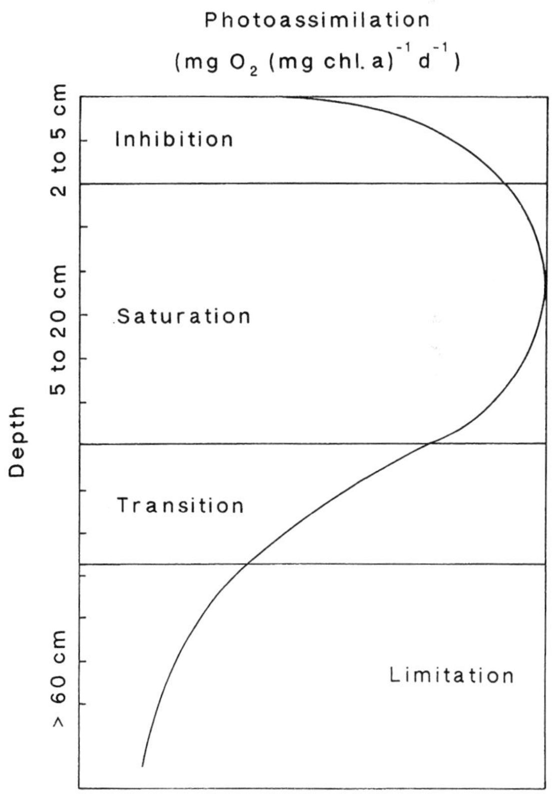

Fig. 35. Generalised diagram of photoassimilation in a homogeneous phytoplankton population according to depth and light penetration within the water mass (from Tilzer, 1980) (with kind permission of Prof. Tilzer, Dr. LeCren & Cambridge Univ. Press).

Oxygen production in a pond is a function of light and depth. The light compensation point is reached at the irradiance value (E_c) where photosynthetic oxygen liberation just equals respiratory oxygen consumption. Above this depth or that irradiance value, net photosynthesis is achieved and photosynthetic rate increases linearly with irradiance, E_d, up to a certain value. After that it curves and eventually levels off. In the range of irradiances where photosynthetic rate does not vary with E_d, photosynthesis is said to be light-saturated. The rate has the value of P_m, the maximum specific photosynthetic rate, sometimes referred to as the photosynthetic capacity. With further increases in irradiance, P begins to decrease again, a phenomenon referred to as photoinhibition.

The light intensity required to saturate photosynthesis, and the compensation point vary markedly from one species to another. These parameters also depend on CO_2 concentration and temperature. The light compensation point changes seasonally, lower in winter or spring than in summer or autumn; whether this is due to differences in temperature or whether other factors are involved is uncertain (Kirk, 1986).

Light absorption or attenuation and trophic loading

The incident irradiance is attenuated underwater, by absorption and scattering (Westlake, 1980). The absorbing and scattering components are the water itself, the plankton, the seston and dissolved substances. Their relative importance in the process and how they compete for spectral energy are largely unknown, partly because of technical difficulties (Yentsch, 1980). The degree of absorption and scattering increases with increasing trophic loading of the water. Irradiance in the green, yellow and even red is least attenuated (Westlake, 1980). Detritus primarily absorbs ultraviolet light (Yentsch, 1980); UV and infrared irradiances are both rapidly attenuated. In ponds with high plankton and organic loads, all light is rapidly attenuated and photosynthetically active radiation is reduced to a low value. Minimum attenuation in water with high phytoplankton abundance occurs in the orange-red, while the phytoplankton has its most marked effects on light in the blue spectral region (Westlake, 1980; Yentsch, 1980).

Use of light in photosynthesis

Autotrophs utilize light as an energy source. Two photochemical reactions, occurring in series in photosynthesis harness this energy. Light reaction 1 is associated with the reduction of NADP (photosystem I); light reaction 2 brings about the liberation of oxygen from water (photosystem II). Each photoreaction takes place in a reaction centre in association with specialized light-harvesting pigment-protein and electron transfer agents. The two systems are referred to as the photosynthetic units. Each photosystem has a chlorophyll a/carotenoid-protein which harvests light and is intimately associated with the reaction centre. The pigment-protein represents a funnel through which all the excitation energy collected by other pigment-protein must pass before it is delivered to the reaction centre. Complex I is made of chlorophyll a/ß-carotene-protein and is present in all plants, photosystem II is not yet so well characterized (Kirk, 1986).

For a plant to be able to make efficient use of light quanta absorbed at a given rate, the activity of the electron transfer components must be high enough to ensure that the excitation energy collected by the light-harvesting pigments is utilized as fast as it arrives at the reaction centres. At low light intensity, the autotroph is achieving its highest conversion efficiency and quantum yield. As the incident light intensity increases, the rate of absorption of quanta reaches a point at which excitation energy arrives at the reaction centres faster than it can be used by the electron transfer components and/or the CO_2 fixation enzymes. Some of the additional absorbed quanta are utilized for photosynthesis, and some are not, the energy of the latter being eventually dissipated, mainly as heat. In the light-saturated state, the electron transfer and/or the CO_2 fixation enzymes are working as fast as they are capable, and so any additional absorbed quanta are not used for photosynthesis at all.

The amount of photosynthesis achieved per unit light absorbed is lower at some wavelengths than at others. For *Chroococcus* (blue-green alga), *Chlorella* (green alga), *Navicula* (diatom), a fall in quantum yield is observed in the red absorption peak (680-710 nm) of the chlorophyll and in the region of carotenoid absorption (440-520 nm, blue portion of the spectrum). This is particularly marked for the blue-green alga, significant for the green alga, slight for the diatom. One explanation is that light absorbed by the carotenoid fucoxanthin in diatoms is efficiently used, while light absorbed by carotenoids in blue-green and green algae is used with much lower efficiency. The more likely explanation is now considered to be the fact that the two light reactions must be energized for photosynthesis to occur. Algae have different light-harvesting pigment arrays. Monochromatic light is strongly absorbed by one photosystem, but weakly by the other. Photosynthetic efficiency is low (Kirk, 1986).

Therefore algae have to be exposed simultaneously to both wavebands (the blue and the red bands), which have an synergistic effect. This is referred to as enhancement. When photosystem II is absorbing light faster than photosystem I, some of the excitation energy is transferred to photosystem I: this is referred to as the spill-over hypothesis. The reverse is not possible: light absorbed in system II can drive both light reactions, but that of system I drives only light reaction I. The fall-off in quantum yield in certain parts of the spectrum indicates that those spectral regions absorb mainly due to photosystem I pigments.

The rate of light collection does not only depend on their pigment complement but also on their position and orientation within the cell. In brown algae, chloroplasts moved to the cell walls facing the light at low irradiance and to the side walls parallel to the light direction at high irradiance. The change of position is completed within 1-2 hours.

In the filamentous green algae *Mougeotia* and *Mesotaenium* which have a single, flat centrally-located rectangular chloroplast in each cell, the characteristic movement of the chloroplast in response to a change in the light regime is to rotate along its longitudinal axis, rather than to move around the wall. In planktonic algae, where the cells contain one or a few chloroplasts, the plasts shrink and aggregate in bright light (Kirk, 1986).

Effects of limiting or excess light on phytoplankton

Light deeply affects algal growth, as will be discussed later (Factors affecting growth rates, p. 138).

Reactions to light intensities vary according to species. Light requirements, the ability to adapt to low or changing light intensities, and resistance to high light irradiations determine the outcome of competition between algae (Nicklisch & Kohl, 1989). Some, such as the blue-greens, thrive at high light intensities, which would kill others by photo-oxidation.

In some conditions, reduced growth is caused by self-shading of algae, which can occur independently of lack of nutrients. In a survey on phytoplankton communities in 165 Florida lakes, this caused a modification of the phytoplankton assemblage, in which small diatoms were replaced by blue-greens of large size (Agustí *et al.*, 1990).

Excess light has an inhibitory effect. This effect increases in the course of time. The higher the temperature at a given light intensity, the more rapidly inhibition ensues. Phytoplankton can recover from the inhibitory effect of intense light if transferred to a lower light intensity.

Mechanisms of photoinhibition have been studied in detail in higher plants. The spectrum of photoinhibition of electron transport showed its main activity in the ultraviolet region, with a peak at 250-260 nm. Photoinhibition also occurred in the visible region but with a much lower quantum efficiency. Between 400-700 nm, the action spectrum followed the absorption spectrum of chloroplast pigments, with a distinct chlorophyll peak at 670-680 nm. Lesions appear primarily to affect the light reactions of photosynthesis, possibly by damaging the reaction centre of photosystem II.

Photoinhibition in the surface layer is not as frequent as was originally thought. Except under very still conditions with no wind or waves, there is always circulation of water in the upper layer. This avoids phytoplankton from being exposed to intense light, thus preventing photoinhibition. The transport downward must, however, not reach too low.

Efficiency of the use of solar energy in ponds

In a fish pond under intensive culture using manure and artificial feed, exploited according to traditional Chinese practise, a total efficiency of solar energy utilization by primary production was 1.2 %. The efficiency of transformation of plankton by planktivorous fish reached 4.2 %, which gave an efficiency of utilization of solar energy by filter-feeding fish of 0.05 %. Compared to natural aquatic ecosystems, in which the efficiency lies around 0.1 to 0.5 %, such fish ponds are extremely efficient (Li, 1987).

Photosynthesis determinants and influence of light

1. It is not possible to measure net photosynthetic rate exactly, as part of the organic compounds elaborated by the photosynthesis are directly used (by respiration) for metabolic needs. Such respiration, occurring simultaneously with photosynthesis, is termed photorespiration as opposed to dark respiration, which occurs in darkness. Photoassimilation is an approximate measure of photosynthetic rate.

2. Photosynthesis is determined by the amount of light available, nutrient availability, the volume of autotrophs, and temperature.

3. Light is progressively absorbed in water and by particles in suspension in the water. The availability of light according to depth leads to a stratification with regard to photosynthesis. Three zones can be distinguished: that of light inhibition causing photoxidation of the pigments, that of light saturation, and that of light limitation.

4. An increase in phytoplankton biomass results in a decrease of photosynthetic rate, owing to self-shading and nutrient depletion.

1.2.2. Influence of phytoplankton biomass

Phytoplankton abundance and measurement of phytoplankton density

In a pelagic system, the amount of phytoplankton determines the rate of photosynthesis. The phytoplankton is composed of an assemblage of algal species. Inventorization of such assemblages requires good systematic skills and has to be done regularly as species assemblages change with time and with modifications of water quality. The phytoplankton is grazed upon by the zooplankton and phytoplanktivorous fish and this grazing pressure also modifies phytoplankton composition.

Phytoplankton density is measured by using plankton netting of adequate mesh size and measuring the quantity of phytoplankton filtered from a known volume of water. It can be expressed as dry weight, as chlorophyll or by numbers of taxa per unit volume or surface area. The biomass of phytoplankton recorded is the standing crop and can be compared to the carrying capacity of the aquatic environment, which is the maximum density attainable.

To express photosynthetic capacity of phytoplankton, it is necessary to have a common conventionally, the amount of chlorophyll a is used as the reference unit, because the ratio of chlorophyll a to cellular organic carbon is fairly constant, in the range of 21.3 to 24.4 mg of chlorophyll a per gram of cellular carbon (Westlake, 1980). Use of chlorophyll as a biomass estimator has, however, been questioned by Heyman & Lundgren (1988), who state that the chlorophyll content of algae may vary up to 5 times with temperature, light, nutrient, and species. Westlake (1980) showed a negative parabolic relationship between phytoplanktonic biomass and chlorophyll a (Fig. 36). A fairly constant ratio between chlorophyll a and carotenoids has also been found (Westlake, 1980).

Phytoplankton biomass changes results from the growth of different species composing it, and from environmental factors modifying it.

Phytoplankton growth

Phytoplankton growth is expressed as a rate, i.e. an increase of biomass with time. This increase is proportional to a specific growth rate, μ, and biomass, B:

$$\frac{dB}{dt} = \mu . B$$

The growth of phytoplankton shows a J-shaped curve in unlimiting conditions, and an S-shaped curve in restricting conditions (Fig. 37). Density rises until a factor becomes limiting. Limiting fac-

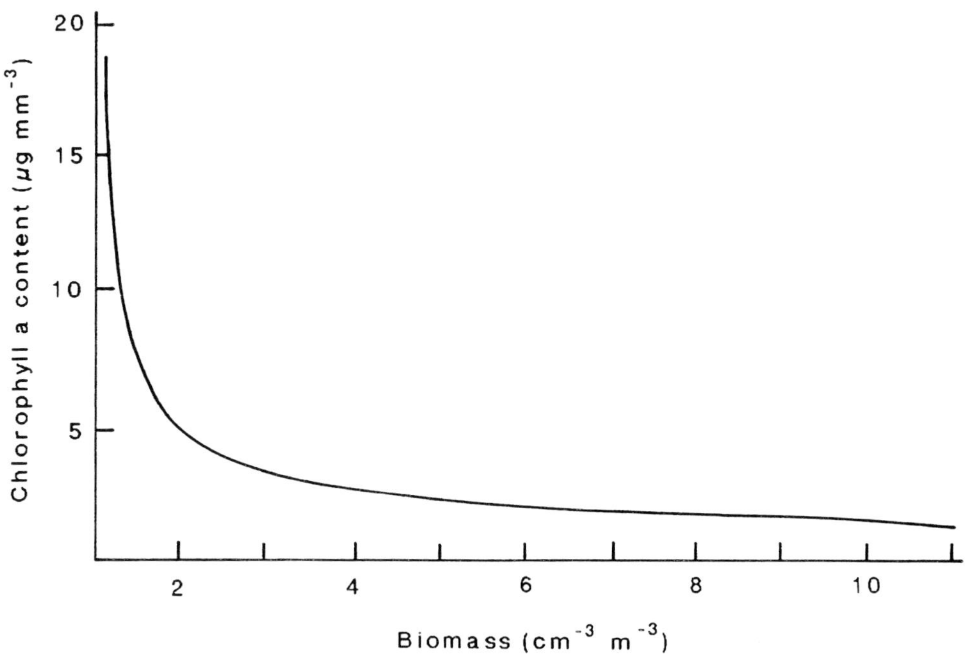

Fig. 36. Chlorophyll a content of phytoplankton (expressed in μg mm^{-3}) against mean biomass concentration (cm m^{-3}) in the euphotic zone of a Czechoslovakian reservoir (modified from Javornický, 1980).

tors are nutrients, or light. The specific growth rate varies and is fastest (the maximum specific growth rate) at the point of inflection of the S-curve (Ahlgren, 1988; Heyman & Lundgren, 1988).

Factors affecting specific growth rates

A common model of restricted growth is the model of Verhulst (1838), where growth rate is plotted against carrying capacity, i.e. maximum attainable biomass. Another group of models considers growth of a population in relation to nutrient concentrations, either extracellular or intracellular. In fact, nutrient concentration determines carrying capacity.

Specific growth rate is also dependent on external nutrients (S) (Monod model (Monod, 1942)) and internal nutrient concentrations (q) (Droop model (Droop, 1974)) (Fig. 38). The relations between the external nutrient concentration and the specific growth rate, and between the internal nutrient content and the specific growth rate are both hyperbolas. In the case of the internal nutrient level, the origin of the curve does not depart from zero but from the minimal cellular nutrient level, q_o. For

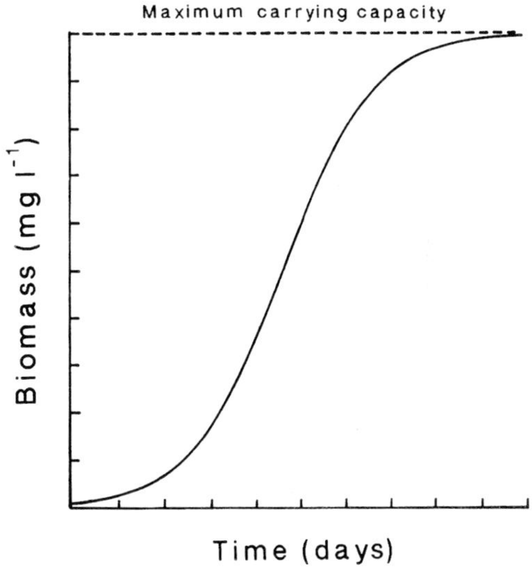

Fig. 37. Generalised growth curve of algal biomass as a function of time (modified from Ahlgren, 1988).

each nutrient, such a relation can be established. Such models consider the role of each nutrient independently, i.e. without interaction between nutrients. It is a widespread opinion that the single nutrient

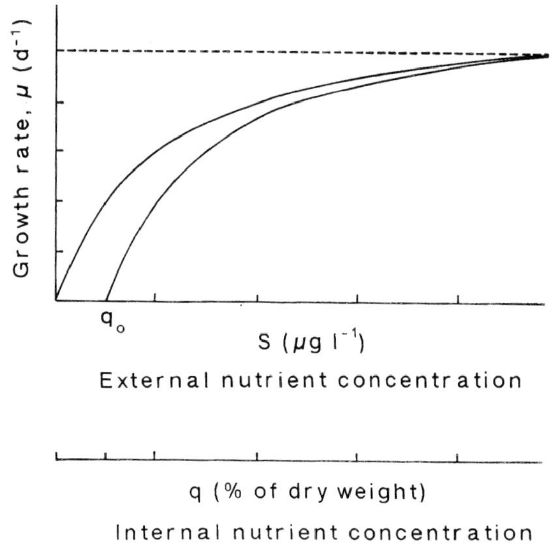

Fig. 38. Generalised specific growth rate curve for algae as a function of (1) external nutrient concentration, S, and (2) internal nutrient concentration, q. q_o is the minimal internal nutrient concentration or minimal cell quota. The dotted line shows the maximum specific growth rate associated with S and q, respectively (modified from Ahlgren, 1988).

which is in short supply, determines the algal growth rate, independently of other nutrients. This is called the threshold or Liebig-Blackman model. Opposed to this is the interactive or Mitscherlich-Baule model, which considers suboptimal nutrient levels to act simultaneously, leading to multiplicative effects (Ahlgren, 1988) (see also p. 140). Clear relations between phytoplankton growth and nutrient concentrations are difficult to show in the field; for example, biomasses for a given level of phosphorus might vary by a factor of 5 to 10. In the case of phosphorus being the limiting nutrient, any supply of phosphorus will quickly be removed, and recorded concentrations will appear low, though production will be high. If a population is heavily grazed upon, biomass can be kept low in comparison with carrying capacity, but specific production will be high (Heyman & Lundgren, 1988). This is common in tropical waters.

The specific growth rate depends also on light and temperature. Interactions between light and nutrient limitation are complex and the respective effect of each factor cannot be singled out. But, light limitation is more severe than nutrient limitation, because nutrients cannot be utilized if sufficient energy is not provided. This in indirectly confirmed by plankton adaptations to changing light regimes (Tilzer, 1990).

Nutrient content increases at low light intensities, as has been shown for silicon and nitrogen (Fig. 39).

In the case of nitrogen, this is an adaptative mechanism, associated with increases in the size of the photosynthetic units. The combined effects of light and nitrate limitation are greater than the sum of the individual effects.

The effect of light on phosphorus metabolism is more complex. Light stimulates phosphorus uptake by algae (Jansson, 1988). In cultures of *Skelotonema costatum* and *Dunaliella tertiolecta*, it was shown that under P limitation the value for light saturation decreased. In both cultures, the respiration rate under P limitation was reduced to a lower value (25 %) than under N limitation (40 %) (McAllister *et al.*, 1964). Phosphorus limitation reduces the availability of phosphate for high-energy phosphate compounds, and affects cellular respiration. Excretion of dissolved organic P increased with light, but N and P limitation reduced excretion. Growth rates were also reduced under such P and N limitations (Kuenzler, 1970).

A three-way interaction between light, P and carbon dioxide has been also shown. To sustain a

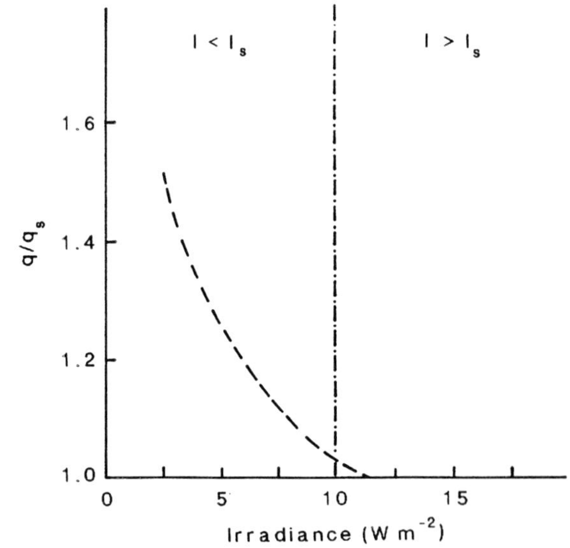

Fig. 39. Ratio between the actual cell quota, q, and the cell quota, q_s, at light-saturation, I_s, as a function of light irradiance, I (modified from Ahlgren, 1988).

given growth rate, the alga needed 10 times higher CO_2 concentrations at both P and light limitation than when the factors were limiting separately. At higher growth rates, light became the primary limiting factor (Young & King, 1980).

The effect of light and nutrient limitation on protein and polysaccharide synthesis was studied in several experiments. Some experiments followed the fate of fixed CO_2 in algal cells (*Merismopedia tenuissima*, a blue-green) under light and nutrient limitation. The percentage of total fixed carbon increased with μ and the increase was stronger under P than under N limitation and is explained because protein synthesis is obstructed by N limitation (Konopka, 1983). The amount of C incorporated into protein decreased with increasing light, but in P-limited conditions, incorporated C was constant under different light conditions. The percentage of total fixed C in polysaccharides decreased with μ in P-limited conditions and was independent of μ in N-limited conditions. In nutrient-sufficient conditions, the percentage of polysaccharides increased with light (Konopka & Schnur, 1980 & 1981). Polysaccharides and protein synthesis should be mirror images of each other, when they are measured as percentage of total fixed carbon (see Hecky & Kilham (1988): 799, Fig. 1) (Ahlgren,1988).

Comparing the percentage of C incorporated into protein against a light gradient for P-limited and nutrient-sufficient algae, (preconditioned at two different light intensities) it was shown that below 20 $\mu E\ m^{-2}\ s^{-1}$ (light limitation) the percentage of protein increased with increasing light. Between 20-100$\mu E\ m^{-2}\ s^{-1}$ (both light and P-limitation) it decreased, and above 100 $\mu E\ m^{-2}\ s^{-1}$ (only P-limitation), it was independent of light. P-limited algae incorporated only half of the C into protein than nutrient-sufficient algae. Algae preconditioned to low light showed smaller (especially at low light intensities, 20 $\mu E\ m^{-2}\ s^{-1}$ under light limitation) differences between non-limited and P-limited conditions, than algae preconditioned in strong light. There may be interaction between light and nutrient limitation, which affects the physiological state of algae (Konopka, 1983). One of the consequences of this observation is that nutrient-sufficient populations cannot be distinguished from nutrient-deficient populations without due consideration of light conditions (Ahlgren,1988).

Interacting effects of light, N, and P was investigated in chemostats on the blue-green *Synechococcus linearis*. In a narrow range of irradiance, both N and light or P and light simultaneously influenced the growth rate of the alga. With decreasing irradiance, the internal content of the limiting nutrient increased. The q_o for N clearly increased with decreasing light, but not the q_o for P. The critical N/P ratio (R_c, where the algae change from N to P limitation) increased with decreasing light (Healey, 1985). Since μ is strongly correlated with light, μ should also be correlated with R_c. With decreasing light or μ, algae need relatively more N than P.

As conditions become adverse, an increasingly higher internal content of the limiting nutrient is required. Three explanations are suggested (1) that in depleted conditions, a physical factor, light, or a major nutrient may be able to substitute each other (Healey, 1985); (2) that in non-optimum conditions algae adapt by increasing the concentrations of enzymes per cell, and as enzymes consist of essential nutrients, their requirements increase (Steemann Nielsen & Hansen, 1959); or (3) that uptake mechanisms of N and P are less influenced by low light than growth metabolism. The metabolic utilization of such nutrients is therefore reduced, hence their accumulation (Ahlgren, 1988).

Ammonium ions are smaller in size than phosphate and should be taken up with less energy expenditure than phosphate. The second explanation is considered the more likely.

Temperature affects growth rate, because it affects the kinetics of enzymatic reactions (Fig. 40) and determines maximum growth rate. Contents of phosphorus, nitrogen and, to a lesser extent, carbon are increased at lower temperature. The minimum cell quota varies with temperature, being minimum at the optimum temperature and increasing at low and high temperatures (Fig. 41). Average cell size decreases with increasing temperature to reach a minimum, before increasing again (U-shaped relation) while biomass remains fairly stable (Fig. 42) (Ahlgren, 1988).

Specific growth rate and biomass

The specific growth rate controls the amount of biomass produced. Growth rate decreases from a maximum at low biomass to an insignificant value

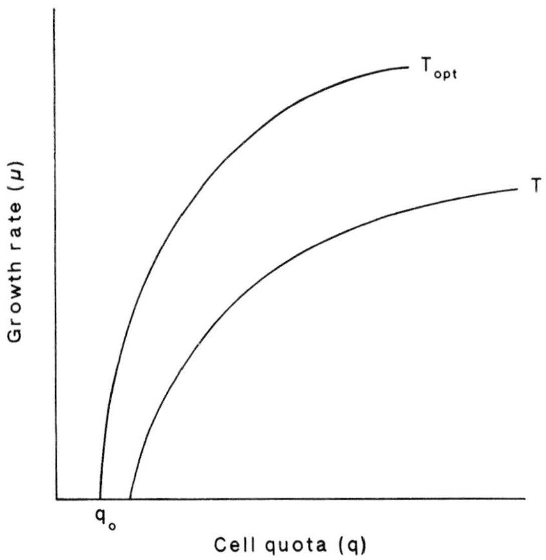

Fig. 40. Generalized specific growth rate curve for algae as a function of the internal nutrient concentration or cell quota, q, at optimum temperature, T_{opt}, and a different temperature, T. The minimum cell content, q_o is minimum at the optimum temperature (absolute minimum cell quota) and the maximum specific growth is maximum at optimum temperature (absolute specific growth rate) (from Ahlgren, 1988) (with kind permission of Prof. Ahlgren).

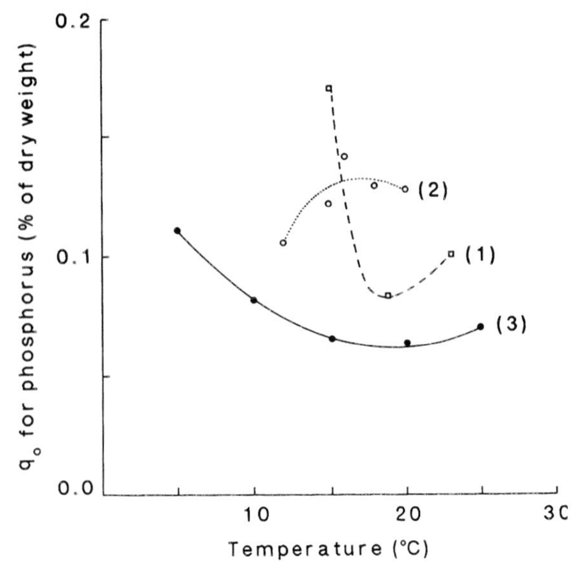

Fig. 41. Minimum cell quota, q_o, for phosphorus in percent of dry weight against temperature. (1) *Monochrysis lutheri*, (2) *Scenedesmus sp.* and (3) *Scenedesmus quadricauda* (from Ahlgren, 1988) (with kind permission of Prof. Ahlgren).

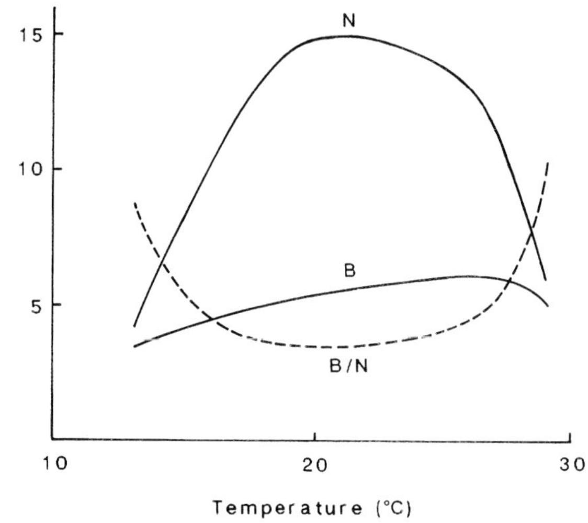

Fig. 42. Schematic representation of the relation between algal cell number, N, biomass, B, and average cell size, B/N as a function of temperature in a chemostat culture The Y-axis is graded in relative units (from Ahlgren, 1988) (with kind permission of Prof. Ahlgren).

close to carrying capacity (Fig. 43). At low growth rates, the same amount of phosphorus will be used more efficiently and produce a higher biomass than at higher growth rates. This applies also for nitrogen, but not in the same proportion. The amount of nutrient consumed to produce a given biomass is defined as the growth yield (Ahlgren, 1988). The relation between yield and growth rate at different temperatures for phosphorus, for instance, shows that yield and growth rate increase with temperature. The higher the temperature, the higher the possible growth rate and yield (Fig. 44) (Ahlgren, 1988).

A shift from nitrogen-limitation to phosphorus-limitation does not appear when algae are exposed to different N/P ratios in culture media. In the field, production of algal biomass in eutrophic lakes was reduced only when phosphorus and nitrogen inputs were both decreased, not when one or the other nutrient was singly decreased (Ahlgren, 1988). This tends to confirm interactive effects of nutrients on phytoplankton growth. Similarly, Yusoff & McNabb (1987) showed in experiments performed in ponds at Serdang, Indonesia, that nitrogen and phosphorus supplies to waters of low nutrient content had an

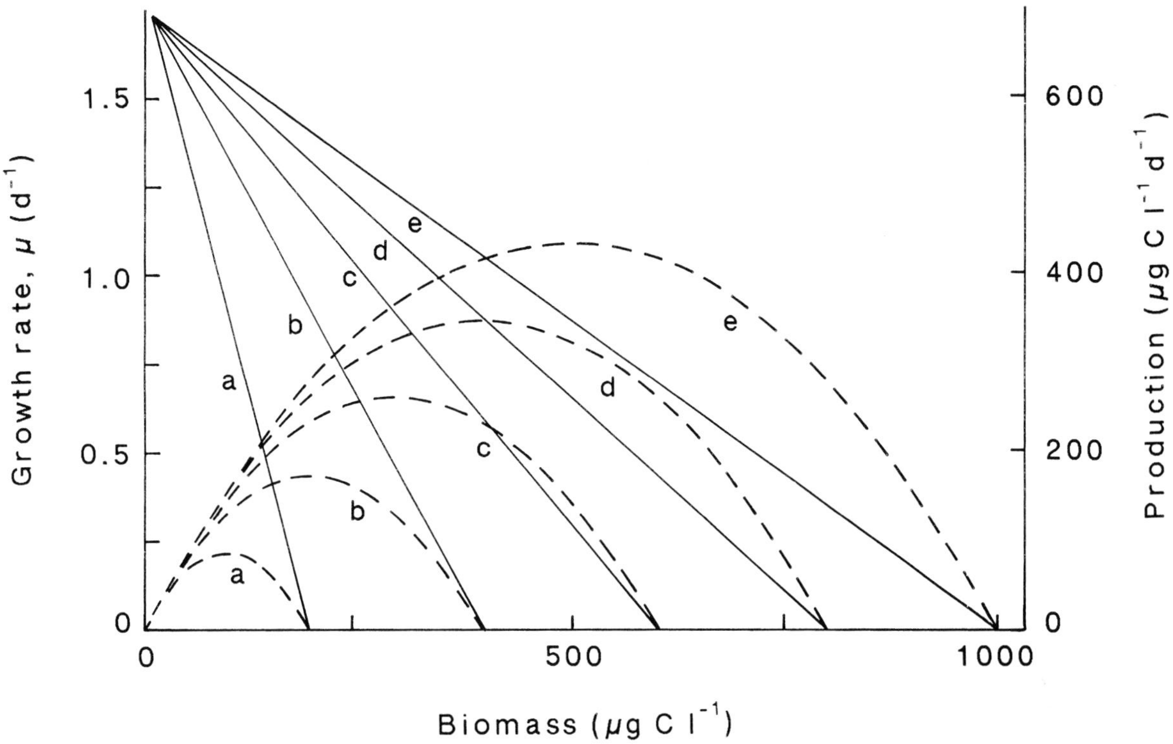

Fig. 43. Schematic relation between specific growth rate, production and biomass. Straight lines indicate specific growth rates and broken lines, production. The letters denote relations at different carrying capacities increasing from (a) 200 μg C l^{-1} to (b) 400 μg C l^{-1}, upto (e) 1000 μg C l^{-1} (after Heyman & Lundgren, 1988).

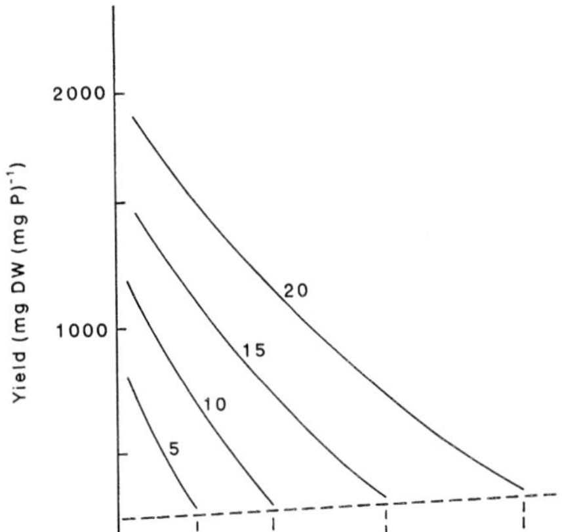

Fig. 44. Generalized diagram of phosphorus yield, i.e. the biomass produced per unit of phosphorus consumed, against specific growth rate, μ, at four different temperatures, 5, 10, 15, and 20 °C. μ_5, μ_{10}, μ_{15}, and μ_{20} indicate maximum specific growth rates at the different temperatures (from Ahlgren, 1988) (with kind permission of Prof. Ahlgren).

interactive effect; applications of one nutrient alone were not effective, as the other nutrient rapidly became limiting.

Community production and biomass

Biomass increase versus biomass (community production divided by biomass, P/B) appears to remain fairly constant within the trophogenic layer, when light is adequate. The main factors determining P/B are biomass level, carrying capacity (Fig. 43) and temperature, while nutrient level does not seriously affect this ratio. This means that all changes to biomass, such as elimination of phytoplankton, modify the P/B ratio. Elimination of phytoplankton biomass is of prime importance in regulating production and biomass, as a low biomass would increase drastically if there was not some kind of elimination (Heyman & Lundgren, 1988).

Three different types of phytoplankton dynamics

can be distinguished: (i) under light-limiting conditions, algae are characterized by a high cellular nutrient content and a low growth rate. The finite source of nutrients determines biomass, and temperature determines maximal growth rate (Fig. 45); (ii) under light-sufficient conditions, temperature determines the maximum growth rate, and nutrients determine carrying capacity. Biomass remains well below carrying capacity. Temperature simultaneously affects grazing, sedimentation, mineralization, and species composition; (iii) under initial conditions of neither nutrient nor light limitation, biomass increases close to carrying capacity, inducing nutrient-deficiency and light limitation due to self-shading, i.e. bloom conditions.

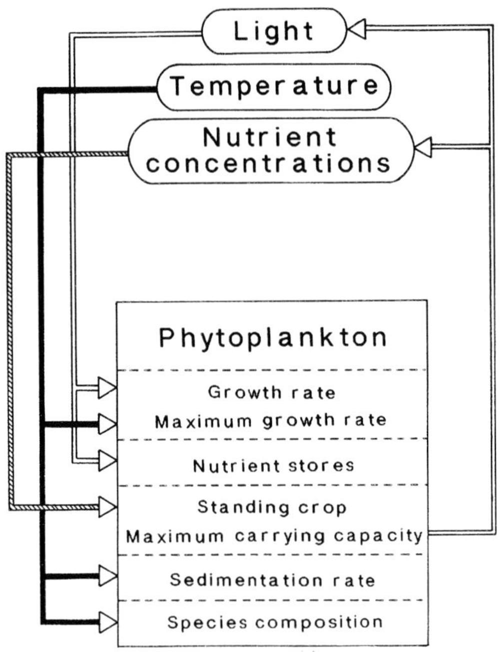

Fig. 45. Effect of different water characteristics on phytoplankton dynamics. The right-hand side arrows indicate a feed-back effect of the algal biomass having reached carrying capacity and causing self-shading and nutrient-limiting conditions. Not included are all the indirect effects, such as of temperature or standing crop on nutrient regeneration, of sedimentation rate on nutrient concentration, etc.

The bloom situation will be discussed later. When biomass is close to carrying capacity, growth rate is slow and hampered by a lack of nutrients and by self-shading (Heyman & Lundgren, 1988).

Steps in the increase of plankton biomass

From the management point of view, it is important to know what is the minimum plankton population density, from which exponential growth of the population will start, as well as the maximum population density which will lead to a reduction in photosynthetic rate by self-shading. The species spectrum which can possibly give rise to an algal bloom is also important.

In the evolution of phytoplankton biomass, four successive periods may be distinguished: a period of low stable biomass, followed by exponential growth, then a high standing biomass (bloom), ending eventually with an algal crash. It is during the period of low biomass that dissolved nutrient concentrations (orthophosphate, silica, nitrate, nitrite) are maximum. Not all species bloom (Krom *et al.*, 1989a); blooms occur in blue-green algae, diatoms, dinoflagellates, and green algae, but not amongst Chrysophyceae (Heyman & Lundgren, 1988). Populations of blue-green algae may reach a maximum of 400 mg DW m^{-3}.

Effects of water circulation on phytoplankton production

Water movements modify phytoplankton production. In stagnant water, nutrients and phytoplanktonic biomass are low at the beginning of production. In the course of production, nutrient concentration in the water and phytoplankton density increase. After some time, biomass reaches a given load, the carrying capacity, determined by light and nutrient availability. At high phytoplankton densities, anaerobic and aerobic patches develop and toxic metabolites accumulate. The conditions for an algal crash and for deterioration of water quality are set. In the case of running water, phytoplankton biomass is determined by nutrient supply. An exchange of water between 1 to 5 times a day allows the daily phytoplankton growth rate to be multiplied up to 5 times (Krom *et al.*, 1989b). According to Erez *et al.* (1990), the dilution rate should not exceed the specific growth rate, and daily replacement of water should be maximum half the volume of the pond. The outflowing water removes metabolites produced in the pond. A periodic crash of algal populations cannot, however, be prevented. If water flow is too fast, phytoplankton is washed away and diluted, which further reduces phytoplankton growth. Only bacteria and algae then grow on the dykes (Krom *et al.*, 1989b).

Water flow has another direct effect on photosynthesis, by fostering carbon dioxide uptake. Immediately next to the surface of any phytoplankton cell, there is an unstirred layer of water across which CO_2 must diffuse. Thickness of this layer is about 5 μm around small cells like *Chlorella*, and around 30-150 μm for larger cells and macrophytes. If the plant is subjected to a water flow, the thickness of this layer decreases and the uptake of carbon dioxide is increased. Morphological adaptations which increase the surface to volume ratio of phytoplankton cells also help to overcome the diffusion problem (Kirk, 1986).

Oxygen release and algal biomass

Algae, besides being food for fish, contribute to the ecosystem by releasing oxygen. Net oxygen production per algal biomass unit decreases, when algal biomass increases, because nutrients and light tend to become limiting and photosynthetis is depressed. Oxygen production is no longer sufficient to satisfy the need of total community production, entailing an oxygen deficit and environmental stress. Paradoxically, excessive phytoplankton biomass may lead to dissolved oxygen depletion. Nutrient concentrations in water are not homogeneous and oxygen depletion is therefore occurring in patches; light is subject to the same heterogeneity, some areas being less illuminated than others (Smith, 1988).

Plankton population composition and trophic state of the water

Phytoplankton composition changes in the course of season and time, adapting to changes in the trophic state of the water body. The relative importance of the different taxonomic groups in a plankton assemblage depends on trophic state, on nutrient supply and on zooplankton grazing. Complex interrelations exist between herbivorous zooplankton, nutrient concentrations, and phytoplankton standing crop (discussed further below) and prevent as yet modelling of these relations (Uehlinger & Bloesch, 1989). Phytoplankton succession (seasonal change) is discussed on p. 158.

Phytoplankton biomass and primary productivity

1. Primary production relies not only on the amount of phytoplankton present in a pond, but also on its rate of increase, i.e. the growth rate of this biomass.

2. Phytoplankton growth depends on the carrying capacity of the environment and on nutrient concentration, both intracellular and external. Light limitation causes an increase in intracellular nutrients and in the size of the photosynthetic units. Light and phosphorus limitation result in a reduction of cell respiration, while light, carbon and nitrogen limitation result in a reduction of organic compound elaboration. Temperature determines maximum growth rate. The factors determining growth under light-saturating conditions are, in order of importance: temperature, nutrient concentrations, and carrying capacity. Growth rate is inversely related to phytoplankton biomass.

3. Nutrient requirements vary with light availability. Under conditions of low light irradiances, and at low growth rates, phosphorus is used more efficiently than at higher growth rates or higher irradiances.

1.2.3. Elimination of phytoplankton biomass

Types of elimination or biomass decreases

Amongst the factors directly involved in determining standing crop, those discussed above are related to biomass increase. Factors related to biomass decrease include predation or grazing by zooplankton and fish, parasitism and diseases caused by viruses and bacteria, sedimentation, and cell lysis, caused by environmental stress.

Elimination of phytoplankton can also result from physical factors: dilution or concentration of nutrients, evaporation and drying up of the pool, and changes in climatic conditions, leading to excessive turbulence capable of damaging cells, excessive temperatures, and harmful light intensities.

If phytoplankton biomass is measured by organic carbon content, algal exudations can also be included among factors contributing to algal elimination.

Plankton sedimentation

Losses by sedimentation of algae from the euphotic zone are a significant process. Sedimentation is not a purely physical phenomenon, but is affected by species composition and by nutrient status of the cells (Heyman & Lundgren, 1988). A number of green algal species, heavy and large-sized, are maintained in the water only by continuous turbulence. During calm periods and when stratification occurs, losses by sedimentation can be sufficiently considerable to cancel out daily production. In an Eastholstein lake, Germany, elimination of algal cells was more important during nutrient deficiency (Meffert, 1989).

Near the surface, sedimentation reduces overall algal biomass, but buoyant blue-green algae accumulate. In Windermere, U.K., densities of colonies of *Anabaena lemmermannii* reached 975 kg m^{-3}, with intact gas vesicle, and 1022 kg m^{-3}, after gas vesicles collapsed (Walsby *et al.*, 1991).

Sedimentation rate increases with a decrease in water density and viscosity, i.e. with increasing temperature, and differs between species. Mean floatation rate of colonies of blue-greens in lake Windermere was 181 $\mu m\ s^{-1}$ and 186 $\mu m\ s^{-1}$, with intact gas and collapsed gas vesicles, respectively (Walsby *et al.*, 1991). Algal biomass is usually greater in the lower part of the epilimnion. The ascendent migration of algae is regulated by changes in buoyancy. In the case of blue-green algae, the density of a cell is modified by the formation and resorption of gas vacuoles (Walsby & Reynolds, 1980). These gas vacuoles ensure vertical movements. When irradiation is too high, the gas-vacuole collapses, and the cell sinks; the vacuole is rebuilt under conditions of low light intensity (Westlake, 1980).

In *Aphanizomenon flos-aquae* from two lakes in Indiana, U.S.A., an irradiation above 100 μmol photons $m^{-2}\ s^{-1}$ reduced buoyancy. In the presence of 2 μmoles of KH_2PO_4, irradiation must be above 200 μmol photons $m^{-2}\ s^{-1}$ to cause cell descent (Konopka, 1989). Carbohydrates are used as a ballast to help cells sinking. Cellular turgidity, linked to carbohydrate synthesis, has been considered responsible for the disappearance of the gas-vacuole, but this seems unlikely. Cell turgor pressure is linked to the presence of a gas-vacuole, modifying the size of the latter, and hence, buoyancy (Walsby & Reynolds, 1980). Gas vacuoles are only present in prokaryotic algae, i.e. blue-greens, while eukaryotic algae posses only fluid vacuoles. In planktonic diatoms, they are well developed, serving as a means of reducing average cell density and in nutrient storage. Other algae have evolved particular shapes, hair-like expansions, or large volume vacuoles, to lessen density (Taylor, 1980). In the case of flagellates, migrations are active.

Algal consumption

A study of consumption of diatoms, filamentous blue-green algae and green algae by cladocerans, copepods and rotifers indicates that herbivores are capable of consuming filamentous algae at the same or higher rates than unicellular algae. Some phytoplankton species are even heterotrophic or phagotrophic and consume other phytoplankton. Parasites, pathogenic bacteria and diseases also affect phytoplankton, as well as growth-inhibiting substances, which are either auto-antagonistic in populations of aged algae or hetero-antagonistic in species which exclude each other, such as *Aphanizonemon* and *Scenedesmus* (Westlake, 1980).

Analysis of grazing of caddis-fly larvae

(*Neophylax automnus*) upon diatoms has shown that almost half of the diatoms are still alive in the larval faeces. Algae of large size are consumed more than those with a small size. Those which adhere better to the substratum escape predation to a greater degree. As a result of lower predatory pressure, their relative number increases (Peterson, 1987) and because of grazing, the structure of the diatom population changes.

The quantitative or relative importance of each of the factors causing elimination of algae is difficult to establish; it appears that grazing cannot be generalized to be the main cause of algal destruction, though it may be important in eutrophic waters (see also the sections hereafter).

Phytoplankton defenses against predation

The importance of zooplankton grazing is controversial, as zooplankton is not always capable, for anatomical reasons, of feeding directly on all algae. A certain number of filamentous or colonial algae reach a volume superior to the maximum particle size that can be ingested by zooplankton. Other algae possess thick gelatinous sheaths (Vanni, 1987a). Some blue-green algae, such as *Anabaena flos-aquae*, are not consumed due to production of toxic or harmful products, and not to the morphology of the filaments. Resistance to digestion as mentioned above is also a defense. Blue-green algae (*Microcystis*, *Aphanizomenon*, *Oscillatoria*, *Anabaena*) Dinoflagellates (*Peridinium*), Chlorococcales, Eugleninae resist well to grazing and digestion by fish, as numerous viable cells are present in the faeces of herbivores, such as silver carp and *Oreochromis aureus* (Lazzaro, 1987; Prowse, 1964; grazing of *O. aureus* and *Sarotherodon galilaeus* in Lake Kinneret on *Peridinium*: Gophen, 1983; Landau, 1983; Ben-Tuvia & Reich, 1983).

Controls of the dynamics of phytoplankton density

Control of phytoplankton biomass is either achieved by grazing (top-down) of herbivorous zooplankton and phytoplanktivorous fish, or bottom-up, by nutrient availability (McQueen *et al.*, 1986). The control of primary productivity by predation is termed the trophic cascade (Carpenter *et al.*, 1987). Determination of the predominant process in overall control of phytoplankton is established from relations between phytoplankton and zooplankton, by comparisons of zooplankton filtration rates to population size, and responses of phytoplankton to nutrient enrichment. Results range from dependency to independency, with high to low grazing pressure, including pressure equal to production, and are seasonally changing. Experiments in limnocorrals, with removal of portions of the plankton, show that zooplankton greatly affect phytoplankton biomass. A negative relation between zooplankton and phytoplankton biomass is indicative of a grazing-controlled system.

Top-down control, when apparent, is always restricted to some period of time. Though no general statement is likely to be true, the type of control appears linked to trophic state of the water body: in eutrophic systems, control of primary production is from below, while in oligotrophic systems, grazing controls phytoplankton.

Importance of elimination of algae

1. Algal elimination is an important process. Phytoplankton is grazed upon by fish and zooplankton. Algae, which are not consumed, supply organic matter to decomposers, which regenerate nutrients for phytoplankton growth.

2. The main fraction of algal elimination is due to sedimentation: grazing is seldom a limiting factor.

3. Phytoplankton growth is controlled by a combination of grazing (top-down control) and nutrient supply (bottom-up control).

1.2.4. Phytoplankton behaviour and life strategies

Importance of life strategies

Phytoplankton species have evolved a variety of life-history strategies. These involve such aspects as nutrient uptake affinities (high versus low), capabilities to withstand grazing by motility or protection, ability to form resting stages, adaptations to low or high irradiations, to different temperature optima, and to living at low or high densities, i.e. close or away from carrying capacity. Production to biomass ratios are fairly constant; this implies that, if production rate is high, natural elimination to maintain a stable biomass will be high and vice versa.

The life-history strategy of phytoplankton can be appraised on the following criteria (Kilham & Kilham, 1980):

- the magnitude of the intrinsic rate of increase
- the amount of resource expenditure used to produce offspring
- the extent of environmental saturation with individuals
- the degree of niche overlap
- the degree to which productivity or efficiency is maximized.

In the context of phytoplankton management, the discussion will focus on species' ability to utilize environmental resources, considered at a patchy, ephemeral and microhabitat scale, species distribution within the water and its modification, and changes induced in the water by algae, such as exudations, blooms and crashes.

r and K strategies of phytoplankton species

The influence of nutrients operates through the nutrient assimilation capacity of the different species. Several strategies can be distinguished: some species specialize in assimilation, i.e. they absorb nutrients quickly in considerable quantities and use them for cell multiplication; others specialize in stocking nutrients in intracellular stocks; still others show high affinities, which allow them to absorb nutrients at very low concentrations.

r-selected species have evolved primarily in temporary habitats, where they cannot maintain their maximum density for any length of time owing to competition from other species, or to rapid changes in the environment. They allocate most resources to reproductive efforts. K-selected species are adapted to operate close to the species' carrying capacity in the environment, being selected for competitive ability and efficient resource utilization. Non-reproductive activities receive more resources, and genetic survival is optimized.

r-selected species are synonymous with fugitive, colonist, weedy or opportunistic species, while K-selected species are equilibrium or autochthonous species (Kilham & Kilham, 1980). r-selected species have rapid growth capacities and can use varying nutrient concentrations, such as pulsed supplies; they have high absorption affinities. When the cell quota are depleted, they show growth rates depending upon nutrient concentrations available in the water; they are usually edible to consumers. K-selected species are unable to react quickly to nutrient enrichment of the water, showing cell quota often close to that of minimal intracellular concentrations, and normal growth rate close to the maximum. They can sustain stable populations for long periods of time; they are usually inedible (Kilham & Hecky, 1988).

The plankton paradox

The paradox of the plankton was formulated by Hutchinson (1961), who asked how a high species diversity in the pelagic of lakes and ponds can be maintained. The water column is a typically homogeneous, unstructured environment; thus, it is puzzling how niche diversification can occur here. Besides a permanent state of non-equilibrium of the communities, the existence of a series of micro-habitats, ephemeral and patchy in nature, not revealed by conventional samplings, but providing spatial and temporal heterogeneity (Smayda, 1980) is advocated as an explanation. The ephemeral character of these patches does not allow one species to develop in such abundance that it can outcompete other species.

Vertical distribution and migration of phytoplankton

The vertical distribution of algae in the water is stratified. This stratification is the result of active migration and of passive sinking (Table 19). In Lake

Table 19. Causes prompting phytoplankton migration

Downward migration	• to take up nutrients/vitamins released by the sediments or produced by microorganisms • to avoid deleterious irradiation causing pigment bleaching • to avoid zooplankton grazing
Upward migration	• to escape light limiting conditions

George, Uganda, phytoplankton distribution is homogeneous at dusk. During daytime, the lake changes from isothermic to stratified, followed by mixing during late afternoon. Algae sink during the day, but begin to distribute uniformly when thermal stratification disappears (Ganf & Horne, 1975).

In deep, aphotic water, nutrients are often more concentrated and some chemical gradients occur around the thermocline and in the epilimnion (Westlake, 1980). Algae move to those areas to take up and stock nutrients, which are assimilated later in euphotic conditions. Aggregations of algae (*Cryptomonas*, for example) in mid-column may be linked to the presence of vitamins at that level. These vitamins are produced by bacteria, whose presence is, in turn, connected to an accumulation of detritus in the deoxygenated hypolimnion (Westlake, 1980). Some algae, such as *Oscillatoria agardhii* var. *isothrix* may begin their growth in the deoxygenated hypolimnion and later move upwards (Saunders, 1972; Westlake, 1980).

In Wellington reservoir, near Pearth, New Zealand, diel migrations were promoted by advective currents, produced by physical processes (Viner, 1989). At mid-day, photosynthesis close to the surface decreases (light inhibition), but increases in deeper waters. In the afternoon, the photosynthetic rate is reduced by nutrient depletion, by the increase of respiration, by pigment bleaching, and by zooplankton grazing. Seasonal variations superimposed on this are changes in light regime, temperature, wind-induced turbidity, supplies of run-off water, nutrient availabilities, water stratification, and species composition. However, in tropical waters, seasonal changes are damped or non-existant (Westlake, 1980).

Algal exudations

Primary productivity is the increase in algal carbon per unit area and time. It is measured by the amount of carbon fixed, but some bias originates from the discrepancy between carbon uptake and carbon assimilation, i.e. its incorporation into organic molecules. Some of the carbon absorbed is reconverted into carbon dioxide by respiration. Some is excreted as glycolate and other organic compounds, such as carbohydrates (hexoses: Chrost, 1989), amino acids and larger molecules; such excretions are called algal exudations.

Glycolate excretion occurs in two circumstances, provided that glycolate production exceeds its rate of consumption: (1) when growth rates are low, due to algal senescence or nutrient depletion; (2) when cells are metabolically stressed or subject to photoinhibiting conditions (Harris, 1980).

Exudations are produced by healthy algae (Bjørnsen, 1988; Hino, 1988, for blue-green algae, in lake Barato, central Hokkaido, Japan). A proper explanation of the physiological mechanism behind these exudations is lacking. One interpretation is that the active release or passive permeation (over-flow) of photosynthetic products is the result of carbon uptake in excess of the transformation or assimilation capacities of the cell. Exudations are nitrogen-rich compounds of low molecular weight, such as amino acids. They are quickly utilized by pelagic bacteria and are a carbon supply for them under limiting conditions. They have been qualified as an 'income tax' or as a 'property tax' (continuous loss) contributed by the algae to the bacteria (Bjørnsen, 1988). In a eutrophic lake, in Michigan, U.S.A., the presence of pelagic bacteria could not, however, be solely explained on the basis of these exudations. Other food sources, for instance, organic carbon from decomposition, were needed by pelagic bacteria (Storch & Saunders, 1978).

Laboratory experiments on the effects of light intensity and temperature on excretion of dissolved organic carbon by three different species of algae (*Chlorella vulgaris*, chlorophyte; *Isochrysis galbana*, prymnesiophyte, *Synechoccus sp.*, blue-green alga) revealed a tight coupling between carbon fixation and net excretion of dissolved organic carbon. Excretions increase under extreme irradiation and at temperatures which depress photosynthesis. The dissolved organic carbon flux towards the bacterial food web is important even when conditions are suitable for photosynthetic activities, though the percentage excreted is low (Zlotnik & Dubinsky, 1989).

Phytoplankton is used as food by zooplankton, which also prey on bacteria, which, in turn, live partly on algal exudations. This intricacy of trophic relations will be further discussed later (p. 173).

Some organic substances are released by algal cells undergoing lysis. Unlike exudations of healthy cells, they are macromolecules, like polysaccharides (Chrost, 1989).

Algal blooms

Algae are said to bloom when a population suddenly develops in such a way as to become dominant. In lake Okaro, New Zealand, high algal biomasses were the result of low and steady growth periods, maintained over long periods of time, during relatively stable environmental conditions, rather than quick responses to exceptional conditions lasting for short periods of time (Vincent & Dryden, 1989). Blooms coincide with stratification, high photosynthetically active radiation, and high supplies of organic matter and nutrients (Paerl, 1988).

The dominance of a species is the expression of its competitive superiority for limiting resources. The appearance of a bloom of nitrogen-fixing blue-green algae often indicates nitrogen limitation, combined with other algal characteristics, such as inedibility (Sterner, 1989). Other characteristics promoting blue-green algae dominance are positive buoyancy of the cells, which increase the average exposure of the algae to light. The interactive effects of diffusion by turbulence, photic depth, mixing depth, and daily migrations on growth rate and production of algae differ fundamentally for algae with positive buoyancy and algae with negative buoyancy. Maximum growth is higher for species with positive buoyancy than for species with negative buoyancy, until self-shading densities are reached. The capacity of blue-green algae to migrate vertically allows them to cope with the spatial separation of nutrients and light (Humphries & Lyne, 1988).

The density of algae with negative buoyancy (naturally sinking) increases with depth, while that of algae with positive buoyancy (naturally rising) is greatest at the surface. Among the characteristics ensuring positive buoyancy are low density of the cells, a low surface/volume ratio and coloniality. Usually these algae have a low growth rate, and are restricted to eutrophic waters. This is confirmed by observations that blooms develop during periods of low nutrient concentrations (Humphries & Lyne, 1988).

Nutrient concentrations do not reflect nutrient supply, as a high turn-over of nutrients can satisfy the demand of a large biomass. The standing crop of algae increases until an equilibrium is reached between demand and supply of nutrients; at that stage, prevailing nutrient concentrations in water reach a minimum. The advantage of species with negative buoyancy compared to species with positive buoyancy depends on the nature of the stratification period; species with positive buoyancy perform better when stratification is destroyed daily. The biomass of species with negative buoyancy is continuously diluted by sedimentation and its increase of biomass is lower than that of species with positive buoyancy (Humphries & Lyne, 1988). Species which are subjected to blooms generally possess resting stages, cysts or akinetes (Paerl, 1988).

Algal crashes

An algal crash is the sudden death of a bloom and the release, from this algal biomass, of considerable amounts of organic matter. There result stressed conditions for the pond: the concentrations of carbon dioxide and ammonium increase (see p. 76) and there can be transformation of ammonium to toxic ammonia, which can lead to fish mortalities. The concentrations of nitrogen and phosphorus increase strongly, owing to decay of organic matter settled on the sediments (ammonification processes). These decompositions remove oxygen from the water, leading to anoxy, and release of phosphate from the sediments (Søndergaard *et al.*, 1990) (see Table 7, p. 39, 40 & p.39).

Causes of algal crashes

Though important for the management of phytoplankton populations, the process whereby biomass increases till populations collapse is not fully understood. A detailed discussion on possible causes is presented by Smith (1988) (Table 20). In most cases, algal crashes happen in communities dominated by blue-greens, with one species dominant. Species most commonly affected by blooms and subsequent crashes are those with cells of large size or colonial forms, which are less vulnerable to grazing. Several causes have been considered: photolysis, attack by pathogenic microorganisms, nutrient deficiencies, and grazing by microzooplankton (ciliates and flagellates). Photolysis is the destruction of cells under the effect of intense light. Blue-green algae, which belong to the group of algae most exposed to high light intensities, grow moderately well under conditions of intense irradiation at the water surface. They produce carotenoids for their photoprotection and possess the specific capacity to form scums of dead algae floating at the surface.

Photolysis does not explain algal crashes, as they frequently appear during overcast periods and sudden temperature drops. The hypothesis of pathogenic viruses or bacteria causing algal death does not seem likely either, as they are omnipresent (e.g. in Lake Erie, U.S.A., Fraleigh & Burnham (1988)) and there is therefore no reason why they should suddenly become harmful. A more probable reason for the death of algae is nutrient deficiency. The appearance of nitrogen-fixing blue-green algae is often promoted by nitrogen-limiting concentrations. The presence of a bloom changes the wind mixing depth of the water. In lakes Rotongaio (eutrophic) and Okaro, New Zealand, they remained at the surface because they were able to sustain varying light conditions, from intense radiations to low light intensities. Lastly, the zooplankton did not graze upon them (Vincent, 1989).

When a dominant species comes close to its carrying capacity, the population is stressed by decreasing nutrient supplies, which could lead to mass destruction. The fate of nutrient-deficient algae has never been clarified: do the algae get senescent or do they die and disappear? There is no convincing evidence that algal crashes are caused by nutrient deficiencies. In Lake Sobygård, Denmark, Jeppesen *et al.* (1990c) observed that a crash was preceded by an increase of sedimentation rate, suggesting that some nutrients were limiting. When algal crashes occur during overcast or cloudy weather, an increase of intracellular nutrient demand, as a result of the lower light intensities or temperature drop, could cause their death. Calm weather, reducing the mixing depth of water and preventing a supply of nutrients to the areas with high phytoplankton densities, enhances stress. A dense population reacts in a more extreme way to changing environmental conditions. Severe nutrient depletion halts or reduces protein synthesis, which induces photolysis, in a way analogous to the action of chloramphenicol (an inhibitor of protein synthesis). Jeppesen *et al.* (1990c) noticed a considerable increase of the filtrating zooplankton population, concurrent with the initiation of an algal crash.

Table 20. Possible causes for algal crashes

photolysis
temperature jumps
attacks by pathogenic microorganisms
nutrient deficiencies due to high standing crop or light-limiting conditions
grazing by microzooplankton

As none of the hypotheses discussed can be considered as the sole cause of algal crashes, a combination of causes acting synergistically seems likely.

Effect of zooplankton on algal blooms

Zooplankton grazing on phytoplankton blooms involves a two-way reaction. Algal blooms promote the abundance of algae of large size and of low edibility. The large-sized zooplankton, which does not find adequate food, disappears progressively. This leads to a decrease in the proportion of nutrients recycled by the zooplankton, which, in turn, affects phytoplankton growth (Krom *et al.*, 1989a). Interactions between algae and zooplankton are discussed in more detail later (p. 162).

The presence of algal blooms affects the whole plankton community. Blooms of blue-green algae modify the behaviour, the composition, and the structure of the zooplankton community. In Rotangaio, New Zealand, for example, *Anabaena minutissima* var. *attenuata*, produces toxins (Forsyth *et al.*, 1990) which decrease the filtration and ingestion rates of limnetic zooplankton and their vertical migration. An inhibition of zooplankton feeding by mechanical interference with algal filaments also occurred.

Algal disappearance or physiological death

In the absence of a clear dominance of particular algal species or of characterized blooms, large portions of the phytoplankton, composed of a range of algae, die, but in an inconspicuous, unnoticed way. Each species lives according to its metabolic requirements and when these cannot be met, the alga disappears progressively (Smith, 1988).

Phytoplankton management

Management practices should be aimed at preventing the appearance of algal blooms and subsequent algal crashes. Ideally, the objective would be to maximize plankton production and optimize quality (Tilzer, 1989) (see also p. 157). In the light of the above discussion, it appears that the way to maintain the phytoplankton standing crop at or just below its carrying capacity, is to supply nutrients in sufficient amounts. Optimum fertilization will reduce the probability of an algal crash. When primary production is at a maximum, ammonium uptake is also at a maximum and 0.15 of a mole of nitrogen is assimilated for each molecule of oxygen released (Smith & Piedrahita, 1988). A diversified algal community is perhaps the best guarantee against algal blooming and subsequent crashes. Biological control of phytoplankton by planktivorous fish or by zooplankton has been attempted. Although it has not yet reached a level of excellence, it is progressing. It effectively reduces the overall volume of phytoplankton. The role of fish and zooplankton in the ecosystem are discussed on p. 162 and p. 180. After an algal crash, a supply of green water, enriched with phytoplankton, from an adjacent pond, will help remove the excess nutrients leached from the dead algal biomass (Erez *et al.*, 1990).

Chemical methods for the control of plankton biomass have been discussed earlier (p. 64).

Algal abundance and off-flavours

Another nuisance caused by excessive abundances of algae is off-flavours, which affect fish in some culture conditions and which are often linked to the presence of blue-green algae. It should, however, be pointed out that off-flavours are not only due to blue-greens, as some other taxa also produce them. The actual flavours originate from the decomposition of organic matter by various fungi; the chemical compounds responsible are geosmin and methylisoborneol. These flavours affect fish for several days to several weeks after their release to the water; this renders preventive measures difficult (Smith, 1988).

Phytoplankton life-strategies and behaviour

1. Phytoplankton species have developed life-strategies analogous to r- and K-selection. r-selected species colonize new environments, have high growth and reproductive capacities and are edible, while K-selected species are equilibrium species, with low growth rates, high biomass and lower edibility.

2. The pelagic environment, though seemingly homogeneous, is heterogeneous, made up of an assemblage of microzones or patches, with ephemeral, space-limited conditions. This explains the diversity of phytoplankton species present in water.

3. Algal exudations are releases of organic compounds, which constitute significant supplies of organic carbon to bacteria.

4. An algal bloom is the development of an algal biomass close to the carrying capacity of the environment. The algae prone to these blooms are often inedible.

5. Algal crashes follow the blooms and are a nuisance in fish ponds. The causes for such crashes are as yet undetermined. Possible causes are light or nutrient limitations.

6. Management of the phytoplankton biomass includes proper fertilization practices, and water circulation to avoid the appearance of anaerobic microzones and large-sized inedible phytoplankton species. After an algal crash, addition of green water may help to restore normal conditions and remove excess nutrients.

1.2.5. Influence of nutrient concentration

Primary productivity and nutrient concentration

Limnologists and aquaculturists have searched for relations between primary productivity and nutrient concentration in water (Strabskaba, 1980). An increase of nutrients in water enhances three functions of the aquatic ecosystem. They are: production, decomposition, and recycling of nutrients. However, none of these processes are sensitive indicators of ecosystem responses (Schindler, 1987). Nutrient concentrations are the sole factors on which action is possible, as the light régime cannot be modified and as the volume of autotrophic organisms depends on proliferation of an initial stock, and, thus, on prevailing limnological conditions.

Trophic loading of a water body

The amount of available nutrients determines the trophic load of a water body. According to trophic level, waters are oligotrophic, mesotrophic or eutrophic. In oligotrophic environments, flagellates and Chrysophyceae dominate; in mesotrophic environments, diatoms are predominant; in eutrophic waters, blue-green algae and dinoflagellates are dominant. In lakes of the north-west of U.S.S.R. (9 lakes of the Karelian Isthmus and 8 Latgalian Elevation lakes), algal biomasses less than 0.5 g m^{-3} (< 1.5 mg chl a m^{-3}) characterized oligotrophic lakes. Values of 1-5 g m^{-3} (1.5-10 mg chl a m^{-3}) were considered mesotrophic, 5-10 g m^{-3} (10-25 mg chl a m^{-3}) eutrophic, and > 10 g m^{-3} (> 25 mg chl a m^{-3}) highly eutrophic (Trifonova, 1989). Quite a few water bodies in tropical areas have been classified as oligotrophic on the basis of their nutrient concentrations. Considering the frequencies of algal blooms, net production rates, the extent of macrophyte development and extent of oxygen depletion with depth, it is clear that they are eutrophic, however.

Relations between nutrient concentration and requirements

Nutrient concentrations determine phytoplankton growth. Phytoplankton biomass increases as a function of light, temperature and nutrient supply. Nutrient-limited growth can be modeled using either the Droop model (internal nutrient concentration)

(Senft, 1978; Brown & Harris, 1978) or the Monod model (external nutrient concentrations) (see p. 137).

Sommer (1988) considers that there is common dissatisfaction with the Monod model as a descriptor of the nutritional status of natural phytoplankton. Unsteady state conditions in the field and intracellular storage permit temporal uncoupling between phytoplankton reproductive rates and ambiant nutrient concentrations. The cellular content reflects the concentrations in the water during the recent past and the volume of uptake during the same period (Heyman & Lundgren, 1988). Under such circumstances the intracellular concentration of limiting nutrients (cell quota) is considered to be a better indicator of the nutritional status than ambient concentrations. Ambient nutrient concentrations do not give a true picture of the recent past, and have only a remote relation with plankton biomass at the time of measurement.

In chemostat culture under limiting nutrients, algae belonging to different taxa (chlorophytes, chrysophytes, cyanophytes and dinoflagellates) were shown to have similar internal nutrient concentrations and physiological responses: N- and P-limitation occurred below 70 $\mu g\ mg^{-1}$(DW) N and 5-10 $\mu g\ mg^{-1}$ (DW) P (Healey, 1978; Hecky & Kilham, 1988: 800, Fig. 2).

However, in a natural assemblage *in situ*, internal nutrient concentrations are not a handy deficiency indicator. Also, at low concentrations (especially for phosphorus), dissolved nutrients are not accurately measurable. Yet, numerous plankton species are still capable of absorbing such analytically undetectable quantities.

In climates with little seasonal variation, like the tropics, phytoplankton biomass is high, despite low ambient concentrations. In these conditions, extremely rapid nutrient turnover times occur. A value for phosphate turnover of less than 10 minutes indicates high P deficiency, while values of days would be found when no P deficiency exists (Lean *et al.*, 1988). From studies involving many large African lakes, extending from L. Abhé, Ethiopia to L. Cheski, Zambia, Kilham & Kilham (1990) showed that recycling of nutrients is indeed high and rapid.

Uptake velocity of nutrients was assumed to provide an estimate of the degree of nutrient deficiency. Uptake kinetics, as an index of competitiveness between species, could, however, not be related to growth rates (Grover, 1989a).

In enclosure experiments conducted in Lake Okaro, New Zealand, Lean *et al.* (1989) evaluated nutrient deficiency against phytoplankton growth, and uptake rates in response to nitrate and phosphate additions. The uptake rates for phosphate, ammonium and nitrate normalized per unit particulate P or N were compared to that of C assimilation per unit C. The remarkable ability of phytoplankton to store phosphorus, as compared to nitrogen, was hereby illustrated (see also p. 108). Under extreme N deficiency, the maximum uptake velocity for ammonium and nitrate was only twice as high as when no N deficiency existed, while for phosphate, the velocity increased by about 100 times when the population became P-deficient (Lean *et al.*, 1989).

Algal growth rate is a function of internal concentration of the limiting nutrient, i.e. the one present in the cell in the smallest amount relative to the need of the alga (Droop, 1975). Synchronous multiple nutrient limitation has not been shown for any unialgal culture. This should not be expected, since macronutrients cannot substitute each other in biochemical functions. However, nutrients can compensate, within limits, for physical limitation, light and temperature, or vice-versa, hence colimitation by a physical factor and a nutrient (Hecky & Kilham, 1988) does occur. In multi-species chemostats, multiple nutrient limitation of phytoplankton growth has been demonstrated: species differ in their optimum nutrient ratios for growth. Different species dominate at varying Si:P ratios. At intermediate ratios, the species coexist since each was either limited by Si or P. Nutrient-loading ratios can have a strong selective effect on natural communities of phytoplankton, and can affect the biomass yield for the nutrient limiting most species.

In non-steady state (pulsed nutrients), even more species can be supported, as the kinetics of nutrient uptake is then uncoupled from growth and allows species that are limited by the same nutrient to persist. In a natural phytoplankton assemblage, one or more species are limited by a common nutrient (for which they compete), different nutrients, light, temperature, light or temperature and a nutrient, or any other combination.

In natural phytoplankton, the focus is on total biomass of phytoplankton rather than on one species. The relevant questions therefore are: can the growth of natural phytoplankton be increased significantly by the addition of one or more nutrients? Which nutrient has the greatest effect? These questions have focused interest on complex variables like primary productivity and algal biomass, and their response to nutrient enrichment (Hecky & Kilham, 1988).

Relations between intracellular and extracellular nutrient concentrations and limiting concentrations

Benthic algae in milkfish ponds in Taiwan contain on average 3.8 times more energy in fertilized ponds than in non-fertilized ponds. For blue-green algae, this ratio is 1 to 8 (Fong & Ju, 1987). The effect of nutrients is obvious when dealing with algal communities, which, previously deprived of nutrients, react to nutrient-enriched water. In biomanipulation experiments using plankton from three small lakes in Michigan, U.S.A., where nutrients are limiting, Elser & MacKay (1989) observed significant increases of growth and photosynthetic capacity following nutrient enrichment. Improvement of the physiological status of algae leads to an increase in herbivorous zooplankton biomass. Loss of algal biomass by grazing is thus compensated by an increase in growth rate, due to nutrient regeneration by animals.

Limiting nutrients, and essential differences between nitrogen, phosphorus and silicon cycling

When the cycling of the macronutrients is compared, essential differences appear (Table 21). Nitrogen and carbon cycles are open to the atmosphere (Smith *et al.*, 1989), which is a sink and a source. Phosphorus is immobilized by the sediments, which are a substantial sink, though under some conditions, they also can release phosphorus. The phosphorus cycle is open to the sediment. The main pool of nutrients associated with living organisms is actively exchanged between producers, consumers and decomposers and is maintained in the active portion of the cycle of the living community.

In a study of four small prairie lakes of different trophic levels in south-western Manitoba, central Canada, Barica (1990) found that nitrogen was supplied more easily, being fixed from the air when required; this is why more was wasted, being denitrified when in excess. The appearance of nitrogen-fixing algae was induced by a nitrogen deficit relative to phosphorus. The inputs of nitrogen into the environment by these nitrogen-fixing algae were, however, only temporary, as the accumulated nitrogen could not be stored. It was removed by denitrification, as soon as the algae perished (Barica, 1990). Phosphorus, which is rarer and more difficult of access, is quickly taken up, when present, as shown by a comparison with the uptake kinetics of nitrogen.

Silicon, supplied by run-off water, is a rather inert nutrient, of which a low number of chemical species occur (see p. 122). It is easily lost to sediments within sedimenting diatoms. It is limiting after a period of intense production only. The role of silica is considered, by some authors, as essential in determining the phytoplankton assemblage, mainly through the silicon-phosphorus proportion (Si:P ratio) (Tilman *et al.*, 1986) or by specific requirements of phytoplankton species for amounts available in the water (Hecky & Kilham, 1988).

Relations between phosphorus and nitrogen requirements

Nutrient requirements are usually assessed on the basis of the Redfield ratio. For the phytoplankton, the ratio for C:N:P is 106:16:1. When the ratio of N:P is below 29:1, nitrogen-fixing blue-green algae are favoured, while, above this value, blue-green algae are outcompeted by other algal groups.

In a study analyzing the nutritional status of phytoplankton in a shallow, hypertrophic, brackish, coastal lake in northern Germany (Großer Binnensee) involving stoichiometric analysis of seston and bio-assay enrichment experiments, Sommer (1989) showed that the most limiting nutrient was nitrogen, then phosphorus, and finally silicon.

In Colorado mountain lakes, Morris & Lewis (1988) showed that the dissolved inorganic nitrogen to total phosphorus ratio represented effectively those fractions of nitrogen and phosphorus really available for phytoplankton and, by using this ratio, possible nutrient limitations were detected. Barica (1990) compared different nitrogen to phosphorus ratios (particulate, total, total dissolved and inorganic dissolved N:P), to establish which forms are representative of systems and may be used to predict system evolution. The ratios were based on data collected over a period of several years in four

Table 21. Comparisons between phosphorus, nitrogen and silicon cycling

	Phosphorus	Nitrogen	Silicon
Intracellular stocks	high	no	low
Natural turn-over	high for the biogenic P, sometimes only a few minutes between uptake and excretion, but millions of years for the geochemical cycle	high, varying between tens to fractions of hours	generally low, but shorter when redissolution of the diatom shells occurs in the euphotic zone. The dissolution of silica is accelerated by the presence of salt.
Regeneration	high recycling, up to 70 % of the inputs derive from P mineralization	low recycling, losses around 50 % of the cycling pool	low regeneration rate
Uptake capacities	high	high	limited uptake, not flexible and restricted to cell division and cell wall formation
Supplies	from the soil geochemistry	from air (by N-fixation)	from run-off water (and upwelling in deep lakes)
Losses	to sediment	to the atmosphere	to the sediment

Central Canadian (south western Manitoba) lakes, and were representative of some 100 lakes of that region. The lakes were eutrophic to hypereutrophic. The total and particulate nitrogen to phosphorus ratios were least variable and closest to intracellular values; the total inorganic N:P ratio had the greatest amplitude of variation, and its use to characterize a water body was not convincing. The evolution of the N:P ratio with time, and with season, showed low values, around 5:1 and less; these, and not the average seasonal values, are most likely liable to lead to the appearance and stimulation of growth of nitrogen-fixing blue-green algae.

Chronic or temporary and changing limitations

The problem of nutrient limitation cannot be pinned down to one major nutrient; it is a subtle balance between varying supplies of different essential nutrients. Successive, opportunistic limitations occur, for the relative abundance of one nutrient renders the nutrient which is at the lowest concentration limiting. This is shown by experiments of Yusoff & McNabb (1987) in Malaysian fish ponds. They calculated ratios between inorganic nitrogen and orthophosphate, before and after fertilization (triple superphosphate (TSP) at 1.4 kg P ha^{-1} $week^{-1}$ and urea at 16.6 kg N ha^{-1} $week^{-1}$), and observed that the ratio shifted from conditions of limiting phosphorus (N:P of 36 in control and 44 in ponds fertilized with TSP and urea) to conditions of limiting nitrogen (N:P of 2 in ponds fertilized with urea only). The evolution of this ratio after fertilization demonstrates the frail balance between the correct amounts of nitrogen and phosphorus required to achieve optimal production.

In conditions of low concentrations of phosphorus

and nitrogen, the supplementation of phosphorus to water quickly entails a limitation by nitrogen. The amounts of phosphorus required are, however, more vital than those of nitrogen, and the effect of nitrogen supplementation is less apparent. Chemostat cultures of *Scenedesmus sp.* with nitrogen and phosphorus additions in a ratio of N:P of 5:1 to 80:1 show successive limitation by nitrogen, when the ratio is below 30:1, and by phosphorus when above 30:1. At the same time, cellular nitrogen content remains constant until the nitrogen-phosphorus ratio reaches an optimum value, and increases proportionally (linearly) afterwards. Cellular phosphorus content remains high at low N:P ratios, when nitrogen is limiting, but decreases quickly when the N:P ratio reaches an optimum. At high N:P ratios, intracellular phosphorus remains low (Rhee, 1978).

Ponds react to changes in phosphorus-nitrogen-carbon ratio by a shift in the phytoplankton species present: an increase in phosphorus supply leads to a community of nitrogen-fixing blue-green algae, while a decrease of carbon dioxide favours algae adapted to low concentrations of carbonate (Schindler, 1987). A simultaneous nitrogen and phosphorus limitation rarely occurs, like in Lake Okaro, New Zealand (White *et al.*, 1989). Additions of the most limiting nutrient lead to only slight modifications in algal growth. The prevailing phytoplankton community is the result of inter-specific competition between species for nutrients. It cannot exactely be predicted in the actual state of our knowledge (see p. 157). Nutrient availability depends on supply, which, in turn, depends on recycling processes, specific utilisation of nutrients, and transfer across trophic levels (Elser *et al.*, 1988).

Control model for carbon, nitrogen and phosphorus

In the controversy whether nitrogen or phosphorus are the prime control of organic production, Smith *et al.* (1989) developed a conceptual model, on evidence gathered in Tomales Bay, California, U.S.A., according to which carbon cycling controls nitrogen cycling, but is itself controlled by phosphorus cycling. It is based on the assumption that assimilation and release of nitrogen and phosphorus follow closely carbon cycling, i.e. production (C fixation) and respiration (C oxidation). In most shallow ecosystems, there is a balance between primary production and respiration, and the sum of carbon fluxes of production and respiratory oxidation tends toward zero. The difference between these fluxes is expressed in nitrogen flux units (Fig. 46). Excess respiration or surplus production is compensated by denitrification or nitrogen fixation. This capacity of adjusting the nitrogen balance is potentially present in most systems, though the actual rates of denitrification or N-fixation may be low. The trophic relations between nitrogen-fixing blue-greens, releasing carbon exudates to bacteria, and bacteria providing phosphorus to algae through phosphate exchange can be viewed as a coupling of the cycling of the three nutrients (Bloem & Bär-Gillisen, 1989).

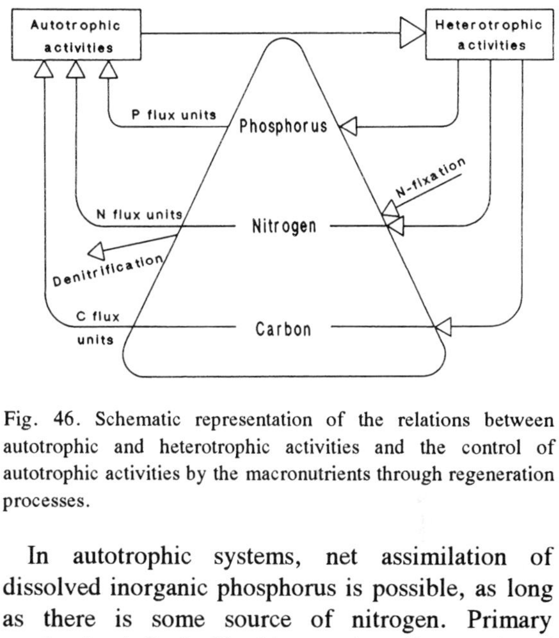

Fig. 46. Schematic representation of the relations between autotrophic and heterotrophic activities and the control of autotrophic activities by the macronutrients through regeneration processes.

In autotrophic systems, net assimilation of dissolved inorganic phosphorus is possible, as long as there is some source of nitrogen. Primary production is limited by biomass (standing crop) and specific growth rate, which in turn is limited by nutrient availability. As long as there is a phosphorus supply, nitrogen-fixing will supply the required amounts of nitrogen for the elaboration of organic matter. Many communities show high primary production, though dissolved inorganic nitrogen concentrations in water may be low. This high primary production results from high rates of nitrogen fixation, which meet the nitrogen needs. Phytoplankton production is significantly correlated to autotroph biomass, but there is only a low correlation with dissolved inorganic nitrogen; this suggests that biomass is ultimately more important than growth rate in the regulation of primary production.

A supply of dissolved inorganic nitrogen will have an immediate stimulatory effect on primary production, provided phosphorus is available. However, along with stimulation of primary production goes an almost proportional increase of respiration; dissolved inorganic nitrogen is released, which stimulates denitrification, bringing the levels of dissolved inorganic nitrogen down to former levels. Hence, inorganic nitrogen loading by a point source enhances production locally and temporarily, but there is no long-term or large-scale effect.

As organic matter is oxidized, nutrients, i.e. phosphorus, nitrogen and carbon, are recycled, thus stimulating primary production. The fate of phosphorus, immobilized in sediment or by primary production, determines that of nitrogen. A nitrogen surplus in relation to phosphorus availability will be denitrified, while a nitrogen deficit relative to available phosphorus is eliminated by nitrogen fixation.

If light supply to a system which relies on autotrophic production increases, autotrophic biomass will increase only as long as phosphorus is available. If nitrogen is required, nitrogen fixation will be stimulated. If available phosphorus is limiting, an increase in organic production can be achieved by a faster recycling and regeneration, but without a change in net production.

Only situations in which phosphorus is supplied externally to the system appear to have long-term impacts on biomass. Systems, with large supplies of organic matter, have an average net heterotrophic metabolism. In these systems, benthic regeneration of organic matter here results in an increase of the relative importance of denitrification, causing a nitrogen-limited growth rate. However, the primary production of such a system remains high, because of the high autotrophic biomass, controlled by phosphorus, regenerated from organic matter.

The control of the production system by nitrogen or phosphorus can be explained by a change in nutrients available or in the N:P ratio. There is a shift from phosphorus to nitrogen limited growth (Smith *et al.*, 1989).

Reaction of the environment to nutrient limitation

Another way to approach the problem of nutrient limitation is to consider how the environment copes with the available amounts of nutrients, or which is the phytoplankton assemblage prevailing under the range of nitrogen and phosphorus fluxes present. A phytoplankton community is composed of a number of species, whose nutrient requirements correspond to the range of nutrients available in the environment. This can be shown by enrichment experiments in limnocorrals with fertilizer applications of various nitrogen-phosphorus ratios. In lake Okaro, New Zealand, the phytoplankton assemblages displayed major changes depending on the nature of fertilizer applications, involving appearance and disappearance of various taxonomic groups (Pick, 1989). Some algae in two lakes (Kennedy and Sproat lakes, Vancouver Island) of British Columbia, Canada, were thriving under limiting nitrogen conditions, while others sustained limiting phosphorus conditions (Suttle & Harrison, 1988). As in Kennedy lake, fertilizer application gave rise to a low N:P ratio, nitrogen limitation promoted the appearance of nitrogen-fixing blue-green algae (Stockner & Shortreed, 1988).

Specific nutrient requirements of algae have been demonstrated by Blomqvist *et al.* (1989), in enclosure experiments in the acidified lake Njuffatet, central Sweden. They experimentally modified the inorganic nitrogen sources and, irrespective of the dominant species at the start of the experiment, the addition of ammonium to the culture medium lead to the dominance of *Merismopedia*, while addition of nitrate lead to dominance of *Peridinium*. Algae will therefore only grow in an environment in which their specific physiological needs are met.

Changes in the supply ratio of silicon and phosphorus also affects the algal community. In semi-continuous cultures of a natural algal community (originated from a Minnesota lake), at Si:P ratios below 20:1, green algae (Chlorophytes) were dominant, while at ratios of around 80:1, diatoms became abundant, and at ratios of around 160:1, siliceous Chrysophytes (silicoflagellates of the genus *Mallomonas*) took over (Grover, 1989b). Diatoms were capable of competitive differentiations, by reducing the size of their cells, when the Si:P ratio decreased. A greater variety of species was associated with variability in phosphorus supply (Grover,

1989b).

Zooplankton affects phosphorus sedimentation; the relation between phosphorus cycling in water and the abundance of zooplankton is discussed on p. 164.

Ways of determining nutrient limitations

Three approaches have been used, with a view to understanding the nutrient requirements of plankton: the first attempts to establish direct correlations between nutrient concentrations and phytoplankton biomasses. Seymour (1980) applied multiple regression (discriminant analysis), to the relation between nutrient concentrations and plankton biomass in a fish pond in the South of France, and obtained significant negative correlations between phytoplankton volume and ammonium and organic nitrogen concentrations in the course of one production season, and between phytoplankton volume and nitrite, nitrate (highly significant) and orthophosphate concentrations, during the following production season. The negative correlation was due to the inverse relation between nutrient abundance and the development of a plankton bloom. At the die-off of the bloom, an important amount of nutrients was released. Jana & De (1983) similarly established a correlation matrix between the physical and chemical variables of a polyculture fish pond in West Bengal, India, and its gross primary productivity. They concluded that productivity was positively correlated with water temperature, and with free carbon dioxide, dissolved oxygen, ammonium, nitrite, nitrate and phosphate concentrations.

A second way to approach the problem is to determine intracellular nutrient concentrations and to compare these with the critical content. These studies are, however, not possible with field populations.

The third type of studies attempts to understand algal nutrient requirements by means of selective enrichment bio-assays, examining phytoplankton responses (especially growth), to additions of one or more nutrients (Sanders *et al.*, 1987). Growth may be an increase in biomass, in the number of cells, in chlorophyll a content, or in particulate nitrogen (White *et al.*, 1989).

Comparative results of these three approaches seem contradictory (Gerhart & Likens, 1975). This is due to differences between field and laboratory studies. Studies of nutrient enrichment have the best potential to identify the limiting nutrient. In the field, there is probably not just one limiting factor, but rather a combination of nutrients, whose concentrations are close to limiting thresholds.

Most mathematical models developed until now consider light, rather than nutrient concentration, as the principal determining factor.

Reliability of predictions on the appearance of algal groups

As indicated previously, predicting the appearance of a given group of phytoplankton, following enrichment experiments with fertilizers, is still unreliable. In simultaneous, comparative trials in adjacent ponds, different phytoplankton associations may appear. For example, in two ponds stocked with the same fish (silver carp, common carp, bighead carp, grass carp, hybrid buffalo) and with similar fertilization (pig dung at 39 and 66 hogs ha^{-1}), different phytoplankton taxa evolved differently. In the first pond (39 hogs ha^{-1}), Euglenophyta were dominant one month after stocking, followed by Chlorophyta the month after, then by Chrysophyta the month after and by blue-green algae a month later. In the other pond (66 hogs ha^{-1}), Euglenophyta were dominant at the same time as in the first pond, but Chlorophyta took over faster, and remained dominant during all subsequent months, while blue-green algae were never present in sizeable amounts. The total amount of phytoplankton in the samples varied over a 10-fold range (Buck *et al.*, 1978). The factors which regulate such succession dynamics are not well understood; the successions observed are probably the result of different nutrient requirements on the part of the various phytoplankton species, and of opportunistic exploitation by the species present. Limnological differences at a micro-environmental scale certainly exists, but are not easily accessible to investigation. These may prove to be major determining factors in the evolution of phytoplankton assemblages.

Seasonal succession is probably less a species replacement than the regular appearance of a sequence of species over the year, related to changes in the environment, which are repetitive (Evans, 1988).

Plankton group-succession or temporal modification of plankton assemblage structure

For a long time, limnologists have been aware of successions in dominance of plankton groups in the course of a year. In temperate waters, this succession has been described by a model (the PEG-model), based on 24 statements about the sequential events which occur in the course of the season (Sommer *et al.*, 1986). Succession is dictated by seasonal imperatives, linked to modifications of the régimes of temperature, nutrient concentration, and of grazing pressure (Vanni & Temte, 1990).

At the end of winter, when temperature becomes favourable, a high availability of nutrients leads to an abundance of small, edible diatoms, with high growth rates. Diatoms require silicon, which is abundant at that timeand of which they exhaust the reserves. In a microcosm experiments on plankton from Schöhsee, northern Germany, nutrient recycling by zooplankton returned proportionally more phosphorus than silicon to the environment and was disadvantageous for diatoms and benefitted green algae (Sommer, 1988). Following this change, herbivores with short generation times appear, for instance rotifers, followed by cladocerans, whose abundance increases quickly and which graze intensely on the algae present. These species are r-selected (Komárková, 1989, observations in the moderately eutrophic Římov reservoir, Czechoslovakia).

Grazing quickly exceeds algal reproduction, and results in a decrease in the number of algal cells, followed quickly by a decline of the grazers. This period is designated as the clearwater phase. Reduced grazing pressure and increasing recycled nutrients lead to a restoration of phytoplankton biomass, the amount of which is inferior to the initial colonial biomass, but richer in species and functionally more diversified. The species, mostly green algae, are now more resistant to grazing, being larger in size or colonial, or protected by gelatinous sheaths or thickened cellular walls. "Canopy species" appear. These summer species are K-selected (Komárková, 1989).

Differential recycling of phosphorus and nitrogen leads to an imbalance in nutrient supplies. A nitrogen depletion promotes the appearance of nitrogen-fixing blue-green algae, at a period when mixing depth is reduced. Large-sized herbivorous zooplankters are replaced by small-sized species, less vulnerable to predation by fish. Autogenic succession is mainly determined by the mixing depth.

The maintenance of a phytoplankton population is dependent on two prime factors: solar light, as the energy source for photosynthesis, and nutrient concentrations in the water. Nutrient supplies during the year come from regeneration and from microbial activities, which are at maximum at a certain depth below the main phytoplankton biomass.

Algae which sink and sediment are colonized by decomposers and release nutrients there. Some algae migrate vertically to gain access to such nutrients, while others depend on physical mixing to carry nutrients upwards from deeper water (Gawler *et al.*, 1988, on Lake Geneva, Switzerland; Vincent & Dryden, 1989, on lake Okaro, New Zealand). The reduction of the amount of light in autumn restricts photosynthetic activities, which limits algal volume.

The population at any given moment reflects a balance between instantaneous conditions, the recent past, and the phytoplankton species best adapted to these conditions, species being selected from a spectrum of available types (Houk, 1989, on Zelivka reservoir, Czechoslovakia).

In marine environments (Mediterranean and Caribbean Sea), phytoplankton succession has been divided into three main stages in relation to changes in the environment (Smayda, 1980). The first stage is one of nutrient rich water, with algal cells characterized by a high ratio of cell surface area to cell volume, i.e. high nutrient uptake capacities, high rates of potential increase, with 1 to 2 cell divisions per day, and individual population densities reaching 100 to 1000 cells per ml. Such algae belong to the diatoms, various flagellates and Chrysophytes. The second stage is a more mixed community, with larger cell sizes, lower growth rates, cell divisions occurring every second day or even less, and population densities of 20 to 100 cells per ml. These algae are diatoms and dinoflagellates. The third stage appears when nutrient concentrations have reached a minimum, and the water is well stratified. The algae now have lower cell surface area to cell volume ratios, and population densities are less than 10 cells per ml. The most common species are nitrogen-fixing blue-greens.

The evolution of plankton communities in fish ponds grossly follows the "marine" pattern. There is considerable availability of nutrients at the start of production, soon after impoundment (Fig. 47). Nutrients are released from the sediments and by the fertilizers applied at the beginning of production. Ponds are usually drained at harvest time and left dry for some time. During this time, accumulated

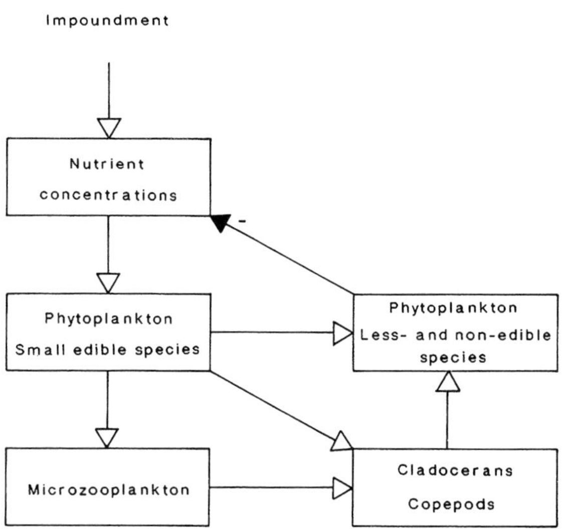

Fig. 47. Schematic representation of the sequence of events occurring within a pond after impoundment. The black arrow indicates a feed-back effect.

organic matter decomposes and the resulting nutrients readily dissolve, when the pond is impounded again (see also p. 24).

Indirect evidence of the relation between nutrient concentrations and primary production was provided by Legendre *et al.* (1987), who examined the evolution of plankton populations and biomass in coastal fish ponds in Côte d'Ivoire. During the first days after impoundment, preceded by a drying period and by liming, they noticed a considerable increase of nutrients in the water. These nutrients stimulated the growth of a large phytoplankton population, leading to the appearance, after 8 to 10 days, of a zooplankton population, mainly composed of rotifers, reaching 34000 ind. l^{-1} or 2.5 g (DW) m^{-3}. The phytoplankton biomass varied widely (up to 15 g C m^{-3}) during the initial phase, but stabilized after a period of 20 to 30 days at a level less than the initial one (around 3 g C m^{-3}). This decrease was due to a drop in average concentration of nutrients, especially ammonium, and also to zooplankton grazing. A second zooplankton population (700 ind. l^{-1} or 1.4 g (DW) m^{-3}) appeared after 13 to 20 days, consisting of crustaceans. The decrease of the phytoplankton was accompanied by an increase in bacterial biomass. The zooplankton biomass stabilized after 25 to 30 days around 0.3 g (DW) m^{-3}.

Spatial heterogeneity of nutrient concentrations

In all observations of physical and chemical variables in water, measurements are made on samples, but results are considered to be representative of the whole lake or pond. However, within a water mass, there is a continuum of conditions, with microzones of varying nutrient concentrations. Samples collected less than 2 m and few seconds apart can show differences in dissolved inorganic nitrogen concentrations of three orders of magnitude. The higher the volume of a water sample, the more such local variations are masked (Smith, 1986).

The development and proliferation of a given phytoplankton species was earlier explained by the presence of such micropatches. This implies that there is no hope to explain the relationship between major nutrient concentrations in water and observed primary production, by conventional sampling procedures.

The presence of a heterogeneous nutrient distribution, resulting from zooplankton excretion and distribution, and, in particular, its effect on phytoplankton growth is another controversial subject (Currie, 1984; Lehman & Scavia, 1984). Still another form of heterogeneity comes from pulsed nutrient supplies, varying in the course of time. Trials of phytoplankton cultured in conditions of pulsed supplies have resulted in the coexistence of various species, competing for the same resource (Sommer, 1984).

In attempts at mapping horizontal, vertical and diel variations of particulate phosphorus, carbon and nitrogen concentrations in ponds in Oahu, Hawaii, Wei & Laws (1989) suggested mechanisms which modify the distributions. About 75 % of the variance of a recorded value of particulate carbon and nitrogen was explained by position within a pond; for dissolved reactive and total dissolved phosphorus, position explained 35 % of the variance, and for particulate N:C ratio 7 %.

Horizontal variability was reduced in conditions of vertical homogeneous mixing (isothermy); determination of temperature difference between surface and bottom of a pond was recommended as an index of mixing conditions. During conditions of homogeneous mixing, depth explained not more than 12 % of the variance among measurements. Time of day was responsible for more than 50 % of total variance; this was ascribed to active migrations and assimilation of plankton, which modified nutrient concentrations in solution. The increase of particulate carbon at the surface between morning and afternoon was the result of photosynthesis. Simultaneously, nutrient concentrations decreased, as a consequence of uptake by phytoplankton.

Total dissolved phosphorus was at a maximum in the morning, and decreased about 7 % in the afternoon, but dissolved reactive phosphorus did not show such a change.

Wind had an appreciable effect on horizontal distribution (see p. 54). Microscale variability of particulate material in water at the bottom of the pond was due to patchy distribution of microzones with intense degradation or to activity of benthic macroinvertebrates. Spatial and temporal variations are not easily distinguished, and the processes have phases of different amplitudes and frequencies. Small scale, high frequency processes sometimes hide those of large scale and low frequency.

In lake Rotongaio, New Zealand, spatial heterogeneity was considered the rule in planktonic life and was the result of the interaction of various factors, including water movement induced by wind action, and zooplankton grazing (Horne & Commins, 1989a). This heterogeneity was temporally unstable and generally lasted from few hours to one day. It was related to the distribution and vertical migration of algae. These migrations were induced by photosynthesis and depended on carbohydrate ballast of the cells. Algae in bad physiological condition did not migrate vertically (Horne & Commins, 1989b).

Temporal heterogeneity of nutrient concentrations

Uptake of nutrients occurs rapidly. This has been shown, for example, by homogeneity tests made near a point source supplying water to a marine fish pond in Israel. At 1 m beyond the water supply, the concentration of nutrients in the water was similar to that of the pond and considerably different of that of the water supply. This homogenization could only result from immediate uptake of nutrients by phytoplankton (Krom *et al.*, 1989a).

Control of excess nutrients

The control of excess nutrients, deleterious to water quality, like in sewage, must also be mentioned. In Israel, Avnimelech & Lacher (1979) studied the recycling of nutrients from surplus feed, which accumulates on pond bottoms. They observed a bottom accumulation of 25 %, 75 % and 80 % of organic carbon, nitrogen and phosphorus respectively, the origin of which was the feed, either directly or indirectly after transformation by algae. When they established the budgets of various nutrients, they found that carbon was lost in the process, through release of carbon dioxide. Small amounts of nitrogen were also lost, but phosphorus was gained, probably through resuspension of particulate material from the sediment. Deterioration of water quality brings about environmental stress. Organisms with high reproductive rates and high dispersion capacities, such as some phytoplankton species, respond quickly to this stress. The most vulnerable species disappear first. Among animals, benthic invertebrates are particularly affected, because accumulation of polluting and harmful substances is most severe in sediments. Anatomical malformations amongst invertebrates are indicators of pollution (Schindler, 1987).

Inorganic nitrogen in excess can be controlled, for instance, by production of unicellular algae (single cell protein), themselves consumed by fish. However, this production needs the presence of other nutrients in adequate amounts, especially inorganic carbon (Avnimelech *et al.*, 1989). In experiments conducted at an aquaculture farm in Singapore, the macrophyte *Elodea densa* removed nitrogen compounds from water and reduced turbidity. A total amount of 25 % of the inorganic nitrogen was found to be removed, 26 % being nitrate, 50 % being nitrite and 17 % ammonium, i.e. 61.36 mg, 2.27 mg and 9.09 mg of N m^{-2} d^{-1} of nitrate, nitrite and ammonium, respectively. At the same time, total alkalinity and total organic carbon increased by 5 %, while dissolved phosphate increased by almost 300 %. The phosphate released originated from the sediment (Ng *et al.*, 1990). The water hyacinth (*Eichornia crassipes*) is a macrophyte which can also be used for the same purpose (Liu & Seki, 1988).

Influence of nutrient concentrations

1. Primary production depends heavily on nutrient availability, but the relation between primary production and nutrient concentrations is complex. Nutrient fluxes and intracellular nutrient contents are more important than the actual concentrations. Algal growth depends on specific uptake capacities of the algal assemblage present.

2. Phosphorus uptake capacities of algae are usually larger than nitrogen uptake capacities and vary to a wider extent between deficiency and normal conditions.

3. Differences in the cycling of nutrients are apparent: phosphorus is readily stored and kept in active cycling, while nitrogen in excess of the immediate needs is wasted by denitrification. Silicon is not recycled and is progressively lost to the sediment.

4. Nutrients control primary production sequentially according to the following model: phosphorus controls carbon, which, in turn, controls nitrogen. Nitrogen concentrations in the environment are adjusted by nitrogen-fixation when in deficit or denitrification when in excess. Phosphorus determines long term modifications of biomass, as it can be stocked; supplies result in a gradual increase in biomass. Nitrogen cannot be stored and any supply immediately increases respiratory activities.

5. Phytoplankton adapts to prevailing N:P ratios in the water through changes in species assemblage, which promote species with specific nutrient requirements. Minimum N:P ratios appearing in the course of time may be more important in governing the evolution of the assemblage than average ratios.

6. Phytoplankton succession is essentially an adaptation of the plankton community to changing water quality. Appearance and disappearance of algae depends on major modifications in nutrient availabilities. The zooplankton community changes accordingly. The phases of the succession which are of most interest to the fish farmer are those which involve a large growth of small edible algae, in a nutrient-rich, changing environment.

7. The major obstacle to an understanding of the dynamics of nutrient cycling in water is the heterogeneity, at a microenvironmental scale, of nutrient distribution in water. Furthermore, this patchiness in distribution is often ephemeral.

1.2.6. Influence of zooplankton

Zooplankton composition

Zooplankton is subdivided in microzooplankton, composed of flagellates, ciliates and rotifers, and macrozooplankton, which consists of planktonic crustaceans, cladocerans, copepods, and less frequent groups, such as typhloplanid flatworms, Acari, and Medusae, and insect larvae (mainly *Chaoborus*).

Microzooplankters are mostly detritivores and bacterivores. They develop abundantly when water is loaded with suspended organic matter, and are consumed by rotifers and small crustaceans (Hul, 1987).

Most but not all cladocerans (also classified as mesozooplankton) are herbivores. In fact they are the most important group of herbivorous crustaceans. Small- and large-bodied species are found, all promptly consumed by planktivorous fish. They are sensitive to predatory pressure by planktivorous fish. As they are numerous, of large size and easily cultured, they have been widely studied and used in biomanipulation experiments. Their relations with phytoplankton are complex and not completely unraveled, which makes predictions about them difficult (McQueen & Post, 1988).

The food of cladocerans is not restricted to planktonic algae. They also consume bacteria, fungi, detritus, organic aggregates, and sometimes even ciliates and rotifers (Burns & Gilbert, 1986).

Cladocerans select their food on the basis of size and by chemoreception. However, great differences in the feeding of even closely related species exist (Matveev & Balseiro, 1990), some being selective, some not (Fulton, 1988).

Copepods (classified as macrozooplankton *sensu stricto)* are subdivided in calanoids, cyclopoids and harpacticoids and comprise species with various feeding regimes, often modified in the course of development. Calanoids are microphages, herbivores or predators. Cyclopoids are omnivorous, herbivorous, or carnivorous. Fish consume cyclopoids (late-stage copepodites and adults), though not so eagerly as other prey. All life stages of harpacticoids are detritivorous or herbivorous (Dussart, 1980).

Copepods in general and calanoids in particular are less preyed upon by fish, as they have better swimming, hence evasion, capabilities than Cladocera (O'Brien, 1987). Cyclopoids have an erratic jumping motion, which makes then more conspicous than the gliding motion of calanoids. The calanoid/cyclopoid ratio is considered to give a reflection of the predation pressure by visual planktivorous fish. The ratio decreases with increasing predation pressure (Hrbáček, 1962; Brooks & Dodson, 1965).

Swimming behaviour of copepods has evolved in close relation to their feeding behaviour. Carnivorous copepods possess mechanoreceptors enabling them to detect vibrations and water displacements created by swimming prey (Strickler, 1970 & 1975; Kerfoot *et al.*, 1980), while herbivorous copepods possess chemoreceptors enabling them to discriminate among different algal prey (Stemberger & Gilbert, 1987). They are considered to be selective feeders (Lampert, 1987).

Zooplankton distribution within ponds

Zooplankton distribution is non-homogeneous. In Dutch fish ponds, some species were mainly found in the littoral, while others selected limnetic waters (Verreth, 1990). Such distributions are probably related to food availability and avoidance of predators.

Distribution varies with time, due, in particular, to vertical migration, which is an almost universal phenomenon. The archetypical form of vertical migration is a swimming movement to deep water during the day, and an active ascent towards the surface during the night. However, there are so many exceptions, including pronounced reverse migration, that generalizations are hazardous. This diel movement from the warm, food-rich surface layers to the cold nutritionally poor deep waters has evolved in response to benefits to the community: (1) vertical migration results in predator avoidance by zooplankton; (2) fluctuations of the grazing pressure owing to vertical migration enables unimpeded growth of the algae during daytime. This results in higher rates of algal production, compared to those obtained in conditions where zooplankton graze continously on the same algal biomass. The disavantages of migration for the zooplankton are: (1) a reduction of energy intake; (2) an increase of energy expenditure required for swimming up and down; (3) a reduction of the rate of development of the eggs carried by migrating animals (Lampert, 1987).

For cladocerans, vertical migration is a compromise between avoiding predatory fish, by staying in deeper water, and the need to feed at the surface. When the amount of available food is insufficient, the degree of migration is reduced. The animals stay longer at the surface and use more time to satisfy their food requirements, defying predation (Gliwicz, 1985; Johnsen & Jakobsen, 1987). The deep waters act as a refuge for zooplankton: low visibility impedes visual prey pursuit and prevailing low oxygen concentrations are avoided by predatory fish. In lake Rotangaio, New Zealand, some zooplankters even survived in deeper water at dissolved oxygen concentrations of less than 0.2 mg l^{-1} (Forsyth *et al.*, 1990).

Influences of zooplankton on the plankton community

Zooplankton is a link between lower trophic levels and fish. Interactions between different plankton groups have been investigated and attempts have been made to establish a general relationship between zooplankton community structure, especially the relative importance of microzooplankton, and the trophic level of water bodies. In a study involving 12 sites in Quebec, Canada, significant relationships between ciliates, rotifers, cladocerans and cyclopoids, but not calanoids, and a range of trophic levels of lakes was shown. Also a clear relationship between microzooplankton and macrozooplankton biomass versus total phosphorus was established. The relative microzooplankton biomass increased with increasing trophy (Pace, 1986).

Though quantification of the relationships between different plankton groups and nutrient availability is the ultimate aim of the ecologist and the fish farmer, qualitative interactions are the first to be investigated. The interactions between species in a community are direct or indirect, involving trophic linkage, behavioural, and chemical responses (Miller & Kerfoot, 1987). Effects of zooplankton on the plankton community are as follows:

a) Grazing

Zooplankton influences phytoplankton community structure, through grazing. Biomanipulation experiments with zooplankton in limnocorrals have clarified the role of zooplankton in ecosystems. In an oligotrophic Swiss lake, the removal of large-bodied zooplankters (above 95 μm in size), consisting mainly of cladocerans (*Daphnia*), lead to the conclusion that herbivorous zooplankton reduced, by grazing, algal biomass, but did not affect the size structure of the phytoplankton and the relative percentage of size classes contributing to total primary production (Uehlinger & Bloesch, 1987). Similar experiments in meso-oligotrophic Lake Lucerne, Switzerland, where removal of zooplankton (with a plankton net of 95 μm net size) was followed by fertilization with phosphorus, showed an increase in chlorophyll a, organic particulate carbon and particulate phosphorus (in fact, of total algal biomass). The proportion of nanoplankton decreased to the benefit of larger cells and the settling rate increased. This was, however, not attributed only to selective grazing, but also to differences in algal strategies (Bloesch & Bürger, 1989).

In enclosures in a eutrophic farm pond and an oligo-mesotrophic lake in Illinois, the impact of Cladocera on phytoplankton depended on the body size of the grazer. The replacement of small-sized species of cladocerans by large-bodied species was found to reduce the abundance of phytoplankton by one order of magnitude, and also decreased the density of herbivorous cyclopoids and rotifers (Vanni, 1986).

In enclosure experiments in the moderately eutrophic lake Ros (northern Poland), the impact of large-sized zooplankton, such as *Daphnia magna*, and zooplankton of normal size, on phytoplankton were different: the species of large size controlled the density of phytoplankton of all size classes, while the zooplankters of normal size had no effect on net plankton (above 50 μm). Algae of large size increased, because they were subjected to less competition from small sized algae, in conditions of limiting phosphorus concentrations. The efficiency of control of phytoplankton density by large cladocerans decreased with increasing dominance of phytoplankton of large size (Dawidowicz, 1990b).

The impact of large-bodied cladocerans is sometimes so important that it drastically changes the structure of the algal community. For example, in fish ponds in Regensburg, Germany, the introduction of *Daphnia magna* resulted in disappearance of the phytoplankton and appearance of a population of *Elodea canadensis* (Theiss *et al.*, 1990).

In the moderately eutrophic Schöhsee, northern Germany, the clearwater phase, which appeared in the course of phytoplankton succession in temperate waters, and which was characterized by an important increase in Secchi disc transparency, was due, at least partly, to Cladocera and other herbivores grazing on the phytoplankton (Lampert *et al.*, 1986).

b) Phytoplankton defenses against grazing (see also p. 145 & p.150)

In laboratory experiments, large-bodied cladocerans have been found to control the proliferation of filamentous algae. These observations cannot, however, be transposed to nature: here, the stocking of such cladocerans does not prevent the appearance of algal blooms; the animals may even disappear when algal density becomes high, probably because the Cladocera cannot reproduce sufficiently fast (Gliwicz, 1990).

Observations on high and low densities of *Daphnia pulex* and of rotifers, respectively, during a bloom of the filamentous alga, *Anabaena affinis*, were followed by trials on interactions of various species of cladocerans and rotifers with filamentous blue-green algae. The presence of filamentous blue-green algae influenced cladoceran populations directly and this effect was proportional to algal density. Rotifers, however, were not affected. The sensitivity of cladocerans varied with species and with size of the individuals (Gilbert, 1990).

Similarly blooms of *Microcystis aeruginosa*

negatively influence large-bodied cladocerans; feeding is aselective in large cladoceran individuals, so that toxic, non-edible algae or algae without nutritive value are readily ingested (Fulton & Paerl, 1987).

Another reason why cladocerans do not easily coexist with filamentous blue-green algae is due to mechanical interference by the filaments with the feeding appendages. Their thoracic trunk limbs are easily clogged by algal filaments. Blue-green algae also produce endotoxins (Gilbert, 1990). All species do not have the same sensitivity these toxins; *Bosmina* is less. sensitive than *Daphnia* (Burns *et al.*, 1989). By fragmenting non-toxic blue-green algal filaments into pieces, large-sized species of cladoceran, make these edible for smaller cladocerans (Dawidowicz, 1990a).

c) Changes in sedimentation rate caused by zooplankton (see also p. 144)

Zooplankton modifies the amount of phytoplankton lost via sedimentation, by changing the average size of the algal population. Grazing results in the appearance of algae of large size, which are less edible and sink more quickly. Grazing by Cladocera transforms small algae into a slurry of partially digested gut fluids that dissipate rapidly. Particulate matter in cladoceran faeces is likely smaller on average than the food particles, with sinking characteristics comparable to those of small moribund algae (Kitchell & Carpenter, 1987). Observations in limnocorrals showed that small algae (> 12 μm) contributed to 80 % of primary production. Removal of the large-sized herbivores from the limnocorral increased sedimentation rate of particulate phosphorus and particulate organic carbon. From this, Uehlinger & Bloesch (1987) concluded that small-sized algae contribute in a more important proportion to sedimentation when consumed than when unconsumed. The portion of primary production ingested by zooplankton follows two paths: part is incorporated in the food web, being transformed into zooplankton biomass, while the rest becomes excrement, which settles and/or is decomposed. The supply of particulate matter to the seston pool thus generated is important, as comparatively more material is rejected than assimilated (Gulati, 1990).

d) Zooplankton effect on nutrient cycling and supply to phytoplankton

By feeding on algae, zooplankton recycles part of the nutrients (P and C) contained in the algal food, and reduces sedimentation. (1) Herbivores reject in their excrements about half the phosphorus contained in the algae ingested. Dominance of large-bodied herbivores in a Polish lake reduced the recycling of phosphorus contained within phytoplankton and zooplankton biomass and detritus (Esjmont-Karabin *et al.*, 1989). In the absence of zooplankton, there is no orthophosphate excretion, the inorganic phosphorus pool decreases, and the abundance of algae decreases. (2) Detritivorous zooplankton has a consumption rate of detritus similar to the consumption rate of algae by herbivores. Therefore, the cycling of phosphorus within the trophic foodweb of the water column is always maintained (Esjmont-Karabin *et al.*, 1989).

The specific release rate of phosphorus by cladocerans increased as the P:C ratio of the food increased. For a ratio of 6 to 8 μg P mg^{-1} C, the release rate of phosphorus was nil. At that concentration, all ingested phosphorus was used for growth and reproduction. In eutrophic lake Nesjøvatn, central Norway, phosphorus limitation occurred when the P:C ratio of phosphorus-limited algal cells and of detritus was well below this value (Olsen *et al.*, 1986).

Comparisons between 49 lakes in north-east Poland have shown that the phosphorus regeneration rate by zooplankton increases when phosphorus concentration increases, in such a way that the turnover time of phosphorus is maintained fairly constant, even if the total phosphorus pool is variable. The uptake of phosphorus by algae is directly affected by the supply rate of available phosphorus, zooplankton regeneration rate being more important than total phosphorus concentration. The rate of phosphorus regeneration determines the abundance of algae; the specific composition of the assemblage is more related to specific characteristics of the species present. In some cases, this regeneration favours blue-green algae, partly because they are not consumed (Esjmont-Karabin & Spodniewska, 1990).

While zooplankton graze on phytoplankton, exudations of organic carbon are released into the water (sloppy feeding). This algal carbon is put at the disposal of the bacteria and stimulates their growth (Güde, 1988, on Lake Constance, Switzerland).

e) Predation on zooplankton

The zooplankton community of Mondsee, Austria, was affected in its structure by predation of invertebrates and fish (Dokulil *et al.*, 1990). These aspects will be discussed in paragraph 1.2.8., Influence of fish production.

Influence of macrozooplankton on microzooplankton

In a small lake of the Spanish Sierra Nevada, the removal of zooplankton above 45 μm in size from limnocorrals has been found to result in the disappearance of flagellates and in an increase in chlorophyll a concentration to 4 to 6 times that of the control. The number of phytoplankton cells increased to about 3 to 4 times that of the control. This not only indicates the importance of zooplankton grazing pressure on algae, but also on microzooplankton. The net growth of phytoplankton increases when zooplankton increases in abundance, probably due to regeneration of nutrients, which partly make up for the loss by grazing. The phosphate released by grazers is assimilated by algae or recovered by bacteria. The disappearance of flagellates reduces the internal regeneration rate of nutrients (Carrillo *et al.*, 1990).

Cladocerans influence rotifer and ciliate communities. In laboratory experiments, five species of cladocerans isolated from lakes in Vermont, U.S.A., were tested on their capacity to kill, during their normal feeding behaviour, individuals of the rotifer species *Keratella cochlearis*. It appeared that mortalities inflicted were independent of species and positively correlated with size of the cladocerans (Burns & Gilbert, 1986). The suppression of rotifer species by cladocerans is thus often attributed to mechanical interference. Rotifers and ciliates are sucked into the branchial cavity, where they can get fatally injured (Gilbert, 1989).

Similarly, the selective removal of zooplankton from water causes an increase of the relative importance of ciliates, which are apparently more reduced in number by predators than by their resources. Bacterivorous microflagellates, as well as ciliates, are important for the regeneration of nutrients (Hamilton & Taylor, 1987).

Controversy about the diversity of tropical zooplankton compared to that of temperate zooplankton

The species assemblage of tropical zooplankton is different from that of temperate region. Tropical zooplankton is characterized by the loss of several elements. Among the Cladocera, no large-sized herbivorous or raptorial species are found in tropical areas; cyclopoid and calanoids copepods are also less numerous. Only the number of predatory insects is more important than in temperate waters (Fernando *et al.*, 1990). For instance, the larvae of *Chaoborus* (Diptera Nematocera) are important predators of zooplankton in Lake George (Moriarty *et al.*, 1973). From these observations, it has been concluded that tropical zooplankton is less diverse than temperate one, though a range of pelagic invertebrates uncommon in temperate water are present: medusae, typhloplanid flatworms (Mesostominae), water mites or Acari. Their presence often remains undetected, but they may be a typical feature of tropical zooplankton (Dumont, 1992).

One of the reasons invoked for this allegedly lower diversity is that the efficiency of food capture does not increase in proportion to the increase of metabolic needs with rising temperature (Fernando, 1980). Crustaceans of smaller size are promoted because of their lower volume-surface ratios. In the case of rotifers, parthogenetic reproduction and resting stages are not suitable to environments with low seasonal differences (Lehman, 1988).

Cladocerans which do well in the tropics are small and bilaterally compressed and show morphological or chemical defense features, such as spines, spinules, predator repellents. Pelagic invertebrate are not visible, equipped with nettle cells, distastefull, produce toxic substances or are so jelly-like that they are not worth being eaten by fish (Dumont, 1991).

Considering the differences between tropical and temperate zooplankton, Dumont (1992) advocated that predation is probably the most important driving force shaping zooplankton communities in the tropics. Warmwater fish are important predators of zooplankton (Crisman & Beaver, 1990), as a large fraction of them remains planktivorous throughout life. The role of herbivorous macrozooplankton is partly taken over by phytoplanktivorous fish. The number of invertebrate predators is also increased by novel elements such as medusae, typhloplanid flatworms and water mites. Top-down control overrides the effects of resource competition or bottom-up control, as shown by numerous defense adaptations (Dumont, 1992).

The phytoplankton-zooplankton interface

The phytoplankton-zooplankton interface is the crucial point where changes in the predators at the top of the food web are translated into changes in the properties of the ecosystem, such as primary producer biomass, productivity and nutrient recycling (Elser *et al.*, 1990).

In ecosystems of high or low productivity, the coupling between phytoplankton and zooplankton is weak. The quantitative response of algae to grazing pressure is often not a direct function of grazing, but involves secondary compensatory effects, such as nutrient recycling. The qualitative responses of algae are specific and are, for instance, the appearance of inedible algal species, large-sized or colonial.

A study of three Californian lakes, varying in trophic level, showed that in ultra-oligotrophic environments, nutrient concentrations limit primary production and herbivore populations, while in hyper-eutrophic waters, the proliferation of inedible algae hampers the development of herbivores. In strongly eutrophic lakes, the corresponding zooplankton was unable to produce a detectable grazing effect, even when increased to eight times its natural density. Thus, the strength of coupling between phytoplankton and zooplankton is greatest in waters of intermediate trophic status (Elser *et al.*, 1990).

Types of control in plankton populations

The concept of interface implies an interaction of plankton with its environment. In experiments of selective removal of large-sized zooplankton, followed by fertilization of the water, increases in total algal biomass occurred. This indicates control of the phytoplankton not only by the grazers (top-down), but also by the resources (bottom-up), as the increase in resource availability stimulated phytoplankton production (Fig. 48). The structure of the zooplankton community depends on the specific characteristics of the phytoplankton assemblage with regard to edibility, and of the morphological capabilities of grazers to make use of this food source.

Analysis of data from three north-temperate lakes, bioassays, nutrient enrichment experiment and mesocosm experiments lead Elser *et al.* (1988) to conclude that the sequence of complex interactions

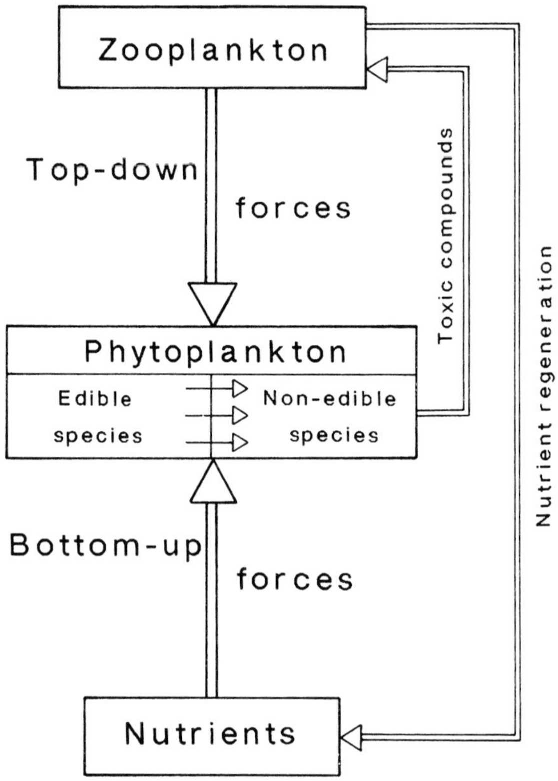

Fig. 48. Schematic representation of the types of controls exercised on the phytoplankton. Right-hand arrows indicate feed-back controls of zooplankton on phytoplankton.

between nutrient concentrations and the phytoplankton assemblage, through interspecific competition for nutrients, is another type of bottom-up force (producer controlled). The edibility of the phytoplankton assemblage affects the zooplankton community, especially the selective zooplankton grazers. This creates a feed-back whereby selective grazing promotes less edible and better protected algal species, and finally a change in nutrient recycling, mediated by zooplankton. The various direct and feed-back effects are not easily separated, which makes it difficult to establish a hierarchy of causes.

One of the specific mechanisms considered in connection with the modifications driven by the zooplankton is that the P:C ratios of nutrient requirements by zooplankton are lower than those of algae; zooplankton recycles greater amounts of nitrogen than phosphorus. This results in changes in relative availability of nitrogen versus phosphorus, and modifications in zooplankton biomass and/or size structure occur. Also, differences exist between the P:C ratios of substances which are ingested and those which are excreted or metabolized by zooplankton; nutrients present in excrements are more directly available for uptake than those present in algae or in zooplankton. The path followed for the reincorporation of nutrients is differential, being a function of the transformation and uptake they have been subjected to (Elser *et al.*, 1988).

In microcosm experiments on a natural plankton suspension enriched with netphytoplankton from the epilimnion of the moderately eutrophic Schöhsee, northern Germany, the Si:P ratio was modified by grazing and subsequent recycling of nutrients. As proportionally more phosphorus than silicon was returned to the water, species requiring more silicon, such as diatoms were outcompeted by green algae (Sommer, 1988).

Population controls do not only temporarily shape community structure, but also trigger micro-evolutionary changes. An important feature of the bottom-up and top-down forces is that they induce genetic changes within a community. The species evolve in response to availability of resources, but also with respect to predator-triggered defenses (environmental cues originating from predators rather than from prey (Havel, 1987)) by developing antipredator traits (Sih, 1987). The selection of traits within a population acts upon genotypes, with identical morphological phenotypes, as in the case of vertical migration (Dumont & De Meester, 1990). This ensures genetic heterogeneity within populations. Conservation of genetic heterogeneity is needed, if a population is to adapt to future changes in the environment.

Controls and regulations on plankton populations

It is important to understand the regulatory mechanisms, including feed-back actions, which promote or inhibit the development of various communities.

Do buffer mechanisms exist, which enable the maintenance of an assemblage of plankton communities (Moss, 1990)? Or are some equilibrium or successive equilibrium stages reached, which are maintained within some limits of changes in external factors, after which there is a sudden shift towards another equilibrium (Levitan, 1987; Scheffer, 1990)? Are trophic interdependence and stability of ecosystems imperatively linked (Paine, 1988)? A change in light regime in a shallow eutrophic lake, which resulted in a radical transformation of the structure of the ecosystem (Lammens *et al.*, 1990), would tend to confirm this contention.

Some uncoupling of the links between zooplankton and phytoplankton has also been observed (McQueen & Post, 1988), indicating that the controls of these trophic levels are numerous and their respective effects sometimes balanced and compensating each other. The effects observed in experiments with single species are seldom additive when species are in polyculture under natural conditions.

Interaction and exploitative competition

Interactions between organisms of similar status in the food web are manifold, but can be summarized as competition for the exploitation of common resources. A resource is here understood as all that is needed by an organism to satisfy its requirements of space, food, oxygen, etc. The tactics of organisms for obtaining such resources can be reduced to r- and K-strategies, i.e. species which have high adaptation potentialities, and high growth and reproduction rates, occurring in temporary habitats, and specialized species, with low growth rates, efficiently using all resources, and living in stable environments.

At the physiological level, this results in different adaptations for the uptake of nutrients or food, based either on velocity of uptake (fast versus slow) or on the affinity of uptake (high versus low concentration thresholds). Species can be distinguished on the basis of these properties. Calanoid copepods exploit aquatic environments with low levels of food, while some rotifers with high food requirements, along with high growth rates, have an advantage over other organisms in conditions of high food abundance. The concepts of transient dominance and mutual competitive exclusion apply to the case. They partly explain seasonal succession between species with fast growth rates, dominant when resources are plentiful, and species with lower growth rates, successful when resources are scarce (Rothhaupt, 1990).

Other hypotheses which explain competition between species have been discussed by Rothhaupt (1990) and are related to such characteristics as metabolism, feeding behaviour and predator escape responses. It is clear that a better understanding of the ecophysiology of the various plankton species will reveal a range of capabilities which allow species to respond to the various constraints and opportunities of the environment.

Interspecific competition acts at different levels: on minimum threshold concentrations required for a given resource, on partitioning of the resource within the environment, on abilities of organisms to substitute or take up alternative food, and on their responses to fluctuations in resource availability. The responses of species to changes in the food availability are variable and even include temporary halts of growth (resting stages) during life history (Rothhaupt, 1990).

Control by limitation of resources

From 25 whole-lake experiments in different countries of Europe, Benndorf (1990) concluded that top-down control on lower trophic levels decreased with increasing distance from the level on which it directly acted. The maximum potential biomass which could be reached by each trophic level was determined by the availability of resources, but the actual biomass reached was the combined effect of control mechanisms which were not only a function of resource availability but also of predation. The potential productivity of a trophic level is determined by nutrient supply, the action of predators, or a combination of these two factors. Actual productivity depends on nutrient recycling and nutrient partitioning between various populations with different growth rates (Gophen, 1990).

An effect on a given trophic level or on a group of organisms is quickly dissipated within the food web. Feed-back actions, which are not always directly apparent, must also be taken into consideration. For example, a reduction of predatory pressure of fish on zooplankton leads to an increase in zooplankton biomass. Beyond a certain zooplankton biomass, their living conditions deteriorate, however. The increase of zooplankton biomass amplifies the availability of nutrients to the surviving phytoplankton, not only by a reduction of phytoplankton biomass, but also by increased mineralization of nutrients by the zooplankton. This nutrient availability benefits phytoplankton species with high nutrient requirements associated with high growth rates, or species which can escape zooplankton grazing. However, increased grazing on the phytoplankton usually leads to higher water transparency, which stimulates photosynthetic activity. This results in some cases in colonization of the environment by macrophytes (Benndorf, 1990).

Some predatory effects are independent of or poorly dependent upon mechanisms of resource control. Reduction of top-down control rebalances the phytoplankton-zooplankton relationship. Its leads to the dominance of large-bodied herbivorous zooplankton and induces changes in the structure of the phytoplankton community, with selection of algae protected against grazing or which have characteristic high growth rates. Other effects are largely dependent on mechanisms acting on the resources. Average long-term biomasses are reduced by predation only if a certain level of availability of the resources is not exceeded. The probability of a mass development of inedible algae increases when resource (for instance P) limitation decreases. The mechanisms of control of the resources rely for their action on the availability of phosphorus. A phytoplankton standing crop can only be reduced, if the availability of phosphorus is decreased.

Herbivore tactics to escape predation

Zooplankton adaptations to escape predation are manifold, ranging over morphological, physiological, chemical and behavioural features (Sih, 1987). In some cases, defenses are predator-induced (Havel, 1987). Amongst crustacean prey, defenses include hardened integuments, spines on different parts of the body, for instance around the thoracic and cephalic zone (Williamson, 1987), size reduction, as in some cladocerans (Stenson, 1987), or unwieldy sizes or shapes rendering the handling process of the prey difficult for the predator (Sih, 1987). Among rotifers, the presence of spines, of a rigid lorica, of mucus sheats, protect them from predation (Stemberger & Gilbert, 1987). A small size entails a reduction of visibility and retention by gill rakers of fish, which is even more reduced by a lateral compression of the body as in cladocerans. Absence of pigmentation or hemoglobin (transparency), and reduction of the size of the eye spot lessen the visibility and enhance survival, though several disavantages go along with these features. Camouflage, mimicry, warning coloration have also evolved as a means of defense. Some prey, such as mites, are distasteful or toxic to predators. Behavioural modifications include reduced mobility, capacity to detect water currents (rheotaxis of rotifers) and escape them, vertical migration (O'Brien, 1987), and aggregation (Folt, 1987) or coloniality (Stemberger & Gilbert, 1987). The evolution of antipredator traits were divided by Sih (1987) into two main categories based on whether they act before or after the encounter with the predator. This choice affects the life-style of the species.

Avoidance of predators results either from features enabling the non-detection by the predator (avoidance of entering the sensory field of the predator) or from a hiding behaviour. Cladocerans benefit from the presence of refuges and provision of artificial refuges was found to increase their abundance (Irvine *et al.*, 1990). Species living in temporary or stressful habitats are considered to have evolved traits to decrease encounter rates with predators and competitors. The difference between species based on resource utilization, which has lead to the distinction between species with r and K strategies, can be extended to antipredators traits (Sih, 1987).

Evolving defenses results in costs for the prey, such as lower feeding, growing or reproductive rates than prey without these defenses. Sometimes they are constraining the prey's overall life-style. On the whole, behavioural traits appear to less constraintful than morphological ones, as they inflict lesser costs on feeding and mating (Sih, 1987). Riessen & Sprules (1990) have quantified the costs of these defenses for *Daphnia pulex*: individuals with spines in the neck region reached sexual maturity 8.4 to 14.6 % later than individuals not developing neck spines.

Zooplankton predator-to-prey proportions

Based on data from a range of freshwater habitats in Britain and North America, Jeffries & Lawton (1985) considered that an invertebrate predator to prey species ratio of 1:2 to 1:3.5 seems to be optimum. In a small enclosed ecosystem in Villefranche, France, the predation of copepods on ciliates and of ciliates on flagellates reached an equilibrium stage, whereby biomass increase compensated predation (Sheldon *et al.*, 1986).

In the Řimov reservoir, Czechoslovakia, the size structure of zooplankton was related to fish biomass and its predation, indicating the importance of their planktivory (Seďa *et al.*, 1989). The predatory pressure of fish is far more intense than that of all other groups.

Influence of zooplankton on the plankton community

1. Zooplankton is subdivided, according to size, into microzooplankton, mesozooplankton and macrozooplankton. Microzooplankton are the smallest organisms, consisting of flagellates, ciliates and rotifers, mostly detritivorous and bacterivorous, though rotifers also consume small algae. Mesozooplankton is comprised of cladocerans, and macrozooplankton of copepods (or cladocerans and copepods). The dietary range of zooplankton organisms is often wide.

2. The influence of zooplankton on phytoplankton is manifold: (1) zooplankton herbivores graze the algae and reduce the overall volume of the algal biomass. The algal community reacts to grazing pressure by an increase in proportion of large-sized and less edible algae. This, in turn, reduces the stock of grazers; (2) the zooplankton modifies the sedimentation rate of algae, by its grazing pressure. It releases faecal pellets of the algae consumed, which would otherwise have been kept in suspension. They also promote algae of greater size, which sediment faster; (3) they promote the recycling of nutrients in the water column. Nutrients are regenerated, recycled, and kept in constant turn-over, owing to the presence of zooplankton. This results from grazing of the zooplankton, whereby organic compounds (via sloppy feeding) and faecal pellets are released into the water, and subsequently decomposed, these nutrients thereby again becoming available for phytoplankton.

3. The diversity of tropical zooplankton compared to that of temperate regions is a controversial subject. The role which zooplankton plays in temperate regions may be taken over in the tropics by phytoplanktivorous fish, but predation pressure by zooplanktivores is another factor, and tropical zooplankton is enriched in unconventional elements with defenses against vertebrate and invertebrate predators.

4. The control of phytoplankton by zooplankton is quite loose and there is often no strong coupling between phytoplankton and zooplankton. The forces controlling such populations dynamics are a combination of bottom-up and top-down controls.

5. The relations between communities of organisms depend on trophic links, which command the establishment of equilibria between groups of organisms. But probably other types of relations besides trophic levels must be taken into consideration, such as competition for space, partitioning of resources within the environment (spatial heterogeneity in the distribution of resources), and resource limitation. Where grazing is not a limiting factor to the development of phytoplankton, limitation of available resources (especially phosphorus) must be taken into account.

1.2.7. Contributions to secondary production

Limitations of the concept of primary production

Primary production is conceptually confined to the autotrophic production of microphytes and macrophytes and designates the elaboration of particulate organic material at the lowest level of the trophic pyramid. However, in fact, this definition also includes heterotrophic bacterial production, unless the definition of the elaboration of organic material is limited to carbon fixation. The extension of primary production to microbial activities allows the inclusion of algal 'losses', such as organic carbon exudations and secretions, and a better assessement of primary production. However, the differences in significance of the processes of autotrophic and heterotrophic production are too wide to allow us to collect them under the same label (Flynn, 1988).

The microbial food loop

Many aquatic environments are characterized by a trophic dependency of much of the animal biomass on dead plant material. Sometimes, consumers directly ingest large amounts of plant detritus, but often there is no such direct link. Heterotrophic bacteria and fungi are then the intermediates between detritus and primary consumers (Miller, 1987).

Transfer of primary production to secondary production is also possible, and involves the microbial loop (Fig. 49). This designates the energy and nutrient flow from algae through bacteria and flagellates to zooplankton (Heyman & Lundgren, 1988); it constitutes a rechannelling of carbon and energy through the microbes into the planktonic food web (Bloem & Bär-Gillissen, 1989, in lake Vechten, Netherlands). The term bacterial food loop was coined by Azam *et al.* (1983) to designate trophic links between bacteria, heterotrophic flagellates and the microzooplankton in marine water. This loop should, however, be considered as an integral part of the food web of any ecosystem (Sherr & Sherr, 1988), and not as a source or sink of carbon. Bacteria and protozoa channel dissolved organic matter back into the food web by repackaging parcels of fixed carbon or recovering fixed carbon. The repackaging designates the uptake of bacteria by mixotrophic algae, while recovering is done by heterotrophic flagellates ingesting and assimilating polysaccharides of high molecular weight at relatively low concentrations (Sherr & Sherr, 1988).

Relations between bacterial populations, living plankton and dead cells

Bacteria recover organic matter with energetic value and release nutrients. Bacteria are present wherever there is accumulation of planktonic cells, which sink and decompose, and where algae exudations occur. In stratified lake Vechten, Netherlands, Bloem & Bär-Gilissen (1989) observed that important amounts of carbon and energy released as organic material byphytoplankton was processed by bacteria and returned to the 'main' food web through the microbial

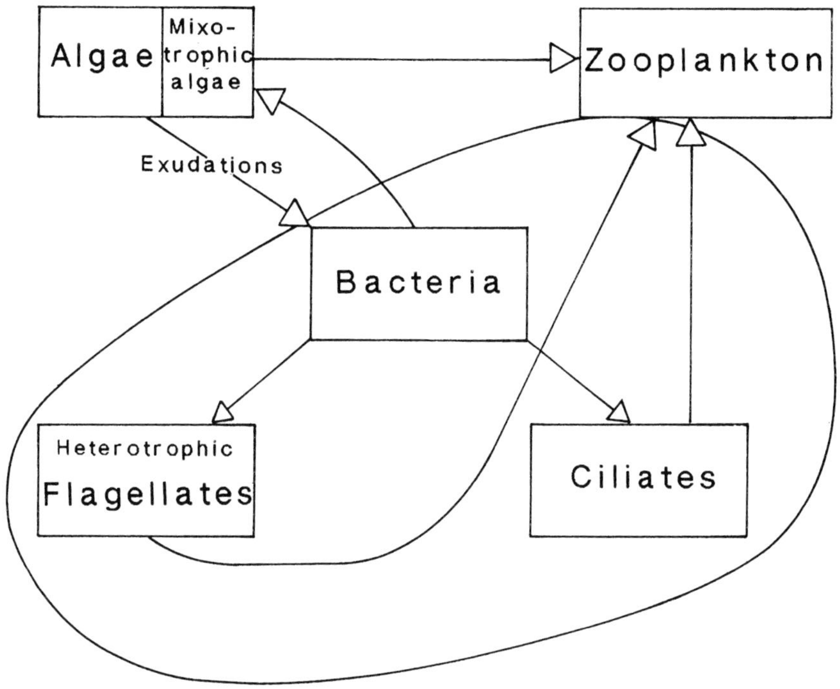

Fig. 49. Schematic representation of the bacterial food loop (encircled portion).

loop. Coupling between bacteria and heterotrophic nanoflagellates was such that about 10 to 70 % of the water column was daily cleared of bacteria. Efficiency of the process depended on the yield, i.e. the gross growth efficiency, of the grazers.

In the metalimnon of the lake, chlorophyll concentration and primary production rate were high, because of deep penetration of light and an upward flux of nutrients. Primary production along with detritus accumulation also promoted bacterial production. Bacteria were consumed by heterotrophic flagellates and ciliates, which themselves were consumed by microzooplankton. Ciliates (two *Coleps* species and one unidentified spherical ciliate) were found in great numbers at the oxic-anoxic boundary, where they were feeding on bacteria. There, ciliate predators were avoiding these oxygen-poor zones, while ciliates were either shifting to nitrate respiration, as an alternative energy source, or having a microaerophilic metabolism. High bacterial activity coinciding with high chlorophyll concentration indicated a link between phytoplankton and bacteria, phytoplankton exudations and organic material from decomposing algae cells, providing carbon and energy to the bacteria.

In eutrophic lake Nesjøvatn, central Norway, an algal die-off caused an increase in bacterial abundance, which constituted up to 30 % of primary production (Vadstein *et al.*, 1989). Methods of measuring microbial production are still controversial. Two main methods are used, namely thymidine and Karl's adenine, which give results varying by orders of magnitude (see, for example, the discussion between Moriarty and Costa-Pierce, 1990). The biomass and production of bacterial populations increase with nutrient concentrations. This makes the relation between primary production and nutrient concentration difficult to grasp, as nutrient supply occurs partly through the decomposition cycle.

Differences in organic material

Organic material in the water column is either dissolved (DOM) or particulate (POM). DOM is usually material freshly released by organisms, of greater nutritional value, consisting, for example, of amino acids which can be directly taken up (Mann, 1988). The organic substances released by algal cells during their active growth are photosynthetic products, made of small organic molecules, easily assimilated by bacteria, such as hexoses. Algal decomposition products are macromolecules, probably polysaccharides. The products exuded during active algae growth inhibit and repress the synthesis of bacterial decomposition enzymes, while the decomposition products induce their synthesis (Chrost, 1989).

Organic particles or aggregates (POM) are also released by organisms, as part of dead organisms, and become substrata for decomposer microorganisms. In the process, nutrients get regenerated and transferred between living communities on and around particles in an almost closed circuit. This is what Mann (1988) called the aggregate spinning wheel concept. Bacteria also aggregate around phytoplankton cells, using their DOM excretions and regenerating nutrients.

The organic material in suspension or solution is regenerated as far as it releases nutrients or energy for the organisms which use it as food, as a source of nutrients, or as substrate for extracellular enzymatic attack. The importance of regeneration is therefore function of the nutritional value of the organic material. The complexity of the organic compounds present in water is great and has been discussed in the section on organic phosphorus compounds (p. 97). Most particulate organic carbon in the detritus of a humic lake, nearby Oslo, Norway, was of low nutritional value for bacteria (Hessen *et al.*, 1990).

In laboratory experiments with humic water, and in a study involving eleven oligotrophic Swedish lakes with differing humic contents, Transvik & Sieburth (1988) showed that bacteria use humic material, when it is partially flocculated or dissolved. As such, humic material could contribute positively to secondary production as the flocs of organic aggregates were colonized by attached or free-swimming bacteria, on which free-swimming or attached flagellates fed.

Bacterial types and decomposition in water

Dissolved organic matter is used as a carbon and energy source by heterotrophic bacteria. The amounts of organic compounds transferred from algae to the bacterial loop are largely undetermined; it has been estimated that 10 to 50 % of the carbon fixed by photosynthesis might pass into the bacterial loop (Berman, 1988; Pomeroy & Wiebe, 1988). Largely undetermined are also the types and populations of bacteria occurring in the water column. In lake Constance, Switzerland, bacteria were separated into small (0.2 to 1 μm) and large sizes (1.0 to 3.0 μm), and into free swimming and attached types. Small free-swimming bacteria were estimated to represent two-thirds of secondary bacterial production. Attached bacteria are usually of less importance. The bacterial biomass and production were predominantly controlled by supplies of organic matter for decomposition (Simon, 1987).

The attraction of bacteria to material to be decomposed is chemosensitive and allows pelagic bacteria to remain in contact with particles (Jackson, 1989). Bacterial activity results in the presence of enzymes in water. The activity of glucosidase is linked to the presence of heterotrophic bacteria, forming part of a coupled system of hydrolysis and assimilation. This enzymatic activity is more important in the 0.2 to 3.0 μm size fraction. Synthesis and exocellular excretion of ß-glucosidase is induced by the presence of macromolecules stemming from the decomposition and autolysis of algal cells. Other exudates, such as hexoses, are directly assimilated by algae (Chrost, 1989).

The microbial food loop : source or sink of energy?

Pomeroy & Wiebe (1988) consider the bacterial food web as an energy sink, because detritivorous metazoans and bacteria compete for the same resource. However, through this loop, dissolved organic carbon, mainly polysaccharides of high molecular weight at low concentrations, oozing out of algae, is recovered by bacteria and is returned to the main food web. These organic compounds are repackaged by bacteria which are themselves consumed by mixotrophic algae and heterotrophic flagellates (Sherr & Sherr, 1988).

Nutrient regeneration by bacteria is in essence a repackaging of nutrients released by organic material into new bacterial tissues. But bacteria do not use carbon, nitrogen and phosphorus in the same ratio as phytoplankton, i.e. the Redfield ratio. The value of organic material, assessed on the C:N:P ratio, therefore depends on the nutritional requirements of the organisms making use of it. Different organisms have different nutrient requirement ratios. Nutrients taken up are used for cell respiration or growth and each process releases or keeps nutrients in specific ratios. Therefore, nutrient composition of organisms does not reflect nutrient intake. From this, it can be concluded that at each trophic step, some nutrients are wasted, being in surplus of the ratio required. For instance, Goldman *et al.* (1987) showed in a batch culture that marine bacteria do not use nitrogen efficiently.

If the efficiency of energy transformation across different trophic levels is considered, losses appear important. In Israeli fish ponds, Schroeder (1987) assessed the loss factor of carbon by microbial transformation to about 3; this, nevertheless, is much less than if the organic material was not recovered.

Trophic links between organisms

Between different groups of organisms, links are established. Bacterial populations appear there where organic detritus and dead cell accumulate. Bacterial production is mainly grazed upon by heterotrophic nanoflagellates (2 to 20 μm in size), which are themselves consumed by microzooplankton, especially ciliates (20 to 200 μm in size). Populations of ciliates, flagellates and bacteria therefore develop accordingly to the trophic links binding them.

In lake Vechten, Netherlands, a high density of blue-green algae induced high densities of bacteria and of heterotrophic nanoflagellates, the main consumers of these bacteria (Bloem & Bär-Gillisen, 1989).

The predators of bacteria are flagellates (almost exclusively at some well-defined depth), rotifers, and ciliates. Heterotrophic flagellates in fresh water are able to feed solely on bacteria. Ciliates are important predators of flagellates, consuming bacteria less readily. In Lake Constance, flagellates and ciliates feed on picoplankton, nanoplankton and detritus (Weisse, 1990).

The impact of feeding of some mixotrophic chrysophytes, such as *Dinobryon*, on bacteria is important and at some depths, even exceeds that of zooplankton. In fact, these algae depend more on bacteria for their metabolism than on photosynthesis (Bird & Kalff, 1987). Phagotrophic protozoans form an interface in the regulation of carbon and energy fluxes of aquatic systems. Their rapid growth, the grazing they are subjected to and which they exert on bacteria, and their nutrient regeneration capabilities are responsible for this. Fast feed-back mechanisms regulate these transfers (Sherr *et al.*, 1988). The mixotrophy of flagellates is an opportunistic response, controlled by environmental conditions. They may use either particulate and dissolved organic material, or the photosynthetic process (Porter, 1988).

In a study involving zooplankton of an eutrophic lake of Georgia, U.S.A., Sanders *et al.* (1989) showed that cladocerans inflicted only little mortality to bacterial populations, but actively consumed heterotrophic flagellates. They ingested particles of 2 to 8 μm of size, but also consumed ciliates of 100 to 200 μm in size.

The trophic pyramid or trophic food web concept

The trophic interrelations between the species in the pelagic and benthic assemblages show that the ideal image of a trophic pyramid is unrealistic. The diversity and flexibility of feeding regimes of the various species making up such communities do not warrant simple pigeonholing of species or groups of species into producers, consumers, and decomposers. In fact, the situation ressembles a web, where the entanglement of production and consumption does not allow a distinction of separate trophic levels. Organisms, such as cladocerans, considered as strict herbivores, appear, at close scrutiny, unclassifiable as to their feeding regime. In laboratory experiments, they feed on planktonic algae, detritus, bacteria, heterotrophic flagellates (Sanders & Porter, 1990), ciliates, and rotifers, thus crisscrossing the traditional trophic levels and short-circuiting them.

In the same context, the transfer of material and energy through a food web should not be viewed too rigidly or too schematically. A detailed knowledge of the feeding regimes of all organisms present in an ecosystem is required to understand the functioning of the food web (Paine, 1988).

The complexity and diversity of the trophic relations in this part of microbial production is so complex that some do not hesitate to call it a bacterial maze rather than a bacterial loop (Berman, 1988). The energy and nutrient fluxes in the microbial loop are still largely unexplored; in the case of intensive culture systems, they are quite important: they constitute a nutrient regeneration system benefitting production.

The bacterial food loop: contribution to production

1. The microbial food loop is the feeding-back of nutrients into the trophic food web via the bacteria-flagellates-ciliates pathway.

2. The organic material in solution and suspension in water, produced by phytoplankton (algal exudations and lysis) is decomposed by bacteria, which support a population of flagellates and ciliates. Even some mixotrophic algae feed directly on bacteria. The energy and nutrient content of these organic compounds are thus recovered for natural productivity.

1.2.8. Influence of fish production

Effects of fish production on primary production and on water quality

Lastly, the role of fish in the process of primary production must be examined. Fish release various waste products, such as carbon dioxide, ammonium, and organic material containing nutrients, while they remove oxygen and particulate organic material (plankton and bacterioplankton) from the system.

According to the effect of fish on the plankton community and on water quality, their overall influence will be positive, i.e. stabilize the food web and maintain a steady production, or negative, leading with increasing fish biomass, to a deterioration of the production factors. Examples are polyculture manure-treated ponds, like the traditional ponds in China, as opposed to high-density artificially fed monoculture crops, which show deteriorated water quality, at the end of the culture period.

Effects of fish on artificial environments

In Israel, Van Rijn *et al.* (1986) compared physical, chemical and biological conditions in cement tanks used for intensive culture of tilapia and eel (Fig. 50). Daily dissolved oxygen was low in tanks with tilapia (Fig. 50a), a consequence of the consumption of phytoplankton by the fish (Fig. 50e). As a result, the water became enriched with ammonia (Fig. 50g). The relative increase in carbon dioxide enhanced the buffer capacity of the water and stabilized pH (Fig. 50c). Bacterial activities became more important (Fig. 50b, d & f) owing to the presence of more detritus and the relative decrease of photosynthesis. The influence of feeding also clearly appears (Fig. 51). The effect of a temporary cessation of feeding was reflected in the physical and chemical condition in the tanks. Feed polluted the water directly and indirectly: left-overs of feed, as well as excrements produced by fish, supplied organic matter to the water, which, on decomposition, consumed oxygen (Fig. 51a) and released ammonia (Fig. 51e). Bacterial activities paralleled the amount of organic material present. Over a period of six weeks, a progressive deterioration of the culture conditions was clear (Fig. 52), with an increase in ammonia content (Fig. 52e) and a progressive alkalinization of the pH, which rendered the proportion of toxic free ammonia threatening. Bacterial activities increased in a similar way.

The effects of two planktivorous fish, the filter-feeding gizzard shad (*Dorosoma cepedianum*) and the visual-feeding blue-gill (*Lepomis macrochira*) were investigated in tanks for one month by Lazzaro *et al.* (1991a & b). Gizzard shad is an omnivorous planktivore, whereas blue-gill is carnivorous planktivorous and benthivorous. An increase of fish biomass lead to increasing primary production: higher chlorophyll a, turbidity, more unicellular green algae and pennate diatoms. The two planktivores were affecting the water independently. Fish biomass reduced particulate phosphorus in the size fractions > 200 μm (macroplankton), but increased it in the size fraction of 20-200 μm (microplankton)

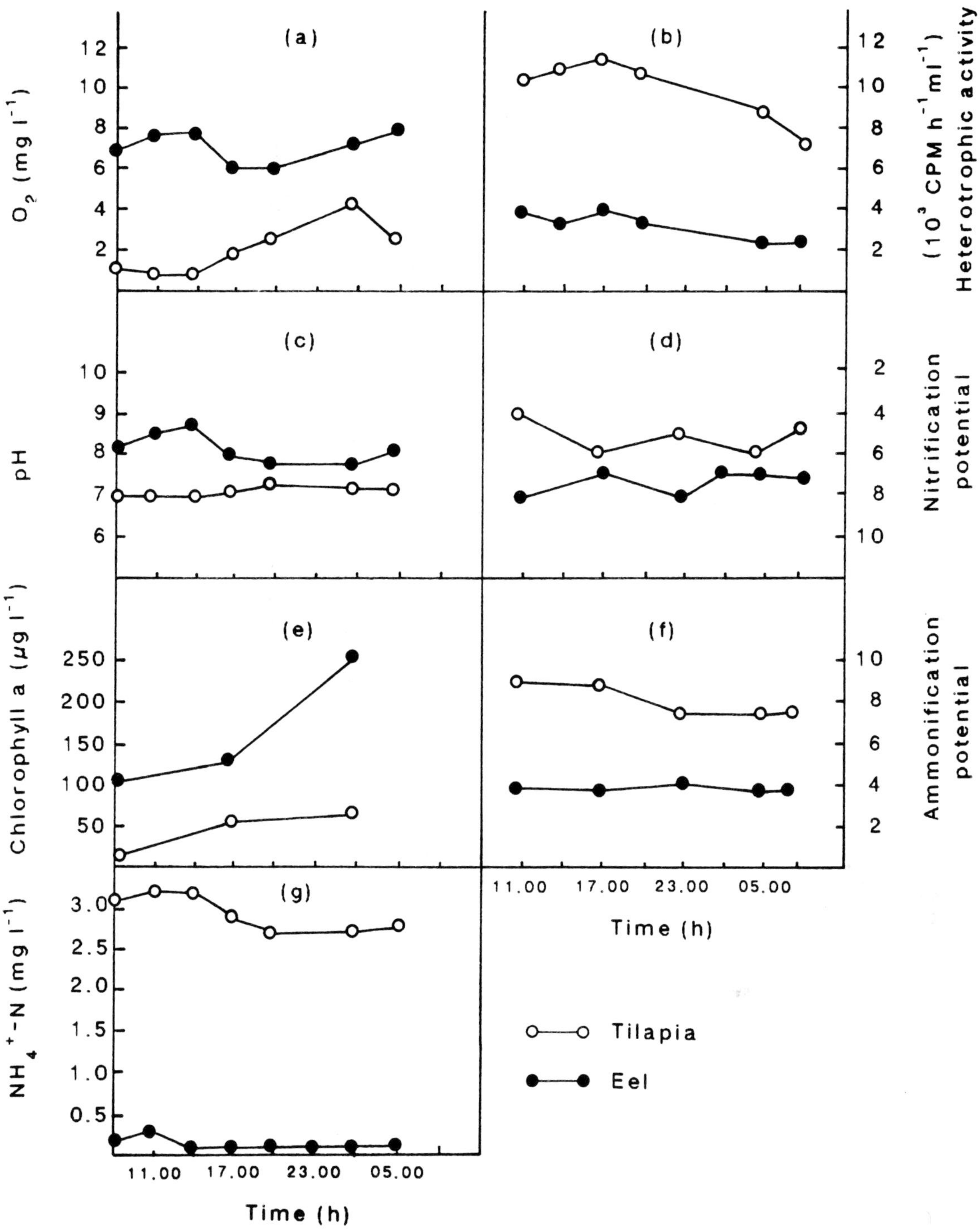

Fig. 50. Comparison of diel changes in physical, chemical and biological production parameters in two concrete ponds (50 m^2 x 0.8 m), stocked with tilapia (625 kg and 1250 kg of total biomass) and eel (less than 200 kg) respectively. Water was supplied at a continuous rate of 1 l s^{-1}. Ammonification potential is expressed as daily ammonium production (mg NH_4^+ d^{-1}). Nitrification potential is expressed as the number of days needed to convert half of the added ammonia to nitrite (d^{-1}). Heterotrophic activity is expressed in counts of radioactivity by a scintillation counter (from Van Rijn *et al.*, 1984) (with kind permission of Prof. Van Rijn on behalf of authors).

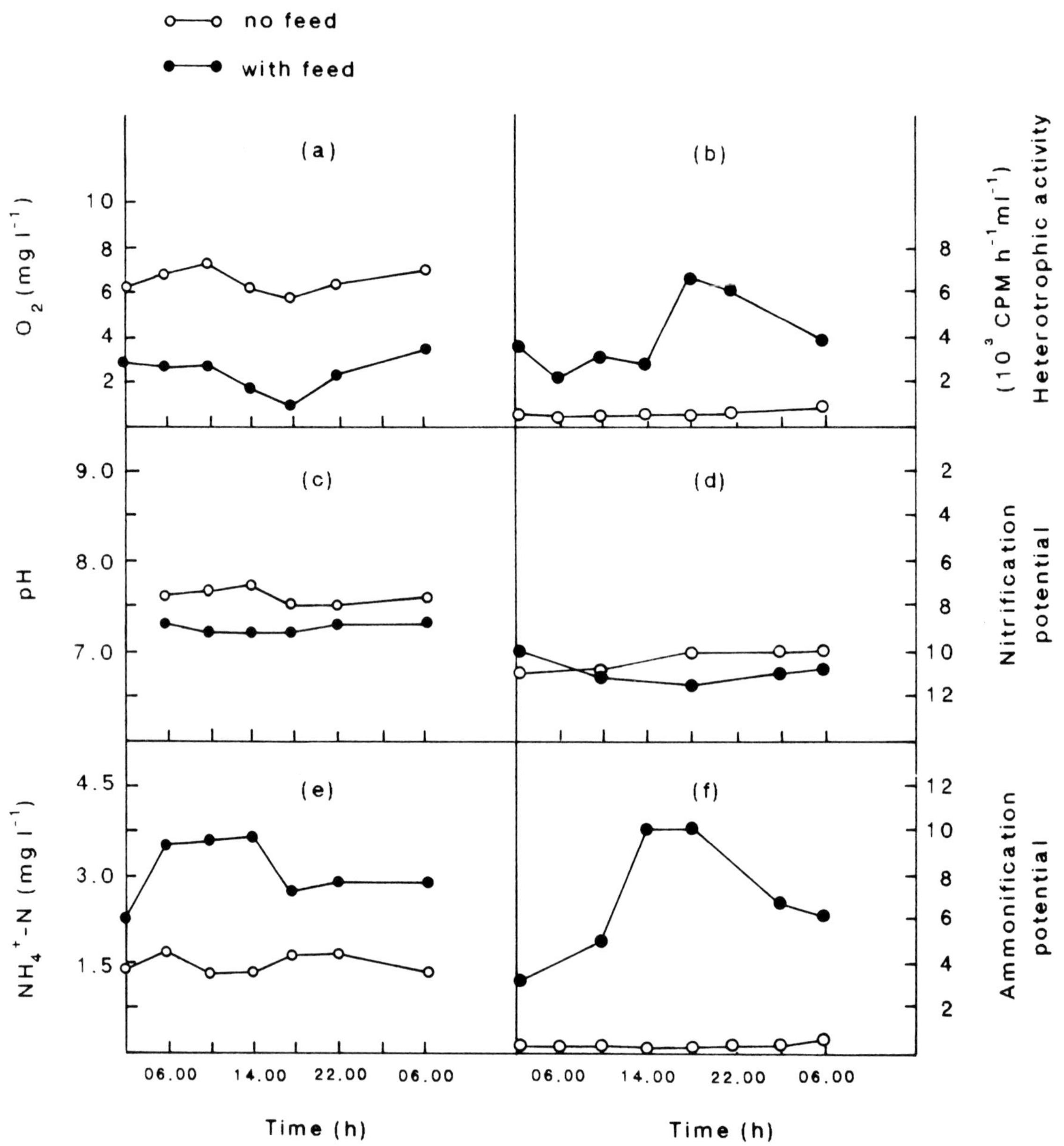

Fig. 51. Comparison of diel changes in physical, chemical and biological production parameters in two concrete ponds, stocked with tilapia, fed (3 % of the biomass, daily) and non-fed (ca. 1200 kg of total fish biomass). The water supply is the same as in Fig. 50, so are as well the units of ammonification, nitrification potentials and heterotrophic activity (from Van Rijn *et al.*, 1984) (with kind permission of Prof. Van Rijn on behalf of authors).

and 12-20 μm (large nanoplankton), which is consistent with the consumption of zooplankton and the resulting increase in phytoplankton. Total phosphorus increased with increasing fish load, while total nitrogen remained unchanged.

Effects of fish on ponds

Tucker *et al.* (1979) monitored changes in physical, chemical and biological factors in ponds used for culture of American catfish. The culture period extended over 10 months and three different stocking densities were used. The ichthyomass was important

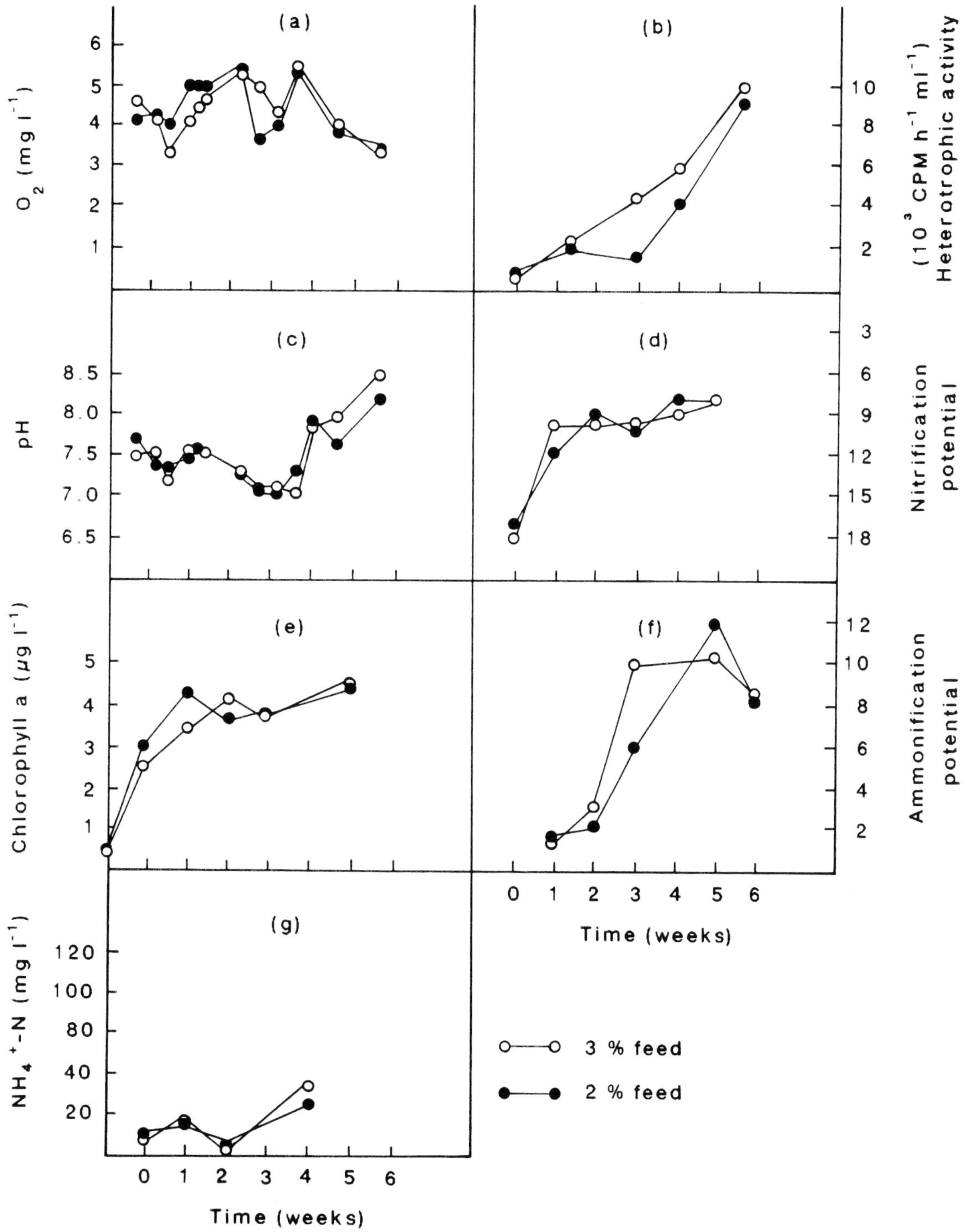

Fig. 52. Effects of two different feeding rates, 2 and 3 %, on the modification of physical, chemical and biological production parameters in two concrete ponds, stocked with tilapia (ca. 1000 to 1400 kg of total fish biomass) during a period of 6 weeks. The water supply is the same as in Fig. 50, and so are as well the units of ammonification, nitrification potentials and heterotrophic activity (from Van Rijn *et al.*, 1984) (with kind permission of Prof. Van Rijn on behalf of authors).

and increased in the course of time, stimulating decomposition of organic material, as well as photosynthesis. This is illustrated by increases in ammonium concentration (Fig. 53e), in COD (Fig. 53a), in chlorophyll a (Fig. 53c), and in turbidity (Fig. 53a). The increase of photosynthesis was a consequence of the decomposition and recycling of nutrients, boosted by fish excrements and feed left-overs.

Cole & Boyd (1986) confirmed these results, by establishing correlations between physical and chemical features of water and amounts of feed added to ponds, which varied between 0 to 200 kg ha^{-1} d^{-1}. The ponds were used for the culture of American catfish. Dissolved oxygen was negatively correlated, while chlorophyll a, chemical oxygen demand, carbon dioxide and ammonium concentrations were all positively correlated with the amount of food added. The use of feed at a rate above 112 kg ha^{-1} d^{-1} resulted in a decline in fish

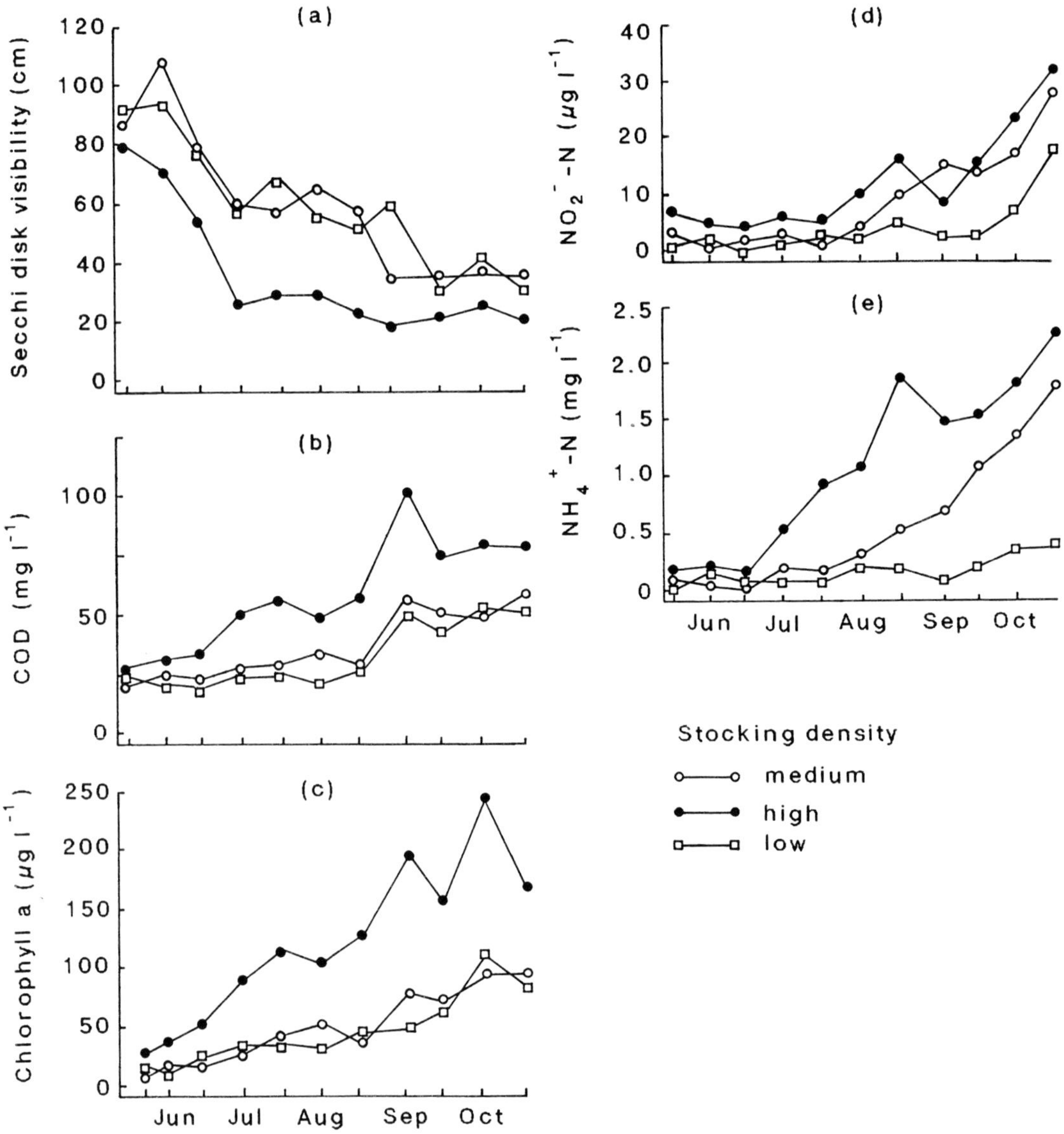

Fig. 53. Effects of three different stocking rates on the physical, chemical and biological production parameters of channel catfish ponds. The three different stocking rates are: 4942 fish ha^{-1} (low), 10007 fish ha^{-1} (medium), and 20385 fish ha^{-1} (high). (COD = chemical oxygen demand) (from Tucker *et al.*, 1979) (with kind permission of Prof. Boyd on behalf of the authors, & Am. Fish. Soc.).

production, which could not be alleviated by the use of an aerator. The authors did not find an explanation for this decline, but it was probably associated with accumulation of metabolites.

Specific effects on ponds of some groups of fish

Planktivorous fish have a major influence on the structure of the whole plankton: they modify the density and size structure of communities (Hrbáček, 1962; Brooks & Dodson, 1965), and the nature of detrital seston, as the sizes of settling particles are determined by the zooplankton (Søndergaard *et al.*, 1990).

The most obvious effects of fish on primary production are direct consequences of their feeding, i.e. the removal of organisms from the pond. Besides these, other effects appear in cascade. They diffuse through the production system and reveal the intricacy of the interrelations. But do links between organisms exist other than trophic ones?

Main effects of fish on plankton communities result from direct and indirect impacts of predation, depending on feeding mode. Fish show three different modes of feeding: (1) particulate or visual feeders visually select their prey; (2) pump filter feeders use rhythmic suction to engulf water straining off the plankton, while swimming slowly or remaining stationary; (3) tow-net feeders swimm rapidly with their mouth fully agape, engulfing water and capturing prey. In each case, capture depends on retention ability of prey by the predator. In the case of filter-feeding fish, the sieving structures are the gill rakers, the microbranchiospines, the pharyngeal teeth, assisted by the secretion of mucus. The mode of feeding of a species can change within life-history or be seasonal, due to changes in prey abundance. With increasing abundance, fish shift from visual feeding to filter feeding. Visual feeding is also modified by learning processes (Lazzaro, 1987).

Main direct and indirect effects on the environment are listed hereunder.

Effects of benthophagous fish, especially common carp

The impact of benthophagous fish on the trophic structure of temperate ecosystems is quite important (Lammens *et al.*, 1990) (Fig. 54). Their effects on the sediments are direct and indirect. They establish a contact between the benthic and pelagic compartments, which are otherwise fully separated (Richardson *et al.*, 1990):

a) predation modifies the structure of the benthic community:

The common carp feeds on benthic macroinvertebrates, such as Chironomidae, Oligochaeta, and Mollusca. Its mode of feeding is mainly tactile and olfactory; it ploughs through sediments in search of food, ingesting sizeable amounts of detritus. Its degree of selection is low and practically all organisms are suppressed. Other benthophagous species, which feed relying on visual stimuli, select prey based either on size or on apparent visibility. They modify the structure of the benthic community towards organisms of smaller size and cryptic behaviour.

b) predation on benthic organisms reduces bioturbation:

By reducing the number of benthic organisms, fish cause a reduction in bioturbation, which is of particular importance for aeration of the sediment.

c) transfer of benthic resources to water is promoted:

Fish transfer particles of sediment to the water column. Some become incorporated to fish biomass or are released as excrement and made available to pelagic bacteria.

d) by resuspension of sediments, fish modify the cycling of nutrients, especially of phosphorus:

The resuspension of sediments makes it possible for nutrients, accumulated in pond mud, to become dissolved in the water (see p. 50). This mainly concerns soluble substances from interstitial water. Sedimentary phosphorus, soluble or labile, is thus transferred to water. The mechanical mixing of sediments with water does not dissolve phosphorus (Qin & Threlkeld, 1990). The importance of phosphorus transfer during sediment disturbance is dependent on sorption properties of soil, amounts of labile phosphorus, and concentrations of interstitial dissolved phosphorus. The influence of benthophagous fish, especially common carp, on planktonic production has been shown by Lamarra (1975).

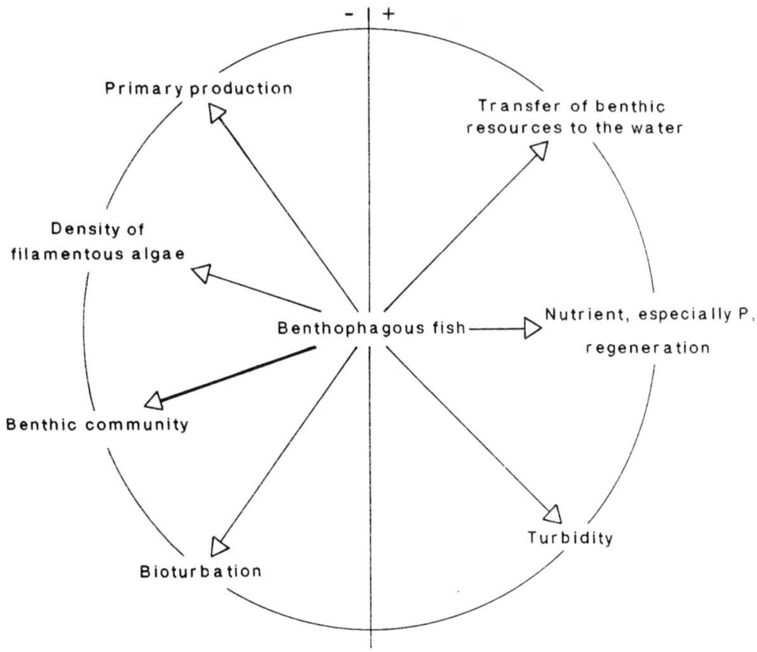

Fig. 54. Schematic diagram of the effects of benthophagous fish on a pond. The positive or enhancing effects are on the left-hand side, while the negative or decreasing effects are on the right-hand side. The bold line indicates the main effect.

Similarly, herbivorous fish (*Scardinius erythrophthalmus*), feeding on aquatic plants removed and transferred phosphorus from sediments to the pelagic compartment, as shown in aquaria (Hansson *et al.*, 1987).

e) the resuspension of sediments affects turbidity:
The common carp, feeding on the bottom of a pond, brings mud into suspension and makes the water turbid. Carp ponds are notorious for their turbidity, especially if pond soil contains clay. A reduction in density of benthophagous fish has resulted in an increase in water transparency in Dutch fish ponds (Meijer *et al.*, 1990b).

f) benthophagous fish cause a decrease in primary production by modifying the light regime in the water:
The increase in mineral turbidity in the water results in a decrease of primary production, entailing a decrease in dissolved oxygen, sometimes leading to oxygen depletion (Szumiec, 1989).

g) the density of filamentous algae is reduced:
Benthophagous fish reduce the density of filamentous algae, directly by feeding on them, and indirectly by changing turbidity and light penetration.

Effects of zooplanktivorous fish on ponds

There is an abundant literature on effects of planktivorous fish on plankton communities, initiated in the thirties by research in fish ponds (Lazzaro, 1987). The main effects are here summarized using a selection of recent papers. Richardson *et al.* (1990) conducted experiments in 18 large outdoor tanks at Kingston, Oklahoma, U.S.A.; Mazumder *et al.* (1990) performed *in situ* experiments, in enclosures (8 m in diameter and 15 m deep), in Lake St. George, Toronto, Canada. The presence of zooplanktivorous fish changed the plankton community as follows (Fig. 55):

a) a decrease in density and biomass of large filtering herbivores (Brooks & Dodson, 1965):
In experiments using a combination of planktivorous and benthophagous species (Richardson *et al.*, 1990), or only planktivores (Mazumder *et al.*, 1990), a decrease in density of large and medium-

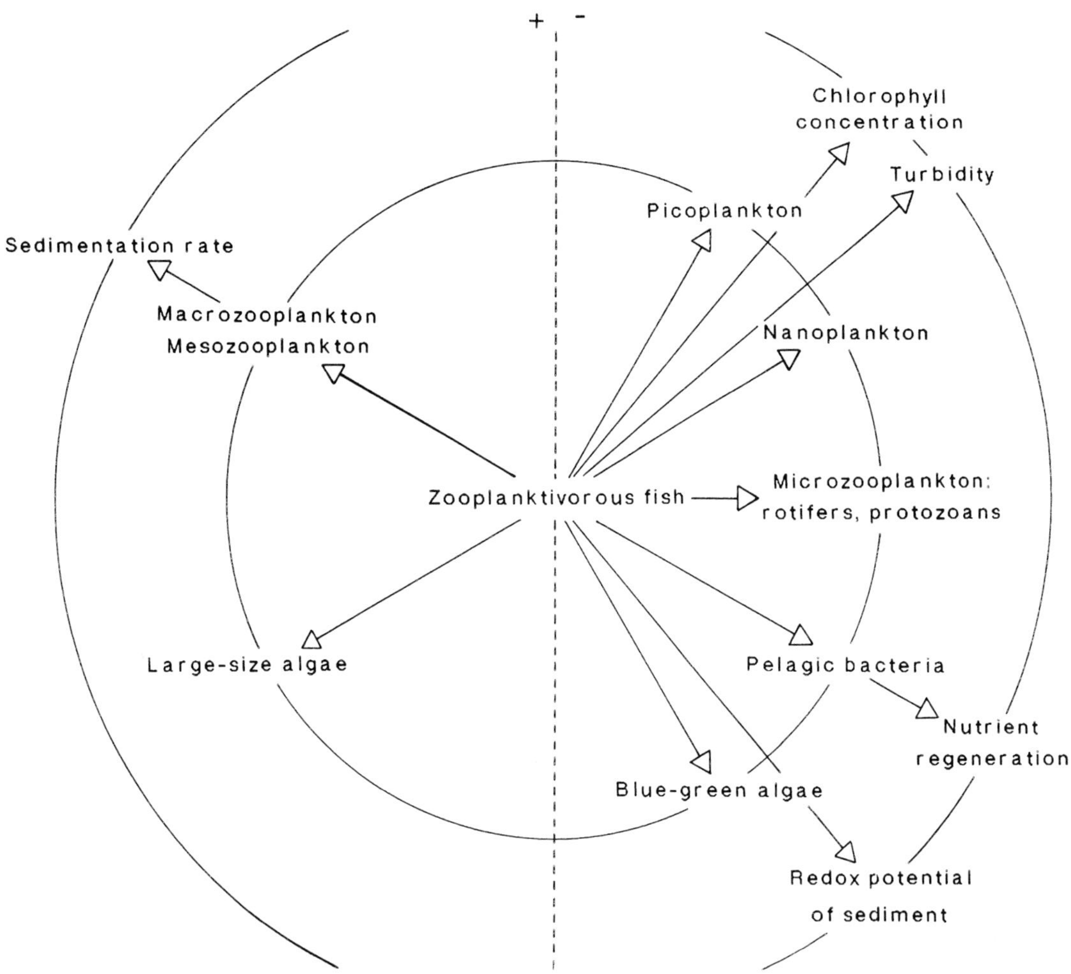

Fig. 55. Schematic diagram of the effects of zooplanktivorous fish on a pond. The positive or enhancing effects are on the left-hand side, while the negative or decreasing effects are on the right-hand side. The bold line indicates the main effect. The inner circle groups the direct effects on the living organisms, the outer circle the indirect effects.

sized zooplankton (copepods and Cladocera) was observed, which themselves prey on microzooplankton and on nanoplankton. In lakes in southern Norway, removal of larger zooplankton resulted in an increase of microzooplankton, in particular of rotifers and protozoans. An increase in density of zooplankton, especially herbivores, and of grazing pressure on algae, following the removal of planktivorous fish, was found by Lyche *et al.* (1990). A decrease in average size and size at maturity of Cladocera has also been observed (Gliwicz *et al.*, 1981). In some cases, predatory cyclopoids increase in density, fecundity, biomass and production as a result of the increase of microzooplankton (Hillbricht-Ilkowska & Weglenska, 1973; Kerfoot, 1987).

b) increase of the picoplankton and nanoplankton: The presence of planktivorous fish alters the structure of the planktonic assemblage, increasing the relative importance of picoplankton and nanoplankton. Their losses, by grazing of zooplankton, are reduced and their biomass increases (Richardson *et al.*, 1990; Mazumder *et al.*, 1990). This results in an increase of chlorophyll and of turbidity (biotic turbidity) (Threlkeld, 1988; experiment in 5 outdoor tanks in Kingston, Oklahoma). This increase of nanoplankton, composed of green algae of less than 10 μm of size, is also observed in the presence of silver carp (phytoplanktivorous) and bighead carp (omnivorous).

c) this increase in algal biomass is observed in the presence or absence of sediment:

The increased algal biomass has been considered to result from nutrient enrichment, following resuspension of sediment by benthophagous fish. The impact of a possible transfer of nutrients from sediment to water has been tested by comparing the increase of algal biomass in the presence and absence of sediment. In a study involving indoor and outdoor tanks of different capacities (900 l and 7000 l) in Kingston, Oklahoma, Qin & Threlkeld (1990) observed that, in mixed cultures of common carp and planktivorous fish, an increase of phytoplankton occurred in the presence and in the absence of sediment, regardless of whether fish had access to the sediment or not. Their interpretation is that the increase of phytoplankton resulted from the presence of the planktivorous fish, which fed on large-sized herbivorous zooplankton, thus causing a reduction in the grazing pressure of the zooplankton on the phytoplankton. In eight enclosures in Lake St. George, Toronto, Canada, the grazing of zooplankton and predation of fish on zooplankton promoted growth of small algae, which would otherwise be subjected to competition with large-sized algae and to nutrient limitation (Mazumder *et al.*, 1986). Increased nutrient regeneration may also be caused by smaller zooplankton, which excrete nutrients at a greater rate (per unit biomass) than large zooplankton (Bartell, 1981; Vanni, 1987b).

d) planktivorous fish reduce the density of large-sized algae:

The removal of about 50 % of planktivorous fish was found to alter the planktonic community towards an increase in large-sized diatoms, green algae and cryptophytes in Lake Søbygård, Denmark (Søndergaard *et al.*, 1990).

e) increase in microzooplankton:

Microzooplankton (rotifers, nauplius larvae, ciliates, and heterotrophic flagellates) develops better when predatory pressure by zooplankton crustaceans is reduced (Richardson *et al.*, 1990; Mazumder *et al.*, 1990). Such an increase in zooplankton has been observed in a eutrophic lake near Wuhan city, China, in the presence of silver carp and bighead carp (Miura, 1990). In enclosures in a small humic lake, nearby Oslo, south-eastern Norway, Hessen *et al.* (1989) noticed an inverse relation between heterotrophic ciliates and crustacean zooplankton, the abundance of the ciliates increasing as that of crustaceans declined. Zooplankton filter-feeding on small suspended particles and bacteria also benefits from the increasing density of detritus resulting from fish feeding. In such conditions, fish may thus actually cause an increased zooplankton production (Hillbricht-Ilkowska & Weglenska, 1973).

f) specific increase in rotifers:

Rotifers increase in the presence of fish, if fish increase rotifer resources; the particles filtered by herbivorous rotifers are partly similar to those consumed by cladocerans. This leads to exploitative competition between both groups, in which cladocerans have the advantage. Therefore, suppression of cladocerans and an increase of algae results in an increase of rotifers (Richardson *et al.*, 1990; Mazumder *et al.*, 1990). The reverse is also true. In Lake Søbygård, Denmark, the removal of about 50 % of the planktivorous fish shifted the plankton community from a dominance of rotifers to large-sized cladocerans (Søndergaard *et al.*, 1990).

g) increased release of organic waste products in the presence of fish:

Organic matter released to the water from excrements, residual feed and sloppy feeding of fish, act as substrata for pelagic bacteria (Richardson *et al.*, 1990; Mazumder *et al.*, 1990). This enrichment of the environment by fish has been called ichthyeutrophication.

h) increase in pelagic bacteria:

As a consequence of increased dissolved organic load, pelagic bacteria increase in biomass; this boosts the microzooplankton, which is mainly bacterivorous. Bacteria are important in the bacterial food loop owing to their role in organic matter decomposition and regeneration of nutrients. Their activity controls the quality of water (Zweig, 1989). The predation pressure on bacteria is not reflected in their population size, because of their rapid growth, especially if availability of substrata is non-limiting (Hessen *et al.*, 1989).

i) increased nutrient regeneration:

The increase of suspended organic material entails a proliferation of decomposer bacteria, hence an increase of nutrient recycling rates and nutrient availability. This increase of nutrient availability favours efficient nutrient absorbers such as the picoplankton and nanoplankton. At the same time, bacterial regeneration of nutrients supplied by the feed and excretion leads to an increase in productivity (Richardson *et al.*, 1990; Mazumder *et al.*, 1990). Qualitative modifications of nutrient concentrations in the water, resulting from changes in grazing pressure in Norwegian lakes, were as follows: the selective retention of nutrients by grazing organisms changed the nutrient proportions in the water; the supply of nutrients was modified from a pulsed mode, when fish were dominant, to a steady mode, where zooplankton was dominant. The phytoplankton assemblage was therewith modified (Lyche *et al.*, 1990). In conditions of phosphorus limitation, increased grazing pressure results in a decreased algal biomass, which is accompanied either by a reduction of the phosphorus pool within the phytoplankton, or by a reduction of the intracellular C:P ratio of the phytoplankton. The removal of fish from the small eutrophic lake Mosvatn, Norway, reduced its internal phosphorus loading (Sanni & Waervagen, 1990), while the presence, for example, of yellow perch reduced phosphorus turn-over time by a factor 3 and increased phytoplankton biomass by a factor 1.5.

j) decreased sedimentation:
Small-sized algal cells settle less quickly than large-sized algae. They decompose in the water column and the nutrients released remain in the pelagic compartment (Richardson *et al.*, 1990; Mazumder *et al.*, 1990).

k) effect on the seston:
The size structure of the seston depends on its mode of formation. Part of the particles are zooplankton excrements. Modifications in species assemblage lead to changes in the types of excrements released and entail modifications in the cycling of nutrients within the water column. The dynamics of particulate and dissolved carbon and phosphorus in the water column and the dynamics of nutrients from the sediments are thereby also affected. In Lake Søbygård, Denmark, the removal of about 50 % of the planktivorous fish was found to reduce the internal loading of phosphorus, enhance the redox conditions of the sediment, and decrease sedimentation rate (Søndergaard *et al.*, 1990).

l) modifications at the sediment level
The decrease of the sedimentation rate enhances redox conditions and decreases the oxygen demand of the sediment, with a favourable effect on the benthic fauna. In Lake Søbygård, Denmark, the removal of 50 % of the planktivorous fish was found to stimulate the production of microbenthos and macrophytes (Søndergaard *et al.*, 1990).

m) long-term dominance of blue-green algae
Uncontrolled development of nanoplankton is considered to be responsible for long-term dominance of blue-green algae in lake Ringsjön, Sweden (Bergstrand, 1990). However, it seems that generalized conclusions cannot yet be drawn. Experiments involving a reduction of planktivorous fish populations in eutrophic lakes have resulted in effects differing between lakes: blue-green algae developed at different densities and showed various tendencies to form blooms. This disparity stems from initial differences in nutrient concentrations, mixing depths and water renewal rates (residence time) between lakes. Such parameters induce differences in the plankton community prevailing after fish reduction, e.g. through discrepancies in specific loss rates of algae.

The impact of planktivorous fish also decreases with increasing mean depth. Lastly, lakes show differences in their ability to create a reduction in dissolved phosphorus, relative to their phosphorus loading (Jeppesen *et al.*, 1990b), e.g. by the establishment of a macrophyte belt (Jeppesen *et al.*, 1990a).

Impact of phytoplanktivorous fish

The effect of phytoplanktivores on plankton communities is less well documented than that of zooplanktivores. Phytoplanktivorous fish are filter feeding, and tilapia are essentially pump filter feeders. Their food selection is passive, size-selective, retaining all material, larger than the interraker spacing, contained in the engulfed water. (A discussion on retention capacities of the filtering apparatus is found in Lazarro, 1987:116). Their effect on plankton is therefore different of that of visual feeders. Filter-feeding fish consume net phytoplankton, enhance nanoplankton and thus induce better water quality (Lazarro, 1987). Their impact on phytoplankton depends on fish feeding rate and algal growth rate. By removing large-sized algae, with slower growth rates, they favour small algae. However, environmental conditions affecting decomposition rates and nutrient regeneration, such as water temperature, mixing pattern, residence time, surface level fluctuation, and initial trophic state, tend to obscure the impact of filter-feeding planktivores.

As mentioned earlier (p. 165), the role of herbivorous invertebrates is less in tropical waters, and the importance of phytoplanktivorous fish on plankton communities (and on trophic state of the water body) is greater here.

a) on zooplankton

In pond experiments conducted in U.S.A., *Oreochromis aureus* reduced rotifers and increased copepodids and adult stages of a large calanoid copepod. The effect on plankton communities was to enhance large evasive zooplankters (Drenner *et al.*, 1984b). Similar effects were obtained with gizzard shad, which is a pump filter-feeding omnivore (Drenner *et al.*,1982 & 1984a)

b) on phytoplankton

The omnivorous silver carp was used in Poland to improve water quality and overcome algal blooms in Lake Warniak (Mazurian lakeland, Northern Poland) (Kajak *et al.*, 1975). Silver carp kept in pens within the lake reduced by 4.5 times the average biomass of phytoplankton regardless of density (450 or 1350 kg ha^{-1}). At the highest densities (1350 kg ha^{-1}), it reduced blue-green algae, but enhanced dinoflagellates and nanophytoplankton. The grazing of plankton cleared the water and the resulting fish faeces enriched the sediment benefitting the benthos. This effect was stronger in stratified lakes than in well-mixed or shallow waters. Silver carp resorts to feeding on bottom sediment during periods of low plankton or at high abundance of blue-greens. Filamentous blue-green algae, of the species *Oscillatoria agardhii*, or *Aphanizomenon flos-aquae* are not digested and if these species dominate, silver carp suffer high mortalities, if access to the lake bottom is denied. The increase in nanoplankton results from the silver carp's consumption of herbivorous zooplankton; the fish do not feed on nanoplankton, having access to particles larger than 20 μm only (Lazarro, 1987).

O. aureus (in the size range of 12.5 to 19.0 cm SL) reduced standing crops of the largest algae (> 150 μm), but did not affect smaller algae (> 15 μm), probably because of their higher growth rate which masked the impact of grazing. Smallest algae (< 8 μm) were enhanced, owing to inefficient filtration, nutrient regeneration by fish, higher growth rates, algal enhancement during gut passage, and modification of the herbivorous zooplankton. Tilapia in the size range of the experiment were feeding on particles larger than 25 μm (Drenner *et al.*, 1984b).

Plankton consumption by fish

Our knowledge of the consumption of plankton by fish is still fragmentary.

Trophic links between fish and plankton are often loose, (1) because plankton is a changing assemblage of pelagic organisms of variable nutritional value. Some of the changes occur as sequels to the grazing or the predation, while other are linked to environmental abiotic changes (seasonality and nutrient availability). (2) The predatory pressure of fish on plankton changes (for a species) with age, with density, and with the presence of other species with similar feeding regimes.

As mentioned earlier (p. 145 & 168), some plankton species are inedible or of poor nutritional value, while some other species are readily consumed.

Blue-green algae are often regarded as a nuisance in ponds, because indigestible to fish and reference was made to observations of Fish (1955) that *Microcystis* is not digested. It has been shown that indigestibility of the alga by fish resulted from the fact that diel rhythms of digestion were not appreciated (Trewawas, 1983). The non-consumption of blue-green algae has also been contradicted with regard to *O. niloticus* by observations of Moriarty (1973), Moriarty & Moriarty (1973), Moriarty *et al.* (1973), Bowen (1982). Tilapia and silver carp are indeed capable of digesting and assimilating all but some toxic blue-green algae (Gophen, 1990, in Lake Kinneret, Israel; Leventer & Teltsch, 1990; Miura, 1990).

Moreover, how selective is fish feeding on plankton? The impact of fish on plankton depends on the selectivity of its feeding, which in turn depends

the left-overs of feed, is limited. This process of nutrient supply to ponds and natural environments, with a certain autotrophic capacity, leads to ichthyoeutrophication.

2. The impact of fish on the environment depends on its mode of feeding: benthophagous, phytoplanktivorous or zooplanktivorous. The effect on a group of prey organisms diffuses and becomes dissipated in a cascade of effects through the network of interrelations within the foodweb.

3. The effects of fish on the environment are, in first instance, the result of feeding activities. These effects are difficult to investigate because the diets of fish change over time, are often plastic, a great variety of food being consumed, and because predation or grazing induces changes in the dynamics of the prey populations.

4. The interactions of communities of organisms within the environment reveal the existence of positive synergisms between groups of organisms, such as the grazing on blue-green algae by fish, benefitting overall water quality. Fish species with similar feeding regimes have to share the available food resources and their feeding regimes will adapt to reduce direct competition.

2. Fish production in a pond

2.1. Stocking, management practices and production

Types of management operations in a fish pond

To conduct fish production in a pond, several operations must be performed: stocking with fingerlings, feeding the fish, and fertilizing the pond, monitoring fish growth, and harvesting the fish. The farmer must decide on the choice of the species of fish to culture, the stocking density, the type of feed and feeding rates, the type of fertilizers and fertilization rates, and the length of the culture period.

Fertilization has already been discussed; the choice of artificial feed and the feeding rates will not be discussed here. Selection of fish for fish farming includes scrutinizing available species for an array of desirable characteristics, including fast growth rates at high stocking densities leading to high fish yields, resistance to manipulation and to diseases, ease of spawning in captivity, high fecundity, wide feeding regime and ready acceptance of artificial feed, good food conversion rates, good market value, etc. The species currently most cultured in Africa are tilapia and catfish (Table 22).

The best tilapia species in terms of growth rate is *Oreochromis niloticus*. It is also the most commonly cultured species, though in some countries it is not available or some other species are cultured, because they are better adapted to local farming conditions, such as brackish water (*Tilapia guineensis*, *Oreochromis mossambicus*). Information on distribution of tilapia species in Africa is found in Philipppart & Ruwet (1982) and in Trewawas (1983).

Tilapia hybrids are used to avoid some inherent problems of tilapia culture. Unselected strains of tilapia in high density culture conditions show precocious maturation and prolific breeding. This results in reduced individual growth rate and a production

Table 22. Species names of most common farmed fish in Africa.

Tilapia	Catfish	Other potential species	Introduced species
Oreochromis niloticus	*Clarias gariepinus*	*Heterotis niloticus*	*Cyprinus carpio* (common carp)
O. aureus	*Heterobranchus longifilis*	some species of *Labeo*,	*Ctenopharyngodon idella* (grass carp)
O. mossambicus	*Chrysichthys nigrodigitatus*	*Citharinus*, and others.	*Micropterus salmoides* (black bass)
O. hornorum			*Salmo trutta* (trout)
O. macrochir			
Sarotherodon galilaeus			
Tilapia guineensis			

of fish of low average weight. Food ingested by female fish is diverted to production of gametes and this energy is lost to weight increase (resource allocation principle). To curb the breeding activity of the fish, different techniques exist: (1) fish are are selected on the basis of fast growth rates and late maturity. It has been shown that all the traits of gonadal development and body development of *O. niloticus* are genetically controlled and mutually independent. Selection for late maturity can therefore be efficiently combined with selection for fast growth (Kronert *et al.*, 1989; Oldorf *et al.*, 1989); (2) All-male tilapia are produced essentially by three techniques: hand-sexing of adult fish (> 30 g of body weight), hormonal treatment, and hybridisation. (Castration or neutering of fry by chemicals or radiation has also been attempted). The hormonal treatment is a sex-inversion, successfully achieved with various androgens (methyltestosterone) and estrogens and to which fry are submitted after yolk-sack resorption. Hybridisation is achieved by crossing two pure species which produce 90-100 % male hybrids. *O. niloticus* x *O. aureus* combines fast growth and hardiness, but only 80 % are males; *O. niloticus* x *O. hornorum* and *O.mossambicus* x *O. hornorum* gives 100 % males, of which the *hornorum-niloticus* cross gives better growth performances. The number of hybrid crosses available for all-male hybrid production is limited. It depends on obtaining pure species and on an ability to maintain this purity. The sophistication of this technique is adding costs to the culture as well (McAndrew, 1988).

Tilapia hybrids are currently produced in some private farms in Africa, e.g. in Ibadan, Nigeria, in Mombassa, Kenya, but to a much larger extent in farms in Israel, in the Philippines, in Taiwan, in Hong Kong, and in U.S.A. To start with this technique, a batch of pure strain tilapia of the species mentioned as well as holding facilities which will ensure isolation of the fish are required. A live collection of pure strains of tilapia are held at the Aquaculture Institute of the University of Sterling, in Scotland. Farming hybrid tilapia in Africa considering local facilities, will need the production, within the farm itself, of the hybrid fish, as purchase of the fish from the outside will imply high costs.

Other species with high production potential are the catfish species, *Clarias gariepinus* (formerly known as *Clarias lazera*) *Heterobranchus longifilis* and *Chrysichthys nigridigitatus*, more adapted to brackish water farming. All these species were difficult to breed in captivity, but breakthroughs for mass breeding these fish have been achieved in Bangui, C.A.R, for *C. gariepinus*, and in Layo, near Abidjan, Côte d'Ivoire for the two other species.

Species with high potential include some *Labeo*, *Citharinus*, *Bagrus*, and others and more research on their real growth rates is needed. Among the introduced species, the common carp is the most commonly cultured. In some countries, Chinese carp have been also introduced.

The monitoring operations of the farmer must be directed at maintaining a good water quality, while maximizing fish biomass in the pond. Control of water quality aims at keeping the fish in optimum ecological conditions, i.e. meeting its oxygen requirements, avoiding exposure to harmful ammonia, and ensuring optimum limnological conditions for maximum primary productivity. Control of fish growth involves the adjustment of feeding rates and the length of the culture period. To be monitored are the increase of individual weight of fish (growth rate), its increase in biomass, and the utilization of feed by the fish.

The role of the fish farmer is primarily to adjust the conditions to his aim of obtaining a high production in the most profitable way. Each intervention of the fish farmer must carefully weigh the supplementary cost involved against the extra yield or benefit gained.

Importance of growth rate and production rate

The absorption of food leads to an increase in weight of individual fish, measured by the growth rate, the increase in weight per unit time (usually per day). The specific growth rate is the daily increase divided by the weight of the fish. Growth is a specific characteristic of fish and varies within a given range. Growth rate is usually greater in young animals and decreases with age. Fish of low initial average weight grow proportionally faster than fish of larger size, but, in absolute terms, their increase in weight is lower. Differences in growth also exist between strains of the same species. Growth can thus be improved by genetic selection. Graphically, growth is represented by an S-shaped curve: at the beginning, the rate is low, then steeply increases until it reaches an inflexion point, and levels off afterwards.

At the population level, the biomass stocked represents a given weight per unit surface area, and the increase of cumulative individual weight represents fish production. It appears that, at low loading, the increase is slow, increases rapidly for a given biomass/volume ratio and levels off when fish biomass gets too large for the available space. The maximum fish biomass for a given space is called the carrying capacity (Fig. 56). The graphical representation of the increase of biomass is also an S-shaped curve. The carrying capacity concept was brought forward by Chen & Prowse (1964) and appears to be the limiting factor at high fish density (Edwards *et al.*, 1981). This phenomenon is controlled by a so-called crowding factor, which combines different physiological and behavioural factors affecting growth, defying strict physiological interpretation (Brett, 1979). Growth rate is affected by the rate of exchange of oxygen and carbon dioxide and by feeding rates, but also by behavioural factors. In territorial species, they include social interaction involving dominance, restricting free movement and access to food. In schooling species, fish feeding is promoted by social stimulation, which is lacking at low densities, resulting in poor growth.

2.1.1. Feeding operations

Importance of feeding

Fish production at or below carrying capacity of the pond can be considered a function of the food available to the fish. This food can be subdivided into three categories: the natural food, produced by the natural productivity of the pond, the extra food produced by fertilization of the pond, and the artificial feed cast into the pond.

The fish farmer has a variety of artificial feeds at his disposal. The efficiency of its transformation into fish depends on the qualitative value of the food, i.e. the quantity of food consumed per unit of increase expressed by a feed conversion ratio (FCR), i.e. the

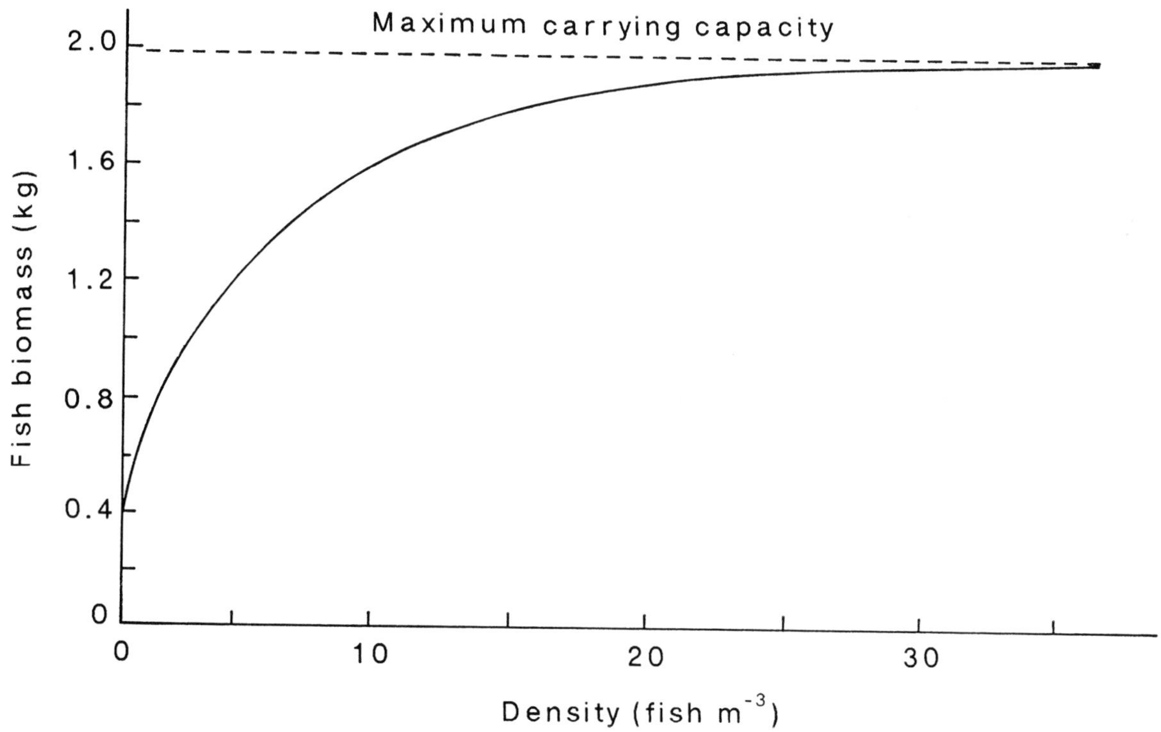

Fig. 56. Tilapia biomass at harvest as a function of fish density in concrete ponds. The fish biomass reached at the highest densities is close to the carrying capacity of the culture system (modified from Edwards *et al.*, 1981).

ratio of the amount of feed offered over the increase of biomass. The food should not only be available in sufficient amounts, but should also suit the nutritional requirements of the species cultured. Usually some amount of natural food is available and artificial feeding is thus supplementary. The artificial feed is never completely consumed and the left-overs fertilize or even pollute the pond. Abundant consumption of feed by fish produces an important volume of excreta, which also fertilize the pond.

Influence of feed on growth rate

The influence of feed on the growth of different fish species is illustrated by experiments of Moav *et al.* (1977) in Israel. In polyculture experiments, they fed silver carp, tilapia (*O. aureus* or hybrids) and common carp with high- and low-level protein feed, at varying stocking densities. The results (Fig. 57) are measures of the growth potential of the three species in the different culture conditions. The growth of common carp was significantly increased by high-protein feed, while tilapia and silver carp met their protein requirements using the natural food available in the pond. A negative effect of density (crowding) was more evident in common carp than in silver carp or tilapia.

The fish communities increased in weight in the course of the culture period and the daily weight increases were regular (Fig. 58). The end of the growing period was reached when a decrease in the average daily increase occurred. The study of Moav *et al.* (1977) showed a sharp increase in FCR at the end of the culture period (Fig. 59). This was explained by a deterioration of culture conditions and an arrestation of growth. In the context of intensive feeding, other aspects, such as frequency, nature and form of feed, i.e. pellets, dough, meal, must also be taken into account.

In experiments conducted in small containers, Opuszyński *et al.* (1989) noted that fingerlings of various species reacted differently to food composed mainly of zooplankton: the common carp and the bighead carp grew quickly when fed zooplankton *ad libitum*, while in the same conditions silver carp and grass carp showed low growth rates.

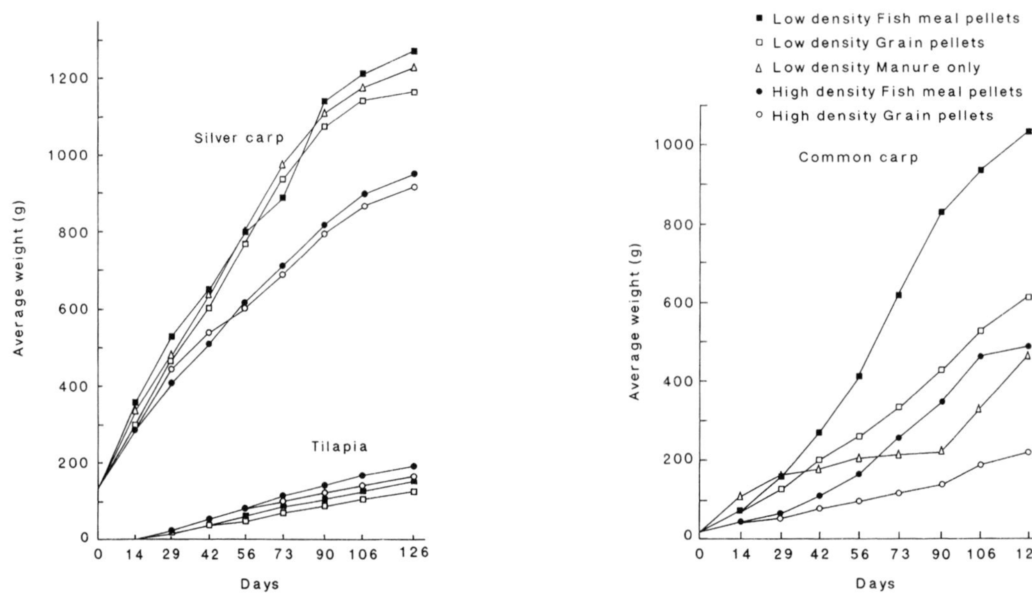

Fig. 57. Comparative growth curves of common carp, silver carp and tilapia submitted to five different treatments. The average size of the ponds was 400 m^2, total stocking rates were 19700 fish ha^{-1} (high) and 6380 or 6130 fish ha^{-1} (low), i.e. 11450 fish ha^{-1} (high) or 3300 fish ha^{-1} (low) for common carp, 2500 fish ha^{-1} (high) or 1250 fish ha^{-1} (low) for silver carp, 5000 fish ha^{-1} (high) or 1500 fish ha^{-1} (low) for *Oreochromis aureus*. The remaining fish were grass carp (from Moav *et al.*, 1977) (with kind permission of Dr. Wohlfarth, on behalf of the authors, & Aquaculture).

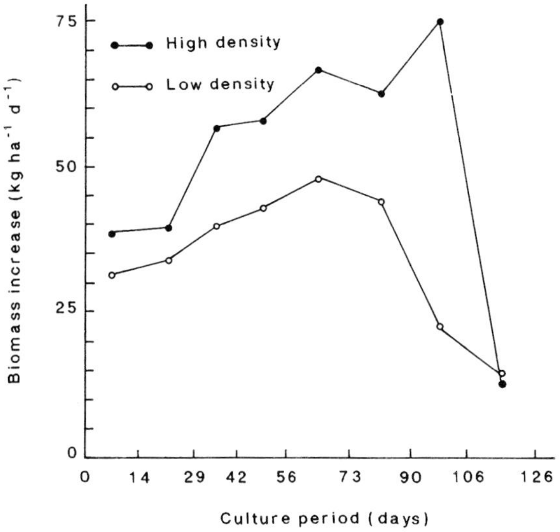

Fig. 58. Calculated average daily yield increments at high and low fish densities in the course of a production period. The stocking rates and other production parameters were the same as in the experiment of Fig. 57. The fish were fed predetermined feed ratios with high-protein fish meal pellets (from Moav *et al.*, 1977) (with kind permission of Dr. Wohlfarth, on behalf of the authors, & Aquaculture).

Relations between feeding rate and production rate

Production depends directly on the amount of available feed, as shown by the experiments of Edwards *et al.* (1981). In trials conducted in cement tanks using *O. niloticus* at high densities (25 fish m^{-2}), the weight of the fish harvested after 12 weeks of culture was directly dependent on the amount of phytoplankton available (Fig. 60). The fish were fed this phytoplankton at concentrations ranging between 20 to 70 mg l^{-1}. At low fish densities, when average plankton concentration increased from 50 mg l^{-1} to 80 mg l^{-1}, the amount of fish harvested did not change much, indicating an excess of available food. At higher densities, the relationship was more obvious. When feed was abundant, fish production became more or less constant; when the fish density was low, individual growth was high, and, with increasing fish density, the decrease of individual growth was compensated by an increase of the biomass at harvest. This range of phytoplankton, 20 to 70 mg l^{-1}, corresponded to a range in increase in fish biomass of 1 to 3 (Fig. 60). Continuous changes in phytoplankton composition, but without appearance of blue-greens (*Microcystis*) euglenoids (*Euglena*) or

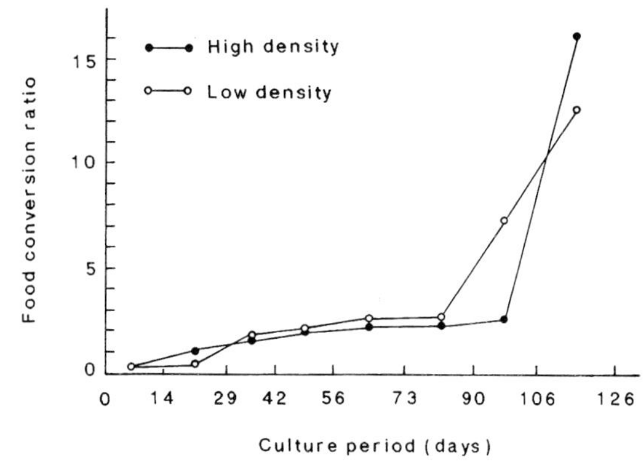

Fig. 59. Evolution of feed conversion ratios at high and low fish densities in the course of a production period. The pond conditions and feeding rates are the same as in Fig. 58 (from Moav *et al.*, 1977) (with kind permission of Dr. Wohlfarth, on behalf of the authors, & Aquaculture).

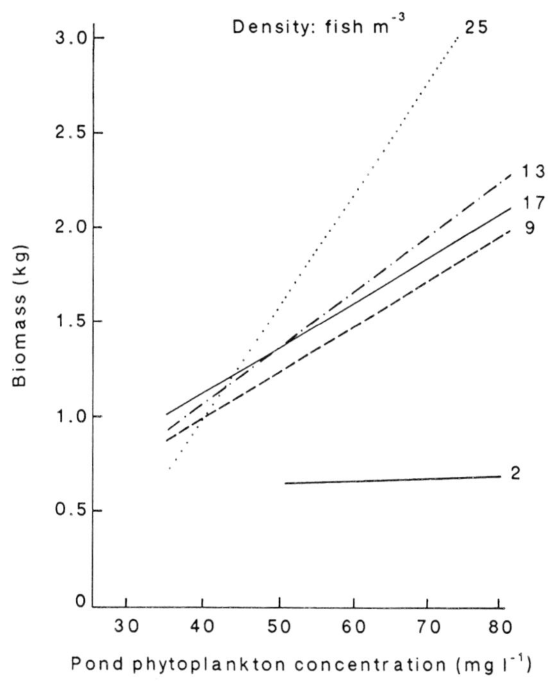

Fig. 60. Tilapia (*Oreochromis niloticus*) biomass at harvest as a function of different pond phytoplankton concentrations and different fish densities. The tilapia were held in concrete ponds (4 m^3), fed with algal-laden high-rate stabilization pond effluents, for a period of 12 weeks (from Edwards *et al.*, 1981) (with kind permission of Prof. Edwards, on behalf of authors, & Aquaculture).

cryptophyte species (*Cryptomonas*), did not affect the feeding of tilapia. Phytoplankton composition comprised 32 genera, 17 green algae, 6 blue-green algae, three euglenoid algae, three diatoms, one dinoflagellate and one cryptophyte.

The use of feed was more efficient at high densities. The estimated feed conversion ratio (FCR) varied between 1.3 and 2.5 (Edwards *et al.*, 1981).

2.1.2. Stocking density and growth

Definition of stocking density

The optimum stocking of a fish pond is that amount of fish released into the pond at the start of the production period, which guarantees the highest possible yield of fish on harvesting. This depends on pond productivity and on the specific potential of the fish species used.

Density is expressed as the number of fish per unit surface area or volume in a pond (m^{-2}, ha^{-1} or m^{-3}). Some authors distinguish between stocking density and population density. By definition, the first designates the initial density, while the second considers the actual density at any time of the culture period. Differences between these two densities result from mortalities, which occur during the course of the culture or are a consequence of stocking manipulations, and reproduction occurring during the culture period. The farmer is concerned with the stocking density, which he decides on, and the density at the harvest.

Relations between density, growth rate and final mean weight

Edwards *et al.* (1981) quantified the effect of fish density on mean average weight at harvest of *O. niloticus* cultured in cement tanks of 4 m^3 and fed phytoplankton originating from sewage stabilization ponds. Average individual weight of the fish after 12 weeks of culture at 1, 10, 20, and 30 m^{-3}, decreased with density (Fig. 61). At a density of 1 fish m^{-3}, the average weight at harvest (after 12 weeks) was little above 120 g, while at 10 fish m^{-3} the average weight remained around 20 g and less for higher densities. Grouping the data according to increasing density illustrates the exponential decrease of mean weight at harvest (Fig. 62) which results.

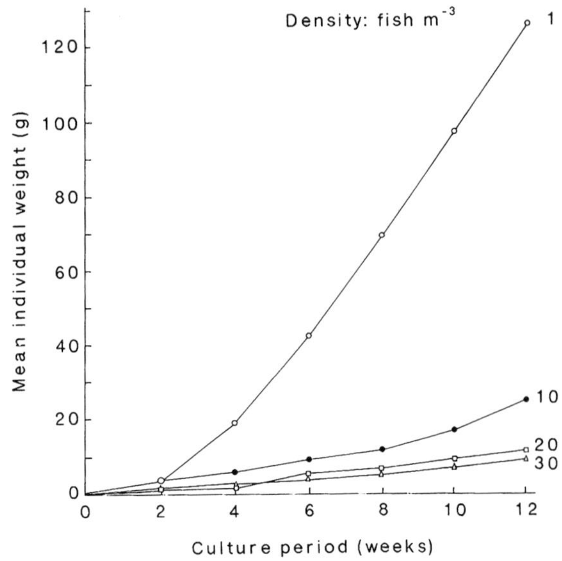

Fig. 61. Evolution of average weight of tilapia (*Oreochromis niloticus*) at two weeks intervals in the course of a culture period as a function of different fish densities and under pond phytoplankton concentrations corresponding to ca. 40 mg l^{-1}. The tilapia were held in concrete ponds (4 m^3) (modified from Edwards *et al.*, 1981).

Individual growth and fish production are in direct relation with population density (Fig. 61, 56, 62 and 63). At low density, the individual growth rate is high and production low, while at high density, the individual growth is low, but production is high. The fish farmer must compromise between a good production and an acceptable individual growth rate, resulting in larger amounts of marketable table-size fish.

Table 23 lists various data on growth of tilapia obtained at different densities in production ponds. It aims at showing the influence of density on growth in real production conditions. The influence of an increase in density can be counterbalanced by artificial aeration, as shown by experiments of Sin & Chiu (1983) obtained in Hong Konk, but growth rates varied widely between experiments indicating a strong influence of food. In indoor tanks experiments, in Belgium, Mélard (1986) explained 80 % of the variance in growth of inbred *O. niloticus* by the following intrinsic and extrinsic variables: initial average weight, gender of the fish, food, feeding frequency, dissolved oxygen concentration, water temperature, and illumination.

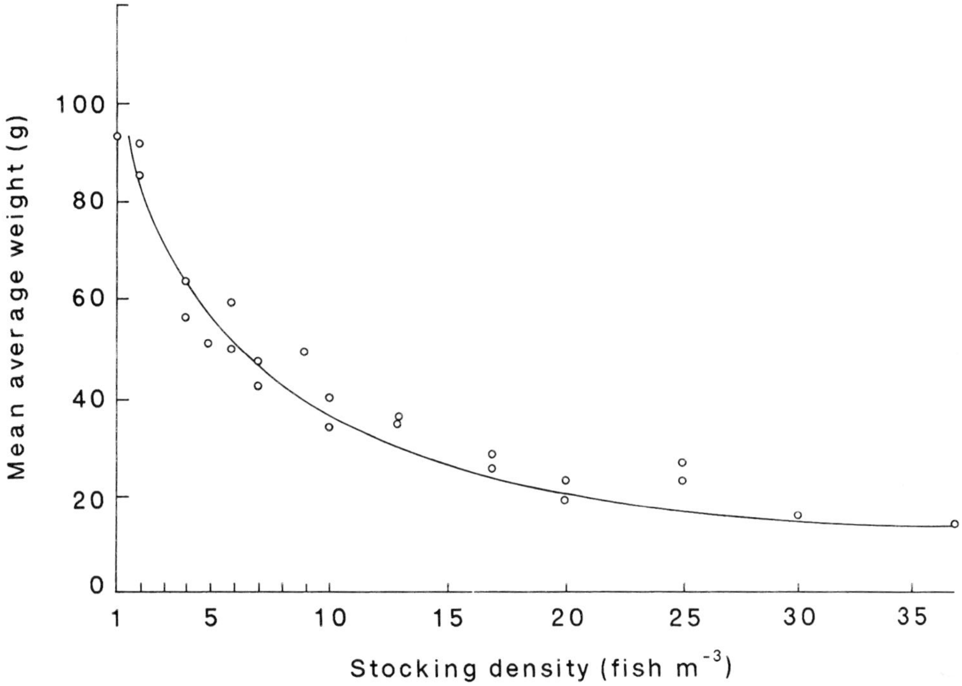

Fig. 62. Mean average weight of tilapia (*Oreochromis niloticus*) at harvest as a function of stocking density. The fish are harvested after a culture period of 12 weeks in concrete tanks (4 m^3) and held under pond phytoplankton concentrations varying in a range of 55 to 60 mg l^{-1} (modified from Edwards *et al.*, 1981).

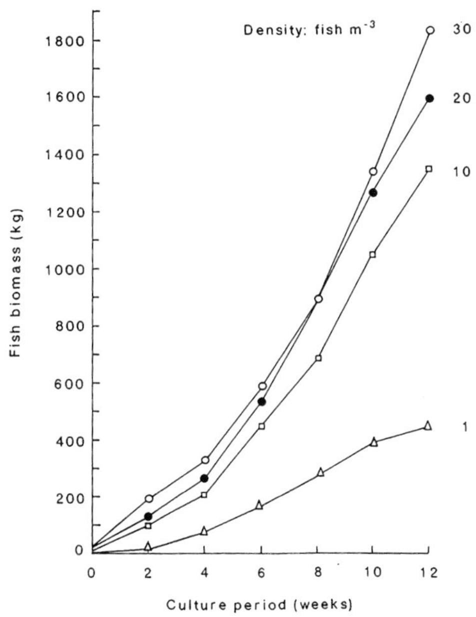

Fig. 63. Evolution of total weight of tilapia (*Oreochromis niloticus*) stocked in concrete ponds (4 m^3) and sampled at two weeks intervals during a culture period as a function of different fish densities and under pond phytoplankton concentrations corresponding to ca. 55 mg l^{-1} (modified from Edwards *et al.*, 1981).

Density and reproduction of tilapia

In tilapia culture, farmers hope that an increase of stocking density will lead to a reduction in reproductive rate of the fish, and thus solve the problem of uncontrolled reproduction (see p. 188). Yet, even at very high densities, this fish continues to reproduce. Edwards *et al.* (1981) observed spawning in tanks of 4 m^3 at densities of 2 to 25 *O. niloticus* m^{-3}, for culture periods of 12 to 18 weeks. Mélard (1986) noticed spawning in growth trials of *O. niloticus* in indoor tanks, outdoor cages and small ponds, in Tihange, Belgium at every density below 500 fish m^{-3}. Densities above that value are not used in farming.

2.1.3. Stocking density and production

Influence of density on production rate

An increase of total biomass with increasing density

Table 23. Daily average growth rate of tilapia, *O. aureus* and hybrids, at different initial densities and stocking weights.

Initial density (ha^{-1})	Average initial weight	Daily weight increase ($g\ d^{-1}$)	Species *Oreochromis*	Remarks and references
59000	6.6	1.01	hybrids:	Sin & Chiu (1983)
122000	5.0	0.69	*mossambicus* x *niloticus*	
30000	111.7	1.40	hybrids:	no aeration
111000	97.7	1.30	-idem-	aeration same authors
20000	16.7	1.12	hybrids:	Verani *et al.*
30000	14.6	0.85	*niloticus* x	(1983)
40000	15.0	1.00	*hornorum*	
50000	14.0	0.80		
20000	fingerlings	2.25	hybrids:	Sarig & Arieli
100000	fingerlings	1.19	*niloticus* x *aureus*	(1980)
		At harvest		
460000	0.8	0.32	hybrids:	Sarig & Arieli
700000	0.8	0.15	-idem-	(1980)
20000	279	2.40	males of *O.*	Sarig & Arieli
30000	265	1.96	*aureus*	(1980)
60000	240	3.39		
100000	243	1.48		
1800	100	3.8	*O. aureus* in	Sarig & Arieli
3000	190	2.8	polyculture with	(1980)
4000	175	3.3	carps	
4900	150	2.4		

has been shown in experiments by Edwards *et al.* (1981). *O. niloticus* were cultured in cement tanks of 4 m^3 and fed phytoplankton. The densities tested were 1, 10, 20, and 30 m^{-3} (Fig. 63). A correlation between total biomass at harvest and density was established, shown in Fig. 56. Total biomass increased exponentially with increasing density until the carrying capacity of the cement tanks (2.0 kg) was reached. In the experiment, under apparently non-limiting food conditions, a density of 10 fish m^{-3} permitted an extrapolated production of 11043 kg ha^{-1} yr^{-1} of fish to be obtained. Higher densities (20 and 30 fish m^{-3}) increased the fish yield to 12240 kg ha^{-1} yr^{-1} and 13679 kg ha^{-1} yr^{-1}, respectively.

Details of pond production trials conducted in Israel and Taiwan at various densities are listed in Table 24.

Fish used were hybrid tilapias or *O. aureus*, in monoculture or polyculture, at densities varying between 0.3 and 10 fish m^{-2}. Sin & Chiu (1980) also tested the influence of artificial aeration, which was beneficial and allowed, in fact, a doubling of production. Information with regard to artificial aeration of ponds in Israel is not clear.

Table 24. Daily average weight increase, per fish and per surface area, of tilapia, *O. aureus* and hybrids, at different densities.

Initial population density or at harvest (ha^{-1})	Daily weight increase per fish (g d-1)	Daily weight increase per ha (kg d^{-1})	Species cultured *Oreochromis*	References and remarks
59000	1.01	52	hybrid tilapia:	without aeration
122000	0.69	67	*mossambicus* x	with aeration
			niloticus	Sin & Chiu (1983)
30000	1.40	41.7	-idem-	without aeration
110000	1.30	133.4		with aeration
				same authors
20000	2.25	45	hybrid tilapia:	Sarig & Arieli (1980)
100000	1.29	129	*niloticus* x	
			aureus	
18000	2.75	50.3	*O. aureus*	male monoculture
20000	2.40	46.0		Sarig & Arieli (1980)
30000	2.50	75.6		
30000	1.96	59.0		
59000	1.75	103.0		
60000	2.39	143.4		
80000	2.00	163.0		
100000	1.48	146.0		
3000	2.80	8.5	*O. aureus*	polyculture with
4000	3.30	13.2		common carp
4900	2.40	11.1		Sarig & Arieli (1980)
5740	2.53	14.3		
7000	3.10	22.0		
8720	2.87	24.6		
9000	3.30	29.0		
10000	5.50	55.0		
20000	3.30	66.0		

For Israel, yield versus stocking density is analyzed graphically on data from Table 24. The relationship is an exponential increase (Fig. 64).

Appropriate densities

In the past, a density of 2 m^{-2} or 20,000 ha^{-1} was recommended for *O. niloticus*. The results obtained in Israel, Hong Kong, and other places, indicate that this density can be notably increased, with intensification of production conditions. Under a regime of abundant feeding, with high-protein pellets, regular fertilization, and intensive organic fertilization, the stocking densities to be recommended are 3 to 6 fish m^{-2}. With supplementary aeration of the ponds, at least at night, 6 to 10 fish m^{-2} can be stocked. The intensification of production conditions should be accompanied by a selection of either monosex male fish, hybrids or, at least, superior strains.

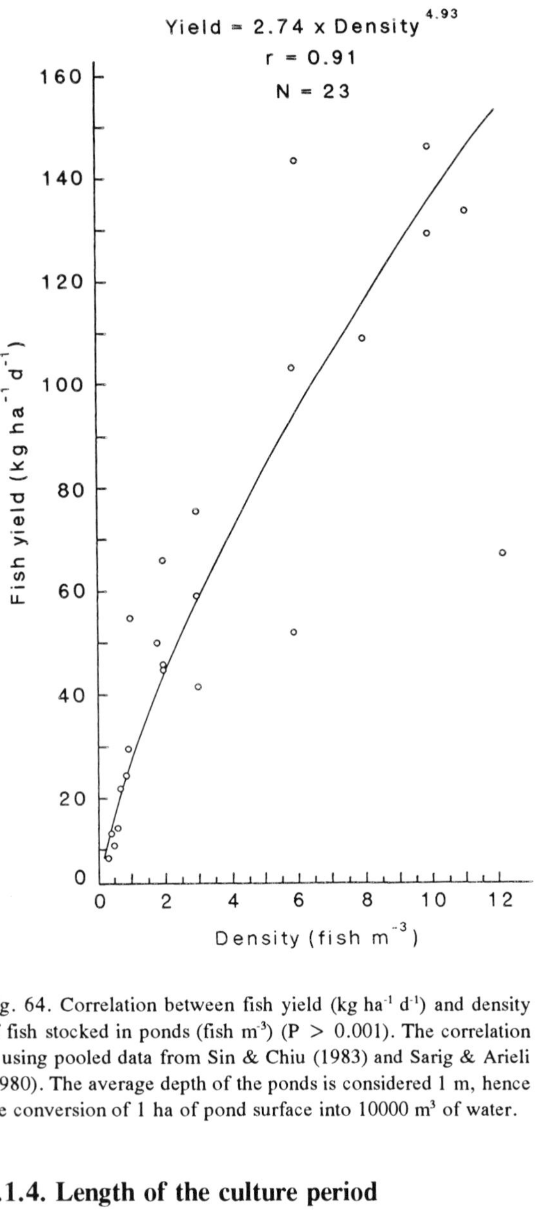

Fig. 64. Correlation between fish yield (kg ha^{-1} d^{-1}) and density of fish stocked in ponds (fish m^{-3}) ($P > 0.001$). The correlation is using pooled data from Sin & Chiu (1983) and Sarig & Arieli (1980). The average depth of the ponds is considered 1 m, hence the conversion of 1 ha of pond surface into 10000 m^3 of water.

2.1.4. Length of the culture period

The length of the culture period in a pond is determined by initial stocking density and the growth rate of the fish. The stocking density should be situated within the range of fast growth rates. Harvest should occur before growth levels off, close to the maximum carrying capacity. The culture period should only be as long as significant increases in biomass occur. If the average size (or weight) of the fish should not have reached a marketable size at maximum carrying capacity, the farmer will have to choose between two possibilities: reduce the fish biomass by selective shinning-out, which will remove the largest fish (intermediate cropping), or transfer the fish to a larger pond. The first solution has the disadvantage of removing the fish with the highest growth rates, leaving only the slow-growers in the pond. Also, if culture conditions have started to deteriorate, the fish remaining in the pond will not fully benefit from this increase in 'breathing space'. The second solution is adopted in the Philippines in the culture of shrimp and milkfish. Here, growout ponds are constructed in modules of three ponds of increasing size, the areas of which vary in a ratio of 1:2:4 or 1:3:9. The fish are transferred from one pond to the other in the course of the culture period by opening sluice-gates and allowing smaller ponds to drain into larger ones (dela Cruz, 1983).

2.2. Fish production and production factors

2.2.1. Production and production factors

Definition of fish production

The fish production of a pond is the amount of fish gained at harvest. Gross yield is the total weight of fish harvested from the pond, and net yield the difference between total weight of fish harvested minus stocking weight. It is conventionally expressed as weight unit per hectare and per day or per year, though it is understood that seldom a production lasts for a whole year. The calculation of yield per ha and per year is extrapolated linearly from harvest size, culture period, and culture surface area.

Benefits to the producer will be maximized when the production of table-size fish is greatest. Therefore, total production should be distinguished from market-size production. Total production includes the production of fingerlings, as most fish, especially tilapia, breed within the pond. Total production and table-size production coincide when the fish do not breed.

Factors determining production

Fish yields are a function of intrinsic factors, related to the genetics of the cultured fish population, and extrinsic factors, peculiar to the culture regime and the nature of the pond. The dependence of individual growth on population density, already mentioned on the preceding pages, operates through competition for space and food. To determine what part of the daily feed is available to each fish of the population, the rules which govern the social behaviour of food procurement should be known. A high level of competition causes an important decrease in individual growth rate, which, in turn, affects final fish biomass and thus yield.

The fish production in a pond depends on farmer's management, including fish care measures. This covers a wide array of factors. Some attempts have been made to quantify these (Piedhrahita, 1988 & 1990), with a view to determining their respective impacts and making recommendations to the farmer as to how he should intervene most efficiently to increase production, but too few publications with a comprehensive description of the culture factors are available. This is one of the greatest handicaps to modelling the production of tilapia. In Africa, for example, the sad reality is that stocking densities, feeding frequencies, and amount of feed cast, for example, often depend on what is locally and momentarily available, and even so, the exact procedures which were followed are not reported. In situations of high yield, it is therefore impossible to reconstruct the culture conditions of the production. One of the few exceptions to this rule is described hereafter.

Production factors in fish farming in CAR

Thirty-five pond production cycles of *O. niloticus*, in monoculture, at the governmental farms in Central African Republic (CAR) (Delincé, original data). were performed between 1974 and 1977, mainly at the Landjia fish farm, Bangui, and yields ranged between 400 to 9300 kg ha^{-1} yr^{-1}.

The feeds were classified on an arbitrary scale, ranging from 1 to 5, according to their nutritional value, 1 standing for no feed, 2 for the least nutritive feed and 5 for the fish-cum-pig culture, which is widely considered the most productive treatment (Fig. 65).

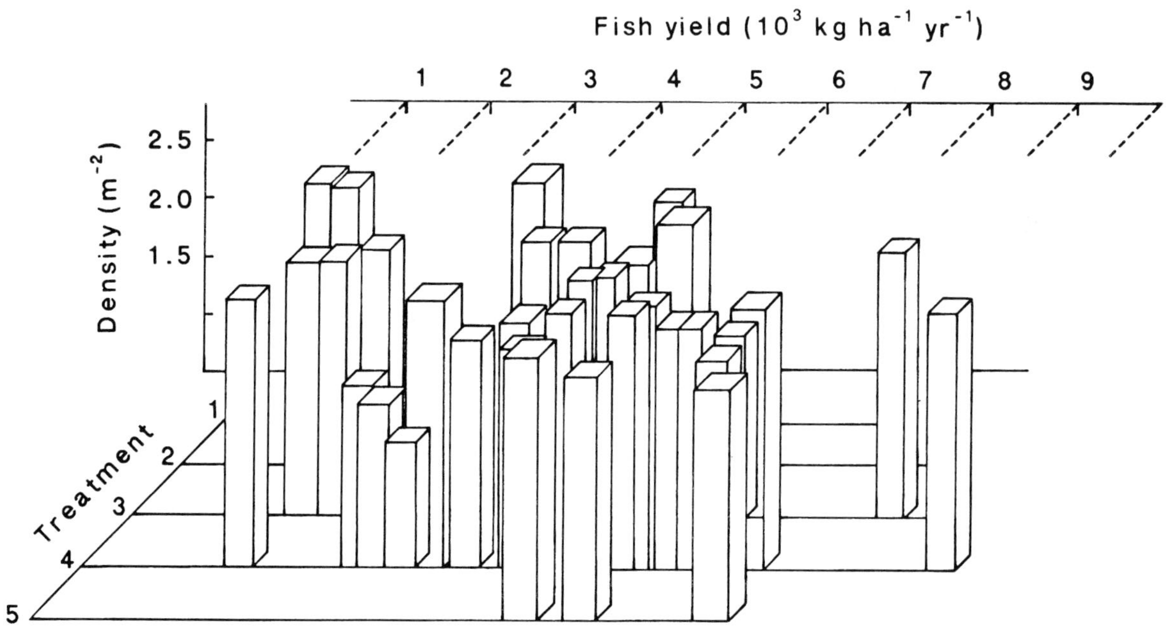

Fig. 65. Fish yield in a monoculture of tilapia (*Oreochromis niloticus*) in earthen ponds in Central African Republic as a function of density and pond treatment. The Y-axis is graded in a relative scale. Detailed explanation in the text.

The various feeds (and/or fertilizers) used were the following: brewery waste, brewery waste and cow dung, brewery waste and stale flours, crushed cotton seeds, crushed cotton seeds and rumen content, pellets, mixed feed unpelleted (same composition as the pellets), duck manure (duck-cum-fish farming), pig dung (pig-cum-fish farming). Production under control conditions, i.e. without feeding or fertilization, was also carried out.

Stocking densities ranged from 1 to 2.6 fish m^{-2}, and mean stocking weight between 8 and 60 g. The culture period extended from 53 to 188 days. Two sizes of ponds were used, the first ca. 350 m^2 in area (16 trials), and the second ca. 3200 m^2 in area (19 trials).

A multiple regression analysis was carried out between production (dependent variable) and the following factors (independent variables): initially stocked biomass, initial average individual weight and density, the feed used, and pond surface area. The multiple correlation coefficient, 0.86, was significant. Individual regressor reports for the initially stocked biomass, the treatment intensity, the initial density, the average individual weight stocked, the surface area of the pond, and the culture period were 0.87, 0.92, 0.90, 0.79, 0.73, and 0.69, respectively. The dependency of fish yield on intrinsic and extrinsic factors of production was therewith confirmed. The relation between production and the explicative variables is expressed by the following equation:

Total production = - 4.00 x stocked biomass (in kg ha^{-1})
+ 862.59 x treatment intensity (scale ranging between 1 to 5)
- 13.44 x culture period (in days)
- 1554.67 x stocking density (m^{-2})
+ 67.53 x average initial stocking weight (in g)
+ 0.14 x pond area (in m^2)

2.2.2. Critical factors limiting production

Deterioration of production conditions

In production trials of increasing intensity, the fish biomass ceases growing at the end of production as a consequence of a deterioration of production conditions. This deterioration has been observed by several authors (Moav *et al.*, 1977; Edwards *et al.*, 1981; Lazard, 1980, in Côte d'Ivoire), who tried to determine its causes. The following factors, all of which were discussed in detail earlier were advocated: a decrease in dissolved oxygen, combined with accumulation of waste products (for example, association of ammonia with suboptimal oxygen concentration in early morning hours); accumulation of inhibitory substances; a decrease in number of organisms consumable by tilapia, and a paralled increase in inedible or toxic organisms, such as euglenoids and blue-green algae (Moav *et al.*, 1977).

Problems with oxygen consumption in ponds

When the biomass of the fish increases, dissolved oxygen concentrations may become limiting. Fish respiration increases, and the environment becomes depleted of oxygen. The fish are not the only oxygen consumers in the pond, however. If the fish load is significant, the amount of excrements released by them is important and the mineralization of this organic matter alone consumes much oxygen.

Jana & De (1983) studied the oxygen consumption of the plankton community in a pond used for polyculture in West Bengal, India. The oxygen consumption of bacteria reached up to 26 g O_2 m^{-2} d^{-1}, with an average of 7-8 g O_2 m^{-2} d^{-1}. This represented about 65 to 100 % of total planktonic consumption of oxygen. In comparison, tilapia (*O. niloticus*) consume between 3.96-4.40 g O_2 kg^{-1} d^{-1} when feeding (fish between 200-400 g, at 25-30° C) (Mélard, 1986). African catfish (*Clarias gariepinus*) consumption was between 3.3 and 7.5 g O_2 kg^{-1} d^{-1} according to feeding level, at 25° C (Hogendoorn, 1983). The oxygen demand of fish ponds depending on organic matter supply amounts to a maximum of 3.5 g m^{-2} d^{-1} (Oláh *et al.*, 1987). Bacterial population sizes of ponds vary little, being about 10^6 cells per ml. In the case of large phytoplankton populations, as found in ponds with high organic fertilization, the autotrophic dark respiration must also be added. The result is that dissolved oxygen levels vary cyclically, and that the concentration often drops to critical values at the end of the night.

Measures to avoid anoxic conditions

Aeration of ponds involves the use of such devices as aerators, diffusers, paddle wheels and air blowers. Extensive discussion on their use can be found in Boyd (1979). The paddle wheel is considered by Boyd & Tucker (1979) as the most efficient device available. In the absence of the above-mentioned devices, aeration by an out-board engine can be a solution. Aeration induces an increase in net photosynthetic activity of the phytoplankton, and alleviates the respiratory demand for oxygen of the pond (Segupta & Jana, 1987).

The use of aerators or other machines to supply air or oxygen to a pond is particularly useful to combat oxygen depletion in the early morning hours. In culture experiments using tanks, aerators are frequently used too. However, their cost is high and their use thereby limited. According to some authors, the solution is palliative, because only a temporary oxygen-enriched refuge for fish is created, and the fundamental problem of oxygen supply to the pond is not solved.

The lack of oxygen also affects the plankton community and the decomposers of the pond; this leads to a die-off of plankton, an additional supply of organic matter, and the establishment of even more acute anoxic conditions. The production conditions deteriorate further.

The replacement of oxygen-depleted stagnant water by fresh oxygen-enriched water is another solution. Unfortunately, such an exchange of water leads to a loss of water fertility and of the natural food present, which is adverse to production.

An uncommon practice in Africa, but which has proved useful in China, is the use of agitators. These devices mix the pond water, like by the movement of a paddle-wheel. This agitation is done during the hot hours of the day, when photosynthesis is active, and helps to oppose thermal stratification, and the establishment of dissolved oxygen gradients. Surface oxygen supersaturation is reduced, as the entire water mass comes into contact with the oxygen-rich upper layers, resulting in an homogenization of dissolved oxygen. This is important, because it builds up a storage of oxygen in deeper strata, to be used during nocturnal oxygen consumption.

In China, in the provinces of Kiangsu and Shanghai, aerators are commonly used. Types found are either blowers powered by a turbine engine (Tapiador *et al.*, 1978) or electrically powered paddle wheels (FAO, 1981). These devices are floating in the center of the pond and are operated 2 to 3 hours a day, around noon, or in the morning by overcast weather. Fish harvest is increased by 14 % by use of these aerators (Tapiador *et al.*, 1978).

Biological control of oxygen production and consumption

Oxygen supply to a pond is dependent on a phytoplankton population of adequate size. One of the most important managemental problems therefore is the control of phytoplankton. Often, an excessive increase of algae results in wide diel fluctuations of dissolved oxygen, which often lead to a sudden die-off of the plankton bloom responsible for these fluctuations (see p. 148). The fluctuations are such that supersaturation of oxygen in the upper layers builds up during the day, followed by a decrease to full depletion during nighttime, when all living organisms, autotrophs and heterotrophs, consume oxygen.

In view of controlling phytoplankton populations by biological methods, Costa-Pierce *et al.* (1985) studied the effect of a polyspecific fish population on primary production in a *Macrobrachium rosenbergii* pond in Hawaii. Six different treatments were analyzed. The ponds were stocked with shrimps at the usual density (11.5 postlarvae m^{-2}), and fed an artificial feed, or fertilized with organic manure, at a rate of 700 kg per hectare and per week. Water hyacinths were added to two ponds, in view of determining their role in removing excess nutrients and controlling phytoplankton. The ponds containing the hyacinths (covering about 25 % of the water surface) were normally fed and fertilized. In two other ponds, grass carp (500 fish ha^{-1}) and silver carp (2000 and 25000 fish ha^{-1}) were stocked, fed, and the ponds fertilized. The water quality of these six types of ponds was compared. Chlorophyll a concentration was more important in the fertilized polyculture ponds, with primary productivity being heavily stimulated and diel oxygen fluctuations reduced. In the other ponds, however, the oxygen reached critical values, potentially dangerous to production. Water hyacinth were of little use in controlling water quality in ponds.

The salient point was that fish polyculture

markedly improved water quality. Turnover time of nitrogen and phosphorus in water was reduced and primary productivity increased. Fish grazing stabilized phytoplankton biomass (lowest coefficient of variation for chlorophyll a).

Means of eliminating deleterious decomposition products

A novel way of removing part of the sediments stemming from the accumulation of left-over feed and settled algae, and thus improve water quality, has been introduced by Israeli farmers (Zohar *et al.*, 1984). They observed that detritus accumulates on the anaerobic sediment and takes much time to decompose and be recycled. Evidently, such material contains metabolites harmful to fish growth. Removal of this detritus is done by a quick flushing of the water of the sediment-water interface, using a pipe laid at the sediment level, and draining the accumulated detritus (Fig. 66). This flush of water from the bottom of the pond is done three times daily for 20 seconds, which is sufficient to remove all the sediment of dark colour.

Analysis of the sludge drained showed that it was rich in organic matter (80 mg l^{-1}) and thus a significant source of oxygen consumption and of metabolites had been removed from the pond.

In flushed fish ponds, with stockings of 15 tilapia per m^2 and with aeration, production amounted to some 50 tons per ha. The FCR of the feed was even decreased, which attests to good production conditions (Zohar *et al.*, 1984). Aeration alone would be detrimental to production, since it would mix sedimented organic material to the water mass, thus even increasing oxygen demand and metabolite concentrations.

In a similar way, bivalves may be used to improve water quality near pond bottoms. In fish ponds in Illinois, Buttner (1986) observed that populations of *Corbicula* filter and remove a great part of the material in suspension. He found significant differences in dissolved oxygen concentrations, in the number of days when dissolved oxygen dropped below 3 mg l^{-1} at dawn, in primary production, turbidity, and ammonium concentration in the water, between ponds stocked with bivalves and those without. The stocking rates ranged around 1000 kg ha^{-1}. However, in some cases, nitrogen

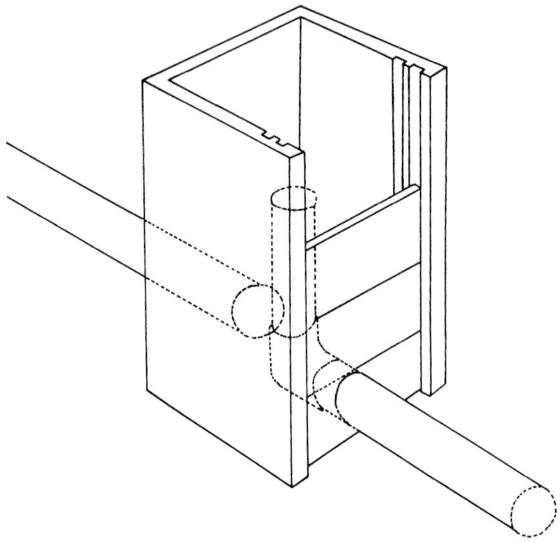

Fig.66. Modified pond monk (drainage device) enabling the draining of the water at the water-sediment interface level. A 15-cm diameter pipe is piercing through the lowest shutter and connected behind the shutter to a removable stand-up pipe, the height of which regulates the water level in the pond. The removal of the stand-up pipe allows the draining of the water at the level of the pipe in the course of the culture period. The extremity of the pipe is screened off to avoid the escape of fish (modified from Zohar *et al.*, 1984).

detritus was more important in the ponds with bivalves.

Successive limiting factors in fish farming

The different components of a pond ecosystem were seen to be complementary and to contribute synergistically to production. Primary productivity acts on fish production, through the quality of the water, primarily via the diel cycling of dissolved oxygen concentration, and through direct use of the phytoplankton as food.

Amongst all environmental variates, water temperature is, of course, a critical factor. Temperature directly affects fish metabolism, but since it influences all other biological and chemical activities within the pond, it is implicitly present in the effects of all other variables.

The factors which influence fish production really act within a network of interrelations. There is not just one limiting factor but a suite of them, each of which is linked to a major increase in production.

This implies that, at each production increment, the bottleneck of that particular limiting factor must be overcome. The three main limiting factors are (1) availability of feed (*ad libitum*), (2) dissolved oxygen concentration, and (3) ammonia concentration of the water. In cultures in artificial environments, these variables can be constantly monitored.

Mélard (1986), in a study of the biological basis of intensive tilapia production in indoor facilities in Belgium, found that the main effect of fish density on the culture environment was to reduce the dissolved oxygen level, so that the effect of fish density on production could be completely masked. Increasing density reduces either the amount of feed available to each fish or reduces dissolved oxygen, which in turn reduces feed intake and consequently growth. Jana & De (1983), in their study of production factors, also found a correlation between gross daily photosynthetic rate and fish production.

Cuenco *et al.* (1985) in a study of environmental factors affecting the population growth of American catfish, showed that, besides available food and oxygen, non-ionized ammonia in the water is an important factor (see Ammonia toxicity, on p. 89). In a culture system with high stocking and feeding rates, daily $N\text{-}NH_3$ inputs are high due to fish excretion and decomposition of excess feed. At alkaline pH, the ammonium/ammonia equilibrium shifts towards unionized ammonia, which is toxic for fish even at low concentrations. Increased fish production was achieved by successive optimalization of growth factors. Stocking densities of catfish were 7000 fish ha^{-1} in all cases. In natural conditions, the yield was 394 kg ha^{-1} (1.1 kg ha^{-1} d^{-1}); with *ad libitum* feeding, the yield reached 3897 kg ha^{-1} or 15.2 kg ha^{-1} d^{-1} (daily feed ratio was computed considering size, temperature, DO and unionized ammonia rather than only body weight); by aeration of the pond, the yield could be further increased to 6141 kg ha^{-1} (25.0 kg ha^{-1} d^{-1}). By keeping non-ionized ammonia concentration low, finally the yield reached 27659 kg ha^{-1} (118.5 kg ha^{-1} d^{-1}).

In ponds, the farmer's art consists in making best possible use of this ecological wisdom, so that each available resource is utilized and does not accumulate beyond a point where it becomes a nuisance. The autotrophic and heterotrophic production cycles must be complementary to ensure a continuous availability of nutrients for photosynthesis and a supply of organic matter to the decomposers. The natural food available to fish then remains continuously high. The diversity of natural production suggests the use of different species of fish with non-overlapping feeding niches in polyculture, adapted to the array of food resources available. Polyculture of tilapia with carp, or with catfish is commendable. Other polyculture associations, with intensive organic fertilization can, however, be established, using other species to widen the resource utilization. In Asian countries, polyculture often involves shrimp and fish, but except for some experimental ponds at the Malindi fish farm, near Mombassa, Kenya, this type of culture is not yet practised in Africa.

2.2.3. Ranges of fish yields

Yield levels according to production intensity

According to yield obtained, four groups of production intensity have been distinguished (Fig. 67) (Balarin & Haller, 1983; Balarin, 1988; Vincke, 1988). Extensive culture produces, on harvest, between 400 to 2500 kg ha^{-1} yr^{-1}, or 1 to 7 kg ha^{-1} d^{-1}. The degree of fertilization is low, as is stocking density, being about 0.3 to 2 fish m^{-2}. Subsistence ponds in rural areas of Africa (most countries in West, Central, East and Southern Africa, in all about 40 countries) indulge in this type of culture. The ponds are usually of small size (less than 0.5 ha), rain-fed or fed by the water table, lacking a water control structure. They depend on fingerlings provided from public or support services and the fish are fed domestic wastes. The ponds are under water for 6 to 12 months and angling occurs regularly. The fish harvested at the end of the culture period consist often of small-sized tilapia, due to the removal of all large-sized fish in the course of the production and due to overcrowding, resulting of poor feeding, prolific breeding and space competition. Fishfarming is practised as a part-time occupation by communities, families or individuals. In that category fall also most of the brackish water farms, such as the many farms in the mangrove areas of Nigeria. Their production is low due to the acidity of the local soil.

The second group is termed semi-intensive (or semi-extensive) and reaches yields ranging between

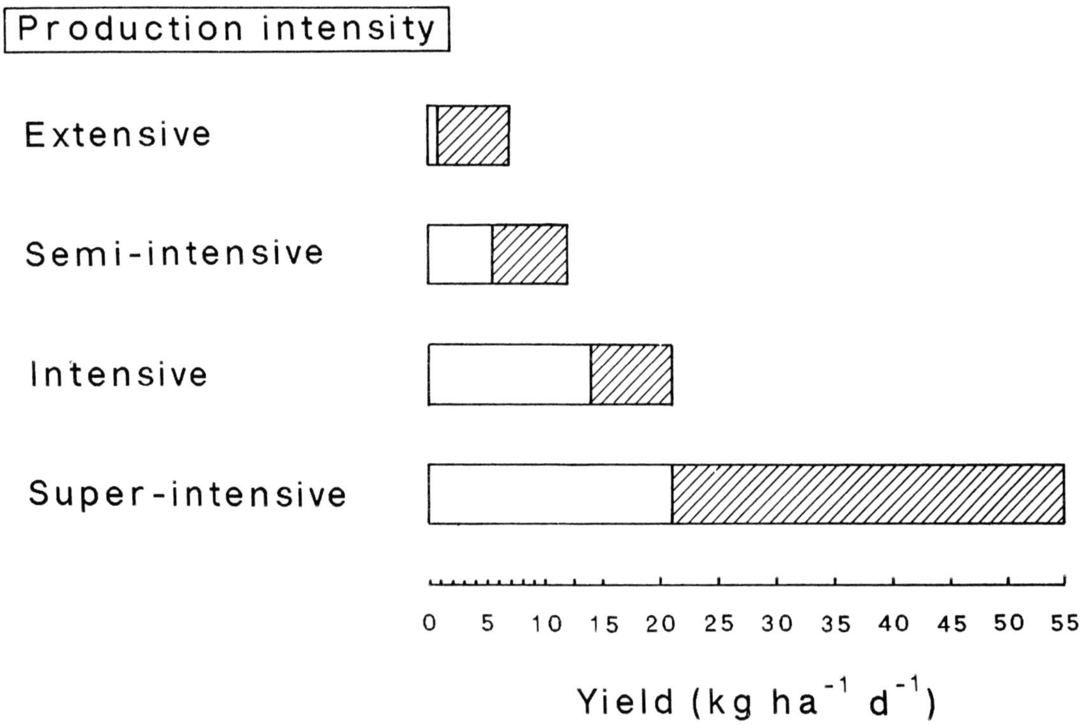

Fig. 67. Ranges of the fish yields obtained at different production intensities.

2000 to 4500 kg ha^{-1} yr^{-1} (or 5.5 to 12 kg ha^{-1} d^{-1}). The layout and the construction of the ponds is properly planned; stocking of the pond is made with tilapia, at 1-3 fish m^{-2}, and the fish are fed and the pond fertilized regularly. The feeds used are single agricultural byproducts or mixed ingredients, with low nutritional value, such as rice or wheat bran, brewery wastes, rumen content, whole or crushed cotton seeds, and molasses. Fertilizers are sometimes used alone. This type of fish farming relies often, but not always on an organised extension service, and is currently practised by individuals or families owning a few ponds from which they draw their living. This semi-intensive culture is found in many African countries, such as in Zimbabwe, Zambia, Malawi, Central African Republic, Kenya, Nigeria, Ivory Coast, Benin, Cameroon, Congo, Madagascar. In some countries, these farms are managed by small entreprises.

Intensive culture is practised in commercial farms. Farm infrastructure comprises sizeable ponds, fingerling production facilities, feed and fertilizer stores, water supply structures. Adequate management of the farm and entrepreneurial skills are required. The ponds are stocked with tilapia in association with catfish, carp or other species, such as *Heterotis*, *Heterobranchus*, *Hemichromis*, Chinese carp. The feed used are composite, nutritionally balanced, in pellet form or not, accompanied with a light organic or inorganic fertilization. The yields of this type of culture range between 5000 to 7000 kg ha^{-1} yr^{-1} or 14 to 21 kg ha^{-1} d^{-1}. This type of culture is practised in the same countries as the semi-intensive type, though not to the same extent. The major limitation to larger extension of this type of culture is the initial investment, and regular supply of feeds. Competition for agricultural byproducts with live-stock farms is impeding the development of these intensive fish farms.

Super-intensive culture is also practised in commercial farms and makes use of intensive organic manuring, resulting from a livestock crop of pigs or ducks, for example, or uses liquified dung, with or without supplementary feeding, and polyculture. Cultured fish are tilapia, the African catfish, *Clarias gariepinus*, common carp. Temporary aeration is carried out. The harvests range from 7500 to 20000 kg ha^{-1} yr^{-1} or 21 to 55 kg ha^{-1} d^{-1}. This type of culture is often part of a commercial farm and is integrated in a larger farm complex, where different agricultural activities are practised. Examples of this type of farms are found, for instance, in Central African Republic, Nigeria, Ivory Coast.

Pond size and length of production period are especially important for the last two production types. If rich artificial feed is used, a homogenous distribution of the feed across the whole surface of the pond is required, to ensure that all fish have access to it. The feed should not be wasted, as its value as fertilizer is much less than as feed. The unconsumed feed transforms into fertilizer. Ponds of large size, i.e. larger than 5000 m^2, are more difficult to feed this way. In ponds where fertilization comes directly from livestock stables, several pens should be constructed at different locations around the pond to avoid accumulation of manure at just one site, resulting in a polluted area. The location of pens at different sites can, however, increase such hazards as theft, a non-negligeable problem in poor third-word countries.

The harvest should take place when fish biomass has come close to the maximum carrying capacity. At the beginning of the production period, an approximate time for harvest is set on the basis of expected daily fish growth. As the production period is reaching its end, regular sampling of the fish should be done, to monitor biomass and FCR. These two items of information will signal when biomass reaches its carrying capacity.

The two intensive types of culture should use selected fish, such as tilapia of a known strain, in male monoculture, or hybrid tilapia (see p. 187).

Maximum yields obtained in Africa

Examples of yields obtained in intensive and super-intensive productions are summarized in Tables 25 and 26. These data come from the governmental farm Landjia, at Bangui, Central African Republic, and are in the same range as the figures of Moav *et al.* (1977), obtained in Israel, and which are considered very high.

The top productions were obtained in polyculture with intensive organic (pig dung) fertilization. The pigsties were located close to or over the ponds. The two fish species used, the tilapia, *O. niloticus*, and the African catfish, *Clarias gariepinus*, have excellent potential, which, in the case of tilapia, is counteracted by its prolific breeding. The advantage of the intense metabolic rate of tilapia is lost when it is diverted from growth to the elaboration of breeding products. In polyculture, catfish recovers part of this energy nutritionally for its own growth,

Table 25. Data on yields obtained in monoculture ponds of *O. niloticus* in the Central African Republic.

Size of the pond (m^2)	3325	2865	3325	3325	2200	484	450	2800
Period (day)	172	53	102	119	146	63	62	142
Initial density (ind m^{-2})	2.32	1.45	2.28	2.20	1.90	2.00	2.00	2.00
Stocking weight (g)	8.00	50.	9.0	15.2	23.0	24.4	23.4	15.3
Total production (kg ha^{-1} yr^{-1})	8168	6395	7167	5644	6818	9348	6163	7261
Table size production (kg ha^{-1} yr^{-1})	3093	3615	4315	3040	2930	4249	3401	1028
Treatment				Artificial feeding				Pig dung

Table 26. Data on intensive production (polyculture) of tilapia, *O. niloticus*, and African catfish, *Clarias gariepinus*, in ponds in Central African Republic.

	Size of the pond (m^2)	3400	3400	250	3400	3400	3400	3400
Tilapia	Initial density (ind m^{-2})	2.6	2.0	2.0	3.88	3.27	2.10	1.98
	Initial weight (g)	24	7.1	8.0	10	10	26	14.8
	Weight at harvest (g)	124	110	93.2	120	-	90.3	71
	Production (kg ha^{-1} yr^{-1})	6465	6003	5856	8609	7152	10317	10461
Clarias	Initial density (ind m^{-2})	0.32	0.25	1.0	1.18	0.2	0.15	0.17
	Initial weight (d)	32.9	8.2	6.4	8.2	6.0	1.17	2.3
	Weight at harvest (g)	471	389	128.4	834	112	1170	1320
	Production (kg ha^{-1} yr^{-1})	667	1595	3824	2126	4035	1552	695
Tilapia (%)		81.4	79.0	60.5	80.2	63.9	86.9	93.8
Clarias (%)		18.6	21.0	39.5	19.8	36.1	13.1	6.2
Period (days)		168	153	90	173	180	148	132
Total production (kg ha^{-1} yr^{-1})		7942	7598	9680	10735	11187	11869	11156
Treatment		Organic manuring (pig excreta)						

by feeding on the tilapia fry and fingerlings. As such, it also alleviates the problem of food and space competition brought about by uncontrolled reproduction of tilapia. With improved strains, selected for late maturity, in all-male culture, or in sterile hybrids culture, even this problem could have been eliminated (see p. 187).

Considering that the production conditions were intensive only from the point of view of organic fertilizer supply, and not from the point of view of stocking densities, which were invariably around 2 fish m^{-2}, or fish used in production, which were unselected tilapia strains, it can be expected that production can be increased even more. Using higher stocking densities, hybrid strains and aeration, it seems reasonable to reach 15 to 20 tons per hectare and per year.

Fish production in ponds

1. Fish production depends on food availability, fish growth rate and density, and available space, which determines carrying capacity.

2. Food, when available at satiation levels, is used more efficiently at higher fish densities. Food requirements are specific, and depending on fish species used. The food conversion ratio indicates whether growing conditions are optimum.

3. Fish production increases, but mean individual weight at harvest decreases, with increasing stocking densities.

4. Production is restricted by dissolved oxygen availability. Measures to increase dissolved oxygen concentrations include the use of aerators, blowers and agitators. The removal of decomposition products which accumulate on the pond bottom is beneficial to production.

5. Successive limiting factors impede the increase of fish production; they are: food, dissolved oxygen and ammonia.

6. The production figures in African culture ponds compare favourably with the maximum production figures obtained in intensive farming outside the region.

References

Adams, D. D., G. Matisoff & W. J.Snodgrass, 1982. Flux of reduced chemical constituents (Fe^{2+}, Mn^{2+}, NH_4^+ and CH_4) and sediment oxygen demand in Lake Erie. Hydrobiologia 91: 405-41.

Agustí, S., C. M. Duarte & D. E. Canfield, Jr., 1990. Phytoplankton abundance in Florida lakes: evidence for the frequent lack of nutrient limitation. Limnol. Oceanogr. 35: 181-188.

Agustí, S., C. M. Duarte & J. Kalff, 1987. Algal cell size and the maximum density and biomass of phytoplankton. Limnol. Oceanogr. 32: 983-986.

Agustí, S. & J. Kalff, 1989. The influence of growth conditions on the size dependence of maximal algal density and biomass. Limnol. Oceanogr. 34: 1104-1108.

Ahlgren, G., 1988. Phosphorus as growth-regulating factor relative to other environmental factors in cultured algae. Hydrobiologia 170: 191-210.

Almazan, G. & C. E. Boyd, 1978. Effects of nitrogen levels on rates of oxygen consumption during decay of aquatic plants. Aquatic Botany 5: 119-126.

Almer, B., W. Dickson, C. Ekström, E. Hornström & U. Miller, 1974. Effects of acidification of Swedish lakes. Ambio 3: 30-36. Quoted by Mulholland *et al.*, 1986.

American Public Health Association, 1976. Standard methods for the examination of water and wastewater, 14th Ed. APHA, Washington, D.C., 874 pp.

Anderson, R. F. & S. L. Schiff, 1987. Alkalinity generation and the fate of sulfur in lake sediments. Can. J. Fish. aquat. Sci 44 (Suppl. 1): 188-193.

Andersson, G., W. Granéli & J. Stenson, 1988. The influence of animals on phosphorus cycling in lake ecosystems. Hydrobiologia 170: 245-266.

Auer, M. T., M. S. Kieser & R. P. Canale, 1986. Identification of critical nutrient levels through field verification of models for phosphorus and phytoplancton growth. Can. J. Fish. aquat. Sci 43: 379-388.

Avnimelech, Y. & M. Lacher, 1979. A tentative nutrient balance for intensive fish pond. Bamidgeh 31: 3-8.

Avnimelech, Y., S. Mokady & G. L. Schroeder, 1989. Circulated ponds as efficient bioreactors for single cell protein production. Israel. J. Aquacult. 41: 58-66.

Azam, F., T. Fenchel, J. G. Field, J. S. Gray, L. A. Meyer-Reil & F. Thingstad, 1983. The ecological role of water-column microbes in the sea. Ecol. Prog. Ser. 10: 257-263.

Balarin, J. D., 1988. Systems in use and species cultivated. In H. R. King & K. H. Ibrahim (eds), Village level aquaculture development in Africa. Proc. Comm. Cons. Workshop on Village Level Aquaculture Development in Africa. Freetown, Sierra Leone, 14-20 February 1985. Commonwealth Secretariat, London: -23.

Balarin, J. D. & R. D. Haller, 1983. Commercial tank culture of Tilapia. In L. Fishelson & Z. Yaron (eds), Tel Aviv University, Tel Aviv, Israel. Proc. Int. Symp. on Tilapia in Aquaculture: 473-483.

Barat, S. & B. B. Jana, 1987. Effects of farming management on the distribution pattern of ammonification rate, protein mineralizing and ammonifying bacterial population in experimental culture tanks. Bamidgeh 39: 120-132.

Barbosa, F. A. R. & J. G. Tundisi, 1989. Diel variations in a shallow tropical Brazilian lake. I. The influence of temperature variations on the distribution of dissolved oxygen and nutrients. Arch. Hydrobiol. 116: 333-349.

Barica, J., 1990. Seasonal variability of N:P ratios in eutrophic lakes. Hydrobiologia 191: 97-103.

Bartell, S. M., 1981. Potential impact of size-selective planktivory on P release by zooplankton. Hydrobiologia 80: 139-146.

Bartell, S. M. & J. F. Kitchell, J. F., 1978. Seasonal impact of planktivory on phosphorus release by Lake Wingra zooplankton. Verh. int. Ver. Limnol. 19: 2478-2484. Quoted by Sternik, 1983.

Barthelmes, D., 1989. Evidence for system dependent interactions of zooplancton-eating fish with phytoplankton. Arch. Hydrobiol. Beih. Ergebn. Limnol.33: 579-586.

Benndorf, J., 1990. Conditions for effective biomanipulation; conclusions derived from whole-lake experiments in Europe. Hydrobiologia 200/201: 187-203.

Benndorf, J., H. Schultz, A. Benndorf, R. Unger, E. Penz, H. Kneschke, K. Kossatz, R. Dumke, U. Hornig, R. Kruspe, S. Reichel & A. Köhler, 1989. Food web manipulation by enhancement of piscivorous stocks: long-term effects in the hypereutrophic Bautzen reservoir. Arch. Hydrobiol. Beih. Ergebn. Limnol. 33: 567-569.

Ben-Tuvia, A. & K. Reich, 1983. Stocking and management of cichlids fishery resources in Lake Kinneret (Lake Tiberias). In L. Fishelson & Z. Yaron (eds), Tel Aviv University, Tel Aviv, Israel. Proc. Int. Symp. on Tilapia in Aquaculture: 18-27.

Bergstrand, E., 1990. Changes in the fish and zooplankton communities of Ringsjön, a Swedish lake undergoing man-made eutrophication. Hydrobiologia 191: 57-66.

Berman, T., 1988. Microbes in a watery world. Hydrobiologia 159: 5-6.

Beveridge, M. C. M., M. Begum, G. N. Frerichs & S. Miller, 1989. The ingestion of bacteria in suspension by tilapia *Oreochromis niloticus*. Aquaculture 81: 373-378.

Bird, D. F. & J. Kalff, 1987. Algal phagotrophy: regulating factors and importance relative to photosynthesis in *Dinobryon* (Chrysophyceae). Limnol. Oceanogr. 33: 277-284.

Björk-Ramberg, S., 1984. Changes in sediment nutrient content in a subartic lake subjected to lake fertilisation. Freshwat. Biol. 14: 157-163.

Björk-Ramberg, S., 1985. Uptake of phosphate and inorganic nitrogen by a sediment- algal system in a subartic lake. Freshwat. Biol. 15: 175-183.

Bjørnsen, P. K., 1988. Phytoplankton exudation of organic matter: why do healthy cell do it? Limnol. Oceanogr. 3: 151-154.

Blauw, T. S., 1988. Nitrogen fixation (acetylene reduction) in the sediments on the Pluß-See. With special attention to the rôle of sedimentation. Arch. Hydrobiol. (Monogr. Beitr.), Suppl. 81 (4): 359-489.

Blažka, P., T. Backiel & F. W. Taub, 1980. Trophic relationships and efficiencies. In E. D. Le Cren & R. H. Lowe-McConnell (eds), The functioning of freshwater ecosystems. Cambridge University Press, Cambridge. International Biological Programme 22 : 393-410.

Bleiwas, A. S. H., P. M. Strokes & M. M. Olaveson, 1984. Six years of plankton studies in the LaCloche region of Ontario, with special reference to acidification. Verh. int. Ver. theor. angew. Limnol. 22: 332-337. Quoted by Mulholland *et al.*, 1986.

Bloem, J. & M.-J. B. Bär-Gilissen, 1989. Bacterial activity and protozoan grazing potential in a stratified lake. Limnol. Oceanogr. 34: 297-309.

Bloesch, J. & H. R. Bürger, 1989. Changes in phytoplankton and zooplankton biomass and composition reflected by sedimentation. Limnol. Oceanogr. 34: 1048-1061.

Blomqvist, B., H. Olsson, H. Olofsson & O. Broberg, 1989. Enclosure experiments with low-dose additions of phosphorus and nitrogen in the acidified lake Njuffatet, Central Sweden. Int. Revue ges. Hydrobiol. 74: 611-631.

Boavida, M. J. & R. T. Heath, 1988. Is alkaline phosphatase always important in phosphate regeneration? Arch. Hydrobiol. 111: 507-518.

Boon, P. I., 1990. Organic matter degradation and nutrient regeneration in Australian freshwaters: II. Spatial and temporal variation, and relation with environmental conditions. Arch. Hydrobiol. 117: 405-436.

Bouldin, D. R., R. J. Johnson, C. Burda & C. W. Kao, 1974. Losses of inorganic nitrogen from aquatic systems. J. envir. Qual. 3: 107-114.

Boström, B., J. M. Andersen, S. Fleischer & M. Jansson, 1988a. Exchange of phosphorus across the sediment-water interface. Hydrobiologia 170: 229-244.

Boström, B., M. Jansson & C. Forsberg, 1982. Phosphorus release from lake sediments. Arch. Hydrobiol. Beih. Ergebn. Limnol. 18: 5-59.

Boström, B., G. Persson & B. Broberg, 1988b. Bioavailability of different phosphorus forms in freshwater systems. Hydrobiologia 170: 133-155.

Boström, B. & K. Pettersson, 1982. Different patterns of phosphorus release from lake sediments in laboratory experiments. Hydrobiologia 91: 415-429.

Bowen, S. H., 1981. Digestion and assimilation of periphytic detrital aggregate by *Tilapia mossambica*. Trans. am. Fish. Soc. 110: 241-247.

Boyd, C. E., 1972. Sources of CO_2 for nuisance blooms of algae. Weed Sci. 20: 492-497.

Boyd, C. E., 1976a. Nitrogen fertilizer effects on production of *Tilapia* in ponds fertilized with phosphorus and potassium. Aquaculture 7: 385-390.

Boyd, C. E., 1976b. Chemical and textural properties of muds from different depths in ponds. Hydrobiologia 48: 141-144.

Boyd, C. E., 1976c. Lime requirements and application in fish ponds. FAO Techn. Conf. on Aquaculture, Fishing News Book, Ltd., Farnham, England: 120-122.

Boyd, C. E., 1979. Water quality in warmwater fish ponds. Auburn University, Craftmaster Printers, Opelika, Alabama, 359 pp.

Boyd, C. E., 1981a. Solubility of granular inorganic fertilizers for fish ponds. Trans. am. Fish. Soc. 110: 451-454.

Boyd, C. E., 1981b. Comparison of five fertilization programs for fish ponds. Trans. am. Fish. Soc. 110: 541-545.

Boyd, C. E., 1981c. Liquid fertilizers for fish ponds. Highlights of Agricultural Research 28 (4): 3.

Boyd, C. E. & M. J. Cuenco, 1980. Refinement of the lime requirement procedure for fish ponds. Aquaculture 21: 293-299.

Boyd, C. E. & Y. Musig, 1981. Orhophosphate uptake by phytoplankton and sediment. Aquaculture 22: 165-173.

Boyd, C. E., Y. Musig & L. Tucker, 1981. Effects of three phosphorus fertilizers on phosphorus concentrations and phytoplankton production. Aquaculture 22: 175-180.

Boyd, C. E., E. E. Prather & R. W. Parks, 1975. Sudden mortality of a massive phytoplankton bloom. Weed Sci. 23: 61-67.

Boyd, C. E., R. P. Romaire & E. Johnston, 1978. Predicting early morning dissolved oxygen concentrations in channel catfish ponds. Trans. am. Fish. Soc. 107: 484-492.

Boyd, C. E. & J. W. Sowles, 1978. Nitrogen fertilisation of ponds. Trans. am. Fish. Soc. 107: 737-741.

Boyd, C. E. & C. S. Tucker, 1979. Emergency aeration of fish ponds. Trans. am. Fish. Soc. 108: 299-306.

Boyd, C. E. & L. Tucker, 1980. Determination of filtrable orthophosphate in water from fish ponds. Trans. am. Fish. Soc. 109: 314-318.

Brady, N. C., 1984. The nature and properties of soil. Ninth Edition. Macmillan Publ. Co., New York, 750 pp.

Brandl, Z., B. Desortova, J. Hrbacek, J. Komárková, V. Vyhnalek, J. Seďa & M. Straškraba, 1989. Seasonal changes of zooplankton and phytoplankton and their mutual relations in some Czechoslovak reservoirs. Arch. Hydrobiol. Beih. Ergebn. Limnol. 33: 597-604.

Brett, J. R., 1979. Environmental factors and growth. In W. S. Hoar, D. J. Randall & J. R. Brett. Fish physiology. Vol. VIII. Bioenergetics and growth. Academic Press, N.Y., San Francisco, London: 599-675.

Broberg O. & G. Persson, 1988. Particulate and dissolved phosphorus forms in freshwater: composition and analysis. Hydrobiologia 170: 61-90.

Broberg, O. & K. Pettersson, 1988. Analytical determination of orthophosphate in water. Hydrobiologia 170: 45-59.

Brooks, J. L. & S. I. Dodson, 1965. Predation, body size and the composition of plankton. Science 150: 26-35.

Brown, E. J. & R. F. Harris, 1978. Kinetics of algal transient phosphate uptake and the cell quota concept. Limnol. Oceanogr. 23: 35-40.

Buck, D. H., R. J. Baur & C. R. Rose, 1978. Utilization of swine manure in a polyculture of Asian and North American fishes. Trans. am. Fish. Soc. 107: 216-222.

Buckman, H. O. & N. C. Brady, 1964. The nature and properties of soil. The MacMillan Company, New York, 516 pp.

Burns, C. W., D. J. Forsyth, J. F. Haney, M. R. James, W. Lampert & R. D. Pridmore, 1989. Coexistence and exclusion of zooplankton by *Anabaena minutissima* var. *attenuata* in Lake Rotongaio, New Zealand. Arch. Hydrobiol. Beih. Ergebn. Limnol. 32: 63-82.

Burns, C. W. & J. J. Gilbert, 1986. Effects of daphnid size and density on interference between *Daphnia* and *Keratella cochlearis*. Limnol. Oceanogr. 31: 848-858.

Burns, R. P. & R. R. Stickney, 1980. Growth of *Tilapia aurea* in ponds receiving poultry wastes. Aquaculture 20: 117-121.

Buttner, J. K., 1986. *Corbicula* as a biological filter and polyculture organism in catfish rearing ponds. Progr. Fish-Cult. 48: 136-139.

Callender E., 1982. Benthic phosphorus regeneration in the Potomac River Estuary. Hydrobiologia 91: 431-446.

Capone, D. G. & R. P. Kiene, 1988. Comparison of microbial dynamics in marine and freshwater sediments: contrasts in anaerobic carbon catabolism. Limnol. Oceanogr. 33: 725-749.

Cappenberg, T. E., 1988. Quantification of aerobic and anaerobic carbon mineralization at the sediment-water interface. Arch. Hydrobiol. Beih. Ergebn. Limnol. 31: 307-317.

Cappenberg, T. E., K. A. Hordijk, G. J. Jonkheer & J. P. M. Lauwen, 1982. Carbon flow across the sediment-water interface in Lake Vechten, The Netherlands. Hydrobiologia 91: 161-168.

Carlton, R. G. & R. G. Wetzel, 1988. Phosphorus flux from lake sediments: effect of epipelic algal oxygen production. Limnol. Oceanogr. 33: 562-570.

Carmouze, J.-P. & P. Caumette, 1985. Les effets de la pollution organique sur les biomassses et activités du phytoplancton et des bactéries hétérotrophes dans la lagune Ebrié (Côte d'Ivoire). Rev. Hydrobiol. trop. 18: 183-211.

Carmouze, J. P., C. Arze & J. Quintamilla, 1984. Le lac Titicaca: stratification physique et métabolisme associé. Rev. Hydrobiol. trop. 7: 3-11.

Carpenter, S. R., J. F. Kitchell, J. R. Hodgson, P. A. Cochran, J. J. Elser, M. M. Elser, D. M. Lodge, D. Kretchmer & X. He, 1987. Regulation of lake primary productivity by food web structure. Ecology 68: 1863-1876.

Carrillo, P., L. Cruz-Pizarro & P. Sanchez-Castillo, 1990. Analysis of phytoplankton-zooplankton relationships in an oligotrophic lake under natural and manipulated conditions. Hydrobiologia 200/201: 49-57.

Cattaneo, A., 1987. Periphyton in lakes of different trophy. Can. J. Fish. aquat. Sci 44: 296-303.

Chakroff, M., 1976. Freshwater fish pond culture and management. VITA Publications, Manuel Series, No. 36 E, Mt. Rainier, USA, 191 pp.

Chang, W. Y. B. & H. Ouyang, 1988. Dynamics of dissolved oxygen and vertical circulation in fish ponds. Aquaculture 74: 263-276.

Chattopadhyay, G. N. & L. N. Mandal, 1980. Effects of different levels of water salinity on the decomposition of organic manures in a brackish water fish pond soil. Hydrobiologia 72: 287-292.

Chen, F. Y. & G. A. Prowse, 1964. The effect of living space on the growth rate of fish. Ichthyologica 3: 11-20. Quoted by Edwards *et al.* (1981)

Chrost, R. J., 1989. Characterization and significance of ß-glucosidase activity in lake water. Limnol. Oceanogr. 34: 660-672.

Chrost, R. J., W. Siuda & G. Z. Halemejko, 1974. Longterm studies on alkaline phosphatase activity (APA) in a lake with fish-aquaculture in relation to lake eutrophication and phosphorus cycle. Arch. Hydrobiol./Suppl. 70: 1-32.

Coche, A., 1986. Pisciculture continentale - Le sol. Méthodes simples pour l'aquaculture. Collection FAO: Formation N° 6. FAO, Rome, 174 pp.

Coche, A. G. & H. Van der Wal, 1981. Water for freshwater fish culture. FAO Training Series No. 4. FAO, Rome, 110 pp.

Cole, B. A. & C. E. Boyd, 1986. Feeding rate, water quality and channel catfish production in ponds. Progr. Fish-Cult. 48: 25-29.

Compton, R. G., K. L. Pritchard & P. R. Umoin, 1989. The dissolution of calcite in acid water: mass transport versus surface control. Freshwat. Biol. 22: 285-288.

Cook, R. B., C. A. Kelly, D. W. Schindler & M. A. Turner, 1986. Mechanisms of hydrogen ion neutralization in an experimentally acidified lake. Limnol. Oceanogr. 31: 134-148.

Cooke, J. G. & R. E. White, 1987a. The effect of nitrate in streamwater on the relationship between denitrification and nitrification in a stream-sediment microcosm. Freshwat. Biol. 18: 213-226.

Cooke, J. G. & R. E. White, 1987b. Spatial distribution of denitrifying activity in a stream draining an agricultural catchment. Freshwat. Biol. 18: 509-519.

Costa-Pierce, B. A., 1990. Letter to the Editor. Aquabyte 3 (2): 18-19.

Costa-Pierce, B. A. & D. B. Craven, 1987. Estimating microbial production and growth rates in aquaculture ponds using rates of RNA and DNA synthesis. Aquaculture 66: 69-78.

Costa-Pierce, B. A., S. P. Malecha & E. A. Laws, 1985. Effects of polyculture and manure fertilization on water quality and heterotrophic productivity in *Macrobrachium rosenbergii* ponds. Trans. am. Fish. Soc. 114: 826-836.

Couillard, D. & Y. Lefebvre, 1985. Analysis of water quality indices. J. environ. Mgt 21: 161-179.

Crisman, T. L. & J. R. Beaver, 1990. Applicability of planktonic biomanipulation for managing eutrophication in the subtropics. Hydrobiologia 200/201: 177-185.

Cuenco, M. L., R. R. Stickney & W. E. Grant, 1985. Fish bioenergetics and growth in aquaculture ponds : III. Effects of intraspecific competition, stocking rate, stocking size and feeding rate on fish productiity. Ecol. Modelling 28: 73-95.

Cuffney, T. F., J. B. Wallace & G. J. Lugthart, 1990. Experimental evidence quantifying the rôle of benthic invertebrates in organic matter dynamics of headwater streams. Freshwat. Biol. 23: 281-299.

Cuker, B. E., 1987. Field experiment on the influence of suspended clay and P on the plankton of a small lake. Limnol. Oceanogr. 32: 840-847.

Currie, D. J., 1984. Microscale nutrient patches: do they matter to the phytoplankton. Limnol. Oceanogr. 29: 211-214.

Currie, D. J., E. Bentzen & J. Kalff, 1986. Does algal-bacterial phosphorus partioning vary among lakes? A comparative study of orthophosphate uptake and alkaline phosphatase activity in freshwater. Can. J. Fish. aquat. Sci 43: 311-318.

Currie, D. J. & J. Kalff, 1984a. A comparison of the abilities of freshwater algae and bacteria to acquire and retain phosphorus. Limnol. Oceanogr. 29: 298-310.

Currie, D. J. & J. Kalff, 1984b. The relative importance of bacterioplankton and phytoplankton in phosphorus uptake in freshwater. Limnol. Oceanogr. 29: 311-321.

Davison, W., 1982. Transport of iron and manganese in relation to the shapes of their concentration depth profiles. Hydrobiologia 91: 463-471.

Dawidowicz, P., 1990a. The effect of *Daphnia* on filament length of blue-green algae. Hydrobiologia 191: 265-268.

Dawidowicz, P., 1990b. Effectiveness of phytoplankton control by large-bodied and small-bodied zooplankton. Hydrobiologia 200/201: 43-47.

Degani, G., C. Dosoretz & D. Levanon, 1984. The influence of cow manure on growth rates of *Oreochromis aureus* and *Clarias lazera* in Israel small outdoor tanks. Bamidgeh 36: 114-120.

De Haan, H. & T. De Boer, 1986. Geochemical aspects of aqueous iron, phosphorus and dissolved organic carbon in the humic Lake Tjeukemeer, The Netherlands. Freshwat. Biol. 16: 661-672.

dela Cruz, C. R., 1983. Fishpond engineering: A technical manual for small- and medium scale coastal fish farms in Southeast Asia. South China Sea Fisheries Development and Coordination Programme, SCS Manual No. 5, Manila, Philippines, 180 pp.

Delmendo, M. N., 1980. A review of integrated live-stock-fowl-fish farming systems. In Pullin, R. S. V. & Z. H. Shehadeh (eds), Integrated agriculture- aquaculture farming system. I.C.L.A.R.M., Manila, Philippines. Conf. Proc. 4: 59-71.

De Pinto, J. V., T. C. Young, J. S. Bonner & P. W. Rodgers, 1986. Microbial recycling of phytoplankton phosphorus. Can. J. Fish. aquat. Sci 43: 336-342.

Dévai, G., 1990. Ecological background and importance of the change of chironomid fauna (Diptera: Chironomidae) in shallow Lake Balaton. Hydrobiologia 191: 189-198.

Diab, S. & M. Shilo, 1986. Transformation of nitrogen in sediments of fish ponds in Israel. Bamidgeh 38: 67-88.

Diab, S. & M. Shilo, 1988. Effects of on the activity and survival of *Nitrosomonas sp.* and *Nitrobacter sp.* isolates from fish ponds. Israel. J. Aquacult. 40: 50-56.

Dokulil, M., A. Herzig & A. Jagsch, 1990. Trophic relationships in the pelagic zone of Mondsee, Austria. Hydrobiologia, 191: 199-212.

Doremus, C. & L. S. Cleresci, 1982. Microbial metabolism in surface sediments and its rôle in the immobilization of phosphorus in oligotrophic lake sediments. Hydrobiologia 91: 261-268.

Drenner, R. W., J. R. Mummert, F. de Noyelles Jr. & D. Kettle, 1984a. Selective particle ingestion by a filter-feeding fish and its impact on phytoplankton community structure. Limnol. Oceanogr. 29: 941-948.

Drenner, R. W., F. de Noyelles Jr. & D. Kettle, 1982a. Selective impact of filter-feeding gizzard shad on zooplankton community structure. Limnol. Oceanogr. 27: 965-968.

Drenner, R. W., W. J. O'Brien & J. R. Mummer, 1982b. Filter-feeding rates of gizzard shad. Trans. am. Fish. Soc. 111: 210-215.

Drenner, R. W., S. B. Taylor, X. Lazzaro & D. Kettle, 1984b. Particle grazing and plankton community impact of an omnivorous cichlid. Trans. am. Fish. Soc. 113: 397-402.

Droop, M. R., 1974. The nutrient status of algal cells in continuous culture. J. Mar. Biol. Assoc. U.K. 54: 825-855.

Dumont, H. J., 1991. The predation-controlled uniqueness of tropical zooplankton: less Cladoceran diversity, but more Limnomedusae, flatworms, mites and numerous prey defenses. Hydrobiologia (in press).

Dumont, H. J. & L. De Meester, 1990. Are contrasting patterns of vertical migration in zooplankton the result of differential natural selection? Rev. Brasil. Biol. 50: 867-874.

Dussart, B. H., 1980. Copépodes. In Durand, J.-R. & C. Lévêque (eds), Flore et faune aquatiques de l'Afrique sahélo-soudanienne, 1. O.R.S.T.O.M., Paris. Documn Techqs 44: 333-356.

Edwards, P., 1980. A review of recycling organic wastes into fish, with emphasis on the tropics. Aquaculture 21: 261-279.

Edwards, P., O.A. Sinchumpasak & M. Tabucanon, 1981. The harvest of microalgae from the effluent of a sewage fed high rate stabilisation pond by *Tilapia nilotica*. Part 2 : Studies of the fish ponds. Aquaculture 23: 107-147.

Eggers, D. M., 1982. Planktivore preference by prey size. Ecology 63: 381-390.

Einsele, W., 1936. Über die Beziehungen des Eisenkreislaufs zum Phosphorkreislauf im eutrophen See. Arch. Hydrobiol. 29: 664-686.

Ejsmont-Karabin, J., M. Bubień & T. Węgleńska, 1989. The effects of zooplankton on phosphorus cycling in man-made Lake Zegrzyński. Arch. Hydrobiol. Beih. Ergebn. Limnol. 31: 493-501.

Ejsmont-Karabin, J. & I. Spodniewska, 1990. Influence on phytoplankton biomass in lakes of different trophy by phosphorus in lake water and its regeneration by zooplankton. Hydrobiologia 191: 123-128.

El Samra, M. I. & J. Oláh, 1979. Significance of nitrogen fixation in fish ponds. Aquaculture 18: 367-372.

Elser, J. J., H. J. Carney & C. R. Goldman, 1990. The zooplankton-phytoplankton interface in lakes of contrasting trophic status: an experimental comparison. Hydrobiologia 200/201: 69-82.

Elser, J. J., M. M. Elser, N. A. MacKay & S. R. Carpenter,

1988. Zooplancton- mediated transitions between N- and P-limited algal growth. Limnol. Oceanogr. 33: 1-14.

Elser, J. J. & N. A. MacKay, 1989. Experimental evaluation of effects of zooplankton biomass and size distribution on algal biomass and productivity in three nutrient-limited lakes. Arch. Hydrobiol. 114: 481-496.

Emerson, K., R. C. Russo, R. E. Lund & R. V. Thurston, 1975. Aqueous ammonia equilibrium calculations : effects of pH and temperature. J. Fish. Res. Bd Can. 32: 2379-2383.

Enell, M. & S. Löfgren, 1988. Phosphorus in interstitial water: methods and dynamics. Hydrobiologia 170: 103-132.

Eren, Y., T. Tsur & Y. Avnimelech, 1977. Phosphorus fertilization of fish ponds in Upper Galilee. Bamidgeh 29: 87-92.

Erez, J., M. D. Krom & T. Neuwirth, 1990. Daily oxygen variations in marine fish ponds, Elat, Israel. Aquaculture 84: 289-305.

Evans, T. G., 1988. A framework for discussing seasonal succession and coexistence of phytoplankton species. Limnol. Oceanogr. 33: 1027-1036.

FAO, 1967. Actes du symposium mondial FAO sur la pisciculture en étang à temperature élevée, Rome (Italie), 18-25 mai 1966. Vol. 1 - Rapport. FAO Fish. Rep. (Fr.), 44, Vol.1: 57 pp.

FAO, 1976. Report of the Symposium on Aquaculture in Africa. Accra, Ghana, 30 September - 2 October 1975. CIFA Tech. Pap., 4: 36 pp.

FAO, 1981. Développement de l'aquaculture continentale en Chine. FAO Doc. Tech. Pêches, 215: 152 pp.

Fairchild, G. W. & A. C. Everett, 1988. Effects of nutrient (N, P, C) enrichment upon periphyton standing crop, species composition and primary production in an oligotrophic softwater lake. Freshwat. Biol. 19: 57-70.

Faybenro, O. A. & D. H. J. Sydenham, 1988. Evaluation of *Clarias isheriensis* (Sydenham) under semi-intensive management in ponds. Aquaculture 74: 287-291.

Fernando, C. H., 1980. The freshwater zooplankton of Sri Lanka with a discussion of tropical freshwater zooplankton composition. Int. Revue ges. Hydrobiol. 65: 85-125.

Fernando, C. H., C. Tudorancea & S. Mengestou, 1990. Invertebrate zooplankton predator composition and diversity in tropical lentic waters. Hydrobiologia 198: 13-31.

Feuillade, M. & J. Feuillade, 1990. Heterotrophic capabilities of the blue-green algae *Oscillatoria rubescens*. Arch. Hydrobiol. 117: 61-76.

Fish, G. R., 1955. The food of Tilapia in East Africa. Uganda J. 19: 85-89. Quoted by Trewawas (1983).

Flynn, K. J., 1988. The concept of "primary production" in aquatic ecology. Limnol. Oceanogr. 33: 1215-1216.

Folt, C. L., 1987. An experimental analysis of costs and benefits of zooplankton aggregation. In W. C. Kerfoot & A. Sih (eds), Predation: direct and indirect impacts on aquatic communities. The University Press of New England, Hanover (N.H.); Lond.: 300-332.

Fong, S.-C. & H. P. Ju, 1987. Energy value of biomass within benthic algae of milkfish ponds. Aquaculture 64: 31-38.

Förstner, U., 1982. Accumulative phases for heavy metals in limnic sediments. Hydrobiologia 91: 269-284.

Forsyth, D. J., M. R. James & M. Cryer, 1990. Alteration of seasonal and diel patterns in vertical migration of zooplankton by *Anabaena* and planktivorous fish. Arch. Hydrobiol. 117: 385-404.

Foy, R. H. & A. G. Fitzsimons, 1987. Phosphorus inactivation in a eutrophic lake by the direct addition of ferric aluminium sulphate: changes in phytoplankton populations. Freshwat. Biol. 17: 1-17.

Fitzgerald, G. P., 1970. Aerobic lake muds for the removal of phosphorus from lake water. Limnol. Oceanogr. 15: 550-555.

Fraleigh, P. C. & J. C. Burnham, 1988. Myxococcal predation on cyanobacterial populations: nutrient effects. Limnol. Oceanogr. 33: 476-486.

Francko, D. A., 1985. Phytoplankton metabolism and cyclic nucleotides. II. Nucleotide-induced perturbations of alkaline phosphatase activity. Arch. Hydrobiol. 100: 409-421.

Franko, D. A., 1986. Epilimnetic phosphorus cycling : influence of humic materials and iron on coexisting major mechanism. Can. J. Fish. aquat. Sci 43: 302-310.

Friday, L. E., 1987. The diversity of macroinvertebrate and macrophyte community in ponds. Freshwat. Biol. 18: 87-104.

Froelich, P. N., 1988. Kinetic control of dissolved phosphate in natural rivers and estuaries: a primer on the phosphate buffer mechanism. Limnol. Oceanogr. 33: 649-668.

Fryer, G., 1987. Quantitative and qualitative: numbers and reality in the study of living organisms. Freshwat. Biol. 17: 177-189.

Fukuhara, H. & M. Sakamoto, 1988. Ecological significance of bioturbation of zoobenthos community in nitrogen release from bottom sediments in a shallow eutrophic lake. Arch. Hydrobiol. 113: 425-445.

Fulton, R. S. III., 1988. Grazing on filamentous algae by herbivorous zooplankton. Freshwat. Biol. 20: 263-271.

Fulton, R. S. III & H. W. Paerl, 1987. Effect of colonial morphology on zooplankton utilization of algal resources during blue-green algal (*Microcystis aeruginosa*) blooms. Limnol. Oceanogr. 32: 634-644.

Gächter, R., J. S. Meyer & A. Mares, 1988. Contribution of bacteria to release and fixation of phosphorus in lake sediments. Limnol. Oceanogr. 33: 1542-1558.

Gage, M. A. & E. Gorham, 1985. Alkaline phosphatase activity and cellular phosphorus as an index of the phosphorus status of phytoplankton in Minnesota lakes. Freshwat. Biol. 15: 227-233.

Ganf, G. G. & A. J. Horne, 1975. Diurnal stratification, photosynthesis and nitrogen-fixation in a shallow equatorial Lake (Lake George, Uganda). Freshwat. Biol. 5: 13-59.

Gaardner, T. & H. H. Gran, 1927. Investigations of the production of planktion in the Oslo fjord. Rapp. Proc.-Verb. Cons. Int. Expl. Mer 42: 3-48.

Gardner, W. S., T. F. Nalepa & J. M. Malczyk, 1987. Nitrogen mineralization and denitrification in Lake Michigan sediments. Limnol. Oceanogr. 32: 1226-1238.

Gawler, M., G. Balvay, P. Blanc, J. C. Druart & J. C. Pelletier, 1988. Plankton ecology of Lake Geneva: a test of the PEG-model. Arch. Hydrobiol. 114: 161-174.

Gerhart, D. Z. & G. E. Likens, 1975. Enrichement experiments for determining nutrient limitation : four methods compared. Limnol. Oceanogr. 20: 649-653.

Gerritsen J. & S. W. Bradley, 1987. Electrophoretic mobility of natural particles and cultural organisms in freshwater. Limnol. Oceanogr. 32: 1049-1058.

Gilbert, J. J., 1989. The effect of *Daphnia* interference on a natural rotifer and ciliate community: short-term bottle experiment. Limnol. Oceanogr. 34: 606-617.

Gilbert, J. J., 1990. Differential effects of *Anabaena affinis* on cladocerans and rotifers: mechanisms and implications. Ecology 71: 1727-1740.

Gliwicz, Z. M., 1985. Predation or food limitation: an ultimate reason for extinction of planktonic cladoceran species. Arch. Hydrobiol. Beih. Ergebn. Limnol. 21: 419-430.

Gliwicz, Z. M., 1990. Why do cladocerans fail to control algal blooms? Hydrobiologia 200/201: 83-97.

Gliwicz, Z. M., A. Ghilarov & J. Pijanowska, 1981. Food and predation as major factors limiting two natural populations of *Daphnia cucullata* Sars. Hydrobiologia 80: 155-177.

Golachowska, J., 1986. Diurnal fluctuations of phosphorus forms in lake water and seston in Pluß-See. Pol. Arch. Hydrobiol. 33: 165-175.

Goldman, J. C., D. A. Caron & M. R. Dennett, 1987. Regulation of gross efficiency and ammonium regeneration in bacteria by substrate C:N ratio. Limnol. Oceanogr. 32: 1239-1252.

Golterman, H. L., 1982. Loading concentration models for phosphate in shallow lakes. Hydrobiologia 91: 169-174.

Golterman, H. L., 1988. Reflections on fractionation and bioavailability of sediment bound phosphate. Arch. Hydrobiol. Beih. Ergebn. Limnol. 30: 1-4.

Golterman, H. L. & F. A. Kouwe, 1980. Chemical budgets and nutrient pathways. In E. D. Le Cren & R. H. Lowe-McConnell (eds), The functioning of freshwater ecosystems. Cambridge University Press, Cambridge. International Biological Programme 22: 85-140.

Gophen, M., 1990. Biomanipulation: retrospective and future development. Hydrobiologia 200/201: 1-11.

Gophen, M., R. Drenner, G. Vinyard & P. Spaturu, 1983. The cichlid fishery of Lake Kinneret: history and management recommendations. In L. Fishelson & Z. Yaron (eds), Tel Aviv University, Tel Aviv, Israel. Proc. Int. Symp. on Tilapia in Aquaculture: 1-7.

Green, B. W., R. P. Phelps & H. R. Alvarenga, 1989. The effect of manures and chemical fertilizers on the production of *Oreochromis niloticus* in earthen ponds. Aquaculture 76: 37-42.

Grover, J. P., 1989a. Phosphorus-dependent growth kinetics of eleven species of freshwater algae. Limnol. Oceanogr. 34: 341-348.

Grover, J. P., 1989b. Effects of Si:P supply ratios, supply variability, and selective grazing in the plankton: an experiment with a natural algal and protistan assemblage. Limnol. Oceanogr. 34: 349-367.

Güde, H., 1988. Direct and indirect influence of crustacean zooplankton on bacterioplankton of Lake Constance. Hydrobiologia 159: 63-73.

Guiral, D., 1984. Devenir de la matière organique particulaire dans un milieu eutrophe tropical. Rev. Hydrobiol. trop. 17: 191-206.

Guiral, D., R. Arfi & J. P. Torréton, 1989. Mécanismes et incidences écologiques de l'homogénéisation annuelle de densité dans un milieu eutrophe stratifié. Hydrobiologia 183: 195-210.

Gulati, R. D., 1990. Structural and grazing responses of zooplankton community to biomanipulation of some Dutch water bodies. Hydrobiologia 200/201: 99-118.

Gunatilaka, A., 1982. Phosphate adsorption kinetics of resuspended sediments in a shallow lake, Neusiedlersee, Austria. Hydrobiologia 91: 293-298.

Hall, P. O. J., L. G. Anderson, M. M. Rutgers van der Loeff, B. Sundby & S. F. G. Westerlund, 1989. Oxygen uptake kinetics in the benthic boundary layer. Limnol. Oceanogr. 34: 734-746.

Hansson, L. A., 1988. Effects of competitive interactions on the biomass development of planktonic and periphytic algae in lakes. Limnol. Oceanogr. 33: 121-128.

Hansson, L. A., 1989. The influence of a periphytic biolayer on phosphorus exchange between substrate and water. Arch. Hydrobiol. 115: 21-26.

Hansson, L. A., L. Johansson & L. Persson, 1987. Effects of fish grazing on nutrient release and succession of primary producers. Limnol. Oceanogr. 32: 723-729.

Hamilton, D. T. & W. D. Taylor, 1987. Short-term effects of zooplankton manipulations on phosphate uptake. Can. J. Fish. aquat. Sci 45: 1038-1044.

Hardy, F. J., K. S. Shortreed & J. G. Stockner, 1986. Bacterioplankton, phytoplankton and zooplankton communities in a British Columbia coastal lake before and after nutrient reduction. Can. J. Fish. aquat. Sci 43: 1504-1514.

Harris, G. H., 1980. The measurement of photosynthesis in natural populations of phytoplankton. In I. Morris (ed.), The physiological ecology of phytoplankton. Blackwell Scientific Publication, Oxford. Studies in Ecology, Volume 7: 129-187.

Harrison, W. G., 1978. Experimental measurements of nitrogen remineralization in coastal waters. Limnol. Oceanogr. 23: 684-694.

Hart, B. T., 1982. Uptake of trace metals by sediments and suspended particulates: a review. Hydrobiologia 91: 299-313.

Hasselrot, B. & H. Hultberg, 1984. Liming of acidified Swedish lakes and streams and its consequences for aquatic ecosystems. Fisheries 9: 4-9.

Havel, J. E., 1987. Predator-induced defenses: a review. In W. C. Kerfoot & A. Sih (eds), Predation: direct and indirect impacts on aquatic communities. The University Press of New England, Hanover (N.H.); Lond.: 263-278.

Hayes, F. R. & J. E. Philips, 1958. Lake water and sediment : IV. Radiophosphorus equilibrium with mud, plants and bacteria under oxidized and reduced conditions. Limnol. Oceanogr. 3: 459-475.

Healey, F. D., 1978. Physiological indicators of nutrient deficiency in algae. Mitt. Int. Ver. Theor. Angew. Limnol. 21: 34-41.

Healey, F. P., 1985. Interacting effects of light and nutrient limitation on the growth rate of *Synechococcus linearis* (Cyanophyceae). J. Phycol. 21: 134-146. Quoted by Jansson *et al.* (1988)

Heath, R. T., 1986. Dissolved organic phosphorus compounds: do they satisfy planktonic phosphate demand in summer? Can. J. Fish. aquat. Sci 43: 343-350.

Hecky, R. E. & P. Kilham, 1988. Nutrient limitation of phytoplankton in freshwater and marine environments: a review of recent evidence on the effects of enrichment. Limnol. Oceanogr. 33: 796-822.

Hecky, R. E., R. H. Spigel & G. W. Coulter, 1991. The nutrient regime. In G. W. Coulter (ed.), Lake Tanganika and its life. Oxford Univ. Press, London, 76-89.

Heinen, J. M., G. C. G. Fornshell & P. J. Foley, 1988. Removal of floating filamentous algae from ponds with a shallow seine. Progr. Fish-Cult. 50: 187.

Hendrey, G. R. & F. R. Wright, 1976. Acid precipitation in Norway: Effects on aquatic fauna. J. Great Lakes Res. 2 (Suppl. 1): 192-207. Quoted by Mulholland *et al.*, 1986.

Hepher, B., 1958. On the dynamics of phosphorus added to fishponds in Israel. Limnol. Oceanogr. 3: 84-100.

Hepher, B., 1962. Primary production in fishponds and its application to fertilization experiments. Limnol. Oceanogr. 6: 131-136.

Hepher, B., 1966. Some aspects of the phosphorus cycle in fishponds. Verh. int. Ver. Limnol. 16: 1293-1297.

Hessen, D. O., T. Andersen & A. Lyche, 1989. Differential grazing and resource utilization of zooplankton in a humic lake. Arch. Hydrobiol. 114: 321-347.

Hessen, D. O., T. Andersen & A. Lyche, 1990. Carbon metabolism in a humic lake: pool sizes and cycling through zooplankton. Limnol. Oceanogr. 35: 84-99.

Heyman, U. & A. Lundgren, 1988. Phytoplankton biomass and production in relation to phosphorus: some conclusions from field studies. Hydrobiologia 170: 211-227.

Hickling, C. F., 1962. Fish Culture. Faber & Faber, London, 295 pp.

Hieltjes, A. H. M. & L. Lijklema, 1980. Fractionation of inorganic phosphates in calcareous sediments. J. environ. Qual. 9: 405-407.

Hillbricht-Ilkowska, A. & T. Weglenska, 1973. Experimentally increased fish stock in the pond type Lake Warniak. VII. Numbers, biomass and production of zooplankton. Ekol. pol. 21: 533-552.

Hino, S., 1988. Extracellular release of organic matter associated with the physiological state of freshwater blue-green alga. Arch. Hydrobiol. 113: 307-317.

Hogendoorn, H., 1983. Growth and production of the African catfish, *Clarias lazera* (C. & V.) III. Bioenergetic relations of body weight and feeding level. Aquaculture 35: 1-17.

Hollerman, W. D. & C. E. Boyd, 1985. Effects of draining on water quality and production of channel catfish in ponds. Aquaculture 46: 45-54.

Hollerman, W. D. & C. E. Boyd, 1986. Effects of annual draining on the limnology of sunfish ponds. Progress. Fish-Cult. 48: 120-124.

Holtan, H., L. Kamp-Nielsen & A. O. Stuanes, 1988. Phosphorus in soil, water and sediment: an overview. Hydrobiologia 170: 19-34.

Horne, A. J. & M. L. Commins, 1989a. Phytoplanktonic monotony. I. Low spatial heterogeneity for nitrogen fixation, algae, nutrients, and temperature, but not for zooplankton, in Lake Rotongaio, New Zealand. Arch. Hydrobiol. Beih. Ergebn. Limnol. 32: 83-97.

Horne, A. J. & M. L. Commins, 1989b. Phytoplanktonic monotony. II. Patterns of temporal stability in phytoplankton patches in Lake Rotongaio, New Zealand. Arch. Hydrobiol. Beih. Ergebn. Limnol. 32: 99-112.

Horne, A. J. & A. B. Viner, 1971. Nitrogen fixation and its significance in tropical lake George, Uganda. Nature 232: 417-418.

Houk, V., 1989. Seasonal changes of phytoplankton development near the dam of the Zelivka-Reservoir (Czechoslovakia). Arch. Hydrobiol. Beih. Ergebn. Limnol. 33: 397-407.

Howard-Williams, C., 1985. Cycling and retention of nitrogen and phosphorus in wetlands: a theoretical and applied perspective. Freshwat. Biol. 15: 391-431.

Howarth, R. W., R. Marino, J. Lane & J. J. Cole, 1988a. Nitrogen fixation in freshwater, estuaries, and marine ecosystems. 1. Rates and importance. Limnol. Oceanogr. 33: 669-687.

Howarth, R. W., R. Marino & J. J. Cole, 1988b. Nitrogen fixation in freshwater, estuaries, and marine ecosystems. 2. Biogeochemical controls. Limnol. Oceanogr. 33: 688-701.

Hrbáček, J., 1962. Species composition and the amount of zooplankton in relation to the fish stock. Rozpr. ČSAV, Ser. mat. nat. sci. 72: 1-117.

Huet, M. & J. A. Timmermans, 1970. Traité de pisciculture. Ed. Ch. de Wijngaert, Bruxelles, 718 pp.

Hul, M., 1987. The effect of domestic sewage on the structure of the microbenthic ciliate communities in the Łyna River. Pol. Arch. Hydrobiol. 34: 567-578.

Humphries, S. E. & V. D. Lyne, 1988. Cyanophyte blooms: the role of cell buoyancy. Limnol. Oceanogr. 33: 79-91.

Hunt, D. & C. E. Boyd, 1981. Alkalinity losses from ammonium fertilizers used in fish ponds. Trans. am. Fish. Soc. 110: 81-85.

Huntsman, S. A. & W. G. Sunda, 1980. The role of trace metals in regulating phytoplankton growth. In I. Morris (ed.), The physiological ecology of phytoplankton. Blackwell Scientific Publication, Oxford. Studies in Ecology Volume 7: 285-328.

Hutchinson, G. E., 1961. The paradox of the plankton. Am. Nat. 95: 137-145.

Iltis, A., 1980. Les algues. In J.-R. Durand & C. Lévêque (eds), Flore et faune aquatiques de l'Afrique sahélo-soudanienne. O.R.S.T.O.M., Paris. Documn Techqs 44: 9-61.

Irvine, K., B. Moss & J. Stansfield, 1990. The potential of artificial refugia for maintaining a community of large-bodied cladocera against fish predation in a shallow eutrophic lake. Hydrobiologia 200/201: 379-389.

Jackson, G. A., 1989. Simulation of bacterial attraction and adhesion to falling particles in an aquatic environment. Limnol. Oceanogr. 34: 514-530.

Jacquet, J.-M., E. Davaud, F. Rapin & J.-P. Vernet, 1982. Basic concepts and associated statistical methodology in the geochemical study of lake sediments. Hydrobiologia 91: 139-146.

Jana, B. B., S. K. Roy & G. N. Patel, 1983. Relationship between the nitrification and denitrification and the fish growth in the fish ponds under polyculture system. Pol.

Arch. Hydrobiol. 30: 65-68.

Jana, B. B. & U. K. De, 1983. Primary productivity of phytoplankton, environmental covariates and fish growth in West Bengal ponds under a polyculture system. Int. Revue ges. Hydrobiol. 68: 45-58.

Jansson, M., 1976. Phosphatases in lake water: Characterization of enzymes from phytoplankton and zooplankton by gel filtration. Science 194: 320-321. Quoted by Jansson *et al.*, 1988.

Jansson, M., 1988. Phosphate uptake and utilization by bacteria and algae. Hydrobiologia 170: 177-189.

Jansson, M., H. Olsson & K. Pettersson, 1988. Phosphatases: origin, characteristics and function in lakes. Hydrobiologia 170: 157-175.

Jeffries, H. J. & J. H. Lawton, 1985. Predator-prey ratios in communities of freshwater invertebrates: the role of enemy free space. Freshwat. Biol. 15: 105-112.

Jeppesen, E., 1982. Diurnal variation in the oxygen uptake of river sediments *in* by use of continuous flow-through systems. Hydrobiologia 91: 189-195.

Jeppesen, E., J. P. Jensen, P. Kristensen, M. Søndergaard, E. Mortensen, O. Sortkjær & K. Obrik, 1990a. Fish manipulation as a lake restoration tool in shallow, eutrophic temperate lakes. 2: Threshold levels, long-term stability and conclusions. Hydrobiologia 200/201: 219-227.

Jeppesen, E., M. Søndergaard, E. Mortensen, P. Kristensen, B. Riemann, H. J. Jensen, J. P. Müller, O. Sortkjær, J. P. Jensen, K. Christoffersen, S. Bosselmann & E. Dall, 1990b. Fish manipulation as a lake restoration tool in shallow, eutrophic temperate lakes. 1: Cross-analysis of three Danish case-studies. Hydrobiologia 200/201: 205-218.

Jeppesen, E., M. Søndergaard, O. Sortkjær, E. Mortensen & P. Kristensen, 1990c. Interactions between phytoplankton, zooplankton and fish in a shallow, hypertrophic lake: a study on phytoplankton collapses in Lake Sobygård, Denmark. Hydrobiologia 191: 139-148.

John, D. M., 1986. The inland waters of tropical West Africa. Arch. Hydrobiol. Beih. Ergebn. Limnol. 23: 1-244.

Johnsen, G. H. & P. J. Jakobsen, 1987. The effect of food limitation on vertical migration in *Daphnia longispina*. Limnol. Oceanogr. 32: 873-880.

Johnson, R. K., B. Boström & W. van de Bund, 1989. Interactions between *Chironomus plumosus* (L.) and the microbial community in the surficial sediments of a shallow, eutrophic lake. Limnol. Oceanogr. 34: 992-1003.

Jones, R. I., 1990. Phosphorus transformations in the epilimnion of humic lakes: biological uptake of phosphate. Freshwat. Biol. 23: 323-337.

Jones, R. I., K. Salonen & H. De Haan, 1988. Phosphorus transformations in the epilimnion of humic lakes: abiotic interactions between dissolved humic materials and phosphate. Freshwat. Biol. 19: 357-369.

Jørgensen, B. B. & J. Sørensen, 1985. Seasonal cycles of O_2, NO_3 and SO_4^{2-} reduction maximum in spring. Mar. Ecol. Prog. Ser. 24: 65-74.

Jurkowska, J., 1987. Estimation of phytoplankton biomass and production, and organic matter decomposition in three small field-ponds. Pol. Arch. Hydrobiol. 34: 35-49.

Kaiser, G. E. & F. W. Wheaton, 1983. Nitrification filters for aquatic culture systems : State of art. J. World Maricul. Soc. 14: 302-324.

Kajak, Z., 1988. Considerations on benthos abundance in freshwaters, its factors and mechanism. Int. Revue ges. Hydrobiol. 73: 5-19.

Kajak, Z., J. I. Rybak, I. Spodniewska & W. A. Godlewska-Lipowa, 1975. Influence of the planktivorous fish, *Hypothalmichthys molitrix* (Val.) on the plankton and benthos of the eutrophic lake. Pol. Arch. Hydrobiol. 22: 301-310.

Kamp-Nielsen, L., H. Meyer & S. E. Jørgensen, 1982. Modelling the influence of bioturbation on the vertical distribution of sedimentary phosphorus in L. Esrom. Hydrobiologia 91: 197-206.

Kaspar, H. F., 1985. The denitrification capacity of sediment from a hypereutrophic lake. Freshwat. Biol. 15: 449-453.

Kelderman, P. & A. M. van de Repe, 1982. Temperature dependence of sediment- water exchange in Lake Grevelingen, S.W. Netherlands. Hydrobiologia 91: 489-490.

Kelly, C. A. & J. W. M. Rudd, 1989. Reply to comment by Urban and Baker. Limnol. Oceanogr. 34: 1146-1148.

Kelly, C. A., J. W. M. Rudd & D. W. Schindler, 1988. Carbon and electron flow via methanogenesis, SO_4^{2-}, NO_3^-, Fe^{3+}, and Mn^{4+} reduction in the anoxic hypolimnia of three lakes. Arch. Hydrobiol. Beih. Ergebn. Limnol. 31: 333-344.

Kerfoot, W. C., 1987. Cascading effects and indirect pathways. In W. C. Kerfoot & A. Sih (eds), Predation: direct and indirect impacts on aquatic communities. The University Press of New England, Hanover (N.H.); Lond.: 57-70.

Kerfoot, W. C., D. L. Kellogg, Jr & J. R. Strickler, 1980. Visual observations of live zooplankton: evasion, escape, and chemical defenses. In W. C. Kerfoot (ed.), Evolution and ecology of zooplankton communities. Special Symp. Vol. 3, Am. Soc. Limnol. Oceanogr., University Press of New England, Hanover, New Hampshire, London: 10-27.

Kerekes, J. J., A. C. Blouin & S. T. Beauchamp, 1990. Trophic response to phosphorus in acidic and non-acidic lakes in Nova Scotia, Canada. Hydrobiologia 191: 105-110.

Ketchum, B. & A. C. Redfield, 1949. Some physical and chemical characteristics of algal growth in mass culture. J. cell. comp. Physiol. 33: 281-299.

Kilham, P., 1990. Mechanisms controlling the chemical composition of lakes and rivers: data from Africa. Limnol. Oceanogr. 35: 80-83.

Kilham, P. & R. E. Hecky, 1988. Comparative ecology of marine and freshwater phytoplankton. Limnol. Oceanogr. 33: 776-795.

Kilham, P. & S. S. Kilham, 1980. The evolutionary ecology of phytoplankton. In I. Morris (ed.), The physiological ecology of phytoplankton. Blackwell Scientific Publication, Oxford. Studies in Ecology Volume 7: 571-597.

Kilham, P. & S. S. Kilham, 1990. Endless summer: internal loading processes dominate nutrient cycling in tropical lakes. Freshwat. Biol. 23: 379-389.

Kilham, P., S. S. Kilham & R. E. Hecky, 1986. Hypothesized resource relationships among African planktonic diatoms. Limnol. Oceanogr. 31: 1169-1181.

Kilham, S. S., 1986. Dynamics of lake Michigan natural

phytoplankton communities in continuous cultures along a Si:P loading gradient. Can. J. Fish. aquat. Sci. 43: 351-360.

Kirk, J. T. O., 1986. Light and photosynthesis in aquatic ecosystems. Cambridge University Press, Cambridge, London, N.Y., 401 pp.

Kirk, K. L. & J. J. Gilbert, 1990. Suspended clay and the population dynamics of planktonic rotifers and cladocerans. Ecology 71: 1741-1755.

Kitchell, J. F. & S. R. Carpenter, 1987. Piscivores, planktivores, fossils, and phorbins. In W. C. Kerfoot & A. Sih (eds), Predation: direct and indirect impacts on aquatic communities. The University Press of New England, Hanover (N.H.); Lond.: 132-146.

Klapwijk, A. & W. J. Snodgrass, 1982. Experimental measurement of sediment nitrification and denitrification in Hamilton Harbour, Canada. Hydrobiologia 91: 207-216.

Klapwijk, S. P., J. W. Kroon & M.-L. Meijer, 1982. Available phosphorus in lake sediments in the Netherlands. Hydrobiologia 91: 491-500.

Kling, G. W., 1988. Comparative transparency, depth of mixing, and stability of stratification in lakes of Cameroon, West Africa. Limnol. Oceanogr. 33: 27-40.

Komákorvá, J., 1989. Changes in phytoplankton assemblage during the spring period in the moderately eutrophic Římov reservoir (Czechoslovakia). Arch. Hydrobiol. Beih. Ergebn. Limnol. 33 : 419-433.

Konopka, A., 1983. The effect of nutrient limitations and its interaction with light upon the products of photosynthesis in *Merismopedia tenuissima* (Cyanophyceaea). J. Phycol. 19: 403-409. Quoted by Jansson *et al.* (1988)

Konopka, A., 1989. Metalimnetic cyanobacteria in hard-water lakes: buoyancy regulation and physiological state. Limnol. Oceanogr. 34: 1174-1184.

Konopka, A. & M. Schnur, 1980. Effect of light intensity upon macromolecular synthesis in cyanobacteria. Microb. Ecol. 6: 291-301. Quoted by Jansson *et al.* (1988)

Konopka, A. & M. Schnur, 1981. Biochemical composition and photosynthetic carbon metabolism of nutrient limited cultures of *Merismopedia tenuissima* (Cyanophyceae). J. Phycol. 17: 118-122. Quoted by Jansson *et al.* (1988)

Korzeniewski, K. & A. Moczulska, 1985. Sorption of cations by bottom sediments of Szczytno Małe and Łętowo Lakes, with cage trout culture. Pol. Arch. Hydrobiol. 32: 131-144.

Krom, M. D., J. Erez, C. B. Porter & S. Ellner, 1989a. Phytoplancton nutrient uptake dynamics in earthen marine fishponds under winter and summer conditions. Aquaculture 76: 237-253.

Krom, M. D., A. Neori & J. van Rijn, 1989b. Importance of water flow rate in controlling water quality processes in marine and freshwater fish ponds. Israel. J. Aquacult. 41: 23-33.

Kronert, U., G. Hörstgen-Schwark & H.-J. Langholz, 1989. Prospect of selecting for late maturity in tilapia (*Oreochromis niloticus*). I. Family studies under laboratory conditions. Aquaculture 77: 113-121.

Kuenzler, E., 1970. Dissolved organic phosphorus excretion by marine phytoplankton. J. Phycol. 6: 7-13. Quoted by Jansson *et al.* (1988).

Kurihara, Y., 1989. Ecology of some ricefields in Japan as exemplified by some benthic fauna, with notes on management. Int. Revue ges. Hydrobiol. 74: 507-548.

Kwiatkiwski, R. E. & J. C. Roff, 1976. Effects of acidity on the phytoplankton and primary productivity of selected northern Ontario lakes. Can. J. Bot. 54: 2546-2561. Quoted by Mulholland *et al.*, 1986.

Lam, D. C. L., W. M. Schertzer & A. S. Fraser, 1982. Mass balance models of phosphorus in sediments and water. Hydrobiologia 91: 217-225.

Lamarra, V. A., 1975. Digestive activities of carp as a major contributor to the nutrient loading of lakes. Verh. int. Ver. Limnol. 19: 2461-2468.

Lammens, E. H. R. R., R. D. Gulati, M.-L. Meijer & E. van Donk, 1990. The first biomanipulation conference: a synthesis. Hydrobiologia 200/201: 619-627.

Lampert, W., 1987. Vertical migration of freshwater zooplankton: indirect effects of vertebrate predators on algal communities. In W. C. Kerfoot & A. Sih (eds), Predation: direct and indirect impacts on aquatic communities. The University Press of New England, Hanover (N.H.); Lond.: 291-299.

Lampert, W., W. Fleckner, H. Rai & B. E. Taylor, 1986. Phytoplankton control by grazing zooplankton: a study on the spring clear-water phase. Limnol. Oceanogr. 31: 478-490.

Landau, R., 1983. Factors influencing tilapia biomass in Lake Kinneret. In L. Fishelson & Z. Yaron (eds), Tel Aviv University, Tel Aviv, Israel. Proc. Int. Symp. on Tilapia in Aquaculture: 8-17.

Lazard, J., 1980. Le développement de la pisciculture intensive en Côte d'Ivoire : Exemple de la ferme piscicole pilote de Natio-Kobadara (Korhogo). C.T.F.T., Notes et Documents, Nouvelle Série 21: 1-44.

Lazzaro, X., 1987. A review of planktivorous fishes: their evolution, feeding behaviour, selectivities, and impacts. Hydrobiologia 146: 97-167.

Lazzaro, X., R. W. Drenner, R. A. Stein & J. D. Smith, 1991a. Planktivores and plankton dynamics: effects of fish biomass and planktivore type. Can. J. Fish. aquat. Sci. (in press).

Lazzaro, X., R. W. Drenner, R. A. Stein & J. D. Smith, 1991b. Interactions between gizzard shad and bluegill as regulators of aquatic communities: a mesocosm experiment. Trans. am. Fish. Soc. (in press).

Lean, D. R. S., F. R. Pick, S. F. Mitchell, M. T. Downes, P. H. Woods & E. White, 1989. Lake Ontaro enclosure experiments: test ecosystems to evaluate plankton phosphorus and nitrogen deficiency. Arch. Hydrobiol. Beih. Ergebn. Limnol. 32: 195-211.

Leavitt, P. R., S. R. Carpenter & J. F. Kitchell, 1989. Whole-lake experiments: the annual record of fossil pigments and zooplankton. Limnol. Oceanogr. 34: 700-717.

Legendre, M., M. Pagano & L. Saint-Jean, 1987. Peuplements et biomasse zooplanctonique dans les étangs de pisciculture lagunaire (Layo, Côte d'Ivoire). Etude de la recolonisation après la mise en eau. Aquaculture 67: 321-341.

Lehman, J. T., 1988. Ecological principles affecting community structure and secondary production by zooplankton in marine and freshwater environments.

Limnol. Oceanogr. 33: 931-945.

Lehman, J. T. & D. Scavia, 1984. Measuring the ecological significance of microscale nutrient patches. Limnol. Oceanogr. 29: 214-216.

Lehninger, A. L., 1965. Bioenergetics. The molecular basis of biological energy transformations. W. A. Benjamin, Inc., N.Y., Amsterdam, 258 pp.

Lennox, L. J., 1984. Lough Ennell : laboratory studies on sediment phosphorus release under varying mixing, aerobic and anaerobic conditions. Freshwat. Biol. 14: 183-187.

Leventer, H. & B. Teltsch, 1990. The contribution of silver carp (*Hypophthalmichthys molitrix*) to the biological control of Netofa reservoirs. Hydrobiologia 191 : 47-55.

Levitan, C., 1987. Formal stability analysis of a planktonic freshwater community. In W. C. Kerfoot & A. Sih (eds), Predation: direct and indirect impacts on aquatic communities. The University Press of New England, Hanover (N.H.); Lond.: 71-100.

Li, S., 1987. Energy structure and efficiency of a typical Chinese integrated fish farm. Aquaculture 65: 105-118.

Liu, Q. & H. Seki, 1988. The effect of a "Bio-filter" aquatic treatment plan on Doh-Hoh-Numa Bog, Tsukuba City, Japan: rapid oligotrophication without damage to the phytoplancton community. Arch. Hydrobiol. 114: 63-82.

Lijklema, L. & A. H. M. Hieltjes, 1982. A dynamic phosphate budget model for a eutrophic lake. Hydrobiologia 91: 227-233.

Lindmark, G. K., 1982. Acidified lakes: sediment treatment with sodium carbonate - a remedy? Hydrobiologia 91: 537-547.

Lipschultz, F., S. C. Wofsy & L. E. Fox, 1985. The effects of light and nutrients on rates of ammonium transformation in an eutrophic river. Mar. Chem. 16 : 329-335.

Little, D. & J. Muir, 1987. A guide to integrated warmwater aquaculture. Publ.Inst. Aquaculture, Univ. of Sterling, Sterling, UK, 224 pp.

Logan, T. J., 1982. Mechanisms for release of sediment-bound phosphate to water and the effect of agricultural land management on fluvial transport of particulate and dissolved phosphate. Hydrobiologia 91: 519-530.

Lund, J. W. G., 1965. The ecology of the freshwater phytoplankton. Biol. Rev. 40 : 231-293.

Lyche, A., B. A. Faafeng & A. Braband, 1990. Predictability and possible mechanisms of plankton response to reduction of planktivorous fish. Hydrobiologia 200/201 : 251-261.

Mann, K. H., 1988. Production and use of detritus in various freshwater, estuarine, and coastal marine ecosystems. Limnol. Oceanogr. 33: 910-930.

Marsden, M. W., 1989. Lake restoration by reducing external phosphorus loading: the influence of sediment phosphorus release. Freshwat. Biol. 21: 139-162.

Marvalin, O., L. Aluya & C. Amblard, 1989. Importance relative des fractions bactériennes libres et fixes en milieu lacustre eutrophe. Arch. Hydrobiol. 115: 371-390.

Matveev, V. E. & E. G. Balseiro, 1990. Contrasting responses of two Cladocerans to changes in the nutritional value of nannoplankton. Freshwat. Biol. 23: 197-204.

Mazumder, A. & M. D. Dickman, 1989. Factors affecting the spatial and temporal distribution of phototrophic sulfur bacteria. Arch. Hydrobiol. 116: 209-226.

Mazumder, A., D. J. McQueen, W. D. Taylor & D. R. S. Lean, 1986. Effects of fertilization and planktivorous fish (yellow perch) predation on size distribution of particulate phosphorus and assimilated phosphate: large enclosure experiments. Limnol. Oceanogr. 33: 421-430.

Mazumder, A., D. J. McQueen, W. D. Taylor, D. R. S. Lean & M. D. Dickman, 1990. Micro- and mesoplankton grazing on natural pico- and nanoplankton in contrasting plankton communities produced by planktivore manipulation and fertilization. Arch. Hydrobiol. 118: 257-282.

McAllister, C. D., N. Shah & J. D. H. Strickland, 1964. Some interactions of temperature, light intensity and nutrient concentration during the continuous culture of *Nitzschia closterum* and *Tetraselmis sp.* Limnol. Oceanogr. 9: 79-86. Quoted by Jansson *et al.* (1988)

McAndrew, B. J., 1988. Tilapia hatchery systems. In H. R. King & K. H. Ibrahim (eds), Village level aquaculture development in Africa. Proc. Comm. Cons. Workshop on Village Level Aquaculture Development in Africa. Freetown, Sierra Leone, 14-20 February 1985. Commonwealth Secretariat, London: 137-146.

McCarthy, J. J., 1980. Nitrogen. In I. Morris (ed.), The physiological ecology of phytoplankton. Blackwell Scientific Publication, Oxford. Studies in Ecology Volume 7: 191-233.

McFarland, M. A. & D. W. Toetz, 1988. Nitrogen fixation (acetylene reduction) in Lake Hefner, Oklahoma. Arch. Hydrobiol. 114: 213-230.

McGeachin R. & R. R. Stickney, 1982. Manuring rates for production of blue tilapia in simulated sewage lagoons receiving laying hen waste. Prog. Fish-Cult. 44: 25-28.

McKnight, D. M. & F. M. M. Morel, 1980. Copper complexation by siderophores from filamentous blue-green algae. Limnol. Oceanogr. 25: 62-71.

McQueen, D. J. & J. R. Post, 1988. Cascading trophic interactions: uncoupling at the zooplankton-phytoplankton link. Hydrobiologia 159: 277-296.

McQueen, D. G., J. R. Post & E. L. Mills, 1986. Trophic relations in freshwater pelagic ecosystems. Can. J. Fish. aquat. Sci. 43: 1571-1581.

Meade, J. W., 1985. Allowable ammonia for fishculture. Prog. Fish-Cult. 47: 135-145.

Meffert, M.-E., 1989. Planktic unsheathed filaments (Cyanophyceae) with polar and central gas vacuoles. II. Biology, population dynamics and biotopes of *Limnothrix redekei* (Van Goor) Meffert. Arch. Hydrobiol. 116: 257-282.

Meijer, M.-L., E. H. R. R. Lammens, A. J. P. Raat, M. P. Grimm & S. H. Hosper, 1990a. Impact of cyprinids on zooplankton and algae in ten drainable ponds. Hydrobiologia 191 : 275-284.

Meijer, M.-L., M. W. de Haan, A. W. Breukelaar & H. Buiteveld, 1990b. Is reduction of the benthivorous fish an important cause of high transparency following biomanipulation in shallow lakes? Hydrobiologia 200/201: 303-315.

Mélard, Ch., 1986. Les bases biologiques de l'élevage intensif du tilapia du Nil. Cahiers Ethol. appl. 6 (3): 1-224.

Mesfin, M., C. Tudorancea & R. M. Baxter, 1988. Some

limnological observations on two Ethiopian hydroelectric reservoirs: Koka (Shewa administrative district) and Finchaa (Walega administrative district). Hydrobiologia 157: 47-55.

Miller, J. C., 1987. Evidence for the use of non-detrital dissolved organic matter by microheterotrophs on plant detritus in a woodland stream. Freshwat. Biol. 18: 483-494.

Miller, J. W., 1976. Fertilization and feeding practices in warmwater pond fishculture in Africa. FAO/CIFA symposium on aquaculture in Africa, 1975. Accra, Ghana. CIFA Techn. Pap. 4, suppl.1: 512-541.

Miller, T. E. & W. C. Kerfoot, 1987. Redefining indirect effects. In W. C. Kerfoot & A. Sih (eds), Predation: direct and indirect impacts on aquatic communities. The University Press of New England, Hanover (N.H.); Lond.: 33-37.

Milstein, A., M. Feldlite, N. Mozes & Y. Avnimelech, 1989. Limnology of reservoirs used for fish farming and crop irrigation, with integrated free and cage fish culture. Israel. J. Aquacult. 41: 12-22.

Miura, T., 1990. The effects of planktivorous fishes on the plankton community in a eutrophic lake. Hydrobiologia 200/201: 567-579.

Moav, R., G. Wohlfarth, G. L. Schroeder, G. Hulata & H. Barash, 1977. Intensive polyculture of fish in freshwater ponds. I. Substitution of expensive feeds by liquid cow manure. Aquaculture 10: 25-43.

Monod, J., 1942. Recherches sur la croissance des cultures bactériennes. Hermann, Paris, 210 pp.

Moore, W. J., 1956. Physical chemistry. Longmans, Greens & Co., London. 633 p.

Moriarty, D. J. W., 1973. The physiology of digestion of blue-green algae in the cichlid fish *Tilapia nilotica*. J. Zool. 171: 25-39.

Moriarty, D. J. W., 1990. Letter to the Editor. Aquabyte 3 (2): 18.

Moriarty, D. J. W., J. P. E. C. Darlington, I. G. Dunn, C. M. Moriarty & M. P. Tevlin, 1973. Feeding and grazing in Lake George, Uganda. Proc. R. Soc., Lond. B 184: 299-319.

Moriarty, D. J. W. & C. M. Moriarty, 1973. The assimilation of carbon from phytoplankton by two herbivorous fishes: *Tilapia nilotica* and *Haplochromis nigripinnis*. J. Zool. 171: 41-56.

Morris, D. P. & W. M. Lewis Jr., 1988. Phytoplankton nutrient limitation in Colorado mountain lakes. Freshwat. Biol. 20: 315-327.

Mortimer, C. H., 1941. The exchange of dissolved substances between mud and water in lakes. Part I & II. J. Ecol. 29: 280-329.

Mortimer, C. H., 1942. The exchange of dissolved substances between mud and water in lakes. Part III & IV. J. Ecol. 30: 147-201.

Moss, B., 1990. Engineering and biological approaches to the restoration from eutrophication of shallow-lakes in which aquatic plant communities are important components. Hydrobiologia 200/201: 367-377.

Msiska, O. V., 1981. Quality and nutrient release of poultry manure immersed in water. J. Limnol. Soc. Sth. Afr. 7: 59-62.

Msiska, O. V., 1983. Yields of *Sarotherodon mossambicus* using inorganic fertilizers and rice bran in Southern Malawi. Aquaculture 32: 201-205.

Msiska, O. V. & M. A. Cantrell, 1985. Influence of poultry manure on growth of *Oreochromis shiranus chilwae*. Aquaculture 44: 67-73.

Mulholland, P. J., Elwood, J. W., A. V. Palumbo & R. J. Stevenson, 1986. Effect of stream acidification on periphyton composition, chlorophyll, and productivity. Can. J. Fish. aquat. Sci. 43: 1846-1858.

Murad, A. & C. E. Boyd, 1987. Experiment on fertilization of sport-fish ponds. Progr. Fish-Cult. 49: 100-107.

Murphy, T. P., Lean, D. R. S. & C. Nalewajko, 1976. Blue-green algae: their excretion of iron-selective chelators enables them to dominate other algae. Science 198: 900-902.

Nalewajko, C. & D. R. S. Lean, 1980. Phosphorus. In I. Morris (ed.), The physiological ecology of phytoplankton. Blackwell Scientific Publication, Oxford. Studies in Ecology Volume 7: 235-258.

Nalewajko, C., B. Paul, K. Lee & H. Shear, 1986. Light history, phosphorus status and the occurence of light stimulation or inhibition of phosphate uptake in Lake Superior phytoplankton and bacteria. Can. J. Fish. aquat. Sci. 43: 329-335.

Nembrini, G., J. A. Capobianco, J. Garcia & J. M. Jaquet, 1982. Interaction between interstitial water and sediment in two cores of Lake Léman, Switzerland. Hydrobiologia 91: 363-375.

Ng, W. J., T. S. Sim, S. L. Ong, K. Kho, L. M. Ho, S. H. Tay & C. C. Goh, 1990. The effect of *Elodea densa* on aquaculture water quality. Aquaculture 84 (3/4): 267-276.

Nicklisch, A. & J.-G. Kohl, 1989. The influence of light on the primary production of two planktic blue-green algae. Arch. Hydrobiol. Beih. Ergebn. Limnol. 33: 451-455.

Nilssen, J.P., 1984. Tropical lakes - functional ecology and future development: the need for a process-oriented approach. Hydrobiologia 113: 231-242.

Noriega-Curtis, P., 1979. Primary productivity and related fish yield in intensely manured fishponds. Aquaculture 17: 335-344.

Nriagu, J. O., 1978. Dissolved silica in pore water of Lake Ontario, Erie, and Superior sediments. Limnol. Oceanogr. 23: 53-67.

O'Brien, W.J., 1987. Planktivory by freshwater fish: thrust and parry in the pelagia. In W. C. Kerfoot & A. Sih (eds), Predation: direct and indirect impacts on aquatic communities. The University Press of New England, Hanover (N.H.); Lond.: 3-16.

Oláh, J., V. R. P. Sinha, S. Ayyappan, C. S. Purushothaman & S. Radhayshyam, 1987. Sediment oxygen consumption in tropical undrainable fishponds. Int. Revue ges. Hydrobiol. 72: 297-305.

Oldorf, W., W. Kronert, J. Balarin, R. Haller, G. Hörstgen-Schwark & H. J. Langholz, 1989. Prospect of selecting for late maturity in tilapia (*Oreochromis niloticus*). II. Strain comparisons under laboratory and field conditions. Aquaculture 77: 123-133.

Olsen, Y., A. Jensen, H. Reinertsen, K. Y. Borsheim, M. Heldal & A. Langeland, 1986. Dependence of the rate of release of phosphorus on the P:C ratio in the food supply, as calculated by a recycling model. Limnol. Oceanogr. 31: 36-44.

Olsen, Y., G. Knutsen & T. Lien, 1983. Characteristics of phosphorus limitation in *Chlamydomonas reinhardtii* (Chlorophyceae) and its palmelloids. J. Phycol. 19: 313-319. Quoted by Jansson *et al.*, 1988.

Olsson, H., 1990. Phosphatase activity in relation to phytoplankton composition and pH in Swedish lakes. Freshwat. Biol. 23: 353-362.

Ongley, E. D., M. C. Bynoe & J. B. Percival, 1982. Physical and geochemical characteristics of suspended solids, Wilton Creek, Ontario. Hydrobiologia 91: 41-57.

Opuszyński, K., L. Myskowski, G. Okoniewska, W. Opuszyńska, M. Szlamińska, J. Wolnicki & M. Wożniewski, 1989. Rearing of common carp, grass carp, silver carp and bighead carp larvae using zooplancton and/or different dry feeds. Pol. Arch. Hydrobiol. 36: 217-230.

Ostrofsky, M. L., 1987. Phosphorus species in the surficial sediments of lakes of Eastern North America. Can. J. Fish. aquat. Sci. 44: 960-966.

Paasche, E., 1980. Silicon. In I. Morris (ed.), The physiological ecology of phytoplankton. Blackwell Scientific Publication, Oxford. Studies in Ecology Volume 7: 259-284.

Pace, M. L., 1986. An empirical analysis of zooplankton community size structure across lake trophic gradients. Limnol. Oceanogr. 31: 45-55.

Paerl, H. W., 1985. Microzone formation: its role in the enhancement of aquatic N_2 fixation. Limnol. Oceanogr. 30: 1246-1252.

Paerl, H. W., 1988. Nuisance phytoplankton blooms in coastal, estuarine, and inland waters. Limnol. Oceanogr. 33: 823-847.

Paine, R. T., 1988. Food webs: road maps of interactions or grist for theoretical development. Ecology 69: 1641-1654.

Pańczakowa, J. & T. Szyszka, 1986. Concentrations of nitrogen and phosphorus in waters of two hypertrophic lakes and their effects on the quantity of phytoplankton biomass. Pol. Arch. Hydrobiol. 33: 151-164.

Paterson, D. M., 1989. Short-term changes in the erodibility of intertidal cohesive sediments related to the migratory behavior of epipelic diatoms. Limnol. Oceanogr. 34: 223-296.

Pearshall, W. H., 1924. Phytoplankton and environment in the English Lake District. Revue Algologique 1: 54-67. Quoted by Shoesmith & Brook, 1983.

Pearsall, W. H. & C. H. Mortimer, 1939. Oxidation and reduction potentials in water logged soils, natural waters and muds. J. Ecol. 27: 483-501.

Peters G. T., J. R. Webster & E. F. Benfield, 1987. Microbial activity associated with seston in headwater streams: effects of nitrogen, phosphorus and temperature. Freshwat. Biol. 18: 405-413.

Peterson, C. G., 1987. Gut passage and insect grazer selectivity of lotic diatoms. Freshwat. Biol. 18: 455-460.

Pettersson K., B. Boström & O.-S. Jacobsen, 1988. Phosphorus in sediments - speciation and analysis. Hydrobiologia 170: 91-101.

Philippart, J.-C. & J.-C. Ruwet, 1982. Ecology and distribution of tilapias. In R. S. V. Pullin & R. H. Lowe-McConnell (eds), The biology and culture of tilapias. ICLARM Conf. Proc. 7, International Centre for Living Aquatic Resources Management, Manila, Philippines: 15-59.

Pick, F. R., 1989. Species-specific phytoplankton response to nutrient enrichment in limnetic enclosures. Arch. Hydrobiol. Beih. Ergebn. Limnol. 32: 177-187.

Piedrahita, R. H., 1988. Introduction to computer modelling of aquaculture pond ecosystems. Aquac. Fish. Manage. (Aquacultural and Fisheries Management) 19:1-12.

Piedrahita, R. H., 1990. Calibration and validation of TAP, an aquaculture pond water quality model. Aquac. Engin. (Aquacultural Engineering) 9:75-96.

Piedrahita, R. H., D. E. Brune, G. Tchobanoglous & G. T. Orlob, 1984. A general model of the aquaculture pond ecosystem. J. World. Maricul. Soc. 15: 355-366.

Pomeroy, L. W., E. E. Smith & C. M. Grant, 1965. The exchange of phosphate between estuarine water and sediments. Limnol. Oceanogr. 10: 167-172.

Pomeroy, L. R. & W. J. Wiebe, 1988. Energetics of microbial food webs. Hydrobiologia 159: 7-18.

Porter, K. G., 1988. Phagotrophic phytoflagellates in microbial food webs. Hydrobiologia 159: 89-97.

Post, J. R. & D. J. McQueen, 1987. The impact of planktivorous fish on the structure of a plankton community. Freshwat. Biol. 17: 79-89.

Preisendorfer, R. W., 1986. Secchi disk science: visual optics of natural waters. Limnol. Oceanogr. 31: 909-926.

Prowse, G. A., 1964. Some limnological problems in tropical fish ponds. Verh. int. Ver. Limnol. 15: 480-484.

Psenner, R., R. Pucsko & M. Sager, 1985. Fraktionierung organischer und anorganischer Phosphorverbindungen von Sedimenten. Versuch einer Definition ökologisch wichtiger Fraktionen. Arch. Hydrobiol. Suppl. 70: 11-115.

Qin, J. & S. T. Threskeld, 1990. Experimental comparison of the effects of benthivorous fish and planktivorous fish on plankton community structure. Arch. Hydrobiol. 119: 121-141.

Rappaport, U. & S. Sarig, 1978. The results of manuring on intensive growth fish farming at the Ginosar Station ponds in 1977. Bamidgeh 30: 27-36.

Rai, H. & T. R. Jakobsen, 1990. Phosphate (^{32}P)-uptake capabilities of natural picoplankton and ultraplankton communities in lakes of differing degrees of eutrophication. Arch. Hydrobiol. 118: 421-435.

Rhee, G.-Y., 1978. Effects of N:P atomic ratios and nitrate limitation on algal growth, cell composition and nitrate uptake. Limnol. Oceanogr. 23: 10-25.

Richardson, W. B., S. A. Wickham & S. T. Threlkeld, 1990. Foodweb response to the experimental manipulation of a benthivore (*Cyprinus carpio*), zooplanktivore (*Menidia beryllina*) and benthic insects. Arch. Hydrobiol. 119: 143-165.

Riessen, H. P. & W. G. Sprules, 1990. Demographic costs of antipredator defenses in *Daphnia pulex*. Ecology 71: 1536-1546.

Rippey, B. & D. H. Jewson, 1982. The rates of sediment-water exchange of oxygen and sediment bioturbation in Lough Neagh, Northern Irland. Hydrobiologia 91: 377-382.

Romaire, R. P. & C. E. Boyd, 1979. Effects of solar

radiation on the dynamics of dissolved oxygen in channel catfish ponds. Trans. am. Fish. Soc. 108: 473-478.

Romaire, R. P., C. E. Boyd & W. J. Collins, 1978. Predicting nighttime dissolved oxygen decline in ponds used for *Tilapia* culture. Trans. am. Fish. Soc. 107: 804-808.

Rosseland, B. O. & O. K. Skogheim, 1984. Attempts to reduce effects of acidification on fishes in Norway by different mitigation techniques. Fisheries 9: 10-16.

Rothhaupt, K. O., 1990. Ressource competition of herbivorous zooplankton: a review of approaches and perspectives. Arch. Hydrobiol. 118: 1-29.

Sanchez, P. A., 1976. Properties and management of soils in the tropics. John Wiley & Sons, N.Y., 618 pp.

Sanders, J. G., S. J. Cibik, C. F. D'Elia & W. R. Boynton, 1987. Nutrient enrichment studies in a coastal plain estuary: changes in phytoplankton species composition. Can. J. Fish. aquat. Sci. 43: 83-90.

Sanders, R. W. & R. G. Porter, 1990. Bacterivorous flagellates as food resource for the freshwater crustacean zooplankter *Daphnia ambigua*. Limnol. Oceanogr. 35: 188-191.

Sanders, R. W., K. G. Porter, S. J. Bennett & A. E. DeBiase, 1989. Seasonal pattern of bacterivory by flagellates, ciliates, rotifers, and cladocerans in a freshwater planktonic community. Limnol. Oceanogr. 34: 673-687.

Sanni, S. & S. B. Waervagen, 1990. Oligotrophication as a result of planktivorous fish removal with rotenone in the small, eutrophic, Lake Mosvatn, Norway. Hydrobiologia 200/201: 263-274.

Sarig, S. & Y. Arieli, 1980. Growth capacity of tilapia in intensive culture. Bamidgeh 32: 57-65.

Saunders, G. W., 1972. Potential heterotrophy in a natural population of *Oscillatoria agardhii* var. *isothrix* Skuja. Limnol. Oceanogr. 17: 704-711.

Saunders, G. W. (Coord.), 1980. Organic matter and decomposers. In E. D. Le Cren & R. H. Lowe-McConnell (eds), The functioning of freshwater ecosystems. Cambridge University Press, Cambridge. International Biological Programme 22: 341-392.

Scheda, S. M. & B. C. Cowell, 1988. Rotifer grazers and phytoplankton: seasonal experiments on natural communities. Arch. Hydrobiol. 114: 31-44.

Scheffer, M., 1990. Multiplicity of stable states in freshwater systems. Hydrobiologia 200/201: 475-486.

Schindler, D. W., 1978. Factors regulating phytoplankton production and standing crop in the world's freshwaters. Limnol. Oceanogr. 23: 478-486.

Schindler, D. W., 1987. Detecting ecosystem responses to anthropogenic stress. Can. J. Fish. aquat. Sci. 44 (Suppl. 1): 6-25.

Schindler, D. W., R. H. Hesslein & M. A. Turner, 1987. Exchange of nutrients between sediments and water after 15 years of experimental eutrophisation. Can. J. Fish. aquat. Sci. 44: 26-33.

Schindler, D. W., K. H. Mills, D. F. Malley, D. L. Findlay, J. A. Shearer, I. J. Davies, M. A. Turner, G. A. Linsey & D. R. Cruikshank, 1985. Long-term ecosystem stress: the effects of years of experimental acidification on a small lake. Science 228: 1395-1401.

Schroeder, G. L., 1980. Fish farming in manure-loaded ponds. In R. S. V. Pullin & Z. H. Shehadeh (eds.), Integrated agriculture-aquaculture farming systems. I.C.L.A.R.M., Manila, Philippines. I.C.L.A.R.M. Conference Proceedings 4: 73-86.

Schroeder, G. L., 1987. Carbon and nitrogen budgets in manured fish ponds on Israel's coastal plain. Aquaculture 62: 259-279.

Schuurkes, J. A. A. R. & C. J. Kok, 1988. In vitro studies on sulphate reduction and acidification in sediments of shallow soft water lakes. Freshwat. Biol. 19: 417-426.

Seďa, J., J. Kubečka & Z. Brandl, 1989. Zooplankton structure and fish population development in the Římov Reservoir, Czechoslovakia. Arch. Hydrobiol. Beih. Ergebn. Limnol. 33: 605-609.

Ségalen, P., 1964. Le fer dans le sol. O.R.S.T.O.M., Paris. Initiations - Documn Techqs 4: 150 pp.

Ségalen, P., 1973. L'aluminium dans les sols. O.R.S.T.O.M., Paris. Initiations - Documn Techqs 22: 281 pp.

Segupta, S. & B. B. Jana, 1987. Effect of aeration on the primary productivity of the phytoplancton in experimental tanks. Aquaculture 62: 131-142.

Seitzinger, S. P., 1988. Denitrification in freshwater and coastal marine ecosystems: ecological and geochemical significance. Limnol. Oceanogr. 33: 702-724.

Senft, W. H., 1978. Dependance of light-saturated rates of algal photosynthesis on intracellular concentrations of phosphorus. Limnol. Oceanogr. 23: 709-718.

Seymour, E. A., 1980. The effects and control of algal blooms in fish ponds. Aquaculture 19: 55-74.

Shapiro, J., 1988. Introductory lecture at the international symposium "Phosphorus in Freshwater Ecosystems" Uppsala, Sweden in October 1985. Hydrobiologia 170: 9-17.

Shapiro, J., 1990. Biomanipulation: the next phase - making it stable. Hydrobiologia 200/201: 13-27.

Shapiro, J., V. Lamarra & M. Lynch, 1975. Biomanipulation: an ecosystem approach to lake restoration. In P. L. Brezonik & J. L. Fox (eds), Proc. Symp. on Water Quality Management through Biological Control. University of Florida, Gainesville, 85-96.

Shearer, J. A., E. J. Fee, E. R. DeBruyn & D. R. DeClercq, 1987. Phytoplancton productivity changes in a small, double-basin lake in response to termination of experimental fertilization. Can. J. Fish. aquat. Sci. 44 (Suppl. 1): 47-54.

Sheldon, R. W., P. Nival & F. Rassoulzadegan, 1986. An experimental investigation of a flagellate-ciliate-copepod food chain with some observations relevant to the linear biomass hypothesis. Limnol. Oceanogr. 31: 184-188.

Sherr, B. F., E. B. Sherr & C. S. Hopkinson, 1988. Trophic interactions within pelagic microbial communities: indications of feedback regulation of carbon flow. Hydrobiologia 159: 19-26.

Sherr, E. & B. Sherr, 1988. Role of microbes in pelagic food webs: a revised concept. Limnol. Oceanogr. 33: 1225-1227.

Shoesmith, E. A. & A. J. Brook, 1983. Monovalent-divalent cation ratios and the occurence of phytoplankton, with special reference to the desmids. Freshwat. Biol. 13: 151-155.

Sieburth, J. McN., V. Smetacek & J. Lenz, 1978. Pelagic ecosystem structure: heterotrophic compartments of the plankton and their relationship to plankton size fractions. Limnol. Oceanogr. 23: 1256-1263.

Sigler, J. W., T. C. Bjornn & F. H. Everest, 1984. Effects of chronic turbidity on density and growth of steelheads and coho salmon. Trans. am. Fish. Soc. 113: 142-150.

Sih, A., 1987. Predators and prey lifestyles: an evolutionary and ecological overview. In W. C. Kerfoot & A. Sih (eds), Predation: direct and indirect impacts on aquatic communities. The University Press of New England, Hanover (N.H.); Lond.: 203-224.

Simon, M., 1987. Biomass and production of small and large free-living and attached bacteria in Lake Constance. Limnol. Oceanogr. 32: 591-607.

Sin, A. W. & M. T. Chiu, 1983. The intensive monoculture of the tilapia hybrid *Sarotherodon nilotica* (male) x *S. mossambica* (female) in Hong-Kong. In L. Fishelson & Z. Yaron (eds), Tel Aviv University, Tel Aviv, Israel. Proc. Int. Symp. on Tilapia in Aquaculture: 506-516.

Sinke, A. J. C. & T. E. Cappenberg, 1988. Influence of bacterial processes on the phosphorus release from sediments in the eutrophic Loosdrecht Lakes, The Netherlands. Arch. Hydrobiol. Beih. Ergebn. Limnol. 30: 5-13.

Sinke, A. J., A. A. Cornelese, P. Keizer, O. F. R. Van Tongeren & T. E. Cappenberg, 1990. Mineralization, pore water chemistry and phosphorus release from peaty sediments in the eutrophic Loosdrecht lakes, The Netherlands. Freshwat. Biol. 23: 587-599.

Smayda, T. J., 1980. Phytoplankton species succession. In: I. Morris (ed.), The physiological ecology of phytoplankton. Blackwell Scientific Publication, Oxford. Studies in Ecology Volume 7: 493-570.

Smayda, T., 1990. The influence of lime and biological activity on sediment pH, redox and phosphorus dynamics. Hydrobiologia 192: 191-203.

Smith, D. F., 1986. Small-scale spatial heterogeneity in dissolved nutrient concentrations. Limnol. Oceanogr. 31: 167-171.

Smith, D. W., 1988. Phytoplankton and catfish culture: a review. Aquaculture 74: 167-189.

Smith, D. W. & R. H. Piedrahita, 1988. The relation between phytoplankton and dissolved oxygen in fish ponds. Aquaculture 68: 249-265.

Smith, R. E. & J. Kalff, 1985. Phosphorus competition among phytoplankton: a reply. Limnol. Oceanogr. 30: 440-444.

Smith, S. V., J. T. Hollibaugh, S. T. Dollar & S. Vink, 1989. Tomales Bay, California: a case for carbon-controlled nitrogen cycling. Limnol. Oceanogr. 34: 37-52.

Smith, S. V., S. Serruya, Y. Geifman & T. Berman, 1989. Internal sources and sinks of water, P, N, Ca, and Cl in Lake Kinneret, Israel. Limnol. Oceanogr. 34: 1202-1213.

Sommer, U., 1984. The paradox of the plankton: fluctuations of phosphorus availability maintain diversity of phytoplankton in flow-through culture. Limnol. Oceanogr. 29: 633-636.

Sommer, U., 1988. Phytoplankton succession in microcosm experiments under simultaneous grazing pressure and resource limitation. Limnol. Oceanogr. 33: 1037-1054.

Sommer, U., 1989. Nutrient status and nutrient competition of phytoplankton in a shallow, hypertrophic lake. Limnol. Oceanogr. 34: 1162-1173.

Sommer, U., Z. M. Gliwicz, W. Lampert & A. Duncan, 1986. The PEG-model of seasonal succession of planktonic events in fresh waters. Arch. Hydrobiol. 106: 433-471.

Søndergaard, M., 1989. Phosphorus release from a hypertrophic lake sediment: experiments with intact sediment cores in a continuous flow system. Arch. Hydrobiol. 116: 45-59.

Søndergaard, M., E. Jeppesen, P. Kristensen & O. Sortkjær, 1990. Interactions between sediment and water in shallow and hypertrophic lake: a study on phytoplankton collapses in Lake Søbygård, Denmark. Hydrobiologia 191: 129-138.

Spotte, S., 1979. Seawater aquarium : the captive environment. John Wiley & Sons, N.Y., 413 pp.

Stabel, H.-H., 1986. Calcite precipitation in Lake Constance: chemical equilibrium, sedimentation, and nucleation by algae. Limnol. Oceanogr. 31: 1081-1093.

Stack, V. T., Jr., 1973. Stabilization oxygen demand. Biological methods for the assessment of water quality. ASTM STP 528, American Society for Testing and Materials: 209-220.

Starling, F. L. R. M. & A. J. A. Rocha, 1990. Experimental study of the impacts of planktivorous fishes on plankton community and eutrophication of a tropical Brazilian reservoir. Hydrobiologia 200/201: 581-591.

Steemann Nielsen, E., 1952. The use of radioactive carbon (^{14}C) for measuring organic production in the sea. J. Cons. perm. Int. Expl. Mer 18: 117-140.

Steemann Nielsen, E. & V. K. Hansen, 1959. Light adaptation in marine phytoplankton populations and its interrelation with temperature. Physiol. Pl. 12: 353-370. Quoted by Jansson *et al.* (1988)

Stemberger, R. S. & J. J. Gilbert, 1987. Defenses of planktonic rotifers against predators. In W. C. Kerfoot & A. Sih (eds), Predation: direct and indirect impacts on aquatic communities. The University Press of New England, Hanover (N.H.); Lond.: 227-239.

Stenson, J. A. E., 1987. Variation in capsule size of *Holopedium gibberum* (Zaddach): a response to invertebrate predation. Ecology 68 (4): 928-934.

Sterner, R. W., 1989. Resource competition during seasonal succession toward dominance by cyanobacteria. Ecology 70: 229-245.

Sternik, K. H., 1983. Untersuchungen zur Phosphor-Abgabe und insbesondere zur Orthophosphat-Exkretion junger Karpfen (*Cyprinus carpio*). Ein Beitrag zur Phosphat-Remobilisierung in Gewässern. Arch. Hydrobiol./Suppl. 66: 1-82.

Stickney, R. R., J. H. Hesby, R. B. McGeachin & W. A. Isbell, 1979. Growth of *Tilapia nilotica* in ponds with different histories of organic fertilization. Aquaculture 17: 189-194.

Stockner, J. G., 1988. Phototrophic picoplankton: an overview from marine and freshwater ecosystems. Limnol. Oceanogr. 33: 765-775.

Stockner, J. G. & N. J. Antia, 1986. Algal picoplankton from marine and freshwater ecosystems : a multidisciplinary perspective. Can. J. Fish. aquat. Sci. 43: 2472-2503.

Stockner, J. G. & K. S. Shortreed, 1988. Responses of *Anabaena* and *Synechococcus* to manipulation of nitrogen: phosphorus ratios in a lake fertilization experiment.

Limnol. Oceanogr. 33: 1348-1361.

Storch, T. A. & G. W. Saunders, 1978. Phytoplankton extracellular release and its relation to the seasonal cycle of dissolved organic carbon in a eutrophic lake. Limnol. Oceanogr. 23: 112-119.

Stumm, W. & J. J. Morgan, 1981. Aquatic chemistry. An introduction emphasizing chemical equilibria in natural waters. 2nd Ed., John Wiley & Sons, N.Y., 780 pp.

Straškraba, M., 1980. The effects of physical variables on freshwater production : analyses based on models. In E. D. Le Cren & R. H. Lowe-McConnell (eds), The functioning of freshwater ecosystems. Cambridge University Press, Cambridge. International Biological Programme 22: 13-84.

Strickler, J. R., 1970. Über das Schwimmverhalten von Cyclopoiden bei Verminderungen der Bestrahlungsstärke. Schweiz. Z. Hydrol. 32: 150-180.

Strickler, J. R., 1975. Swimming of planktonic *Cyclops* species (Copepoda, Crustacea): pattern, movements and their controls. In T. Y.-T. Wu *et al.* (eds), Swimming and flying in nature, Vol. 2, Plenum: 599-613.

Sugita, H., K. Oshima, T. Fushino & Y. Deguchi, 1987. Substrate specificity of heterotrophic bacteria in the water and sediment of a carp culture pond. Aquaculture 64: 39-46.

Suttle, C. A., J. A. Fuhrman & D. G. Capone, 1990. Rapid ammonium cycling and concentration-dependent partitioning of ammonium and phosphate: implications for carbon transfer in planktonic communities. Limnol. Oceanogr. 35: 424-433.

Suttle, C. A. & P. J. Harrison, 1988. Ammonium and phosphate uptake rates, N:P supply ratios, and evidence for N and P limitation in some oligotrophic lakes. Limnol. Oceanogr. 33: 186-202.

Sweerts, J. P., J. W. M. Rudd & C. A. Kelly, 1986. Metabolic activities in flocculent surface sediments and underlying sandy littoral sediments. Limnol. Oceanogr. 31: 330-338.

Szumiec, M. A., 1989. The rôle of carp, temperature and morphometry in solar radiation transmission in ponds. Arch. Hydrobiol. Beih. Ergebn. Limnol. 31: 571-577.

Talling, J. F., 1969. The incidence of vertical mixing, and some biological and chemical consequences, in tropical African lakes. Verh. Internat. Verein. Limnol. 17: 998-1012.

Talling, J. F. & I. B. Talling, 1965. The chemical composition of African lake waters. Int. Revue ges. Hydrobiol. 50: 421-463.

Tapiador, D. D., H. F. Henderson, M. N. Delmendo & H. Tsutsui, 1978. Pêche et aquaculture continentales en Chine. FAO Doc. Tech. Pêches, 168: 86 pp.

Tarapchak, S. J. & L. R. Herche, 1986. Phosphate uptake by microorganisms in lake water : deviations from simple Michaelis-Menten kinetics. Can. J. Fish. aquat. Sci. 43: 319-328.

Tarapchak, S. J. & C. Nalewajko, 1986a. Introduction : Phosphorus - plankton dynamics symposium. Can. J. Fish. aquat. Sci. 43: 293-301.

Tarapchak, S. J. & C. Nalewajko, 1986b. Synopsis : Phosphorus - plankton symposium. Can. J. Fish. aquat. Sci. 43: 416-419.

Taylor, F. J. R., 1980. Basic biological features of phytoplankton cells. In In I. Morris (ed.), The physiological ecology of phytoplankton. Blackwell Scientific Publication, Oxford. Studies in Ecology, Volume 7: 3-55.

Theiss, J., K. Zielinski & H. Lang, 1990. Biomanipulation by introduction of herbivorous zooplankton. A helpful shock for eutrophic lakes? Hydrobiologia 200/201: 58-68.

Thompson, P. A., M. E. Levasseur & P. J. Harrison, 1989. Light-limited growth on ammonium vs. nitrate: what is the avantage for marine phytoplankton. Limnol. Oceanogr. 34: 1048-1061.

Threlkeld, S. T., 1988. Planktivory and planktivore biomass effects on zooplankton, phytoplankton and the trophic cascade. Limnol. Oceanogr. 33: 1362-1375.

Tilman, D., R. Kiesling, R. Sterner, S. S. Kilham & F. A. Johnson, 1986. Green, blue-green and diatom algae: taxonomic differences in competitive ability for phosphorus, silicon and nitrogen. Arch. Hydrobiol. 106: 473-485.

Tilzer, M. M., 1989. The productivity of phytoplankton and its control by resource availability. A review. In P. D. Kumar (ed.), Phycotalk, Varansi Hindi University, Varansi: 1-40.

Tilzer, M. M., 1990. Environmental and physiological control of phytoplankton productivity in large lakes. In M. M. Tilzer & C. Serruya (eds), Large lakes: ecological structure and function. Springer.

Tipping, E., C. Woof & M. Ohnstad, 1982. Forms of iron in the oxygenated waters of Eastwaite Water, U. K. Hydrobiologia 91: 383-393.

Tranvik, L. J. & J. McN. Sieburth, 1989. Effects of flocculated humic matter on free and attached pelagic microorganisms. Limnol. Oceanogr. 34: 688-699.

Trewawas, E., 1983. Tilappine fishes of the genera *Sarotherodon*, *Oreochromis*, and *Danakilia*. British Museum, London.

Trifonova, I. S., 1989. Changes in community structure and productivity of phytoplankton as indicator of lake reservoir eutrophication. Arch. Hydrobiol. Beih. Ergebn. Limnol. 33: 363-371.

Tucker, C. S., 1984. Potassium permanganate demand of pond waters. Prog. Fish- Cult. 46: 24-28.

Tucker, C. S., 1989. Method for estimating potassium permanganate disease treatment rate for channel catfish in ponds. Progr. Fish-Cult. 51: 24-26.

Tucker, C. S. & C. E. Boyd, 1977. Relationships between potassium permanganate treatment and water quality. Trans. am. Fish. Soc. 106: 481-488.

Tucker, C. S. & C. E. Boyd, 1978. Consequences of periodic applications of copper sulfate and simazine for phytoplankton control in catfish ponds. Trans. am. Fish. Soc. 107: 316-320.

Tucker, L., C. E. Boyd & E. W. McCoy, 1979. Effects on feeding rate on water quality, production of channel catfish, and economic returns. Trans. am. Fish. Soc. 108: 389-396.

Tudorancea, C., C. H. Fernando & J. C. Paggi, 1988. Food and feeding ecology of *Oreochromis niloticus* (Linneaus, 1758) juveniles in Lake Awassa (Ethiopia). Arch. Hydrobiol. (Monogr. Beitr.), Suppl. 79: 267-289.

Tudorancea, C. & A. D. Harrison, 1988. The benthic community of the saline lakes Abijata and Shala (Ethiopia).

Hydrobiologia 158: 117-123.

Tudorancea, C. & A. Zullini, 1989. Associations and distribution of benthic nematodes in the Ethiopian Rift Valley lakes. Hydrobiologia 179: 81-96.

Twinch, A. L. & C. M. Breen, 1982. Vertical stratification in sediments from a young oligotrophic South African impoundment: implication in phosphorus cycling. Hydrobiologia 91: 395-404.

Uehlinger, U. & J. Bloesch, 1987. The influence of crustacean zooplankton on the size structure of the algal biomass and suspended and settling seston (biomanipulation in limnocorral II). Int. Revue ges. Hydrobiol. 72: 473-486.

Uehlinger, U. & J. Bloesch, 1989. Primary production of different phytoplankton size classes in an oligo-mesotrophic Swiss lake. Arch. Hydrobiol. 116: 1-21.

Urban, N. R. & L. A. Baker, 1989. Alkalinity generation associated with the formation of ester sulfates. Limnol. Oceanogr. 34: 1144-1146.

Vadstein, O., B. O. Harkjerr, A. Jensen, Y. Olsen & H. Reinertsen, 1989. Cycling of organic carbon in the photic zone of a eutrophic lake with special reference to the heterotrophic bacteria. Limnol. Oceanogr. 34: 840-855.

Vadstein, O., A. Jensen, Y. Olsen & H. Reinertsen, 1988. Growth and phosphorus status of limnetic phytoplankton and bacteria. Limnol. Oceanogr. 33: 489-503.

van Eck, G. T. M., 1982. Forms of phosphorus in particulate matter from the Hollands Diep/Haringvliet, The Netherlands. Hydrobiologia 91: 665-681.

van Liere, L. & L. R. Mur, 1982. The influence of groundwater-movement on the phosphorus release from sediments, as measured in a continuous flow system. Hydrobiologia 91: 511-518.

Vanni, M. J., 1986. Competition in zooplankton communities: suppression of small species by *Daphnia pulex*. Limnol. Oceanogr. 31: 1039-1056.

Vanni, M. J., 1987a. Effects of nutrients and zooplankton size on the structure of a phytoplankton community. Ecology 68: 624-635.

Vanni, M. J., 1987b. Indirect effect of predators on age-structured prey populations: planktivorous fish and zooplankton. In W. C. Kerfoot & A. Sih (eds), Predation: direct and indirect impacts on aquatic communities. The University Press of New England, Hanover (N.H.); Lond.: 149-160.

Vanni, M. J. & D. L. Findlay, 1990. Trophic cascades and phytoplankton community structure. Ecology 71: 921-937.

Vanni, M. J. & J. Temte, 1990. Seasonal patterns of grazing and nutrient limitation of phytoplankton in a eutrophic lake. Limnol. Oceanogr. 35: 697-709.

Van Rijn, J., S. Diab & M. Shilo, 1987. Phytoplankton succession in relation to the nitrogen régime in shallow, brackish-water fishponds. Arch. Hydrobiol. 111: 183-196.

Van Rijn, J., S. R. Stutz, S. Diab & M. Shilo, 1986. Chemical, physical and biological parameters of superintensive concrete fish ponds. Bamidgeh 38: 35-43.

Verani, J. R., C. S. R. M. Pinto, P. Paiva & Y. A. Tabata, 1983. Experimental studies on intensive fish culture of the all male hybrid of *Sarotherodon niloticus* (female) x *S. hornorum* (male) stocked various levels. In L. Fishelson & Z. Yaron (eds), Tel Aviv University, Tel Aviv, Israel. Proc. Int. Symph. on Tilapia in Aquaculture: 517-523.

Verhulst, P. F., 1838. Notice sur la loi que la population suit dans son accroissement. Corresp. Math. Phys. 10: 113-121.

Verreth, J., 1990. The accuracy of population density estimates of a horizontally distributed zooplankton community in Dutch fish ponds. Hydrobiologia 203: 53-61.

Vieillefon, J., 1977. Les sols des mangroves et des tannes de Basse Casamance (Sénégal). Importance du comportement géochimique du soufre dans leur pédogenèse. O.R.S.T.O.M., Paris, Mémoires 83: 291 pp.

Vincent, W. F., 1989. Cyanobacterial growth and dominance in two eutrophic lakes: review and synthesis. Arch. Hydrobiol. Beih. Ergebn. Limnol. 32: 239-254.

Vincent, W. F. & S. J. Dryden, 1989. Phytoplankton succession and cyanobacterial dominance in a eutrophic lake of the mid-temperate zone (Lake Okaro, New Zealand). Arch. Hydrobiol. Beih. Ergebn. Limnol. 32: 137-163.

Viner, A. B., 1989. Fluorescence and carbon fixation by phosphorus deficient phytoplankton in a Western Australian reservoir. Arch. Hydrobiol. 115: 1-19.

Viner, A. B. & I. R. Smith, 1973. Geographical, historical and physical aspects of Lake George. Proc. R. Soc. Lond. B 184: 235-270.

Vincke, M. M. J., 1988. Developing productive systems under village conditions. In H. R. King & K. H. Ibrahim (eds), Village level aquaculture development in Africa. Proc. Comm. Cons. Workshop on Village Level Aquaculture Development in Africa. Freetown, Sierra Leone, 14-20 February 1985. Commonwealth Secretariat, London: 39-57.

Vostradovský, J., J. Křížek, O. Albertová, L. Růžička & M. Vostradovská, 1989. The changes in fish communities and biomanipulation in water supply reservoirs. Arch. Hydrobiol. Beih. Ergebn. Limnol. 33: 587-594.

Walsby, A. E., R. Kinsman, B. W. Ibelings & C. S. Reynolds, 1991. Highly buoyant colonies of the cyanobacterium *Anabaena lemmermannii* from persistent surface waterblooms. Arch. Hydrobiol. 121: 261-280.

Walsby, A. E. & C. S. Reynolds, 1980. Sinking and floating. In I. Morris (ed.), The physiological ecology of phytoplankton. Blackwell Scientific Publication, Oxford. Studies in Ecology Volume 7: 371-412.

Wei, S. L. & E. A. Laws, 1989. Spatial and temporal variation of water column measurements in aquaculture ponds. Aquaculture 78: 253-266.

Weilenmann, U., C. R. O'Melia & W. Stumm, 1989. Particle transport in lakes: models and measurements. Limnol. Oceanogr. 34: 1-18.

Weisse, T., 1990. Trophic interactions among heterotrophic microplankton, nanoplankton, and bacteria in Lake Constance. Hydrobiologia 191: 111-122.

Welcomme, R. L., 1972. The inland waters of Africa. Les eaux intérieures d'Afrique. FAO CIFA Techn. Pap. 1, 117 pp.

Welcomme, R. L., 1983. River basins. FAO Fish. Techn. Pap. 202, 60 pp.

Weninger, G., 1985. Principal freshwater-types and comparative hydrochemistry of tropical running water systems. Rev. Hydrobiol. trop. 18: 79-110.

Westlake, D. F. (Coordinator), 1980. Primary production. In

E. D. Le Cren & R. H. Lowe-McConnell (eds), The functioning of freshwater ecosystems. Cambridge University Press, Cambridge. International Biological Programme 22: 141-246.

Wetzel, R. G., 1983. Limnology. 2nd Ed., Saunders, Philadelphia.

White, E., M. T. Downes, G. W. Payne, S. Pickmere, P. Woods & J. Rueter, 1989. Community responses of the late summer phytoplankton of Lake Okaro (New Zealand) to experimental changes in nutrient abundance: evidence of nitrogen limitation in the presence of phosphorus deficiency. Arch. Hydrobiol. Beih. Ergebn. Limnol. 32: 165-175.

Williamson, C. E., 1987. Predator-prey interactions between omnivorous diaptomid copepods and rotifers: the rôle of prey morphology and behavior. Limnol. Oceanogr. 33: 167-177.

Williamson, C. E. & N. M. Butler, 1986. Predation on rotifers by the suspension-feeding calanoid copepod *Diaptomus pallidus*. Limnol. Oceanogr. 31: 393-402.

Wurtsbaugh, W. A., W. F. Vincent, R. Alfaro Tapia, C. L. Vincent & P. J. Richerson, 1985. Nutrient limitation of algal growth and nitrogen fixation in a tropical alpine lake, Lake Titicaca. Freshwat. Biol. 15: 185-195.

Wynne, D. & T. Berman, 1990. The influence of environmental factors on nitrate reductase activity in freshwater phytoplankton. Hydrobiologia 194: 235-245.

Yan, N. D., 1979. Phytoplankton community of an acidified, heavy metal-contaminated lake near Sudbury, Ontario: 1973-1977. Water Air Soil Pollut. 11: 43-55. Quoted by Mulholland *et al.*, 1986.

Yentsch, C. E., 1980. Light attenuation and phytoplankton photosynthesis. In I. Morris (ed.), The physiological ecology of phytoplankton. Blackwell Scientific Publication, Oxford. Studies in Ecology Volume 7: 95-127.

Young, T. C. & J. V. DePinto, 1982. Algal-availability of particulate phosphorus from diffuse and point sources in the lower Great Lakes basin. Hydrobiologia 91: 111-119.

Young, T. C. & D. L. King, 1980. Interacting limits to algal growth: light, phosphorus, and carbon dioxide availability. Water Res. 14: 409-412. Quoted by Jansson *et al.* (1988)

Yusoff, F. M. & C. D. McNabb, 1987. Effects of nutrient availability on primary productivity and fish production in fertilized tropical ponds. Aquaculture 78: 303-319.

Zaret, T. M., 1980. Predation and freshwater communities. Yale Univ. Press, New Haven, 187 pp.

Zhang, F. L., Y. Zhu & X. Y. Zhou, 1987. Studies on the ecological effects of varying the size of fish ponds loaded with manures and feeds. Aquaculture 60: 107-116.

Zlotnik, I. & Z. Dubinsky, 1989. The effect of light and temperature on DOC excretion by phytoplankton. Limnol. Oceanogr. 34: 831-839.

Zohar, G., U. Rappaport, Y. Avnimelech & S. Sarig, 1984. Results of the experiments carried out in the Genosar experimental station in 1983. Cultivation of tilapia in high densities and with periodic flushing of the pond water. Bamidgeh 36: 63-69.

Zurbuch, P. E., 1984. Neutralization of acidified streams in West Virginia. Fisheries 9: 42-47.

Zweig, R. D., 1989. Evolving water quality in a common carp and blue tilapia high production pond. Hydrobiologia 171: 11-21.

Index